HANDBOOK ON THE ECONOMICS OF ECOSYSTEM SERVICES AND BIODIVERSITY

Handbook on the Economics of Ecosystem Services and Biodiversity

Edited by

Paulo A.L.D. Nunes

Scientific Coordinator, Policy and Technical Experts Committee, WAVES – Wealth Accounting and the Valuation of Ecosystem Services, The World Bank, USA

Pushpam Kumar

Chief, Ecosystem Services Economics Unit, United Nations Environment Programme (UNEP), Kenya

Tom Dedeurwaerdere

Director, BIOGOV Unit, Université Catholique de Louvain (UCL), Belgium

Edward Elgar
Cheltenham, UK • Northampton, MA, USA

Published by
Edward Elgar Publishing Limited
The Lypiatts
15 Lansdown Road
Cheltenham
Glos GL50 2JA
UK

Edward Elgar Publishing, Inc.
William Pratt House
9 Dewey Court
Northampton
Massachusetts 01060
USA

A catalogue record for this book
is available from the British Library

Library of Congress Control Number: 2013957778

This book is available electronically in the ElgarOnline.com
Economics Subject Collection, E-ISBN 978 1 78195 151 4

MIX
Paper from
responsible sources
FSC
www.fsc.org
FSC® C013056

ISBN 978 1 78195 150 7

Typeset by Servis Filmsetting Ltd, Stockport, Cheshire
Printed and bound in Great Britain by T.J. International Ltd, Padstow

Contents

Contributors

Pertti Ala-aho is Associate Researcher at the Department of Process and Environmental Engineering, Water Resources and Environmental Engineering Laboratory, University of Oulu, Finland. His research interests include hydrological modelling in boreal regions, the interaction between groundwater and surface water, lake hydrology and lake water balance and the use of environmental tracers.

Ioannis Anastasiou is an Associate Researcher with ReSEES (Research tEam on Socio--Economic and Environmental Sustainability) at Athens University of Economics and Business.

Jorge Angulo-Valdés is a professor at the University of Havana, Cuba. He also is the Director of the Center for Marine Research at the University of Havana and Director of the International Ocean Institute Operational Center in Cuba. His research interests include marine management effectiveness of marine protected areas, ecology of reef fish, ecology of manatees and sharks, natural resources conservation and bioeconomics.

Vassilis Babalos is a lecturer at the Kavala Institute of Technology (Greece).

Tomáš Badura is a PhD candidate at the Centre for Social and Economic Research on the Global Environment at the University of East Anglia and holds an MSc in Theoretical Economics from the University of Amsterdam. He specializes in the economic dimension of ecosystem valuation, focusing in his current research on the methodological issues associated with the use of valuation for policy and decision-making purposes.

Kenneth J. Bagstad is a research economist working with the US Geological Survey's Geosciences and Environmental Change Science Center in Denver, Colorado. For this work, he uses GIS and modelling to map and value ecosystem service flows and social values for ecosystem services at multiple sites in the United States. He has previously worked with the US Department of Interior–Bureau of Land Management on testing alternative ecosystem services tools, including the ARIES (Artificial Intelligence for Ecosystem Services) and InVEST (Integrated Valuation of Environmental Services and Tradeoffs) models, for their value in decision-making for public land management.

Harold E. Balbach is Senior Research Biologist with the US Army Engineer Research and Development Center, Champaign, IL. He earned his Bachelor's degree from Chicago State University and his MS and PhD from the University of Illinois. His research has emphasized environmental impact analysis, natural resource demand modelling, and many aspects of military installation land management, focusing recently on endangered species and climate change concerns.

Edward B. Barbier is John S. Bugas Professor of Economics, Department of Economics and Finance, University of Wyoming. His main expertise is natural resource and development economics as well as the interface between economics and ecology. He has served as a consultant and policy analyst for a variety of national, international and non-governmental agencies, including many UN organizations and The World Bank.

Amitrajeet A. Batabyal is Arthur J. Gosnell Professor of Economics at the Rochester Institute of Technology, New York. His research interests include natural resource and environmental economics, regional economics, international trade theory and issues at the interface of economics and political science.

Amos Bien is a population ecologist by academic training and administrator, and works in the field of the sustainability of tourism and standardization. He was the lead consultant in developing the 'Global Sustainable Tourism Criteria', author of the 'Tourism Sustainability Scorecard' for the Inter-American Development Bank (IDB).

Luke M. Brander is a freelance environmental economist based in Hong Kong. He is affiliated to the Institute for Environmental Studies (VU University Amsterdam) as Associate Researcher and to the Division of Environment, Hong Kong University of Science and Technology (HKUST) as Adjunct Assistant Professor. Luke's main research interests are in the design of economic instruments to control environmental problems and the valuation of natural resources and environmental impacts.

Adele Catzim-Sanchez is lead social scientist for the Belize ISIS Enterprises Ltd. She has worked on several research projects on the socioeconomic and governance effects of marine protected areas.

Hongyan Chen is Research Fellow at the School of Environmental Sciences, Liverpool University. Her research interests include economics of ecosystems and biodiversity, natural resource management, spatial statistical analysis and modelling, model integration and spatial decision-making support.

William W.L. Cheung has been Assistant Professor at the Fisheries Centre, University of British Columbia (UBC) since 2011 and is the Principal Investigator of the Changing Ocean Research Unit. His main research area is on assessing the biophysical and socioeconomic vulnerabilities and impacts of marine climate change and other human stressors, and identifying mitigation and adaptation options.

Joseph C. Cooper is chief of the Agricultural Policy and Models Branch and Deputy Director for Staff Analysis (Acting) for the Market and Trade Economics Division of the Economic Research Service of the US Department of Agriculture. Joe is an expert on farm policy and associated research, and has worked on topic areas ranging from commodity support to agri-environmental policy. Joe has also served as an economist at the Food and Agricultural Organization of the United Nations.

Jessica Coria is Associate Professor at the Environmental Economics Unit, Department of Economics, University of Gothenburg. Her work lies in the effects of the choice of market-based policy instruments on the rate of adoption of environmentally friendly technologies and compliance with environmental regulations. Her research deals with the optimal design of environmental policies that considers the multiple complexities of environmental problems, such as for instance, the temporal and spatial patterns of pollution as well as the interaction across pollutants and joint environmental damages.

Giuseppe Cucuzza is Associate Professor of Environmental Economics at the University of Catania (Italy). His research focuses on policy and economics of agricultural externalities, and biodiversity.

Arianne T. de Blaeij is a research scholar at the Research Unit on Regional Economy and Land Use of LEI Wageningen UR. She is a specialist in monetary valuation of and payment mechanisms for conservation of and improving ecosystem services, and in cost–benefit analysis.

Tom Dedeurwaerdere is Director of the Biodiversity Governance Unit of the Centre for the Philosophy of Law, Professor of Philosophy of Science at the Université Catholique de Louvain and Senior Research Associate at the National Research Foundation, Belgium (F.R.S.-FNRS).

Maria De Salvo is a Research Assistant at the University of Verona (Italy). Her research focuses on non-market valuation methods, and assessment of climate changes impact on agricultural and natural ecosystems.

Salvatore Di Falco is Professor of Environmental Economics at the University of Geneva. His research focuses on the intersection between environmental and development economics using econometric models. He has analysed the contribution of natural resources such as biodiversity on agricultural productivity, food security and weather risk in arid environments. His work also focuses on the role of institutional structures (common property forests) on natural resource conservation, rural development and poverty reduction. More recent work includes the impact of social capital and traditional sharing norms on consumption, saving accumulation, investment decisions and adaptation to climate change.

Sahan T.M. Dissanayake is Assistant Professor in Economics at Colby College, Maine, USA and an Assistant Professor (by courtesy) in Economics at Portland State University. His research uses tools from economics and mathematical programming to understand consumer preferences and solve resource allocation problems. His work focuses on theoretical reserve design models, choice experiment surveys of preferences for conservation, application of land use models to military installations, and spatially explicit renewable energy policy.

Anantha Kumar Duraiappah is Executive Director of the International Human Dimensions Programme on Global Environmental Change (UNU-IHDP). He is an experienced environmental development economist whose work largely focuses on the equity of access and use of ecosystem services. Professor Duraiappah helped to initiate the Intergovernmental Platform for Biodiversity and Ecosystem Services (IPBES) and has since then played a pivotal role in its recent approval. He additionally served on the scientific committee of DIVERSITAS, one of the Earth System Science Partnership (ESSP) partners. Professor Duraiappah continues to successfully incorporate his expertise in science–policy interaction, economics, development and ecosystem services into his work at the United Nations University and IHDP.

William H. Durham, is the co-founder and Co-Director of the Center for Responsible Tourism (CREST) at Stanford University. He is also the Bing Professor in Human

Biology in the Department of Anthropology, and the Yang and Yamazaki University Fellow. Bill has worked in Peru, Brazil, Ecuador, El Salvador, Honduras, Costa Rica and Panama, including a year with the Native American Kuna.

Riku Eskelinen is an Associate Researcher at the Department of Process and Environmental Engineering, and Water Resources and Environmental Engineering Laboratory, University of Oulu, Finland. He is mainly interested in peatland hydrology and peatland restoration. His Bachelor's degree in Civil Engineering was from Kotka University, Finland and his MSc in Environmental Engineering from Oulu.

Tamara Figueredo Martín is Researcher for the Center for Coastal Ecosystems, also Professor of Economy and Environment at the Center for Marine Research of the University of Havana, Cuba. Her research interests include natural resource and environmental economics, marine protected areas design and performance, integrated coastal zone management, status of fisheries, environmental issues of recreational fishing and SCUBA diving, socioeconomics issues and environment planning.

Patrick Fong is Senior Scientific Officer (Environment) at the Institute of Applied Science (IAS) of the University of the South Pacific, Fiji. Patrick's research focuses on resource management and poverty reduction; livelihood impacts of adaptive management; leadership and governance; management effectiveness; resilience and well-being.

Masahiko Gemma is Professor of Economics, Waseda University in Tokyo His research interests include economic development issues related to productivity and production efficiency problems for the transition economies in Europe and Asia. He is currently serving as director of the Institute of Japan-US Studies at Waseda University.

John M. Gowdy is Rittenhouse Professor of Humanities and Social Science, Department of Economics, Rensselaer Polytechnic Institute in Troy, New York. He is past President of the International Society for Ecological Economics. His current research interests include climate change, biodiversity valuation, behavioural economics, and evolutionary economics.

Martha Honey is co-founder and Co-Director of CREST, the Center for Responsible Travel and head of its Washington, DC office. She was Executive Director of The International Ecotourism Society (TIES) from 2003 to 2006.

Gary W. Johnson is a computer scientist and ecological modeller at the University of Vermont (UVM) who has worked for years in ecosystem services flow modelling. His research interests include neural and evolutionary computation, knowledge representation and reasoning under uncertainty, decision theory, statistical modelling, functional programming, and pattern classification.

Timo Karjalainen is Senior Research Fellow at Thule Institute, Adjunct Professor (Environmental Sociology, University of Oulu, Finland). His research interests are environmental impact assessment and social impact assessment, natural resource governance, analytic-deliberative methods, transdisciplinary and collaborative processes, and ecosystem services.

Marianne Kettunen is Senior Policy Analyst at the Institute for European Environmental Policy (IEEP), currently based as a guest researcher at the Finnish Environment Institute in Helsinki. She coordinated the assessment of socioeconomic benefits of protected areas for the book *The Economics of Ecosystems and Biodiversity (TEEB) in National and International Policy Making* in 2011 and is the lead editor of the Earthscan book *Social and Economic Benefits of Protected Areas* released in 2013.

Björn Klöve is Professor and Director of the Laboratory of Water Resources and Environmental Engineering at the University of Oulu (Finland). He has published extensively in the following areas: hydrology of boreal systems, tracer hydrology, peatland hydrology, peat hydraulic properties, sediment transport, groundwater hydrology, groundwater-dependent ecosystems, groundwater directives, environmental impacts of peat harvesting, water pollution control, treatment wetlands, diffuse loading, land use and water quality.

Eva Kougea is Associate Researcher with ReSEES (Research tEam on Socio-Economic and Environmental Sustainability) at Athens University of Economics and Business.

Phoebe Koundouri is Associate Professor at the Department of International and European Economic Studies, Athens University of Economics and Business and the Director of the Research tEam on Socio-Economic and Environmental Sustainability (ReSEES). She is also Senior Research Fellow at the London School of Economics and Political Science, Grantham Research Institute on Climate Change and the Environment. She is also the Vice-President of the European Association of Environmental and Resource Economics.

Pushpam Kumar is Chief of the Ecosystem Services Economics Unit, Division of Environment Programme Implementation, UNEP, where he works on mainstreaming of ecosystem services into development policy. He is also on the Faculty of School of Environmental Sciences, University of Liverpool, UK.

Vicky W.Y. Lam is a fisheries economist and a postdoc at the Fisheries Economics Research Unit (FERU) at the University of British Columbia. Vicky's research is focused on the effect of climate change on the economics of major commercial marine fisheries on the global scale.

Glenn-Marie Lange leads the environmental policy and economics work of The World Bank's Department of Agriculture and Environmental Services, which is responsible for The World Bank's work on wealth accounting and adjusted net savings, including the recent report, *The Changing Wealth of Nations* (World Bank, 2011). She also leads the Global Partnership for Wealth Accounting and the Valuation of Ecosystem Services (WAVES), which is working in more than ten countries to mainstream natural capital in national economic accounts and development planning.

Vincent Linderhof is a researcher at the Agricultural Economics Research Institute (LEI Wageningen UR) in The Hague. His research interests focus on the economic aspects of water allocation and the quality of water in low-income and high-income countries, including the valuation of water, water policy assessment and climate change impact assessment.

Anil Markandya is Director of the Basque Centre for Climate Change in the Basque Country, Spain and Honorary Professor of Economics at the University of Bath, UK.

He has acted as a consultant to a number of national and international organizations and served as Lead Economist at The World Bank.

Juan Maté is Coastal and Marine Science Advisor of the Smithsonian Tropical Research Institute (STRI) in Panama City.

Leonardo Mazza is a Senior Policy Analyst in IEEP's Environmental Economics Programme. He has led or contributed to work on mainstreaming biodiversity and ecosystem service values across relevant policy areas, the transition to the green economy, phasing out and reform of environmentally harmful subsidies (EHS), environmental tax and fiscal reform, policies for resource efficiency, indicators to complement GDP as a progress indicator and natural capital accounting.

Carlos Mena is Professor of Geography and Ecology at the School of Biological and Environmental Sciences in the Universidad San Francisco de Quito (USFQ), Ecuador. Dr Mena is Co-Director of the Galapagos Science Center and Director of the Unit for Socio-environmental Systems Research at USFQ.

Yohei Mitani is Assistant Professor in the Division of Natural Resource Economics, Graduate School of Agriculture at Kyoto University. He is also associated with the School of Economics and Business at the Norwegian University of Life Sciences (HH, NMBU).

Esther Naikal is a research analyst in the environmental policy and economics team of The World Bank's Agriculture and Environmental Services Department. For the past three years, she has been responsible for the database update and management of the wealth accounting work, which includes indicators published in the *World Development Indicators* such as adjusted net saving, adjusted net national income, and resource rents. She previously worked on The World Bank's Climate Change Knowledge and Learning initiative, leading the module on the Economics of Climate Change.

Daiju Narita is a researcher at the Kiel Institute for the World Economy, Germany. His research field is environmental economics, and his research interests include climate policy and uncertainty, climate change adaptation, and economic assessment of climate change impacts such as those associated with tropical cyclones and ocean acidification. Prior to his training in environmental economics, he studied chemistry and holds an MSc from the University of Tokyo.

Ståle Navrud is Professor of Environmental and Resource Economics at the School of Economics and Business at the Norwegian University of Life Sciences (HH, NMBU). He has been a Visiting Fulbright Research Fellow at the University of California (UC), Berkeley, and a Visiting Researcher at UC-San Diego and University of New Mexico, Albuquerque. His interests include cost–benefit analysis and economic valuation of terrestrial, aquatic and marine ecosystem services, biodiversity, landscape aesthetics, transportation noise, animal welfare, external effects of renewable energy, public health and cultural heritage, and spatial and temporal benefit transfer of these values.

Peter Nijkamp is Professor of Regional and Urban Economics at the VU University, Amsterdam. His main research interests include regional growth, quantitative policy evaluation, regional and urban modelling, transport systems analysis, environmental and resource management, and sustainable development. He has broad expertise in the areas of public policy, services planning, infrastructure management, and environmental protection.

Paulo A.L.D. Nunes is the Scientific Coordinator of the Policy and Technical Expert Committee, WAVES – Wealth Accounting and the Valuation of Ecosystem Services, World Bank and guest professor at the University of Padova, Italy. Dr Nunes has contributed technical expertise to a wide set of international initiatives, including 'The Economics of Ecosystems and Biodiversity – TEEB', 'Ocean Health Index – OHI', and the Global Partnership for the Oceans, World Bank. His research interests include environmental policy and sustainable development, marine economics and governance, and economic valuation of ecosystem services for national accounting.

Hayri Önal is Professor of Operations Research (2005–present) in the Department of Agricultural and Consumer Economics, University of Illinois.

Ruslana Rachel Palatnik is a Senior Lecturer of Economics at the academic college of Emek Yezreel, and a Deputy Director of the Natural Resources and Environmental Research Center, University of Haifa, Israel. She developed a computable general equilibrium model of the Israeli economy and has been conducting climate change adaptation and mitigation policy analyses. In addition, she carried out econometric researches on external costs and economic incentives in solid waste management.

Charles Palmer is Lecturer in the Department of Geography and Environment of the London School of Economics.

Sarah Parks is president and founder of New York-based Amala Consulting, an interdisciplinary consulting company specializing in geographic information systems (GIS) and ecological economics. Her consulting work has focused on the development and analysis of ecological, social and economic data. Notable projects have included the mapping and analysis of water quality monitoring and hydraulic fracturing sites in New York and Pennsylvania, and the creation of ecological features maps and the valuation of ecosystem services on the New York Rensselaer Plateau.

Marta Pascual is a Postdoctoral Researcher at the Basque Centre for Climate Change (BC3). Her current research interests are focused on ecosystem services mapping, modelling and evaluation; as well as in marine spatial planning management, marine socio-economics, marine governance and the implementation of Marine Strategy Framework Directives (MSFDs). Her work has also been published in international peer-reviewed journals.

Marta Pérez-Soba is Senior Researcher in the team Earth Informatics at Alterra Wageningen University and Research Centre (the Netherlands). Her current research is focused on the interaction of land science with society and policy, involving the development of tools for integrated impact assessment of land use change, mapping of ecosystem services, and scenarios for future regional developments in Europe.

Fabián Pina-Amargós is Senior Researcher at the Center for Coastal Ecosystems Research and a Professor of Protected Areas of the Center for Marine Research of the University of Havana, Cuba. His research interests include fish community structure, marine protected areas design and performance, movement and spillover of fish, status of tropical ecosystems, integrated coastal zone management, hurricane impacts on tropical ecosystems, status of fisheries, environmental issues of recreational fishing and SCUBA diving, fish biodiversity and environmental economics.

Nico B.P. Polman is Senior Researcher responsible for investigations regarding (institutional) economics, economic governance, and linking ecological and economic system perspectives, including agent-based modelling. His research interests focus on the economic aspects of agriculture and freshwater supply, regional development and rural policy.

Lawrence Pratt is Director of the Latin American Center for Competitiveness and Sustainable Development (CLACDS) at the INCAE Business School (Alajuela, Costa Rica).

Manuel Pulido-Velazquez is Associate Professor of Hydrology and Water Resources Engineering at the Technical University of Valencia (UPVLC), Spain (since 2001), and a senior researcher within the Research Institute of Water Engineering and Environment.

Maarten J. Punt is currently working at the Technische Universität München as a Postdoctoral Researcher.

Diego Quiroga is Professor of Anthropology and Environmental Sciences at the Universidad San Francisco de Quito, Ecuador. His research interests focus on the human environment interaction and he is doing research both in the Amazon Forest and in the Galapagos, analysing the impact of tourism, conservation and extractive economies on the local population.

Katrin Rehdanz is Associate Professor of Environmental and Resource Economics at the University of Kiel associated with the Kiel Institute for the World Economy. She has a strong background in environmental valuation and environment-economy modelling.

Stijn Reinhard of the Dutch Agricultural Economics Research Institute (LEI part of Wageningen UR) is currently head of both the Department of Regional Economy and Land Use and the Department of Natural Resources.

Kalle Reinikainen is Associate Researcher at Pöyry Finland Oy, Tutkijantie, Oulu. He was Assistant Professor at the University of Oulu from 1995 to 2007.

Elizabeth Robinson is Reader in Environmental Economics at the University of Reading's School of Agriculture, Policy and Development, UK and a member of the UK Department for Environment, Food and Rural Affairs Economic Advisory Panel. She was a lead author for the Millennium Ecosystem Assessment, and coordinating lead author for the International Assessment of Agricultural Knowledge, Science and Technology for Development (IAASTD) Sub-Saharan Africa report, member of the global and Sub--Saharan Africa design teams, and co-author of the global synthesis report.

Pekka M. Rossi holds an MSc in Environmental Engineering and is a doctoral candidate at the University of Oulu, Finland. His research interests are groundwater management, groundwater flow modelling, groundwater–surface water interactions and natural tracers in hydrological studies.

Giselle Samonte is an economist for ERT (Earth Resources Technology, Inc.); she leads the Office of Habitat Conservation's efforts at the United States National Oceanic and Atmospheric Administration (NOAA)–Fisheries, in demonstrating the societal benefits of habitat conservation as related to marine and coastal ecosystem services.

Andrew Seidl is a Professor and Public Policy Specialist in the Department of Agricultural and Resource Economics at Colorado State University. Andy often works in communities and countries that feature unique or valuable natural wealth to identify strategies for local people to use economic tools to improve management decisions, capture economic benefits and encourage environmental stewardship.

Darius Semmens is with the US Geological Survey in 2008 as a research physical scientist. His current research interests include the development of new methods and tools for the assessment and valuation of ecosystem goods and services – the specific benefits that we derive from nature. More specifically, his work involves developing methods and tools that can account for the spatial and temporal dynamics of service production and incorporate that and other information into more rigorous analyses of the trade-offs associated with landscape management.

Mordechai Shechter currently serves as the Founding Dean of the School of Sustainability at the Interdisciplinary Center Herzlyia (IDC) in Israel. He is Professor Emeritus of the Department of Economics and of the Department of Natural Resource and Environmental Management at the University of Haifa, Israel.

Benyamin Shitovitz is a Professor of Economics at the University of Haifa since 1980. His main research areas and interest are mathematical economics, game theory and the theory of social situations, oligopoly theory, public economics and information economics.

Giovanni Signorello is a Professor of Environmental Valuation at the University of Catania (Italy). He leads the Master's programme on Conservation of Land, Environment and Landscape, the University Center for the Conservation and Management of Environmental Resources and Agroecosystems (CUTGANA), the Environmental Valuation Laboratory, the international programmes of the Department of Agri-food and Environmental Systems Management, the Belpasso International Summer School on Environmental and Resource Economics.

R. David Simpson, US Environmental Protection Agency, USA, is Director of Ecosystem Economic Studies in the United States Environmental Protection Agency's National Center for Environmental Economics. Simpson was a coordinating lead author for the Responses volume of the Millennium Ecosystem Assessment, a review editor for The Economics of Ecosystems and Biodiversity (TEEB) initiative, and now serves on the Policy and Technical Experts Committee to The World Bank's Wealth Accounting and the Valuation of Ecosystem Services project. He has written extensively on the economics of ecosystems, biological diversity and conservation policy, as well as industrial policy, technological change and green accounting.

Geraldine Slean was the Coordinator for CESD (Center on Ecotourism and Sustainable Development) at Stanford University. She has since switched careers and is currently finishing up medical school at the University of California, San Francisco.

Henrik G. Smith is Professor in Animal Ecology at the Department of Biology, Lund University. He is also the director of Lund University Centre for Environmental and Climate Research that aims at gathering, strengthening and visualizing education and research within the area of environment and climate. He coordinates several research initiatives, including the strategic research area Biodiversity and Ecosystem Services in a Changing Climate, which engage researchers in multiple disciplines at Lund and Gothenburg Universities.

Rodney B.W. Smith is Associate Professor in the Department of Applied Economics at the University of Minnesota. His current research focus is on macroeconomic growth and economic development, applied production economics, natural resource economics and the economics of climate change. He has international research experience in Brazil, China, India, Japan and South Africa, and has lectured, presented seminars and participated in workshops in Africa, and throughout Asia, Europe and the USA.

Thomas Sterner is Visiting Chief Economist of the Environmental Defense Fund in New York and Professor of Environmental Economics at the University of Gothenburg, Sweden. He has built up the Unit for Environmental Economics, a prominent European centre that provides a unique PhD programme in climate economics for students from developing countries.

Mavra Stithou is Associate Researcher with ReSEES (Research tEam on Socio--Economic and Environmental Sustainability) at the Athens University of Economics and Business . . . She specializes in water and marine policy, integrated socioeconomic assessment of marine planning, the application of environmental valuation techniques (non-market valuation methods), cost–benefit analysis, and the use of benefits transfer.

U. Rashid Sumaila is Professor and Director of the Fisheries Economics Research Unit at UBC Fisheries Centre. He specializes in bioeconomics, marine ecosystem valuation and the analysis of global issues such as fisheries subsidies, IUU (illegal, unreported and unregulated) fishing and the economics of high and deep seas fisheries. Sumaila has experience working in fisheries and natural resource projects in Norway, Canada and the North Atlantic region, Namibia and the Southern African region, Ghana and the West African region and Hong Kong and the South China Sea.

Daniel Suman is Professor in Marine Policy at the Rosenstiel School of Marine and Atmospheric Science of the University of Miami. His research and project areas focus on coastal management, governance of marine resources and space, and marine protected areas – particularly in Latin America.

Rady T. Tawfik is an Assistant Professor in King Faisal University in Saudi Arabia. He also works for the Egyptian Environmental Affairs Agency. He has been involved in the European Union project for the development of South Sinai protectorates and quickly moved up to manage the Nature Conservation Training Centre.

Patrick ten Brink is Senior Fellow at IEEP and Head of the Brussels Office. He also leads IEEP's environmental economics programme, which focuses on the values of nature, green economy, environmental tax and fiscal reform including environmental harmful subsidies, beyond GDP indicators and environmental accounting. Patrick led The

Economics of Ecosystems and Biodiversity (TEEB) in National and International Policy Making work and is editor of the associated Earthscan book released in 2011.

Richard S.J. Tol is Professor at the Department of Economics, University of Sussex, UK and the Professor of the Economics of Climate Change, Institute for Environmental Studies and Department of Spatial Economics, Vrije Universiteit, Amsterdam.

R. Kerry Turner is Professorial Research Fellow in the School of Environmental Sciences at the University of East Anglia (UEA), UK and was the Director of the Centre for Social and Economic Research on the Global Environment from 1996 to 2010. Kerry specializes in environmental economics, coastal zone and wetland management, conservation economics and waste management research. He is currently a contributor to Defra's National Ecosystem Assessment Initiative and several POST reports. In 2000, Kerry Turner was awarded a CBE for his services to sustainable development.

C. Martijn van der Heide has been working as a research scholar at LEI Wageningen UR, The Hague, where his main areas of research are environmental cost–benefit analysis, economics (and valuation) of biodiversity conservation, ecological economics and landscape economics.

Ekko C. van Ierland Currently he is Full Professor of Environmental Economics and Natural Resources at Wageningen University and Head of the Environmental Economics and Natural Resources Group.

Peter Verweij is an environmental informatics engineer at Alterra, Wageningen-UR. His current research is on discussion support systems in participatory settings and ontology--based clearinghouses. He has been lead developer in several international service contracts and large European research projects that include the design and development of environmental modelling, integrated sustainability assessment tools, ecosystem service assessments and climate change adaptation tools. Peter is a member of the International Environmental Modelling and Software Society (iEMSs).

Ferdinando Villa is leading, the ARIES project, aims to redefine the integrated assessment of ecosystem services and to provide new approaches to address the challenge of environmental decision-making in the twenty-first century.

Sissel Waage, PhD, is Director of Biodiversity and Ecosystem Services for BSR.

Xuanwen Wang is a Research Associate at CPWR – The Center for Construction Research and Training at Silver Spring, Maryland, USA. Her expertise is in environmental and natural resource economics, health economics and labour economics.

Hans-Peter Weikard has been Associate Professor at the Environmental Economics and Natural Resources Group of Wageningen University, the Netherlands, His interests are in natural resource economics, climate change economics, risk management and game theory applied to environmental problems. Jointly with Ekko van Ierland he has been coordinating the Stability of Coalitions project (STACO) examining drivers of stability of climate coalitions. In recent years this work has been extended to other types of international environmental agreements.

James D. Westervelt is a Research Scientist with the Engineer Research and Development Center, part of the United States Corps of Engineers. His research has focused on the sustainable management of lands supporting military training and testing, with special interests in endangered species simulation modelling, regional urban growth, climate change impacts on natural systems and modelling of social systems.

Manuel Winograd is consultant and advisor for Alterra (WUR, Wageningen, the Netherlands) and ETC-LUSI (Spain) in decision support tools. His applied research examines the assessment methods for vulnerability, adaptation, ecosystem services and sustainability evaluation, the use of scenarios and models for participatory assessment and indicator development as well as strengthening institutional capacities for decision support tools and modelling.

Sirini Withana is Senior Policy Analyst in IEEP's Environment and Climate Governance programme, specializing in environmental governance and environmental fiscal reform. She has led or contributed to work on reforming environmentally harmful subsidies (EHS), environmental tax reform, financial instruments and greening the EU budget.

Shiri Zemah-Shamir is a research fellow at NRERC (Natural Resource and Environmental Research Center), University of Haifa, where her main areas of research are economics (and valuation) of biodiversity conservation and ecosystem services and ecological economics.

Introduction

Paulo A.L.D. Nunes, Pushpam Kumar and Tom Dedeurwaerdere

SETTING THE SCENE

Economic analysis of causes and impact of biophysical changes like climate change, ecosystem services and biodiversity loss has commanded special attention in academic research as well as policy discourse. While the economic dimension of climate change has entered into a stage of being embraced and accepted by nearly everyone, it would be presumptuous to say this for ecosystem and biodiversity changes. How society is affected as a result of changes in the status and condition of ecosystems like forest and wetlands is clearly understood, at least to a great extent; the question remains how to capture those changes? One of the possible approaches typically suggested by the economics profession (admitting that economics is not a monolithic discipline) is to value the changes based on consumers' and producers' surplus-based concepts. However, the outcomes obtained through these approaches also create confusion and propel the need for credibility. In fact it may not be an exaggeration to state here that the bulk of the literature in economics of ecosystems and biodiversity is devoted to methodological nuances.

Economic valuation of ecosystem services helps to identify and resolve the trade-offs between different stakeholders engaged in managing ecosystems. Ecosystem management plans often result in net gains for some sections of society and net losses for others. For example, forest conservation might increase carbon sequestration (a global benefit), but as a result, local populations might be deprived of access to forest and services like timber and non-timber forest products. Valuation of ecosystem services is a tool that can help to ensure that the decision-making process incorporates considerations of equity and sustainability.

Economic valuation of ecosystem services also helps to link the conservation strategy with mainstream policies at national and regional levels. For any ecosystem service, its social value must be equated with the discounted net present value of the flow of that service (Hanley and Barbier, 2009). In theory, decision-makers can then see how the marginal benefits, for example the value of urban or coastal wetland conservation, equates with the marginal costs of conservation. Estimating the economic value of services like timber and fish, known as provisioning services, is relatively easy because they enter the domain of the market. However, this is not the case for regulating or indirect services, which can be defined as the benefits people obtain from the regulation of ecosystem processes, including, for example, the regulation of climate, water and some human diseases (Heal et al., 2005; Kumar and Wood, 2010). Typically, these regulating services are ignored as they are outside the conventional market. As a result, the marginal cost of conservation exceeds the marginal benefit, which sends the wrong signal to policy-makers.

One area of confusion in the valuation of regulating services has been the decision on what should be valued. Biogeochemical processes and subsequent functions of an ecosystem create services, but not all of these services are appropriated by society. Only those benefits that people obtain from ecosystems should be considered as services (MA, 2003, 2005). Thus, valuation should target final rather than intermediate services (Fisher and Turner, 2008). Most of the regulating services are public goods and intermediate in character, but some of the services, like groundwater flow maintained by forests, could be used by lowland people for drinking (consumption) or industrial use (production). Economic valuation of ecosystem services is instrumental, anthropocentric, individual-based, subjective, context-dependent, marginal and state-dependent (Goulder and Kennedy, 1997; Nunes and van den Bergh, 2001; Freeman, 2003; Baumgartner et al., 2006; Barbier et al., 2009; Dasgupta, 2009, 2013). The value of ecosystem services is essentially a marginal concept arising out of scarcity and depends on the ecosystem condition and the social-cultural context in which people make choices (Heal, 2000; Barbier and Heal, 2006; Kumar, 2012). Thus, those who undertake ecosystem valuations should focus on ecosystems that are socially important, evaluate ecological responses in economic value-relevant terms, and consider the possible use of a broad range of valuation methodologies to estimate values (EPA, 2009).

While this discourse on economic valuation of ecosystem services and biodiversity still continues, there are other equally important issues like estimating natural wealth and income, trade affecting natural capital and ecosystems, role of governance and property rights in better management of ecosystems and biodiversity, and an entire range of issues like distribution and ownership of ecosystems and biodiversity, pushing the discussion into the political economy arena. They are equally critical in understanding the complexity of economics of ecosystems and biodiversity.

In the last ten years or so we have observed an interesting and probably encouraging phenomenon of the emergence of a need for a global consensus on the issue of economics of ecosystems and biodiversity. The global community and policy-makers have clearly expressed and articulated their demand for synthesis on valuation, accounting and mainstreaming of ecosystem services into conventional decision-making frameworks of fiscal and monetary policies, trade and investment measures and use of economic instruments in response to policies for biodiversity, ecosystems and climate change.

The United Nations supported probably the largest ever scientific assessment – the Millennium Ecosystem Assessment (MA, 2005) – which aimed to assess the capacity of ecosystems to support human well-being, and showed that over the last half-century humans have changed ecosystems more rapidly and extensively, which has resulted in a substantial and largely irreversible loss in the diversity of life on earth. These problems of changes, unless addressed, will considerably reduce the benefits of ecosystems and increase the risks of non-linear changes, and the aggravation of poverty. The MA's prominence on ecosystem services and their role in human well-being is widely acknowledged and is considered one of the key influential reports on clarifying the linkage between biodiversity conservation and poverty alleviation. The MA also recommended a greater use of economic principles while developing intervention strategies for management.

Although the MA created public awareness about how ecosystem services influence the constituents and determinants of human well-being, on economic aspects of ecosystems it was not as vocal as The Economics of Ecosystems and Biodiversity (TEEB) report, which came after the Stern Review (2006) on climate change. TEEB (2010) was a global initiative aiming to increase awareness on the economic paybacks of ecosystem services and biodiversity. In all, five volumes of compiled evidence and practical case examples have highlighted that value of the natural capital is systematically underestimated and thus goes unnoticed in economic decisions. The TEEB guided policy-makers on how to incorporate use and conservation of natural resources and the services they provide into decision-making. The TEEB report also provided economic evidence on why, how and where ecosystems and biodiversity are essential for human well-being and play an important role for economic development. The report substantiates that valuing ecosystem services makes economic sense and argues that economic growth could be increasingly compromised by the continued reduction of natural capital. It was very logical that after the MA and TEEB, multilateral environmental agreements like the Strategic Plan for Biodiversity 2011–2020 with its Aichi Biodiversity Targets, adopted by the Convention on Biological Diversity (CBD, 2010), aim to address the underlying causes of biodiversity loss by mainstreaming biodiversity across government and society; to reduce the direct pressures on biodiversity and promote sustainable use; to improve the status of biodiversity by safeguarding ecosystems, species and genetic diversity; to enhance the benefits to all from biodiversity and ecosystem services; and finally to enhance implementation through participatory planning, knowledge management and capacity building. One of the Aichi Targets is to substantially increase financial resources from current levels. In support of the implementation of the Strategic Plan for Biodiversity 2011–2020, adequate, predictable and timely new and additional financial resources have to be provided.

After the completion of the MA in 2005, a plethora of national assessments and sub national assessments followed the economic assessments of ecosystems and biodiversity with greater vigour and confidence as reliable estimates and studies clearly showed the relationship between changing ecosystems and their impacts on people, especially the poor. The outcomes and recommendations following from the MA paved the way for the creation of the Intergovernmental Platform on Biodiversity and Ecosystem Services (IPBES), which aims to assess the condition of the world's ecosystems, synthesize knowledge on the issue to be used by policy-makers, and develop capacity to assess and use scientific information for better decisions leading to enhanced human well-being. The UK's Ecosystem Services for Poverty Alleviation (ESPA) programme is also producing impactful research to manage ecosystems sustainably to reduce poverty. The evidence and tools for decision-makers are aimed to improve understanding of how ecosystems operate, the services humans acquire, and their connection with the economy and sustainable development.

Ecosystem and poverty dynamics also became the centre of attention after UNEP--UNDP Poverty-Environment Initiative (UNPEI) was launched in 2007. In brief, PEI supports developing countries' policy-makers and various stakeholders in managing the natural capital by focusing on improving livelihoods. The programme has provided financial and technical support to nine countries across Africa – Botswana, Burkina Faso, Kenya, Malawi, Mali, Mauritania, Mozambique, Rwanda and Tanzania; five countries in Asia – Bangladesh, Bhutan, Lao PDR, Nepal and Thailand; and four countries in Europe and Latin America – Kyrgyzstan, Tajikistan, Uruguay and Dominican Republic – to build capacity for mainstreaming poverty–environment linkages into national development planning processes.

Similarly, the United Nations Environment Programme (UNEP) Global Environment Facility-funded Project for Ecosystem Services (ProEcoServ) is being implemented in five countries – Chile, South Africa, Lesotho, Trinidad and Tobago, and Vietnam. The overall goal of the project is to better integrate ecosystem assessment, scenario development and economic valuation of ecosystem services into national sustainable development planning. Country teams have developed very strong relationships with the relevant governmental organizations in implementing the project; and the teams have significantly advanced in the area of providing support and inputs to the targeted policy development mechanisms in pilot sites.

The Ecosystems Services for Poverty Alleviation (ESPA) primarily focused on research excellence and capacity development in the area of ecosystem management and poverty alleviation in the UK. The national ecosystem assessment furnishes a complete synopsis of the state of the natural capital. The assessment successfully reveals that natural wealth has been under-valued. Five hundred experts from the natural sciences, economics and the social sciences have contributed to this first ever national-scale assessment. The team collected, analysed and synthesized a significant body of peer-reviewed empirical evidence and information about the UK's environment. Within this process new tools for valuing the worth of the natural capital were developed. Valuing the ecosystem service properly may lead to better-informed decisions on investment, job creation and enhancement of human well-being.

UNEP's Green Economy Initiative defines the green economy as an economy whose growth in income and employment is driven by the investments that also reduce carbon emissions and pollution, improve resource efficiency, and ensure a continuous flow of ecosystem services by preventing the loss of biodiversity and ecosystem services. This path of creating economic prosperity depends on maintaining, restoring and enhancing natural capital. In most economic decisions, natural capital should be considered a critical economic asset and source of public benefits. First, the ecosystem is the main provider of various raw materials that serve as an engine for economic development. Second, this would also link poverty alleviation policies with economic development policies since livelihoods and security of poor people depend strongly on nature.

The degradation of ecosystems and loss of biodiversity have led some communities to experience a decline in their well-being. If the degradation continues, ecosystems may lose ability to regulate the climate and this could lead to further, unanticipated, and potentially permanent changes in the system and reductions in life-supporting ecosystem services. Therefore, the protection and restoration of ecosystems is at the core of the 'green economy'. Governments can play a key role in influencing and catalysing the green investment by targeted public expenditure policy reforms. For instance, investment in green forestry can preserve the economic livelihoods of more than 1 billion people who live from timber, paper and fibre products.

It is evident that the national-level economic assessment of ecosystems has gained momentum after the MA and has also led to serious questions regarding use of GDP as a measure of progress on the one hand, and framing approaches and methodologies involving ecosystem accounts on the other. The UNEP-led Inclusive Wealth Index emphasizes the need to estimate wealth of all types including natural capital, in order to say something on the sustainability of economy and society. The United Nations Statistics Division (UNSD) through its System of Environmental-Economic Accounting (SEEA) and Experimental Ecosystem Accounting (EEA) takes the System of National Accounts (SNA) to its logical culmination where Statistics recognizes the need for indicators to better capture the global and national sustainability discourse. Countries willing to implement natural capital accounting (NCA) are also backed up by The World Bank's Wealth Accounting and the Valuation of Ecosystem Services (WAVES) programme. This partnership comprises several UN agencies, governments, NGOs and scholars. The programme is helping Botswana, Colombia, Costa Rica, Madagascar and the Philippines to establish natural capital accounts and integrate them within the national accounts. The UNEP led new initiative Valuation and Accounting of Natural Capital for Green Economy known as VANTAGE lays down the strategy to embrace and integrate contribution of natural capital especially in poorer parts of the world to attain the elements of Green Economy. This also includes the UNEP-IHDP led Inclusive Wealth Index (IWI) which advocates the need to account wealth-natural, produced and human to say something meaning on the direction and sustainability of Economy.

These initiatives fill the gap in environment–economy literature but create a void too where new paradigms of analysis and approaches seek identity and credibility. Indeed, the growing interest of the researchers and practitioners of ecosystem conservation and use over the last two decades has led to the proliferation of new modelling approaches and management tools. As a result, there is an increasing need for clarifying the scope and the limits of the various models, for improving our understanding of their use in decision-making, and for building tools for model choice in interaction with the key stakeholders. The chapters in this volume address these challenges by presenting the latest developments in the economic analysis of ecosystem services and by identifying the key research needs that have to be addressed in order to move the field forward. Nearly all the chapters in the volume present original in-depth case study research to substantiate their argument. Therefore, in addition, the book contains a wealth of case study material on the assessment of ecosystem services and biodiversity conservation of unique breadth and policy relevance.

STRUCTURE OF THIS BOOK

Innovative approaches have been developed by economists to analyse various types of services and natural ecosystems. Part I of the handbook 'Setting the scene: the need for ecosystem service valuation' lays the basis for the further analysis in the book by a series of chapters on the usefulness and the limits of innovative approaches for improving and motivating policy action. Chapter 1 presents the general framework of natural capital accounting as part of a more comprehensive approach for measuring long-term economic well-being. The analysis of natural wealth accounts over the decade from 1995 to 2008 shows how economic development can be understood as a process towards building wealth. In this process, the composition of wealth has changed over the last decade: shifting away from natural capital and toward produced capital and, increasingly, intangible capital. This changing composition has in return an impact on long-term human welfare. The accounting exercises provide the basic framework to better assess such impacts and the many trade-offs between the various forms of capital – for example, decay in natural capital can have an impact through the moral and aesthetic satisfactions afforded by preserving wild areas and the biodiversity they shelter. As highlighted by the authors, the main challenges in these accounting exercises is data availability and the development of more fine-grained indicators that can range from the building of single composite indexes to complex multi-criteria indicators that do not presuppose substitutability among the different forms of capital.

The subsequent chapters in Part I address precisely this question of data contribution and modelling of various forms of capital by discussing specific case studies of biodiversity conservation and ecosystem management. Chapter 2 addresses the question of the protective value of estuarine and coastal ecosystems, mainly through their ability to attenuate waves or buffer winds in the case of storm and coastal floods. Chapter 3 extends this analysis of ecosystem services by addressing the long-term recreational value of ecosystems, through an analysis of the ecological and social footprint of cruise tourism in Belize. Chapters 4 to 6 add to the analysis by integrating the indirect climate-related human impacts on ecosystem services and biodiversity conservation into the model. These chapters respectively address the issue of marine fisheries, coral reefs and other marine ecosystems and show that when climate change has an adverse impact on ecosystem services, economic losses related to ecosystems decay can dramatically increase. All in all, the chapters in Part I show the need to integrate contributions by researchers from multiple disciplines, including economists, natural scientists and sociologists, to address the complex linkages between natural and socioeconomic processes in the flow of the various ecosystem-related services.

One of the core problems in the economic valuation of ecosystem services is to account for the evolution of the actor's preferences regarding the value of the services and the variation of these preferences according to social and cultural factors that lead to different outcomes at various spatial scales of analysis. Part II on 'Emerging economic valuation methods' presents recent work on valuation addressing this issue, by broadening the conventional toolbox and integrating deliberative, spatially heterogeneous and macro-level assessment techniques.

Chapter 7 presents the general limits of the conventional Walrasian welfare model for understanding human motivations for conservation and management choices regarding ecosystem services. The cornerstone of the Walrasian welfare model is an approach of human behaviour that characterizes consumer preferences as stable, consistent, insatiable and independent of the preferences of others. Recent work in economics has exposed flaws in this conventional model, in particular by showing that many preferences are 'other' regarding, as well as 'nature' regarding, and that these preferences vary significantly according to cultural conditioning, relative positioning and other reference points. Therefore, understanding the social process of preference formation and the integration of welfare modelling in broader spatial macroeconomic frameworks has become critical to formulating sound economic policies. Deliberative methods can contribute to this goal, along with the mapping of trade-offs and values at a macro level of analysis. This broadened toolbox should facilitate clearer communication to decision-makers and the public and thereby could contribute to the formation of more collective preferences on biodiversity conservation and management of ecosystem services.

Chapters 8 and 9 show that it is possible to use a general equilibrium framework to assess micro-level welfare impacts of macro-level evolutions such as climate change and agricultural production. Results highlight some well-known aspects of these macro–micro interactions such as the contrast between regions that would benefit from climate-change-induced impacts, such as Mediterranean Europe, and regions that would suffer, such as many developing countries. Chapters 10 to 12 contribute further to the toolkit of methods that can ensure better knowledge generation for concerned stakeholders (including public officials, scientific experts and decision-makers). More specifically, Chapter 10 analyses the use of an important software tool, Artificial Intelligence for Ecosystem Services (ARIES) by the US Bureau of Land Management. This software tool integrates a set of agent-based algorithms that allow accounting for the spatial dynamics of ecosystem service flows. Chapters 11 and 12 present respectively a selection algorithm for land allocation for the conservation of keystone species (Chapter 11) and a decision support tool, QUICKS for exploring various scenarios of land cover/use, which is applied to the case of the EU agricultural and environmental policy (Chapter 12). The common point highlighted by the authors of these various chapters is the importance of a proper design for the process of model choice and use. In particular, proper attention should be given to ensuring transparency in tool development, capacity strengthening to use a diversified toolbox and communicating results and improving exchange between developers and users (e.g., technicians, decision-makers, etc.).

The extent to which the choices of protecting biodiversity versus promoting biodiversity-related ecosystem services are likely to coincide depends on complex and yet little understood interactions between biodiversity and ecosystem services. However, biodiversity and ecosystem services are in serious jeopardy and the best hope to protect them is to create and align diverse incentives for conservation wherever possible and to integrate these into the larger policy-maker arena. Part III on 'Ecosystem services and conservation policy' applies the insights on natural capital accounting, modelling and economic valuation developed in Parts I and II to this complex nexus between protection of biodiversity and management of biodiversity-related ecosystem services.

Chapter 13 sets the stage of this analysis by presenting a simple model of land allocation that integrates the economic opportunity costs of biodiversity conservation and provision of ecosystem services on the same land, the effectiveness of conservation and management efforts and population density. Some themes that emerge from the stylized model is that greater reliance on ecosystem services is not always attractive to a country – as alternative land uses might be preferred – and that the potential for synergies between biodiversity conservation and ecosystem services will increase with the number of ecosystem services that are considered (see also Chapter 19). These points underscore the importance of relying on a wide diversity of mechanisms to address conservation, some of which can be based on marketable ecosystem services, while others might address non-marketable services such as moral and aesthetic satisfaction through subsidy schemes such as payments for ecosystem services.

Chapters 14 and 15 analyse the impact on conservation decisions of the availability of marketable products from ecosystems, such as the valuable research derived from biodiversity prospecting or the presence of fisheries adjacent to marine protected areas. The analysis in these chapters shows the temptation of actors to only consider the private benefit from the use of the marketable products, while such strategic or short-term self-interested behaviour does not always lead to the most socially desirable outcome. The further development of such studies at the intersection of conservation policy and management of ecosystem services, however, depends on the availability of richer data and more sophisticated data aggregation tools, as also stressed in Part I. Chapters 16 and 17 contrast the data needs and the role of uncertainty in the modelling in the case of valuation of ecosystem services and the case of the assessment of optimal biodiversity preservation policies respectively. Particularly in the first case, lack of high-quality data is a major barrier. As shown in Chapter 16, the building of a data portal for accessing all the existing valuation studies and a more systematic use of spatial data in these studies could contribute to alleviating this problem. Chapters 18 and 19 finally broaden the analysis of the synergies between ecosystem services and biodiversity conservation by reviewing the literature on biodiversity, poverty and development on the one hand (Chapter 18) and the ecological-oriented literature on irreversibility and scale-dependency of certain services on the other (Chapter 19). In order to take into account this broader social context, both chapters advise looking at payments for ecosystem services (PES) as an important tool to provide the necessary incentives for local conservation activities that yield wider social benefits.

When looking at the types of valuation methods used in those studies that link valuation to policy and management approaches, there is a clear bias in favour of market-based methods in most of them. As also highlighted in Part III, this is unfortunate as it means that some important services are excluded, such as existence values or aesthetic values, leading to the possibility of placing zero value on such services. Part IV on 'Shedding light on non-market values of ecosystem services' presents a wealth of original case study research showing the importance of the non-market values within the ecosystem services framework.

Chapters 20 to 22 use choice experiments and contingent valuation techniques to value the non-use values in wetlands and coastal areas. The findings of these studies indicate the preference of users for more environmentally sound management scenarios, whether it is in the case of visits to coral reefs (Chapter 20) or restoration of wetlands (Chapters 21 and 22). However, even though these studies provide unambiguous evidence of the importance of the non-use values, further research is needed to better assess the social costs of investment in the management and restoration of these areas. Chapters 23 and 24 aim at better integrating this cost dimension by using a total economic value method for integrating the market price in the evaluation of the benefits, cost evaluation methods and data on non-use values gathered by choice experiments and contingent valuation. The resulting analysis of the value of unextracted groundwater in an aquifer in Greece (Chapter 23) shows that the incorporation of the groundwater's indirect ecosystem value and the non-use value increase the overall value of the groundwater-related ecosystem services in the assessment exercise. In a similar way, the non-use values estimated in the study of Jardines de la Reina National Park in Cuba (Chapter 24) are higher than the use values in the two management scenarios that are analysed in the case study. Chapter 25, finally, shows the interesting result that, although respondents to a contingent valuation survey show a clear interest in and demand for man-made wetlands, they indicate a zero willingness to pay. Indeed, the respondents would prefer to allocate public payments for the construction of man-made wetlands, exploited on a commercial basis. The latter result provides an interesting alternative to the more conventional choice for a public management scenario, such as envisioned in the study of the management of groundwater in a groundwater-dependent ecosystem in Northern Finland (Chapter 26).

While the debate on the economic tools and the synergies between biodiversity conservation and ecosystem service management is clearly at the heart of the major innovations in the field over the last decade, there are other equally important issues that determine conservation policies, which were already mentioned above, like the role of governance and property rights, and the entire range of issues of distribution of ownership of ecosystems and biodiversity. Part V on 'The role of governance and science–policy–business interface in bringing visible ecosystem values' aims to provide an outlook on this broader set of issues by following the same combination of original case study material and literature review used in the other parts of the handbook.

Chapters 27 and 28 show the importance of broad stakeholder participation both in ecosystem management and in producing the science that supports such management. Chapter 27 analyses extensive survey data on governance of marine management areas based on different governance arrangements, such as national governance, co-management and community governance. In all these different forms increased participation leads to increased sustainable use and information gathering on use of resources in marine protected areas. However, as shown in the chapter, for effective participation to occur, one must critically examine the transparency of the processes involved in the development of the management plans and search for mechanisms to more effectively include the social actors that are really interested and affected by the creation of the marine managed areas. A similar caveat applies to the analysis of participation by stakeholders and policy-makers in the Intergovernmental Platform on Biodiversity and Ecosystem Services (IPBES), which is analysed in Chapter 28. Indeed, as stated in the chapter, the success of IPBES will to a large extent be dependent on the governance structures put in place, especially the participation of government representatives with a strong science background.

Chapter 29 provides an extensive review of the emerging discourse on the transition to the green economy, which allows better understanding of the various distributional issues that have to be considered in moving towards effective ecosystem management. Indeed, while the transition to the green economy will lead to many win-wins, it may also mean losses for certain groups and trade-offs across sectors and countries over time. These impacts need to be duly accounted for in transition plans. Some of the necessary steps that are essential components of such an effort are analysed in more depth in this chapter. First, discussions on the green economy need to build on a sound understanding of the value of nature and the role it plays in crucial issues such as poverty alleviation, the promotion of economic growth and the realization of multiple policy objectives simultaneously. Second, addressing these multiple values will require a combination of market and non-market tools, the most important of which are payment for ecosystem services, regulation and spatial planning.

Finally, Chapter 30 closes with an industry perspective on the use of some of the models that have been presented in the various chapters of the handbook. Indeed, if the interest in the economics of ecosystem services continues to grow, then it is likely that companies will need to employ new ecosystem-services-related risk and impact assessment protocols to identify potential effects of new projects and possible disruptions to supply chains based on changes in ecosystem services flows. The chapter discusses a case study to compare the various models – the location of a hypothetical residential housing project in the US San Pedro Watershed in Arizona. As shown through this analysis, it is both possible and desirable to combine a set of approaches to modelling and assessment. A possible scenario presented in the chapter is, for example, to offer a structure for priority setting through approaches that integrate stakeholder preferences, then conduct a landscape-level assessment through approaches that are primarily focused on ecological data (such as those analysed in Part II), and finally use a tool that can assist with site--level analysis. Therefore, as highlighted throughout the contributions of the handbook, further progress in the design of the economic tools for assessing ecosystem services and the synergies between biodiversity conservation and ecosystem service management is to be found in multi-method research and integration of stakeholder preferences in assessment and modelling exercises.

We sincerely hope that broader constituencies of policy-makers and researchers in economics of ecosystems and biodiversity will find this volume useful not only in their daily work but coax them to take the discourse further, which we think has a long way to go.

CONCLUDING REMARKS

How can we use the ideas presented here to formulate an integrated, effective framework to assess the value of ecosystem services and biodiversity? And what can we learn from current valuation studies in terms of their role in the design of economic policy? The answers to these questions require, inter alia, that a clear ecosystem service be chosen, that a concrete policy change scenario be formulated, that the relevant ecosystem services changes be described, and within certain boundaries, and that the particular perspective on value be made explicit.

So far, most studies lack a uniform and clear perspective on ecosystem services as a distinct, univocal concept. In addition, at present we have a rather poor knowledge about, for example, how biodiversity is affecting the overall ecosystem performance, including the provision of ecosystem services, as well as the level of resilience that this system, and relative ecosystem service supply capacity, has. For this reason alone, it is very difficult to assess the economic value of ecosystem services. To completely answer the question, 'What is the value of ecosystem services?' we have to include the value of the variety of interrelationships in which species exist in different ecosystems, the functions among ecosystems, and all the respective interactions, and impact, in terms of provision of ecosystem services. Therefore, the range and degree of ecosystem functioning needs be estimated, especially in terms of ecosystem-functional relationships and the respective outputs, including ecosystem services. Furthermore, we should have a clear understanding of beneficiaries, that is, the people and communities, who indeed receive welfare from the consumption or experience of the ecosystem service(s) under consideration. Without any doubt, full monetary assessment is impossible or would be subject to much debate. An additional problem is that, at the global level, ecosystem values can differ significantly, even for similar entities, due to unequal international income distribution.

The physical assessment of the functions performed by ecosystems is an essential prerequisite of any ecosystem services evaluation. However, simply identifying these functions is insufficient if we want to present resource managers and policy-makers with relevant policy response options. It is necessary to develop criteria for the expression of the functions in a form that allows for evaluation. For example, one can identify the range of management strategies by exploring the use of spatial modelling, including data such as the Red Data Species List, biological diversity indexes and ecosystem productivity. Computer models have become available to help decision-making by simulating different policy scenarios. Models have been applied to calculate minimum dynamic areas that support the minimum viable population of a certain species. In addition, computer models have been used for ecosystem services evaluation, predicting conservation values under different development scenarios. This approach to ecological evaluation allows for a direct comparison of management or conservation strategies.

From an economic perspective, certain aspects of ecosystem services and biodiversity are scarce and highly desirable, which is the reason why they have economic value. The concept of economic value is founded in welfare economics, which developed around the theory of consumer behaviour. Economic valuation assumes interaction between a subject – a human being – and an object – for example, ecosystem services. As a result, economic value is distinct from the notion of intrinsic value, which assumes that an object has or can have value in the absence of any (human) subject. It is important to recognize that economists do not pursue absolute value assessment of environmental systems or all the ecosystem services they contain, but always focus attention on valuing environmental system changes. This means that the terms 'economic value' and 'welfare change' are two sides of the same coin. Economics can thus assess the human welfare significance of ecosystem changes, namely through the determination of changes in the provision of biodiversity-related goods and services – including ecosystem services – and their consequent impacts on the well-being of humans who derive – use or non-use – benefits from their provision.

Integrated economic-ecological modelling can contribute to, and may even be essential for, a thorough understanding of the intricate relationship between ecosystem functioning and ecosystem services and economic dynamics. Although integrated modelling has something of a tradition, both at the ecosystem level and at the global level, applications to ecosystem valuation and economic policy are still in their infancy. Integrated modelling can be linked to ecosystem performance and evaluation of ecosystem services in various ways. Integrated models can generate a set of biological and economic, possibly monetary indicators that can be further aggregated through multi-criteria analysis techniques. In addition, it is possible to provide for a closer, innovative connection between modelling and valuation, among other methods, by generating conditional values for specific environment–economic scenarios, using scenario modelling outcomes such as tables and graphs in valuation studies, and using spatial models to aggregate monetary values related to specific areas.

Finally, one needs to be aware of the limitations of economic valuation and analysis. Ecosystem services and biodiversity are rather complex concepts. They can be associated with a wide range of benefits to human society, most of them still poorly understood. In general terms, the value of ecosystem services can be assessed in terms of its impact on the provision of inputs to production processes, on human welfare, and on the regulation of ecological functions. A complete understanding of these and their integration into multidisciplinary studies provides a great challenge for future research, in which economists, ecologists and others, including policy-makers, will have to work closely together. Only then can one expect to offer an insightfuly policy. There is no doubt that the economics of ecosystem services and biodiversity will face many research and policy challenges in the years ahead and that is another way to ensure the growth and importance of this important transdisciplinary theme.

REFERENCES

Barbier, E.B. and G.M. Heal (2006), 'Valuing ecosystem services', *The Economists' Voice*, **3**(3), Article 2.

Barbier, E.B., S. Naeem, D.E. Bunker, A. Hector, M. Loreau and C. Perrings (2009), *Biodiversity, Ecosystem Functioning, and Human Well-Being*, Oxford: Oxford University Press.

Baumgartner, S., C. Becker, R. Manstetten and M. Faber (2006), 'Relative and absolute scarcity of nature: assessing the roles of economics and ecology for biodiversity conservation', *Ecological Economics*, **59**(4) 487–98.

CBD (2010), *Strategic Plan for Biodiversity 2011–2020 and the Aichi Targets*, accessed 30 January 2014 at https://www.cbd.int/doc/strategic-plan/2011–2020/Aichi-Targets-EN.pdf.

Dasgupta, P. (2009), 'The welfare economic theory of green national accounts', *ERE*, **42**(1), 3–38.

Dasgupta, P. (2013), 'The nature of economic development and the economic development of nature', October (unpublished manuscript), University of Cambridge.

Environmental Protection Agency (EPA) (2009), *Valuing the Protection of Ecological Systems and Services*, Washington, DC: United States Environmental Protection Agency.

Fisher, B. and K. Turner (2008), 'Ecosystem services: classification for valuation', *Biological Conservation*, **141**(5), 1167–9.

Freeman, A.M. (2003), *The Measurement of Environmental and Resource Values: Theory and Methods*, Washington, DC: Resources for the Future.

Goulder, L.H. and D. Kennedy (1997), 'Valuing ecosystem services: philosophical bases and empirical methods', in G.C. Daily (ed.), *Nature's Services: Societal Dependence on Natural Ecosystems*, Washington, DC: Island Press, pp. 23–47.

Hanley, N. and E.B. Barbier (2009), *Pricing Nature Cost Benefit Analysis and Environmental Policy*, Cheltenham, UK and Northampton, MA, USA: Edward Elgar Publishing.

Heal, G. (2000), 'Valuing ecosystem services', *Ecosystems*, **3**(1), 24–30.

Heal, G., E.B. Barbier, K.J. Boyle, A.P. Covich, S.P. Gloss and C.H. Hershner et al. (eds) (2005), *Valuing Ecosystem Services: Toward Better Environmental Decision Making*, Washington, DC: The National Academies Press.

Kumar, P. (2012), 'Impact of economic drivers on mangroves of Indian Sundarbans: an exploration of missing links', *Environment, Development and Sustainability*, **14**(1), 939–53.

Kumar, P. and M. Wood (eds) (2010), *Valuation of Regulating Services*, Oxford: Routledge.

Millennium Ecosystem Assessment (MA) (2003), *People and Ecosystems: A Framework for Assessment*, Washington, DC: Island Press.

Millennium Ecosystem Assessment (MA) (2005), *Findings from Responses Working Group*, Washington, DC: Island Press.

Nunes, P.A.L.D. and J.C.J.M. van den Bergh (2001), 'Economic valuation of biodiversity: sense or nonsense', *Ecological Economics*, **39**(2), 203–22.

Stern, N. (2006), *The Economics of Climate Change*, London: Cabinet Office, HM Treasury.

TEEB (2010), *The Economics of Ecosystems and Biodiversity: Ecological and Economic Foundations*, London and Washington, DC: Earthscan.

PART I

SETTING THE SCENE: THE NEED FOR ECOSYSTEM SERVICE VALUATION

1. Comprehensive wealth accounting: measuring sustainable development

Glenn-Marie Lange and Esther Naikal

What we measure affects what we do; and if our measurements are flawed, decisions may be distorted.
(Stiglitz et al., 2009)

1.1 INTRODUCTION: PROGRESS WITH NATURAL CAPITAL ACCOUNTING

The Conceptual Framework

The goal of development policy should be to maximize social welfare, and the measure of current progress would be the change in social welfare. A seminal paper by Samuelson (1961, p. 50) highlighted that income, or for that matter consumption, does not have a direct welfare connotation. The choice of a welfare measure has to be made 'in the space of all present and future consumption . . . the only valid approximation to a measure of welfare comes from computing wealth-like magnitudes not income magnitudes'. Pearce et al. (1989, p. 34) argued that sustainable well-being is possible if next generations inherit 'a stock of wealth . . . no less than the stock inherited by the previous generation'.

A considerable body of theoretical work (for example, Hamilton and Clemens, 1999; Dasgupta and Mäler, 2000, 2004; Asheim and Weitzman, 2001; Arrow et al., 2003) has established the link between social welfare and a nation's total wealth – wealth defined comprehensively to include its stock of manufactured capital, natural capital and human and social capital. If wealth is decreasing, for example from depletion or degradation of natural capital, then a country will not be able to sustain its current level of income and well-being. The theory has been tested empirically by Ferreira and Vincent (2005) and Ferreira et al. (2008).

More recently, the Stiglitz et al. (2009) report summarized the widely recognized criticisms of gross domestic product (GDP), the most commonly used indicator of economic progress. With respect to natural capital, that report notes that GDP growth is not a comprehensive measure of development due to important inter-temporal factors; long-term economic well-being can only be assessed by considering both income and wealth.

Progress with Implementation

The World Bank took up the challenge of environmental accounting early on, and in the 1990s began constructing a global database for 'comprehensive wealth accounts' to monitor country economic progress using comprehensive wealth accounts and the related macroeconomic indicators, adjusted net savings (ANS) and adjusted net national income

(ANNI). Development is conceived as a process of building and managing a portfolio of assets. The challenge of development is to manage not just the total volume of assets – how much to save versus how much to consume – but also the composition of the asset portfolio, that is, how much to invest in different types of capital, including the institutions and governance that constitute social capital.

Two reports on wealth accounting have now been published: the wealth accounts were first reported in *Where is the Wealth of Nations?: Measuring Capital for the 21st Century* (World Bank, 2006) and updated in 2010 for 150 countries in *The Changing Wealth of Nations: Measuring Sustainable Development for the New Millennium* (World Bank, 2010). The natural capital component of the wealth accounts includes agricultural land, forest land, protected areas and subsoil assets. ANS, or genuine savings, and ANNI, reported annually in World Development Indicators, are related indicators that measure whether a country is building its wealth or running it down. ANS is intended to be used alongside traditional macroeconomic indicators such as GDP: GDP indicates whether an economy is growing; ANS indicates whether that growth is sustainable.

A new initiative in wealth accounting, undertaken by the United Nations University and UNEP, published *The Inclusive Wealth Report 2012*, which estimated wealth for 20 countries from 1990 to 2008. While some of the data sources and methodology are quite similar to those used by The World Bank, some aspects are quite different. For example, *The Inclusive Wealth Report* estimates health as a separate form of capital (rather than as an aspect of human capital); 'health capital' dwarfs all other assets, accounting for an average of 95 per cent of total wealth.

The Statistical Framework for Implementation

While the 'comprehensive wealth' approach to sustainable development provides the conceptual framework, a parallel effort has been underway over the past 25 years to develop the statistical methodology to expand the System of National Accounts (SNA; EC et al., 2009) for natural capital.[1] The System of Environmental-Economic Accounting (SEEA), developed under the auspices of the UN Statistical Commission, provides a comprehensive framework for incorporating the role of the environment and natural capital into national accounts through satellite accounts for the environment. In 2012 the UN Statistical Commission adopted the SEEA Central Framework (EC et al., 2012) as an international statistical standard, paving the way for implementation by national statistical agencies in all countries, like the SNA. The SEEA Central Framework includes accounts for flows and stocks, the latter forming the basis for natural capital in expanded national balance sheets, or wealth accounts.

The construction of wealth accounts by national statistical agencies has largely been limited to Organisation for Economic Co-operation and Development (OECD) countries, and most of these countries do not include natural capital. A review of country experiences with the SEEA for natural capital accounting found that 14 countries regularly compile asset accounts, mainly for energy and minerals, timber and land (World Bank, 2010, Chapter 8). Only six of these countries have expanded their balance sheets to include some natural capital. The most comprehensive balance sheets are produced by Australia.

Ecosystem Services in Environmental Accounting

There has been great progress in accounting for material natural resources like minerals and timber, which often have market prices or 'near market' prices that can be used to estimate asset value (most of the provisioning and recreational services). The challenge lies with accounting for the regulating ecosystem services, whose value derives indirectly from their use as intermediate inputs to economic production or consumption, like other intermediate inputs to production in the national accounts. The value of many of these services is already included, implicitly, in the value of other assets for which we have markets – for example, agricultural land includes the value of natural pollination or groundwater.

But because the value of regulating services is not explicit, their values are hidden and we don't have the information to make informed decisions about management of natural capital. Ecosystem services are especially important for developing countries, where the livelihoods of many communities depend directly on healthy ecosystems. In addition, these countries contain most of the world's biodiversity, and ecosystem services such as water provisioning and regulation, or soil protection are under greatest threat, while they often have fewer resources to cope with loss of ecosystem services. Following the adage, 'what we do not measure, we cannot manage', an objective of ecosystem accounting is to identify and make explicit the value of regulating services and the assets that provide them.

The World Bank's wealth accounts do not explicitly identify the regulating ecosystem services, but a new programme is underway to address this gap, Wealth Accounting and the Valuation of Ecosystem Services, or WAVES (to be discussed at the end of the chapter). A theoretical approach to ecosystem accounting was provided by Mäler et al. (2008) and was further developed in *The Inclusive Wealth Report 2012* (UNU-IHDP, 2012). Several countries have research programmes to develop ecosystem accounts, including Australia, Canada and the UK. The SEEA includes a draft experimental volume on ecosystem accounting, but a large number of issues remain unresolved and extensive testing of concepts is needed before clear guidelines can be recommended.

This chapter uses The World Bank's comprehensive wealth accounts to describe how far we have been able to go with wealth accounting, and what we can learn from them about development and wealth. Although the accounts do not, at present, adequately represent ecosystems, they provide a sound basis for further development of these accounts under the new World Bank initiative, WAVES. The chapter includes new data for 2008, providing for wealth accounts for more than 150 countries over a longer period of time, from 1995 to 2008, and deepening the inter-temporal assessment of global, regional and country performance in building wealth and achieving sustainable development.

The next section provides a brief overview of the methodology for the comprehensive wealth approach, followed by an analysis of wealth over the period 1995 to 2008. Finally, the chapter concludes with a discussion of new initiatives to incorporate ecosystems in wealth accounting and some major research challenges.

1.2 METHODOLOGY AND DATA

Given space constraints, only a very brief summary of the main features of the wealth calculations can be given here. The reader is directed to Hamilton and Hartwick (2005) for the theoretical model, and World Bank (2010) for a discussion of the methodology and data for implementation of the theoretical model.

The comprehensive wealth accounts presented here include the following categories of assets:

- *Total wealth*: the measure of total (or comprehensive) wealth is built upon the intuitive notion that current wealth must constrain future consumption.
- *Produced capital*: machinery, structures and equipment.
- *Natural capital*: agricultural land (crop and pastureland), protected areas, forests, minerals and energy resources.
- *Intangible capital*: this asset is measured as a residual, the difference between total wealth and the sum of produced and natural capital. It implicitly includes measures of human, social and institutional capital, which includes factors such as the rule of law and governance that contribute to an efficient economy. Net foreign financial assets, the balance of a country's total financial assets and financial liabilities, are included as part of intangible capital.[2]

Each of these is briefly described below.

Total wealth

Total wealth can be calculated as:

$$W_t = \int_t^{\infty} C(s) \cdot e^{-r(s-t)} ds$$

where W_t is the total value of wealth, or capital, in year t; $C(s)$ is consumption in year s; and r is the social rate of return to investment. The social rate of return to investment is expressed as:

$$r = \rho + \eta \frac{\dot{C}}{C}$$

where ρ is the pure rate of time preference and η is the elasticity of utility with respect to consumption. Under the assumption that $\eta = 1$ and that consumption grows at a constant rate, the total wealth can be expressed as:

$$W_t = \int_t^{\infty} C(t) \cdot e^{-\rho(s-t)} ds \qquad (1.1)$$

The current value of total wealth at time t is a function of the consumption at time t and the pure rate of time preference. Expression (1.1) implicitly assumes that consumption is on a sustainable path, that is, the level of saving is enough to offset the depletion of natural resources.

For computation purposes, we assumed the pure rate of time preference to be 1.5 per cent, and we limited the time horizon to 25 years. This time horizon roughly corresponds to a generation. We adopted the 25-year truncation throughout the calculation of wealth, in particular, of natural capital.

Produced capital

For the calculation of produced capital stocks, the well-established perpetual inventory method is used, deriving stock estimates for machinery, equipment and structures from the accumulation of investment in gross capital formation, adjusted for depreciation. It is assumed that the average service life is 20 years. The value of urban land is added as a fixed proportion (24 per cent) of the value of machinery, equipment and structures.

Natural capital

Accounts for natural capital include:

- *energy and mineral resources*: oil, gas, hard coal, soft coal, bauxite, copper, gold, iron ore, lead, nickel, phosphate, silver, tin and zinc;
- *crop land*;
- *pasture land*;
- *forest* (timber and non-timber forest products);
- *protected areas*.

The conceptual approach used in the estimation of natural capital is based on the well-established economic principle that asset values should be measured as the present discounted value of economic profits over the life of the resource. This value, for a particular country and resource, is given by the following expression:

$$V_t = \sum_{i=t}^{t+T-1} \pi_i q_i / (1 + r)^{(i-t)}$$

where $\pi_i q_i$ is the economic profit or total rent at time i, (π_i denoting unit rent and q_i denoting production), r is the social discount rate, and T is the lifetime of the resource.

This formula is implemented in various ways for the different kinds of natural capital, depending on data availability. The reader is referred to World Bank (2010) for a detailed description.

Intangible capital

Intangible capital implicitly includes human and social capital, net foreign financial assets and any missing or mismeasured wealth. It is calculated as the residual of total wealth minus the sum of produced and natural capital.

1.3 RESULTS: THE CHANGING WEALTH OF NATIONS 1995 TO 2008

Trends in Global Wealth

Table 1.1 shows wealth and per capita wealth by type of capital for countries grouped by income over the years 1995 and 2008. Grouping countries by income is useful because it reveals the relationship between wealth, income and development. Trends for low-income countries are of particular interest because of the concentration of the world's poor in these countries. Middle-income countries are important because they shed light on the process of wealth creation during the transition from low to high income, and high-income countries provide insight into the volume and composition of wealth in those countries that have achieved high material standards of living.

Between 1995 and 2008, global wealth increased in per capita terms by 23 per cent from US$115 884 to US$142 763 in constant 2008 US dollars. Wealth grew fastest in the lower-middle-income countries, which are dominated by the economies of China and India; per capita wealth in this group increased by 80 per cent. High-income OECD countries continue to hold most of the world's wealth (79 per cent), while the world's poorest countries, accounting for 10 per cent of the global population, hold less than 1 per cent of global wealth.

By 2008, intangible capital is the largest single component of wealth in all income groups, and the fastest-growing component for most countries. The growth of intangible capital reflects improvements in human capital, but also institutions, governance and technology that support more efficient use of all assets.

But natural capital is especially important for low-income countries; in 1995, natural capital was the largest component of wealth (44 per cent), surpassing intangible capital. It was still important in 2008 at 31 per cent. Natural capital's share of total wealth has decreased or remained the same for all income groups from 1995 to 2008, with the exception of upper-middle-income countries where its share actually increased from 17 per cent to 26 per cent. This trend likely captures the effect of the commodity boom in the mid-2000s, where record-high prices in minerals and energy commodities drove up asset values for countries that rely heavily on their extractive industries. Generally, between 1995 and 2008, not only did total wealth grow, but there was a significant shift in the composition of wealth as well. In most countries, produced and intangible assets grew faster than natural capital, increasing their shares, a topic explored further in the next section.

Wealth in Developing Nations

Taking a closer look at low- and middle-income countries by geographic region,[3] human and social capital dominate the wealth of most regions, except the Middle East and North Africa, where subsoil assets are particularly important. Natural capital is larger than produced capital in four out of six regions.

For developing countries, especially low-income countries, the challenge for development is to leverage natural capital for growth. Table 1.2 shows the relative importance of various natural resources for developing countries by region. Subsoil assets greatly dominate the natural capital of two regions: Middle East and North Africa and Europe and

Table 1.1 *Wealth and per capita wealth by type of capital and income group, 1995 and 2008*

Income Group	1995					2008				
	Total wealth (2008 US$ billions)	Per capita wealth (2008 US$)	Intangible capital (%)	Produced capital (%)	Natural capital (%)	Total wealth (2008 US$ billions)	Per capita wealth (2008 US$)	Intangible capital (%)	Produced capital (%)	Natural capital (%)
Low income	2725	6 269	43	12	44	4 590	7 670	54	15	31
Lower-middle income	38 695	13 071	42	24	34	82 428	23 481	48	29	23
Upper-middle income	49 409	88 527	64	19	17	70 543	105 113	55	19	26
High income: OECD	465 055	524 546	79	19	2	655 080	680 071	80	18	2
World	565 526	115 884	74	19	7	827 578	142 763	73	19	7

Note: Figures are based on the set of countries for which wealth accounts are available from 1995 to 2008. Data in this table do not include high-income oil exporters, which are a special case.

Source: Authors' calculations based on World Bank data.

Table 1.2 Composition of assets in developing regions, 2008

Region	Per Capita Wealth (2008 US$)				Composition of Natural Capital (%)			
	Total wealth	Intangible capital	Produced capital	Natural capital	Crop land	Pasture land	Forest and protected areas	Subsoil assets
East Asia & Pacific	30730	12760	10828	7141	34	18	13	35
Europe & Central Asia	110193	59529	22903	27761	8	6	7	79
Latin America & Caribbean	112447	65447	19603	27397	52	7	10	31
Middle East & North Africa	40755	11587	11255	17913	11	15	2	82
South Asia	12970	8277	2841	1852	46	19	9	26
Sub-Saharan Africa	17504	9500	2363	5641	34	10	12	44

Note: Based on a balanced sample of 123 countries for 1995–2008 and 17 Europe and Central Asia countries for 2000–08.

Source: Authors' calculations based on World Bank data.

Central Asia, and account for the largest share of natural capital in Africa. For countries dependent on subsoil assets, transforming this non-renewable capital into other forms of wealth is the path to sustainable development.

Agricultural land, both crop and pasture land, is important for Latin America and the Caribbean and South Asia. Forests are most important in Latin America, and agricultural land is most important for the other regions. Where natural capital is potentially renewable, such as forest land, appropriate property rights and efficient management regimes are essential if the country is to develop sustainably.

Among developing countries, wealth creation, both total and per capita, was highest in East Asia, followed by South Asia (Figure 1.1). Surprisingly, total wealth creation in Sub-Saharan Africa was relatively high, and higher than several other regions. However, on a per capita basis all regions outperformed Sub-Saharan Africa, in part because of much smaller and more slowly growing populations compared to Sub-Saharan Africa. But the regional results for Sub-Saharan Africa combine two very different trends: a small group of oil exporters, led by Nigeria, where per capita wealth declined, and a much larger number of countries, dominated by South Africa, where wealth increased.

Most of the growth in wealth was in human and social capital. Investment in human capital, but especially in good institutions, the rule of law and governance, is critical for wealth creation.

Whether one compares wealth across income groups in a single year or looks at a single income group over time, the comprehensive wealth accounts tell a clear story about the relationship between development and wealth: development entails building total wealth, but also changing the composition of wealth. The challenge is to manage the total volume of assets – how much to save versus how much to consume – and the composition of the asset

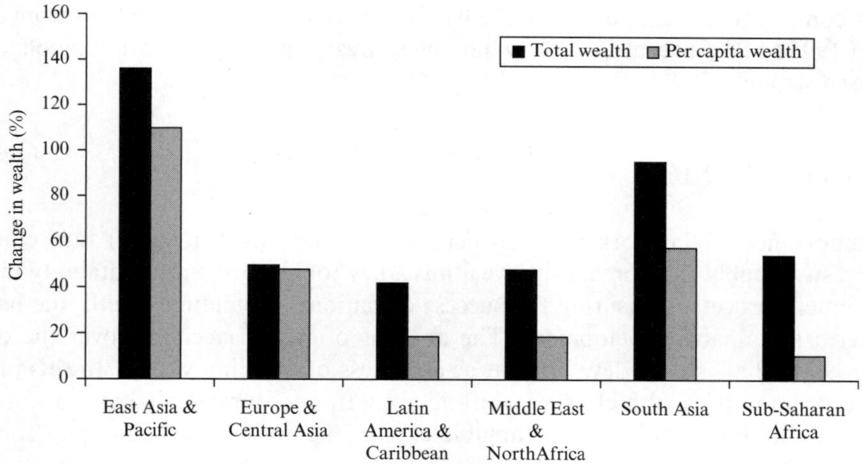

Source: Authors' calculations based on World Bank data.

Figure 1.1 Change in wealth and per capita wealth in developing countries, 1995–2008

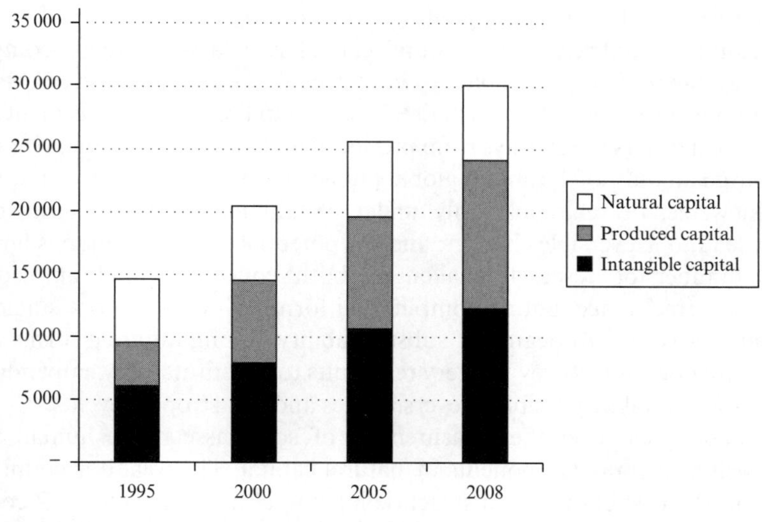

Source: Authors' calculations based on World Bank data.

Figure 1.2 Changing volume and composition of wealth in East Asia and Pacific countries, 1995–2008 (2008 US$ per capita)

portfolio – how much to invest in different types of capital. Most countries start out with a relatively high dependence on natural capital, such as agricultural land or subsoil assets, but they use these assets to build more wealth, especially produced and intangible capital.

This relationship between development and capital is clearly seen in the East Asia and Pacific countries, where the economy of China dominates. As Figure 1.2 shows, per capita wealth has increased dramatically from 1995 to 2008, and just as importantly, the wealth composition has changed markedly. The share of natural capital fell from 34 per cent in 1995 to 23 per cent in 2008, while the shares of produced and intangible capital increased strongly.

1.4 CONCLUSION

The importance of the work reported here is in the lessons it holds for how countries develop sustainably. Comprehensive wealth is a key tool to bring accountability into the development process: measuring the success of nations in building wealth, the basis of long-term sustainable development. The analysis of wealth accounts over the decade from 1995 to 2008 shows development as a process of building wealth. In this process, the composition of wealth changes, shifting away from natural capital and toward produced capital and, increasingly, intangible capital. The role of the changing composition of wealth in the development process points to the need for comprehensive wealth accounting.

Intangible capital plays the key role in building wealth in all regions. Investing in human capital is important in this process, but building good institutions and governance is equally important because this provides the basis for more efficient use of, and higher economic returns to, all forms of capital.

For developing countries, where natural capital is a large share of comprehensive wealth, management of natural capital to build wealth is critical. However, even when its share of comprehensive wealth is small, it is essential to focus on management of natural capital because it differs in key ways from produced and intangible capital. Natural capital includes a large amount of local and global public goods; many forms of natural capital are non-renewable, or renewable only under certain management; losses of natural capital may lead to irreversible changes; and the potential for substitution is limited.

A large number of research challenges, both conceptual and applied, remain. Comprehensive wealth accounting combines all forms of wealth into a single measure, which assumes a very high degree of substitutability among different forms of capital. Such a measure does not convey the very real limits to substitutability, impending thresholds for natural capital or possible irreversibilities and catastrophic events.

Data availability impedes the measurement of some assets, like human and social capital, as well as certain components of natural capital. The wealth accounts reported here represent the best effort to construct detailed wealth accounts for 152 countries on a regular basis using publicly available data that are comparable both across countries and over time. Specific natural resources are included in the database when they meet two criteria: (1) reliable data on price and volume are available on a regular basis, not from occasional or one-off studies, and (2) data are available for a large number of countries, if not for all. There are some natural resources where the available data do not meet these criteria, notably fisheries, certain minerals, and certain water services such as hydropower. As a result, the value of natural capital is underestimated, and for specific countries this omission can be significant (for example, see Sumaila et al., 2012 for estimate of global fisheries).

There is a renewed global momentum for natural capital accounting. The adoption of the SEEA as a statistical standard has put the methodology on a solid footing. The 2012

World Development Summit in Brazil featured widespread discussion of this topic, for both national accounting and private sector, corporate accounting. One of the outcomes of the Summit was a call for broader measures of well-being, going beyond GDP. Several global programmes – including TEEB, Green Growth programme, and the UN's Green Economy – call for natural capital accounting and provide platforms for rapid implementation. Botswana hosted the African Sustainability Summit in May 2012 and issued a communiqué calling for implementation of natural capital accounting that was signed by heads of state and ministers from ten countries. The OECD is considering a programme to construct expanded national balance sheets that go beyond produced capital to include natural and human assets.

The World Bank initiated a global partnership, WAVES, that calls on countries to: (1) implement natural capital accounting where we have internationally agreed methodology (for example, the SEEA Central Framework), (2) work to develop methodology for ecosystem accounting, and (3) demonstrate the policy relevance of natural capital accounting. WAVES is currently supporting programmes in seven developing countries in partnership with ten developed countries, international agencies, academics and NGOs. The invitation to join this initiative was extended to other interested countries and 62 countries have joined so far.

Wealth accounts will continue to improve as countries themselves implement them, using more accurate country-specific information only available to national statistical institutions. The exploration of different concepts and methods through The World Bank's programme, the *Inclusive Wealth Report* and various country initiatives, is a welcome step toward the development of internationally agreed statistical standards – a process that is necessary before wealth accounting is routinely included in official statistics.

NOTES

1. The SNA provides guidelines for national balance sheets, but these do not include human or social capital and until recently had a limited representation of natural capital.
2. As a residual, intangible capital also includes any 'missing' assets, as discussed in the concluding section.
3. Due to data constraints for Europe and Central Asia, wealth in this region is reported from 2000 to 2008, while results for the other regions compare wealth in 1995 and 2008.

REFERENCES

Arrow, K., P. Dasgupta and K.-G. Mäler (2003), 'Evaluating projects and assessing sustainable development in imperfect economies', *Environmental and Resource Economics*, **26**(4), 647–85.

Asheim, G.B. and M.L. Weitzman (2001), 'Does NNP growth indicate welfare improvement?', *Economics Letters*, **73**(2), 233–9.

Dasgupta, P. and K.-G. Mäler (2000), 'Net national product, wealth and social well-being', *Environment and Development Economics*, **5**(1) 69–93.

Dasgupta, P. and K.-G. Mäler (eds) (2004), *The Economics of Non-Convex Ecosystems*, Dordrecht: Kluwer Academic Publishers.

European Commission, International Monetary Fund, Organisation for Economic Co-operation and Development, United Nations and World Bank (2009), *System of National Accounts 2008*, New York: United Nations.

European Commission, International Monetary Fund, Organisation for Economic Co-operation and Development, United Nations and World Bank (2012), *System of Environmental-Economic Accounting*, New York: United Nations.
Ferreira, S. and J. Vincent (2005), 'Genuine savings: leading indicator of sustainable development?', *Economic Development and Cultural Change*, **53**(3), 737–54.
Ferreira, S., K. Hamilton and J. Vincent (2008), 'Comprehensive wealth and future consumption: accounting for population growth', *World Bank Economic Review*, **22**(2), 233–48.
Hamilton, K. and M. Clemens (1999), 'Genuine savings rates in developing countries', *World Bank Economic Review*, **13**(2), 333–56.
Hamilton, K. and J. Hartwick (2005), 'Investing exhaustible resource rents and the path of consumption', *Canadian Journal of Economics*, **38**(2), 615–21.
Pearce, D., A. Markandya and E. Barbier (1989), *Blueprint for a Green Economy*, London: Earthscan.
Samuelson, P.A. (1961), 'The evaluation of "social income": capital formation and wealth', in F.A. Lutz and D.C. Hague (eds), *The Theory of Capital*, London: Macmillan, pp. 32–57.
Stiglitz, J.E., A. Sen and J.-P. Fitoussi (2009), *Report by the Commission on the Measurement of Economic Performance and Social Progress*, Paris: Commission on the Measurement of Economic Performance and Social Progress.
Sumaila, R., G. Lange, E. Naikal and U. Narain (2012), 'Marine fisheries as a source of sustainable wealth of nations', paper presented at the AAAS conference in Vancouver, British Columbia, 16–20 February 2012.
United Nations University and United Nations Environment Programme (2012), *The Inclusive Wealth Report 2012*, Bonn: UNU-IHDP.
World Bank (2006), *Where is the Wealth of Nations? Measuring Capital for the 21st Century*, Washington, DC: World Bank.
World Bank (2010), *The Changing Wealth of Nations: Measuring Sustainable Development in the New Millennium*, Washington, DC: World Bank.

2. The protective value of estuarine and coastal ecosystems

Edward B. Barbier

2.1 INTRODUCTION

Estuarine and coastal ecosystems (ECEs) are some of the most heavily used and threatened natural systems globally (Lotze et al., 2006; Worm et al., 2006; Halpern et al., 2008). Their deterioration due to human activities is intense and increasing; 50 per cent of salt marshes, 35 per cent of mangroves, 30 per cent of coral reefs and 29 per cent of seagrasses are either lost or degraded worldwide (MA, 2005; Orth et al., 2006; UNEP, 2006; FAO, 2007; Waycott et al. 2009; Spalding et al., 2010; Barbier, 2011). Since Hurricanes Katrina and Rita in 2005 and the Indian Ocean tsunami in 2004, attention has focused on how the continuing worldwide loss of ECEs is making coastlines and coastal communities vulnerable to flooding and storm events (Braatz et al., 2007; Day et al., 2007; Cochard et al., 2008).

There is mounting evidence that a variety of ECEs, including marshes, mangroves, near-shore coral reefs, seagrass beds, and sand beaches and dunes, provide some type of protection against storms and coastal floods, mainly through their ability to attenuate waves or buffer winds (Barbier et al., 2008, 2011; Koch et al., 2009; Barbier, 2011; Gedan et al., 2011; Paul and Amos, 2011; Shephard et al., 2012). However, to date, there are few economic studies that estimate the protective value of many systems, although some estimates are beginning to emerge for marsh and mangroves (see Barbier et al., 2011 and Table 2.1). Although many more studies exist than those indicated in Table 2.1, there are problems of reliability in the estimates of protection value produced by some of these earlier studies because of the arbitrary valuation methods often employed (Barbier, 2007, 2011).

The following chapter reviews and discusses the current economic methods of valuing the protective service of key ECEs. As indicated in Table 2.1, this service is directly related to how these ecosystems attenuate and/or dissipate waves and buffer winds from storms. Here, we will focus mainly on the protection value that arises from the ability of ECEs to attenuate storm surge and waves, as most current valuation studies focus on this benefit (see Table 2.1). The next section reviews current ecological evidence on how various ECEs, such as marshes, mangroves, near-shore coral reefs, seagrass beds, and sand beaches and dunes, are able to attenuate waves. Understanding this ecological function is important, as it underlies the correct approach to valuing the protective service provided by these ecosystems. The subsequent section then provides an overview of existing valuation studies of valuing the protective benefits of ECEs. As two common methods appear to be in use – the replacement cost and expected damage function – these two valuation approaches are described and then illustrated with the example of mangroves in Thailand. The chapter concludes by discussing key challenges and new directions for research for improving estimating the protective value of ECEs.

Table 2.1 Examples of studies of the protective value of estuarine and coastal ecosystems

Ecosystem Structure and Function	Ecosystem Service	Valuation Examples
Attenuates and/or dissipates waves, buffers wind	Protection of coastal communities against properties damages, loss of life and/or injuries	Badola and Hussain (2005), mangroves, India Barbier (2007), mangroves, Thailand Barbier et al. (2008), mangroves, Thailand Chong (2005), mangroves and coral reefs, various regions Costanza et al. (2008), marsh, US Atlantic and Gulf Coasts Das and Vincent (2009), mangroves, India Farber (1987), marsh, Louisiana, United States King and Lester (1995), marsh, United Kingdom Laso Bayas et al. (2011), mangroves, Aceh, Indonesia Sathirathai and Barbier (2001), mangroves, Thailand Wilkinson et al. (1999), coral reefs, Indian Ocean

2.2 WAVE ATTENUATION BY ESTUARINE AND COASTAL ECOSYSTEMS

As Table 2.1 indicates, the protective value of estuarine and coastal ecosystems (ECEs) is directly related to their ability to attenuate, or reduce the height of the storm surges and waves as they approach shorelines. This wave attenuation function derives from the amount of plant and sedentary animal material obstructing the water column and the bathymetry, or underwater depth, of the area (Koch et al., 2009). The vegetation contained in some ECEs, such as marsh, seagrass beds and mangroves, are an important source of friction to moving water (Massel et al., 1999; Koch et al., 2009; Gedan et al., 2011; Paul and Amos, 2011; Shephard et al., 2012). In the case of coral reefs and sand dunes, it is their reticulated structure that acts as a natural barrier to storm waves, although the presence of grasses on dunes enhances wave attenuation (Madin and Connolly, 2006; Stockdon et al., 2007; Barbier et al., 2008).

For example, the protection against storms provided by mangroves depends on this critical ecological function in terms of 'attenuating', or reducing the height, of storm waves (Barbier et al., 2008; Koch et al., 2009; Gedan et al., 2011). Ecological and hydrological field studies suggest that mangroves are unlikely to stop storm-induced waves that are greater than 6 m (Forbes and Broadhead, 2007; Wolanski, 2007; Alongi, 2008; Cochard et al., 2008). Mangroves are effective in reducing storm-induced waves less than 6 m in

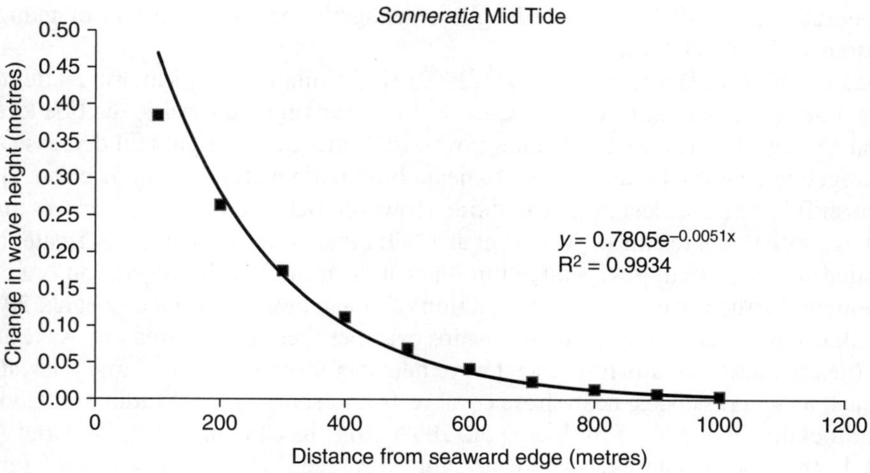

Source: Koch et al. (2009) based on data in Mazda et al. (1997).

Figure 2.1 Wave attenuation across a mangrove landscape

height, and studies suggest that the wave height decreases non-linearly for each 100 m that a mangrove forest extends out to sea (Mazda et al., 1997; Barbier et al., 2008). In other words, wave attenuation is greatest for the first 100 m of mangroves but declines as more mangroves are added to the seaward edge. For salt marshes, wave attenuation also diminishes with increasing habitat distance inland from the shoreline (Barbier et al., 2008; Koch et al., 2009; Gedan et al., 2011). For example, for five mangrove and ten salt marsh sites, the seaward margin of all the wetlands exhibited greater wave attenuation than equivalent landward distances, and the non-linear decline in wave attenuation was similar for marsh and mangrove landscapes (Gedan et al., 2011). A meta-analysis of field studies of the wave attenuation function of marshes found that salt marsh vegetation had a significant positive effect in reducing wave height per unit distance across a marsh (Shephard et al., 2012).

Figure 2.1 illustrates the non-linear wave attenuation function of mangroves based on field study data by Mazda et al. (1997) from a coastal site in Vietnam where *Kandelia candel* and *Sonneratia caseolaris* mangrove plantations have been created over a wide intertidal shoal as a coastal defence against typhoon waves. Wave data was measured in situ at the seaward edge of the forest up to a distance inland of approximately1000 m. Koch et al. (2009) employ these data to construct a wave attenuation relationship as a function of 100 m inshore mangrove distance, assuming a mangrove forest extending 1000 m seaward along a 10 km coastline (i.e., a 10 km^2 mangrove landscape). Figure 2.1 plots the wave attenuation relationship for *Sonneratia* spp. at mid-level tide, showing the change in wave height corresponding to every 100 m that the 10 km^2 mangrove landscape extends inshore from its seaward boundary. Without any mangroves (distance 0 m) waves have a maximum height of 1.1 m, but the presence of mangroves over the first 100 m from the seaward boundary reduces wave height significantly (0.38 m). However, as shown in the figure, this wave attenuation effect is non-linear across the 1000 m mangrove landscape. The change in wave height due to the presence of subsequent mangroves declines exponentially, until the fall in wave height is negligible for the last 100 m of mangroves (for example only 0.004 m).

Gedan et al. (2011) and Koch et al. (2009) find similar non-linear wave attenuation across other mangrove landscapes, regardless of the mangrove species, the tide level and coastal geology. In the absence of mangroves, the wave height might still decrease across an unvegetated coastal landscape due to near-shore bathymetry, bottom friction, and the abrupt shift in bottom elevation near shore. However, field studies of mangroves and salt marshes confirm the results of Mazda et al. (1997) that wave attenuation is greater across vegetated wetlands than unvegetated mudflats, indicating that the vegetation is a critical component for the wave attenuation function of coastal wetlands (Gedan et al., 2011).

Similar non-linear landscape relationships exist between habitat area and wave attenuation (i.e., reduction of storm-induced wave height) for other estuarine and coastal habitats, such as seagrass beds, near-shore coral reefs and sand dunes (Madin and Connolly, 2006; Stockdon et al., 2007; Barbier et al., 2008, 2011; Koch et al., 2009; Paul and Amos, 2011). In the case of seagrasses and near-shore coral reefs, wave attenuation is a function of the water depth above the grass bed or reef, and these relationships are also non-linear. There is also a spatial relationship between the percentage cover of dune grasses and the size of oceanic waves blocked by the sand dunes produced by the grasses.

2.3 VALUATION STUDIES

The growing evidence indicating that a number of estuarine and coastal ecosystems (ECEs) have a significant wave attenuation function has led to interest in valuing their storm protection benefit. But despite the importance of this coastal protection service, very few economic studies have estimated a value for it. Those studies that have been conducted tend to use benefit transfer and replacement cost methods of valuation in an ad hoc manner, which undermine the reliability of the value estimates (see Chong, 2005; Barbier, 2007; and further discussion below). Here, we review the handful of more reliable storm protection valuation studies indicated in Table 2.1. Most of these studies of the protective value of ECEs have focused on marshes and mangroves.

Widespread reef destruction caused by catastrophic events and global change, such as hurricanes, typhoons and coral bleaching, gives some indication of the value of the lost storm protection services. For example, as a result of the 1998 bleaching event in the Indian Ocean, the expected loss in property values from declining reef protection is estimated to be US\$174 ha^{-1} $year^{-1}$ (Wilkinson et al., 1999). Evidence from the Seychelles documents how rising coral reef mortality and deterioration have significantly increased the wave energy reaching shores that are normally protected from erosion and storm surges by these reefs (Sheppard et al., 2005). However, to date, this effect has not been valued explicitly.

Although field studies indicate that seagrass meadows and sand dunes may have a significant impact on reducing storm waves, no valuation studies currently exist of the resulting protection benefit. For seagrasses, one problem is that coastal protection can vary significantly if damaging storm events occur when plant biomass and/or density are low (Koch et al., 2009; Paul and Amos, 2011). This is particularly important in temperate regions, where seasonal fluctuations of biomass may differ from the seasonal occurrence of storms. For example, along the US Atlantic Coast, the biomass of seagrass peaks in the summer (April–June) yet decreases in the fall (July–September) when storm events usually strike (Koch et al., 2009). In tropical areas, seagrass beds have relatively constant biomass throughout the year, so the coastal protection service is relatively unaffected by seasonal or temporal variability.

An analysis of the economic damages associated with 34 major hurricanes striking the United States coast since 1980 found that coastal wetlands explained 60 per cent of the variation in relative damages inflicted on coastal communities (Costanza et al., 2008). The additional storm protection value per unit area of coastal wetlands from a specific hurricane ranged from a minimum of US\$23 ha^{-1} for Hurricane Bill to a maximum of US\$463 730 ha^{-1} for Hurricane Opal, with a median value of just under US\$5000 ha^{-1}. Recent hydrodynamic storm surge models developed for southern Louisiana also show how the attenuation of surge by wetlands is affected by the bottom friction caused by vegetation, the surrounding coastal landscape and the strength and duration of the storm forcing (Resio and Westerink, 2008; Loder et al., 2009; Wamsley et al., 2010).

Marsh wetlands may also act as a buffer against the wind damages from hurricanes and other storms. Using historical storm frequencies for the Louisiana Gulf Coast, Farber (1987) estimates the expected wind damage to property from the loss of intervening marsh. The present value of the loss of a mile-long strip of wetlands amounts to between US\$1.1 and US\$3.7 million (1980 dollars). The increased cost of property damage amounted to between US\$7 and US\$23 $acre^{-1}$.

A different approach was taken to estimating the protective value of marsh as a sea defence in East Anglia, United Kingdom (King and Lester, 1995). In this region, existing marsh areas in front of constructed sea walls together provide protection against storms, and less marsh means that higher sea walls have to be built. The authors therefore estimate the value of marsh as sea defence by calculating the additional capital and maintenance costs that would needed to build higher walls as the marsh disappears. Evaluation of an 80 m width of salt marsh in front of a sea wall yields a value over the whole area of between £30 and £60 m^{-2}.

Mangroves significantly reduced the number of deaths and damages to property, livestock, agriculture, fisheries and other assets during the 1999 cyclone that struck Orissa, India (Badola and Hussain, 2005; Das and Vincent, 2009). Statistical analysis indicates that there would have been 1.72 additional deaths per village within 10 km of the coast if mangroves had been absent (Das and Vincent, 2009). Economic losses incurred per household were greater (US\$154) in a village that was protected by a constructed embankment compared to those (US\$33) in a village protected by mangrove forests (Badola and Hussain, 2005).

Since the 2004 Indian Ocean tsunami, there has been considerable debate as to whether the presence of mangroves reduced the impacts of the extremely large storm surges associated with this event, thus protecting lives and property (see Cochard, 2011 for a review). In a definitive study for one of the worst affected regions, Aceh, Indonesia, Laso Bayas et al. (2011) confirm that not only coastal topography and near-shore bathymetry, but also vegetation including the presence of mangroves, plantations and other coastal forests, were effective in reducing the deaths and damage caused by the tsunami. Mangroves, forests and plantations situated between villages and the coastline may have decreased loss of life by 3–8 per cent, as the trees appear to have slowed or diverted the waves. If these natural barriers were located behind the villages, casualties increased by 3–6 per cent, because the debris from the trees increased the risk of death.

A series of studies for Thailand also confirm the protective value of mangroves against the damages caused by frequent storm events (Sathirathai and Barbier, 2001; Barbier, 2007; Barbier et al., 2008). Sathirathai and Barbier (2001) employed the replacement cost method to estimate the value of coastal protection and stabilization provided by mangroves in Surat Thani Province, Thailand. Using the cost of constructing breakwaters to replace protection by mangroves, the authors calculate that the present value over 20 years of mangrove protection and stabilization service is \$12 263 ha^{-1}.The contribution of mangrove deforestation to economic damages of storms was estimated for 39 coastal storm events affecting Southern Thailand from 1975 to 2004 (Barbier, 2007). Over 1979 to 1996, the marginal effect of a 1 km^2 loss of mangrove area was an increase in expected storm damages of about US\$585 000 km^{-2}, and from 1996 to 2004, the expected increase in damages from a 1 km^2 loss in mangroves was around US\$187 898 km^{-2} (US\$1879 ha^{-1}). Barbier et al. (2008) further show how variation in this protective value of mangroves across a 10 km^2 landscape could lead to substantial change in land use decisions, including the conversion of mangroves to shrimp farms. Barbier (2012) further shows how the type of declining wave attenuation function as depicted in Figure 2.1 affects the mangrove conversion decision, including the optimal location of shrimp ponds in the mangrove ecosystem, as well as the risk of ecological collapse.

2.4 VALUATION METHODS

The above review of selective valuation studies suggests that an important development has occurred in the methods used to estimate the protective value of estuarine and coastal ecosystems (ECEs). Previously, many studies that have attempted to value the storm prevention and flood mitigation services of the 'natural' storm barrier function of mangrove and other ECEs have employed the replacement cost method by simply estimating the costs of replacing coastal habitat by constructing physical barriers to perform the same services (King and Lester, 1995; Sathirathai and Barbier, 2001; Chong, 2005). However, economists recommend that the replacement cost approach should be used with caution in estimating value of ecosystem services such as storm protection because, first, one is essentially estimating a benefit (e.g., storm protection) by a cost (e.g., the costs of constructing sea walls, groins and other structures), and second, the human-built alternative is rarely the most cost-effective means of providing the service (Shabman and Batie, 1978; Ellis and Fisher, 1987; Freeman, 2003; Barbier, 2007).

Figure 2.2 illustrates the limitation of using the replacement cost method to estimate the protective value of an ECE. Assume that the ecosystem comprises a coastal wetland, such as a marsh or mangrove, of initial landscape area S_0.The cost of the storm protection service provided by the ecosystem is 'free' and thus corresponds to the horizontal axis, $0S_0$. However, suppose part of the wetland is lost or converted, and so the ecological landscape decreases to S_1.The replacement cost method would suggest that the value of this loss in wetland area could be estimated by the cost of 'replacing' the lost wetlands with sea walls, breakwaters, levies and other human-built structures to reduce storm surge and waves. In Figure 2.2, the marginal cost of an alternative, human-built coastal storm

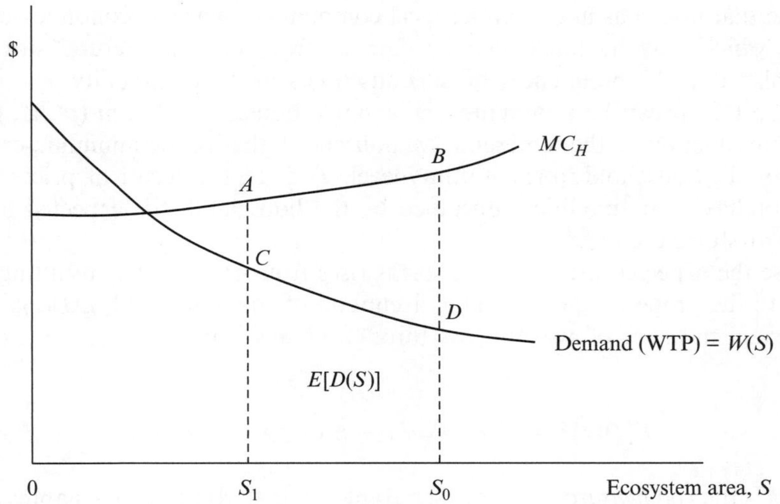

Figure 2.2 Replacement cost vs expected damage function estimation of protective value

barrier is MC_H. Thus, the 'replacement cost' of using the human-built barrier to provide the same storm protection service as the S_0S_1 amount of wetlands lost is the difference between the two supply curves, or area S_0ABS_1. However, this overestimates the benefit of having the wetlands provide the storm protection service. The true benefit of this ecosystem service is the demand curve, or total willingness to pay, for the service provided by S_0S_1 amount of wetlands less the costs of providing protection. In Figure 2.2, this net benefit corresponds to area S_0CDS_1.Thus, the replacement cost method overestimates the net benefits of the storm protection service by area *ABCD*.

As an alternative to the replacement cost method, some valuation studies of the protective value of estuarine and coastal ecosystems (ECEs) have used the expected damage function approach (Farber, 1987; Barbier, 2007; Hanley and Barbier, 2009, Chapter 6). In such cases, the ECE may be thought of as producing a non-marketed service, such as 'protection' of economic activity, property and even human lives, which benefits individuals through limiting damages. As a result, the expected damage function approach is an adaptation of the production function methodology of valuing the environment as an input into a final benefit (Barbier, 2007; Hanley and Barbier, 2009, Chapter 6). Utilizing this approach requires modelling the 'production' of this protection service and estimating its value as an environmental input in terms of the expected damages avoided.

The following example illustrates how the expected damage function (EDF) methodology can be applied to value the storm protection service provided by an ECE, such as a marsh or mangrove ecosystem. The starting point is the standard 'compensating surplus' approach to valuing a quantity or quality change in a non-market environmental good or service (Hanley and Barbier, 2009, Chapter 2). As the example illustrates, unlike the replacement cost method, the expected damage function approach can yield an exact measure of the willingness to pay for the loss or gain in wetland area that results in storm protection.

Assume that in a coastal region the local community owns all economic activity and property, which may be threatened by damage from periodic natural storm events. Assume also that the preferences of all households in the community are sufficiently identical so that it can be represented by a single household. Let $m(p^x, z^0, u^0)$ be the expenditure function of the representative household, that is, the minimum expenditure required by the household to reach utility level, u^0, given the vector of prices, p^x, for all market-purchased commodities consumed by the household, the expected number or incidence of storm events, z^0.

Suppose the expected incidence of storms rises from z^0 to z^1. The resulting expected damages to the property and economic livelihood of the household, $E[D(z)]$, translates into an exact measure of welfare loss through changes in the minimum expenditure function:

$$E[D(z)] = m(p^x, z^1, u^0) - m(p^x, z^0, u^0) = c(z), \qquad (2.1)$$

where $c(z)$ is the *compensating surplus* (see Hanley and Barbier, 2009, Chapter 2). In this example, compensating surplus is the minimum income compensation that the household requires to maintain it at the utility level u^0, despite the expected increase in damaging storm events. Alternatively, $c(z)$ can be viewed as the minimum income that the household needs to avoid the increase in expected storm damages.

However, the presence of coastal wetlands could mitigate the expected incidence of damaging storm events. Because of this storm protection service, the area of coastal wetlands, S, may have a direct effect on reducing the 'production' of natural disasters, in terms of their ability to inflict damages locally. Thus the 'production function' for the incidence of potentially damaging natural disasters can be represented as:

$$z = z(S),\ z' < 0,\ z'' > 0. \qquad (2.2)$$

It follows from Equations (2.1) and (2.2) that $\partial c(z)/\partial S = \partial E[D(z)]/\partial S < 0$. An increase in wetland area reduces expected storm damages and therefore also reduces the minimum income compensation needed to maintain the household at its original utility level. Alternatively, a loss in wetland area would increase expected storm damages and raises the minimum compensation required by the household to maintain its welfare. Thus, we can define the marginal willingness to pay, $W(S)$, for the protection services of the wetland in terms of the marginal impact of a change in wetland area on expected storm damages:

$$W(S) = -\frac{\partial E[D(z(S))]}{\partial S} = -E\left[\frac{\partial D}{\partial z}z'\right], W' < 0. \qquad (2.3)$$

The 'marginal valuation function', $W(S)$, is analogous to the Hicksian compensated demand function for marketed goods. The minus sign on the right-hand sign of Equation (2.3) allows this 'demand' function to be represented in the usual quadrant, and it has the normal downward-sloping property (see Figure 2.2). Although an increase in S reduces z and thus enables the household to avoid expected damages from storms, the additional value of this storm protection service to the household will fall as wetland area increases in size. This relationship should hold across all households in the coastal community. Consequently, as indicated in Figure 2.2, the marginal willingness to pay by the community for more storm protection declines with S.

The value of a non-marginal change in wetland area, from S_0 to S_1, can be measured as:

$$-\int_{S_0}^{S_1} W(S)dS = E[D(z(S))] = c(S). \tag{2.4}$$

If there is an increase in wetland area, then the value of this change is the total amount of expected damage costs avoided. If there is a reduction in wetland area, as shown in Figure 2.2, then the welfare loss is the total expected damages resulting from the increased incidence of storm events. As indicated in Equation (2.4), in both instances the valuation would be a compensation surplus measure of a change in the area of wetlands and the storm protection service that they provide.

A comparison of using an expected damage function approach and replacement cost method of estimating the welfare impacts of a loss of the storm protection service due to mangrove deforestation in Thailand confirms that the latter method tends to produce extremely high estimates compared to the EDF approach (Barbier, 2007). The comparison of annual and net present values produced by the two methods are depicted in Table 2.2. But the expected damage function has its own limitations, especially when households are risk averse, and in such circumstances can be a poor proxy for the *ex ante* willingness to pay to reduce or avoid the risk from storm damages (Freeman, 2003,

Table 2.2 Valuation of storm protection service of mangroves, Thailand, 1996–2004

Annual Deforestation Rate	FAO[a] 18.0 km^2	Thailand[b] 3.44 km^2
Valuation approach (US$)		
Replacement cost method[c]		
Annual welfare loss	25 504 821	4 869 720
Net present value (10% discount rate)	146 882 870	28 044 836
Net present value (12% discount rate)	135 896 056	25 947 087
Net present value (15% discount rate)	121 698 392	23 236 280
Expected damage function approach		
Annual welfare loss	3 382 169	645 769
Net present value (10% discount rate)	19 477 994	3 718 998
Net present value (12% discount rate)	18 021 043	3 440 818
Net present value (15% discount rate)	16 138 305	3 081 340

Notes:

a. FAO (Food and Agriculture Organization) estimates from Wilkie and Fortuna (2003). 2000 and 2004 data are estimated from 1990–2000 annual average mangrove loss of 18.0 km^2.

b. Thailand estimates from various Royal Thailand Forestry Department sources reported in Aksornkoae and Tokrisna (2004). 2000 and 2004 data are estimated from 1993–96 annual average mangrove loss of 3.44 km^2.

c. Based on replacement cost method assumptions of Sathirathai and Barbier (2001).

Source: Adapted from Barbier (2007).

pp. 243–7; Barbier, 2007). Nevertheless, because the EDF approach is a direct compensation surplus measure of a change in the area of ECEs and the storm protection service that they provide, it is a promising method of estimating the protective value of these ecosystems.

2.5 DISCUSSION AND CONCLUSIONS

Given the growing interest in the protective value of estuarine and coastal ecosystems (ECEs), there will be continual progress in the valuation methods employed to estimate this benefit. Improvements in the hydrodynamic modelling of storm surges, accounting for the influence of coastal topography of near-shore bathymetry, and allowing for the varying attributes of storms will also lead to better estimates of the protective value of ECEs.

For example, recent storm surge models developed for southern Louisiana along the US Gulf Coast show how the attenuation of surge by wetlands is affected by the bottom friction caused by vegetation, the surrounding coastal landscape, and the strength and duration of the storm forcing (Resio and Westerink, 2008; Loder et al., 2009; Wamsley et al., 2010). Although existing studies of the protective value of Gulf Coast wetlands do not incorporate such factors (Costanza et al., 2008), more accurate determination of this value will require allowing for the hydrodynamic properties of storm surges as well as the effects of varying wetland landscape and vegetation across coastal systems. Similarly, one of the most important innovations in recent assessments of the role of coastal forests, including mangroves, in protecting against the damages and casualties caused by the 2004 Indian Ocean tsunami has been separating out the influence of coastal topography, such as shoreline slope, distance of villages to shore and other coastal features, from the protection provided by forests (Cochard, 2011). For example, while the analysis by Laso Bayas et al. (2011) confirms that the presence of coastal vegetation significantly reduced the casualties caused by the tsunami in Aceh, Indonesia, distance to coast was the dominant determinant of casualties and infrastructure damage.

As discussed in Section 2.2, a growing number of field studies and experiments are showing that the wave attenuation function of ECEs, which is critical to their protective value, may vary spatially and temporally. For example, wave attenuation by coral reefs, seagrass beds, salt marshes, mangroves and sand dunes provides protection against wind and wave damage caused by coastal storm and surge events, but the magnitude of protection will vary spatially across the extent of these habitats (Madin and Connolly, 2006; Stockdon et al., 2007; Barbier et al., 2008; Koch et al., 2009; Gedan et al., 2011; Shephard et al., 2012). Only recently are valuation studies taking into account spatial and temporal variability of wave attenuation by ECEs in estimating their potential protective value (Barbier et al., 2008; Koch et al., 2009; Barbier, 2012).

Another unique feature of ECEs is that they occur at the interface between the coast, land and watersheds. The location of these ECEs in the land–sea interface suggests a high degree of 'interconnectedness' or 'connectivity' across these systems, which could lead to the linked provision of the storm protection service by more than one ECE. For example, Alongi (2008) suggests that the extent to which mangroves offer protection against catastrophic storm events, such as tsunamis, may depend not only on the relevant features and conditions within the mangrove ecosystem, such as width of forest, slope of forest floor, forest density, tree diameter and height, proportion of above-ground biomass in the roots, soil texture and forest location (open coast versus lagoon), but also on the presence of foreshore habitats, such as coral reefs, seagrass beds and dunes. Similar cumulative effects of wave attenuation are noted for seascapes containing coral reefs, seagrasses and marshes (Koch et al., 2009). For instance, evidence from the Seychelles documents how rising coral reef mortality and deterioration have increased significantly the wave energy reaching shores that are normally protected from erosion and storm surges by these reefs (Sheppard et al., 2005). In the Caribbean, mangroves appear not only to protect shorelines from coastal storms but may also enhance the recovery of coral reef fish populations from disturbances due to hurricanes and other violent storms (Mumby and Hastings, 2008). Modelling simulations for an interconnected reef–seagrass–mangrove seascape confirm that the storm protection service of the whole system is greater than for a single coastal habitat on its own (Sanchirico and Springborn, 2011).

As the world's estuarine and coastal ecosystems continue to disappear due to human population and development pressures, it becomes increasingly essential to assess the values of these important systems. Existing valuation studies suggest that the protective value of ECEs may be one of the more significant benefits sacrificed when these habitats are lost or degraded. As we improve our understanding of how various ECEs attenuate waves and buffer winds, we must also develop better methods of assessing the protective benefits of these ecosystems. Understanding the role of vegetation and other ECE attributes in storm protection compared to coastal topography and near-shore bathymetry is also essential, as is better hydrodynamic modelling of the storm surge and wind characteristics of various storm events. Finally, perhaps the biggest but most interesting challenge lies in allowing for the connectivity across ECE habitats to assess the wave attenuation and wind buffering functions underlying coastal protection. Only recently have valuation studies begun to model this connectivity and assess the cumulative implications for protective values across various ECEs.

REFERENCES

Aksornkoae, S. and R. Tokrisna (2004), 'Overview of shrimp farming and mangrove loss in Thailand', in E.B. Barbier and S. Sathirathai (eds), *Shrimp Farming and Mangrove Loss in Thailand*, Cheltenham, UK and Northampton, MA, USA: Edward Elgar Publishing.

Alongi, D.M. (2008), 'Mangrove forests: resilience, protection from tsunamis, and responses to global climate change', *Estuarine, Coastal and Shelf Science*, **76**(1), 1–13.

Badola, R. and S.A. Hussain (2005), 'Valuing ecosystems functions: an empirical study on the storm protection function of Bhitarkanika mangrove ecosystem, India', *Environmental Conservation*, **32**(01), 85–92.

Barbier, E.B. (2007), 'Valuing ecosystem services as productive inputs', *Economic Policy*, **22**(49), 177–229.

Barbier, E.B. (2011). *Capitalizing on Nature: Ecosystems as Natural Assets*, Cambridge, UK: Cambridge University Press.

Barbier, E.B. (2012), 'A spatial model of coastal ecosystem services', *Ecological Economics*, **78**, 70–79.

Barbier, E.B., E.W. Koch, B.R. Silliman, S.D. Hacker, E. Wolanski and J.H. Primavera et al. (2008), 'Coastal ecosystem-based management with non-linear ecological functions and values', *Science*, **319**(5861), 321–3.

Barbier, E.B., S.D. Hacker, C. Kennedy, E.W. Koch, A.C. Stier and B.R. Silliman (2011), 'The value of estuarine and coastal ecosystem services', *Ecological Monographs*, **81**(2), 169–83.

Braatz, S., S. Fortuna, J. Broadhead and R. Leslie (2007), *Coastal Protection in the Aftermath of the Indian Ocean Tsunami. What Role for Forests and Trees?* Bangkok: FAO, pp. 157–79.

Chong, J. (2005), *Protective Values of Mangrove and Coral Ecosystems: A Review of Methods and Evidence*, Gland, Switzerland: IUCN.
Cochard, R. (2011), 'On the strengths and drawbacks of tsunami-buffer forests', *Proceedings of the National Academy of Sciences*, **108**(46), 18571–2.
Cochard, R., S.L. Ranamukhaarachchi, G.P. Shivakotib, O.V. Shipin, P.J. Edwards and K.T. Seeland (2008), 'The 2004 tsunami in Aceh and Southern Thailand: a review on coastal ecosystems, wave hazards and vulnerability', *Perspectives in Plant Ecology. Evolution and Systematics*, **10**, 3–40.
Costanza, R., O. Pérez-Maqueo, M.L. Martinez, P. Sutton, S.J. Anderson and K. Mulder (2008), 'The value of coastal wetlands for hurricane protection', *Ambio*, **37**(4), 241–8.
Das, S. and J.R. Vincent (2009), 'Mangroves protected villages and reduced death toll during Indian super cyclone', *Proceedings of the National Academy of Sciences*, **106**(18), 7357–60.
Day, J.W. Jr., D.F. Boesch, E.J. Clairain, G.P. Kemp, S.B. Laska and W.J. Mitsch et al. (2007), 'Restoration of the Mississippi Delta: lessons from Hurricanes Katrina and Rita', *Science*, **315**(5819), 1679–84.
Ellis, G.M. and A.C. Fisher (1987), 'Valuing the environment as input', *Journal of Environmental Management*, **25**, 149–56.
Farber, S. (1987), 'The value of coastal wetlands for protection of property against hurricane wind damage', *Journal of Environmental Economics and Management*, **14**(2), 143–51.
Food and Agricultural Organization (FAO) of the United Nations (2007), 'The world's mangroves 1980–2005', FAO Forestry Paper No. 153. Rome: FAO.
Forbes, K. and J. Broadhead (2007), *The Role of Coastal Forests in the Mitigation of Tsunami Impacts*, RAP Publication 2007/1, Bangkok: Food and Agricultural Organization of the United Nations, Regional Office for Asia and the Pacific.
Freeman, A.M. III (2003), *The Measurement of Environmental and Resource Values: Theory and Methods*, 2nd edition, Washington, DC: Resources for the Future.
Gedan, K.B., M.J. Kirwan, E. Wolanski, E.B. Barbier and B.R. Silliman (2011), 'The present and future role of coastal vegetation in protecting shorelines: answering recent challenges to the paradigm', *Climatic Change*, **106**(1), 7–29.
Halpern, B.S., S. Walbridge, K.A. Selkoe, C.V. Kappel, F. Micheli and C. D'Agrosa et al. (2008), 'A global map of human impacts on marine ecosystems', *Science*, **319**(5865), 948–52.
Hanley, N. and E.B. Barbier (2009), *Pricing Nature: Cost–Benefit Analysis and Environmental Policy*, Cheltenham, UK and Northampton, MA, USA: Edward Elgar Publishing.
King, S.E. and J.N. Lester (1995), 'The value of salt marsh as a sea defence', *Marine Pollution Bulletin*, **30**(3), 180–89.
Koch, E.W., E.B. Barbier, B.R. Silliman, D.J. Reed, G.M.E. Perillo and S.D. Hacker et al. (2009), 'Non-linearity in ecosystem services: temporal and spatial variability in coastal protection', *Frontiers in Ecology and the Environment*, **7**(1), 29–37.
Laso Bayas, J.C., C. Marohn, G. Dercon, S. Dewi, H.P. Piepho and L. Joshi et al. (2011), 'Influence of coastal vegetation on the 2004 tsunami wave impact in West Aceh', *Proceedings of the National Academy of Sciences*, **108**(46), 18612–17.
Loder, N.M., J.L. Irish, M.A. Cialone and T.V. Wamsley (2009), 'Sensitivity of hurricane surge to morphological parameters of coastal wetlands', *Estuarine, Coastal and Shelf Science*, **84**(4), 625–36.
Lotze, H.K., H.S. Lenihan, B.J. Bourque, R.H. Bradbury, R.G. Cooke and M.C. Kay et al. (2006), 'Depletion, degradation, and recovery potential of estuaries and coastal seas', *Science*, **312**(5781), 1806–9.
Madin, J.S. and S.R. Connolly (2006), 'Ecological consequences of major hydrodynamic disturbances on coral reefs', *Nature*, **444**(7118), 477–80.
Massel, S.R., K. Furukawa and R.M. Brinkman (1999), 'Surface wave propagation in mangrove forests', *Fluid Dynamics Research*, **24**(4), 219–49.
Mazda, Y., M. Magi, M. Kogo and P.N. Hong (1997), 'Mangroves as a coastal protection from waves in the Tong King Delta, Vietnam', *Mangroves and Salt Marshes*, **1**(2), 127–35.
Millennium Ecosystem Assessment (MA) (2005), *Ecosystems and Human Well-being: Synthesis*, Washington, DC: Island Press.
Mumby, P.J. and A. Hastings (2008), 'The impact of ecosystem connectivity on coral reef resilience', *Journal of Applied Ecology*, **45**(3), 854–62.
Orth, R.J., T.J.B. Carruthers, W.C. Dennison, C.M. Duarte, J.W. Fourqurean and K.L. Heck Jr. et al. (2006), 'A global crisis for seagrass ecosystems', *BioScience*, **56**(12), 987–96.
Paul, M. and C.L. Amos (2011), 'Spatial and seasonal variation in wave attenuation over Zostera noltii', *Journal of Geophysical Research*, **116**(C8), C08019.
Resio, D.T. and J.J. Westerink (2008), 'Hurricanes and the physics of surges', *Physics Today*, **61**(9), 33–8.
Sanchirico, J.N. and M. Springborn (2011), 'How to get there from here: ecological and economic dynamics of ecosystem service provisions', *Environmental and Resource Economics*, **48**(2), 243–67.

Sathirathai, S. and E.B. Barbier (2001), 'Valuing mangrove conservation, Southern Thailand', *Contemporary Economic Policy*, **19**(2), 109–22.
Shabman, L.A. and S.S. Batie (1978), 'Economic value of natural coastal wetlands: a critique', *Coastal Zone Management Journal*, **4**(3), 231–47.
Shephard, C.C., C.M. Crain and M.W. Beck (2012), 'The protective role of coastal marshes: a systematic review and meta-analysis', *PLoS ONE*, **6**(11), e27374.
Sheppard, C., D.J. Dixon, M. Gourlay, A. Sheppard and R. Payet (2005), 'Coral mortality increases wave energy reaching shores protected by reef flats: examples from the Seychelles', *Estuarine, Coastal and Shelf Science*, **64**(2–3), 223–34.
Spalding, M., M. Kainuma and L. Collins (2010), *World Atlas of Mangroves*, London: Earthscan.
Stockdon, H.F., H.A. Sallenger, R.A. Holman and P.A. Howd (2007), 'A simple model for the spatially-variable coastal response to hurricanes', *Marine Geology*, **238**(1–4), 1–20.
United Nations Environment Programme (UNEP) (2006), *Marine and Coastal Ecosystems and Human Wellbeing: A Synthesis Report Based on the Findings of the Millennium Ecosystem Assessment*, Nairobi: UNEP.
Wamsley, T.V., M.A. Cialone, J.M. Smith, J.H. Atkinson and J.D. Rosati (2010), 'The potential of wetlands in reducing storm surge', *Ocean Engineering*, **37**(1), 59–68.
Waycott, M., C.M. Duarte, T.J.B. Carruthers, R.J. Orth, W.C. Dennison and S. Olyarnik et al. (2009), 'Accelerating loss of seagrasses across the globe threatens coastal ecosystems', *Proceedings of the National Academy of Sciences*, **106**(30), 12377–81.
Wilkie, M.L and S. Fortuna (2003), 'Status and trends in mangrove area extent worldwide', Forest Resources Assessment Working Paper No. 63. Rome: FAO.
Wilkinson, C., O. Linden, H. Cesar, G. Hodgson, J. Rubens and A.E. Strong (1999), 'Ecological and socio-economic impacts of 1998 coral mortality in the Indian Ocean: an ENSO impact and a warning of future change?', *Ambio*, **28**(2), 188–96.
Wolanski, E. (2007), *Estuarine Ecohydrology*, Amsterdam: Elsevier.
Worm, B., E.B. Barbier, N. Beaumont, J.E. Duffy, C. Folke and B.S. Halpern et al. (2006), 'Impacts of biodiversity loss on ocean ecosystem services', *Science*, **314**(5800), 787–90.

3. Cruising for a bruising: challenges in sustainable capture of ecosystem service values from cruise ship tourism in Belize

Andrew Seidl, Lawrence Pratt, Martha Honey, William H. Durham, Geraldine Slean and Amos Bien

3.1 INTRODUCTION

Over the last several decades, Belize has built an international reputation for small-scale, nature and cultural tourism, widely known as 'ecotourism'. Ecotourism holds the promise of providing a nature-based solution to a country's economic development challenges by facilitating local capture of the valuable ecosystem services it nurtures. Since 2000, the cruise industry has also put down roots in Belize, and today cruise tourism is widely viewed as a permanent part of the country's tourism landscape. Beginning in 2002, cruise passenger numbers surpassed stayover visitors and, in 2010, more than 3.2 cruise passengers arrived for every stayover visitor (TEEB, 2010). From 2000 to 2005 in fact, Belize was the fastest growing cruise market in the Caribbean. Today, there are indications that cruise visitor numbers may have peaked and tapered off somewhat, but the sector remains vitally important to ongoing development efforts. The Belize government, like others in Central America and the Caribbean, is faced with choices about how best to use resources in the service of the country's tourism development. This study is intended to provide data and analysis to inform local decision-making and to illustrate an approach to facilitating better management of ecosystem services through their improved measurement.

Globally, both 'experiential' forms of tourism (including ecotourism) and cruise tourism are growing rapidly (UNWTO, 2001). In Belize, perhaps more than anywhere else in either region, cruise tourism is competing with and in some instances colliding with ecotourism. Belize's national tourism motto – 'Nature's Best Kept Secret' – and its strategic vision of promoting 'responsible tourism [that encourages] a strong "eco-ethic" to ensure environmental and socio-cultural sustainability' (BTB, 2004a) are challenged by the rapid growth of cruise tourism in the last decade-plus.. Balancing cruise tourism, where visitors bring their hotel with them, and various forms of 'stayover' tourism, where they typically sleep in local, land-based accommodation, has proved an enormous challenge. Fears have been raised that 'Belize is killing its golden goose' of ecotourism with far too many cruise passenger 'day trippers'.

Cruise tourism has brought new revenue, more employment and improved infrastructure, particularly to port cities. On the other hand, the size and scale of cruise ships, their resource consumption, waste generation and volume of visitors has produced increasingly visible impacts on the region, highlighting the need for much firmer controls and regulations. Governments welcome the increased revenue and jobs that cruise tourism generates, but are also concerned about negotiating the best contracts and balancing the numbers and needs of cruise and stayover visitors. While some businesses vie for a piece of the cruise tourism pie, others opt out or are shut out of the cruise economy. Environmentalists are concerned about visitor impacts on parks, preserves and biodiversity. Local communities are interested in improving their livelihoods and their attractiveness to future tourists. All sectors are vitally interested in measuring and weighing the comparative advantages of land-based resort tourism, ecotourism and cruise tourism. According to the World Travel and Tourism Council (WTTC): 'there is widespread concern within the Caribbean tourism industry that there has been a lack of balance until now between cruise tourism development and that of land-based tourism' (WTTC, 2003 in CESD, 2006a, p. 15).

To date, much of the attention in cruise research and policy discussion has focused on wastewater discharges and other shipboard operations (for example, Wood, 2004). In contrast, there has been little systematic analysis of the economic, environmental and socio-cultural impacts of cruise tourism on host communities and visitor sites, and on the way that those impacts are perceived locally. National governments have often lacked the data needed to make sound policy decisions. The Barbados Tourism Minister, Noel Lynch, told 700 delegates attending the annual Caribbean Tourism Conference: 'There needs to be an independent study of the impact of cruise tourism and land-based tourism and how they can work together. I believe the jury is still out on what the real impact is from the cruise ships'. He urged 'that we have one study collectively commissioned together to get rid of these myths' (CESD, 2006a, p. 16).

This study examines the terrestrial economic effects of cruise tourism in Belize. Through analysis of cruise passenger surveys, the study compares spending patterns, activities, perceptions and preferences of cruise visitors with independently collected information regarding stayover tourists.

3.2 METHODOLOGY

The cruise tourist survey instrument consisted of a structured passenger questionnaire that was pre-tested in Costa Rica, adapted for Belize, and then implemented in and around Belize City. The 10- to 15-minute survey consisted of a four-page questionnaire with 65 questions divided into four sections. Inquiries were made concerning trip characteristics and activities, trip satisfaction, expenditure patterns, tour purchases, willingness to spend and demographic information.

The research team in Belize worked under the director of the Belize Tourism Board. The research design in Belize included a mix of quantitative and qualitative interviews with various stakeholders regarding the impacts of cruise tourism. Economic effects comprised the bulk of the quantitative research. Economic issues were examined by tracking expenditures of disembarking passengers and detailing the effects of direct and indirect impacts on port communities and port authorities. Results from the passenger surveys were compared with airport exit surveys of departing stayover tourists (Visitor Expenditure and Motivation Survey or VEMS) conducted by the Belize Tourism Board and Central Bank of Belize in 2003 (BTB, 2004b; Central Bank of Belize, 2005). We used these two surveys to analyse a range of preferences, spending patterns and impacts of cruise passengers and stayover tourists.

A travel cost method (TCM) approach facilitated the ability to study patterns in activity choice, spending, preferences and willingness to pay among cruise passengers. Commonly employed to analyse demand for tourism services, TCM relies on tourist surveys to obtain a profile of visitors' actual expenditures and to elicit sensitivity to exogenous changes in travel costs. In addition to basic demographics, TCM also investigates trip characteristics in order to derive a demand curve for tourism visitation. In general, TCM allows researchers to extrapolate survey results to broader populations, infer visitors' willingness to pay for tourism services, and explore the effect of local, national, or industry policy changes on tourism behaviour.

Surveying resulted in 609 useable surveys (of 623 completed surveys). Most of the 14 rejected surveys were due to non-target populations (for example, minors, business travellers, people visiting family or friends). The sampling plan originally called for random sampling of cruise tourists within the Belize City Tourist Village just prior to their return to the ship. However, as most tourists participate in a tour upon disembarking from the ship, and few subjects were willing and able to complete the surveys while they were within the village limits, arrangements were made with Bel-Cruise Limited to administer the surveys during bus rides for Carnival cruise ship passengers on their return from a local attraction known as cave tubing at Caves Branch. In 2008, over 50 per cent of all cruise passengers to Belize came by Carnival. Since Bel-Cruise is the primary supplier of tours offered through Carnival shore excursions, its tours were identified as a convenient way to survey a representative majority of cruise visitors to Belize. As a consequence of interviewing during pre-arranged Bel-Cruise rides, practically all respondents were on a cruise ship from among Carnival Cruise Lines' fleet (for example, *Elation*, *Valor*, *Glory* and *Miracle*). Moreover, according to the 2003 VEM survey data of cruise passengers, cave tubing at Caves Branch is the most commonly selected tour package (Central Bank of Belize, 2005). In 2005, over 84 000 cruise visitors participated in this activity, representing 82 per cent of all visitors to Caves Branch.[1] In 2008, over 94 000 tourists visited Caves Branch, of whom cruise visitors were the most important group.[2]

Royal Caribbean's ownership of the Village is unfortunate from a research perspective due to the potential to invite bias in data obtained there from too many Royal Caribbean customers. We are sensitive to this potential source of bias. However, our surveyors were only able to secure permission to interview on the buses run by tour companies with contractual arrangements with a second cruise company, Carnival, and this arrangement potentially counteracts any bias created by surveying shoppers in a Royal Caribbean--owned tourist village. Since Royal Caribbean and Carnival lines represented 72 per cent of all cruise ship arrivals in Belize, we feel our data still remain broadly representative.

For comparison with our field data and contextual understanding, general profiles of tourists and tourism in Belize were collected from the published literature (including NGO studies, government reports and destination site reports). Additional information was compiled on the political and economic history of cruise tourism, the history of tourism in Belize and the Caribbean, national tourism policy, cruise impacts and environmental and scientific studies of various marine and terrestrial visitor sites. In addition, researchers collected the latest statistics on tourism growth, passenger spending (by both cruise and stayover guests), departure taxes, cruise head taxes and other fees. As shown below, these aggregate measures provide valuable perspectives on the data collected specifically for this study.

3.3 CARIBBEAN CRUISE TOURISM

In the Caribbean, one of the most tourism-intensive regions of the world, tourism is a major source of income for most countries. Within that sector, cruise tourism has recently expanded as one of the most dynamic components, having benefited from both the depreciation of the US dollar and the region's image as a travel destination still safe from terrorism. Following 9/11, many cruise lines pulled their voyages out of the Mediterranean, placed them closer to their home ports of Fort Lauderdale or Miami, increased and diversified marketing campaigns and offered discounts to attract a wider clientele (Mahler, 2003). Cruise tourism was expanding at a time when the traditional staple of Caribbean tourism – namely sun-and-sand resort tourism – appeared to be losing its lustre. According to the United Nations World Tourism Organization (UNWTO), by 2001 sun-and-sand resort tourism had 'matured as a market' and its growth was projected to remain flat (UNWTO, 2001).

At the same time, cruise tourism began moving into locations in the Caribbean (Dominica, Grenada) and Central America (Belize, Honduras) where small-scale nature-based tourism – ecotourism – and small-scale 'pocket' cruises and yachts have so far defined the country's image. In the Caribbean, the total number of cruise ship passenger and stayover arrivals were currently about equal in 2005 – some 15 million each – and the markets for both were dynamic and growing. The UNWTO ranks 'experiential' tourism – which encompasses ecotourism, nature, heritage, cultural and soft adventure tourism, as well as sub-sectors such as rural and community tourism – as among the sectors expected to grow most quickly over the next two decades. It also predicts that cruise tourism will continue to be one of the top products worldwide (ibid.).

About 50 per cent of the global cruise market operates in the Caribbean, and cruise tourism in the region has generally grown faster than land-based tourism (WTTC, 2004). Not only has Caribbean cruise travel's image as a safe and secure holiday risen in the post-9/11 era, but the cruise industry's tax-free status and its 'token' port charges have supported its rapid development, in contrast to the 'significant' departure taxes paid by stayover tourists (ibid.). The number of cruise passengers globally has more than doubled since 1990. Ship size has grown exponentially, and the industry has consolidated so that today three lines – Carnival, Royal Caribbean and Norwegian – control 90 per cent of the North American market and account for almost 75 per cent of total capacity deployed in the Caribbean.[3] Between 2004 and 2009, 21 new liners were launched. These newer ships accommodate, on average, 3000 passengers and 1000 crew. This represents dramatic growth from earlier cruise liners that had capacities for only 500 to 800 passengers (Mintel International Group, 2005). Large ships are also likely to lead to increased levels of 'bunching' – with port congestion on some days and little business on others – so that facilities are not utilized as efficiently as they might be.

Most of these trends and concerns have played out in Belize, a country that uniquely carries both a Caribbean and a Central American identity. The extraordinarily rapid growth of cruise arrivals in Belize between 2000 and 2004 yielded benefits in terms of revenue and job creation, but it also presented enormous challenges and scant time for reflection and readjustment. The government and private sector responded to the new demand with heavy investment into cruise tourism and plans for future expansions.

Today, as Belizean experts perceive that cruise tourism likely peaked in 2004 and is now in decline, new strategies and policies are required.

3.4 CRUISE VS STAYOVER TOURISM IN BELIZE

Belize is a prime example of a country that is trying to protect its international reputation for ecotourism in the wake of the rapid rise of mass market cruise tourism. Encompassing 22 966 km^2 – 386 of which are coastline – Belize is best known as a nature-based tourist destination. Its spectacular coral reefs, white beaches, dense rainforests and Mayan archaeological sites attract visitors from all over the world. Marketed as 'Mother Nature's Best Kept Secret', Belize was a relatively unknown tourist destination until 20 years ago. Today, over 41 per cent of all land in Belize is protected under an extensive network of national parks, preserves, sanctuaries and archaeological sites, such as the Caye Caulker Marine Reserve, Hol Chan Marine Reserve, Half Moon Caye, Crooked Tree Wildlife Sanctuary, Cockscomb Wildlife Sanctuary, Altun Ha, and Lamanai. These areas offer tourists numerous natural, cultural and adventure opportunities. For these reasons, Belize is regarded as a showcase site of ecotourism (Blackstone, 1998; Eltringham, 2001).

Stayover tourism in Belize has grown steadily but gradually over the past quarter-century. Despite having many of the right ingredients – outstanding natural and archaeological sites; a friendly English-speaking population; a network of small lodges and hotels; an international reputation for high-quality ecotourism; political stability; and proximity to North America – Belize lags far behind other countries in Central America and a number of smaller Caribbean islands. In fact, Belize has had one of the lowest non-cruise tourism arrivals in Central America, and one of the slowest growth rates in stayover tourism arrivals of any country in the region (a reduction of 3.7 per cent from 2006 to 2010) (SITCA and CCT, 2011).

In the meantime, perhaps not coincidentally, cruise tourism in Belize was experiencing a boom. In 2003, Belize had become 'the fastest growing cruise destination' (WTTC, 2004). By 2004, Belize ranked first in growth rate and eighth in total highest annual cruise arrival rates in the Caribbean and Central America; it also ranked 22nd in stayover tourism arrivals (UNWTO, 2005). According to former Tourism Minister Mark Espat, until 1999 cruise tourism was an insignificant sub-sector to what is now popularly called overnight tourism (Espat, 2004).

Since then, cruise arrival numbers have increased dramatically. By 2002, the number of cruise tourists exceeded stayover tourists. In 2004, over 850 000 cruise passengers made up 78.7 per cent of all visitors to Belize – a 25-fold increase since 1999. In 2005, despite expectations of reaching more than 1 million cruise passenger arrivals, numbers declined by 6 per cent that year due to the frequency and intensity of hurricanes (for example, Katrina, Rita, Wilma, Dennis and Emily), which caused cruise ships to alter their itineraries (BTO, 2005b). In contrast with nature-based, stayover tourism, which grew steadily from the 1980s until 2007, cruise tourism has had a very short but influential history in Belize. It is noteworthy that while cruise tourism grew by 18 per cent from 2007 to 2010, stayover tourism decreased by 6 per cent.

Cruise ship arrivals are not evenly dispersed throughout the year or the week. Cruise ships arrive most frequently during the peak tourism season from November through April, sometimes bringing over twice as many monthly visitors as in other months. In 2002, for instance, when cruise passenger numbers jumped six-fold to nearly 320000, most arrived on Wednesdays (Mahler, 2003). More recently, multiple cruise ships – each carrying over 2000 guests – may be seen docking on any given day. In 2004, Belize received 399 port calls from Carnival, Royal Caribbean and Norwegian, while in 2005, the total dropped slightly to 370 arrivals. Challenges of uncoordinated management of cruise ship arrivals persist, creating periodic inefficiencies of both over- and under-capacity (ibid.). This uneven distribution unnecessarily puts added burdens on the country's attractions, services and infrastructure.

Typically, ships arrive in Belize City in the early morning and leave in the late afternoon. Cruise days turn the sleepy, steamy Belize City port into a beehive of activity. Dozens of modern buses drive along Belize City's narrow streets to line up by Memorial Park, taxis queue down side streets, vendors station themselves alongside roads to the docks. In the city streets, hundreds of small vendors materialize, smartly dressed and ready to offer cruise passengers a range of local foods, crafts, clothing, tours and trips, hair braiding and other services. In addition, a handful of big Belizean and international companies with stores at the port or in Belize City advertise on board the cruises, and pay commissions to the cruise company in typical arrangements.

A select number of 'preferred' tour operators have negotiated contracts with each cruise line, resulting in 100 per cent mark-ups for cruise ship visitor purchases of their tour services over their standard fares. The majority of the cruise ship tours and excursions are handled by a relatively small number of companies. About 60 per cent of disembarking passengers have already purchased tours through shore excursions; only 18.2 per cent of individuals that disembark pay a local onshore operator for their tour experience (Mahler, personal communication).

Belize's cruise tourism infrastructure investments have been supported (in part) by the cruise passenger head tax. As part of a 15-year agreement signed in 2001, the government pledged US$4 (BZ$8) of each head tax to the Tourism Village. Of the US$5 (BZ$10) head tax, US$1 went to the government. In January 2005, the head tax was raised to US$7, with US$3 divided almost equally between PACT (Belize's Protected Areas Conservation Trust) to support national conservation and the Belize Tourism Board to support tourism activities. The number of cruise passengers who arrived in Belize between September 2001 (when US$4 payments to the owners of the Tourism Village are said to have started) and the end of 2005 was over 2.5 million. Based on this, the government has paid the Tourism Village owners (first the Feinstein Group and then Royal Caribbean and Diamonds International) over US$10 million during these years, while the government agencies received about US$4 million.[4]

According to the 2003 Belize Cruise Ship Policy (BTB, 2003), between 2000 and 2003 both the government and the private sector 'invested significantly in preparation for the growth of cruise tourism', including the purchase of tenders and buses and the opening of new attractions. This also included an extensive landfill project along the water front to permit better movement of cruise buses and taxis. The Belize government invested 'significantly in improving. . .major archaeological sites and human resources' as part of an InterAmerican Development Bank (IDB)-funded project (BTB, 2003, p. 2).

Table 3.1 Cruise tourists in Belize: demographic information

	Mean	Median	Mode	Max	Min
Male (%)	41	0	0	1	0
Age (years)	40.26	42	42	75	18
US citizen (%)	96	1	1	1	0
US or Canadian citizen (%)	97	1	1	1	0
Employed (% yes)	83	1	1	1	0
Paid vacation (weeks)	3.66	3	2	18	0
Household size	3.12	3	2	9	0
Household income earners	1.78	2	2	5	1
Household income (US$)	91 422	90 000	140 000	140 000	30 000

3.5 CRUISE SHIP PASSENGER SURVEYS RESULTS

3.5.1 Sample Demographics

In our 2005 survey of Belize cruise passengers, 41 per cent of respondents were male and their average age was just over 40 years, somewhat younger and less gender balanced than expected. Respondents were also somewhat more educated than expected, with most having completed a four-year college or university education. (These findings may reflect the fact that the surveys were undertaken with passengers going on land excursions who may well have been a younger and more active sample than ship passengers who either did not disembark or stayed only within the dock-side 'Village' or Belize City.) About 8 per cent of respondents were retired, somewhat lower than expectations; and about 83 per cent of respondents were employed. Employed respondents received an average of three to four weeks of paid vacation per year, though there was substantial variation in response to this question (Table 3.1).

An overwhelming 96 per cent of respondents were US citizens, and more than 1 per cent was Canadian. If Mexican and Puerto Rican nationals were included in the calculations, some 99 per cent of cruise passengers to Belize would have been North Americans. Reported household sizes of two to three people, including two income earners, are typical in the United States. However, an average household income of about US$90 000 indicates a wealthier demographic than was expected based on other studies conducted on this tourism sector (Table 3.1).

3.5.2 Cruise Vacation Characteristics

Our sample of cruise passengers found that the average length of Carnival cruises that include Belize is seven days, a figure that accords nicely with company websites. Carnival itineraries include several additional ports of call, including the Cayman Islands; Costa Rica; Panama; Roatán, Honduras; Cancun or Cozumel, Mexico; and Key West, Florida. Royal Caribbean and Norwegian Cruise Lines also travel to Belize on five- to seven-day tours, with stops at either Cozumel or Costa Maya, Mexico. Thus, Carnival tends to visit

Table 3.2 General cruise vacation travel expenses, 2005, in US$

	N	Mean	Median	Mode	Max	Min	St Dev
Cost of cruise	574	1 335.83	1200	1 000	5 000	75	671.27
Transportation to/from home port	509	370.18	240	200	8 000	0	632.37
Daily onboard spending	554	103.77	50	100	2 000	0	164.12
Total cruise-related expenditures	526	2 345.22	1 922.5	1 900	11 750	35	1 451.99
Total cruise-related expenditures per diem	521	342.08	271.43	271.43	2 850	5	247.13

more ports per itinerary than its competitors. Port visits, pricing and capacity are nearly equivalent for all three companies.[5]

Respondents were asked to report the per person cost of their cruise, the transportation costs to and from the ship, and their daily onboard expenditures. These figures can be used to estimate the total cruise-related expenditures of the cruise vacation, excluding expenditures for purchases at the ports of call themselves or for tours in the ports of call/host countries. Some port and tour expenditures accrue to the cruise ship, while some go to the host port community and country. The cruise costs and onboard expenditures clearly accrue to the cruise company, while some of the transportation, tour, shop and restaurant expenditures may find their way to the cruise company, depending on the contractual agreements between the cruise lines and local businesses.

According to people surveyed, the average cruise price for itineraries that include a Belizean port of call was estimated at US$1336 per person, excluding transportation to and from the cruise's home port. Substantial variation in cruise rates was observed, but the majority of responses fall between US$1000 and US$2000 per person (Table 3.2). The average per person expenditure on transportation to and from the home port was US$370. However, there was a great deal of variation in response, with some people reporting essentially no costs and most respondents reporting transportation costs of under US$1000 per person (Table 3.2). The average daily per person expenditure while on the ship was US$104. Many people estimated onboard expenditures at substantially less (median US$50 per day), but some big spenders reported very high onboard purchases, thus skewing the average (Table 3.2). The average total cruise-related expenditure, including transportation, was US$2345 per person, with the median and mode expenditure calculated at about US$400–450 less per person. This indicates a substantial spread in spending practices beyond the basic cruise package price. Survey results also found that the average daily expenditures for a cruise vacation including a Belizean port of call was US$342 per person (Table 3.2). In total, cruise tourists whose vacation included a port call in Belize spent about US$1.9 billion in 2005, excluding their in-port expenditures.

Although cruises have the reputation of being a 'bargain vacation', the survey findings from the cruise passenger surveys and the airport surveys (analysed in more detail below) showed that cruises to Belize were more expensive than pre-packaged stayover trips in Belize of equal length. To compare, the average cruise to Belize, including airfare, cost US$2345 per passenger, while the average pre-packaged stayover tour to Belize costs US$1522, according to the 2003 VEM survey (BTB, 2004b). These findings demonstrate that, on average, week-long stayover holidays to Belize cost less than the average seven-day cruise vacation stopping in Belize (ibid.). However, as we will show, only a fraction of the average cruise vacation costs remain in Belize.

3.5.3 Cruise Tourist In-port Activities

The rate of disembarkation in Belize was 85 per cent, according to BTB statistics, making it 'higher than the rates in other Caribbean destinations' (BTB, 2005a). Several factors appear to contribute to this. Under Belize's 2000 Cruise Ship Policy, cruise lines must shut down all activities on board the vessels – casinos, restaurants and entertainment – while in Belize, which encourages people to disembark. In addition, Belize offers a relatively wide variety of tours compared than other cruise destinations (BTB, 2005a). For instance, Carnival's website listed 25 different shore excursions in Belize that passengers could choose to purchase on board or online. They ranged in duration from two-and-a-half to seven hours and in cost from US$40 to a bit under US$200. In comparison, other ports of call on itineraries with Belize offered 16 shore excursions in Limón, Costa Rica; nine excursions in Colón, Panama; 35 in Roatán, Honduras; and 26 in Costa Maya, Mexico.[6]

Of those surveyed for this study, some 86 per cent indicated that they disembarked at all ports of call, while an additional 8 per cent indicated they left the ship at only some ports of call. Practically all (93 per cent) respondents purchased land tours, with more than half of respondents purchasing tours at all ports of call on the ship's itinerary. Again, the interview pool was made up largely of people on excursion buses and is therefore biased towards passengers who take excursions. According to the BTB, approximately 85 per cent of all cruise passengers choose to disembark; and 60 per cent of these individuals had purchased tours while onboard the cruise ship.[7] Using this information, it can be estimated that of the 800 331 cruise passengers who arrived in Belize in 2005, approximately 408 169 purchased shore excursions from cruise lines. About 18.2 per cent of the passengers who disembarked in Belize (or 123 811 passengers) purchased their tours onshore either through operators in the Village or independent guides outside the Village. The remaining 21.8 per cent of disembarked passengers (or 148 301 individuals) walked around the Tourism Village and Belize City.

When cruise tourists visit Belize they engage in a variety of recreational activities, and many engage in multiple activities while onshore. The most popular activities for cruise visitors to Belize were shopping, a city tour, a visit to Mayan archaeological sites, cave tubing, a trip to the beach, hiking or walking in the rainforest, and snorkelling (Table 3.3). While shopping nearly always ranks as the top activity for cruise passengers in any port, cruise tourists in Belize also showed preference for activities for which the country is internationally known: more than one-third of cruise tourists visited archaeological sites, went cave tubing and to the beach, and about one-quarter went snorkelling, on a river cruise or boat tour (Table 3.3).

Most of these onshore activities involved spending money in the Belizean economy. Community economic development is driven by the amount of money spent in the local economy (referred to as 'drop'), the proportion of local content in the goods and services purchased, and the distribution of those expenditures through the local economy (or 'multiplier'). Respondents were asked about their expenditures in Belize, including tour expenditures, local arts and crafts, duty-free shopping, transportation, food and drink,

Table 3.3 Cruise tourist activities and excursions in Belize

Activity	% Participating	Rank	Activity	% Participating	Rank
Shopping	47	1	Horseback ride	5	13
Visit archaeological sites	42	2	Drive for pleasure	3	14
			Attend local music/dance	3	15
City tour	39	3			
Cave tubing	38	4	Scuba dive	3	16
Beach	35	5	Canopy tour	3	16
Hike/walk	30	6	Farm/ranch	3	17
Snorkel	28	7	Bicycle/mountain bike	3	18
Boat/boat tour	22	8			
River tour	21	9	Sport fishing	2	19
Museum/zoo visit	17	10	Surf/boogie board	2	20
Wildlife watching	15	11		0	21
National Park visit	15	12	Windsurf	0	22

event tickets, entrance and/or licence fees. An analysis of the impact of various expenditures on Belize's economic development follows.

3.5.4 Tour Expenditures

In analysing the data, questions about tour expenditures were separated from other expenditures because of the strong possibility that tours are purchased on board the ship. When tours are purchased onshore the tour company may capture 100 per cent of the purchase price (or somewhat less if they are required to pay a concession to the port authority). Purchase prices for tours sold onshore ranged from US$25 to US$70. Cave tubing, for example, costs US$65 if purchased through tour operators in the Village. The average tour price paid by cruise passengers to independent onshore tour operators was US$47.50 – or 76 per cent of what they might be expected to pay to Carnival for similar tours.

In our survey sample, 93 per cent of the tours to visit Belize were purchased on board the ship, at an average cost of US$78. Based on the aforementioned data, the ship captured about 56 per cent of the price of tours purchased on board, and these expenditures were not equal to direct local economic impact. Instead, we expected the amount that the local tour provider earned (that is, 44 per cent of the cost of tours) – approximately US$34 per cruise tourist – as contribution toward the Belizean economy. Given that 85 per cent of cruise tourists chose to disembark in Belize, and 60 per cent of these passengers purchased their tour on board, the contribution of these onboard tour purchases to the local Belizean economy could be estimated at US$14 million, based on 800 331 cruise tourist arrivals registered by the Belize Tourism Board in 2005.[8] Some US$17.8 million, therefore, accrued to the cruise lines for brokering these services.[9]

Onshore tour purchases, however, were 76 per cent of the cost of onboard tour purchases. Our survey data revealed an average onboard tour purchase price of US$78 and an onshore tour purchase price of US$59. The difference in local impact between onboard tour purchases and onshore tour purchases is US$25, or 32 per cent of the US$78 average

Table 3.4 Amount spent in US $ (per person) while off the ship, in 2005 (excluding tour purchases) – non-zero responses

	% Reporting	Mean	Median	Mode	Max	Min
Total	89	54.36	26	20	940	1
Other gifts and souvenirs	41	30.63	20	10	300	1
Food and drink	64	16.60	10	10	300	1
Local arts and crafts	32	29.13	20	10	300	1
Duty-free shopping	19	45.87	25	20	600	1
Events tickets	6	61.85	15	10	400	1
Local transportation	3	16.9	10	10	150	5

tour price. For those 123 811 individuals who purchased tours onshore – estimated at an average cost of US$59 – their contribution of US$7.3 million remained primarily within the local economy. The total direct economic impact from tour purchases was estimated at US$21.3 million,[10] and on average each (680 281) disembarked passenger spent US$31 in direct contribution to the Belizean economy through tour purchases.[11] Alternatively, on average each (800 331) cruise passenger contributed US $27.[12]

3.5.5 Non-tour Expenditures

Non-tour local expenditures were reported in two ways. An overall calculation is provided, including all zero and positive value responses. Excluding tour expenditures, a cruise tourist could be expected to spend between US$20 (median) and US$48 (mean) in Belize. In addition, a calculation based only on positive value responses is presented (Table 3.4). Approximately 89 per cent of respondents spent some money apart from any tour expenditures they might have had, thus reporting a non-zero response to the total expenditures category; among these respondents, the average expenditure could be expected to be between US$26 (median) and US$54 (mean).

According to the survey results, cruise tourists' gift and souvenir purchases averaged US$13 overall, and the 41 per cent of respondents who purchased gifts and souvenirs reported an average expenditure of US$31. Food and drink expenditures while in Belize were US$11 per tourist, on average, and US$17 among the 64 per cent of cruise tourists who reported expenditures in this category. About one-third of cruise tourists reported purchasing an average of US$29 in local arts and crafts, creating an overall average expenditure of about US$9 on arts and crafts.

Duty-free shopping was separated from other shopping expenditures due to the extremely high proportion of non-local content in most duty-free items. For instance, a tourist's purchase of a US$150 Swiss watch at a duty-free store does not contribute anywhere near US$150 to the Belizean economy. In contrast, a large portion of a US$150 painting by a local artist would logically return to benefit the local economy.

Only about one-fifth of cruise tourists to Belize reported duty-free shopping expenses. The average among those reporting duty-free expenditures was US$46 and the median was US$19 (Table 3.4). Our overall reported average duty-free purchases (US$9) may appear low, given that BTB data show the minimum per passenger expenditure on duty-free items as US$17.[13] Since our survey respondents were likely to engage in tour activities, it may be that individuals experienced constraints on their time that prevented them from purchasing duty-free items. However, there was no clear evidence to indicate that passengers who decided not to participate in a tour were likely to spend more or less than their counterparts who did purchase a tour. Thus, it seems reasonable to consider expenditure findings as representative of all disembarked passengers for the purposes of this study.

Compared with many Caribbean cruise ship destinations, Belize is not a significant port for duty-free shopping. As detailed below, passengers list duty-free shopping as 14th in a long list of activities in Belize, although shopping in general remains their number one activity. Since so much of the money from duty-free shops goes out of the country, Belize has been wise not to promote this type of tourism activity.

Very few (6 per cent) respondents reported expenditures in event tickets, entrance fees and licence fees – bringing the overall average down to US$4. One important reason was that tour packages generally include these fees. However, those who did report purchases in this area averaged some US$62 per person. Similarly, very few (3 per cent) cruise tourists reported local transportation costs, accounting for US$0.45 of the overall average expenditure but US$16.9 among those who reported purchases in this category (Table 3.4). Again, this figure makes sense because local transportation was included in tour excursion packages.

An admittedly simplified, but still useful, number is the average total individual expenditures in Belize during a cruise-based visit. To the estimated amount spent while off the ship – US$20 – we added the average local contribution of tours per disembarked passenger – US$31 – to obtain an approximate figure: US$51 per disembarked cruise visitor in total expenditures in Belize; or US$44 in total expenditures per cruise visitor.[14] This figure represents the portion of cruise passenger expenditures that contributed to the Belize economy (Figure 3.1).

In an independent study, the Central Bank of Belize estimated the average expenditure of BZ$85 (or US$42.50) per disembarked cruise visitor to Belize remained in the local economy (Central Bank of Belize, 2005).[15] While remembering its very approximate nature, this figure can reasonably be compared with total expenditures in Belize for other types of tourists.

Assuming conservative expenditures of US$20[16] (excluding tours) from our study, cruise tourists would appear to have contributed approximately US$13.6 million in additional direct expenditures to the Belizean economy. In addition to an estimated US$23 million in direct expenditures for tours, total cruise tourist expenditures are calculated at approximately US$36.6 million for Belize. Based upon multiplier estimates (1.40) from Costa Rica, the total annual economic benefit from cruise tourists to Belize was estimated to be US$51.2 million in direct, indirect and induced economic activity.

3.5.6 Factors Influencing Cruise Tourists' Decision to Disembark in Belize

From both marketing and economic development perspectives, it is useful to know what draws people to disembark at a particular port of call. Since such a high proportion of cruise tourists were first-time visitors to Belize, their decision to leave the ship must be based on word of mouth, marketing materials, reputation and additional visitor research.

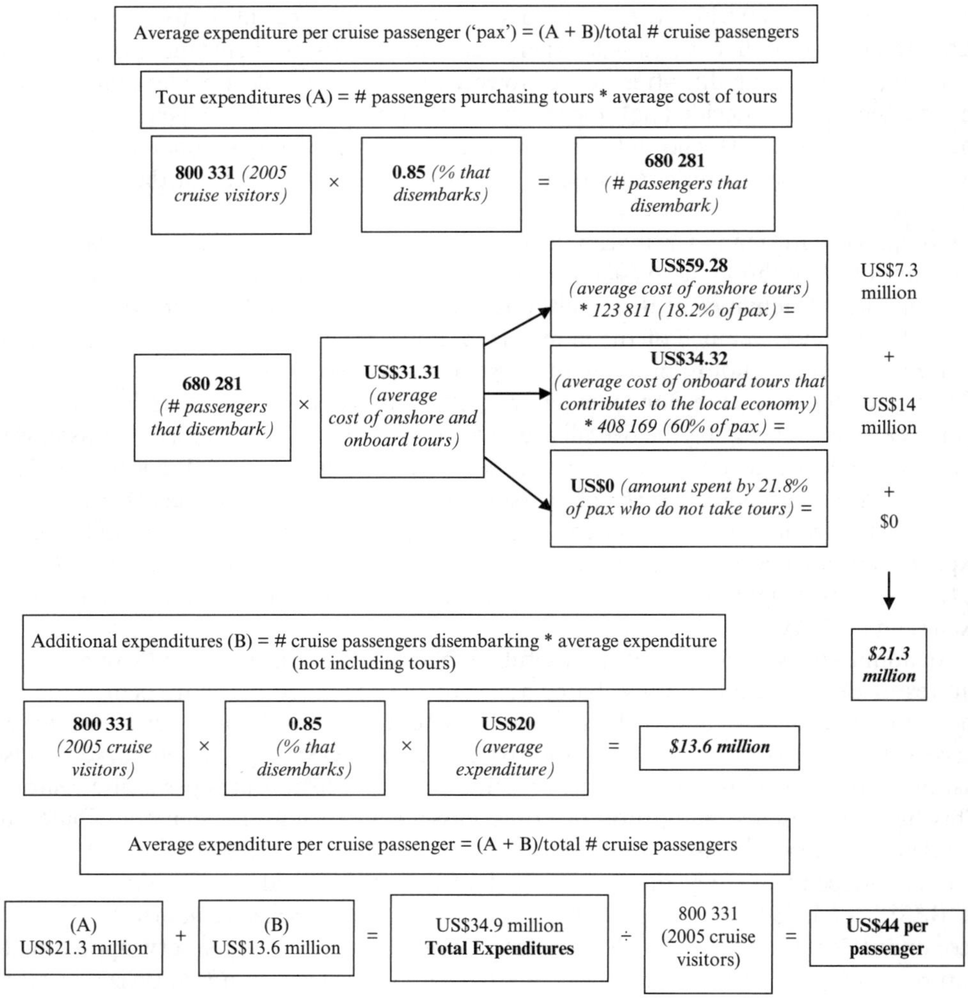

Figure 3.1 Illustration of cruise tourist expenditure calculations

Since the surveys were carried out when tourists were returning to the ship, results were more likely to measure visitor experiences mixed with visitor preconceptions.[17]
Using a five-point Likert scale, respondents were asked to evaluate the relative importance of natural features, traditional culture and historical features, contemporary services, and factors that contribute more generally to the ease and enjoyment of the travel experience anywhere. Cruise tourists indicated that friendly people, personal safety, scenic landscapes, cleanliness and general affordability were the most important reasons for disembarking in Belize, with each item ranking between 'important' and 'very important' on the five-point scale. Thus, four of the top five reasons to disembark in Belize are features that all destinations can and should aspire to, with the possible exception of affordability (Table 3.5).

Four of the top ten reasons to disembark in Belize were features associated with the

Table 3.5 Factors contributing to cruise tourists' decision to disembark in Belize

Feature	Mean Score	Rank	Feature	Mean Score	Rank
Friendly people	4.40	1	High quality restaurants	3.61	10
Personal safety	4.31	2	Local art and crafts	3.53	11
Scenic landscapes	4.21	3	Interesting architecture/ built infrastructure	3.52	12
Cleanliness	4.20	4			
General affordability	4.14	5	Local music, dance or customs	3.48	13
Quality of national	3.89	6			
parks and protected areas	3.89	6	'Duty-free' shopping	3.45	14
			Solitude/lack of crowds	3.43	15
Quality of beaches	3.80	7	High-quality services (medicine, dentistry)	3.16	16
Interesting/high-quality food	3.77	8	Entertainment/nightlife	2.93	17
Quality of coral reefs	3.75	9	Farms and ranches	2.79	18

natural environment: (1) scenic landscapes, (2) national parks and protected areas, (3) beaches and (4) coral reefs each ranked in the 'important' range. On the other hand, farms and ranches, entertainment and nightlife, the quality of local medical or dental services, solitude, and duty-free shopping were the least important features motivating tourists to disembark from the ship. Each item ranked in the neutral range of the five-point scale (Table 3.5).

These results are interesting if viewed in conjunction with the information that tourists provided about their activities in Belize. For example, although many people indicated that they went shopping and toured the city, duty-free shopping was not a particularly common activity (Table 3.4) and it ranked only 14th among factors contributing to their decision to disembark. Belize had the highest percentage age of passenger disembarkation in the Caribbean. It is the quality of the natural environment and other features that seemed to strongly influence tourists' decision to disembark in Belize. Thus, the attraction of the Belizean tourism experience to cruise visitors was largely in keeping with Belize's tourism image and stated goals.

3.5.7 Cruise Tourists' Opinion of Belize

Tourists were asked to respond to five questions intended to assess their opinion of their experience in Belize. It has been argued that relatively brief visits to a country on a cruise ship will generate longer-term, and more financially beneficial, future stayover visits. As a result, tourists were asked to rate the following: their general level of satisfaction with their visit; the likelihood that they would return to Belize either on a cruise ship or on a non-cruise vacation; and the likelihood that they would recommend a visit to Belize to their friends.

Overall, the cruise visitors in our sample were very pleased with their experience in Belize, and many indicated that they were quite likely to return on a future cruise vacation. Moreover, they would recommend to their friends that they also visit Belize on a cruise. Some three-quarters indicated that they were either 'satisfied' or 'very satisfied' with their visit to Belize. About two-thirds indicated that they were either 'likely' or 'very likely' to return on a cruise vacation, and almost 80 per cent indicated they would recommend a cruise including Belize as a port of call to their friends.

However, respondents were more divided about a potential visit to Belize on a non-cruise vacation for themselves or for their friends. Less than one-third of visitors indicated they were 'likely' or 'highly likely' to return to Belize on a non-cruise vacation, and only a similar number indicated they would recommend Belize to their friends for a non-cruise vacation. On the other hand, more than one-third indicated that they were 'unlikely' or 'highly unlikely' to return to Belize on a non-cruise vacation, and about one-third were 'unlikely' or 'highly unlikely' to recommend Belize to their friends for a non-cruise vacation.

These numbers are important, since both the cruise industry and government have argued that one of the benefits of cruise tourism is that it leads to passengers returning as stayover tourists. Officials stated that it was hoped that a sizeable percentage of cruise tourists would sufficiently enjoy their day visit to Belize so as to return as stayover guests, injecting more foreign capital into the economy. If the likelihood is low that cruise tourists will be converted to longer-term stayover tourists in the future, cruise visits must not be evaluated as investments, but as one-offs, and must withstand a higher level of scrutiny to be considered the preferred development path as a result.

3.5.8 Sensitivity to Changes in Travel Costs

In the survey, cruise passengers were also asked to evaluate their relative sensitivity to changes in the costs of cruise vacations. According to common practice, respondents were presented with a feasible reason or reasons for costs to increase without changing the quality of the experience (for instance, fuel costs) as well as 13 feasible bid categories from which to choose (a 'payment card' approach). Cruise tourists were asked to choose the maximum amount they would have been willing to pay for this cruise vacation from bids ranging from zero to US$2300 per person. They were then asked the same question in regard to their decision to disembark in Belize, but with bids ranging from zero to US$500 per person (which roughly reflected the current expenditure ranges in Belize).

Responses to the question regarding tourists' sensitivity to increases in travel costs for this cruise vacation were appropriately arrayed throughout the range of potential bid amounts on the payment card. Some 70 per cent of responses fell in the four bid categories between US$50 and US$250, fewer than 10 per cent of responses were zero (providing little indication of substantial 'protest' bids), and fewer than 2 per cent of responses were in the upper three bid categories (providing evidence that we were capturing a substantial amount of the remaining value of this cruise experience to respondents).

It was expected that cruise tourist responses to a parallel question on their sensitivity to increases in costs to disembark in Belize would be similar but lower in magnitude. The responses to this question may be important in establishing policies with regard to per head charges assessed on cruise ships by port authorities. Here again, about two-thirds of responses were within the middle bid ranges, a relatively modest number of zero bids, and very few bids at the very high end of the bid range. Most port-cost sensitivity bids fell between US$10 and US$50. Given what we now know about how much cruise passengers spend in Belize, these bid amounts appear appropriate.

Table 3.6 Measures of sensitivity to changes in cruise costs and willingness to pay for conservation in Belize

	N	Mean	Median	Mode
Max WTP for cruise vacation	582	201.43	100	100
Max WTP for Belize	584	47.22	25	25
Max WTP for Belizean culture and nature	588	42.92	25	25
% for culture	528	143.78	50	50
% for nature	528	56.22	50	50

The average willingness to pay for increases in cruise costs and costs of disembarking can be derived from these bids. The potential impact of policies designed to capture this surplus value can then be estimated. Cruise passengers indicated they were willing to absorb an average of US$100–200 per person, or a 5–10 per cent cost increase, and still choose to take the cruise. Analogously, they were willing to consider an increase in cost of US$25–50 per person to disembark in Belize (Table 3.6), approximately doubling their current expenditures in Belize. A policy that effectively extracts this additional cost from cruise tourists was potentially worth some US$17 to US$34 million based on 2005 disembarkation rates.

3.5.9 Willingness to Support Local Conservation

In addition to evaluating their sensitivity to increases in travel costs and costs of disembarking in Belize, cruise passengers were asked about their willingness to financially protect the natural and cultural environment of Belize. Again, cruise tourists were confronted with a reasonable policy (local government investment in environment and culture), a payment vehicle (a mandatory per head tourist tax), and a feasible range of potential bid amounts (US$0–500 per person). In addition, respondents were asked to divide their total bid into the proportion to be dedicated to nature and the proportion to be dedicated to local culture. Responses were logically expected to be lower in magnitude than either of the previous two bid amounts. The degree to which the willingness to pay for the conservation of Belizean culture and nature was similar to the willingness to pay to disembark in Belize indicates the extent to which the decision to visit Belize was driven by extant natural environmental and cultural factors.

Here again, the frequency distribution of bid amounts provided by respondents is encouraging. Some 75 per cent of the bids fall between US$5 and US$50, zero bids are relatively few, and fewer than 2 per cent of responses were in the upper three bid categories. Tourist responses were very similar to their willingness to absorb additional costs to disembark in Belize: ‘nature and culture’ responses averaged between US$25 and US$45 per person per trip and amounted to some 90 per cent of their average overall bid to visit Belize. One potential interpretation of this information is that the cultural and natural environment of Belize was the principal reason that cruise tourists got off the ship at all. Respondents allocated their willingness to pay somewhat more strongly toward nature (56 per cent of US$43 or US$24) than to culture (44 per cent or US$19). Given what is now known about cruise tourist behaviour while in Belize, these allocations appear to appropriately reflect their values. Using observed disembarkation and visitation rates, cruise tourists would have been willing to contribute some US$29.2 million toward environmental (US$16.4 million) and cultural (US$12.8 million) investments in Belize.

3.6 CONCUSIONS AND RECOMMENDATIONS

Globally, both cruise tourism and 'experiential' tourism (including ecotourism) has been growing rapidly. Since 2000, the cruise industry has really blossomed in Belize, and virtually everyone now views it as a permanent part of the country's tourism landscape. By 2005, Belize was both the fastest-growing cruise market in the Caribbean, and a bellwether for Central America, where it led the pack in terms of growth rate and size of cruise tourism. Since 2002, the number of cruise tourist arrivals has significantly surpassed the number of stayover visitors to Belize. This creates both opportunities and concerns. Belize, as an English-speaking destination with outstanding natural, cultural and archaeological offerings, political stability and relative proximity to the United States, is likely to remain a staple on Caribbean cruise line itineraries.

The central focus of our work has been to examine the economic dimensions of cruise tourism, a topic that has received surprisingly little study to date. Most other aspects of cruise tourism's impact on Belizean society and environment – most importantly its effects on archaeological sites, marine parks and coastal waters – have not been addressed here. Careful direct measurement of those impacts on ruins, reefs, parks and other sites, of course, involves far more extensive, longitudinal fieldwork than we had the time or funds to undertake. Our data allowed us to reach a number of conclusions and recommendations regarding the challenge of balancing the needs of stayover and cruise tourism. The Belize government, like others in Central America and the Caribbean, is faced with choices about how to best use their resources to promote these very different types of tourism.

In Belize, perhaps more than anywhere else in either region, cruise tourism has competed with – in some instances colliding with – ecotourism, that is, small-scale nature and cultural stayover tourism. Belize was the first country to develop a national Cruise Ship Policy and to foster multi-stakeholder discussions around both cruise and stayover tourism. It also hosted the first ecotourism summit back in 1991, has over 41 per cent of its territory under conservation, has a vibrant and extensive ecotourism sector, and has a host of NGOs actively supporting and monitoring ecotourism, protected areas, and cultural sites.

While more than 3.4 cruise passengers arrived for every stayover visitor in 2005, cruise tourism generated only 17.5 per cent of the total tourism revenue. In dollar terms, cruise tourism generated US$30.6 million in 2005 compared to US$144.1 million for stayover tourism (BTB, 2006). Moreover, the difference in average amount spent per day by stayover visitors (US$96) was more than double that of cruise passengers (US$44). Given that stayover visitors spent an average of 6.8 days in Belize, this difference translated into an average of US$653 that the stayover tourist spent in the country per visit. Put another way, the average stayover visitor spent 15 times more than the average cruise passenger in Belize.

That different types of tourists (or other types of economic activity) generate different financial flows is not problematic unless they are in competition for scarce investments, services, or other resources. However, if cruise tourists crowd out or otherwise discourage stayover visitors, then significant financial losses result from welcoming cruise tourists. Many observers recognize the economic contributions of cruise tourism but worry that massive cruise tourism is undermining Belize's secluded, eco-friendly image, and thus killing the 'golden goose' of ecotourism. It has been argued that cruise tourism motivates the passengers to return to the destination as stayover tourists. However, the evidence of increasing cruise tourism and decreasing stayover tourists from 2007 through 2010 suggests that this supposition may have little basis in fact.

Based on the analysis of our field data, combined with a broad-scale review of pertinent policies, reports and studies, our conclusion was that Belize's cruise tourism was on an untenable course. While the official policy continues to be to promote 'responsible tourism' that ensures environmental and socio-cultural sustainability, in practice the country's tourism policies support cruise tourism often at the expense of stayover tourism that: (1) has a lighter environmental and social footprint, (2) brings higher economic returns per person, per visit, and in total revenue, and (3) has been the historic mainstay of Belize's tourism reputation.

The rationale for Belize or any country to receive cruise passengers is its potential to create value for the society – through income and jobs for local residents, generation of capital investment, improved quality of life, and other means. To date, the evidence from Belize is that cruise tourism creates relatively little value for local communities, contributes relatively little to public revenues, and consumes resources that could support the development of higher value uses.

Belize's current US$7 per passenger tax should be increased. It is below the Caribbean average (US$8.66) and is a fraction of the airport exit tax (US$36.15). Our data show that cruise passengers are willing to pay US$25–50 more to disembark in Belize. Therefore, in negotiations with the cruise lines, Belize should insist on a higher head tax. At the same time, the government of Belize should dialogue with other governments in the region about the establishment of head tax levels that meet a consistent minimum level from one country to the next. Establishing a minimum cruise head tax level that is consistent across Central America and (ideally) the Caribbean would help avoid a race to the bottom as countries compete with each other to attract the cruise industry.

Belize (and other countries) should examine the fee structure of commissions paid by local businesses in return for access to cruise passengers. These commissions should be publicly known, based on agreed upon criteria, and the higher prices should be presented to cruise passengers in terms of assuring reliability, safety standards, quality guides and good service from those tour operators selected to work with the cruise lines. If commissions are paid to the cruise lines, then they should bear some of the responsibility for raising standards and ensuring quality.

One of the key issues relating to the overall cost–benefit and value creation for a society are the opportunity costs of public funds. Belize, like other governments, has a right and obligation to ensure that scarce public funds be dedicated to activities that provide the greatest public benefit. Decisions on potential investments to support cruise tourism must be compared with: (1) realistic estimations of the benefits they will generate, (2) other potential investments within the tourism sector, and (3) other investments outside the sector that could achieve similar development outcomes.

The empirical evidence from Belize suggests that the benefits from cruise tourism do not justify substantial public investment. At present, US$4 of the US$7 head tax goes to the privately owned Tourism Village and is therefore subsidizing Royal Caribbean and Diamonds International's operations in Belize. The head tax is widely recognized as a tax the cruise industry pays to host countries and therefore these funds should be used to manage and improve public sites and protected areas and to further improve infrastructure.

Belize has earned an international reputation for its outstanding coral reef, national parks and archaeological sites and for its small hotels catering to a discriminating clientele. Large groups of cruise passengers are now using the same resources, during the same peak seasons, and this is deteriorating the infrastructure or quality of tourism experience. One of the most consistent complaints our data revealed was that large groups of cruise tour passengers were interfering with the experience of ecotourists who come to Belize's reefs, parks and archaeological sites singly or in small groups. Higher-value stayover tourists create substantial economic and social benefits for the country, visitor sites, the tourism industry and local communities. If the sites or areas the stayovers visit are overcrowded or overused, show signs of poor maintenance, or otherwise deteriorate, their willingness to pay for these sites and willingness to return to the country will be reduced. To the extent that mass cruise tourism is reducing the quality of the experience of stayover tourists, the country suffers.

While some progress has been made, our surveys and press reports revealed that cruise passengers often complained they found little to do in Belize City, while many worried about crime and other social ills. The government, together with the tourism industry and civic associations, could invest more in upgrading and expanding the urban facilities and offerings. Among the potential strategies are the following.

The passenger surveys show a stronger preference for shopping for local handicrafts and souvenirs, and that duty-free shopping had low priority for those disembarking in Belize. Rather than expanding the duty-free shops (and adding a casino), the range of local products and local cultural activities should be expanded for those staying in and around Belize City. Investments should be made by government, the private sector and NGOs in raising the standard and increasing the offerings of handicrafts and souvenirs and highlighting local customs and festivals. This would both serve to increase expenditures in Belize and to meet passengers' desire for local crafts.

Globally, food is an area of increasing interest and concern to all types of tourists. Restaurants and food stalls should be encouraged to promote Belizean cuisine, locally grown fruits, vegetables, seafood and other local products. Restaurants and hotels could demonstrate local recipes, local festivals could feature in-season produce. A national network of restaurants featuring local cuisine could be developed and promoted to both cruise and stayover passengers. Some designated hotels in or near Belize City could offer new products that both capture current market trends and showcase the best of Belizean culture and climate. Excursions could include a spa/hotel offering massages, yoga, mineral baths; language (local dialect) lessons; local dance classes; or tours of organic coffee and other farms.

In summary, we see the challenge ahead for Belize as one of finding a constructive, dynamic balance between the relative 'newcomer', cruise tourism, and the country's well established mainstay and reputation-builder, ecotourism. Because there is a substantial time-depth to ecotourism in Belize, this balance is not an abstract or 'in principal' concept. We propose that cruise tourism in the country should be managed by government and stakeholders so that its positive impacts can be maximized, and it does not constrain or interfere with the continued development of bona fide ecotourism in the country. Any indication that consumer demand for ecotourism in Belize is unambiguously hindered by cruise activity, whether because of direct effects (overcrowded sites, environmental damage, etc.) or indirect effects (damage to its eco-friendly image, overexposure among consumers), is an indication that regulation of the cruise industry needs tightening or adjusting. Plans to build a second cruise port in southern Belize are unwise. Further, our recommendations include increasing the economic contribution of cruise tourism, mitigating environmental impacts of cruise tourism, adhering to realistic limits on cruise visitor volume per year, and separating cruise tourists and ecotourists in space and time. Because ecotourism actively and deliberately promotes both environmental conservation and local livelihoods, we believe its dynamics can be helpful and effective in creating policies to effectively monitor the trajectory of cruise tourism in Belize. The time has come to chart a sustainable course for cruises to Belize.

ACKNOWLEDGEMENTS

This study was made possible through a partnership among four institutions: (1) Center for Responsible Travel (CREST; formerly the Center on Ecotourism and Sustainable Development/CESD), which is based at Stanford University and in Washington, DC; (2) INCAE Business School of Costa Rica; (3) Belize Tourism Board (BTB) and (4) Belize's Protected Areas Conservation Trust (PACT). Many people contributed to this effort. Among them, the authors would like to acknowledge: Dr. Vincent Palacio, Anthony Mahler, Raymond Mossiah, Vernita Ediger, Jackie Crespo, Christina Cairns, Gale Malic, Carolie Verde, Molly Chapman, Zoe Chafe, Claire Nelson, Katherine Roubos, David Krantz and David Russell. We are grateful to Carlos Saavedra and the Summit Foundation, BTB, PACT, INCAE, Stanford, and Fulbright Foundation for their support. Errors remain the responsibility of the authors.

NOTES

1. Tourism and Leisure Europraxis Consulting (2011, p. 4).
2. BTB (2009), p. 59.
3. Carnival Corporation is the largest cruise business in the world, acting as an umbrella corporation for 12 other cruise operations including Princess Cruises and Carnival Corporation and Plc Revises.
4. Personal interviews with BTB, Tourism Village, and tourism industry officials, Belize City, 1–3 August 2006.
5. See http://www.carnival.com; http://www.royalcaribbean.com; http://www.norwegiancruises.com.
6. Retrieved from http://www.carnival.com/cruise-to/Caribbean/Belize.aspx; http://www.carnival.com/Activities/ (accessed 20 February 2014).
7. Personal communication, Anthony Mahler, BTB Director of Product Development, 2–9 August, 2006.
8. This figure is calculated by multiplying total cruise visitors to Belize (800 331 visitors) by the rate of disembarkation (85 per cent) by the percentage age of disembarkers purchasing tours on board (60 per cent) by the amount of the tour purchase price that remains with tour companies (US$34.32).
9. This figure is calculated by multiplying total cruise visitors to Belize (800 331 visitors) by the rate of disembarkation (85 per cent) by the percentage age of disembarkers purchasing tours onboard (60 per cent) by the amount of the tour purchase price that is absorbed by cruise lines (US$43.68).
10. This figure is calculated by adding approximate direct contribution of onboard tour purchases (US$14 million) with approximate expenditures on onshore tour purchases (US$7.3 million).

11. This number is calculated by dividing total direct contribution (US$21.3 million) by all passengers that disembarked (680 281).
12. This number is calculated by dividing total direct contribution (US$21.3 million) by all cruise visitors (800 331).
13. Personal communication, Vincent Palacio, University of Belize, January 2006.
14. This figure is calculated by multiplying estimated total expenditures for disembarked passengers (US$51) by the number of disembarked passengers for 2005 (680 281) and then dividing by all cruise visitors for 2005 (800 331).
15. This survey was administered to 4206 cruise passengers throughout 2003 and assumed that 80 per cent of onboard tour purchase prices remain with the cruise lines.
16. This figure represents the median of all responses to the question on total onshore expenditures.
17. In order to accurately measure reasons for disembarking in Belize, it would be necessary to interview passengers on board the cruise before they get off the ship in order to find out their preconceived notions and motivations.

REFERENCES

Belize Tourism Board (BTB) (2003), *Belize Cruise Ship Policy* (revised), Belize City: Belize Tourism Board.
Belize Tourism Board (BTB) (2004a), *Strategic Vision for Belize Tourism in the New Millennium*, Belize City: Belize Tourism Board.
Belize Tourism Board (BTB) (2004b), *2003 Visitor Expenditure and Motivation Survey*, Belize City: Belize Tourism Board.
Belize Tourism Board (BTB) (2005a), *Belize Tourism Policy 2005*, Belize City: Belize Tourism Board.
Belize Tourism Board (BTB) (2005b), 'Tourism statistics for the third quarter: July through September', accessed 1 December 2005 at http://www.belizetourism.org/press/375.htm.
Belize Tourism Board (BTB) (2006), 'Cruise visits', accessed 16 January 2006 at http://www.travelbelize.org/cruise_belize.html.
Belize Tourism Board (BTB) (2009), *Travel and Tourism Statistics 2008*, accessed 12 February 2014 at http://www.seaaroundus.org/conference/belize/belize_scientific_papers/btb_2008.pdf.
Blackstone Corporation and Resource Management and Tourism Consultants (1998), *A Tourism Strategy Plan for Belize*, prepared for the Ministry of Tourism and the Environment.
Central Bank of Belize (2005), *2003 Visitor Expenditure Survey*, Belize: Central Bank.
Eltringham, P. (2001), *The Rough Guide to Belize*, London: Rough Guides, p. 276.
Espat, M. (2004), 'Keynote address', paper presented at the Cruise Tourism Impact Forum in Belize City, 30–31 September 2004.
Mahler, A. (2003), 'Belize draft cruise ship policy document', *Proceedings of the 1st National Coastal Symposium Belize*, 27 May 2003, accessed 29 January 2014 at http://www.coastalzonebelize.org/wp-content/uploads/2010/04/proceedings_1st_national_coastal_symposium.pdf.
Mintel International Group (2005), *Cruise Market Report*, accessed 22 June 2006 at www.lexisnexis.com.
SITCA and CCT (2011), *Boletín de Estadísticas Turísticas de Centroamérica 2010* [*Bulletin of Tourism Statistics in Central America*], El Salvador, accessed 12 February 2014 at http://www.sidturismo.org/index.php/estudios/file/11-boletines.
TEEB (2010), *The Economics of Ecosystems and Biodiversity: Ecological and Economic Foundations*, edited by P. Kumar, London and Washington, DC: Earthscan.
United Nations World Tourism Organization (UNWTO) (2001), *Tourism 2020 Vision: Global Forecasts and Profiles of Market Segments*, Madrid: World Tourism Organization.
United Nations World Tourism Organization (UNWTO) (2005), *Tourism Market Trends, 2005 Edition*, Annex, p. 5.
Wood, R.E. (2004), 'Global currents: cruise ships in the Caribbean Sea', in D.T. Duval (ed.), *Tourism in the Caribbean: Trends, Development, Prospects*, London: Routledge, pp. 152–71.
World Travel and Tourism Council (WTTC) (2003), 'Blueprint for new tourism', accessed 29 January 2014 at http://www.hospitalitynet.org/file/152001179.pdf.
World Travel and Tourism Council (WTTC) (2004), *The Caribbean: The Impact of Travel and Tourism on Jobs and the Economy*, London: World Travel and Tourism Council.

4. Climate change effects on the economics and management of marine fisheries

U. Rashid Sumaila, William W.L. Cheung and Vicky W.Y. Lam

4.1 INTRODUCTION

Global fisheries contribute tens and hundreds of billions of gross revenues and economic impact, respectively, each year (Sumaila et al., 2007; Dyck and Sumaila, 2010; FAO, 2011). Additionally, marine fisheries provide nearly 3.0 billion people with 15 per cent of their animal protein needs (FAO, 2011), and not only to people who reside in maritime countries of the world because international fish trade has made the contribution of fisheries to food security truly global. Recent estimates put the number of people worldwide who derive employment income from marine fisheries at about 260 million (Teh and Sumaila, 2011). So, global marine fisheries are clearly important to people.

But these benefits of fisheries and many others (e.g., recreational, non-market) are currently under threat from a host of stressors, including overfishing, ocean warming, acidification and hypoxia. Climate change, in particular, will complicate the challenges currently facing global fisheries, as it has begun to alter ocean conditions, particularly water temperature and biogeochemistry. These changes are expected to affect the productivity of marine fisheries (Cheung et al., 2010). Recent studies estimate that climate change will lead to losses in revenues, earnings to fishing companies and household incomes in many regions, although some countries and/or regions may realize increases in fisheries benefits, for example, Greenland (Arnason, 2007).

Effects of global ocean-atmospheric changes act at multiple levels of organization of marine ecosystems and human society, including individual organisms, populations of organisms, ecosystems and communities of organisms, the economics of fisheries, and larger global issues, such as global food security, energy supply and food prices (Sumaila et al., 2011).

Here, we first review existing knowledge on the responses of marine ecosystems to climate change and warming oceans, and how these changes are expected to affect the economics of global marine fisheries. Next, we present an analysis of potential impacts of climate on landings and landed values from the Mexican Exclusive Economic Zone (EEZ). Finally, we conclude.

4.2 BIOPHYSICAL CHANGES TO FISHERIES IMPACTS

Sea surface temperature is increasing (Figure 4.1), which would likely result in the expansion of oxygen minimum zones, changes in primary productivity, loss of sea ice, changes

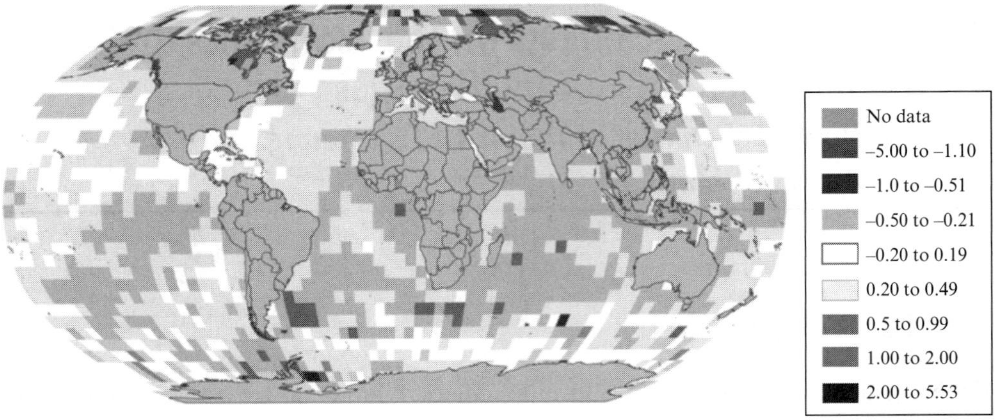

Source: Rayner et al. (2003).

Figure 4.1 Sea surface temperature (SST) changes between the 1960s (average 1950–69) and 2000s (1988–2007)

in ocean circulation patterns, sea level rise, and increase in extreme weather events (IPCC, 2007).

Marine fishes and invertebrates are strongly dependent on oceanographic conditions to stay alive (Pörtner and Knust, 2007). Hence, changes in temperature and ocean chemistry directly affect the physiology, growth and reproduction of these organisms. For example, fishes in warmer temperatures are expected to have a smaller maximum body size (W_∞), and smaller size at first maturity. These changes would have significant impacts on things such as natural mortality rates, population dynamics and productivity (Cheung et al., 2010). Also, ocean acidification and expansion of oxygen minimum zones would likely have negative impacts on marine organisms and fisheries. Current studies suggest that species' responses to more acidic water are likely to vary between species, with invertebrates being impacted more while the effect on finfish is more uncertain (Doney et al., 2009).

Changes in environmental conditions strongly affect the spatial distributions of marine fishes and invertebrates. Fish and invertebrates have different environmental preferences and limits (upper and lower) where animals cannot survive (e.g., temperature, salinity) (Pauly, 2010). Together, these factors are expected to affect, directly and/or indirectly, the distribution of marine species, including those that are targeted by fisheries. Studies have shown that many marine species have moved towards the poles and into deeper waters under ocean warming, with these kinds of shifts seen, for example, in the Northeast Atlantic (Perry et al., 2005; Dulvy et al., 2008), East Coast of the USA (Lucey and Nye, 2010), the Bering Sea (Mueter and Litzow, 2008), Australia (Last et al., 2011) and West Africa (Lam et al., 2012). Shifts in distributions will result in species gains and losses, and therefore changes in catch potential worldwide (Figure 4.2).

Climate change is expected to affect primary productivity, which most marine animals depend on as a source of energy. Boyce et al. (2010) suggest that global ocean phytoplankton biomass may have declined substantially in the past 50 years, although the

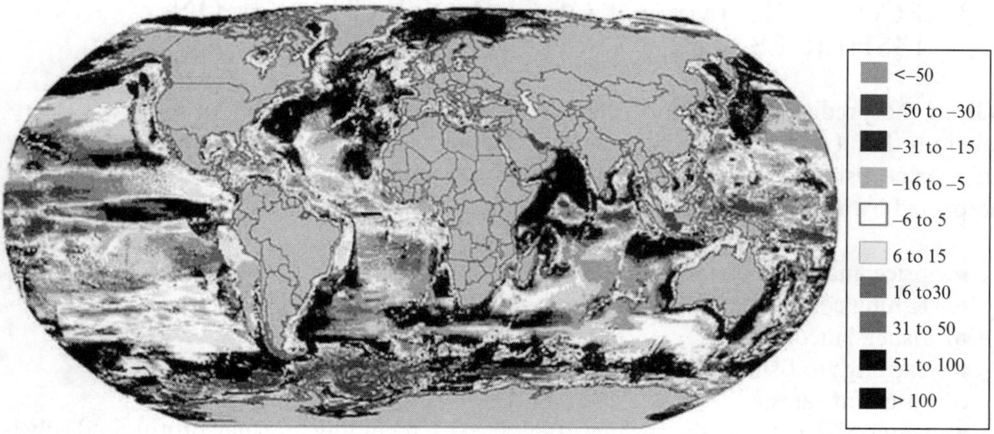

Source: Redrawn from Cheung et al. (2010).

Figure 4.2 Projected changes in maximum potential catch under the Special Report on Emissions Scenarios (SRES A1B) scenario, % relative to 2005

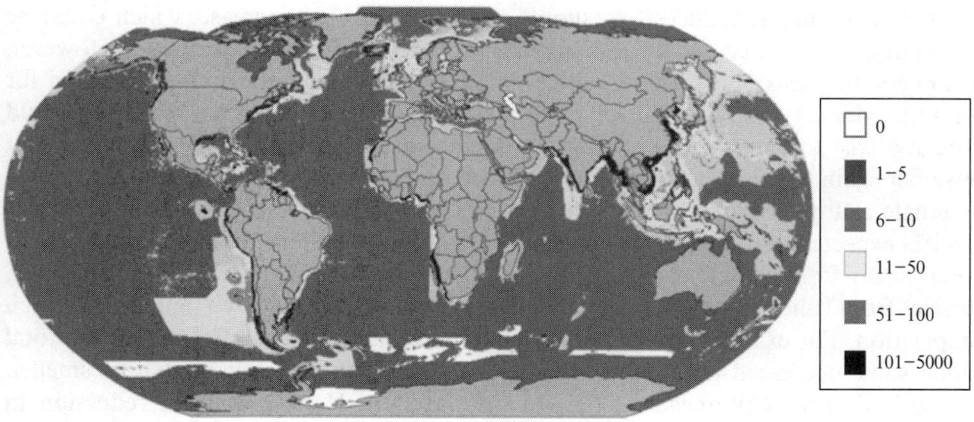

Source: Sea Around Us Project.

Figure 4.3 Estimated global catches (tonnes, average 2000–07)

validity of such results is still being debated. In the Pacific, the oligotrophic zones with low primary production in the Pacific and Atlantic are expanding and such changes are suggested to be attributed to ocean warming (Polovina et al., 2011). Changes in primary productivity and planktonic community structure affect the amount of energy transferred to higher trophic levels and ultimately the productivity of trophic groups that contribute to fisheries catches (Figure 4.3). In particular, analysis of global fisheries landing data suggest that the capacity of global marine capture fisheries production is currently limited by primary production, exacerbating the impacts of climate change on fisheries.

4.3 ECONOMIC IMPACT OF CLIMATE CHANGE ON FISHERIES

Given the predicted impacts of climate change on the biophysics of the ocean, it is obvious that it will affect the economics of fishing because both the quantity and quality of marine fish catch and its distribution within and between nations' EEZs would be impacted by it. Climate change is likely to affect the following (Sumaila et al., 2011):

- price and value of catch;
- fishing costs;
- fishers' incomes;
- earnings to fishing companies;
- discount rates;
- economic rent (i.e., the surplus after all costs, including 'normal' profits, and subsidies have been covered);
- impact on the whole economy.

4.3.1 Price and Value of Catch

When climate change reduces fish supply, fish prices should increase, which could be large enough to balance out the loss in gross revenues due to reduced catches. However, consumers may seek substitutes as prices increase, thereby dampening the demand for fish and reducing the potential for price increases. In addition, price increases would come at a cost to consumers through loss in consumer surplus, that is, the welfare that consumers gain from the consumption of goods and services.

Climate change research predicts that catches in high latitude countries may increase, thus it is expected that fisheries in countries in these regions (e.g., Iceland) would benefit economically from climate change, at least in the short term (Arnason, 2007). However, revenues from fisheries are not only dependent on the amount of catch, but also on catch composition. For example, in spite of the increased catches in the Celtic Sea, the total landed value decreased because a larger proportion of the catch consisted of smaller, lower-priced species (Pinnegar, 2002). In the Southern Hemisphere, the reduction in landings of pelagic fisheries in Peru because of change in sea surface temperature during the 1997–98 El Niño caused more than US$26 million of revenue loss (Badjeck, 2010).

4.3.2 Fishing Costs

Capital costs, that is, the cost of vessels, fishing gear, processing plants, and so on, would be affected by climate change if additional capital for fishing and processing operations is required in order to adapt to the climate change impacts on the quantity, composition and distribution of fishery resources (Pauly et al., 2005). Changes in migratory routes and fish distribution would affect travel time, which can lead to increases or decreases in fuel and ice cost depending on catch levels and patterns, and the management regime in place. It is estimated that under a scenario of a 1.2 C sea surface temperature increase, which corresponds to the ENSO (El Niño-Southern Oscillation) event of 1983, the number of active boats landing sablefish in Monterey Bay, California could decrease by

60 per cent (Dalton, 2001). Decadal oceanographic changes affected the distribution of tuna in the Central Western Pacific, which in turn impacted how the tuna purse seine fleet operated, and thus increased fishing costs (McIlgorm, 2010). In particular, most of the world's large, fuel-consuming fishing vessels are from developed countries, implying that these vessels would face much higher costs of rising fuel and climate mitigation than small fishing vessels. The developed countries may be forced to engage in the expensive business of scrapping their large vessels as climate change impacts intensify.

4.3.3 Fishing Companies and the Resource Rent

With the expected changes in landed values and costs under climate change, earnings to fishing companies and the resource rent generated through fishing will be altered, with the direction and magnitude of change varying across regional fishing zones. For example, earnings to the European sardine fishery are estimated to decrease by up to 1.4 per cent on average per year with rising temperatures (Garza-Gil et al., 2010). The reduction in coral cover and its associated fisheries production due to increased sea temperature is expected to lead to a potential net revenue loss of US$95 to US$140 million (current net revenue is US$310 million) per year in the Caribbean basin by 2015 (Trotman et al., 2009). A World Bank report estimated the annual economic impact of climate change on the coast of Viti Levu, Fiji to be US$0.1 to 2 million for subsistence fisheries, and US$0.05 to 0.8 million for commercial coastal fisheries by 2050 (Lal et al., 2009). For a small country like Fiji, these numbers are significant.

In contrast, resource rent from fishing Pacific sardines (*Sardinops sagax*) could increase as sardines are known to be more productive during warm water regimes in the California Current Ecosystem (Herrick et al., 2009). The fisheries impacts of global warming on both Iceland's and Greenland's gross domestic product (GDP) are likely to be positive, with the economy of Greenland projected to benefit substantially (Arnason, 2007).

4.3.4 Flow of Resource Rent and Discounted Value

Another important consideration is how climate change may affect the stream of benefits that fishery resources are expected to generate over time, and how beneficiaries are likely to discount future benefits. Discounting accounts for the compensation individuals require for sacrificing benefits now for potentially greater benefits in the future. The percentage difference between the required future and the current benefit represents the discount rate, and reflects the weight placed on receiving benefits now in contrast to the uncertainty of receiving greater benefits in the future.

The more distant and uncertain the future benefit, the greater the compensation required and correspondingly the higher the discount rate. As a result those preferring to sacrifice future for immediate benefits will have a higher discount rate than those placing a higher value on benefits realized in the future. Therefore, the uncertainty associated with climate change may result in a relatively high discount rate for privately owned commercial fishing firms, driving them to pursue fishing strategies that favour current catches over those in the future. On the other hand, the greater is society's interest in maintaining both market and non-market values from the ocean for the benefit of all generations, the more it would favour a more conservative fishing strategy that defers current use to maintain higher stock levels for the future. In general, society has a longer time horizon, so that the social discount rate would be comparatively lower than the private discount rate. In view of the range of benefits fishery resources are capable of providing, the choice of discount rates becomes an extremely critical issue in formulating and evaluating conservation and management policy to address climate change (Sumaila and Walters, 2005; Dasgupta, 2008).

It is predicted that developing countries will experience relatively more reductions in fish catch with climate change because they are concentrated in tropical and sub-tropical regions of the world. Whether these predicted decreases in catch would result in real negative economic impacts remains to be seen. This is partly because most of these countries have relatively weak fisheries governance and management institutions (Pitcher et al., 2009), implying that their fisheries are already at or near open access equilibrium, where resource rent from the fisheries are zero (Clark, 1990). In such cases, climate change can hardly make resource rent worse. In fact, because climate change could increase the cost of fishing, it is likely to lead to reductions in fishing effort and overcapacity, resulting in potential increases in resource rent (Sumaila et al., 2011). On the other hand, because these countries tend to rely on fish and fisheries as a source of livelihoods and protein, climate change would intensify their socioeconomic and food security problems (Allison et al., 2009; Lam et al., 2012).

4.3.5 Impact on the National Economy

The added value or impact through the fish value chain is the indirect economic effects of fisheries due to their impact on activities such as boat building/maintenance, equipment supply and the restaurant sector (Pontecorvo et al., 1980). Fisheries is a base industry since fish catch impacts several other sectors (e.g., wholesale, retail, processing, restaurants) of the national economy (e.g., Roy et al., 2009). Since the landings and landed values of fisheries are predicted to be affected by climate change and ocean acidification, it would in turn have an impact on whole economies (e.g., Dyck and Sumaila, 2010).

4.4 CLIMATE CHANGE IMPACTS ON MEXICAN FISHERIES

To predict the future change in the distribution of fish in Mexico's EEZ, we used two climate scenarios generated by the Geophysical Fluid Dynamics Laboratory of the US National Oceanic and Atmospheric Administration (GFDL's CM 2.1) (Delworth et al., 2006). The two scenarios included in our model are the 720 parts per million (ppm) stabilization experiment (Special Report on Emissions Scenarios, or SRES, A1B) and the Committed Climate Change Scenario (Commit), representing high-range (severe scenario) and low-range (mild scenario) climate change, respectively. The severe (A1B) scenario defines a world with a very rapid economic growth, low population growth, rapid introduction of new, more efficient technologies and moderate use of resources with a balanced application of technologies. In the mild (Commit) scenario, climate-forcing agents are assumed to stabilize at the end of twentieth-century levels.

Changes in physical environmental variables such as sea temperature, sea ice coverage, advection and salinity under different climate change scenarios are obtained from GFDL's CM 2.1. Then, given changes in these bioclimate attributes, we use the Dynamic Bioclimate Envelope Model (DBEM, described in Cheung et al., 2008a) to project the resulting changes in species distribution. To get a more realistic projection of future distributions, the DBEM incorporates population dynamics, spatial adult migration and larval dispersal that are dependent on the carrying capacity of each spatial cell determined by its habitat suitability for the studied species (ibid.).

Catch potential can be predicted from macroecology theory, which deals with large spatial and temporal-scale relationships between ecology and biogeography (Cheung et al., 2008b). Maximum catch potential of a species is proportional to its trophic level, mean primary productivity within the species' range of exploitation and the distributional range of the species. Changes in primary production are calculated from three empirical models (see Sarmiento et al., 2004 for details) and the ensemble mean of the estimates from the models are used. The range area of the species is projected from the DBEM described above. The empirical model for predicting maximum catch potential is:

$$\log_{10} MSY' = -2.881 + 0.826\log_{10} P' - 0.505\log_{10}(A) - 0.152\lambda + 1.887\log_{10}CT + 0.112\log_{10} HTC' + \varepsilon$$

where MSY' represents the maximum sustainable yield, P' is the annual primary production from a given exploited range, A is the geographic range, ′ denotes the trophic level, CT represents the number of years of records and HTC' represents the catch from higher taxonomic groups.

4.4.1 Current and Predicted Future Landed Values

The current landed values in 2000 dollars of catches extracted from Mexico's EEZ were calculated using the ex-vessel price data reported in Sumaila et al. (2007) and catch data from the global fisheries catch spatial database reported in Watson et al. (2004), both available at www.seaaroundus.org. Catches of 70 marine fish and invertebrate species and Penaeus shrimps expected to be caught in Mexican EEZ in 2050 under the severe and mild climate change scenarios were modelled using the bioclimate dynamic model and the empirical model described in the last sections. In this study, we assumed that the real ex-vessel price of the different fish caught remain constant throughout the modelling period. With the predicted catch and the ex-vessel price, we computed the annual landed values (in 2000 real values) of each species extracted from Mexico's EEZ under different climate change scenarios in 2050.

4.4.2 The Results

From 1991 to 2000, South American pilchard (*Sardinops sagax*) and Penaeus shrimps contributed 43 per cent and 15 per cent, respectively, to the total annual landings from Mexico's EEZ and hence they are the two most important commercially exploited marine species and group in this country (Figure 4.4). From the model, ten of the top 12 highest fished species in Mexico showed a decline in catch in 2050 under the severe climate

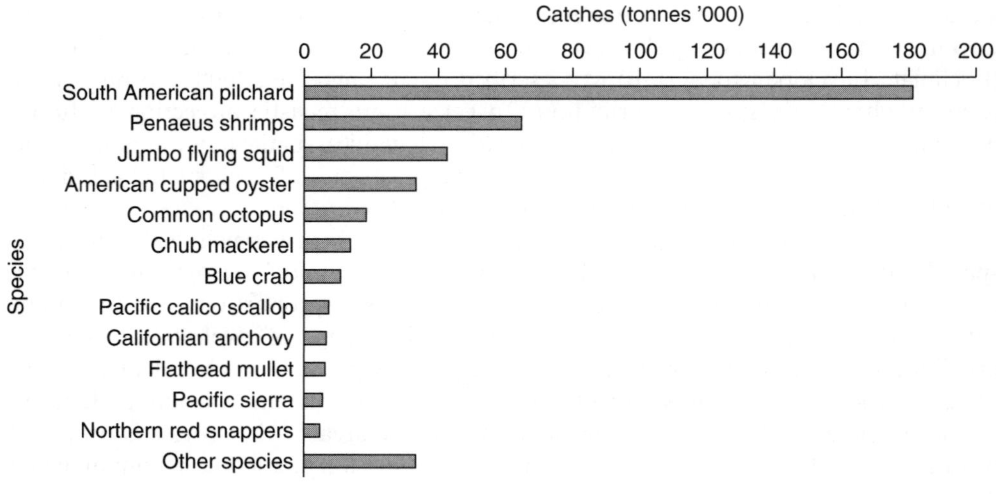

Figure 4.4 Annual average catch (tonnes) for last decade (1991–2000) in Mexican EEZ

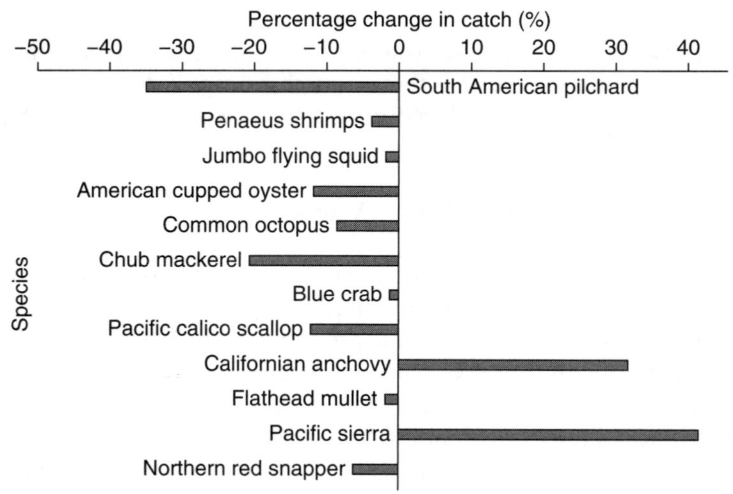

Figure 4.5a Predicted percentage change in 2050 in Mexican EEZ under the SEVERE climate change scenario

change scenario (Figure 4.5a). South America pilchard (*Sardinops sagax*) and club mackerel (*Scomber japonicus*) declined the most and their catch decreased by 35 per cent and 20 per cent, respectively. The two exceptional species are Californian anchovy (*Engraulis mordax*) and Pacific sierra (*Scomberomorus sierra*), which showed an increase in catch in 2050 under this scenario. We obtained similar patterns for the catch of all these species under the mild climate change scenario but the predicted catch of blue crab (*Callinectes sapidus*) increased under this scenario (Figure 4.5b).

In recent decades, Penaeus shrimps have been the most important commercially

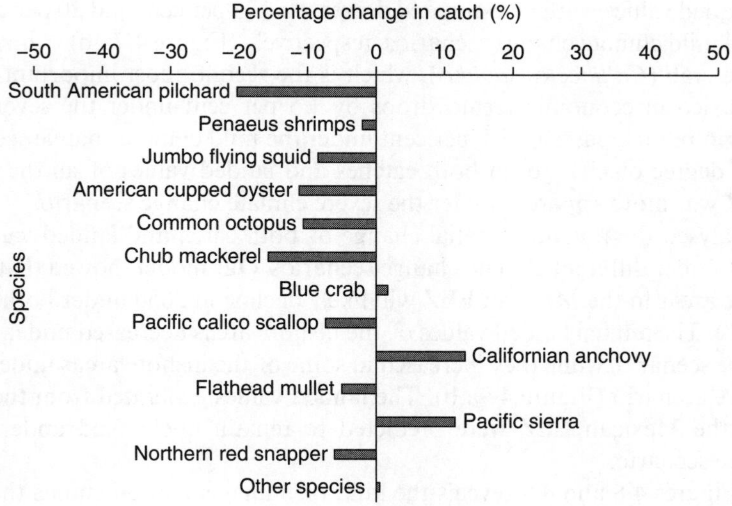

Figure 4.5b Predicted percentage change in catch in 2050 in Mexican EEZ under the MILD climate change scenario

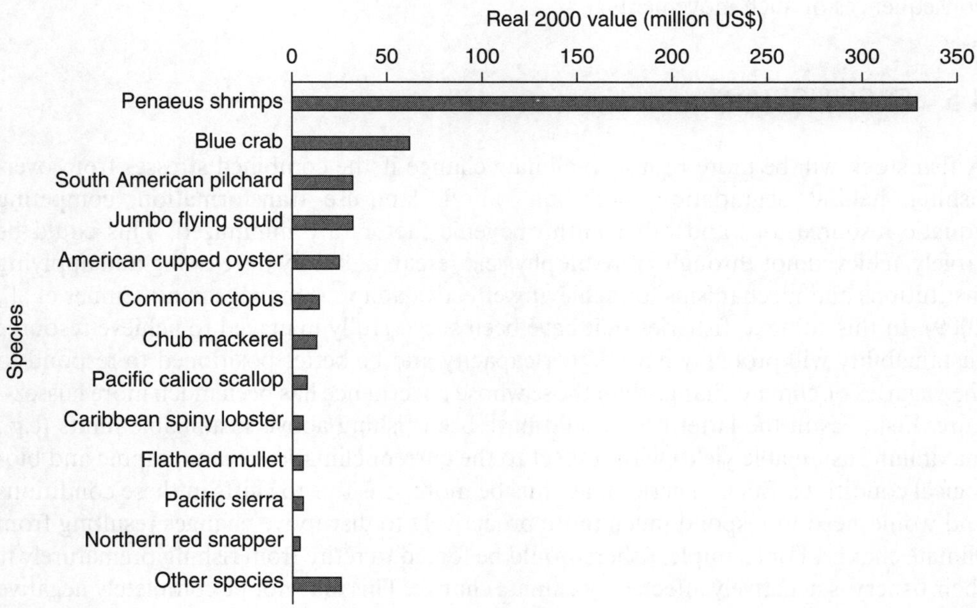

Figure 4.6 Average annual landed values (real 2000 value) (US$) for last decade (1991–2000) in Mexican EEZ

exploited group in terms of annual landed values in Mexico's EEZ and they contributed almost 60 per cent to the total annual landed values (in 2000 real values) in Mexico (Figure 4.6). In 2050, South American pilchard (*Sardinops sagax*), which is the species with the third highest landed value in Mexico currently, is predicted to drop the most in real annual landed values, with an estimated decrease of 35 per cent and 20 per cent under the severe and mild climate change scenarios, respectively (Figure 4.7a,b). Annual landed values of blue crab (*Callinectes sapidus*), which is the second most important exploited species in Mexico in economic terms, drops by 1.5 per cent under the severe climate change scenario but increases by 1.8 per cent under the mild climate change scenario. As expected, the degree of changes in both catches and landed values of all the species in Mexican EEZ was more apparent under the severe climate change scenario.

We also analysed the potential spatial change of both catch and landed values in the Mexican EEZ under different climate change scenarios. Our model showed that the catch in most of the areas in the Mexican EEZ will likely decline in 2050 under both scenarios (Figure 4.8a,b). The annual landed values of the inshore areas decreased under the severe climate change scenario, while they increased in some of the inshore areas under the mild climate change scenario (Figure 4.9a,b). The landed values generated from the offshore areas within the Mexican EEZ were predicted to remain unchanged under the mild climate change scenario.

A look at Figures 4.8 and 4.9 reveals the kind of management dilemmas that climate change is likely to impose on fisheries managers even within a country. For instance, is it the case that fishers whose fishing grounds lose fish (medium shaded areas) are going to be able to simply move into fishing areas that gain fish (black areas)? What are the likely consequences of such movements?

4.5 DISCUSSION

A fish stock will be more robust to climate change if the combined stresses from overfishing, habitat degradation, pollution runoff, land-use transformation, competing aquatic resource uses and other anthropogenic factors are minimized. This could be largely achieved not through more biophysical research, but by developing and applying institutions and mechanisms for achieving effective adaptive management (Cinner et al., 2009). In this context, fisheries that have been successfully managed to achieve resource sustainability will probably have higher capacity and be better positioned to respond to the vagaries of climate change than those whose governance has been much more laissez-faire. Fisheries in the latter case would have been fishing above sustainable limits (e.g., maximum sustainable yield) with respect to the current climatic, oceanographic and biological conditions. Such fisheries may thus be more sensitive to shifts in these conditions and would need to respond much more proactively to disruptive changes resulting from climate change. For example, fishers would be forced to retire from fishing prematurely if their fishery is negatively affected by climate change. This may not be completely negative as labour and capital displaced from fishing might be used more productively in other sectors of the economy.

The global fishing sector has had to adapt to declining fish stocks and catches over time because of overfishing (e.g., by fishing deeper and into the high seas) or because of seasonal, inter-annual and multi-decadal variability (e.g., the warm period in the North Atlantic from 1925 to 1960) or combinations of both. When climate change affects the composition and productivity of exploited species in a region, some fishers can adapt by switching target species or gear type or by moving to marginally productive areas. As

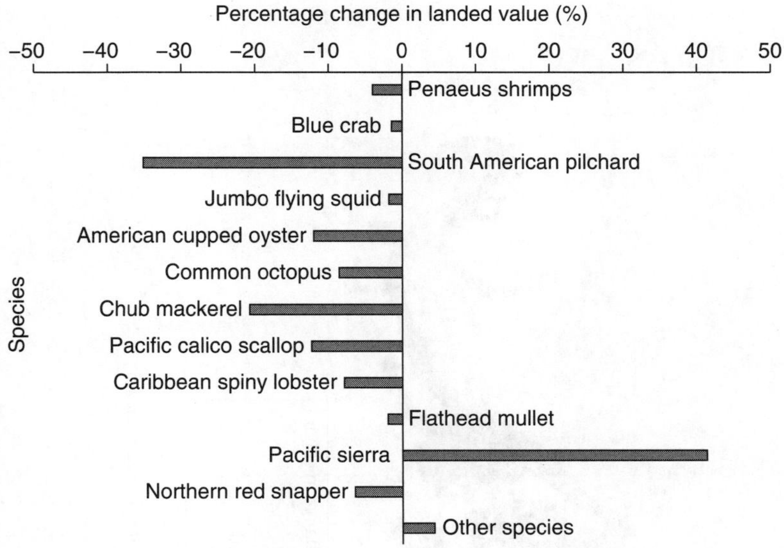

Figure 4.7a *Predicted percentage change in average annual landed values (real 2000 value) in 2050 in Mexican EEZ under a SEVERE climate change scenario*

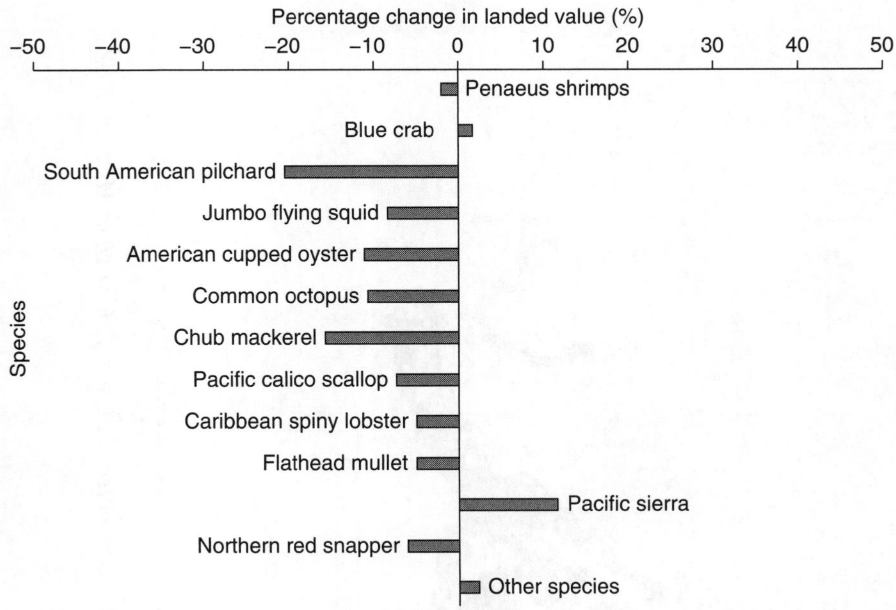

Figure 4.7b *Predicted percentage change in average landed values (real 2000 value) in 2050 in Mexican EEZ under a MILD climate change scenario*

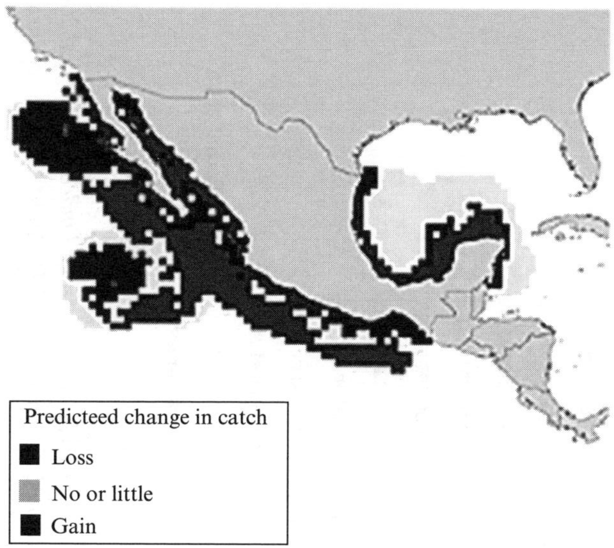

a. SEVERE cclimate change scenario

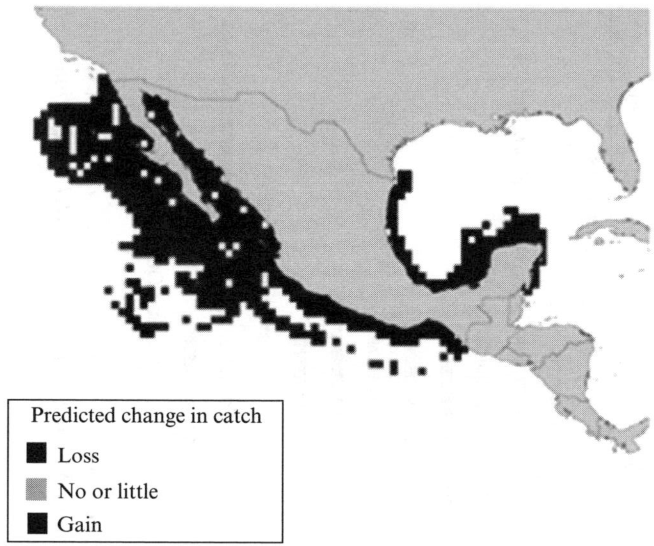

b. MILD climate change scenario

Figure 4.8a,b Predicted change in fish catch in Mexican EEZ

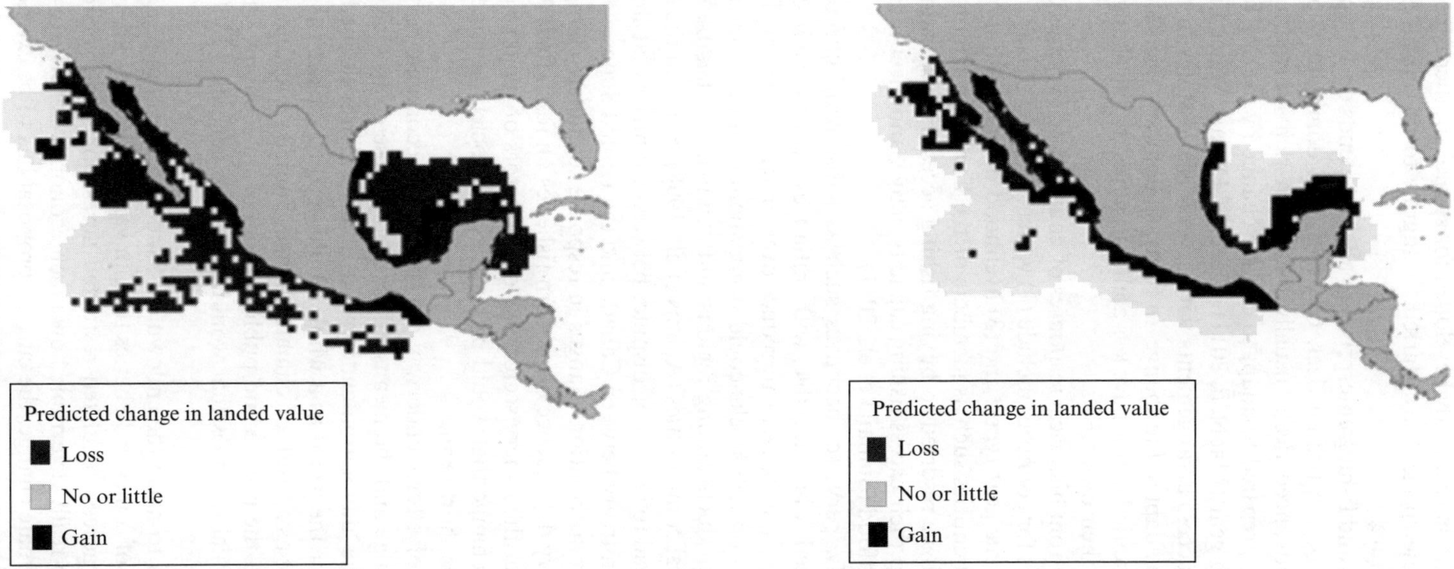

a. SEVERE climate change scenario

b. MILD climate change scenario

Figure 4.9a,b Predicted change in fish landed value in Mexican EEZ

an example, new fisheries have already developed for several southern species in the UK (e.g., red mullet) as these species have started to migrate to the North Sea because of increase in sea temperature.

The ability of fishers and fishing enterprises to adapt depends on a number of factors, including the mobility of the fishing fleet. Currently, the dominant hypothesis is that more technologically advanced fleets, usually located in rich Northern countries, are more likely to be better prepared to adapt to climate change by moving to other fishing grounds and by shifting gear (MacNeil, 2011). Fleets of Distant Water Fishing Nations (DWFN), which have access arrangements with several island states in the Pacific, for example, may be able to adapt to the change. In contrast, domestic fleets and their associated canneries of the Pacific Islands have less ability to adjust to the change because they are usually confined to their own EEZs.

Based on experience from historical responses of countries to fisheries changes, possible adaptation strategies for countries include (1) vessel buybacks (Clark et al., 2005); (2) restricting the use of some gear types; and (3) livelihood diversification measures. Also, large countries/political entities such as members of the European Union, Japan, China and the United States have resorted to buying fishing access rights, mainly from developing countries, to keep their excess fishing capacity active, and meet their populations' growing demand for seafood (Sumaila et al., 2011).

To adapt successfully, perverse incentives such as subsidizing unprofitable fishing fleets need to be replaced, where feasible, with initiatives such as catch shares management and other incentive mechanisms to reduce overcapacity in overexploited fisheries. Climate change impacts on fish stocks could bring about changes in current trade patterns in fish and fish products among regions and countries, as has been shown to be the case with agricultural commodities (Adams et al., 1998). Shifts in the distribution of exploited species may lead to increasing disputes between countries that share fish stocks. For example, the salmon treaty between Canada and the United States would need to be re-negotiated as salmon distribution changes in response to climate change (Miller and Munro, 2004). The heavy dependence of modern fisheries on fossil fuels (Sumaila et al., 2008) would require that the fishery sector, like other sectors of the economy, mitigates its carbon footprint, a change that would be beneficial to society at large, but costly to fishing enterprises in the short term.

There are a number of efforts underway to estimate the economic costs of adapting fisheries to climate change and the means of absorbing these costs. One approach that has been proposed is the 'adaptation endowment fund', which is defined as the capital that a country, region or the world as a whole would need to replace the projected loss in annual gross revenues as a result of climate change (Sumaila and Cheung, 2010). The authors argue that both the private and public sectors of an economy will have to find ways to replace the annual revenues that would have been generated by fisheries in the absence of climate change.

More work is needed to make the endowment fund idea useful in practice. First, the real world is composed of a mix of fisheries that, in varying degrees, are more like open access, regulated open access, regulated restricted access, or some version of optimal management. Hence, to really determine how much economic benefit (or resource rent) would be lost due to climate change, the current proposal has to be extended to include an institutional layer that captures and incorporates how management institutions in a country would help protect fisheries benefits from being eroded by climate change. Second, as per economic theory, using gross revenue loss as a basis for the endowment fund overestimates the compensation needed to mitigate the impact of climate change on fisheries. The appropriate economic indicator to use is resource rent, or more broadly, welfare loss.

How societies deal with climate change will depend largely on their capacity to adapt, which will be strongly influenced by social, economic, political and cultural conditions. A wide range of adaptations and mitigation measures are possible, either carried out in anticipation of future effects, or in response to impacts once they have occurred. Some can be implemented via public institutions, others by private individuals. In general, responses to the direct impacts of extreme events on fisheries infrastructure and communities are likely to be more effective if they are anticipatory, as part of a long-term participatory, broad-based approach to fisheries management. Such an integrated approach has the potential to increase ecosystem and community resilience and provide a valuable framework for dealing with climate change. In any case, preparation should be commensurate with risk, as excessive protective measures could themselves have negative social and economic impacts.

4.6 CONCLUSIONS

We have presented in this chapter recent developments in the biophysical and economic modelling of the effects of climate change on the world's fish and fisheries. Clearly, we are still at the initial stages of model development with much work still to be done before we begin to understand the full economic and management implications of ocean warming. Because of the close linkages between the biophysical components of marine ecosystems and social economics of fisheries, integrated assessments across disciplines are needed to understand climate change impacts on human welfare through marine fisheries.

Given the strong linkages between primary productivity and fishery resources, there is a need for improved understanding on the effects of climate change of primary productivity at the scale that is relevant to fisheries. And, in the absence of published research on the macroeconomic impacts of climate change on fisheries, we can only conjecture that climate change would likely have impacts on national labour markets, industry reorganization and reorientation to changes in export earnings (some negative) due to the predicted declines of fisheries in many maritime countries (Glantz and Thompson, 1981). Further research on the potential macroeconomic effects of climate change on fisheries is needed.

How much consumer surplus is lost under various scenarios? What are the real deadweight losses as a result of climate change impacts on fisheries? These economic questions are yet to be addressed in the literature, and therefore deserve attention. To be certain about the impact of climate change on economic rent globally, we need more information about seafood demand elasticity or the degree of substitution between seafood and other protein sources.

REFERENCES

Adams, R.M., B. Hurd, S. Lenhart and N. Leary (1998), 'Effects of global warming on agriculture: an interpretative review', *Climate Research*, **11**(1), 19–30.

Allison, E.H., A. Perry, M.-C. Badjeck, W.N. Adger, K. Brown and D. Conway et al. (2009), 'Vulnerability of national economies to the impacts of climate change on fisheries', *Fish and Fisheries*, **10**(2), 173–96.

Arnason, R. (2007), 'Climate change and fisheries: assessing the economic impact in Iceland and Greenland', *Natural Resource Modeling*, **20**(2), 163–97.

Badjeck, M.-C., E.H. Allison, A.S. Halls and N.K. Dulvy (2010), 'Impacts of climate variability and change on fishery-based livelihoods', *Marine Policy*, **34**(3), 375–83.

Boyce, D.G., M.R. Lewis and B. Worm (2010), 'Global phytoplankton decline over the past century', *Nature*, **466**(7306), 591–6.

Cheung, W.W.L, V.W.Y. Lam and D. Pauly (2008a), 'Modelling present and climate-shifted distribution of marine fishes and invertebrates', *Fisheries Centre Research Reports*, **16**(3), University of British Columbia, Vancouver.

Cheung, W.W.L., C. Close, V. Lam, R. Watson and D. Pauly (2008b), 'Application of macroecological theory to predict effects on climate change on global fisheries potential', *Marine Ecology Progress Series*, **365**, 187–97.

Cheung, W.W.L., V. Lam, J. Sarmiento, K. Kearney, R. Watson. D. Zeller and D. Pauly (2010), 'Large-scale redistribution of maximum fisheries catch potential in the global ocean under climate change', *Global Change Biology*, **16**(1), 24–35.

Cinner, J.E., T. McClanahan, N. Graham, M. Pratchett, S. Wilson and J.-B. Raina et al. (2009), 'Gear-based fisheries management as a potential adaptive response to climate change and coral mortality', *Journal of Applied Ecology*, **46**(3), 724–32.

Clark, C.W. (1990), *Mathematical Bioeconomics: The Optimal Management of Renewable Resources*, New York: Wiley-Interscience.

Clark, C.W., G.R. Munro and U.R. Sumaila (2005), 'Subsidies, buybacks, and sustainable fisheries', *Journal of Environmental Economics and Management*, **50**(1), 47–58.

Dalton, M.G. (2001), 'El Niño, expectations, and fishing effort in Monterey Bay, California', *Journal of Environmental Economics and Management*, **42**(3), 336–59.

Dasgupta, P. (2008), 'Discounting climate change', *Journal of Risk and Uncertainty*, **37**(2), 141–69.

Delworth, T.L., A. Broccoli, A. Rosati, R.J. Stouffer, V. Balaji and J. Beesley et al. (2006), 'GFDL's CM2 global coupled climate models. Part I: formulation and simulation characteristics', *Journal of Climate*, **19**, 643–74.

Doney, S.C., W.M. Balch, V.J. Fabry and R.A. Feely (2009), 'Ocean acidification: a critical emerging problem for the ocean sciences', *Oceanography*, **22**(4), 16–25.

Dulvy, N.K., S.I. Rogers, S. Jennings, V. Stelzenmüller, S.R. Dye and H.R. Skjoldal (2008), 'Climate change and deepening of the North Sea fish assemblage: a biotic indicator of warming seas', *Journal of Applied Ecology*, **45**(4), 1029–39.

Dyck, A.J. and U.R. Sumaila (2010), 'Economic impact of ocean fish populations in the global fishery', *Journal of Bioeconomics*, **12**(3), 227–43.

FAO (2011), *The State of World Fisheries and Aquaculture, 2010*, Rome: FAO.

Garza-Gil, A.D., J. Torralba-Cano and M.M. Varela-Lafuente (2010), 'Evaluating the economic effects of climate change on the European sardine fishery', *Regional Environmental Change*, **11**(1), 87–95.

Glantz, M.H. and J.D. Thompson (1981), *Resource Management and Environmental Uncertainty: Lessons from Coastal Upwelling Fisheries*, New York: Wiley Interscience.

Herrick, S.F., R. Hannesson, U.R. Sumaila, M. Ahmed and J.P. Torres (2009), 'Global production and economics of small, pelagic fish', in A. Checkley, J. Alheit, Y. Oozeki and C. Roy (eds), *Climate Change and Small Pelagic Fish*, Cambridge, UK: Cambridge University Press, pp. 256–74.

IPCC (2007), *Climate Change Synthesis Report*, contribution of Working Groups I, II and III to the fourth assessment report of the Intergovernmental Panel on Climate Change, Geneva: IPCC.

Lal, P.N., J. Kinch and F. Wickham (2009), 'Review of economic and livelihood impact assessments of, and adaptation to, climate change in Melanesia', paper, accessed 27 March 2011 at http://www2.bishopmuseum.org/ccbm/Areas/Melanesia/Papers/EconomicImpactsMelanesia_Lal.pdf.

Lam, V.W.Y., W.W.L. Cheung, W. Swartz and R. Sumaila (2012), 'Climate change impacts on fisheries in West Africa: implications for economic, food and nutritional security', *African Journal of Marine Science*, **34**(1), 103–17.

Last, P.R., W.T. White, D.C. Gledhill, A.J. Hobday, R. Brown, G.J. Edgar and G. Pecl (2011), 'Long-term shifts in abundance and distribution of a temperature fish fauna: a response to climate change and fishing practices', *Global Ecology and Biogeography*, **20**, 58–72.

Lucey, S.M. and J.A. Nye (2010), 'Shifting species assemblages in the Northeast US Continental Shelf large marine ecosystem', *Marine Ecology Progress Series*, **415**, 23–33.

MacNeil, M.A., N. Graham, J. Cinner, N. Dulvy, P. Loring and S. Jennings et al. (2011), 'Transitional states in marine fisheries: adapting to predicted global change', *Philosophical Transactions of the Royal Society B*, **365**(1558), 3753–63.

McIlgorm, A. (2010), 'Economic impacts of climate change on sustainable tuna and billfish management: insights from the Western Pacific', *Progress in Oceanography*, **86**, 187–91.

Miller, K.A. and G.R. Munro (2004), 'Climate and cooperation: a new perspective on the management of shared fish stocks', *Marine Resource Economics*, **19**, 367–93.

Mueter, F.J. and M.A. Litzow (2008), 'Sea ice retreat alters the biogeography of the Bering Sea Continental Shelf', *Ecological Applications*, **18**(2), 309–20.

Pauly, D. (2010), *Gasping Fish and Panting Squids: Oxygen, Temperature and the Growth of Water-breathing Animals, Excellence in Ecology: Book 22*, Oldendorf/Luhe: International Ecology Institute.

Pauly, D., R. Watson and J. Alder (2005), 'Global trends in world fisheries: impacts on marine ecosystems and food security', *Philosophical Transactions of the Royal Society*, **360**(1453), 5–12.

Perry, A.L., P.J. Low, J.R. Ellis and J.D. Reynolds (2005), 'Climate change and distribution shifts in marine fishes', *Science*, **308**(5730), 1912–15.

Pinnegar, J.K., S. Jennings, C.M.O. O'Brien and N.V.C. Polumin (2002), 'Long-term changes in the trophic level of the Celtic Sea fish community and fish market price distribution', *Journal of Applied Ecology*, **39**(3), 377–90.

Pitcher, T., D. Kalikoski, G. Pramod and K. Short (2009), 'Not honouring the code', *Nature*, **457**(7230), 658–9.

Polovina, J.J., J.P. Dunne, P.A. Woodworth and E.A. Howell (2011), 'Projected expansion of the subtropical biome and contraction of the temperate and equatorial upwelling biomes in the North Pacific under global warming', *ICES Journal of Marine Science*, **68**(6) 986–98.

Pontecorvo, G., M. Wilkinson, R. Anderson and M. Holdowsky (1980), 'Contribution of the ocean sector to the United States economy', *Science*, **208**(4447), 1000–1006.

Pörtner, H.O. and R. Knust (2007), 'Climate change affects marine fishes through the oxygen limitation of thermal tolerance', *Science*, **315**(5808), 95–7.

Rayner, N.A., D. Parker, E. Horton, C. Folland, L. Alexander and D. Rowell et al. (2003), 'Global analyses of sea surface temperature, sea ice, and night marine air temperature since the late nineteenth century', *Journal of Geophysical Research*, **108**(D14), 1–37.

Roy, N., R. Arnason and W.E. Shrank (2009), 'The identification of economic base industries, with an application to the Newfoundland fishing industry', *Land Economics*, **85**(4), 675–91.

Sarmiento, J.L., R. Slater, R. Barber, L. Bopp, S. Doney and A. Hirst et al. (2004), 'Response of ocean ecosystems to climate warming', *Global Biogeochemical Cycles*, **18**(3).

Sumaila, U.R. and W.W.L. Cheung (2010), 'Cost of adapting fisheries to climate change', World Bank Discussion Paper No. 5, Washington, DC: The International Bank for Reconstruction and Development/The World Bank.

Sumaila, U.R. and C. Walters (2005), 'Intergenerational discounting: a new intuitive approach', *Ecological Economics*, **52**, 135–42.

Sumaila, U.R., A.D. Marsden, R. Watson and D. Pauly (2007), 'A global ex-vessel fish price database: construction and applications', *Journal of Bioeconomics*, **9**(1), 39–51.

Sumaila, U.R., L. Teh, R. Watson, P. Tyedmers and D. Pauly (2008), 'Fuel price increase, subsidies, overcapacity and resource sustainability', *ICES Journal of Marine Science*, **65**(6), 832–40.

Sumaila, U.R., W.W.L. Cheung, V.W.Y. Lam, D. Pauly and S. Herrick (2011), 'Climate change impacts on the biophysics and economics of world fisheries', *Nature Climate Change*, **1**, 449–56.

Teh, L.C. and U.R. Sumaila (2011), 'Contribution of marine fisheries to worldwide employment', *Fish and Fisheries*, **14**(1), 77–88.

Trotman, A., R.M. Gordon, S.D. Hutchinson, R. Singh and D. McRae-Smith (2009), 'Policy responses to GEC impacts on food availability and affordability in the Caribbean community', *Environmental Science & Policy*, **12**, 529–41.

Watson, R., A. Kitchingman, A. Gelchu and D. Pauly (2004), 'Mapping global fisheries: sharpening our focus', *Fish and Fisheries*, **5**(2), 168–77.

5. The economic impacts of ocean acidification

Luke M. Brander, Daiju Narita, Katrin Rehdanz and Richard S.J. Tol

5.1 INTRODUCTION

Carbon dioxide dissolves in seawater to form carbonic acid. As the atmospheric concentration of carbon dioxide increases, so does the oceanic concentration in order to maintain the chemical equilibrium between seawater and the atmosphere. Carbon emissions from fossil fuel combustion and land use change thus make ocean water more acidic.[1]

Ocean acidification reduces the availability of calcium carbonate in the oceans. When carbon dioxide bonds with seawater to form carbonic acid, a bicarbonate ion and hydrogen ion are released. The free hydrogen ions then bond with free carbonate ions and thereby reduce the availability of carbonate ions for marine animals that make calcium carbonate shells and skeletons. Ocean acidification therefore spells potential trouble for species with an exoskeleton. Shellfish and corals spring immediately to mind, but many micro-organisms in the ocean crucially depend on calcium carbonate as well. Besides the direct impact on vulnerable species, ocean acidification also affects the food chain for animals in the oceans and elsewhere.

This affects humans too. Fisheries and aquaculture are an important source of income and food, particularly proteins. Coral reefs protect coasts from storm damage and erosion, supply sediments to form beaches and support the livelihoods of entire small island nations. Coral reefs harbour valuable fish and provide excellent opportunities for recreation. Marine species make up a major part of global biodiversity, and play a role in the global carbon cycle. Prima facie, ocean acidification is a reason for concern.

However, despite their great importance, the services provided by marine and coastal ecosystems have received far less attention than those provided by terrestrial ecosystems – possibly due to the difference in access and direct experience (TEEB, 2009). The same is true for the economic impact of ocean acidification. This chapter reviews the available literature on this issue, which is nascent and sparse.

Economic impact estimates of ocean acidification are needed for a number of reasons. Ocean acidification is an externality of carbon dioxide emission, presumably negative, which requires correcting. Damage cost estimates are required in order to determine the optimal level of mitigation. In the likely case that mitigation is insufficient and delayed, and that atmospheric concentrations of carbon dioxide continue to increase over the next century, there is a need for impact estimates to guide adaptation efforts. Adaptation options that reduce the local biogeochemical impacts of ocean acidification may be limited (e.g., spreading pulverized olivine or calcium hydroxide) and so adaptation is limited to replacing impacted ecosystem services and finding alternative sources of food and livelihood.

Ocean acidification is different from, but related to, climate change. Climate change is caused by a range of greenhouse gases, whereas ocean acidification is caused by carbon dioxide only (and has been described as 'the other CO_2 problem'). Climate engineering is a possible solution for climate change, but not necessarily for ocean acidification. In fact, ocean acidification may accelerate if sulphur particles (which acidify too) are used to deflect sunlight. Climate change is slow because of the slow uptake of energy by the ocean, whereas ocean acidification is expected to occur more rapidly. Ocean acidification is therefore not just an extra reason to reduce carbon dioxide emissions but has qualitative impacts on optimal mitigation policy. In addition, a potential feedback effect between ocean acidification and climate change is possible. Ocean acidification is likely to result in a decline in the rate at which the oceans absorb carbon dioxide since less carbon is stored in calcium carbonate shells and skeletons. This means that a greater proportion of carbon dioxide emissions remains in the atmosphere and exacerbates climate change.

The chapter continues as follows. Section 5.2 presents the framework for assessment. Section 5.3 discusses the available estimates. Section 5.4 reviews the gaps in knowledge. Section 5.5 concludes.

5.2 FRAMEWORK FOR ASSESSING THE ECONOMIC IMPACTS OF OA

The assessment of the economic impacts of ocean acidification (OA) requires the linking of biophysical processes with human benefits derived from the marine environment. This link can usefully be made through the concept of ecosystem services (TEEB, 2010). Ecosystem services can be defined as the outputs of ecosystems from which people derive benefits (UK NEA, 2011). Ecosystem services can either be 'final' goods and services that people benefit from (e.g., coastal protection, non-use biodiversity values) or be used as inputs in the production of 'final' goods and services (e.g., wild caught and aquaculture fish). It is the value of 'final' goods and services that need to be quantified in an economic assessment of ecological impacts. Applying the concept of ecosystem services is useful in making a distinction between ecosystem functions (processes) and benefits to humans, and ensures that assessments of ecological impacts are more accessible to economic valuation. Any ecosystem assessment still needs to determine the change in service provision in biophysical terms to give a solid ecological basis to the economic valuation (TEEB, 2010).

The assessment of the impacts of ocean acidification on economic welfare requires that the full impact pathway is understood and modelled. This demands the coupling or integration of models that explain each step in the pathway linking (1) socioeconomic activities, CO_2 emissions, ocean acidification, (2) impacts on marine ecosystems, (3) changes in the provision of ecosystem services, and finally (4) impacts on human welfare (see Figure 5.1). Each of these components in themselves is a complex research topic that is characterized by varying degrees of scientific understanding and uncertainty.

The biogeochemistry (1) of increasing atmospheric concentrations of CO_2, uptake of CO_2 by the oceans, resulting in a decrease in seawater pH and availability of calcium carbonate is relatively well understood, albeit with spatial variation depending on physical determinants of CO_2 solubility such as depth, temperature and circulation (Doney et al., 2009).

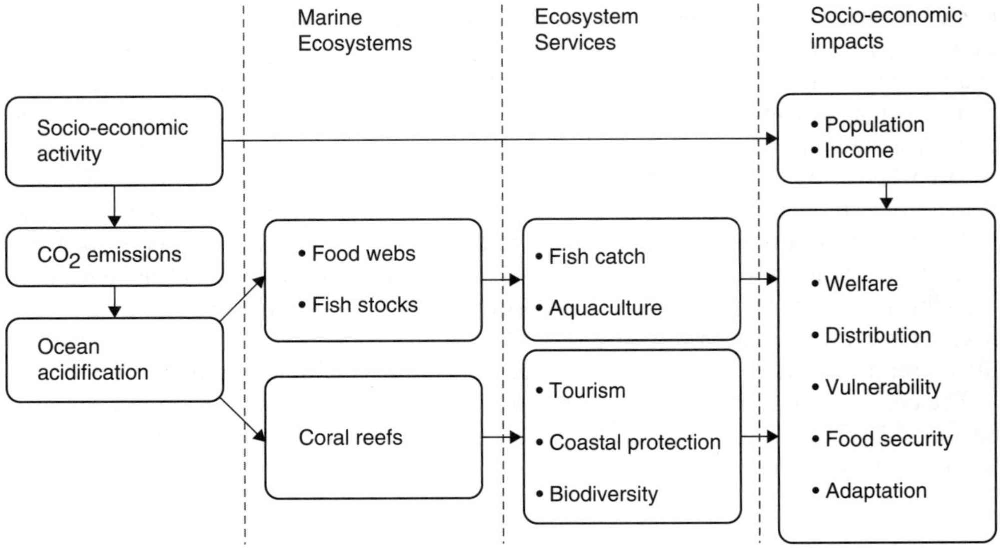

Figure 5.1 Impact pathway for ocean acidification

The impacts of ocean acidification (2) on individual species and marine ecosystems are now the subject of considerable research attention. For a comprehensive review see Gattuso and Hansson (2011). In general the impacts are still not well understood, particularly as the focus moves from individual species that are directly impacted by calcium carbonate availability (e.g., molluscs and reef-building corals) to species that are more distant in the food web (e.g., commercially important fish species and 'charismatic' mammals) and to the impact of multiple stressors on the marine environment (e.g., temperature change, pollution, overfishing).

The next step (3) in the impact pathway, namely the change in provision of ecosystem services, is still less well understood and also less well studied. Although potential impacts on commercial mollusc and fish harvests have been addressed in a number of studies, there remains no modelled or empirical quantification of the effect of species-level biophysical changes on harvests. As yet there has been little research on the impacts of ocean acidification on the provision of ecosystem services from coral reefs (e.g., recreational opportunities, beach formation, coastal protection, and non-use values for biodiversity).

The estimation of economic welfare impacts (4) of changes in the provision of ecosystem services requires information on the value of these services and how values change with the level of provision. It is important to note that changes in the value of marine ecosystem services are determined by changes in both supply and demand, the latter being determined by the characteristics of the beneficiaries that receive those services (population, income, preferences etc.). It is therefore necessary to also account for changes in demand for marine ecosystem services over the course of the time frame of ocean acidification impacts (i.e., the populations that will be impacted will be very different from the present). This point is indicated by the arrow at the top of Figure 5.1. The valuation of ecosystem services from the marine environment is further complicated by the fact that some services are not traded, directly or indirectly, in markets and therefore do not have market prices (e.g., coastal protection provided by coral reefs and non-use biodiversity values). The valuation and distribution of economic impacts have received little research attention and we review the few existing studies in the following section.

It is important to note that the linkages between each component step in the ocean acidification impact pathway are perhaps still less well understood since they fall between disciplines. The assessment of OA impacts therefore requires an integration of research findings that bridges disciplinary boundaries, which involves challenges of information compatibility and communication.

5.3 THE CURRENT EVIDENCE ON ECONOMIC IMPACTS OF OCEAN ACIDFICATION

Despite the potentially severe impacts of ocean acidification on marine ecosystem services, there has been relatively little research on the economic costs involved. In this section we summarize and review the existing literature that addresses the economic impacts of ocean acidification. The purpose of this review is to take stock of the current understanding of this issue, the methods employed, the ecosystem services addressed and the geographic scope of analysis. Summary information for each study is presented in Table 5.1.

Kite-Powell (2009) provides a discussion of the main economic impacts of acidification, namely to fisheries and the ecosystem services provided by coral reefs. The paper gives rough estimates of the total value of these ecosystem services (in gross revenue terms) but argues that currently available knowledge and models are not sufficient to quantify the impacts of ocean acidification with any certainty. The point is made that the adaptive response of marine ecosystems and humans to ocean acidification is not known.

Finnoff (2010) sets out the economic approach to valuing the impacts of ocean acidification on the provisioning of ecosystem services. The paper argues that the welfare implications of ocean acidification need to be measured in terms of changes in consumer and producer surplus rather than changes in gross revenues (as is done in a number of studies). Finnoff makes the point that the standard economic framework for assessing material damages, such as the degradation of an environmental input into production or consumption, has been available for some time but that the implementation remains challenging. A particular challenge is the integration of highly complex ecological processes into economic models. On one hand, the use of a reduced form representation of the natural system may be overly simplified and miss non-convexities in the ecological system. On the other hand, the use of detailed structural models may better capture the complexities of the system but become intractable. Using an existing model of the Bering Sea ecosystem, the paper provides a simulation of ocean acidification impacts in order to highlight the complexities of assessing impacts in systems characterized by non-convexities. The paper does not attempt to estimate values for ocean acidification impacts.

Brander et al. (2012a) provide a first estimate of the global economic impact of ocean acidification on coral reefs. The paper constructs and calibrates simple models of ocean acidification and coral reef area loss, driven by the atmospheric concentration of carbon dioxide. Carbon dioxide emissions and concentrations are modelled for the four SRESs (Special Report on Emissions Scenarios) using the FUND (Climate Framework

Table 5.1 Summary of studies that examine the economic impacts of ocean acidification

Study	Impacts	Geographic scope	Emissions scenario	Period of analysis	Welfare measure[a]	Annual value (US$; billions)[b]
Armstrong et al. (2012)	Fisheries	Norway	pH decrease 0.5	2010–2110	Revenue	0.01
	Carbon storage	Norway	pH decrease 0.5	2010–2110	Damage Cost	3
Brander et al. (2012)	Coral reefs	Global	SRES A1B	2000–2100	Mixed	1,093
Cheung et al. (2011)	Fish and invertebrates	N-E Atlantic	SRES A1B	2005–2050	–	–
Cooley and Doney (2009)	Molluscs	United States	IPCC A1F1	2007–2060	Revenue	0.07
Cooley et al. (2012)	Molluscs	Global	CCSM3	2010–2060	–	–
Finnoff (2010)	Fisheries; non-use values	Baring Sea	–	–	–	–
Harrould-Kolieb et al. (2009)	Coral reefs; fisheries	Global	SRES A1B	2009–2050	–	–
Hilmi et al. (2012)	All	Global	–	–	–	–
Kite-Powell (2009)	Coral reefs; fisheries	Global	IS92a	–	–	–
Moore (2011)	Molluscs	United States	RCP8.5; RCP6	2010–2100	CV	0.31
Narita et al. (2012)	Molluscs	Global	IS92a	2000–2100	CS, PS	139
Rodrigues et al. (2013)	Use and non-use values	Mediterranean	–	–	–	–
Sumaila et al. (2011)	Capture fisheries	Global	–	–	–	–

Notes:
a. CV: compensating variation; CS: consumer surplus; PS: producer surplus.
b. Impact estimates are standardised to annual values for the terminal year in each analysis (i.e., 2060 for Cooley and Doney (2009), and 2100 otherwise) in US$ 2010 price levels.

for Uncertainty, Negotiation and Distribution) integrated assessment model. A meta-analysis of coral reef values is used to estimate a value transfer function for coral reef ecosystem services. An existing model of tourist numbers is applied to estimate future visitor numbers to coral reef locations. Combining these models, the annual value of lost ecosystem services due to ocean acidification-induced coral reef loss is estimated. The economic value of lost coral reef ecosystem services varies across scenarios due to (1) differing rates of CO_2 emissions, ocean acidification and loss of coral cover; (2) differing rates of population and income growth that determine the value of coral reef services per unit area of coral cover. The results show that the annual economic impact (loss of coral reef service value) escalates rapidly over time, essentially because the scenarios have high economic growth in countries with coral reefs and because demand for coral reef services increases more than proportionately with income. Nonetheless, the annual value of foregone ecosystem services from coral reefs in 2100 is still only estimated to be a small fraction of total global income (0.14 per cent or US$870 billion in 2100; 2000 price levels; SRES A1B). The estimated impacts are, however, considered to be partial since the underlying value data is largely focused on recreational values and includes limited information on the value of other services such as coastal protection or non-use values for biodiversity. The study reports the results of a sensitivity analysis and shows that the estimated impact is highly uncertain, with a confidence interval spanning one order of magnitude. It is important to note that other threats to the health of coral reefs and the provision of reef services are not included in this analysis (e.g., overfishing, sedimentation, eutrophication, sea level and temperature rise). The impact of ocean acidification on coral reef services may therefore be overstated in the case that coral reefs are more rapidly degraded by the combined stress of multiple threats. The projection of coral reef functioning under combined stressors adds a considerable degree of complexity, as does the attribution of coral reef damage to specific stressors.

Cooley and Doney (2009) estimate the impact of ocean acidification on gross revenues for US mollusc fisheries up to 2060. They combine experiment-level information on the impact of ocean acidification on growth rates of molluscs with data on US fisheries harvests and prices. Baseline future revenues are projected to 2060 assuming no changes in ecological and economic conditions prevail in 2007 (i.e., catch, prices and revenues remain constant). Under an ocean acidification scenario, the time profile of increasing impacts on mollusc growth is assumed to be linear and proportionately related to revenue for the period 2007–60. The estimated present value of losses in revenue are shown to be sensitive to CO_2 emission trajectories, impacts on mollusc growth and the discount rated used in calculating present values. Under the IPCC A1F1 scenario, the present value of lost revenue is estimated to be US$2557 million (25 per cent reduction in mollusc growth at 740 ppm CO_2; 2 per cent discount rate).

Narita et al (2012) estimate the value of global and regional loss of mollusc production due to ocean acidification over the period 2000–2100. A partial equilibrium analysis is used to quantify both producer and consumer surplus and accounts for two determinants of welfare change, namely reduced production/consumption and increased prices. Following Cooley and Doney (2009), the rate of shellfish harvest loss is assumed to be equal to the decrease in calcification rate due to ocean acidification. Narita et al. (2012) use an estimate of the decrease in calcification rate from a different meta-study (Kroeker et al., 2010), which is higher than that used by Cooley and Doney (2009). The results show that the annual global costs in 2100 could be over 100 billion US$ under a business--as-usual emission trend and assuming that demand for molluscs increases with income, the trend for which is based on the IPCC projections. The major determinants of this cost estimate are the impacts on Chinese production, which is projected to dominate global production, and the expected increase in demand for molluscs in developing countries, including China, in accordance with future income rise. The analysis also indicates that in key regions such as China and the USA, the economic losses are roughly evenly divided between producers and consumers, with slightly greater relative consumer losses for China as a result of relatively inelastic demand of molluscs in that country.

Moore (2011) develops an integrated biogeochemical-economic model to estimate the potential impacts of ocean acidification on the US market for oysters, scallops, clams and mussels for the period 2010–2100. The welfare measure that is estimated is the compensating variation of US households. Compensating variation reflects the change in consumer welfare following a change in prices and is defined as the amount of additional income that a household would need in order to obtain their original level of utility prior to a price increase. The estimated impact therefore represents the loss in consumer welfare due to increased mollusc prices caused by ocean acidification. The change in mollusc prices is modelled using a Cobb-Douglas production function with environmental quality as an input. Changes in household consumption of molluscs and alternative meats with respect to income and food prices is modelled using a two-stage demand system to estimate the parameters of a representative household's expenditure function. The estimated expenditure function is then used to calculate the additional income that a representative household would require to obtain their utility level under a 'medium' ocean acidification scenario (Representative Concentration Pathways RCP8.5) versus a 'high' ocean acidification scenario (RCP6). The present value of aggregated reduced consumer welfare is estimated to be US$735 million for the period 2010–2100 using a discount rate of 5 per cent. The author identifies the most tenuous link in the integrated model to be the relationship between changes in mollusc growth rates and prices. The Cobb-Douglas production function that is used is an assumed relationship (unitary price elasticity with respect to mollusc growth rates) rather than being empirically determined.

Cheung et al. (2011) simulate the impact of ocean acidification on the distributions and estimated catch potentials for commercially exploited demersal fish and invertebrates in the North-East Atlantic for the period 2005–50. They construct a dynamic bioclimatic envelope model that accounts for the effects of changes in ocean biogeochemistry on the distribution, productivity and maximum catch potential of 120 fish and invertebrate species. Under the SRES A1B, they show that the maximum catch potentials decline by 20–30 per cent (ten-year average for 2050 relative to 2005) relative to model simulations that do not account for ocean acidification and reduced oxygen availability. The paper discusses a number of sources of uncertainty in the analysis and shows that the distribution of physiological impacts of ocean acidification is highly dispersed.

Sumaila et al. (2011) provide a review of the global impacts of climate change on capture fisheries, including the impacts of ocean acidification. The paper gives a qualitative discussion of the expected impacts of ocean acidification and argues that the likely effects are for potential catch to decrease, fish prices to increase, costs of fishing and adaptation to increase, and resource rents to increase. The reasoning for an increase in fishery resource rents is not explained. The paper concludes that there are substantial knowledge gaps in disciplinary and interdisciplinary understanding of the full range of impacts and that current research is not necessarily focused on regions that are likely to experience the heaviest impacts.

Armstrong et al. (2012) describe a preliminary analysis of the potential impacts of ocean acidification in Norwegian waters. The study identifies the marine ecosystem services that are likely to be affected by ocean acidification, the economic methods to assess the impacts, and the present knowledge gaps. The study also produces a preliminary analysis of the scale of possible damage costs from ocean acidification with a focus on provisioning and regulating services. The results of this analysis show that ocean acidification may have positive as well as negative effects on provisioning services of fisheries and aquaculture. The largest estimated damage cost, however, is due to the reduced regulating service of carbon storage and the associated increased impacts of climate change.

Rodrigues et al. (2013) provide a discussion of the possible impacts of ocean acidification in the Mediterranean Sea. The sectors for which substantial impacts are considered likely are tourism and recreation, red coral extraction, capture fisheries and aquaculture production. In addition the study considers the effects of ocean acidification on carbon sequestration and non-use values. The framework and methods for conducting an economic assessment are set out and a preliminary qualitative assessment of potential impacts is made. This study is part of the 'European Mediterranean Sea Acidification in a Changing Climate' (MedSeA) project funded by the European Commission and lays the basis for future quantitative assessments within this project.

Hilmi et al. (2012)[2] summarize the current understanding of potential impacts of ocean acidification and set out what further information is required to enable social welfare analyses of ocean acidification. The paper does not review existing assessment results as we do in this chapter but aims to provide clear directions for future multidisciplinary collaborative research. The recommendations for data collection to inform socio-economic studies of ocean acidification include the collection of observational data on ocean chemistry, determination of the biological relevance of chemical changes, long-term multi-generation biological experiments, analysis of synergies between multiple stressors, assessment of ecosystem responses, and understanding the adaptation potential of marine ecosystems. The authors also argue for integrated quantitative assessments of the likely biological, social and economic impacts of ocean acidification, and emphasize the need for information on the vulnerabilities of both ecosystems and human communities.

Harrould-Kolieb et al. (2009) address the likely distribution of economic impacts and vulnerability to ocean acidification. Vulnerability is evaluated at the national level based on four equally weighted criteria, namely fish and shellfish catch, the level of seafood consumption, the extent of coral reefs as a percentage of Exclusive Economic Zones (EEZs) and the projected level of ocean acidification in coastal waters in 2050. Countries are then ranked according to their acidification vulnerability score. The results of this ranking show that it is mainly developed countries that score highly in terms of vulnerability, with Japan, France, the United Kingdom, the Netherlands and Australia making up the top five. This analysis of vulnerability is arguably oversimplified but points to the need for further analysis of this issue. In particular, there is a need for an analysis of vulnerability below the national scale since it is likely that the impacts of ocean acidification will be highly spatially variable with severe impacts in specific locations, for example, coastal communities that are highly dependent on fisheries and coral reef ecosystem services, and that have low capacity to adapt.

Cooley et al. (2012) also address the issue of vulnerability to the impacts of ocean acidification and focus their analysis on mollusc fisheries. They develop a national-level vulnerability scale that first groups countries by net import status and then assigns a point for each of the following five characteristics: (1) mollusc fisheries account for more than 0.001 per cent of GDP; (2) the country is protein insufficient; (3) molluscs account for more than 1 per cent of dietary protein; (4) the increase in mollusc fishery production required to maintain constant per capita volume is greater than 100 per cent; (5) the country does not currently have a mollusc aquaculture industry. Points are also assigned based on each country's adaptability ranking (based on characteristics 4 and 5 above) and on the estimated number of years until each country's EEZ is expected to experience substantial change in mean and variability of calcium carbonate availability. The expected 'transition decade' in which a substantial change in calcium carbonate availability occurs is projected to vary across national EEZs within the range of 10–50 years from 2010. Countries with low adaptive capacity, high nutritional or economic dependence, rapidly growing populations, and rapidly approaching transition decades are therefore identified as the most vulnerable to ocean acidification impacts on molluscs. The highest-ranked vulnerable countries are mainly least developed countries in Sub-Saharan Africa. Again the scale of analysis at the national level may conceal severe impacts at a local or regional scale.

Huelsenbeck (2012) produces a similar analysis of vulnerability that includes the following factors: (1) exposure (aragonite saturation state in EEZ by 2050); (2) dependence (coral reef fishers as a proportion of the national population); (3) dependence (mollusc consumption as a percentage of available protein); (4) adaptive capacity indicators (GDP per capita, population growth rate 2012–50, percentage of the population undernourished). In this analysis the countries at the top of the vulnerability ranking are again least developed countries but include some small island nations.

The studies reviewed here have assessed several of the expected impacts of ocean acidification on ecosystem services. Most have focused on impacts to commercial fisheries and particularly mollusc fisheries. This is understandable given the relatively well-understood relationship between ocean biogeochemistry and mollusc growth, albeit largely at the laboratory experimental level. The paucity of value estimates for the impact of ocean acidification on fisheries reflects the lack of current understanding of the biological/ecological processes involved.

The biophysical impact of OA on coral reefs is also relatively well understood (although recent research points to potentially complex feedbacks – Anthony et al., 2011; Kleypas et al., 2011) but the economic impacts have only been valued by one study to date. The economic valuation of coral reef ecosystem services is characterized by high complexity and uncertainty. This is largely due to the scarcity of information on the value of coral reef ecosystem services (particularly non-recreation services such as support to fisheries, coastal protection and non-use biodiversity values) and the non-market nature of many of these services. Since most coral reef services are not traded directly in markets, values are not readily observable. In response, there are a growing number of studies that apply non-market valuation techniques to estimate values for coral reef services but these are prone to biases and inaccuracies. Moreover, there is a substantial methodological challenge in transferring existing value information that is inherently spatially variable to estimate values for impacted ecosystems in other (future) contexts (Brander et al., 2012b).

The geographic scale of analysis of studies examining the economic impacts of ocean acidification has tended to be very large. Most studies cover a global scope with a level of analysis at either a regional or national scale. National assessments are only available for Norway and the USA. Regional analyses have focused on broad marine ecosystems (i.e., the Bering Sea) or FAO statistical regions (i.e., the North-East Atlantic, FAO Area 27). Cooley and Doney (2009) report results for specific subnational regions, namely New Bedford, New England and the Pacific Coast. The existing research has attempted to provide first estimates of the scale of the ocean acidification problem but it has not determined a spatial disaggregation of which human communities are most at risk. It is likely that ocean acidification will have highly localized ecological and societal impacts. The impact on economic welfare can be expected to vary across locations depending on the localized degree of acidification, the sensitivity of ecosystems to acidification and the extent to which they are already under pressure, the dependence of the population on impacted ecosystem services (e.g., fisheries, coastal tourism), and the capacity to adapt to losses in the provision of those services. All of these factors along the ocean acidification impact pathway vary by location and so it is important to have an understanding of their spatial convergence in order to identify where the impacts of ocean acidification will be most severe. Information on the spatial distribution of impacts is useful in order to target adaptation responses such as additional marine management, reduction in other environmental pressures and the development of alternative sources of protein and income.

Regarding the methodologies employed in the reviewed studies, a broad range of approaches have been used, particularly for the valuation of impacts. The welfare measures that have been estimated are therefore not consistent or directly comparable. Armstrong et al. (2012) estimate gross revenue from fisheries and damage costs from enhanced climate change; Cooley and Doney (2009) estimate loss of gross revenue; Narita et al. (2012) estimate changes in both consumer and producer surplus; Moore (2011) estimates the compensating variation measure of consumer welfare; and Brander et al. (2012a) use a meta-analytic value function that is derived from a mix of underlying welfare estimates (including both consumer and producer surplus). A common methodological weakness in the reviewed studies is the link between biophysical changes resulting from ocean acidification and changes in the provision of ecosystem services. Cooley and Doney (2009), Narita et al. (2012) and Moore (2011) all use an assumed relationship between mollusc growth and harvest. Brander et al. (2012a) use a very simple model of the impact of OA on coral cover and implicitly model the provision of ecosystem services through the meta-analytic value function.

Of the 14 studies reviewed only five provide monetary estimates of the costs of ocean acidification. Three of these are for impacts on mollusc fisheries (two for the USA and one global estimate); one covers impacts on fisheries and carbon storage; and one is for impacts on coral reef services. Central estimates from each study presented in Table 5.1 are standardized to annual values in the terminal year of each analysis in US$ at 2010 price levels. From the limited information that is currently available, it appears that impacts to coral reef services dominate. The global annual loss in value of coral reef services in 2100 is estimated to be an order of magnitude higher than that of mollusc fisheries. The estimated increased damage cost of climate change associated with reduced carbon storage due to ocean acidification suggests that this is also a potentially important impact category. The cost of this feedback effect to climate change has currently only been roughly estimated for the Norwegian Exclusive Economic Zone. The current information on the damage costs of ocean acidification only provides a partial assessment of total impacts given that other impact categories, particularly fin fisheries, are yet to be widely assessed. Gaps in the current knowledge are discussed in the following section.

5.4 GAPS IN CURRENT KNOWLEDGE

The literature overview demonstrates that there are still considerable gaps in knowledge. These gaps refer to (1) understanding the relation between changes in the marine environment and socio-economic impacts, (2) the ecosystem services that have been assessed, (3) the distribution of impacts and (4) the vulnerability of different populations. These individual gaps are discussed in more detail in this section.

The first gap relates to difficulties in linking impacts of ocean acidification on biophysical processes with changes in human benefits. Though natural science is starting to shed light on the functioning of marine ecosystems and the provision of ecosystem services, important links between ecosystem functioning, ecosystem services and human benefits are still poorly understood. However, it needs to be noted that even on the first level, the understanding of biophysical processes, important research gaps exist. The majority of studies look at direct impacts (growth, survival, etc.) of ocean acidification on single species, mostly calcifying organisms (Fabry, 2008). Studies focusing on other organisms and indirect effects, such as habitat changes, adaptation of marine species to changing conditions or food web effects, are very limited. Most existing studies are based on short-term perturbation experiments that do not account for regional differences in the coastal or marine environment. Another limitation is the type of information generated by natural science studies that can be used in economic impact analysis. Studies on the impact of ocean acidification on calcifying organisms generally analyse changes in calcification rates. However, in economic studies information on the potential change in harvest, for molluscs, or loss in reef area, for coral reefs, are required. Due to this lack of information, the studies reviewed in Section 5.3 that estimate the value of impacts are based on changes in calcification rates.

As biologists and ecologists still grapple with the first-order impact of ocean acidification, little (quantitative) attention has been paid to more complex issues – such as variability, heterogeneity, non-linearity, threshold effects and irreversibility – and higher-order effects – such as predator–prey relations and resource scarcity.

The second gap identified regards the types of ecosystem services and values that have been analysed so far. The estimates that are currently available focus on particular aspects of direct use values (mollusc fisheries and coral reef recreation). Other ecosystem services with direct use values (e.g., fin fish) have not yet been considered. Likewise, other value categories including indirect use values (e.g., regulating services) and non-use values (e.g., existence and bequest values for marine biodiversity) have not been addressed. The existing estimates of the economic impact of ocean acidification are, therefore, far from complete. It should be noted that, in general, goods and services provided by marine and coastal ecosystems have received far less attention in the economic valuation literature than those provided by terrestrial ecosystems. Since many marine ecosystem services are much less visible, valuation is a much greater challenge, which puts existing valuation approaches to the test.

Another important gap (3) in the current knowledge is on the distribution of impacts of ocean acidification. So far, research is practically absent as to the question of which groups of people will suffer economically from ocean acidification by how much. The importance of the distribution of impacts is twofold. First, better information about the distribution of impacts would enable governments to make plans and actions to support the coping efforts and adaptation by communities that are particularly affected by acidification. Second, information on how the impacts of ocean acidification are distributed influences our evaluation of the social welfare implications of ocean acidification, and in turn, of desirable stringency of general carbon policy to reduce the problem. For example, a deep reduction of carbon dioxide emissions might be warranted if the damage of ocean acidification disproportionately falls on the poorest regions of the world, since there are no effective public mechanisms that facilitate substantial international transfers of wealth.

For the understanding of impact distributions, both detailed local-level studies and synthetic global or regional studies need to be conducted and compared. Local-level studies are essential for the understanding of impact distributions for two reasons. First, as discussed earlier, the magnitude of biogeochemical impacts of ocean acidification are spatially heterogeneous because of local patterns of ocean circulations, differences in coupling effects with sea temperature or other local geological characteristics, and characteristics of local ecosystems. Only detailed local-level studies can take account of such local differences. Second, the importance of the marine environment to human welfare is spatially heterogeneous depending on local socio-economic conditions and cultural factors. For example, the significance of fisheries for people is different depending on the availability of species, income levels, dietary habits and other cultural traits, and the existence of alternative income sources. Often, a change in the marine environment brings about acute effects on the livelihoods of specific groups within a community, such as fishers or those engaged in the tourism sector. Local-level studies can highlight such differences in impacts affecting different social groups.

Global or regional-scale studies are also important and are complementary to the local-level studies mentioned above. The values of marine ecosystem services are partly determined by their demand, which is determined by the availability of substitutes, the income level of consumers, and also how closely the economy is connected to the global market. To account for all these factors necessitates a general equilibrium analysis. In addition, the analysis of ocean acidification entails a long time horizon, and in this sense it is important to take account of expected economic growth and demographic changes in the future.

A related aspect to the distribution of impacts is vulnerability (4). Some populations will be relatively more vulnerable than others to the negative impacts of ocean acidification. In other words, the biophysical impacts of ocean acidification could incur varied monetary and non-monetary costs, especially in terms of long-term costs such as nutritional deficiency or low educational attainment as a result of losses in income and food availability. Vulnerability is determined both by a population's dependence on ecosystem services that are likely to be impacted by ocean acidification and by its adaptive capacity. Dependence is determined by the extent to which a community is reliant on fisheries and coral reef services for nutrition and income. Adaptive capacity is determined by the availability of substitute production and consumption options. For ocean acidification, there are a number of potential adaptation options. For example, aquaculture may be able to insulate itself from or mitigate the effects of acidification by relocating farms to closed waters or adopting varieties resistant to acidity (for a discussion, see Narita et al., 2012). Adaptation, however, generally incurs costs, and actions are warranted only if the benefits outweigh the costs. At present, little analysis has been done with regard to the costs of adaptation to ocean acidification, and more research is needed on this subject. In evaluating adaptation costs, it is important to consider the coupling of ocean acidification with other factors that affect the marine environment, such as more frequent storm surges or sea level rise under climate change.

5.5 DISCUSSION AND CONCLUSIONS

The assessment of the economic impacts of ocean acidification requires the understanding and quantification of each step in the impact pathway from carbon dioxide emissions through biogeochemical effects, impacts on species- and ecosystem-level processes, provision of ecosystem services, and finally to the value and distribution of economic impacts. Such assessments require the integration or linkage of multiple research disciplines. Environmental economics has all the tools available to analyse the economic impact of ocean acidification, but estimates are few because natural scientists have only recently been able to quantify the impact on organisms and ecosystems.

According to Table 5.1, the annual impact of ocean acidification is measured in tenths of a percentage of global GDP. The impact of climate change on the other hand is measured in per cent (Tol, 2009). Ocean acidification is thus an additional reason for concern about carbon dioxide emissions, but given the large uncertainty about the total impacts, it is not a reason to drastically raise the level of concern.

Although the exact calculations are yet to be made, ocean acidification could have a much larger effect on the optimal Pigou tax on carbon dioxide emissions. The total impact of climate change is an order of magnitude larger than that of ocean acidification, but occurs some 50 years later than the onset of significant acidification in the ocean. With a 5 per cent discount rate and assuming a linear relationship between the degree of ocean acidification and its impact, the net present marginal value of ocean acidification and climate change are of the same order of magnitude (since $1.05^{-50} = 0.1$). Based on this approximate calculation, the estimated impacts reported in Table 5.1 suggest that the optimal tax on carbon dioxide emissions – but not on other greenhouse gases – should be doubled.

These quantitative insights are preliminary and incomplete. Many impacts remain unquantified and the currently available estimates have yet to be replicated. It is arguably even too early to structure or prioritize a research agenda. More work needs to be done on all economic aspects – impacts, adaptation, mitigation – of ocean acidification.

NOTES

1. Seawater is historically slightly alkaline with a global mean of approximately pH 8.6. Ocean acidification describes a relative decrease in seawater pH toward acidity but it is not expected that seawater pH will reach the acid side of the scale.
2. This review paper was initiated following the first international workshop of the Monaco Environment and Economic Group (MEEG) on bridging the gap between ocean acidification impacts and economic valuation, jointly organized by the Centre Scientifique de Monaco and the International Atomic Energy Agency.

REFERENCES

Anthony, K.R., J.A. Kleypas and J.-P. Gattuso (2011), 'Coral reefs modify their seawater carbon chemistry – implications for impacts of ocean acidification', *Global Change Biology*, **17**(12), 3655–66.

Armstrong, C., S. Holen, S. Navrud and I. Seifert (2012), *The Economics of Ocean Acidification – A Scoping Study*, Norway: Fram Centre et al.

Brander, L.M., K. Rehdanz, R.S.J. Tol and P. van Beukering (2012a), 'The economic impact of ocean acidification on coral reefs', *Climate Change Economics*, **3**(1), 1250002.

Brander, L.M., I. Brauer, H. Gerdes, A. Ghermandi, O. Kuik and A. Markandya et al. (2012b), 'Using meta-analysis and GIS for value transfer and scaling up: valuing climate change induced losses of European wetlands', *Environmental and Resource Economics*, **52**(3), 395–413.

Cheung, W.W.L., J. Dunne, J.L. Sarmiento and D. Pauly (2011), 'Integrating ecophysiology and plankton dynamics into projected maximum fisheries catch potential under climate change in the Northeast Atlantic', *ICES Journal of Marine Science*, **68**(6), 1008–18.

Cooley, S.R. and S.C. Doney (2009), 'Anticipating ocean acidification's economic consequences for commercial fisheries', *Environment Resource Letters*, **4**(June), 024007.

Cooley, S.R., N. Lucey, H. Kite-Powell and S.C. Doney (2012), 'Nutrition and income from molluscs today imply vulnerability to OA tomorrow', *Fish and Fisheries*, **13**(2), 182–215.

Doney, S.C., V.J. Fabry, R.A. Feely and J.A. Kleypas (2009), 'Ocean acidification: the other CO_2 problem', *Annual Review of Marine Science*, **1**, 169–92.

Fabry, V.J. (2008), 'Marine calcifiers in a high-CO_2 ocean', *Science*, **320**(5879), 1020–22.

Finnoff, D. (2010), 'Modeling economic impacts of climate change and ocean acidification to fisheries', Working Paper, US Environmental Protection Agency, National Center for Environmental Economics, accessed 31 January 2014 at http://yosemite.epa.gov/ee/epa/eerm.nsf/vwAN/EE-0566-115.pdf/$file/EE--0566-115.pdf.

Gattuso, J.-P. and L. Hansson (2011), *Ocean Acidification*, New York: Oxford University Press.

Harrould-Kolieb, E., M. Hirshfield and A. Brosius (2009), *Major Emitters Among Hardest Hit by OA: An Analysis of the Impacts of Acidification on the Countries of the World*, Oceana Report, accessed 31 January 2014 at http://oceana.org/sites/default/files/Acidity_Vulnerability_Risk_report_2.pdf.

Hilmi, N., D. Allemand, S. Dupont, A. Safa, G. Haraldsson and P.A.L.D. Nunes et al. (2012), 'Towards improved socio-economic assessments of ocean acidification's impacts', *Marine Biology*, **160**(8), 1773–87.

Huelsenbeck, M. et al. (2012), *Ocean-based Food Security Threatened in a High CO_2 World*, accessed 24 February 2014 at http://www.oceanacidification.org.uk/pdf/Ocean-Based_Food_Security_Threatened_in_a_High_CO2_World.pdf.

Kite-Powell, H. (2009), 'A global perspective on the economics of ocean acidification', *Current, The Journal of Marine Education*, **25**, 25–9.

Kleypas, J.A., K.R.N. Anthony and J.-P. Gattuso (2011), 'Coral reefs modify their seawater carbon chemistry – case study from a barrier reef (Moorea, French Polynesia)', *Global Change Biology*, **17**(12), 3667–78.

Kroeker, K.J., R.L. Kordas, R.N. Crim and G.G. Singh (2010), 'Meta-analysis reveals negative yet variable effects of ocean acidification on marine organisms', *Ecology Letters*, **13**(11), 1419–34.

Moore, C. (2011), 'Welfare impacts of ocean acidification: an integrated assessment model of the US mollusk fishery', Working Paper No. 11–06, US Environmental Protection Agency, National Center for Environmental Economics.

Narita, D., R. Rehdanz and R.S.J. Tol (2012), 'Economic costs of ocean acidification: a look into the impacts on global shellfish production', *Climatic Change*, **113**, 1049–63.

Rodrigues, L.C., J.C.J.M. van den Bergh and A. Ghermandi (2013), 'Socio-economic impacts of ocean acidification in the Mediterranean Sea', *Marine Policy*, **38**, 447–56.

Sumaila, U.R., W.W.L. Cheung, V.W.Y. Lam, D. Pauly and S. Herrick (2011), 'Climate change impacts on the biophysics and economics of world fisheries', *Nature Climate Change*, **1**(9), 449–56.
TEEB (The Economics of Ecosystems and Biodiversity) (2009), *The Economics of Ecosystems and Biodiversity for National and International Policy Makers – Summary: Responding to the Value of Nature*, accessed 31 January 2014 at http://www.teebweb.org/publication/teeb-for-policy-makers-summary-responding-to-the-value-of-nature/.
TEEB (The Economics of Ecosystems and Biodiversity) (2010), *The Economics of Ecosystems and Biodiversity: Ecological and Economic Foundations*, edited by Pushpam Kumar, London and Washington, DC: Earthscan.
Tol, R.S.J. (2009), 'The economic effects of climate change', *The Journal of Economic Perspectives*, **23**(2), 29–51.
UK National Ecosystem Assessment (NEA) (2011), *The UK National Ecosystem Assessment: Synthesis of the Key Findings*, Cambridge, UK: UNEP-WCMC, accessed 31 January 2014 at http://archive.defra.gov.uk/environment/natural/documents/UKNEA_SynthesisReport.pdf.

6. Estimating the welfare loss of climate change impact on corals

Pushpam Kumar and Hongyan Chen

6.1 INTRODUCTION

In this chapter, first we highlight the ecosystem services from corals. Second we show the economic significance of corals by showcasing the examples from across the globe. Third, we bring the issue of climate change and its impact on corals. Fourth, we estimate the economic value of the world's corals by using value transfer methods. This has been built upon the database collated under The Economics of Ecosystems and Biodiversity (TEEB)[1] project. Fifth, we analyse the state of coral reefs under dominant climate change scenarios. We estimate the economic value of the corals under different climate scenarios and then compare it with as-usual status and show how much welfare loss would be there due to climate change.

Coral reefs, which are three-dimensional, shallow water structures dominated by scleractinian corals and are the most biologically diverse of shallow water marine ecosystems, provide various ecosystem services. These services are appropriated by humans for different consumptive and productive purposes. It is believed that although coral reefs represent less than 0.2 per cent of total ocean area, they contain more species per unit area than any other ecosystem (Ahmed et al., 2004). In shaping the tropical marine systems, coral reefs have a crucial role, which despite surviving in very low nutrient conditions, are actually highly productive systems (Odum and Odum, 1955). Consequently, coral reefs are often likened to 'oases' within marine nutrient deserts. The Millennium Ecosystems Assessment (MA, 2005) provides a glimpse of the goods and services provided by corals in Table 6.1.

As is evident, most of the services have economic implications in terms of their effect on human well-being. Coral reefs contribute various services to society, which directly and indirectly strengthen the constituents of human well-being. Some of the illustrative services emanating from corals have been identified in various studies (Spurgeon, 1998; Moberg and Folke, 1999; Cesar, 2002). Coral reefs represent crucial sources of income and resources through their role in tourism, fishing, building materials, coastal protection and providing new drugs and biochemicals. Globally, many people depend in part or wholly on coral reefs for their livelihood and around 8 per cent (0.5 billion people) of the world's population lives within 100 kilometres of coral reef ecosystems (Pomerance, 1999). Estimates of the total number of people reliant on coral reefs for their food resources range from 500 million (Wilkinson, 2004) to over 1 billion (Whittingham et al., 2003). Some 30 million of the world's poorest and most vulnerable people in coastal and island communities are entirely reliant on reef-based resources as their primary means of food production, sources of income and livelihoods (Gomez et al., 1994; Wilkinson, 2004). Due to increasing population size, the reliance on reef resources is set to increase over the coming decades.

Table 6.1 Goods and services provided by coral reefs

Provisioning Services (Products Obtained from Ecosystems)	Regulating Services (Benefits Obtained from Regulation of Ecosystem Processes)	Cultural Services (Non-material Benefits Obtained from Ecosystems)	Supporting Services (Natural Processes that Maintain the Other Services)
Food (fish and shellfish) Genetic resources Natural medicines and pharmaceuticals Ornamental resources Building materials	Erosion control Storm protection	Spiritual and religious values Knowledge systems/ educational values Inspiration Aesthetic values Social traditions Sense of place Recreation and ecotourism	Sand formation Primary production

Sources: Adapted from MA (2005).

6.2 DRIVERS OF CLIMATE CHANGE FOR CORALS

Various scientific studies also show that we are losing the corals at an alarming rate. Wilkinson (2004) suggests that 20 per cent of reefs have been destroyed during 1970–2000. MA (2005) also confirms that more than 20 per cent of the corals are badly degraded or under imminent risk of collapse. Detailed studies made at the World Resources Institute (WRI, 1998) suggest that while 58 per cent of the world's reef are potentially threatened by human activities at the global scale, 80 per cent of the corals in South East Asia having the richest biodiversity are under medium to high potential threat. In terms of coral biodiversity, 11 per cent of the world's reefs are at severe risk. The best form of coral reefs are found in developing and tropical regions, which are hotspots of poverty too. The degradation of corals and the slowing down of the flow of ecosystem services from them would throw a significant number of people into the poverty trap. Decaying corals cannot deliver the critically important ecosystem services providing sustenance to the people, especially the poor.

Corals are under serious threat because several direct and indirect drivers are in operation in different forms. Subsidies for fishing, lack of coastal regulation including pollution control, and expansion of unplanned urbanization are some of the indirect drivers. Climate changes, over-exploitation for fishing and recreation, inland pollution and dynamite fishing are some of the direct drivers causing damages to corals across the world. The phenomenon of coral bleaching due to a rise in water temperature arising from climate change has become a serious challenge.

CO_2 coral bleaching starts at 320 ppm (parts per million) CO_2; the current concentration is at 387 ppm CO_2, and the politically 'acceptable target' from the perspective of climate policy is 450 ppm. However, the long-term viability of coral reefs requires a level of atmospheric carbon dioxide that should be reduced significantly below 350 ppm. Current emissions are 370 ppm. Accepting a 450 ppm target is effectively a death sentence on many coral reefs. The demise of coral reefs is an extinction event of never seen before proportions. Such a situation will impoverish 500 million people dependent on coral reefs for livelihoods, damage global fisheries productivity and chances of stock survival, contribute to future food crises and price shocks, and as fisheries are the main source of animal protein for a billion people in the developing world, it will be a cost to their future health.

Imminent coral reef destruction is both a problem built up by historic emissions of the past (that is, the stock of carbon in the atmosphere and influence on sea temperatures) and a problem of new emissions in the present from both the developed and developing world (the 'flow' of annual global GHG emissions).

6.3 ECONOMIC VALUATION OF IMPACTS ON CORALS

The impact of climate change on bleaching of corals and subsequent ecosystem services has significant economic implications. The economic analysis of corals and their decline is critical for designing an effective response option. The analysis through economic valuation of damages or improvement in the condition would help decision-makers in the following ways:

- to prioritize the option where expenditure is best targeted at sustainable utilization;
- to help to justify additional management costs and expenditure;
- to determine appropriate compensation after a damage assessment;
- to help control people's behaviour and utilization of resources;
- to enhance revenue generation;
- to highlight the winners and losers and facilitate equitable distribution.

As discussed above, coral reefs provide a wide range of commercial and non-commercial benefits to human society. Many of these benefits, or 'ecosystem goods and services' are of high value and critical importance to local and national economies. Coral reefs provide habitats for commercially valuable fish, are a magnet for coastal recreation, and reduce the impact of waves on the shore, slowing erosion and beach loss, and lessening damages from storms. Valuation of some of these benefits would greatly help the decision-maker. Valuation of goods and services generated by the coral reef system has widely been attempted by researchers in different parts of the world. Thus, according to an estimate, the total net benefit per year of the world's coral reefs is US$29.8 billion. Tourism and recreation account for US$9.6 billion of this amount, coastal protection for US$9.0 billion, fisheries for US$5.7 billion and biodiversity for US$5.5 billion (Cesar et al., 2003). However, at regional levels the values range widely, mainly due to the variations in type and extent of the coral reef systems and also the underlying socioeconomic state of the beneficiaries. For example, in South East Asia, the total potential sustainable annual economic net benefits per km^2 of healthy coral reef is estimated to range from US$23 100 to US$270 000 arising from fisheries, shoreline protection, tourism, recreation, and aesthetic value (Burke et al., 2002). In the Caribbean, the annual net benefits provided by coral reefs in terms of fisheries were estimated to be about US$300 million (Burke and Maidens, 2004).

In recent years, there have been several attempts to value the services accruing to society from corals. Table 6.2 summarizes the key approaches and findings from some of the credible economic valuation.

Economic valuation has taken different routes and followed a range of methodologies depending upon the purpose of valuation and availability of the data.

6.4 METHODOLOGICAL APPROACH

In order to estimate the economic value of the corals, we have used value transfer method. Value transfer can be used for transfer of figures from a similar study known as a 'study site' to the site in question known as a 'policy site'. Technically, transfer of value is done by estimating the value of an ecosystem or ecosystem services by borrowing the existing valuation for a similar system. Transfer of unit value uses the value of ecosystem services per unit of spatial scale. Other forms of transfer of value could be adjusted unit value transfer and value function (Brander et al., 2007). In this chapter, we have used meta-analytic function transfer. Under meta-analytic analysis, primary valuation studies are collated and analysed in the group and the results from each study are treated as single observations in the new analysis of the combined dataset. This will enable us to evaluate the influence of the characteristics of ecosystem services, the features of the valuation method and other related assumptions. The resulting regression equations explaining variations in unit values can then be used together with socioeconomic contextual data gathered on the independent variables in the model that describes the policy site to construct new value. There have been successful examples of the application of benefit transfer methods by Eshet et al. (2007) and Johnston and Rosenberger (2009).

The general meta-analytic regression model is expressed as:

$$y_i = a + b_S X_{Si} + b_E X_{Ei} + b_C X_{Ci} + \mu_i$$

where y_i measures the benefit of ecosystem site i and is a dependent variable. The explanatory variables include:

X_{Si} (the characteristics of valuation studies, such as valuation method);
X_{Ei} (the characteristics of the valued ecosystem, such as ecosystem service);
X_{Ci} (the socioeconomic data, such as income per capita, and geographical context);
b_S, b_E and b_C are the vectors containing the coefficients of the explanatory variables;
a is a constant and μ_i is an error term.

In this meta-analysis, the natural logarithms of the dependent variable, socioeconomic variables, gross national income per capita and population density are used to improve the model fit and mitigate the heteroskedasticity.

Many experiments have been conducted to fit the regression model. Due to the variation of the values and limited number of samples, a statistically significant model, including the overall model and all the coefficients, is difficult to achieve when all the characteristics of the studies and sites and other factors have to be considered. However,

Table 6.2 Overarching approaches for coral reef valuation

Approach	Fundamental Objective	Methods Practised	Examples
Welfare economics (encompassing total economic value)	How to allocate the resources optimally among the competing uses for maximization of human welfare	Market and constructed market valuation techniques Benefit (value) transfers method Total economic value Cost–benefit analysis	Polunin et al. (2004) calculated an annual total economic value (TEV) of around US$10 million for American Samoa based on fishery, recreation, coastal protection and non-use values and compared future TEV under different management scenarios
Economic impact analysis	To assess contribution to, and/or the effect on, local, regional and national economies (e.g., in terms of expenditures and jobs)	Input–output models Expenditure surveys Benefit transfer methods	Hazen and Sawyer (2001) estimated that the coral reefs of southeast Florida generate US$4.4 billion worth of local sales and US$2 billion of income, and support 71 300 jobs
Socioeconomic analysis	To understand and quantify the social, cultural, economic and political aspects of individuals, organizations and communities	Qualitative and quantitative Focus groups surveys Interviews Visualization techniques Stakeholder analysis	Hoon (2003) identified 24 different types of socioeconomic benefits from coral reefs on Agatti Island (west of India). She also found that 12% of poor households on the island depended on coral reefs for 100% of their incomes, and 59% of poor households relied on reefs for 70% of their incomes
Financial analysis	To determine the financial viability and sustainability of enterprises and organizations, by focusing on transaction-/market--based costs and benefits	Budget forecasts Profit and loss accounts Cash flow analysis Balance sheets Business plans	Belize Coastal Zone Management Authority and The Institute of Belize (2003) undertook a financial analysis to assess financing options for planned marine park and coastal management in Belize. They estimated potential revenues from Belizeans and non-Belizeans of over BZ$5 million and identified a financing gap of BZ$322 000
Other non--monetary'value'--based approaches	To highlight the relative importance of biodiversity and of natural and man-made assets and features	Environmental and social impact assessments Sustainability indicators Index of captured ecosystem value multi-criteria analysis scoring and weighting techniques Energy-based approaches	Fernandes et al. (1999) used multi-criteria analysis to determine the relative importance of various ecological, economic, social and global objectives and indicators amongst different stakeholders for Saba Marine Park. The approach also highlighted the fact that enhanced education and enforcement were commonly agreed by the stakeholders to be the best means of improving upon all four objectives

Table 6.3 Meta-regression results

Explanatory Variable		Variable Definition	Coefficient
Constant			−10.702*
	DExtreme	Dummy: 1 = extreme event protection service; 0 = other service	2.501
	DFood	Dummy: 1 = food provision service; 0 = other service	0.480
X_{Ei}	*DGenetic*	Dummy: 1 = genetic diversity service; 0 = other service	2.715*
	DRecreation	Dummy: 1 = recreation service; 0 = other service	2.675**
	DRawM	Dummy: 1 = material provision service; 0 = other service	−2.027
X_{Si}	*MDirectMP*	Dummy: 1= direct market price method; 0 = other method	1.335
	MReplaceC	Dummy: 1 = replacement cost method; 0 = other service	4.243*
X_{Ci}	*CAfrica*	Dummy: 1 = Africa; 0 = other continent	4.597**
	CAmerica	Dummy: 1 = America; 0 = other continent	3.151*
	CAsia	Dummy: 1 = Asia; 0 = other continent	1.304
	CLatinA	Dummy: 1 = Caribbean; 0 = other continent	2.282*
	LNPopuD2005	Natural log of population density	0.978**
	LNGNIPerCapita2006	Natural log of gross national income per capita	0.838

Note: Significance is indicated with ***, ** and * for the 1, 5 and 10 per cent level, respectively; adjusted $R^2 = 0.378$.

some meaningful conclusions can be drawn from the result in Table 6.3, where the adjusted R^2 is significant, which means the data can be better explained by this model than by other models.

It can be seen from Table 6.3 that the values for services such as genetic biodiversity, extreme events protection and recreation are higher than the values of food and raw material. That means that coral reefs' invisible values are greater than visible provisional ones. Among these coefficients, the one for recreation service is statistically significant at a 5 per cent level and the one for genetic diversity is significant at a 10 per cent level. Of the valuation methods, the direct market pricing and replacement cost methods have higher values than the average of other valuation methods, the replacement cost method is the highest and its coefficient is statistically significant at a 10 per cent level. However, we cannot say that the coral reefs' value in Africa is the highest even if its coefficient is the largest of all the continents. The reason is that the income per capita and population density also influence the values. These several factors work together and produce a composite effect. The coefficients for all the continents are statistically significant, except for Asia, and one of the socioeconomic factors, population density, is statistically significant at a 5 per cent level.

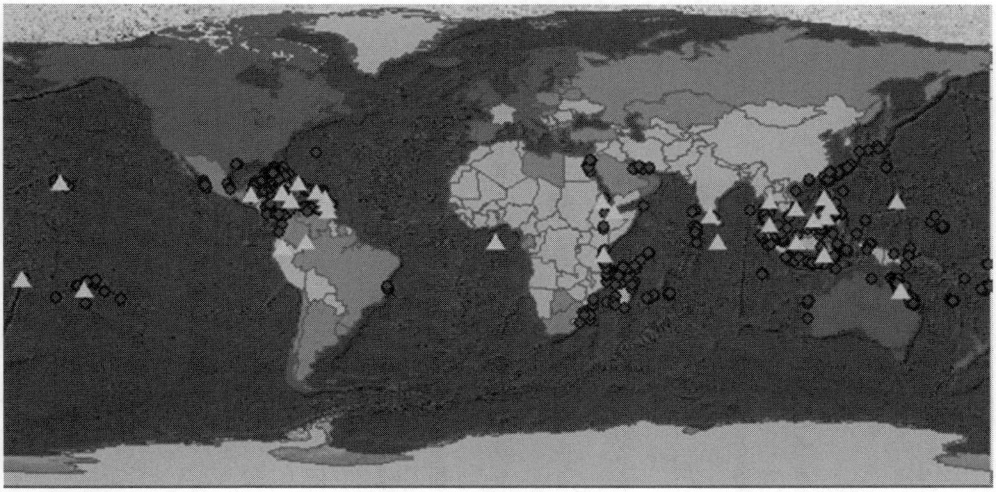

Sources: a. TEEB database; b. ReefBase.org.

Figure 6.1 *Study sites (triangles) for ecosystem service valuation*[a] *and global monitoring sites (dots) of coral reefs*[b]

6.4.1 Description of Primary Studies and Database

The data for the meta-analysis of coral reef valuation are extracted from the TEEB Valuation Database. There are in total 152 data points from 43 publications, for which the desired information including coral reef value, services being valued, location, year of valuation and valuation method, is available.

The study sites are marked on the map of the nations as triangles with global monitoring sites of coral reefs as dots on the map (see Figure 6.1). As we can see in Figure 6.1, the study sites are distributed globally and, to some extension, representative. Since these studies are conducted at different time points for different places and with different purposes, a wide variety of units are used for the coral reef valuation. We converted these values to a common metric, a unit value in US dollars at 2007 prices per hectare. In addition, the primary studies that address one individual service of coral reefs at a specific geographic point with a known valuation method, but not the total valuation value or a world average or with the valuation method unknown, are used. For our dataset, the statistical mean of the coral reef values for different services by region and valuation method are shown in Figures 6.2 and 6.3.

Figure 6.2 shows that the means of the estimated values for different services of coral reefs are different across the continents. Not all services have been addressed for each continent: 'Aesthetic', 'BioControl', 'Climate' and 'Ornamental' services are missing for Asia, and the Extreme Events and Food services are missing for America, for example. The only service that has been valued for all the continents is the Recreation service and the recreation values estimated for all the continents are relatively high, even if not all of them reach the highest among the services. The Genetic Diversity service has a biggest varying range across all the continents, with the highest value occurring for America and

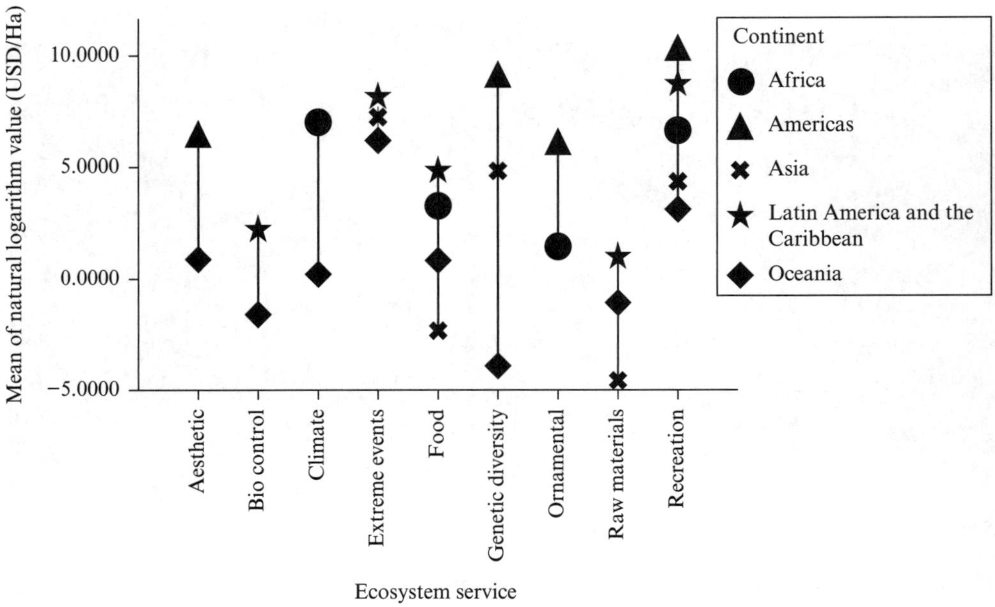

Figure 6.2 Mean of values of coral reef services distinguished among continents

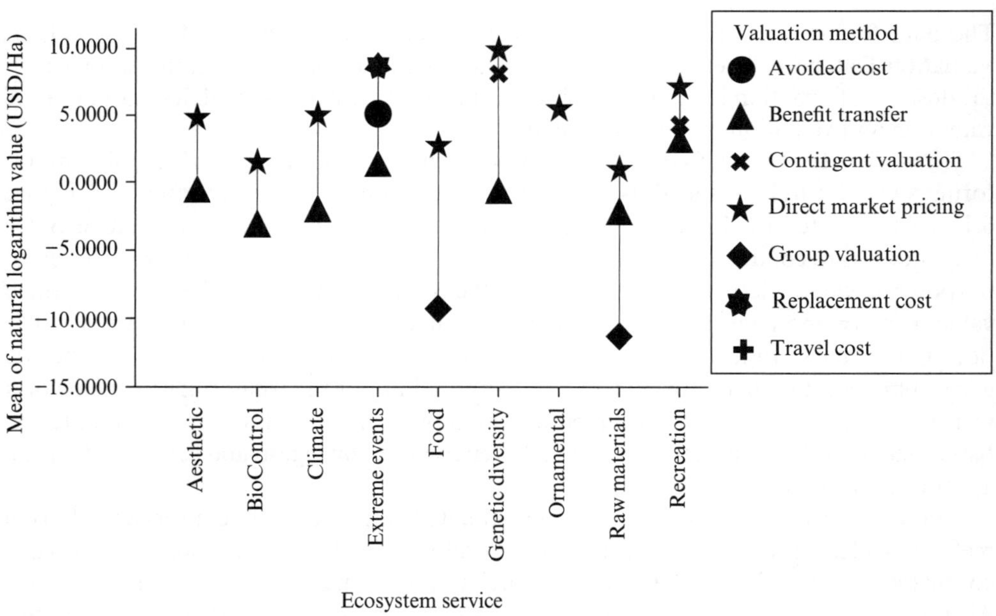

Figure 6.3 Mean of values of coral reef services distinguished across valuation methods

the lowest for Australia and the range spans from about 10^{-1} to 10^{8} or so. Generally, the estimated highest values mostly occur in the Americas and Latin America and the lowest ones in Oceania and Asia.

Figure 6.3 shows the means among different valuation methods. The values estimated by the direct market pricing are usually high while the values by the contingent valuation method are relatively low. The values achieved by benefit transfer are almost always the lowest. The biggest difference occurring among different methods is reflected in the Estimation of Food and Raw Materials services. Among these methods, the group valuation methods give the lowest estimates for Food and Raw Materials services, where the benefit transfer method is not used. Since the values shown on the figure are in a log scale, the real values for coral reefs services vary enormously when different methods are adopted.

6.4.2 Estimations of the Values of Coral Reefs: Value Transfer Method

As discussed earlier, meta-analytic value transfer is the procedure of estimating the benefit of an ecosystem (goods and services from an ecosystem) by borrowing an existing valuation estimate for a similar ecosystem from multiple study results. Rosenberger and Phipps (2007) identify the important assumptions underlying the use of meta-analytic functions for value transfer.

A statistically significant regression model, including the overall model and all the coefficients, is used to calculate the values of the coral reefs at different marine regions. First the model is applied to all the monitoring sites as shown in Figure 6.1 to derive an average unit value for each marine region and then the total value of the region can be worked out by multiplying the total coral reef area of the region. The adopted regression model is given as follows:

$$Ln(Y_i) = -9.495 + 6.542{*}CAfrica + 3.133CAmerica + 5.108CAsi + 4.469CLatin + 1.26LnGNI$$

Figure 6.4 shows the result of the calculation of the unit values of the monitoring sites over the world. The unit values are in natural log and the legend is on the left panel of Figure 6.4.

For each region, the average unit value is calculated through summing all the estimated values of all the coral reef monitoring sites located in the region and then dividing it by the number of the sites. The marine area has been divided into eight regions and the unit values for them are given in Table 6.4. The coral reef areas of the marine regions (Table 6.4) are to be found in Spalding and Grenfell (1997). The total values are the production of these two factors (Table 6.4).

However, the values estimated in the primary studies are somehow the values for a relatively healthy and quality coral reef ecosystem of more than ten years ago. For example, the estimated value by Ruitenbeek and Cartier (1998) for the Montego Bay, Jamaica is US\$893 000/ha/year. However, after several times of coral bleaching during the recent decade, the unit values have to be discounted. According to the status of the coral reef bleaching, we calculate the proportion of the coral reef sites where different levels of

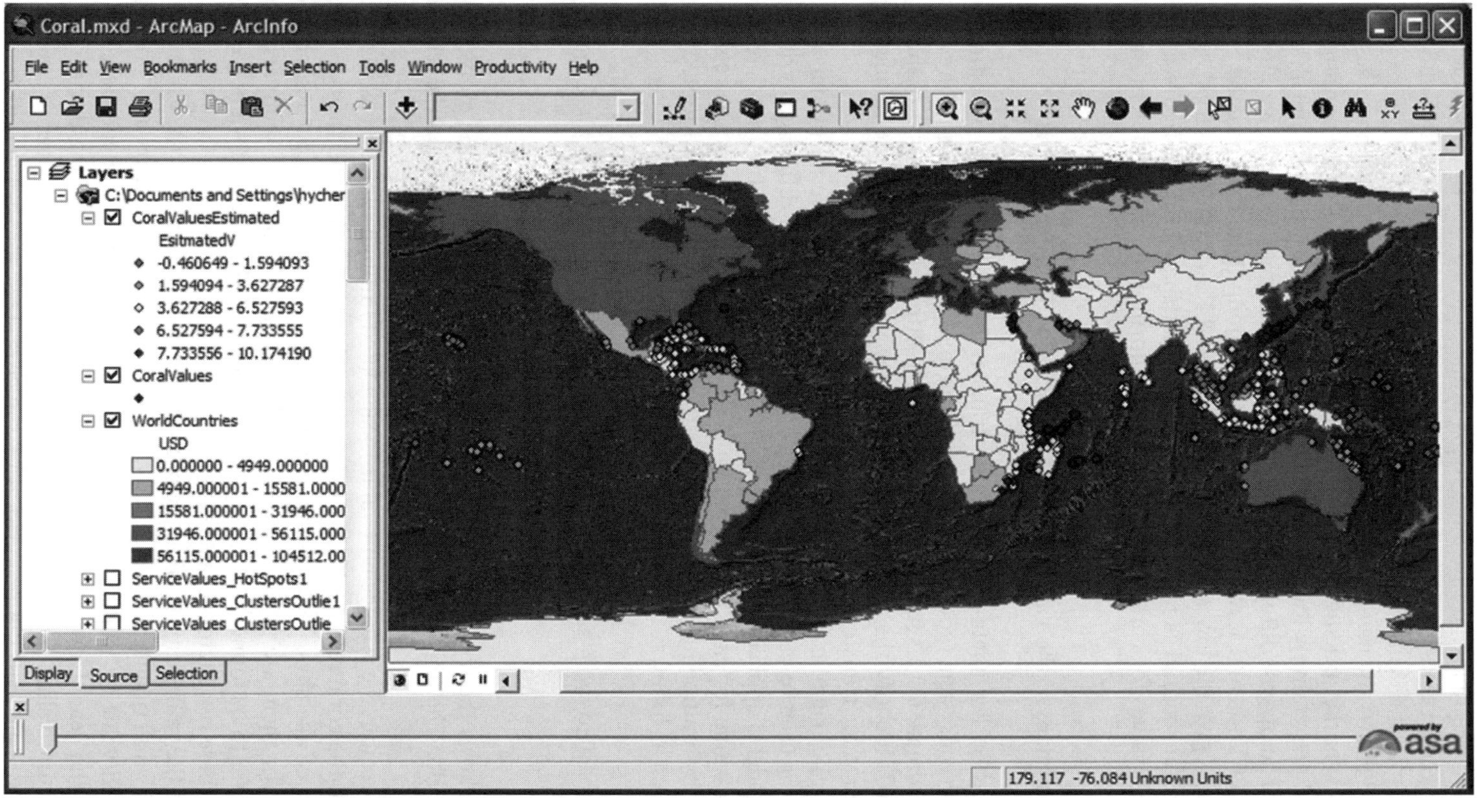

Figure 6.4 The distribution of the estimated unit value of coral reef ecosystem across the world

Table 6.4 Total values of the coral reefs in different marine regions

	North Pacific	Caribbean	South Pacific	India Ocean	Atlantic	South East Asia	Red Sea	Persian Gulf
Unit value (dollars/ha)	2656.6	1097.5	23.7	2912.9	611.3	798.3	2591.6	4649.3
Area (1000 km^2)[a]	17	20	91	36	3	68	17	3
Total (million dollars)	4516	2195	215	10487	1834	5428	4406	1395

Source: a. The source of the estimation is Spalding and Grenfell (1997).

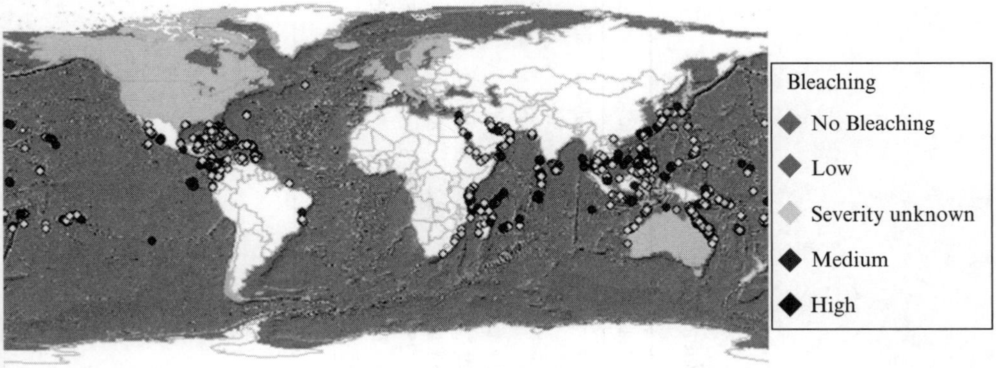

Source: ReefBase.org.

Figure 6.5 Global states of coral reef bleaching

bleaching happen for each marine region. Figure 6.5 gives the bleaching states of the coral reefs in the world.

Table 6.5 gives the numbers and proportions of different bleaching levels for all the eight marine regions. Assume that the values of the highly bleached coral reefs will lose 50 per cent of their original values and the medially bleached will lose 20 per cent, then the total value loss rate of each region is worked out (see the last row in Table 6.5 for the result).

6.5 VALUE OF IMPACTS ON CORALS UNDER DIFFERENT CLIMATE CHANGE SCENARIOS

The climate change scenarios are plausible pathways of temperature rise based upon mix of socioeconomic and natural factors. In this chapter we consider most dominant scenarios and how they might affect the corals and their ecosystem services. The realistic scenarios coming from the IPCC work impacting the corals are:

Table 6.5 The severity of the coral reef bleaching and the value loss rate

Severity	NPacific		Caribbean		SPacific		Indian		Atlantic		SEAsia		RedSea		Persian Gulf	
High	67	35%	179	11%	540	24%	109	41%	62	19%	75	34%	8	40%	31	61%
Medium	37	19%	432	26%	282	12%	71	27%	91	28%	66	30%	4	20%	8	16%
Low	49	26%	582	36%	404	18%	57	21%	119	37%	48	22%	2	10%	5	10%
No bleaching	16	8%	394	24%	962	42%	10	4%	20	6%	13	6%	6	30%	4	8%
Unknown	21	11%	52	3%	79	3%	19	7%	30	9%	16	7%	0	0%	3	6%
Total	190	100%	1639	100%	2267	100%	266	100%	322	100%	218	100%	20	100%	51	100%
Loss rate	21.5%		10.7%		14.4%		25.8%		15.3%		15.3%		24.0%		33.5%	
Value left (million dollars)	3544		1959		184		7779		1554		4599		3349		927	

Table 6.6 Projected frequency that the world's coral reefs experience annual DHM>1 and DHM>2 during 2030–39 and 2050–59[a]

Time Duration	Frequency	0.1	0.2	0.3	0.4	0.5	0.6	0.7	0.8	0.9	1
	DHM>1										
2030–39	HadCM3 A2	100%	100%	100%	97%	94%	89%	70%	50%	32%	16%
	HadCM3 B2	100%	100%	99.5%	98%	94%	90%	72%	60%	38%	20%
2050–59	HadCM3 A2	100%	100%	99%	98%	98%	97%	95%	94%	85%	76%
	HadCM3 B2	100%	100%	99%	98%	98%	95%	90%	84%	75%	50%
	DHM>2										
2030–39	HadCM3 A2	97%	93%	86%	72%	56%	42%	24%	11%	5%	2%
	HadCM3 B2	98%	96%	88%	75%	60%	47%	32%	24%	11%	6%
2050–59	HadCM3 A2	99%	98%	97%	96%	95%	94%	90%	83%	70%	50%
	HadCM3 B2	99%	98%	96%	94%	88%	78%	70%	59%	40%	22%

Note: a. The results are the fraction of 36 km grid cells containing coral reefs with a given frequency of exceeding threshold.

Source: Extracted from Figures 3 and 4 in Donner et al. (2005).

HadCM3 SRES A2: emission scenario is commonly used for 'business as usual' impact studies, projecting a 3°C increase in surface air temperature.
HadCM3 SRES B2: a low emission path, projecting a 2.2°C temperature increase on average across all the models.

The durations considered are the years 2030–39 and 2050–59. Deterioration of the coral reefs under different scenarios (summarized from Donner et al., 2005) is *a monthly bleaching index*: degree heating month (DHM) is equal to one month of sea surface temperature (SST) that is 1°C greater than the maximum in the monthly climatology. The annual DHM total of 1°C was the best proxy for the lower-intensity bleaching threshold and an annual total of 2°C as the higher threshold, for severe coral bleaching with more associated coral mortality.

The predicted fraction of coral reefs for these levels of severity and different frequencies by Donner et al. (2005) are as shown in Table 6.6.

Table 6.6 gives the fraction of corals at ten frequency points. Among them, 0.1 means that the bleaching will happen one year (once) in the decade of 2030–39, 0.5 is every two years, while 1.0 means that it happens every year. During 2030–39, the fraction of the corals with the lower-intensity bleaching occurring every year will reach 16 per cent in HadCM3 A2 and 20 per cent in HadCM3 B2, and the fraction of the corals with the severe coral bleaching will reach 2 per cent in HadCM3 A2 and 6 per cent in HadCM3 B2. During 2050–59, the fraction of the corals with the lower-intensity bleaching occurring every year will reach 76 per cent in HadCM3 A2 and 50 per cent in HadCM3 B2, and the fraction of the corals with the severe coral bleaching will reach 50 per cent in HadCM3 A2 and 22 per cent in HadCM3 B2.

From Table 6.6, we can also see that during 2030–39, the fraction of the corals with

Table 6.7 The fraction of coral reefs that experience a given bleaching frequency during two time durations under different climate scenarios

Time Duration	Frequency	0.0	0.1	0.2	0.3	0.4	0.5	0.6	0.7	0.8	0.9	1
	DHM>1											
2030–39	HadCM3 A2	0%	0%	0%	3%	3%	5%	19%	20%	18%	16%	16%
	HadCM3 B2	0%	0%	1%	2%	4%	4%	18%	12%	22%	18%	20%
2050–59	HadCM3 A2	0%	0%	1%	1%	0%	1%	2%	1%	9%	9%	76%
	HadCM3 B2	0%	0%	1%	1%	0%	3%	5%	6%	9%	25%	50%
	DHM>2											
2030–39	HadCM3 A2	3%	4%	7%	14%	16%	14%	18%	13%	6%	3%	2%
	HadCM3 B2	2%	2%	8%	13%	15%	13%	15%	8%	13%	5%	6%
2050–59	HadCM3 A2	1%	1%	1%	1%	1%	1%	4%	7%	13%	20%	50%
	HadCM3 B2	1%	1%	2%	2%	6%	10%	8%	11%	19%	18%	22%

the lower-intensity bleaching occurring at least every two years will reach 94 per cent in HadCM3 A2 and 94 per cent in HadCM3 B2, and the fraction of the corals with the severe coral bleaching will reach 56 per cent in HadCM3 A2 and 60 per cent in HadCM3 B2. During 2050–59, the fraction of the corals with the lower-intensity bleaching occurring at least every two years will reach 98 per cent in HadCM3 A2 and 98 per cent in HadCM3 B2, and the fraction of the corals with the severe coral bleaching will reach 95 per cent in HadCM3 A2 and 88 per cent in HadCM3 B2. For other frequencies of bleaching, there are other corresponding fractions of coral reefs.

According to Table 6.6, we derive another table, Table 6.7 to reflect a fraction for each frequency (not accumulated frequency) for two bleaching severity degrees.

Assumption of loss rates under different situations led by climate change: rise in sea temperatures by 1°C warmer than the usual summer maxima can cause the coral bleaching and the bleaching occurring at a different frequency with a different severity will lead to a different degree of degradation of coral reefs. For example, frequent low-intensity bleaching even with no coral mortality can lead to long-term degradation of the coral reef ecosystem by slowing coral growth, reducing coral recruitment and reducing resiliency to other disturbances and frequent severe coral bleaching may lead to coral mortality (Hoegh-Guldberg, 2007; Donner et al., 2005). Accordingly, we assume the loss rates under different frequencies of mass coral bleaching as follows (see Table 6.8) – the loss rate for other omitted frequencies can be derived through linear interpolation, for

Table 6.8 Loss rate assumed according to the frequency and severity of coral reef bleaching

Probability	0.2	0.4	0.6	0.8	1
Loss rate	DHM>1				
	5%	10%	15%	20%	25%
	DHM>2				
	20%	40%	60%	80%	100%

Table 6.9 Loss rate of coral reef value under different climate change scenarios

Duration	Climate Scenario	Value of Loss	
		DHM>1	DHM>2
2030–39	HadCM3 A2	0.19	0.49
	HadCM3 B2	0.19	0.54
2050–59	HadCM3 A2	0.24	0.87
	HadCM3 B2	0.22	0.74

instance for frequency F = 0.1 for DHM>1, the loss rate = 2.5 per cent; for frequency F = 0.9 for DHM>2, the loss rate = 90 per cent.

Monetary loss of coral reef ecosystem value changes under different scenarios. According to the predicted fraction of the bleached coral reefs at different bleaching frequencies and the loss rate defined above, the total loss rate for the two severity degrees can be expressed as the following:

$$R_i = \sum_{n=1}^{10} f_{ni} * r_{ni}$$

Where *Ri* is the total loss rate of coral reefs for *i* (= 2) degrees of severity, represented by monthly bleaching index DHM>1 and DHM>2; f_n represents the fraction of the coral reefs bleached at the *n*th frequency: the first one is once during the decade, the fifth is every two years and the tenth is bleaching every year.

According to this formula, the total loss rate of global coral reef value for the two degrees of severity under different climate change scenarios can be calculated and the result is given in Table 6.9.

The monetary loss for each marine area can be derived according to their current values and the loss rate due to climate change and the results are given in Table 6.10.

Since the situation when DHM>1 includes the cases when DHM>2, it is not reasonable to achieve an overall loss by adding together the losses of these two degrees of severity. If we do so, it will raise a double-counting issue. In order to avoid double-counting, we assume a 50 per cent discount for the DHM>1 situation, which means that about half of the coral reefs belonging to the situation also belong to the situation when DHM>2, and their loss has been calculated for when DHM>2 and therefore needs to be excluded from the loss calculation for when DHM>1. Based on this assumption, we achieve the overall annual monetary loss as 13.9 and 15.14 billion dollars respectively for climate change scenarios HadCM3 A2 and HadCM3 B2 (see Table 6.11).

The present value of loss arising from coral loss would depend upon the kind of discount rate we choose and the type of scenario we value. Even for a 20-year time horizon and A2 scenario the present value of the loss at a 2 per cent rate of discount would be approximately US$6.19 billion while at a 3 per cent rate of discount, it would be US$4.11 billion. For an infinite time horizon, the present value of the loss would be US$13.90 billion under the A2 scenario and US$15.14 under the B2 scenario.

Table 6.10 Monetary loss in different marine regions

	Duration	Scenario	Loss Rate	North Pacific	Caribbean	South Pacific	India Ocean	Atlantic	South East Asia	Red Sea	Persian Gulf	Total
Current value				3.54	1.96	0.18	7.78	1.55	4.60	3.35	0.93	23.90
DHM>1	2030–39	HadCM3 A2	0.19	0.66	0.37	0.03	1.45	0.29	0.86	0.63	0.17	4.47
		HadCM3 B2	0.19	0.68	0.38	0.03	1.50	0.30	0.89	0.65	0.18	4.61
	2050–59	HadCM3 A2	0.24	0.83	0.46	0.04	1.83	0.37	1.08	0.79	0.22	5.63
		HadCM3 B2	0.22	0.79	0.44	0.04	1.73	0.34	1.02	0.74	0.21	5.31
DHM>2	2030–39	HadCM3 A2	0.49	1.73	0.96	0.09	3.80	0.76	2.24	1.63	0.45	11.66
		HadCM3 B2	0.54	1.90	1.0	0.10	4.18	0.83	2.47	1.80	0.50	12.83
	2050–59	HadCM3 A2	0.87	3.09	1.71	0.16	6.78	1.35	4.01	2.92	0.81	20.83
		HadCM3 B2	0.74	2.63	1.46	0.13	5.79	1.15	3.42	2.49	0.69	17.77

Note: Units are billion dollars.

Table 6.11 Loss in future value with the interest rate at 2% and %

Climate Scenario	Annuity Value of Loss Infinite Time	Loss with 40 Years Time Horizon		Loss in 20 Years (in Future Value)	
Current total value (23.90)		2%	3%	2%	3%
HadCM3 A2	13.90	9.28	7.56	6.19	4.11
HadCM3 B2	15.14	10.11	8.23	6.75	4.48

Note: Units are billion dollars.

6.6 DISCUSSION AND CONCLUSIONS

The economic value of the ecosystem as they exist has no meaning. Economic valuation of ecosystem services must have a purpose. The valuation must be seen in the context of either alternate scenario or conflicting choices (Kumar and Wood, 2010). Moreover, the economic value of ecosystem services is instrumental, anthropocentric, individual-based, subjective, context-dependent, marginal and state-dependent (Goulder and Kennedy, 1997; Barbier et al., 2009). So, economic value of corals might not make much sense especially to the policy-makers who always like to see the value of going one step further. Therefore, in the chapter we have provided the value of corals as per scenario. The relative difference in the value in one scenario over another should be taken into account. For example, the difference of losses in B2 and A2 is 1.24 billion if the time line is 2030. This value should be seen in the context of societal welfare loss and provide the rationale for effective action to combat climate change. This value should be analysed as the cost of inaction arising just from coral ecosystems. Various economic values can also be derived under different scenarios projected by the Intergovernmental Panel on Climate Change (IPCC) and the meta-value derived on the basis of site-specific studies is relevant to guiding the global community for additional resource allocation to save this critical ecosystem of corals.

NOTE

1. See http://www.teebweb.org/Home/tabid/924/Default.aspx; accessed 31 January 2014.

REFERENCES

Ahmed, M., C.K. Chong and H. Cesar (eds) (2004), 'Economic valuation and policy priorities for sustainable management of coral reefs', *World Fish Center Conference Proceedings*, **70**, 108–17.

Barbier, E.B., S. Baumgärtner, K. Chopra, C. Costello, A. Duraiappah and R. Hassan et al. (2009), 'The valuation of ecosystem services', in S. Naeem, D. Bunker, A. Hector, M. Loreau and C. Perring (eds), *Biodiversity, Ecosystem Functioning, and Human Well-Being*, Oxford: Oxford University Press.

Belize Coastal Zone Management Authority (2003), 'Operationalizing a financing system for coastal and marine resource management in Belize', Strategy Paper, Belize Coastal Zone Management Authority.

Brander, L.M., P. van Beukering and H. Cesar (2007), 'The recreational value of coral reefs: a meta-analysis', *Ecological Economics*, **63**(1), 209–18.

Burke, L. and J. Maidens (2004), *Reefs at Risk in the Caribbean*, Washington, DC: World Resources Institute.

Burke L., L. Selig and M. Spalding (2002), *Reefs at Risk in Southeast Asia*, Washington, DC: World Resources Institute.

Cesar, H.J.S. (2002), *The Biodiversity Benefits of Coral Reef Ecosystems: Values and Markets*, ENV/EPOC/GSP/BIO(2001)5/FINAL, accessed 31 January 2014 at http://www.cbd.int/ doc/external/oecd/oecd-coral--reefs-2002-en.pdf.

Cesar, H., L. Burke and L. Pet-Soede (2003), *The Economics of Worldwide Coral Reef Degradation*, Australia: International Coral Reef Action Network.

Donner, S.D., W.J. Skirving, C.M. Little, M. Oppenheimer and O. Hoegh-Guldberg (2005), 'Global assessment of coral bleaching and required rates of adaptation under climate change', *Global Climate Change*, **11**, 2251–65.

Eshet, T., M.G. Baron and M. Shechter (2007), 'Exploring benefit transfer: disamenities of waste transfer stations', *Environmental and Resource Economics*, **37**(3), 521–47.

Fernandes, L., M.A. Ridgley and T. van't Hof (1999), 'Multiple criteria analysis integrates economic, ecological and social objectives for coral reef managers', *Coral Reefs*, **18**(4), 393–402.

Gomez, E.D., P.M. Alino, H.T. Yap and W.Y. Licuanan (1994), 'A review of the status of Philippine reefs', *Marine Pollution Bulletin*, **29**(1–3), 62–8.

Goulder, L.H. and J. Kennedy (1997), 'Valuing ecosystem services: philosophical bases and empirical methods', in G.C. Daily (ed.), *Nature's Services: Societal Dependence on Natural Ecosystems*, Washington, DC: Island Press.

Hazen and Sawyer (2001), *Socioeconomic Study of Reefs in Southeast Florida*, NOAA.

Hoegh-Guldbergh, O., P.J. Mumby, A.J. Hooten, R.S. Steneck, P. Greenfield and E. Gomez et al. (2007), 'Coral reefs under rapid climate change and ocean acidification', *Science*, **318**(5857), 1737–42.

Hoon, V. (2003), 'A case study from Lakshadweep', in E. Whittingham, J. Campbell and P. Townsley (eds), *Poverty and Reefs: Volume 2 Case Studies*, Paris: UNESCO, pp. 87–226.

Johnston, R.J and R.S. Rosenberger (2009), 'Methods, trends and controversies in contemporary benefit transfer', *Journal of Economic Surveys*, **24**(3), 479–510.

Kumar, P. and M.D. Wood (2010), 'Valuation of regulating services: genesis to current evolution', in P. Kumar and M.D Wood (eds), *Valuation of Regulating Services: Methodology and Applications*, London and New York: Routledge.

Millennium Ecosystem Assessment (MA) (2005), *Ecosystems and Human Well-Being: Synthesis*, Washington, DC: Island Press.

Moberg, F. and C. Folke (1999), 'Ecological goods and services of coral reef ecosystems', *Ecological Economics*, **29**(2), 215–33.

Odum, H.T. and E.P. Odum (1955), 'Trophic structure and productivity of a windward coral reef at Eniwetok Atoll, Marshall Islands', *Ecological Monographs*, **25**, 291–320.

Pomerance, R. (1999), 'Coral bleaching, coral mortality, and global climate change', report presented by Deputy Assistant Secretary of State for the Environment and Development to the US Coral Reef Task Force, 5 March 1999, Maui, Hawaii.

Rosenberger, R.S. and T.T. Phipps (2007), 'Correspondence and convergence in benefit transfer accuracy: a meta-analytic review of the literature', in S. Navrud and R. Ready (eds), *Environmental Values Transfer: Issues and Methods*, Dordrecht: Kluwer Academic Publishers.

Ruitenbeek, J. and C. Cartier (1998), *Issues in Applied Coral Reef Biodiversity Valuation: Results for Montego Bay, Jamaica*, World Bank Research Committee Project RPO No. 682–22, Final Report, Washington, DC: World Bank.

Spalding, M.D. and A.M. Grenfell (1997), 'New estimates of global and regional reef areas', *Coral Reefs*, **16**(4), 225–30.

Spurgeon, J. (1998), 'The socio-economic costs and benefits of coastal habitat rehabilitation and creation', *Marine Pollution Bulletin*, **37**(8–12), 373–82.

Polunin, N. with contributions from J. Spurgeon et al. (2004), *Economic Valuation of Coral Reefs and Adjacent Habitats in American Samoa*, compiled for the Department of Commerce by Jacobs Engineering in association with MRAG Americas, National Institution of Water and Atmospheric Research.

Whittingham, E., J. Campbell and P. Townsley (2003), *Poverty and Reefs*, Paris: UNESCO.

Wilkinson, C. (2004), *Status of Coral Reefs of the World: 2004*, Australia: Global Coral Reef Monitoring Network.

World Resources Institute (WRI) (1998), *Report: Reefs at Risk: A Map-based Indicator of Threats to the World's Coral Reefs*, by D. Bryant, L. Burke, J. McManus and M. Spalding.

PART II

EMERGING ECONOMIC VALUATION METHODS, INCLUDING THE USE OF DELIBERATIVE, MACRO AND SPATIALLY EXPLICIT ECONOMIC VALUATION

7. The behavioral argument for an expanded valuation framework for biodiversity and ecosystem services

John M. Gowdy and Sarah Parks

7.1 INTRODUCTION

The cornerstone of Walrasian welfare economics[1] is a model of human behavior characterizing consumer preferences as stable, consistent, insatiable, and independent of the preferences of others. The assumptions embodied in *Homo economicus* underlie the theoretical assumptions and policy recommendations of most economists, whether explicitly acknowledged or not. In this model, economic value comes solely from the preferences of perfectly rational, self-regarding individuals. For example, environmental economics has become predominantly concerned with uncovering the 'true' value of environmental features using techniques such as hedonic pricing and contingent valuation. Preferences revealed by these techniques can then be used in benefit–cost analysis (BCA) to suggest corrections in market prices and/or to assign property rights more completely (Heal, 1998; Stavins, 2008).

Recent work in theoretical welfare economics has exposed flaws in the Walrasian model that have led many theorists to reject a basic principle supporting BCA, that is, that preferences can be characterized as independent and exogenous so that interpersonal comparison of utility can be avoided (Boadway, 1974; Chipman and Moore, 1978; Blackorby and Donaldson, 1990; Ng, 1997; Bowles, 1998; Suzumura, 1999; Gowdy, 2004).[2] Furthermore, it has been well-established in the behavioral literature that preferences are other-regarding, and that they vary significantly according to cultural conditioning, relative position, and other reference points (Henrich et al., 2001; Camerer et al., 2005). Understanding the social process of preference formation is critical to formulating sound economic policies (Kahneman et al., 1999; Knetsch, 2005). It is argued below that current work in economic theory and experimental economics not only calls into question Walrasian-based BCA approaches to valuation and policy, it also legitimizes new approaches that are both theoretically consistent and empirically valid. Deliberative valuation can be one such approach although, as discussed below, some applications of it fall back on the self-regarding rational actor model.

7.2 THEORETICAL CRITIQUES OF BENEFIT–COST ANALYSIS LAY THE FOUNDATION FOR DELIBERATIVE VALUATION

A classic defense of the Walrasian welfare foundation (as presented, for example, by Bator, 1957) of BCA was given by Harberger in 1971. He argued that (1) interpersonal comparisons of utility can and should avoided (2) market prices are the best indicator of value for consumers and producers and (3) policies should be evaluated by adding costs and benefits 'without regard to the individual(s) to whom they accrue' (Harberger, 1971, p. 785). Following Kaldor (1939) and Hicks (1939) he asserts that questions of aesthetics or distribution may be important but they are 'not a part of that package of expertise that distinguishes the professional economist from the rest of humanity' (Harberger, 1971, p. 785; see also the discussion by Persky, 2001). Harberger lists the major theoretical objections raised against the Walrasian approach, namely, (1) the marginal utility of money is not constant across individuals as the Walrasian model assumes, (2) changes in income distribution are not taken into account, and (3) partial equilibrium measures are used to draw general equilibrium conclusions. Harberger's answers to these objections are surprisingly weak:

1 The assumption of constant marginal utility of money – Harberger argues that most applications of BCA involve such small changes that the assumption does not matter (Harberger, 1971, p. 787). Furthermore, this objection is not important if changing the marginal utility of money leaves relative prices unchanged (ibid., p. 788).
2 Ignoring distributional effects – he argues that these are likely to be minimal in most BCA applications (ibid., p. 787). Harberger confirms the opinion of those who assert that Walrasian theory is driven by mathematical tractability rather than economic realism: 'Giving equal weight to all dollars of income is mathematically the simplest rule' (ibid.).
3 Drawing general equilibrium conclusions from partial equilibrium analysis – Harberger's position is that only price changes in the 'distorted' goods need to be considered and 'it is hoped that in most cases the number of elements in it will be of manageable size' (ibid., p. 791).

In spite of Harberger's claims, all three of these issues are critical and none of them can be satisfactorily addressed without stepping outside the core Walrasian model.[3] These three objections go to the heart of BCA claims that preferences are exogenous (independent of the preferences of others). Avoiding interpersonal comparisons of utility was thought to be a major step toward establishing economics as a positive science. It is a necessary condition for the assertions that welfare changes can be consistently measured and that prices are reliable indicators of utility.

Proposition I: Contemporary neoclassical welfare theory has demonstrated that the (classical welfare) effects of economic policies cannot be judged without making interpersonal comparisons of utility.

Work in welfare economic theory over many decades has demonstrated that the assertion that the economic valuation of alternative states (as in cost–benefit analysis) requires no interpersonal comparisons of utility is not supported theoretically. Recent theoretical critiques of the Walrasian system are particularly damaging because it demonstrates that even if one accepts all the basic postulates of the system, including all the assumptions defining *Homo economicus*, perfect competition, and the Kaldor-Hicks compensation principle, there is no theoretically consistent way to compare alternative states of the economy without making interpersonal comparisons of utility (Boadway, 1974; Chipman and Moore, 1978).

Equation (7.1) shows the usual way cost–benefit measures are obtained (Harberger, 1971):

$$\int dU/\lambda = \Delta Y - \int \sum X_i dD_i \qquad (7.1)$$

where the X_is are the goods in the utility function. The change in welfare from a policy change is equal to the change in income resulting from the new policy minus the net effect on consumer and producer surplus of price changes in the 'distorted' goods (D_i). λ represents the marginal utility of income since in constrained maximization equilibrium all MU_i/p_i are equal (the marginal utility of a dollar spent is the same for all X_is. If only one price changes and income remains constant, Equation (7.1) becomes:

$$\Delta Y = \int_{\bar{U}} X_i dp_i = 0 \qquad (7.2)$$

The critical difficulty in moving from Equation (7.1) to Equation (7.2) is that it must be assumed that the marginal utility of income, λ, is the same for all individuals (in the classic utilitarian model). In evaluating policy changes that affect several persons we need a social welfare function of the form:

$$W = W(U^1, U^2, \ldots, U^n) \qquad (7.3)$$

Changes in social welfare may be estimated by Equation (7.4) (see Boadway, 1974, p. 929):

$$\int dW = \int \sum_j \sum_i \beta_j p_i dX_{ij}, \qquad (7.4)$$

where:

$$\beta_j = \lambda^j \, \partial W / \partial U^j \qquad (7.5)$$

Equation (7.5) represents the marginal social utility of income of individual *j*. It will change as the marginal utility of income, λ, changes. This result uncovers a critical flaw in traditional BCA estimates using Equation (7.1) above. A basic result of consumer theory is that the equilibrium demand system is unchanged after a monotonic transformation of the utility function. That is, all that is needed is an ordinal, not a cardinal, measure of consumer ranking of commodity bundles. But another fundamental result is that the value of λ, the marginal utility of money, is not invariant with a monotonic transformation except under the unrealistic assumption that utility functions are homothetic.[4] For this condition to hold, the utility-maximizing proportions of all goods remains the same as income increases. If this is not the case, then to properly evaluate the effect of price distortions, or lump sum transfers, the changes in the outputs of *all* goods must be considered whether distortions exist for them or not. Boadway (1974, pp. 926–7) summarizes the implications of all this:

> [E]ven if we accept compensation tests as welfare criteria, the summation of total money gains and losses (including surpluses) does not in general indicate satisfaction of these tests. In fact, positive net monetary gains may be observed when compensation is not possible and vice versa. Furthermore, when comparing alternative projects or policies, the one with the largest net gain is not necessarily the 'best' one in the compensation test sense. Since we cannot rely on money surplus measures to determine the best course of action, there may be no alternative but to incorporate interpersonal comparison regarding the marginal social utility of incomes to different groups into the cost–benefit analysis.

Equation (7.1) is an accurate measure of compensating variation only if we make the assumption that the marginal social utility of income is constant or that it is unaffected by changes in relative prices and output. But we cannot even make that claim without invoking an interpersonal utility comparison. Assuming that λ is the same for all individuals requires making interpersonal comparisons (Suzamura, 1999).

Proposition II: Contemporary welfare theory has demonstrated that standard cost–benefit measures of value (equivalent and compensating variations) are arbitrary. No unique relationship exists between utility changes and money income changes (Boadway, 1974).

As Burns (1973), Silverberg (1972) and others have shown, the value of the right-hand side of Equation (7.1) depends on the path of integration. A particular path must be chosen and there is no compelling reason to choose one path over another. In order to assert that equivalent variations in income are comparable we must again invoke the assumption that the utility function is homothetic. If it is not homothetic any utility change can generate an infinite number of imputed rents from a policy change (Silverberg, 1972, p. 947). Measures of equivalent variations are path dependent. In cost–benefit analysis the path chosen is either the one that converts Equation (7.1) into a compensating variation measure by moving along a constant indifference surface or one that keeps utility constant following a change in relative prices and/or income (Boadway, 1974, p. 929). But Silverberg shows that indeterminancy is also a problem in the compensating variation case. Compensating gains or losses will vary according to the terminal prices because of Hicksian income effects.

Current BCA measures of welfare changes have been chosen for convenience rather than for realism or logical consistency. Silverberg (1972, pp. 950–51) states this quite strongly:

> [T]o merely substitute the compensating variations for the equivalent variations as a measure of welfare loss or gain (as Harberger, 1954, 1964 and others do) is to use the inappropriate to measure the undefinable . . . It seems strange, at best, to use a measure of consumer's benefits a construct which explicitly assumes that individuals remain at the same level of utility.

We argue below that deliberative valuation is one way to tease out the underlying other-regarding attributes and value judgments embodied in Equation (7.5). This position is strengthened by the current research in behavioral economics and evolutionary game theory showing the widespread prevalence of other-regarding preferences and the superiority of group decision-making.

7.3 THE BEHAVIORAL AND NEUROSCIENCE CASE FOR DELIBERATIVE VALUATION

The classic paper in neoclassical economic methodology is Friedman's (1952) essay on positive economics. He asserts that a sound theory is one that uses a small number of simplifying assumptions to make good predictions. Theories should be judged on the basis of their predictive ability not the realism of their assumptions. Applying the predictability criterion, recent evidence from behavioral and experimental economics has clearly demonstrated the inadequacy of the behavioral model of Walrasian economics. It has been established in the experimental economics literature that preferences are not stable (Bateman and Mawby, 2004), that they are interdependent (Kahneman and Tversky, 1979; Knetsch, 1989), that they are culturally dependent (Henrich et al., 2001), that lexicographic preferences are common (Rekola, 2003), and that the most commonly used welfare measure, per capita income, is an inadequate indicator of well-being (Easterlin, 1974; Frey and Stutzer, 2002; Layard, 2005). These findings confirm the theoretical critiques of Walrasian welfare economics, namely that economic valuation and policy cannot proceed without making explicit value judgments based on interpersonal comparisons of utility.

Proposition III: Empirical work in behavioral economics has demonstrated that preferences are other-regarding. Individuals judge their own well-being by comparing themselves to others in the context of prevailing cultural norms and are willing to punish others, at cost to themselves, to enforce social norms.

Game theory and behavioral experiments have demonstrated the importance of cultural values in predicting economic behavior. Particularly illuminating is a large-scale research project undertaken by economists, anthropologists and other behavioral scientists involving playing the well-known Ultimatum Game (Güth et al., 1982) in a variety of non-western cultures. Offers in the UG cannot be explained without reference to the particular social values of the groups studied. There is considerable behavioral variation across groups and group-level (cultural) differences explain behavior better than do individual characteristics (Henrich et al., 2001).

Other game theory results show that people act to affect the well-being of others, positively or negatively, even at significant cost to themselves (Fehr and Gächter, 2002). A sense of fairness is a critical factor in economic decisions. Results from the Public Goods Game show that participants are willing to impose punishments on non-cooperators, at great cost to themselves, even in the last round of the game (Bowles and Gintis, 2000). So-called altruistic punishment helps explain why cooperative behavior is pervasive in human societies (Henrich et al., 2006). Understanding these kinds of behavior patterns is critical for environmental valuation and policy design.

Income is judged in relation to the income of others. Ng (1997) argues that the relative income effect is more important than the theoretical objections. The key point here is that in judging welfare gains and losses, one individual's payoff depends on the payoffs to others. This is in sharp contrast to the work of Becker (1976) and others who include preferences for altruism in a utility function but do not consider interactive preferences. We may consider the historical formation of preferences, but once they are formed they are fixed and immutable. As Stigler and Becker (1977, p. 76) put it: 'one does not argue over tastes for the same reason that they do not argue over the Rocky Mountains – both are there, will be there next year, too, and are the same to all men'. By contrast, other-regarding preferences are not immutable nor 'the same to all men'. They can only be teased out through interactive negotiation.

A robust finding is that the social process of preference formation is critical to the valuation of alternative payoffs. For example, in the Ultimatum Game, results (mean offers and rejection rates) vary significantly according to the process through which money is obtained and offers are made. Offers are substantially lower if proposers win their position by doing well in a quiz (Hoffman et al., 1994). Offers are substantially higher if it is made clear that the payoff to be shared is a windfall gain. Experimental results indicate clearly that explanations of behavior that do not take into account social processes such as community norms about fairness are poor predictors of economic behavior. In environmental policy the process of arriving at a decision may be as important for public acceptance as the actual outcome itself.

Another consistent empirical finding is that utility changes depend on multiple perspectives and somehow need to be evaluated simultaneously by those affected and the cultural context of the decision-making process. What is considered 'rational' depends on the institutional structure and cultural context (Sahlins, 1996; Vatn, 2004). If knowledge creation is handled differently by each cultural bias, then the only way to create shared understanding and agreement for action is to produce meaning that lies outside the territory of individual cultural biases. Effective policy depends on successful creation of shared meaning among – not within – cultural groups. Two points are particularly relevant here: (1) shared social values as well as shared meanings are created through effective social interaction. Social values do not exist a priori, nor are they merely the intersection of individual values. Instead, they are created through social interaction. (2) The only effective way to achieve this kind of interaction is through open dialogue.

Other-regarding preferences can also include moral preferences. These preferences are not self-regarding but they include both individual and social aspects. They may be part of shared group values (for example, moral responsibility to other species) but they are not necessarily influenced by social processes (Mayer and Franz, 2004; Meier, 2006; Reeson, 2008; Manner and Gowdy, 2010; Videras et al., 2008). People may be concerned about the environment for selfish reasons (own health, for example) or because of moral responsibility (Ferrer-i-Carbonell and Gowdy, 2007). A highly relevant line of research is the search for evolutionary roots of human behavior including the evolution of morality (de Waal, 1996, 2006; Wilson and Gowdy, 2013). In this chapter we focus on the social aspects of decision-making to highlight social preferences and shared meanings.

Proposition IV: Neuroscience has confirmed the uniqueness of the human 'social brain'.[5]

Many mammals are highly social animals with a variety of behavioral attributes that evolved to facilitate social interaction, but humans seem to be unique in their degree of sociability. Recent evidence for the existence of the social brain (Fehr, 2009; Singer, 2009) unequivocally confirms the need for a theory of decision-making that is other-regarding.

A remarkable finding from neuroscience is that most of the neurons in the human brain develop after birth and the way they are configured depends critically on how a child is socialized. It is another way that variability can be introduced into the evolutionary mix. Wexler (2006, p. 3) writes about the evolutionary advantages of brain plasticity:

> [T]he distinctive postnatal shaping of each individual's brain function through interaction with other people, and through his or her own mix of sensory inputs, creates an endless variety of individuals with different functional characteristics. This broadens the range of adaptive and problem-solving capabilities well beyond the variability achieved by sexual reproduction.

The human brain is much larger in proportion to size than the brains of great apes but other than that few structural differences have been identified. For example, the frontal lobe – associated with cognition and executive functions – is about three times larger in humans compared to other apes. But it is not disproportionately larger when scaled to brain size. This has led to speculation that the uniqueness of human cognitive ability might be due to the internal organization of major brain features rather than gross anatomy. A part of the brain of particular interest is Brodmann's Area 10, which differs among humans in the degree of horizontal spacing of neurons (Semendeferi et al., 2010, Table 2). The relatively recent reorganization of the human brain has apparently enhanced social referencing by activating reward-related brain regions (Schilbach et al., 2010).

Another remarkable finding from neuroscience is the presence in the human brain of Von Economo neurons, also called spindle neurons, which apparently evolved to enable people to make rapid decisions in social contexts. Sherwood et al. (2008, p. 433) write:

> It is interesting that these specialized projection neuron types have been identified in cortical areas that are positioned at the interface between emotional and cognitive processing. Given their characteristics, it has been speculated that Von Economo neutrons are designed for quick signaling of an appropriate response in the context of social ambiguity (Allman et al. 2005). Enhancements of this ability would be particularly important in the context of fission-fusion communities, such as those of panids [bonobos and chimpanzees] and possibly the LCA [last common ancestor], with complex networks of social interactions and potential uncertainties at reunions.

Allman et al. (2005, 370) argue that these neurons help humans to adjust quickly to rapidly changing social situations:

> We hypothesize that the VENs and associated circuitry enable us to reduce complex social and cultural dimensions of decision-making into a single dimension that facilitates the rapid execution of decisions. Other animals are not encumbered by such elaborate social and cultural contingencies to their decision-making and thus do not require such a system for rapid intuitive choice.

Von Economo neurons are also found (in much smaller numbers) in great apes and whales and dolphins, other highly intelligent species with complex social systems (Semendeferi et al., 2010). In humans, most Von Economo neurons are formed *after birth*, again pointing to the blurred line between heredity and socialization. Again, the latest neurological evidence suggests that human behavior is uniquely social. Understanding the social nature of decision-making (the importance of reference groups, for example) is critical both to formulating successful social and environmental policies and to gaining public acceptance of these policies.

Proposition V: Behavioral psychology has confirmed the existence of a kind of 'social cognition', that is, group decision-making exhibits emergent properties apart from the characteristics of individuals in the group.

The reason for the evolution of the structure of such features as Brodmann's Area 10 and Von Economo neurons is hotly debated, but neuroscience suggests that a feature distinguishing humans from other apes is the type of social behavior mediated by the structure of the prefrontal cortex. Neuroscience is confirming the importance of an evolutionary perspective in explaining human behavior including group selection (Caporael, 1997; Sober and Wilson, 1998; van den Bergh and Gowdy, 2009). Evidence from psychological studies points to the existence of a kind of collective intelligence of human groups related to group composition but unrelated to the characteristics of individuals within the groups. In a recent study of group decision-making, Woolley et al. (2010) found evidence for what they called a 'collective intelligence factor'. In two different studies, groups of two to five people were assigned a variety of tasks, then the groups were ranked according to their performance of these tasks. The author found that a collective intelligence factor explained the groups' performance and that:

> The 'c-factor' is not strongly correlated with the average or maximum individual intelligence of group members but is correlated with the average social sensitivity of group members, the equality in distribution of conversational turn-taking, and the proportion of females in the group. (Woolley, 2010, p. 686)

This finding suggests that a strong case can be made for decision process allowing for interaction and deliberation among individuals. It also begs for further research into the 'ideal' composition of groups for making critical decisions. For example, is there an ideal mix of selfish individuals and altruists in collective decision-making? Does voting based on isolated individual decisions preclude solutions based on group deliberative valuation that might result in better outcomes?

To summarize Sections 7.2 and 7.3 above: (1) the theoretical literature in welfare economics has established that preferences cannot be characterized or evaluated in a social welfare context without making interpersonal comparisons of utility, (2) the empirical behavioral literature has established that preferences cannot be understood without reference to social context.

One final point is to note that although experimental economic models go beyond the 'armchair theorizing' of traditional models, they still deal with highly artificial situations. Neurological studies provide at best psycho-physical correlates and one can only make sense of these in combination with psychological and behavioral approaches. But as the field matures, neurological and psychological models are being tested in more and more realistic situations. Measham and Barnett (2008) present evidence from a case study of environmental volunteering as an example of cooperative behavior in real world situations. A pioneering example of empirically examining real world cooperative behavior is the work of Elinor Ostrom (1990, 2010). Based on her extensive fieldwork in a variety of societies she proposed a set of principles for successful institutional arrangements for managing common property resources. Wilson et al. (2013) discuss these in detail, showing how they follow from basic evolutionary principles.

7.4 DELIBERATIVE VALUATION

Deliberative valuation (DV) involves a group of selected persons who explore the values that should guide collective decisions through a process of reasoned discourse (Howarth and Wilson, 2006). The basic idea is that through deliberation, people can reach agreement by exploring arguments and developing mutual understanding and trust. The institutional settings range from exploratory workshops such as focus groups to decision-oriented designs such as citizen juries or consensus conferences. The latter type is often supported by decision-aid methods such as multicriteria appraisal. As envisioned by practitioners like Soderholm (2001), and Spash (2002), DV is an alternative to the self-regarding decision-making process adopted in revealed and stated preference approaches.[6] A deliberative group process allows a representative group of individuals to gather scientific information and analyze their pre-existing preferences in a more critical and interactive way. In this regard, deliberative processes can enhance the effectiveness and support for policy decisions by explicitly recognizing the importance of groups and group identity, making the decision process transparent, and allowing a role for collective intelligence in the decision-making process (Howarth and Wilson, 2006). If properly structured, groups can negotiate distributional outcomes that participants can accept as fair and legitimate. This means replacing welfarism with an approach based on establishing deliberative, democratic institutions that can resolve distributional conflicts given a procedural conception of distributional justice (Dewey, 1938; Habermas, 1971; Bromley, 2006).

The results of empirical DV studies provide evidence for the importance of different realms of value beyond self-referential economic values. These results present challenges to the traditional economic framework, including multiple values, incommensurability and lexicographic preferences, social justice, fairness and non-human values (Spash, 2008b). As the use of formal methods of deliberation is relatively new to the field of environmental valuation (Howarth and Wilson, 2006), the methods to obtain these values, the expected outcomes and their roles vary widely (Spash, 2007). But in general, DV goes beyond traditional stated preference approaches to valuation in that it embraces social, other-regarding preference formation.

Recently, economists have applied deliberative monetary valuation (DMV) methodologies, a combination of participatory deliberation techniques and traditional stated preference methods, as an alternative to BCA (Gowdy, 2007). DMV aims to combine theories and practices of economics, social psychology, decision science and politics in an effort to obtain more reliable environmental values (Sagoff, 1998; Wilson and Howarth, 2002; Stagl, 2006; Spash, 2007). There still remains a gap between the practiced empirical approaches and theoretical underpinnings of DMV (Spash, 2008b). Although the DMV process still faces various challenges, it is still highly praised by ecological economists for valuing many ecological services (Vatn, 2009), because it improves upon the traditional contingent valuation method in a variety of ways (Kenyon and Hanley, 2005).

Positions on theoretical construction of the DMV process vary. Spash (2007, 2008b) categorized proposed theoretical approaches and identified their corresponding advocates. Most theorists support the 'arbitrated social willingness to accept/willingness to pay (WTP/WTA)' that individuals decide alone but it is recognized that they make valuation decisions within institution and social contexts. But Sagoff (1998) offers the three

Table 7.1 Deliberative monetary valuation as willingness to pay

Terms in Which WTP Specified		
Value provider	Individual (disaggregated value)	Social (aggregated value)
Individual	Exchange value	Speculative value
Individual in a group setting	Fair price	Expressed social WTP/WTA
Group	Charitable contribution	Arbitrated social WTP/WTA

Source: Adapted from Spash (2007, 2008b).

options in Table 7.1. In the first option individuals act as isolated consumers expressing a social WTP, defined as 'speculative value'. In the second option, 'expressed social WTP/WTA', Sagoff (1998) proposes that individuals, rather than acting as consumers by providing an individual WTP, can participate as citizens who construct a collective or social WTP of an environmental good. Alternatively, citizens could judge the value of an environmental good, not only for the group, but to society as a whole, and provide 'a statement of the "fair share" they would pay as members of the community to protect those goods' (Sagoff, 1998, p. 226), represented by 'fair price' in Table 7.1. Finally, Sagoff's third proposition, categorized as 'charitable contribution' (Spash, 2007), is a group-mediated individual WTP for an environmental good or service representing what society should invest, proposed by individuals participating in a deliberative process (Sagoff, 1998).

Wilson and Howarth (2002) point out that while many environmental goods and services are public goods, in practice, their values are often expressed by individuals, and therefore do not address issues of other-regarding social equity. Instead, they suggest 'discourse-based valuation' in which a group reaches a consensus-based social value of an ecosystem good or service (ibid.). This deliberative process is consistent with political theory of both liberal democrats and discursive democrats, and can theoretically be modeled using techniques of cooperative game theory, as they specifically demonstrate through the Nash bargaining solution. Their theoretical model finds that the aggregate WTP or WTA in typical BCA will overstate a deliberative group's WTP and understate a deliberative group's WTA (Howarth and Wilson, 2006). Brown et al. (1995) also express concern with the assumption that the larger social good is merely an aggregation of individual personal preferences or of well-organized interest groups. This concern leads them to propose the alternative of a 'values jury'. Participants of the jury would act in the interest of society as a whole, and could both assist in decision-making and recommend a payment regarding the public good (ibid.).

Although most theorists advocate the use of 'arbitrated social WTP/WTA', in practice, the majority of empirical studies are generating a fair price value, that is, a modified exchange value, revealing the divergence between theorists and practitioners. As theorists argue for consensus group outcomes, practitioners employ individual choice models; as theorists advocate for the representation of a variety of social positions on a topic, practitioners use random, semi-random, quota or convenience samples; and as theorists expect to address issues such as lexicographic preferences, incommensurability or fairness, many practitioners wish to remove these results from their analysis (Spash, 2008b).

A full review of these empirical studies categorized under the various theoretical approaches in Table 7.1 can be found in Spash (2008b). While many of the studies in the 'charitable contribution' category claim that the DMV process has improved the validity of their results, none identify what type of validity. Most of these DMV practitioners aim to achieve statistical power, and therefore miss the rich data that could be gathered through interpretative activities, and further attempt to force the data into a preconceived model (ibid.). One empirical study, as categorized by Spash (2008b), did, however, generate an 'arbitrated social WTP/WTA' (James and Blamey, 2005). In this study, while information, time and opportunity for deliberation allowed for the formation of robust preferences, the researchers experienced numerous methodological issues, including consensus formation, decision rules, equality of juror impact and provision of information, as well as theoretical issues, including concern framing, representation, non-consensual outcomes, and the economic interpretation of the results. Although faced with these difficulties, they concluded that research into the management and behavior of legal juries will greatly enhance the development and refinement of the methodology (ibid.). In this regard, staff training to improve comfort with analytical techniques, such as eliciting preferences from community stakeholders and evaluating trade-offs between various facets of value, would aid in the success of deliberative processes. These types of efforts will be beneficial, as they will result in the development of more effective, cost-efficient, and broadly supported environmental policies (Gregory and Wellman, 2001; Gowdy, 2008).

Although the DMV process currently faces the issues and challenges mentioned above, mounting evidence from various disciplinary fields continues to justify and support the use of group approaches. Additionally, the research from these fields is providing useful insights regarding the DMV process, such as the structure of the group and the format of the approach. For example, in studies conducted by Woolley et al. (2010) discussed in Section 7.3, groups ranging from two to five members were asked to work on various tasks including solving visual puzzles, brainstorming, making collective moral judgments, and negotiating over limited resources. Results indicated that collective intelligence was found to be a better predictor of group performance than average or maximum individual intelligence. Neither the average group member intelligence nor the maximum group member intelligence was significantly correlated with collective intelligence of the groups. Three unlikely factors, however, were found to be correlated with collective intelligence: average social sensitivity, equality in distribution of conversational turn-taking, and the proportion of females in the group (ibid.). These latter examples, coupled with evidence provided in Section 7.3, support the continued development of deliberative processes in eliciting complex environmental values. In assessing these complex values, it would be beneficial for economists to accept and understand the importance of qualitative, as well as quantitative data: 'If economics is really an empirical science then its method must improve from trying to fit the data into the model to accepting observations as a means of learning, which often challenges accepted theory and requires change in that theory' (Spash, 2008b, p. 486). Humans evolved as members of social groups and understanding group processes is essential to understand how humans place value on objects and situations that give them utility (van den Bergh and Gowdy, 2009; Sheldon et al., 2000).

Deliberative valuation moves beyond the standard economic theory of value. Unlike contingent valuation, deliberative techniques, like citizens juries, can (1) better address the information problem, (2) treat participants as citizens, rather than consumers, therefore addressing general public interest, not only individual interest, (3) be more useful in dealing with equity and distributional issues, (4) include local communities in the decision-making process, allowing for the possibility that these environmental decisions will be more sustainable, and (5) implicitly, rather than directly, aid participants to construct their values on a given topic (Kenyon and Hanley, 2005). Additionally, deliberative methods are more compatible with democratic institutions and processes, by which society actually responds to issues in terms of political trade-offs (Sagoff, 1998). Interactive valuation can allow for results that are based on the norms and rules compatible with the local, social and natural environment; hence, the results can be more readily accepted and put into action (Arzt, 2005).

7.5 SUMMARY AND CONCLUSIONS

The basis for deliberative valuation can be established from the theoretical and empirical criticisms of contemporary welfare theory as embedded in the basic welfare equation (Equation (7.5) above). The basic flaw is the individual-at-a-point-in-time perspective that became the basis for social decision-making. To summarize, these criticisms imply the following:

- In evaluating welfare changes, changes in the marginal utility of money should be taken into account, implying that changes in all goods must be considered, not just changes in those directly affected. The options considered in deliberative valuation are often presented as scenarios, which by definition explore equity impacts in different markets and many different realms of life.
- Changes in social welfare ($\delta W/\delta U$) depend on relative welfare effects. The group setting of deliberative valuation exercises is regularly observed to enable participants to develop an understanding of the impact on the whole and take this into account in their decisions.
- Welfare involves much more than changes in the utility from accumulating goods. Deliberative valuation processes explore explicitly different dimensions (perspectives, criteria) of the problem at hand.
- The time path in moving from one policy to another is critical and discussion of this should be part of the valuation process. The final part of deliberative valuation workshops normally consists of discussions about different ways to achieve the preferred scenarios or options. Issues raised in these sessions are normally reported to the decision-makers together with preferred option(s); sometimes decision--makers are part of the deliberations.
- All feasible outcomes should be considered in the valuation process. Deliberative valuation processes compare by definition different options or scenarios. Whoever devises the deliberative process, will normally start from all feasible outcomes. Because of constraints of time and cognitive capacity of participants for making comparisons, options for discussion need to limited. This choice should be transparent and open for alteration.
- The possibility that the valuation process might change values should be taken into account. Respondents have been shown to be very sensitive to how a valuation

question is framed. The framing of the question works as a hint to help construct responses, implying that preferences are not set in stone, but rather are formed in part during the valuation process (Slovic et al., 1990; Slovic, 1995). Deliberative valuation explicitly aims to give participants the opportunity to revise their preferences after having explored the problem at hand. Researchers support the individuals in the process of clarifying their preferences when learning about the various elements and characteristics of the (unfamiliar) good involved. Additionally, preferences are perceived as social constructs that may change in a group discourse. Instead of defining away the need for devising a fair and legitimate decision process, the challenge is explicitly addressed and made transparent. This idea is in line with Harsanyi's (1997) argument that for normative issues 'informed preferences' should be used.

At a broad level, we do not reject the basic value system or the fundamental approach of classical economic theory. We agree that individuals are the best judges of what is best for themselves and that, in general, people attempt to their limited resources to the best of their ability, and we agree that Bentham's utilitarian principle of the 'greatest good for the great number' is a good starting point for economic policy. But with the ascent of Walrasian welfare economics these common sense observations were turned into parodies of actual human behavior. Contemporary economic experiments show that it is not 'irrational' to be affected by the opinions and behavior of others and that there is more to sound economic policy than static allocative efficiency.

In the hands of Lionel Robbins, J.R. Hicks, Nicholas Kaldor and others, Walrasian general equilibrium was transformed from a descriptive model into a goal-oriented system that was used to define economics as the science of the optimal allocation of scarce resources among alternative ends. Throughout most of the twentieth century, economic policy became less concerned about the details of specific economic problems and more concerned about the establishment of efficiency in a general equilibrium framework. In recent decades the Walrasian system has fallen into disarray under the combined weight of theoretical intractabilites and empirical refutation of its basic behavioral assumptions. The time has come to incorporate modern welfare economics and scientific evidence about how humans make valuation decisions into economic valuation methods and public policy.

NOTES

1. Following Bowles and Gintis (2000) we use the term Walrasian welfare economics to describe the school of economic thought that came to dominate economic theory in the decades during and following World War II. It is also known as 'The New Welfare Economics' (Chipman and Moore, 1978) or simply neoclassical economics. We prefer the term Walrasian because it emphasizes the mathematical general equilibrium framework with its necessary assumption of self-regarding preferences (Gowdy, 2010). It should be pointed out that one can be Walrasian and yet see social choice in mainly libertarian terms that reject welfarism.
2. The ideological purpose of Walrasian welfare economics is to establish that freely operating market economies with prices representing true social costs are Pareto efficient. No reallocation can make one person better off without making another worse off. The starting point for the mathematical proof of efficiency of competitive markets is that the condition that the rate at which consumers are willing to substitute one

good for another (the marginal rate of substitution) must be the same for both consumers. This condition cannot be established if preferences are other-regarding. If one person's utility depends in any way on the utility of others, then the proof of the efficiency of competitive equilibrium cannot be completed (Henderson and Quandt, 1980, p. 297; Gowdy, 2004, 2010). Interdependent preferences can be modeled in the neoclassical, general equilibrium framework but not without losing the basis for proving the First Fundamental Theorem of Welfare Economics, namely, that competitive markets will correct social prices and will achieve Pareto efficiency in allocation.

3. Many economists believe that these three issues are covered by the Second Fundamental Theorem of Welfare Economics: 'Assume that all individuals are selfish price takers. Then almost any Pareto optimal equilibrium can be achieved via the competitive price mechanism, provided appropriate lump-sum taxes and transfers are imposed on individuals and firms' (Feldman, 1987, p. 891). But it turns out that even moving from one point to another on the production possibilities frontier is plagued by inconsistencies and paradoxes. These problems are compounded when the theorem is extended to compare non-Pareto optimal possibilities as in the Kaldor-Hicks compensation principle.
4. The necessity of the homothetic preferences assumption is illustrated by the so-called Cycling Paradox. Under the potential Pareto improvement criterion a movement from point A to point B on a utilities possibilities frontier may be desirable but so may a movement from B back to A. This cycling phenomenon can only be avoided if the underlying utility functions are homothetic. This implies that any change in income will not affect the relative amounts of the goods in the consumer's market basket.
5. Interestingly, the controversy in economics over the usefulness of neuroeconomics revolves around the issue of cognitive individualism. Some economists argue that current research is too concerned with individual agents and not enough with human groups (Wilcox, 2008) while others criticize neuroeconomics for not being well grounded in standard axiomatic economics (Harrison, 2008).
6. We do not discuss the large literature on the contingent valuation method (CVM) and the controversies surrounding it. CVM began as a stated preference method excluding social and institutional context (Spash et al., 2009). Although the method is based on individual survey it can be expanded to include values to partially take into account social norms (Spash, 2008a).

REFERENCES

Allman, J., T. McLaughlin and A. Hakeem (2005), 'Intuition and autism: a possible role for Von Economo neurons', *Trends in Cognitive Science*, **9**(8), 367–73.

Arzt, K. (2005), 'The challenges of stakeholder participation: agri-environmental policy', in M. Getzner, C. Spash and S. Stagl (eds), *Alternatives for Environmental Valuation*, New York: Routledge, pp. 244–62.

Bateman, I. and J. Mawby (2004), 'First impressions count: interviewer appearance and information effects in stated preference studies', *Ecological Economics*, **49**(1), 47–55.

Bator, F. (1957), 'The simple analytics of welfare maximization', *American Economic Review*, **47**(1), 22–59.

Becker, G. (1976), 'Altruism, egoism, and genetic fitness: economics and sociobiology', *Journal of Economic Literature*, **14**(3), 817–26.

Blackorby, C. and D. Donaldson (1990), 'A review article: the case against the use of the sum of compensating variations in cost–benefit analysis', *Canadian Journal of Economics*, **23**(3), 471–94.

Boadway, R. (1974), 'The welfare foundations of cost–benefit analysis', *The Economic Journal*, **84**(336), 926–39.

Bowles, S. (1998), 'Endogenous preferences: the cultural consequences of markets and other economic institutions', *Journal of Economic Literature*, **36**(1), 75–111.

Bowles, S. and H. Gintis (2000), 'Walrasian economics in retrospect', *Quarterly Journal of Economics*, **115**(4), 1411–39.

Bromley, D. (2006), *Sufficient Reason: Volitional Pragmatism and the Meaning of Economic Institutions*, Princeton, NJ: Princeton University Press.

Brown, T., G. Peterson and B. Tonn (1995), 'The values jury to aid natural resource decisions', *Land Economics*, **71**(2), 250–60.

Burns, M. (1973), 'A note on the concept and measurement of consumer surplus', *The American Economic Review*, **63**(3), 335–44.

Camerer, C., G. Loewenstein and D. Prelec (2005), 'Neuroeconomics: how neuroscience can inform economics', *Journal of Economic Literature*, **43**(1), 9–64.

Caporael, L. (1997), 'The evolution of truly social cognition: the core configurations model', *Personality and Social Psychology Review*, **1**(4), 276–98.

Chipman, J. and J. Moore (1978), 'The new welfare economics 1939–1974', *International Economic Review*, **19**(3), 547–84.

de Waal, F. (1996), *Good Natured: The Origins of Right and Wrong and Other Animals*, Cambridge, MA: Harvard University Press.
de Waal, F. (2006), *Primates and Philosophers: How Morality Evolved*, Princeton, NJ: Princeton University Press.
Dewey, J. (1938), *Experience and Education*, New York: Touchstone Books.
Easterlin, R. (1974), 'Does economic growth improve the human lot? Some empirical evidence', in P. David and M. Reder (eds), *Nations and Happiness in Economic Growth: Essays in Honor of Moses Abramowitz*, New York: Academic Press, pp. 89–125.
Fehr, E. (2009), 'Social preferences and the brain', in P. Glimscher, C. Camerer, E. Fehr and R. Poldrack (eds), *Neuroeconomics: Decision Making and the Brain*, London: Academic Press, pp. 215–32.
Fehr, E. and S. Gächter (2002), 'Altruistic punishment in humans', *Nature*, **415**(6868), 137–40.
Feldman, A. (1987), 'Welfare economics', in J. Eatwell, M. Milgate and P. Newman (eds), *The New Palgrave Dictionary of Economics*, London: Macmillan, pp. 889–95.
Ferrer-i-Carbonell, A. and J. Gowdy (2007), 'Environmental awareness and happiness', *Ecological Economics*, **60**(3), 509–16.
Frey, B. and A. Stutzer (2002), *Happiness and Economics: How the Economy and Institutions Affect Well-Being*, Princeton, NJ: Princeton University Press.
Friedman, M. (1952), 'The methodology of positive economics', in M. Friedman (ed.), *Essays in Positive Economics*, Chicago: University of Chicago Press.
Gowdy, J. (2004), 'The revolution in welfare economics and its implications for environmental valuation', *Land Economics*, **80**(2), 239–57.
Gowdy, J. (2007), 'Toward an experimental foundation for benefit–cost analysis,' *Ecological Economics*, **63**(4), 649–55.
Gowdy, J. (2008), 'Behavioral economics and climate change policy', *Journal of Economic Behavior and Organization*, **68**(3–4), 632–44.
Gowdy, J. (2010), *Microeconomic Theory Old and New: A Student's Guide*, Stanford, CA: Stanford University Press.
Gregory, R. and K. Wellman (2001), 'Bringing stakeholder values into environmental policy choices: a community-based estuary case study', *Ecological Economics*, **39**(1), 37–52.
Güth, W., R. Schmittberger and B. Schwarz (1982), 'An experimental analysis of ultimatum bargaining', *Journal of Economic Behavior and Organization*, **3**(4), 83–116.
Habermas, J. (1971), *Knowledge and Human Interests*, Boston: Beacon Press.
Harberger, A. (1971), 'Three basic postulates for applied welfare economics: an interpretive essay', *Journal of Economic Literature*, **9**(3), 785–97.
Harrison, G. (2008), 'Neuroeconomics: a critical reconsideration', *Economics and Philosophy*, **24**(3), 303–34.
Harsanyi, J. (1997), 'Utilities, preferences, and substantive goods', *Social Choice and Welfare*, **14**(1), 129–45.
Heal, G. (1998), *Valuing the Future: Economic Theory and Sustainability*, New York: Columbia University Press.
Henderson, J. and J. Quandt (1980), *Microeconomic Theory: A Mathematical Approach*, New York: McGraw-Hill.
Henrich, J., R. Boyd, S. Bowles, C. Camerer, E. Fehr, H. Gintis and R. McElreath (2001), 'In search of Homo economicus: behavioral experiments in 15 small-scale societies', *American Economic Review*, **91**(2), 73–8.
Henrich, J., R. McElreath, A. Barr, J. Ensminger, C. Barrett and A. Bolyanatz et al. (2006), 'Costly punishment across human societies', *Science*, **312**(5781), 1767–70.
Hicks, J. (1939), 'The foundations of welfare economics', *Economic Journal*, **49**(196), 696–712.
Hoffman, E., K. McCabe, K. Shachat and V. Smith (1994), 'Preferences, property rights, and anonymity in bargaining games', *Games and Economic Behavior*, **7**(3), 346–80.
Howarth, R. and M. Wilson (2006), 'A theoretical approach to deliberative valuation', *Land Economics*, **82**(1), 1–16.
James, R. and R. Blamey (2005), 'Deliberation and economic valuation: national park management', in M. Getzner, C. Spash and S. Stagl (eds), *Alternatives for Environmental Valuation*, New York: Routledge, pp. 225–43.
Kahneman, D. and A. Tversky (1979), 'Prospect theory: an analysis of decision under risk', *Econometrica*, **47**(2), 263–91.
Kahneman, D., E. Diener and N. Schwarz (eds.) (1999), *Foundations of Hedonistic Psychology: Scientific Perspectives on Enjoyment and Suffering*, New York: Russell Sage Foundation.
Kaldor, N. (1939), 'Welfare propositions of economics and interpersonal comparisons of utility', *Economic Journal*, **49**(196), 549–52.
Kenyon, W. and N. Hanley (2005), 'Three approaches to valuing nature: forest floodplain restoration', in M. Getzner, C. Spash and S. Stagl (eds), *Alternatives for Environmental Valuation*, New York: Routledge, pp. 187–224.

Knetsch, J. (1989), 'The endowment effect and evidence of nonreversible indifference curves', *American Economic Review*, **79**(5), 1277–84.
Knetsch, J. (2005), 'Gains, losses, and the US-EPA Economic Analyses Guidelines: a hazardous product?' *Environmental & Resource Economics*, **32**(1), 91–112.
Layard, R. (2005), *Happiness – Lessons from a New Science*, London: Penguin Books.
Manner, M. and J. Gowdy (2010), 'Group selection and the evolution of moral behavior: toward a coevolutionary foundation for public policy', *Ecological Economics*, **69**, 753–69.
Mayer, F.S. and C.M. Franz (2004), 'The connectedness to nature scale: a measure of individuals' feeling in community with nature', *Journal of Environmental Psychology*, **24**(4), 503–15.
Measham, T.G. and G.B. Barnett (2008), 'Environmental volunteering: motivations, modes and outcomes', *Australian Geographer*, **39**(4), 537–52.
Meier, S. (2006), *The Economics of Non-selfish Behaviour*, Cheltenham, UK and Northampton, MA, USA: Edward Elgar Publishing.
Ng, Y.K. (1997), 'A case for happiness, cardinalism, and interpersonal comparability', *Economic Journal*, **107**(445), 1848–58.
Ostrom, E. (1990), *Governing the Commons: The Evolution of Institutions for Collective Action*, Cambridge, UK: Cambridge University Press.
Ostrom, E. (2010), 'Beyond markets and states: polycentric governance of complex economic systems', *American Economic Review*, **100**(3), 1–33.
Persky, J. (2001), 'Cost–benefit analysis and the classical creed', *Journal of Economic Perspectives*, **15**(4), 199–208.
Reeson, A. (2008), 'Institutions, motivations and public goods', *Socio-Economics and the Environment in Discussion (SEED) Working Paper Series*, No. 2008–01.
Rekola, M. (2003), 'Lexicographic preferences in contingent valuation', *Land Economics*, **79**(2), 277–91.
Sagoff, M. (1998), 'Aggregation and deliberation in valuing environmental public goods: a look beyond contingent pricing', *Ecological Economics*, **24**(2/3), 213–30.
Sahlins, M. (1996), 'The sadness of sweetness: the native anthropology of western cosmology', *Current Anthropology*, **37**(3), 395–428.
Schilbach, L., M. Wilms, S. Eickhoff, S. Romanzetti, R. Tepest and G. Bente et al. (2010), 'Minds made for sharing: initiating joint attention recruits reward-related neurocircuitry', *Journal of Cognitive Neuroscience*, **22**(12), 2702–15.
Semendeferi, K., K. Teffer, D. Buxhoeveden, M. Park, S. Bludau and K. Armunts et al. (2010), 'Spatial organization of neurons in the frontal pole sets humans apart from great apes', *Cerebral Cortex*, doi:10.1093/cercor/bhq191.
Sheldon, K., M. Sheldon and R. Osbaldiston (2000), 'Prosocial values and group assortation within an N-person prisoner's dilemma game', *Human Nature*, **11**(4), 387–404.
Sherwood, C., F. Subiaul and T. Zadiszki (2008), 'A natural history of the human mind: tracing evolutionary changes in brain and cognition', *Journal of Anatomy*, **212**(4), 426–54.
Silverberg, E. (1972), 'Duality and the many consumer surpluses', *The American Economic Review*, **62**(5), 942–52.
Singer, T. (2009), 'Understanding others: brain mechanisms of theory of mind and empathy', in P. Glimscher, C. Camerer, E. Fehr and R. Poldrack (eds), *Neuroeconomics: Decision Making and the Brain*, London: Academic Press, pp. 251–68.
Slovic, P. (1995), 'The construction of preference', *American Psychologist*, **50**(5), 364–71.
Slovic, P., N. Kraus and V. Covello (1990), 'What *should* we know about making risk comparisons?' *Risk Analysis*, **10**, 389–92.
Sober, E. and D. Wilson (1998), *Unto Others: The Evolution and Psychology of Unselfish Behavior*, Cambridge, MA: Harvard University Press.
Soderholm, P. (2001), 'The deliberative approach in environmental valuation', *Journal of Economic Issues*, **35**(2), 487–95.
Spash, C. (2002), 'Informing and forming preferences in environmental valuation: coral reef biodiversity', *Journal of Economic Psychology*, **23**(5), 665–87.
Spash, C. (2007), 'Deliberative monetary valuation (DMV): issues in combining economic and political processes to value environmental change', *Ecological Economics*, **63**(4), 690–99.
Spash, C. (2008a), 'Contingent valuation design and data treatment: if you can't shoot the messenger, change the message', *Environment and Planning C*, **26**(1), 34–53.
Spash, C. (2008b), 'Deliberative monetary valuation and the evidence for a new value theory', *Land Economics*, **84**(3), 469–88.
Spash, C. (2009), 'Motives behind willingness to pay for improving biodiversity in a water system: economics, ethics and social psychology', *Ecological Economics*, **68**(4), 955–64.
Stagl, S. (2006), 'Multicriteria evaluation and public participation: the case of UK energy policy', *Land Use Policy*, **23**, 53–62.

Stavins, R. (2008), 'Environmental economics', in S. Durlauf and L. Blume (eds), *The New Palgrave Dictionary of Economics*, 2nd edition, London: Macmillan.

Stigler, G. and G. Becker (1977), 'De gustibus non est disputandum', *American Economic Review*, **67**(2), 76–90.

Suzumura, K. (1999), 'Paretian welfare judgements and Bergsonian social choice', *Economic Journal*, **109**(445), 204–21.

van den Bergh, J. and J. Gowdy (2009), 'A group selection perspective on economic behavior, institutions and organizations', *Journal of Economic Behavior and Organization*, **72**(1), 1–20.

Vatn, A. (2004), 'Environmental valuation and rationality', *Land Economics*, **80**(1), 1–18.

Vatn, A. (2009), 'An institutional analysis of methods for environmental appraisal', *Ecological Economics*, **68**(8–9), 2207–15.

Videras, J., A. Owen, E. Conover and S. Wu (2008), 'The influence of social relationships on pro-environmental behaviors', *Journal of Environmental Economics and Management*, **63**(1), 35–50.

Wexler, B.E. (2006), *Brain and Culture*, Cambridge, MA: MIT Press.

Wilcox, N. (2008), 'Against simplicity and cognitive individualism', *Economics and Philosophy*, **24**(3), 523–32.

Wilson, D.S. and J. Gowdy (2013), 'Evolution as a general theoretical framework for economics and public policy', *Journal of Economic Behavior and Organization*, **90**(S), S3–S10.

Wilson, D.S., E. Ostrom and M. Cox (2013), 'Generalizing the core design principles for the efficacy of groups', *Journal of Economic Behavior and Organization*, **90**(S), S21–S32.

Wilson, M. and R. Howarth (2002), 'Discourse-base valuation of ecosystems services: establishing fair outcomes through group deliberation', *Ecological Economics*, **41**(3), 431–43.

Woolley, A., C. Chabris, A. Pentland, M. Hashmi and T. Malone (2010), 'Evidence for a collective intelligence factor in the performance of human groups', *Science*, **330**(6004), 686–8.

8. Valuing ecosystem services in macroeconomic settings

Rodney B. W. Smith and Masahiko Gemma

8.1 INTRODUCTION

Consider the following questions. In a region where agricultural production is primarily irrigated, how much of its agricultural gross domestic product (GDP) is accounted for by the provisioning services of water? If agriculture in the region is both rain-fed and irrigated, what is the value of rainwater in agricultural production? How much does a coral reef system contribute to a region's GDP? These questions are indicative of those that economists and national account statisticians are beginning to tackle. Answering these questions requires measuring the aggregate flow value of an ecosystem service: water in the first two cases and the tourism services derived from a coral reef system in the last case.

This chapter lays out an approach to numerically calculating the flow (shadow rental) value of an ecosystem service's contribution to aggregate GDP. The approach develops a conceptual framework that is directly linked to an empirical model amenable to numerical solutions. The underlying conceptual model is based on dynamic, general equilibrium theory and accommodates multiple sectors and multiple regions. In addition to predicting shadow rental values for ecosystem services *over time*, the empirical model can also calculate the unit shadow price (discounted shadow rental values) of the ecosystem service(s) over time.

We focus on the ecosystem services of water primarily because it is a relatively easy service to measure, for example, the quantity of water used in agricultural and manufactured good production, and because it is a service whose use cuts across most productive sectors in an economy and one with which most readers can identify. Another reason for using water is that it is a natural asset that is seldom allocated with markets, and hence, seldom accounted for in GDP. Also, although the ecosystem producing water is complex, if the water comes from a river or rain, and if the water availability is relatively stable, an economist might safely ignore the ecosystem dynamics generating the water and take the quantity of water provided over a period of time as exogenous and given. Hence, for this exercise at least, we can ignore ecosystem dynamics. The choice of Japan as the case study region is driven primarily because the authors are familiar with water issues in Japan, and because Japan has readily available national account data and a rich source of land and water use data – requisite data for the empirical model.

As with physical capital assets, an ecosystem service's (or natural asset's) contribution to GDP is given by the value of the flow of services it provides to producers. One major difference between physical (i.e., man-made) capital and a natural asset or ecosystem, is that markets typically allocate physical capital, but seldom allocate ecosystem services. In principle, physical capital's contribution to GDP is equal to the unit flow value of physical capital – for example, the market-determined interest rate paid on a unit of capital – multiplied by the size of the capital stock. Here, the interest rate is equal to the revenue an additional unit of capital generates, ceteris paribus. Hence, with physical capital the market captures the flow value of the asset.

The unit flow value of an ecosystem service is usually not valued by a market. Water, for example, is seldom allocated via market mechanisms, and in such cases the flow value of water is not captured by markets. Even though a market might not capture the flow value of water, producers who use water certainly do – earning rent on the water they employ in production. The unit flow value of an additional unit of the ecosystem service, for example, an additional unit of water, is equal to the additional unit of revenue the producer receives given an additional unit of the ecosystem service. We call this additional revenue the unit 'shadow rent' of the ecosystem service, and contrast this with the unit 'shadow price' of the ecosystem service – defined here as the discounted present value of current and future unit shadow rents.

In what follows, we link the (regional) shadow rental value of agricultural water to national income by first estimating an agricultural production function and using the production function[1] to estimate water's marginal contribution to agricultural GDP. With this information we disentangle water's contribution to GDP by introducing a factor account for water and readjusting Japan's remaining factor account entries accordingly. Some will view this process as an approach to 'greening' national accounts, as it provides analysts with a way to measure part of the ecosystem's direct use value in generating GDP (see World Bank, 2006). Using the re-parameterized agricultural technology we develop an empirical model that predicts a sequence of water and land shadow rental rates for three regions over time. Each sequence of shadow rental values is then used to calculate a sequence of unit shadow prices of the ecosystem service over time.

With a unit shadow price in hand, one is then in a position to calculate the value of a stock of natural assets, or possibly the stock value of an ecosystem. The process of measuring the stock value of a natural asset or ecosystem ventures into the literature on national welfare and sustainability, as aggregate measures of welfare and sustainability are invariably based on a measure of the total value of a nation's assets: physical capital, human capital, natural assets (ecosystems) and institutions (see Dasgupta, 2009).

Human well-being, that is, welfare, and its definitions and measurement, is the topic of interest for a wide spectrum of disciplines (see Stiglitz et al., 2008). For the past few decades, GDP and gross national product (GNP) have been popular indices of welfare, where GNP is GDP plus income earned by domestic citizens abroad, less income earned by foreign citizens in the region. While GNP is a reasonable measure of economic activity, when measuring welfare in a dynamic setting (e.g., an evolving economy) some economists prefer using wealth or net national product (NNP) as a measure of economic well-being or social welfare, where NNP is equal to GNP less depreciation (see Weitzman, 1976; Dasgupta and Mäler, 2000; Heal and Kriström, 2005; Dasgupta, 2009).

Dasgupta and Mäler (2000) note the NNP concept has been around for over 80 years, but renewed interest in the concept emerged as economists began thinking that NNP should be adjusted to reflect the cost of natural resource depletion and degradation, and environmental damages. NNP adjusted to account for such costs is often referred to as '*green NNP*'. They suggest using a wealth-based measure of welfare defined as the summed value of natural asset stocks, man-made capital stocks and human capital, where the value of natural asset stocks is the unit (possibly shadow) price of the asset multiplied by the stock of that asset. Heal and Kriström (2005) examine the merits of the Dasgupta and Mäler measure, as well as an income-based measure of welfare defined as the present value of consumption at shadow prices, and show (p. 1171) that between the wealth and income measures, the income measure 'tracks welfare changes better. Wealth as the value of stocks only tracks welfare if stock prices are constant'.

As hinted above, in addition to examining the practical aspects of measuring the value of green account factor entries, this chapter also presents an approach for estimating (actually, calculating) the stock value of a natural asset – a requisite component of aggregate total wealth. The methodology exploits recent advances in dynamic general equilibrium theory and its empirical implantation to calculate the discounted present value of the stream of flow shadow values, hence yielding unit values of a natural asset.

We should make one qualification on the unit values we estimate: the shadow value as measured here is only a value of the natural asset or ecosystem as an economically productive factor. If the ecosystem service is associated with externalities, for example, river flow and habitat preservation valued by society – the approach here will not measure the value of the externality unless the externality is explicitly modeled. Hence, the approach outlined here *does not* try to uncover the economic values associated with externalities or natural vistas. Although such exercises are important, we feel they distract from a more fundamental challenge: uncovering a clearer picture of the economic value of natural assets or ecosystem services embedded in GDP – a value typically assigned incorrectly to the factor accounts for labor or capital.

Section 8.2 discusses input-output data and green accounts, and the role production technologies can play in backing out natural assets' contribution to GDP. Section 8.3 gives the reader an overview of the economic environment upon which the empirical model and its results are based, and provides a formal definition of the shadow rental value and its corresponding shadow price. Aside from these two definitions and their relationship, all other details of the mathematical model are relegated to the Appendix to this chapter. Section 8.4 presents the production, consumption and technical change parameter values for Japan, and the empirical model results. Not surprisingly, the empirical simulations reveal that shadow rental values can vary widely across regions and over time. More interestingly, the results also suggest that factor intensities affect the rate at which shadow rental values evolve over time: the more labor intensive the water-using sector is, the slower is the rate of growth in its water shadow rental rate. The results also show the importance of accounting for technical change when deriving shadow prices: ignoring technical change leads to significant underestimates of natural asset shadow prices. Section 8.4 also presents a measure of the stock value of water in Japan, using 2010 base year prices. Section 8.5 concludes.

8.2 INPUT-OUTPUT DATA, GREEN ACCOUNTS AND PRODUCTION TECHNOLOGIES – PRACTICAL CONSIDERATIONS

Input-Output Data and Production Technologies

GDP as measured by value-added is a residual measure, and is the difference between (1) the gross value of all final goods and services produced over a period of time for a given region or economy, and (2) the value of all intermediate goods used in producing the final goods and services. In the United Nations' System of National Accounts (SNA), final good receipts are summed for a sector and the value of intermediate goods used to produce the final goods are subtracted from it to give sector GDP, sometimes referred to as sector value-added. The SNA then decomposes sector value-added into at least two factor account categories: payments to capital and payments to labor.[2]

In most countries, wage income is relatively straightforward to measure, as are some payments to capital. In Japan, values that cannot be clearly assigned to either a labor or capital account end up in a 'mixed income' account – measures of the return to entrepreneurial efforts are included in mixed income, but land rental payments are not. Table 8.1 presents aggregated 'factor accounts' for Tokyo agriculture, manufacturing and services. As suggested above, Table 8.1 has three aggregate factor accounts: labor income, capital income and mixed income. The sum of payments to the labor, capital and mixed income factor accounts is sector value-added. In Table 8.1, the value 15 576 is equal to wL_a, where w is the wage rate and L_a is labor demanded: it is the value of wages the agricultural sector paid to labor. The value 6 310 is equal to rK_a, where r is the capital rental rate and K_a represents the capital stock level in Tokyo agriculture: it is the value of capital stock rent the agricultural sector paid to owners of capital.

When constructing an empirical general equilibrium model (static or dynamic), economists often take the data in the input-output (I-O) table and combine it with data on estimates of labor force and capital stock levels to specify sectoral production technologies. The functional forms for these technologies are typically Cobb-Douglas or constant elasticity of substitution (CES) production functions. The Cobb-Douglas function is the simplest functional form to work with because its parameters relate to I-O data in a very straightforward way. For example, the Cobb-Douglas analog to the agricultural sector in Table 8.11 is:

$$Y_a = \Psi_a K_a^{\alpha_1} L_a^{\alpha_2} O_a^{\alpha_3} \tag{8.1}$$

Table 8.1 Input-output table for Tokyo (2008 billion yen)

Factor Accounts	Sector		
	Agriculture	Manufacture	Service
Capital income	6310	848946	18106355
Labor income	15576	5321070	49052716
Mixed (other) income	20777	874329	19086345
Value-added	42663	7044345	86245416

Here Y_a represents agricultural value-added, L_a is the amount of labor employed in agricultural production, while K_a and O_a are indices of the capital stock and other factors, respectively, used in agricultural production.

Under constant returns to scale, the parameters α_1, α_2 and α_3 are a factor's share in the cost of producing the final good: for example, $\alpha_1 = rK_a / Y_a$. Hence, the parameter α_1 is calculated by dividing the payment to capital, 6310, by agricultural value-added, 42633. Similarly, we get the coefficients for α_2 and α_3 by the respective factor payment by value--added. Then, from Table 8.1, the share of labor costs in producing Tokyo region agriculture is 36.5 percent, the share of capital is 14.8 percent, and the share of mixed income is 48.3 percent. The *scaling parameter*, Ψ_a, is an index of total factor productivity, defined as $\Psi_a = Y / K_a^{\alpha_1} L_a^{\alpha_2} O_a^{\alpha_3}$.

Adjusting Factor Account Entries to Reflect Land and Water Rent

Having discussed the link between I-O data and production technologies, we now turn attention to the role of natural assets in such a framework. For the discussion at hand, the major problem with Equation (8.1) is that it omits the primary factors (natural assets) land and water. A more desirable specification is:

$$Y_a = \Psi_a K_a^{\alpha_1} L_a^{\alpha_2} H_a^{\alpha_3} Z_a^{\alpha_4} \tag{8.2}$$

where H_a and Z_a represent the level of water and land used in agricultural production respectively. Let π represent the unit shadow rent of land and τ represent the unit shadow rental value of water. Then, given the structure of agriculture as reflected in Equation (8.2), a more appropriate factor account breakdown is shown in Table 8.2.

The principal challenge, now, is to reapportion agricultural value-added over the new factor account mix. One approach to reapportioning GDP over the factor accounts in Table 8.2 is to estimate the parameters of the Cobb-Douglas function as represented in Equation (8.2).[3] In the empirical example that follows, we constructed a panel of data on water use from 1980 through 2006 from Japan Ministry of Agriculture data sources. This data was combined with panel data on cultivated land area, agricultural capital asset values and agricultural labor in man-hours and used to econometrically estimate the factor share parameters in Equation (8.2). Table 8.3 presents the summary statistics of the data.

To estimate Equation (8.3), we added regional dummy variables to Equation (8.2) to

Table 8.2 I-O table

Factor Accounts	Sector		
	Agriculture	Manufacture	Service
Labor income	wL_a	–	–
Capital income	rK_a	–	–
Water rent	τH_a	–	–
Land rent	πZ_a	–	–
Value-added	42663		

Table 8.3 Summary statistics of regression variables – by region (all values in 1000s)

Region		Real GDP (yen)	Capital Stock	Labor (man-hours)	Land (hectares)	Water (m^3)
Chugoku	Mean	498748359	4744558394	120886	3891	3136308
	St dev	50975545	1907292245	23724	820	429551
Hokkaido	Mean	1089496847	2259327481	150120	3349	3299005
	St dev	74413876	814582390	9227	704	455029
Hokuriku	Mean	619986542	5815517495	64433	2524	5215874
	St dev	53159214	2590964811	8211	490	465180
Kanto	Mean	2191137998	37683532501	145114	13189	7715504
	St dev	183160543	17380582383	15087	2060	667786
Kinki	Mean	563958471	11650957632	52263	3492	2917188
	St dev	64886850	5697336070	8538	469	325167
Kyushu	Mean	1708996737	9396824376	120788	10020	5168906
	St dev	160177522	4692235473	10775	1522	603380
Shikoku	Mean	506930618	3719117281	56374	3569	1507816
	St dev	58638174	1851387748	5179	646	166844
Tohoku	Mean	1752925545	11181298556	233162	9949	10790373
	St dev	120504985	3150428473	12850	1457	1028029
Tokai	Mean	991999346	14913015784	92746	5430	2763287
	St dev	102544609	6113912699	17990	813	306212

control for regional differences in productivity (and output mix) and added time to control for technical change effects. We also imposed constant returns to scale on the technology by dividing each region's agricultural value-added, capital stock level, man-hour level and water use by its cultivated area, Z_i The normalized, log-linear version of Equation (8.2) plus the time and dummy variables is presented in Equation (8.3):

$$\ln(y_t) = \Psi + \sum_{i=1}^{9} \beta_i D_i + \alpha_0 \,\text{time} + \alpha_1 \ln(k_t) + \alpha_2 \ln(l_t) + \alpha_3 \ln(h_t) + \varepsilon_t \qquad (8.3)$$

Here Ψ is a constant, α_0 is a technical change parameter, D_i is a regional dummy variable, y_t is output per unit of land, k_t is capital per unit of land, l_t is labor per unit of land, h_t is water per unit of land, and ε_t is the error term.

Equation (8.3) was estimated using ordinary least squares (OLS). The coefficient estimates for capital, labor, water and time were all significant at the 99 percent level and had the desired sign. The OLS estimator, however, yielded an R-square of 0.97, suggesting the presence of serial correlation or endogeneity. To correct for potential correlation problems, we estimated both a random-effect and a fixed-effect model, using the nine region groups as the panel (18 observations per region). The estimated coefficients for both models were very close to those obtained using OLS, while the R-square value dropped to 0.64. Also, the estimated coefficients were very similar regardless of the method used to correct the variance–covariance matrix. Table 8.4 presents results for the fixed-effect estimation.[4]

To reapportion the factor payments in Table 8.1 across the Table 8.2 factor categories, simply multiply each factor's share coefficient from Table 8.4 by Tokyo agriculture's

Table 8.4 Regression results for Cobb-Douglas function fixed effects (with nine regional clusters)

ln(y)	Coefficient	Std. Error	t-statistic
α_1	0.0519	0.0118	4.39
α_2	0.5848	0.1341	4.36
α_3	0.3094	0.0731	4.23
α_4	0.0539	–	–
Time	0.0235	0.0022	10.59
Ψ	−47.090	4.5770	−10.29
R^2 (overall) = 0.656, F(4,8) = 49.15			

Table 8.5 Natural asset adjusted I-O accounts for Tokyo agriculture

Factor Accounts	Original I-O Data		'Shadow' I-O Data	
	Level	Share	Level	Share
Capital income	6310	0.148	2214	0.052
Labor income	15576	0.365	24949	0.585
Other income	20777	0.487		
Water rent			13200	0.309
Land rent			2 300	0.054
Value-added	42663		42663	

value-added. For example, capital's payment to agriculture is 0.0519 * 42 663 = 2214. Call this new factor payment value the 'shadow factor payment' and the corresponding I-O table the 'shadow I-O table'.

Table 8.5 presents the original and shadow I-O data for Tokyo agriculture. The shadow results suggest we reapportion all 'Other income' and two-thirds of 'Capital income' to labor, land and water. After the suggested reapportionment, on average, labor costs account for about 58 percent of agricultural production value, while (physical) capital rental payments account for a little over 5 percent of production value. The natural asset, land, accounts for another 5 percent of production value, while water accounts for about 31 percent of production value. Hence, about 36 percent of agricultural value-added accrues to the natural assets (or ecosystem services provided by), land and water, with 86 percent of this being a shadow water rent.

Reconciling the original I-O data with the econometric results raises measurement-related questions. For instance, does the 'Capital income' account include land rent, and some water rent? If so, then using the econometric results directly could be an acceptable decision. If not, then a better choice would be to leave the capital account alone and reapportion 5 percentage points of 'Other income' rent to land, 31 percentage points to water and the remaining 13 percentage points to labor (e.g., as returns to entrepreneurial ability). National account measurement is a challenging task, and the econometric results in Table 8.4 along with the original data in Table 8.2

hint at some of the challenges the System of Environmental-Economic Accounting could face when trying to measure natural assets' (or ecosystem services') contribution to sector GDP.

8.3 SHADOW RENTAL VALUES AND SHADOW PRICES

Below, we provide a brief description of the modeled economy and relegate the model details to the Appendix. Japan is modeled as a small open economy divided into three regions: Tokyo, the rest of Kanto (ROK), and the rest of Japan (ROJ). The country is endowed with four productive factors: capital, labor, land and water, with water and land being regionally specific. Water and land combine to form an ecosystem whose provisioning services contribute to agricultural production. Water is also used in producing manufacturing, residential water and services in Tokyo and the rest of Kanto. The productive factors are used in various combinations across sectors and regions to produce four final goods: agriculture, manufacturing, services in each region, and residential water in Tokyo and the rest of Kanto. Firms produce using constant return to scale technologies, and households make consumption and savings decisions that maximize utility over time. All economic agents interact in a competitive world, that is, an economy in which there are many buyers and sellers of goods and no one agent can influence prices. Finally, the manufactured and agricultural goods are traded, while the service good and residential water are non-traded.

A complete presentation of the conceptual model is presented in the Appendix, and the numerical simulations that follow are based directly upon that model. Before presenting the empirical results, however, we digress to discuss two concepts: shadow rental values and shadow prices.

Shadow Rental Values

Economic theory suggests that producers will use a factor up to the point where the market price (or market rental rate) of the factor is equal to the marginal value product of that factor. If a market does not exist for a factor, for example, water in Japanese agriculture and manufacturing, the unit (shadow) rental value of water is estimated by calculating the marginal value product of water. Use Equation (8.2) and let p_a represent the unit price of agricultural output. Then the shadow rental value of water, denoted SV_H^a is defined as:

$$SV_H^a(p_a, K_a, L_a, H_a, Z_a) = p_a \frac{\partial Y_a}{\partial H_a} = p_a \alpha_3 \Psi_a K_a^{\alpha_1} L_a^{\alpha_2} H_a^{\alpha_3 - 1} Z_a^{\alpha_4} = \frac{\alpha_3 Y_a}{H_a}$$

while the shadow rental value of land, denoted SV_Z^a, is defined as:

$$SV_Z^a(p_a, K_a, L_a, H_a, Z_a) = p_a \frac{\partial Y_a}{\partial Z_a} = p_a \alpha_4 \Psi_a K_a^{\alpha_1} L_a^{\alpha_2} H_a^{\alpha_3} Z_a^{\alpha_4 - 1} = \frac{\alpha_4 Y_a}{Z_a}$$

Equations (A8.1)–(A8.3) in the Appendix show another way to derive the shadow rental value of water and land using value-added (restricted profit) functions whose arguments are own output price, factor prices and the level of natural assets used in production.

Shadow Prices

The shadow price of land and water is linked directly to their corresponding shadow rental rates through no-arbitrage conditions. To develop the no-arbitrage condition for land and water, let Z represent land and H represent water, and define agricultural value-added as:

$$\Pi(p_a, r, w, H, Z) \equiv \max_{K_a, L_a}\{p_a Y_a - rK_a - wL_a : Y_a = F(K_a, L_a, H, Z)\}$$

Assume an economy only has physical capital, labor, land and water, and denote the economy's endowment of capital and labor by K and L, respectively. Given the natural assets H and Z, the total value of physical and natural capital holdings is expressed as:

$$A(t) = K(t) + P_Z(t)Z + P_H(t)H \tag{8.4}$$

Accounting for the two natural assets, the flow budget constraint for the economy is given by:

$$\dot{K}(t) = r(t)K(t) + w(t)L(t) + \Pi_Z(t) + \Pi_H(t) - E(t) \tag{8.5}$$

in terms of changes in the capital stock, and:

$$\dot{A}(t) = r(t)A(t) + w(t)L(t) - E(t) = rK(t) + rP_Z(t)Z + rP_H(t)H + w(t)L(t) - E(t) \tag{8.6}$$

in terms of asset values. Here, $E(t)$ is the value of consumption expenditures, and $\Pi_Z = \frac{\partial}{\partial Z_a}\Pi(\cdot)$ and $\Pi_H = \frac{\partial}{\partial Z_H}\Pi(\cdot)$ are the shadow values of land and water. Next, observe from Equations (8.5) and (8.6):

$$rK + wL - E = \dot{K} - \Pi_Z - \Pi_H = \dot{A} - rP_Z Z - rP_H H$$

then take the total derivative of Equation (8.4) with respect to time, and substitute the result into the above equation:

$$\dot{K} - \Pi_Z - \Pi_H = \dot{K} + \dot{P}_Z Z + \dot{P}_H H - rP_Z Z - rP_H H$$

Rearrange terms and simplify to get the following arbitrage conditions:

$$r = \frac{\Pi_Z}{P_Z Z} + \frac{\dot{P}_Z}{P_Z} = \frac{\Pi_H}{P_H H} + \frac{\dot{P}_H}{P_H} \tag{8.7}$$

In the above expression, r represents the return to the household from investing a unit of income in physical capital. The same unit of income can also buy $1/(P_Z Z_a)$ units of land, generating, at time $t + dt$, a rental income of $\Pi_Z/P_Z Z$ plus the rate of change in the land price. If this condition did not hold, optimizing investors could exploit the arbitrage opportunity and move investments out of land and into capital. Given the no-arbitrage conditions hold across natural and physical assets, the time t shadow price of land per unit of labor is given by (see Roe et al., 2010):

$$p_{\varsigma}(t) = \int_{t}^{\infty} e^{-\int_{t}^{\tau} [r(v) - n - \frac{\dot{P}_k(v)}{P_k(v)}]dv} \frac{\Pi_Z(\tau)}{P_k(\tau)} d\tau \tag{8.8}$$

Here, $p_{\zeta} = p_Z/p_k$ is the unit shadow price of land relative to the unit price of capital. Equation (8.8) is a solution to the differential equation defined by (8.7), and is the discounted present value of all future shadow rents, where the discount factor depends on the rate of return to capital adjusted for depreciation, the rate of growth in the labor force, the rate of exogenous technical change, and the rate of change in the price of (composite) capital. More labor and more efficient labor lends to increased productivity of the natural asset, hence, the impact of n on the shadow price of a natural asset.

8.4 EMPIRICAL MODEL AND RESULTS

Model Parameters

With the *shadow* factor account data in Table 8.5, one can proceed to construct a dynamic, empirical model to predict shadow water rental rates over time, as well as the unit value of the stock of water over time. The technologies for all but the residential water sector are strictly Cobb-Douglas, while the residential water sector is Cobb-Douglas in capital and labor, but Leontief in water. Preferences are homothetic, with Cobb-Douglas structure. Table 8.6 presents the factor shares per sector for each region, and the consumption shares for the final goods agriculture, manufacturing and the three

Table 8.6 Factor shares by sector, across regions, and consumption shares for all final goods

			Agri-culture	Manu-facture	Service			Water Supply	
Factor Shares	Tokyo	Capital income	0.0519	0.2209	0.4624			0.6050	
		Labor income	0.5848	0.7391	0.5345			0.3950	
		Water shadow	0.3094	0.0400	0.0031			–	
		Land rent	0.0539	–	–			–	
	Rest of Kanto	Capital income	0.0519	0.3453		0.4189			0.624
		Labor income	0.5848	0.6147		0.5762			0.375
		Water shadow	0.3094	0.0400		0.0048			–
		Land rent	0.0539	–		–			–
	Rest of Japan	Capital income	0.0519	0.3692			0.3597		–
		Labor income	0.5848	0.6308			0.6403		–
		Water shadow	0.3094	–			–		–
		Land rent	0.0539	–			–		–
Consumption shares			0.0120	0.1634	0.0206	0.023	0.1199	0.0007	0.0006

Table 8.7 Unit shadow water rental values

Year	Agriculture			Manufacturing		Services	
	Tokyo	ROK	ROJ	Tokyo	ROK	Tokyo	ROK
2008	93.88	41.02	69.52	311.14	52.44	188.76	80.06
2018	97.76	42.71	72.39	147.87	58.94	274.27	116.57
2028	111.04	48.52	82.23	125.00	68.90	353.48	150.36
2038	130.12	56.85	96.36	129.81	81.69	436.21	185.61
2048	154.37	67.45	114.32	146.35	97.40	528.98	225.12
2058	184.10	80.44	136.34	170.78	116.41	636.78	271.01
2068	220.05	96.14	162.96	202.22	139.26	764.17	325.23
2078	263.25	115.02	194.96	240.95	166.67	915.80	389.77
2088	315.07	137.67	233.33	287.88	199.51	1096.90	466.85
2098	377.16	164.79	279.31	344.35	238.85	1313.47	559.03
2108	451.52	197.28	334.38	412.10	285.95	1572.64	669.33

service sectors. Preliminary econometric results suggest the elasticity of water in Tokyo and the rest of Kanto manufacturing is about 5 percent. We used this information and decreased the capital share of manufacturing for both these regions, yielding the values shown in Table 8.6.

As aggregated, the most capital-intensive sectors are the two water supply sectors, followed by the Tokyo and rest of Kanto service sectors. The remaining sectors are quite labor intensive. Growth accounting yields a Harrod neutral rate of technical change equal to $x = 0.018$, and we assume the rate of population and labor force growth equal to $n = 0$. Finally, the felicity function is given by $u(q) = \log \bar{u}$.

Empirical Results: Water Shadow Rental Values and Shadow Prices

Table 8.7 lists the unit water shadow rental values for all sectors except residential water. The values show water rents can vary significantly across regions and sectors, and also can exhibit non-monotonic behavior. Water rent for Tokyo manufacturing falls for about 20 years, and then begins increasing again. This is because Tokyo manufacturing is the most labor intensive of all sectors, and increasing wages combined with falling capital rental rates makes it difficult for the sector to compete for resources. Increased labor productivity eventually dominates the wage and capital rental rate effects, leading to increased water rents.

Figure 8.1 reveals that the more capital intensive is a sector, the larger is the rate of growth in its water shadow value over time.[5] Results in Roe et al. (2010) suggest that, in the process of economic growth, capital deepening occurs and tends to favor the more capital-intensive sectors of an economy as it evolves. In other words, capital deepening makes capital-intensive sectors more competitive in factor markets. They refer to this competitiveness effect as a 'Rybczynski-type effect', suggesting that capital deepening tends to have an effect similar to that expressed in the Rybczynski theorem.[6] With capital deepening, the rate of return on (or unit cost of) capital falls over time, while the wage rate increases over time. This Rybczynski-type effect results in the more capital-intensive

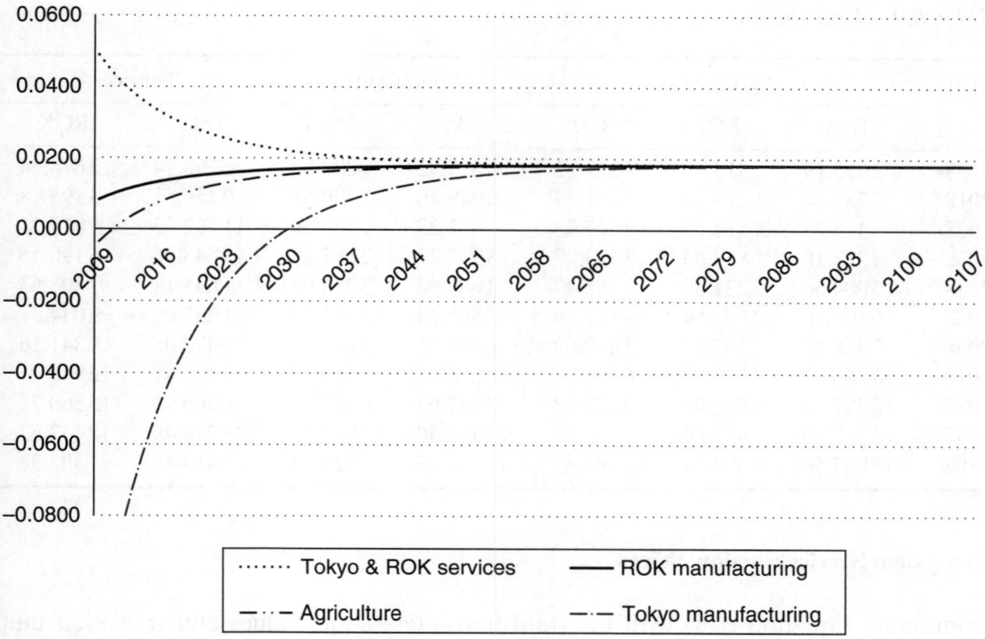

Figure 8.1 *Sector shadow rental values*

Table 8.8 *The share of natural asset value embedded in GDP*

Year	2008	2018	2028	2038	2048	2058	2068	2078
GDP share (%)	0.41	0.33	0.31	0.31	0.30	0.30	0.30	0.30

sectors garnering a larger share of resources over time, and hence, producing a larger share of GDP over time.

The initial question raised in the introduction is how much, or what share, of GDP is attributable to water? With linearly homothetic technologies and sector factor shares, the answer is readily addressed. The Cobb-Douglas technologies tell us the share of GDP coming from land and water in agricultural production is equal to land and water's agricultural cost share values multiplied by agriculture's share of GDP. Table 8.8 presents the trajectory of natural asset shares of GDP over a 60-year period. Natural assets' share of GDP in Japan is very small, and falls slightly over time. These values would increase if we disentangled water's contribution to the rest of Japan's manufacturing and service sectors. The increase, however, is not likely to surpass 1.5 to 2 percent of GDP. Still, the value of this small share is larger than the GDP of over 120 developing countries.

With homothetic technologies, land shadow values will behave analogously to that of water. As such, we do not discuss the land rent dynamics or present their numerical results.

Table 8.9 Unit shadow water price values

Year	Agriculture			Manufacturing		Services	
	Tokyo	ROK	ROJ	Tokyo	ROK	Tokyo	ROK
2008	2095.19	915.46	1551.62	3017.92	1280.23	6298.74	2678.54
2018	2783.68	1216.28	2061.49	2949.86	1739.55	9156.58	3895.78
2028	3466.73	1514.73	2567.33	3352.28	2183.59	11792.33	5018.21
2038	4218.10	1843.03	3123.77	3942.29	2665.25	14545.76	6190.45
2048	5084.60	2221.63	3765.47	4686.96	3216.98	17635.18	7505.53
2058	6105.16	2667.54	4521.26	5595.13	3864.84	21227.06	9034.37
2068	7318.33	3197.62	5419.69	6690.36	4633.95	25472.08	10841.16
2078	8766.27	3830.27	6491.97	8005.53	5551.35	30525.67	12992.05
2088	10497.39	4586.65	7773.98	9582.03	6647.91	36560.95	15560.75
2098	12568.63	5491.64	9307.87	11470.40	7959.76	43778.49	18632.62
2108	15047.56	6574.77	11143.67	13731.55	9529.75	52414.89	22308.38

Ecosystem Service Shadow Prices

Combining Equation (8.8) with the transition interest rate values and transition unit water shadow rental rates allows for the calculation of unit water shadow prices. Table 8.9 lists the unit water shadow prices across sectors and over time.

Except for a ten-year period for Tokyo manufacturing, the water shadow prices for each sector increases over time. Although not reported here, converting each of the prices to present values, however, yields a price trajectory for Tokyo manufacturing that falls over time. Also, not reported are the shadow price growth rates across sectors: as with the water shadow rental values, the more capital intensive is a sector, the higher is the rate of growth in its unit water shadow price.

At this point we investigate the potential impact of deviating from theory when deriving empirical results. We consider two cases. First, we calculate shadow prices using an approach used in most empirical exercises:

$$Price_{asset}(t) = \frac{Rent_{asset}(t)}{r(t)} \tag{8.9}$$

Here, $Price_{asset}(t)$ is the time-t shadow price of an asset, $Rent_{asset}(t)$ is the time-t rental rate, and $r(t) = r^k(t) - \delta$ is the rate of return on capital adjusted for depreciation. Second, we calculate shadow prices while ignoring technical change. Increases in the effective labor force – either by an increase in the number of units of labor or an increase in the productivity of labor via technical change – leads to an increase in the productivity of the asset, and hence its shadow rental rate and its (shadow) price. Also, an increase in the unit value of (composite) capital puts upward pressure on the natural asset (shadow) price, via arbitrage forces. Table 8.10 presents the water shadow price for manufacturing and services calculated using the three different calculations.

The results along the row 'Mod 1' are values calculated using the 'correct' pricing model, as represented by Equation (8.8). The results along the row 'Mod 2' are values

Table 8.10 Unit shadow water price values

Year		Manufacturing				Services			
		Tokyo	% Difference	ROK	% Difference	Tokyo	% Difference	ROK	% Difference
2008	Mod 1	3018		1280		6299		2679	
	Mod 2	1570	−48%	1035	−19%	5664	−10%	2399	−10%
	Mod 3	4057	34%	684	−47%	2461	−61%	1044	−61%
	Mod 1	3942		2665		14546		6190	
2038	Mod 2	700	−82%	1397	−48%	9421	−35%	3995	−35%
	Mod 3	1534	−61%	965	−64%	5154	−65%	2193	−65%
	Mod 1	6690		4634		25472		10841	
2068	Mod 2	667	−90%	1441	−69%	9941	−61%	4216	−61%
	Mod 3	1439	−78%	991	−79%	5439	−79%	2315	−79%

calculated when ignoring technical change. The results along the row 'Mod 3' are values calculated using the 'standard' approximation characterized by Equation (8.9). Both Mod 2 and Mod 3 yield results that deviate from the results one obtains when implementing the theoretical model correctly. In Mod 2, ignoring the impact of improving labor productivity introduces a serious bias in the shadow price predictions. This effect will likely hold for most empirical exercises. The empirical model here, as with the Solow model, predicts that the rate of return to capital will fall as capital deepening occurs. The standard model underestimates shadow prices, as it does not accommodate a fall in interest rates over time. Dividing the current rental rate by the rate of return on capital 50 or 60 years later yields unit water shadow prices closer to those estimated using the correct pricing model. Hence, ignoring technical change and using the standard approximation leads to, what could be, seriously biased estimates of natural asset and ecosystem unit shadow prices.

Natural Assets and Wealth

The last question of interest is: what is the value of the stock of water and land, and how large is this value relative to the value of the stock of physical capital? Since water is ignored in the rest of Japan, we focus attention on the natural and physical asset values in the Kanto region. The price of capital in the initial period, 2008, is numeraire, and the stock of physical capital in Kanto is estimated at $K_{2008} = 4\ 620\ 080$ billion yen. The value of Kanto agricultural land in 2008 is 1272.98 billion yen. Finally, the average unit shadow price of Kanto water in 2008 is equal to 4378 yen per cubic meter, while a lower bound on the estimate of Kanto water stocks is 251.9 billion cubic meters.[7] Then the total value of the ecosystem services produced by land and water in the Kanto Plain is equal to 1 103 256 billion yen, suggesting the stock value of Kanto land and water is about one-fourth that of its physical capital. This ratio is similar to the ratio of natural to man-made capital for the world as measured by The World Bank (see Table 2.3 in World Bank, 2006).

8.5 CONCLUSION

This chapter lays out a methodology for measuring a natural asset's or ecosystem service's contribution to GDP – their shadow rental values – and then projecting those values over time. The approach uses production theory, and its empirical implementation hinges on the ability to identify the parameters of a production technology for the sector using the ecosystem service (see Barbier, 2007, 2009). Using appropriately estimated production technologies allows for spatial and sector differences in water productivity across an economy, which in turn can be used in a dynamic general equilibrium modeling framework having equilibrium factor prices and other endogenous variables that evolve over time.

These sequences of shadow rental values are then used to calculate the unit shadow stock value of the ecosystem service. The approach is perhaps more general than a reader might guess, as it can accommodate non-competitive factors or output market structures, and admits a variety of functional forms for consumers and producers. The approach can also introduce ecosystems dynamics: for example, ecosystem services deriving from a large groundwater aquifer plus surface water sources like that observed in the Punjab, India.

The discussion in this chapter focused on backing out the value of natural assets or ecosystem services embedded in GDP. Using multi-output technologies, one could follow a similar approach and measure the economy-wide or regional value of externalities, for example, the shadow flow value of carbon storage in the event markets for such services do not exist. These values will not enter GDP, but the discounted flow values can be used to calculate the stock value of carbon, and hence, augment the value of natural assets in wealth accounting exercises.

This chapter describes one direction that economists can take in measuring economy-wide flow and stock values of natural assets and ecosystems. Certainly other approaches to measuring these values exist. The integrated global models, MIMES and GUMBO[8] are examples of models that integrate ecosystems and economics and generate values for ecosystem services. Noted shortcomings of both these models are that: (1) neither MIMES nor GUMBO include market allocation mechanisms for allocating capital and labor across competing sectors within a region; and (2) the rate at which capital accumulates – and indirectly, the rate at which the economy grows – is exogenous: the rate at which capital accumulates within a sector is chosen by the modeler (Boumans et al., 2002). Hence, the models have undesirable allocation features, and although touted as dynamic models, the economic side of these models is not dynamic. This appears to be the case with most 'dynamic' models that integrate ecosystems or natural resources, and economics (see, for example, Chen et al., 2002). One exception is the DICE model (Nordhaus, 1993). While dynamic, the DICE model does not accommodate natural capital, and hence, is not designed to yield values of natural capital or ecosystem services.

NOTES

1. See Barbier (2007) for a discussion on using production functions to value the environment.
2. Although the flow of services provided by a natural asset can enter an economy as an intermediate good and/or a primary factor, for the purpose of this chapter we concentrate on its role as a primary factor.

3. A more preferred approach would have been to estimate a production function for each region.
4. We also ran the model using non-normalized data – that is, non-normalized data with land as an additional explanatory variable – and obtained almost identical results.
5. The value for the Tokyo and ROK service sectors are not identical, but close enough to warrant plotting only one of the price series.
6. The Rybczynski theorem uses a two-output, two-input model to show when the level of a factor increases, the output of the sector using that factor most intensively increases, while the output of the competing sector decreases.
7. Here, water stock levels are measured as the quantity of water used in Kanto agriculture, manufacturing, services and residential water. The actual stock of Kanto water is stored in the relatively vast system of dams and aqueducts.
8. See Boumans et al. (2002), for a discussion of the Multiscale Integrated Earth Systems Model (MIMES) and the Global Unified Metamodel of the Biosphere (GUMBO).

REFERENCES

Barbier, E.B. (2007), 'Valuing ecosystem services as productive inputs', *Economic Policy*, **22**(49), 177–229.
Barbier, E.B. (2009), 'Ecosystems as natural assets, foundations and trends in microeconomics', **4**(8), 611–81.
Boumans, R., R. Costanza, J. Farley, M.A. Wilson, R. Portela and J. Rotmans et al. (2002), 'Modeling the dynamics of the integrated earth system and the value of global ecosystem services using the GUMBO model', *Ecological Economics*, **41**(3), 529–60.
Chen, X., D. Zhang and E. Zhang (2002), 'The south to north water diversions in China: review and comments', *Journal of Environmental Planning and Management*, **45**(6), 927–32.
Dasgupta, P. (2009), 'The welfare economic theory of green national accounts', *Environment and Resource Economics*, **42**(1), 3–38.
Dasgupta, P. and K.-G. Mäler (2000), 'Net national product, wealth, and social well-being', *Environment and Development Economics,* **5**(1), 69–93.
Heal, G. and B. Kriström (2005), 'National income and the environment', in K.-G. Mäler and J.R. Vincent (eds), *Handbook of Environmental Economics, Vol. 3*, Amsterdam: Elsevier BV.
Nordhaus, W. (1993), 'Rolling the "DICE": an optimal transition path for controlling greenhouse gases', *Resource and Energy Economics*, **15**(1), 27–50.
Roe, T.L., R.B.W. Smith and S. Saraçoglu (2010). *Multisector Economic Growth Models: Theory and Applications*, New York: Springer.
Stiglitz, J., A. Sen and F. Fitoussi (2008), *Report of the Commission on the Measurement of Economic and Social Progress Submitted to President Sarkozy*, Paris.
Weitzman, M.L. (1976), 'On the welfare significance of national product in a dynamic economy', *Quarterly Journal of Economics*, **90**(1), 156–62.
World Bank (2006), *Where is the Wealth of Nations: Measuring Capital for the 21st Century*, Washington, DC: World Bank.

APPENDIX

The Economic Environment

The economic model takes as its point of departure, the three-sector, small, open economy model in Chapter 4 of Roe et al. (2010). We represent Japan as a small, open economy and divide it into three regions: Tokyo, the rest of Kanto and the rest of Japan – indexed by $i = 1$, 2 and 3 respectively. Each region has an agricultural, manufacturing and service sector, indexed by $j = a$, m and s respectively. In addition, Tokyo and the rest of Kanto have a municipal water authority that provides water to their households and service sector, with sector index $j = h$. The agricultural output from each region is a perfect substitute in consumption, as is the manufacturing output of each region. Service sector outputs are not substitutes for one another in consumption: municipal water also does not serve as substitutes.

Denote the time-t production of residential water, the agricultural, manufacturing and service goods respectively by $Y_{\omega i}(t)$, $Y_{ai}(t)$, $Y_{mi}(t)$and $Y_{si}(t)$. In what follows, the time-t notation is suppressed: for example, we typically represent the time-t production of Kanto-1 agriculture by Y_{a1}, instead of $Y_{a1}(t)$. Agricultural output is a consumption good consumed within the country or traded internationally at given world prices. For each region, residential water is a pure consumption good only consumed within the region with endogenous (regional) final good prices. The service sector output of each region is either consumed or saved, but not traded. Hence, the service good price and the water shadow value(s) are endogenous to each region. Manufacturing output from each region is either consumed or saved, with an exogenously determined (world) final good price. Service sector and manufacturing output are combined to produce a composite capital: an index of aggregate capital. In what follows, composite capital, denoted $Y_k(t)$, is a least cost combination of the three regional manufactured goods, Y_{m1}, Y_{m2} and Y_{m3}, and the three regional service sector goods, Y_{s1}, Y_{s2} and Y_{s3}. The accumulation of composite capital, net of depreciation, yields capital stock, the services of which are employed in producing the agricultural, manufacturing and service goods. The value of water, composite capital and land are assets held by households.

Let H_{ji} denote the water endowment of sector j in region i, and let Z_i denote region i's agricultural land endowment. Assume both the stock of land and available water remains constant over time. Each region's land endowment is a factor specific to agriculture in that region. H_{ji} is also a region/sector-specific factor, while labor and composite capital are mobile across regions and sectors. Subsequent models allow water to move across sectors within a region or across regions. Let L denote the stock of labor, also assumed constant over time. For each region, the services of labor, composite capital and the region's land and water endowment are employed to produce the agricultural, manufacturing, residential water and service goods.

At each instant in time, household income derives from: (1) providing labor services L in exchange for wages $w(t)$, (2) earning interest income at rate $r^k(t)$ on capital assets $K(t)$, (3) receiving rent $\tau_i(t)$ from land Z_i, and (4) capturing rent from the ith region's water resources H_{ji}, with unit water rent denoted $v_{ji}(t)$. The representative agent uses income to invest in composite capital and purchase final consumption goods. Denote the level of total household consumption of the agricultural and manufactured good by the scalar values $Q_a(t)$ and $Q_m(t)$ respectively, the level of service sector output by $Q_s(t)$, and the level of residential water consumption (in region i) by $Q_{hi}(t)$. The initial capital stock, denoted K_0, is given, and the initial endowment of labor, L_0, is normalized to unity, as are the regional land endowments and traded good prices: that is, $z_i = p_a = p_m = 1$. We assume labor force growth is negligible, and hence aggregate labor supply is given by $L(t) = L = L_0$. The non-traded good price for each region i is endogenous and denoted p_{si}, $i = 1, 2, 3$.

Production

Let $K_{ji}(t)$ denote the time-t level of capital stock employed in producing good ji, and firms in each sector employ a constant returns to scale (CRS) technology. The aggregate technology for services in the rest of Japan is represented by:

$$Y_{s3}(t) = F^{j3}(K_{j3}, A(t)L_{j3})$$

The function $A(t)$ represents the exogenous level of growth in labor productivity. The production functions $F^{j3}(\cdot)$ – and all production technologies below – are twice continuously differentiable, non-decreasing and strictly concave in all respective arguments, and satisfy the standard Inada conditions. The cost function corresponding to $F^{s3}(\cdot)$ is:

$$C^{s3}(r^k, w)\, Y_{s3} \equiv \min_{K_{ss}, L_{ss}} \{r^k K_{s3} + wL_{s3}\text{: } Y_{si} = F^{si}(K_{ji}, AL_{ji})\}$$

The technology for municipal water provision in Kanto-1 and Kanto-2 is represented by:

$$H_{hi} \leq \min_{L_{hi}, K_{hi}} \left\{ F^{hi}(K_{hi}, A(t)L_{hi}) \frac{H_{hi}}{\sigma_{hi}} \right\},\ i = 1, 2$$

where the parameter σ_{hi} is the input-output coefficient that determines the amount of 'river' water required to produce one unit of residential water: here, we assume $\sigma_{hi} = 1$. The cost functions for municipal water provision are given by:

$$C^{hi}(r^k, w) H_{hi} \equiv \min_{K_{hi}, L_{hi}} \{r^k K_{hi} + wL_{hi}\text{:} H_{ai} = F^{hi}(K_{hi}, AL_{hi})\}$$

homogeneous in input prices, and satisfy Shephard's lemma.

The aggregate technologies for services in Kanto-1 and Kanto-2, and manufacturing in each region are represented by:

$$Y_{ji}(t) = F^{ji}(K_{ji}, A(t)L_{ji}, \Lambda(t)H_{ji})\text{:}\, j = m \text{ and } i = 1, 2, 3;\ j = s \text{ and } i = 1, 2$$

Here, the function $\Lambda(t)$ represents the exogenous level of growth in water productivity. Thus, in addition to labor-augmenting technological change, technological change in the employment of water in each region also occurs. The dual analog of these manufacturing and service technologies is the value-added function defined as:

$$\prod^{mi}(p_m, r^k, w, H_{mi}) \equiv \max_{K_{mi}, L_{mi}}\{p_m Y_{mi} - r^k K_{mi} - wL_{mi}: Y_{mi} = F^{mi}(K_{mi}, AL_{mi}, \Lambda H_{mi}), i = 1,2,3\} \quad \text{(A8.1)}$$

and

$$\prod^{si}(p_{si}, r^k, w, H_{si}) \equiv \max_{K_{si}, L_{si}}\{p_{si} Y_{si} - r^k K_{si} - wL_{si}: Y_{si} = F^{si}(K_{si}, AL_{si}, \Lambda H_{si}), i = 1, 2\} \quad \text{(A8.2)}$$

The aggregate technologies for agricultural production in each region is represented by:

$$Y_{ai}(t) = F^{ai}(K_{ai}, A(t)L_{ai}, \Lambda(t)H_{ai}, B(t)Z_i), i = 1, 2, 3$$

where the function $B(t)$ represents the exogenous level of growth in land productivity. For each region i, the agricultural value-added function is defined as:

$$\prod^{ai}(p_a, r^k, w, H_{ai}, Z_i) \equiv \max_{K_{ai}, L_{ai}}\{p_a Y_{ai} - r^k K_{ai} - wL_{ai}: Y_{ai} = F^{ai}(K_{ai}, AL_{ai}, \Lambda H_{ai}, BZ_i)\} \quad \text{(A8.3)}$$

Finally, given the properties of $F^{ji}(\cdot)$, each value-added function is concave in the wage rate, the rate of return to capital, and own price, and satisfies Hotelling's lemma. Furthermore, constant returns to scale in the inputs yields value-added functions that are separable in prices and the fixed factors. For example, the value-added function $\prod^{h1}(p_{h1}, r^k, w, H_{h1})$ is concave in prices, and can be written as $\prod^{h1}(p_{h1}, r^k, w)H_{h1}$ where $\prod^{h1}(p_{h1}, r^k, w)$ is the unit shadow value of Kanto-1 municipal water.

We ignore the water endowment for manufacturing and services in the rest of Japan and set the respective values equal to zero. We also implicitly aggregate the municipal water sectors in the rest of Japan with ROJ services.

Composite Capital

In Japan, over 98 percent of savings comes from the manufacturing and service sectors. To accommodate this feature of the economy, we assume the capital stock is created by combining the saved output of the saving sectors, and call the result composite capital, denoted $Y_k(t)$. Composite capital production is governed by the CRS Cobb-Douglas technology:

$$Y_k = Y_{m1}^{\varsigma_{m1}} Y_{m2}^{\varsigma_{m2}} Y_{m3}^{\varsigma_{m3}} Y_{s1}^{\varsigma_{s1}} Y_{s2}^{\varsigma_{s2}} Y_{s3}^{\varsigma_{s3}}, \quad 0 < \varsigma_{ji} < 1$$

where $Y_{ji}(t)$ is the time-t level of manufacturing or service sector output used in producing composite capital. The composite capital good's corresponding cost function is given by:

$$c^k(p_{m1}, p_{m2}, p_{m3}, p_{s1}, p_{s2}, p_{s3})\, Y_k = \prod_{j=m,s}\prod_{i=1}^{3} \varsigma_{ji}^{-\varsigma_{ji}} p_{ji}^{\varsigma_{ji}} Y_k = p_k Y_k,$$

where $c^k(\cdot)$ is with the unit cost of composite capital and is equal to p_k in equilibrium. Since the service sector prices p_{si} are endogenous and evolve over time, it follows that in equilibrium, p_k evolves over time also. The composite capital factor demand function for manufacturing output Y_m is obtained using Shephard's lemma. Note, that if Y_k units of composite capital are produced, the aggregate stock of capital increases by Y_k. Hence, in equilibrium, the instantaneous change in the aggregate stock of capital, denoted K, is given by $\dot{K} = Y_k$ adjusted for losses due to depreciation.

The savings and consumption behavior of households: let $q_{ji}(t) = Q_{ji}(t)/L$, and define the time t per capita consumption vector as:

$$q(t) = (q_a, q_m, q_{h1}, q_{h2}, q_{s1}, q_{s2}, q_{s3})$$

The present value of intertemporal utility is a time-separable weighted sum of all future utility flows:

$$U = \int_0^{\infty} u(q(t))e^{-pt} \quad \text{(A8.4)}$$

where $\rho > 0$ is the discount rate of future consumption. We assume the felicity function $u(\cdot)$ is homothetic, twice continuously differentiable, increasing and strictly concave in each argument.

Given prices $p(t) = (p_a, p_m, p_{h1}, p_{h2}, p_{s1}, p_{s2}, p_{s3})$, the minimum expenditure capable of yielding welfare level $\bar{u}(t)$ per household member is given by:

$$\varepsilon(p, \bar{u}) = x(p)\bar{u} \equiv \min_q\{p \cdot q : \bar{u} \leq u(q)\} \quad \text{(A8.5)}$$

The properties of $u(\cdot)$ imply the expenditure function is increasing and concave in p, increasing in u, and satisfies Shephard's lemma.

A flow budget constraint expresses time-t savings as the difference between income and expenditures. Let τ_i denote rent per effective unit of land BZ_i in region i, and let v_{ji} denote the rents (shadow value) per effective unit of water ΛH_{ji} in sector j of region i. Income is derived from labor income, wL, returns to the capital asset, $r^k K$, returns to land assets in each region, $B\sum_{i=1}^{3}\tau_i Z_i$, and returns to water as measured by the shadow value of water in the two regions, $\Lambda\sum_{i=1}^{2}\sum_{j=a,h,m,s}v_{ji}H_{ji}$. Thus, as modeled, the rents to the government's allocation of water accrue to households. Then the representative household's flow budget constraint in per worker terms is expressed as:

$$\dot{k}(t) = \frac{1}{p_k}\left[w + r^k k + \sum_{i=1}^{3}(\tau_i\tilde{B}Z_i + v_{ai}\tilde{\Lambda}H_{ai}) + \Lambda\sum_{i=1}^{2}\sum_{j=a,h,m,s}v_{ji}H_{ji} - \varepsilon\right] - \delta k \quad \text{(A8.6)}$$

Here $\dot{k}(t) = K(t)/L$ is household saving, while $\tilde{B}(t) = B(t)/L$ and $\tilde{\Lambda}(t) = \Lambda(t)/L$. The representative household chooses the sequence of consumption bundles $\{q(t)\}_{t\in[0,\infty)}$ to maximize intertemporal utility (A8.4) subject to the flow budget constraint (A8.6). A solution to the present value Hamiltonian derived from Equations (A8.4) and (A8.6) is the Euler equation:

$$\frac{\dot{\varepsilon}}{\varepsilon} = \frac{r^k}{p_k} - \delta - \rho + \frac{\dot{p}^k}{p_k} \quad \text{(A8.7)}$$

Given, $L = 1$, the initial capital stock is given by:

$$k(0) = K_0 \tag{A8.8}$$

and the transversality condition satisfies:

$$\lim_{t\to\infty} \lambda(t)k(t) = 0 \tag{A8.9}$$

Here, $\lambda(t)$ is the costate variable for the equation of motion, and is the present value shadow price of income. Together, Equations (A8.6)–(A8.9) characterize the representative household's optimization problem.

The empirical model includes exogenous technical change. Harrod neutral, labor-augmenting technical change is introduced into the model by redefining labor in terms of effective labor, $\eta(t)L(t)$, where $\eta(t) = e^{xt}$, and x is the Harrod neutral rate of technical change. Accompanying the introduction of exogenous technical change, are changes in the Euler equation and flow budget constraint: both expressions need to be specified in units of effective labor. The first step in doing this is to specify expenditure $\varepsilon(\cdot)$ in per-effective-labor units: that is, $\hat{\varepsilon}(\cdot) = \varepsilon(\cdot)e^{xt}$. Then:

$$\frac{\dot{\hat{\varepsilon}}}{\hat{\varepsilon}} = \frac{\dot{\varepsilon}}{\varepsilon} - x$$

and the Euler equation (A8.7) becomes:

$$\frac{\dot{\hat{\varepsilon}}}{\hat{\varepsilon}} = \frac{r^k}{p_k} - \delta - \rho - x + \frac{\dot{p}^k}{p_k} \tag{A8.10}$$

and the flow budget constraint becomes:

$$\dot{\hat{k}}(t) = \frac{1}{p_k}\left[\hat{w} + r^k\hat{k} + \sum_{i=1}^{3}(\tau_i \hat{B}Z_i + v_{ai}\hat{\Lambda}H_{ai}) + \Lambda\sum_{i=1}^{2}\sum_{j=a,h,m,s} v_{ji}\hat{\Lambda}H_{ji} - \varepsilon\right]$$

$$- \delta\hat{k}(\delta + x) \tag{A8.11}$$

where $\hat{w} = we^{xt}$, $\hat{k} = ke^{xt}$, $\hat{B} = \tilde{B}(t)e^{xt}$ and $\hat{\Lambda} = \tilde{\Lambda}(t)e^{xt}$.

Equilibrium

For all intents and purposes, equilibrium is a prediction of how the economy will perform, given the economic environment, and its primitives and factor endowments. We define equilibrium as follows: given water assignment:

$$\{H_{a1}, H_{a2}, H_{a3}, H_{m1}, H_{m2}, H_{s1}, H_{s2}, H_{h1}, H_{h2}\},$$

a competitive equilibrium is an endogenous sequence of capital stock and expenditure levels $\{\hat{k}(t), \hat{\varepsilon}(t)\}_{t\in[0,\infty)}$ and a nine-tuple sequence of positive values:

$$\{\hat{w}\}(t), r^k(t), p_{s1}(t), p_{s2}(t), p_{s3}(t), p_{h1}(t), p_{h2}(t), \hat{y}_{m3}(t), \hat{y}_{s3}(t)\}_{t\in[0,\infty)}$$

such that the representative household's utility is maximized and at each t, the following intra-temporal conditions are satisfied:

1. *Zero profit in ROJ manufacturing and services*:

$$C^{m3}(r^k, \hat{w}) = p_m \tag{A8.12}$$

$$C^{s3}(r^k, \hat{w}) = p_{s3}$$

and zero profit in Tokyo and ROJ municipal water provision:

$$C^{hi}(r^k, \hat{w}) = p_{hi},\ i = 1, 2 \tag{A8.13}$$

2. *Labor market clearing*:

$$-\sum_{i=1}^{3} \frac{\partial \Pi^{ai}(p_a, r^k, \hat{w}, H_{ai}, Z_i)}{\partial \hat{w}} - \sum_{i=1}^{2} \frac{\partial \Pi^{si}(p_{si}, r^k, \hat{w}, H_{si})}{\partial \hat{w}} - \sum_{i=1}^{2} \frac{\partial \Pi^{ai}(p_m, r^k, \hat{w}, H_{mi})}{\partial \hat{w}}$$

$$+ \sum_{i=1}^{2} \frac{\partial C^{hi}(r^k, \hat{w}) H_{hi}}{\partial \hat{w}} + \sum_{j=m,s} \frac{\partial C^{j3}(r^k, \hat{w}) \hat{y}_{j3}}{\partial \hat{w}} = 1 \tag{A8.14}$$

3. *Capital market clearing*:

$$-\sum_{i=1}^{3} \frac{\partial \Pi^{ai}(p_a, r^k, \hat{w}, H_{ai}, Z_i)}{\partial r^k} - \sum_{i=1}^{2} \frac{\partial \Pi^{si}(p_{si}, r^k, \hat{w}, H_{si})}{\partial r^k} - \sum_{i=1}^{2} \frac{\partial \Pi^{ai}(p_m, r^k, \hat{w}, H_{mi})}{\partial r^k}$$

$$+ \sum_{i=1}^{2} \frac{\partial C^{hi}(r^k, \hat{w}) H_{hi}}{\partial r^k} + \sum_{j=m,s} \frac{\partial C^{j3}(r^k, \hat{w}) \hat{y}_{j3}}{\partial r^k} = \hat{k} \tag{A8.15}$$

where $\hat{y}_{ji} = \frac{Y_{ji}}{Le^{xt}}$, $j = m, s$, and $i = 1, 2, 3$.
The non-traded good market clears in each region:

$$\frac{\partial \varepsilon(\cdot)}{\partial p_{si}} + \hat{y}_{ski} = \frac{\partial \Pi^{si}(p_{si}, r^k, \hat{w}, H_{si})}{\partial p_{si}},\ i = 1, 2 \tag{A8.16}$$

$$\frac{\partial \varepsilon(\cdot)}{\partial p_{s3}} + \hat{y}_{sk3} = \hat{y}_{s3} \tag{A8.17}$$

where $\hat{y}_{ski} = \frac{Y_{si}}{Le^{xt}}$, $i = 1, 2, 3$.

If a solution to the system (A8.13)–(A8.17) exists, it will be an eleven-tuple sequence of endogenous variables, with each variable being a function of the exogenous variables:

$$\{pa, pm, Z_1, Z_2, Z_3, H_{a1}, H_{a2}, H_{a3}, H_{m1}, H_{m2}, H_{s1}, H_{s2}, H_{h1}, H_{h2}\}$$

and the remaining endogenous variables $\{\hat{k}, \hat{\varepsilon}\}$. In practice, however, solving the system is facilitated by representing the endogenous variables $\{\hat{w}(t), r^k(t), p_{h1}(t), p_{h2}(t), \hat{y}_{m3}(t), \hat{y}_{s3}(t), \hat{\varepsilon}\}$ as a function of p_{s1}, p_{s2}, p_{s3} and $\hat{k}$. Hence, the solution can be identified with four differential equations. To conserve on space we leave these derivations to the motivated reader.

9. Exploring the use of a macro–micro-based approach to value biodiversity productivity impacts on the agricultural sector

Ruslana Rachel Palatnik and Paulo A.L.D. Nunes

9.1 INTRODUCTION

In the last century the agricultural sector has been profoundly altered by both natural and anthropogenic factors, including climate change, changing world economies, changes in agricultural policies, including the Common Agricultural Policy (CAP), and the subsidies currently paid to European farmers and land managers. All these factors together affected, and still affect, crop productivities and agricultural land use patterns. In this context, various studies have focused their attention on the analysis of agricultural land use under scenarios of global climate change, computing projections on land use changes and respective land productivities induced by fluctuations in temperature and precipitation. However, fluctuations in temperature and precipitation are not the only channel for climatic change to impact agro-ecosystems. Biodiversity is also subject to climatic fluctuations, which, in turn, might impact land productivity. The present chapter aims at analysing the economics of the climate-change-induced impacts on biodiversity in the agricultural sector, including its estimation in terms of changes in agricultural land productivity (and, above all, the geo-climatic impacts, as available from the state-of-the-art literature – see Brown and Rosenberg, 1999; Rounsevell et al., 2005; and Kan et al., 2009 as an illustration of a few representative examples).

The proposed economic valuation of consequences of climate-change-induced impacts on biodiversity is anchored in a three-step approach. The first step refers to the microeconomic assessment of the role of biodiversity in agro-ecosystems, including the estimation of the marginal effect of biodiversity in terms of crop productivity. From the operative/applied viewpoint, we used a microeconomic partial equilibrium approach to single out and measure these direct effects of biodiversity on land productivity. The second step refers to the assessment of the climate-change-induced impacts on crop productivity, including the effect on biodiversity. From the operative/applied viewpoint, we use land use projections as a result of climate change. The third step is the (monetary) assessment of identified crop productivity impacts employing the computable general equilibrium (CGE) model to evaluate the projected impact of climate change/biodiversity combined effects on the agricultural output and ultimately on the national GDPs of the countries under consideration. Our empirical analysis explores the use of four IPCC scenarios, that is, A1FI, A2, B1 and B2,[1] with relation to prognosis for the future scenarios of the agricultural sector in the next 40 years. Moreover, we confront the welfare results with a similar macroeconomic modelling exercise where these climate-change/biodiversity combined effects are not mapped (and are therefore able to explore a comparative static analysis among the two and identify who wins and who loses). To our knowledge, this study represents an original attempt to shed light on the economic impacts of climate-change-induced impacts of biodiversity on agro-ecosystems.

The chapter is organized as follows. Section 9.2 presents a conceptual model for evaluating climate-change-induced impacts on biodiversity in crop productivity, motivating the choice of biodiversity indicators employed in this analysis. Section 9.3 focuses on the use of a microeconomic approach to estimate climate change impacts on crop productivity, including the estimation of climate-change-induced impacts on biodiversity in crop productivity. Section 9.4 focuses on the use of a macroeconomic modelling tool, that is, the CGE model, to scale up micro-based estimation results and value, in monetary terms, the underlying impacts on regional GDP, including loss of human welfare due to climate-change-induced impact of biodiversity on agro-ecosystems. Section 9.5 concludes.

9.2 FRAMEWORK FOR EVALUATING CLIMATE CHANGE IMPACTS IN CROP PRODUCTIVITY

9.2.1 Biodiversity in Agro-ecosystems

Natural and modified ecosystems provide many services and goods that are essential for humankind (Matson et al., 1997). Building upon the Millennium Ecosystem Assessment (MA) we are able to identify the following ecosystem services: food, feed and fibre; soil erosion control; maintenance of biodiversity essential for successful crop and animal breeding; nutrient cycles; biological control of pests and diseases; erosion control and sediment retention; and water regulation (MA, 2005). These are, mainly, local benefits that agro-ecosystems can provide to local communities. In addition, there are also global benefits to human well-being from agro-ecosystems in terms of regulating services such as carbon sequestration (Allen and Vandever, 2003; Swift et al., 2004; MA, 2005). Furthermore, we can also distinguish between croplands and grasslands due to very different types of ecosystem goods and services that these two agro-ecosystems can provide. In terms of provisioning services, croplands provide three kinds of natural products, including food, non-food and bio-energy,[2] whereas grasslands are cultivated primarily for grazing. The distinction between croplands and grasslands is therefore essential to the quantitative projections of agro-ecosystems' productivity under climate change scenarios, and ultimately to the macroeconomic valuation exercise. In the present analysis, we shall be investigating croplands, assessing respective productivity in terms of provision of food services, including wheat. The literature shows that there are many possible indicators that can provide information on biodiversity. Multiple dimensions of biodiversity in cultivated systems make it difficult to categorize production systems into 'high' or 'low' regimes, especially at spatial and temporal scales. In agro-ecosystems a distinction has been made between 'planned' and 'associated' diversity (Walker and Steffen, 1997; Swift et al., 2004). 'Planned' diversity deliberately refers to plants and livestock, which are imported, stocked and managed by farmers. The term 'associated' refers to the nature of the biota (plant, animal and microbial), associated with the planned diversity and influenced by its composition and diversity. Farmers play a dominant role in the context of agricultural biodiversity by the selection of the present biodiversity stock, by the modification of the abiotic environment and by interventions aimed at the regulation of specific populations ('weeds', 'pests', 'diseases' and their vectors, alternate hosts and antagonists). More recent studies have pointed out that permanent grasslands represent 'hotspots' of biodiversity and are therefore very important in the maintenance of associated biodiversity values. Furthermore, the quality of soil is also higher in permanent grasslands, with respect to arable lands, as confirmed by many soil quality indicators, including organic carbon, aggregate stability – see Bignal and McCracken (1996); de Miguel (1999); Anger et al. (2002); Gardi et al. (2002); Nagy (2002); EEA (2007); Baglioni et al. (2009a, 2009b). Notwithstanding, we explore the use of biodiversity indicators following the protocol described in the setting of the European Network of Natura 2000 (EEA, 2010, 2011).[3] These indicators are constructed as a spatial/GIS explicit dataset and were combined with the Corine Land Cover (CLC) dataset, which is also spatial/GIS explicit. We found that all CLC identified as grasslands, when compared to the CLC croplands, are characterized by higher species biodiversity levels, when measured by the N2K indicators. Against this background, the ratio between cropland and grassland is here employed as a proxy indicator for the measurement of the level of biodiversity in agro-ecosystems. Furthermore, we propose to evaluate this link in the context of global climate change through a conceptual framework, which is discussed in the following section.

9.2.2 A Conceptual Model

To understand the impacts of climate change on agro-ecosystems, a graphical presentation is given in Figure 9.1, below. First, we shall be exploring the use of IPCC scenarios and investigating respective induced climate change (CC) impacts on crop productivities and, ultimately, in terms of agricultural provisioning services, including food.

In this context, we propose characterizing the interface between climate change and land productivity by two main components. We refer to geo-climatic-oriented drivers,

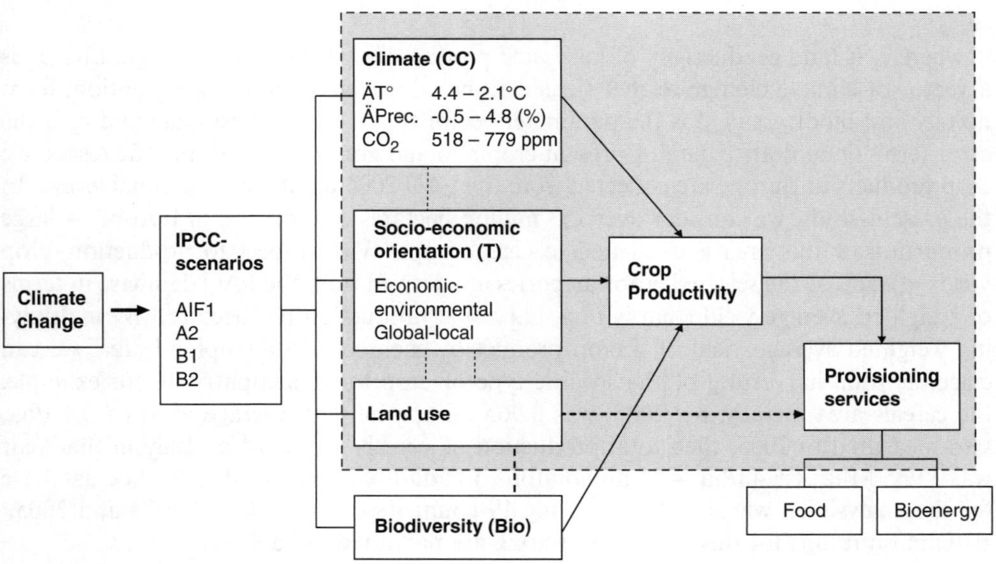

Figure 9.1 Framework for evaluating climate change impacts

which include indicators such as temperature and precipitation, and socio-economic-oriented drivers, which include indicators such as technology and land management practices. Furthermore, we shall be exploring the use of IPCC scenarios to investigate whether or not climate-change-induced impacts on biodiversity affect crop productivity. In other words, we focus on evaluation of welfare effects of induced climate change on biodiversity by assessing the respective impacts on agro-ecosystems (above the well-known geo-climatic and socio-economic drivers). Therefore, we approach such a valuation exercise by pursuing a microeconomic assessment of the role of biodiversity in agro-ecosystems, including the estimation of the marginal effect of biodiversity in terms of crop productivity. Finally, we shall estimate its impacts in monetary terms, reporting these in terms of changes on agricultural output and, ultimately, on the national GDPs of the countries under consideration.

9.3 ASSESSING CLIMATE CHANGE IMPACTS ON CROP BIODIVERSITY

9.3.1 A Microeconometric Approach to Modelling Crop Productivity

Our approach to assessing potential effects of climate change on crop productivity is based on observing how (1) temperature and precipitation (geo-climatic-oriented variables), (2) amount of fertilization and number of tractors (technological-oriented variables) and, (3) proportion of grassland to cropland (biodiversity-oriented variables) evolve in a set of selected countries over a given period of time. Formally, we propose to estimate the following equation:

$$y_{it} = x_{it}\,\beta' + \varepsilon_{it} \tag{9.1}$$

where y_{it} is land productivity of harvested product i in period t, measured in t/ha, x_{it} is a vector of climate-change-related variables, including temperature, precipitation, technology and biodiversity, β is the parameter vector that we aim to estimate and ε_{it} is the error term. Quantitative data of present cropland and grassland areas and the respective crop products in Europe are collected from the FAO 2005 database at national levels.[4] In the present study, we consider over 153 million hectares of cropland in Europe. A large proportion of this area is dedicated to cereal crops. With respect to production, crop yields of each of the selected crop categories are derived from the FAO database in terms of weighted average yield (that is t/ha, harvested production per hectare). By multiplying weighted average yield of a crop product by each country's cropland area, we can calculate total harvesting of the specific type of crop for this country. If, for example, the cereals area in Italy, for 2005, was 3.965 million ha with average yield of 5.4 t/ha, also measured in 2005, then total production of cereals produced by Italy in that year was 3.965 Mha × 5.4 t/ha = 21 million tons. In addition, we created an ad hoc database for the analysis on wheat yields, covering 19 countries over the period 1974 and 2005.[5] Estimation results for this panel data setting are presented in Table 9.1.[6,7]

We can see that all empirical model specifications are statistically significant (p- value < 0.01), as are other variables selected. First of all, land productivity for cereal

Table 9.1 Estimation results

	M1 β (*p*-value)	M2 β (*p*-value)	M3 β (*p*-value)	M4 β (*p*-value)	M5 β (*p*-value)
Bio	(−0.627	0.618	0.549	0.388	0.465
	(0.000)	(0.000)	(0.000)	(0.000)	(0.000)
Avg_t	(−0.512	–	0.469	−0.265	–
	(0.000)		(0.000)	(0.005)	
Avg_t$_2$	(−0.034	–	−0.033	0.107	–
	(0.000)		(0.000)	(0.005)	
Prec	–	0.003	0.004	−0.001	–
		(0.022)	(0.001)	(0.281)	
Prec$_2$	–	0.000	0.000	0.000	–
		(0.018)	(0.006)	(0.983)	
Fert. (t/ha)	(−99.872)	14.937	10.002	–	–
	(0.000)	(0.000)	(0.000)		
Tractor (n/ha)	(−4.188)	−3.810	1.002	–	–
	(0.005)	(0.081)	(0.668)		
With interaction effects					
*Tractor*Prec*	–	–	–	−0.028	–
				(0.000)	
*Fert*Temp*	–	–	–	3.083	–
				(0.000)	
*Fert*Prec*	–	–	–	0.033	–
				(0.000)	
*Clim*Tractor*	–	–	–	–	−0.003
					(0.000)
*Clim*Fert*	–	–	–	–	0.016
					(0.000)
p-value	(−0.000)	0.000	0.000	0.000	0.000
Adjusted R^2	(−0.430)	0.46	0.55	0.41	0.31

Note: With *Clim* is here defined with the *Temp*Prec* cross-effect.

crops is affected by physical climatic variables, including temperature (*t*) and precipitation (*prec*).

In particular, we can see that temperature has a positive impact on crop productivity, however, at a decreasing rate, since the t^2 estimate is negative. On the other hand, precipitation is also associated to a positive impact on yield, but of a much smaller magnitude when compared to temperature estimate. Moreover, marginal productivity of precipitation shows constant returns to scale – see model specification 1 and 2, M1 and M2. Similar estimation results for precipitation and temperature are obtained when bringing these two dimensions together in the same model specification – see M3. In addition, this model specification informs us that the *Bio* parameter is significant (p-value < 0.01) with a coefficient of 0.549. This implies that, when the actual ratio of grassland to cropland is 0.44 for Italy, the contribution of biodiversity to the wheat yield is $0.44 \times 0.549 = 0.24$ t/ha. In addition, the estimate of marginal productivity, associated to biodiversity, is rather robust across these model specifications. In fact, formal testing does not reject the null hypothesis of having the same marginal productivity of biodiversity across the 1, 2 and 3 model specifications. We also tested other empirical model specifications, including the introduction of joint interaction effects between temperature and precipitation and technological-related variables and, here again, the *Bio* parameter estimate is revealed to be rather robust – see M4 and M5 in Table 9.1. At this point, we are in a condition to calculate changes in land productivity due to changes in biodiversity, based on the projections in the variation of the ratio of grasslands to croplands for the IPCC scenarios in 2050. These will be presented and discussed in the next subsection.

9.3.2 Assessing Climate-change-induced Impacts on Crop Productivity, Including Climate-change-induced Impacts on Biodiversity in Europe

In this sub-section we project an effect of climatic change on agricultural yield directly, and also through its impact on biodiversity. In order to project the direct impact of climate change, we use the CC coefficient estimated in Equation (9.1) to introduce variation according to different IPCC scenarios, anchored for 2050. In addition, we create a database of agricultural areas assigned for cropland and grassland in each country for 2050 according to the four different IPCC scenarios. First, we use projections of land use changes as a result of climate change, described by the ATEAM model (Schröter et al., 2004, 2005). These land use changes projections are anchored in the use of the IPCC SRES (Special Report on Emissions Scenarios) circulation model. In addition, we complement the analysis with the use of the IMAGE 2.2 Integrated Assessment Model (IMAGE team, 2001) to calculate the required information on agro-ecosystem land use under consideration for the remaining countries. In this context, the future trends are projected individually for the period of 2050, based on global circulation models, where greenhouse gas concentration and climatic and socio-economic factors are the drivers of land use changes (Nakicenovic and Swart, 2000; Schröter et al., 2004, 2005; Ewert, 2007). Employing this information, we can project the composition and productivity of cereal agro-ecosystems, at country level. All estimated changes in cereal productivity, as projected by changes in climatic conditions (CC), are reported in the central columns of Table 9.2. For example, considering the present Italian cereal productivity (3.2 t/ha) and a CC coefficient value of 0.94 for the scenario A1FI, this country's cereal yield in 2050 will be 3.2 t/ha $\times$ 0.94 = 3.00 t/ha, as a result of the projected changing climatic conditions. These results are validated by the recent study by Ferrise et al. (2010) in which the authors explore the use of crop simulation model (SIRIUS) applied to durum wheat using data from open-field experimentation.

Next, we analyse the additional impact of climate change on agricultural yields through changes in biodiversity, employing the coefficient *Bio* estimated in Equation (9.1). In particular, we calculate changes in land productivity due to changes in biodiversity based on the estimated variation of the ratio *GR/CL* for the IPCC scenarios in 2050, as follows:

$$(GR/CL)_{scenario} \times \beta^{Bio} - (GR/CL)_{2005} \times \beta^{Bio} = Yield_{\_variation} \, [(GR/CL)_{scenario} - (GR/CL)_{2005}] \times \beta^{Bio} = Yield_{\_variation} \, \Delta[GR/CL] \times \beta^{Bio} = Yield_{\_variation} \quad (9.2)$$

where 'scenario' refers to the A1FI, A2, B1 and B2 scenarios as identified by the IPCC. To standardize the wheat yield variation due to biodiversity, we performed the following standardization:

Table 9.2 Estimated changes in cereal productivity as projected by changes in climatic conditions (CC) and as projected by climate-induced changes in biodiversity (Bio), 2050

Country	CC				Bio			
	A1FI	A2	B1	B2	A1FI	A2	B1	B2
Greece	0.91	0.93	0.98	0.96	1.14	0.99	1.20	1.00
Italy	0.94	0.95	0.98	0.97	1.00	0.97	0.99	0.99
Portugal	0.91	0.92	0.98	0.96	0.94	0.87	0.90	0.86
Spain	0.92	0.93	0.98	0.96	1.05	0.97	1.09	1.00
Albania	0.95	0.96	0.99	0.98	0.92	0.94	0.92	0.94
Bosnia and Herz.	1.05	1.04	1.01	1.03	0.91	0.93	0.91	0.93
Bulgaria	1.01	1.01	1.00	1.01	0.94	0.96	0.94	0.96
Serbia and Mont.	1.03	1.03	1.01	1.02	0.94	0.96	0.94	0.95
Turkey	1.03	1.03	1.01	1.02	0.91	0.94	0.92	0.94
TFR of Yugoslavia	1.02	1.02	1.01	1.01	0.88	0.91	0.88	0.90
Austria	1.07	1.06	1.02	1.04	0.94	0.92	0.98	0.93
Belgium	0.98	0.98	0.99	0.99	1.00	0.99	1.00	0.99
France	0.95	0.96	0.99	0.98	0.99	0.99	1.00	0.99
Germany	1.01	1.01	1.00	1.01	0.99	0.99	1.00	0.99
Ireland	0.95	0.96	0.99	0.98	0.98	0.99	1.01	0.99
Luxembourg	0.98	0.99	1.00	0.99	1.00	0.99	1.00	0.99
Netherlands	0.96	0.97	0.99	0.98	1.01	0.98	1.00	0.98
Switzerland	1.08	1.07	1.02	1.04	0.92	0.90	0.96	0.92
Croatia	0.99	0.99	1.00	0.99	0.91	0.94	0.92	0.93
Czech Republic	1.05	1.04	1.01	1.02	0.97	0.98	0.98	0.98
Hungary	1.01	1.01	1.00	1.01	0.98	0.98	0.98	0.98
Poland	1.03	1.03	1.01	1.02	0.97	0.97	0.97	0.97
Romania	1.04	1.03	1.01	1.02	0.92	0.94	0.92	0.94
Slovakia	1.03	1.03	1.01	1.02	0.92	0.93	0.92	0.93
Slovenia	1.08	1.07	1.02	1.04	0.93	0.95	0.93	0.94
Denmark	1.00	1.00	1.00	1.00	0.99	0.99	1.00	0.99
United Kingdom	0.97	0.97	0.99	0.98	0.98	0.98	1.01	0.98
Estonia	1.06	1.05	1.02	1.03	0.98	1.00	0.98	0.99
Latvia	1.04	1.04	1.01	1.02	0.92	0.94	0.92	0.94
Lithuania	1.02	1.01	1.00	1.01	0.99	1.00	0.99	1.00
Finland	1.12	1.10	1.03	1.06	1.03	1.02	1.05	1.02
Norway	1.20	1.17	1.05	1.10	1.00	1.00	1.02	1.00
Sweden	1.12	1.10	1.03	1.06	1.00	1.00	1.01	1.00

$$(Yield_{\_variation}/Yield_{\_2005}) \times 100 = Relative_{\_variation} \quad (9.3)$$

Considering again the example for Italy, we currently have wheat productivity estimated to be at 3.2 t/ha. In addition, according to the FAO dataset, in Italy, the ratio between grassland area and cropland area, that is, GR/CL, is calculated at 0.39. Taking into account the projection of land use changes provided by the ATEAM model this ratio is projected to be at 0.33 in the A2 scenario, by 2050. In this context, the climate-change-induced effect of biodiversity, measured in terms of change in wheat yield as $[(0.33 - 0.39) \times 0.55]/3.2 \times 100 = 0.97$ (or – 3.3 per cent) is expressed as projected final yield values (3.2 t/ha $\times$ 0.97 = 3.10 t/ha). All estimated changes in wheat productivity, as projected by climate-induced changes in biodiversity, are reported on the right-hand side of Table 9.2.

As we can see, climate-induced changes in biodiversity also cause a significant impact on cereal productivity, in addition to the direct impact due to changes in the climate conditions. It is therefore important to estimate the monetary terms of welfare impacts caused by these two factors, that is, climate conditions and biodiversity. Since agricultural products, including cereals, constitute one of the most important international commodity markets, we estimate the overall welfare impacts by exploring the use of a macroeconomic valuation. In this way, we are able to evaluate, in monetary terms, the impact of climate-change-induced variation in biodiversity on the productivity of agro-ecosystems that takes into account that this sector is not in autarchy, but rather in interaction with the other economic sectors, both at domestic and international levels. The next section explores the use of this general equilibrium approach.

9.4 SCALING UP: ASSESSING ECONOMIC VALUATION RESULTS IN MONETARY TERMS

9.4.1 Computable General Equilibrium Model for the Analysis of the Economy–Climate–Biodiversity Interactions

In this section we provide a simple, rigorous, macroeconomic-oriented approach to express the economic valuation impacts in monetary terms, in particular, expressed in terms of changes of the national or regional GDP. We model linkages between climate change, biodiversity and productivity of European agro-ecosystems within an economy-wide framework. In particular, we shall explore the use of computable general equilibrium (CGE) model, which constitutes a common tool for analysis of economy–environment interactions.[8] The present CGE model, underlying algebraic structure and numerical calibration uses the static multi-regional GTAP (Global Trade Analysis Project) model version, as presented by Bigano et al. (2008) and Berrittella et al. (2006). This architecture is calibrated using economic and environmental data, and then solved for the equilibrium values of economic variables in such a way that the architecture can replicate regional GDP growth paths, consistent with the A2 IPCC scenario.

In general, a CGE model is an algebraic representation of the Arrow-Debreu general equilibrium structure that is calibrated on economic data. In this setting, national income is allocated between aggregate household consumption, public consumption and savings. The top-level utility function has a Cobb-Douglas specification. Private consumption is split into a series of alternative composite Armington aggregates. The functional specification used at this level is the constant difference in elasticities form: a non-homothetic function, which is used to account for possible differences in income elasticities for the various consumption goods. In addition, industries are modelled through a representative firm, which maximizes profits in perfectly competitive markets. The production functions are specified via a series of nested, constant elasticity of substitution functions. Domestic and foreign inputs are not perfect substitutes, according to the so-called Armington assumption, which accounts for product heterogeneity. A representative consumer in each region receives income, defined as the service value of national primary factors. We refer to natural resources, land, labour and capital. Capital and labour are perfectly mobile domestically, but immobile internationally. Land (imperfectly mobile) and natural resources are industry specific.

The model operates on the GTAP 6 global database, which combines detailed bilateral trade, transport and protection data characterizing economic linkages among regions, together with individual country input-output databases, which account for inter-sectoral linkages within regions (Dimaranan, 2006). Although regional and industrial disaggregation in the model may vary, the results presented here refer to 19 macro world regions in which several European countries appear disaggregated, as distinct economic entities, whereas the rest of the world is aggregated into four major trading blocks.[9] Regional economies are represented by 19 sectors that can be classified in three major industries, where land-using industries are presented in the broadest disaggregation possible in GTAP database. Table 9.3 depicts regional and sectoral disaggregation.

In this context, we shall be able to illustrate how CGE simulations may be used to analyse the economy-wide impacts of climate-change-induced variation in biodiversity on the productivity of agro-ecosystems, as discussed in the past section.

Table 9.3 GTAP-EF sectoral and regional disaggregation

N	Regions		Sectors
	Code	Description	Description
1	Italy	Italy	Rice
2	Spain	Spain	Wheat
3	France	France	Cereal crops
4	Greece	Greece	Vegetable fruits
5	Malta	Malta	Oil seeds
6	Cyprus	Cyprus	Sugar cane
7	Slovenia	Slovenia	Plant-based fibres
8	Croatia	Croatia	Other crops
9	FYug	Bosnia, Montenegro, Serbia	Animals
10	Albania	Albania	Forestry
11	Turkey	Turkey	Fishing
12	Tunisia	Tunisia	Coal
13	Morocco	Morocco	Oil
14	RoNAfrica	Rest of North Africa	Gas
15	RoMdEast	Rest of Middle East	Oil products
16	RoNME	Non-Med. Europe	Electricity
17	RoA1	Other Annex 1 countries	Other industries
18	ChInd	China and India	Market services
19	ROW	Rest of the World	Non-market services

9.4.2 Scenario Analysis

We proceed with the macroeconomic-based valuation exercise in two steps. The first step refers to the creation of benchmark datasets for the world economy 'without climate change', using the methodology described in Bosello and Zhang (2005). This entails inserting in the GTAP calibration data forecasted values for some key economic variables, to identify a hypothetical general equilibrium state in year 2050. Since we work on the medium to long term, we focus, primarily, on the supply side: projected changes in the national endowments of labour, capital, land, natural resources, as well as variations in factor-specific and multi-factor productivity. We shall interpret this benchmark as the baseline scenario. The second step is imposing over this benchmark equilibrium, climate-change-induced temperature and precipitations (CC), as well as biodiversity (Bio) impacts on land productivity for crops in different regions employing estimations presented in Table 9.2.

We obtained estimates of the regional labour and capital stocks by running the G-Cubed model (McKibbin and Wilcoxen, 1998) and of land endowments and agricultural land productivity. In particular, land, which is an exogenous input fixed at regional level, is allocated among nine land-using sectors, as identified in Table 9.3, in response to changes in relative rental rates. By changing the calibration values for these variables, the CGE model has been used to simulate a general equilibrium state for the future world economy. We run this model for four scenarios about the climate (A1F1, A2, B1, and B2). In this way, we shall be analysing three states of the world: (1) a baseline growth for the world economy, in which climate change impacts are ignored; (2) a climate counterfactual scenario in which temperature and precipitations are imposed; and, finally, (3) a climate counterfactual scenario in which temperature and precipitations and climate-change-induced variation in biodiversity on the productivity of agro-ecosystems impacts are imposed. The results are presented and discussed in the next sub-section.

9.4.3 Monetary Valuation Results of the CGE Model

Table 9.4 presents changes in output of the wheat crop due to climate change direct variations in temperature and precipitations (CC), and the climate counterfactual scenarios in which temperature, precipitations and climate-change-induced variation in biodiversity on the productivity of agro-ecosystems impacts are imposed, all measured for the year 2050 and expressed in terms of changes with respect to the baseline scenario. Estimation results show that significant effects of biodiversity, above direct climatic impact, can be observed. For instance, examining percentage change in wheat output in Italy under the A1F1, A2 and B2 scenarios, it becomes clear that the added effect of biodiversity reversed direct climatic change impact, so that wheat production is projected to increase with climate-change-induced variations in biodiversity (Bio) when compared to benchmark dynamics. Output change was negative when only direct (CC) shock was estimated. In a similar fashion, we shall be able to express the magnitudes in terms of changes in the agricultural output as well as in terms of change in regional GDP. As Figure 9.2 illustrates, for some regions, the added effect of biodiversity operates in the same direction as temperature and precipitation changes. However, there were regions where this effect

Table 9.4 Percentage change in wheat output versus no climate change baseline in 2050

Region	CC				Bio			
	A1F1	A2	B1	B2	A1F1	A2	B1	B2
Italy	−0.067	−0.123	0.150	−0.064	0.333	0.202	0.061	0.108
Spain	−1.683	−1.511	−0.245	−0.821	1.551	−0.522	2.288	0.215
France	−0.436	−0.469	0.478	−0.128	0.609	0.352	0.647	0.173
Greece	−3.331	−2.574	−0.540	−1.432	5.420	−0.536	7.258	0.204
Malta	−1.482	−1.535	0.330	−0.775	2.474	−0.279	3.342	0.468
Cyprus	0.731	0.408	0.775	0.293	1.453	0.577	1.355	0.449
Slovenia	0.419	0.322	0.212	0.198	0.144	0.108	0.026	0.050
Croatia	0.439	0.236	0.432	0.103	−0.595	−0.387	−0.596	−0.615
FYug	0.311	0.255	0.189	0.154	−0.250	−0.193	−0.328	−0.253
Albania	−0.547	−0.443	−0.042	−0.202	−0.703	−0.597	−0.762	−0.594
Turkey	0.317	0.226	0.198	0.146	0.081	0.057	0.024	0.016
Tunisia	0.323	0.235	0.209	0.152	0.101	0.074	0.039	0.035
Morocco	0.322	0.246	0.197	0.156	−0.046	−0.026	−0.072	−0.059
RoNAfrica	0.194	0.145	0.129	0.094	−0.052	−0.030	−0.055	−0.053
RoMdEast	0.984	0.606	0.757	0.396	0.915	0.558	0.708	0.374
RoNME	0.269	0.139	0.209	0.081	0.234	0.145	0.250	0.081
RoA1	0.372	0.250	−0.012	0.159	0.346	0.244	−0.019	0.183
ChInd	−0.612	−0.365	0.184	−0.243	−0.613	−0.366	0.183	−0.243
RoW	−0.630	−0.372	−0.669	−0.251	−0.633	−0.377	−0.666	−0.246

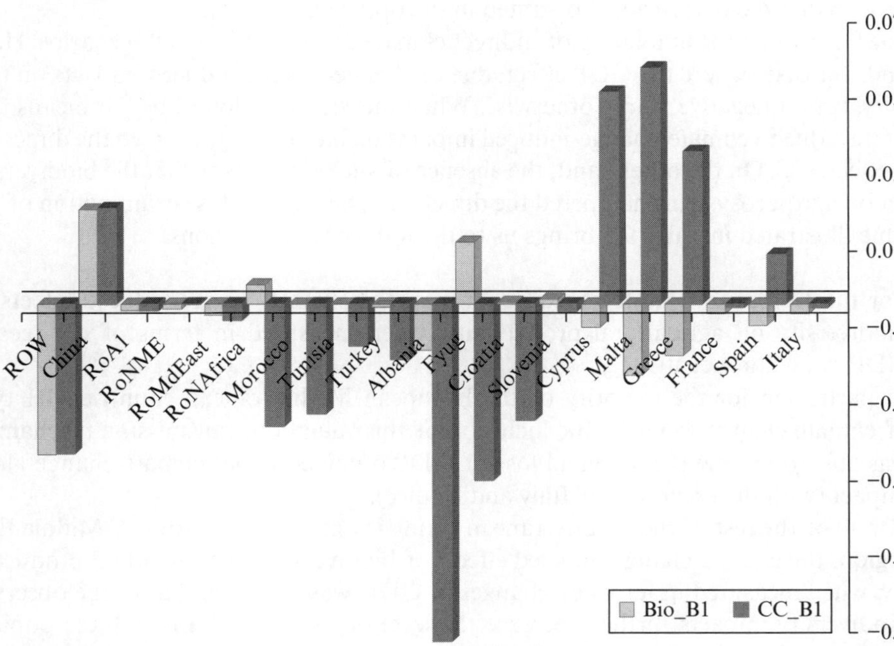

Figure 9.2 Percentage change in regional GDP in 2050 (B1 scenario versus baseline)

Table 9.5 Economic impact of biodiversity above temperature effect on land productivity

Region	A1F1	A2	B1	B2
Italy	+ ↕	+	+	+
Spain	+ ↕	+	+ ↕	+
France	+	+	+	+
Greece	+	+	+ ↕	+
Malta	+ ↕	+	+ ↕	+
Cyprus	+ ↕	+	+ ↕	+
Slovenia	–	–	–	–
Croatia	– ↕	– ↕	–	– ↕
FYug	–	–	–	–
Albania	– ↕	– ↕	– ↕	– ↕
Turkey	–	–	– ↕	–
Tunisia	–	–	–	–
Morocco	–	–	–	–
RoNAfrica	–	–	–	–
RoMdEast	– ↕	– ↕	– ↕	– ↕
RoNME	+	+	+	+
RoA1	+	+	– ↕	+
ChInd	+	+	+	+
RoW	+	+	+	+

was reversed and, in some cases, it was even larger than temperature and precipitation impacts, so that the overall effect operated in the opposite direction.

Table 9.5 reflects the ambiguity of indirect climate change effect in all scenarios. Here, + stands for cases where the GDP effect, due to climate-change-induced impacts on biodiversity, is non-negative, and – otherwise. When this sign is followed by ↕ it means that GDP effect due to climate-change-induced impacts on biodiversity reversed the direct climatic effect, CC. On the other hand, the absence of such sign means that the biodiversity impact on agro-ecosystems hampered the direct climatic effect. Close examination of the outcome illustrated in Table 9.5 brings us to the following conclusions.

1. For the European Mediterranean countries, the climate-change-induced effects of biodiversity on agricultural productivity, when measured in terms of changes in GDP, were non-negative.
2. In particular, for the majority of the European Mediterranean countries, B1 type of climate change scenario, the inclusion of this valuation transmission mechanism was able to reverse the marginal loss of GDP obtained under climate-change-alone impact (with the exception of Italy and France).
3. For all of the rest of the Mediterranean countries, as well as for Rest of Middle East region, the climate-change-induced effects of biodiversity on agricultural productivity, when measured in terms of changes in GDP, was negative; that is, the observed biodiversity impacts further decrease the level of human welfare of these populations, as originally measured by the CGE model.
4. For Albania, the Rest of Middle East countries and Turkey (when analysed at the

B1 scenario) the magnitude of the negative economic impact of biodiversity – above temperature effect on land productivity – was high enough to reverse the original CGE welfare impact.

In short, and despite the fact that, in general, we are assisting a worldwide decrease in the levels of biological diversity, from an economic perspective, which is here approached from the productivity of the agro-ecosystems, this stylized fact is not always corresponding to a similar GDP change pattern. In fact, not only European countries experience diverse GDP changes associated with a decrease in the levels of biological diversity, but also some countries might be more impacted than others. Furthermore, from a monetary, market perspective, as here expressed in terms of changes of the countries' GDP, some countries will be associated with a welfare gain. This effect is not only dependent on the geographical location of the country but, above all, on the structure of the existing markets, overall technology profile, land use patterns as well as its agricultural 'dependency' with respect to biodiversity indicators. Therefore, the world economy as a system of markets interacting through exchanges of inputs, goods and services responding to changes in relative prices, as induced by climate shocks, is solely responsible for a mechanism of adaptation. For this reason, this mechanism is interpreted as market-driven or autonomous social-economic adaptation.

9.5 CONCLUSIONS AND DISCUSSION

Economic valuation results showed that biodiversity contributes to explaining land productivity in the agro-ecosystem sector. Second, economic valuation results of the climate-change-caused impacts on biodiversity and the productivity of European agro-ecosystems are multifaceted and respective monetary magnitudes across the different European countries vary significantly. In some cases, the GDP effect due to climate-change-induced impacts on biodiversity was not only non-negative, but also able to reverse the direct (and negative) climatic effect. Particularly enlightening is the case of Mediterranean Europe, where initial negative impacts were eventually turned into gains. There, negative direct impacts were, in fact, counterbalanced by improvement in terms of trade. In addition, the estimation results showed that, while developed regions lose slightly, or even gain as in the case of Central and Northern Europe, developing regions can lose considerably more. For the majority of the North African economies in the Mediterranean Sea the GDP effect due to climate-change-induced impacts on biodiversity was negative and, hampers the negative effect of the climatic conditions on agro-ecosystems. This highlights the greater economic vulnerability to climatic change of the African countries in the Mediterranean Basin. Even in terms of final impacts on economic activity, the southern basin of the Mediterranean was more severely affected than the northern one. These results reiterate the importance of welfare analyses of climate-change-caused impacts on biodiversity that focus on the redistributive aspects involved with these changes: impacts were not distributed in a uniform way across the Mediterranean countries under consideration; some countries, and respective economies, showed to be less resilient than others; most of the time, the welfare changes involved clearly signal the presence of winners and losers. To our knowledge, this exercise constitutes an original procedure, at a global level of analysis, in the economic welfare assessment of biodiversity impacts induced by climate change.

We recognize, however, that this analysis is (still) looking at the tip of the iceberg. First, from an economic perspective, the welfare effects of climate-change-induced impacts on biodiversity are *not restricted* to market/productivity anchored transmission mechanisms, and the present analysis did not embed a single consideration associated to these non-market aspects. Second, and again from an economic perspective, the market links of biodiversity and human well-being are *not limited* to the agro-ecosystem sector. Other markets will be certainly impacted by climate-change-induced impacts on biodiversity (for example, coastal tourism), and the present analysis did not capture this/these additional market(s) anchored transmission mechanisms.

Finally, and from a natural science perspective, it can always be argued that one could work with additional, and probably more efficient, indicators to measure biodiversity (since our decision was also based in terms of existing data, including projections for the different climate change scenarios among the countries under consideration). Having said that, and since we are not embracing a reductionist approach, we do not have the ambition to provide a clear, unique and indisputable reply to the quantification of the biodiversity loss effects on GDP and, therefore, on human well-being. The crucial point that we raised here is that our exercise, and underlying monetary valuation results, demonstrated that the regional/national economies, which also reflects complex (social) systems, showed different resilience profiles to deal with these types of effects. Some economies, and respective market architectures and trade systems, are able to buffer the climate change impacts, including climate-change-induced impacts on biodiversity, while others are not. Naturally, further research is needed to better understand the ecological–social-systems interactions and the role of biodiversity as a determinant.

ACKNOWLEDGEMENTS

The authors thank M. Bindi and R. Ferrise, Department of Agronomy and Land Management, University of Florence, for their research work in projecting wheat productivity for Italy under a scenario of climate change, and L. Onofri, Department of Economics, University of Venice, for her input and discussions on microeconometric modelling. Finally, the authors also thank S. Silvestri and E. Lugato for their research assistance and data management and B. Racah for the English proof-reading of the full manuscript. Finally, the authors also acknowledge the financial support of the European Union Sixth Framework Programme under the project Climate Change and Impact Research: the Mediterranean Environment – CIRCE, Contract No. 036961, and the European Investment Bank, University Research Sponsorship Programme.

NOTES

1. In the Fourth IPCC Assessment Report (2007) the scenarios are defined as follows: A1F1 describes a future world of very rapid economic growth, global population that peaks in mid-century and declines thereafter, and the rapid introduction of new and more efficient fossil-intensive technologies; B1 is based on a similar

global population growth path, but with rapid change in economic structures and the introduction of clean and resource-efficient technologies; A2 describes a very heterogeneous world with continuously increasing population and slower economic development than other scenarios; B2 describes the world with continuously increasing global population, oriented towards environmental protection and social equity.
2. *Food* includes crops destined for human consumption, such as sugar crops, nuts, cereals, fruits, oils crops, pulses, root and tubers, vegetables. *Non-food* includes provisioning services not destined for human consumption such as latex, pharmaceuticals and agro-chemical products. On the other hand, bio-energy includes crops for energy production such as oilcrop for biodiesel and cereals for ethanol.
3. The same indicators have been used for the economic analysis of the impacts of biodiversity on tourism flows (see Loureiro et al., 2012).
4. Including countries such as Greece, Italy, Portugal, Spain, Albania, Bosnia and Herzegovina, Bulgaria, Serbia and Montenegro, Turkey, TFR of Yugoslavia, Austria, Belgium, France, Germany, Ireland, Luxembourg, Netherlands, Switzerland, Croatia, Czech Republic, Hungary, Poland, Romania, Slovakia, Slovenia, Denmark, United Kingdom, Estonia, Latvia, Lithuania, Finland, Norway and Sweden.
5. Since wheat is the most cultivated crop in Europe, it is considered the most representative of net primary production variation and can therefore be an important crop to be studied in terms of the consequences of changing climatic parameters (such as temperature, and precipitation). Moreover, information regarding wheat yield, grassland and cropland areas, total fertilizers used and total tractors is derived from FAO statistics. Information about the country's temperature and precipitation is derived from the Tyndall database, University of East Anglia.
6. From the empirical estimation viewpoint, we compute the estimator for β in a simple way. We used a regression model in deviations from the individual means. The OLS estimator for β obtained from this transformed model is often called within estimator or fixed effect estimator, and it is exactly identical to the LSDV (least squares dummy variable) estimator (Verbeek, 2002).
7. Estimation results obtained with Stata software.
8. See Wing (2011) for a recent review of CGE models.
9. This regional disaggregation of the world economies was constructed to serve the CIRCE (Climate Change and Impact Research) project's focus on Mediterranean economies.

REFERENCES

Allen, A.W. and M.W. Vandever (eds) (2003), *A National Survey of Conservation Reserve Program (CRP) Participants on Environmental Effects, Wildlife Issues, and Vegetation Management on Program Lands*, Denver, CO: Government Printing Office.

Anger, M., A. Malcharek and W. Kuhbach (2002), 'An evaluation of the fodder values of extensively utilised grasslands in upland areas of Western Germany. I. Botanical composition of the sward and DM yield', *Journal of Applied Ecology*, **76**(1–2), 41–6.

Baglioni, M., G. Macagno, P.A.L.D. Nunes and C. Travisi (2009a), *Natura 2000 Network and Agricultural Activities: Analysing the Potential of the Natura 2000 Database in Developing Agricultural Pressure Indices. Report of the EXIOPOL, Deliverable II.3.b.1. Part 1*, Milan: EXIOPOL.

Baglioni, M., G. Macagno, P.A.L.D. Nunes and C. Travisi (2009b), *Natura 2000 Network, Agricultural Pressures and Biodiversity Conservation: An Italian Outlook: Analysing the Potential of the Natura 2000 Database in Describing the Linkages Between Agricultural Pressures and Biodiversity. Report of the EXIOPOL, Deliverable II.3.b.1. Part 2*, Milan: EXIOPOL.

Berrittella M., A. Bigano, R. Roson and R.S.J. Tol (2006), 'A general equilibrium analysis of climate change impacts on tourism', *Tourism Management*, **27**(5), 913–24.

Bigano, A., F. Bosello, R. Roson and R.S.J. Tol (2008), 'Economy-wide impacts of climate change: a joint analysis for sea level rise and tourism', *Mitigation and Adaptation Strategies for Global Change*, **13**(8), 765–91.

Bignal, E.M. and D.I. McCracken (1996), 'Low-intensity farming systems in the conservation of the countryside', *Journal of Applied Ecology*, **33**, 413–24.

Bosello, F. and J. Zhang (2005), 'Assessing climate change impacts: agriculture', Fondazione Eni Enrico Mattei (FEEM) Working Paper No. 94, Milan: FEEM.

Brown, R.A. and N.J. Rosenberg (1999), 'Climate change impacts on the potential productivity of corn and winter wheat in their primary United States growing regions', *Climatic Change*, **41**(1), 73–107.

De Miguel, J.M. (ed.) (1999), 'Nature and configuration of the agrosilvopastoral landscape in the conservation of biological diversity in Spain', *Revista Chilena de Historia Natural*, **72**(4), 547–57.

Dimaranan, B.V. (ed.) (2006), *Global Trade, Assistance, and Production: The GTAP 6 Data Base*, West Lafayette: Center for Global Trade Analysis, Purdue University.

EEA (European Environmental Agency) (2007), 'Natura 2000 dataset', accessed March 2011 at http://www.eea.europa.eu/data-and-maps/data/.
EEA (European Environmental Agency) (2010), *EU 2010 Biodiversity Baseline*, EEA Report No. 12/2010.
EEA (European Environmental Agency) (2011), 'Natura 2000 dataset', accessed March 2011 at http://www.eea.europa.eu/data-and-maps/data/natura-1.
Ewert, F., J.R. Porter and M.D.A. Rounsevell (2007), 'Crop models, CO_2, and climate change', *Science*, **315**, 459–60.
Ewert, F., I. Rounsevell, I. Reginster, M.J. Metzger and R. Leemans (2005), 'Future scenarios of European agricultural land use. I. Estimating changes in crop productivity', *Agriculture, Ecosystems and Environment*, **107**(2–3), 101–16.
FAO, FAOSTAT, accessed 4 February 2014 at http://faostat.fao.org/.
Ferrise, R., A. Triossi, P. Stratonovitch, M. Bindi and P. Martre (2010), 'Sowing date and nitrogen fertilisation effects on dry matter and nitrogen dynamics for durum wheat: an experimental and simulation study', *Field Crops Research*, **117**(2–3), 245–57.
Gardi, C., M. Tomaselli, V. Parisi, A. Petraglia and C. Santini (2002), 'Soil quality indicators and biodiversity in Northern Italian permanent grasslands', *European Journal of Soil Biology*, **38**(1), 103–10.
IMAGE (2001), *The IMAGE 2.2 Implementation of the SRES Scenarios*, Bilthoven: RIVM.
IPCC (2007), *AR4 – Climate Change 2007*, accessed 4 February 2014 at http://www.ipcc.ch/pdf/assessment--report/ar4/syr/ar4_syr.pdf.
Loureiro, M., P.A.L.D. Nunes, G. Macagno and R. Tol (2012), 'Assessing the impact of biodiversity on tourism flows: an econometric model for tourist behaviour with implications for conservation policy', *Journal of Environmental Economics and Policy*, **1**(2), 174–94.
Kan, I., D. Haim, M. Rapaport-Rom and M. Shechter (2009), 'Environmental amenities and optimal agricultural land use: the case of Israel', *Ecological Economics*, **68**(6), 1893–8.
Macagno, G. and P.A.L.D. Nunes, (2009), *Agricultural Pressures on Biodiversity Conservation: An Analysis of the Effectiveness of Natura 2000 Network in Italy*, Venice: Department of Economics, Ca' Foscari University.
Matson, P.A., W.J. Parton, A.G. Power and M.J. Swift (1997), 'Agricultural intensification and ecosystem properties', *Science*, **277**(5325), 504–9.
McKibbin, W.J. and P.J. Wilcoxen (1998), 'The theoretical and empirical structure of the G-Cubed model', *Economic Modelling*, **16**(1), 123–48.
MA (Millennium Ecosystem Assessment) (2005), *Ecosystems and Human Well-Being: Synthesis*, Washington, DC: Island Press.
Nagy, G. (2002), 'The multifunctionality of grasslands in rural development in a European context', *Acta Agronomica Hungarica*, **50**(2), 209–22.
Nakicenovic, N. and R.J. Swart (eds) (2000), *Special Report on Emissions Scenarios*, Cambridge, UK: Cambridge University Press.
Rounsevell, M.D.A., F. Ewert, I. Reginster, R. Leemnas and T.R. Carter (2005), 'Future scenarios of European agricultural land use. II: Projecting changes in cropland and grassland', *Agriculture, Ecosystems and Environment*, **107**(2–3), 117–35.
Schröter, D. et al. (2004), *ATEAM (Advanced Terrestrial Ecosystem Analyses and Modelling) Final Report*, Potsdam: Potsdam Institute for Climate Impact Research.
Schröter, D. et al. (2005), 'Ecosystem service supply and vulnerability to global change in Europe', *Science*, **310**(3752), 1333–37.
Swift, M.J., A.-M.-N. Izac and M. van Noordwijk (2004), 'Biodiversity and ecosystem services in agricultural landscapes – are we asking the right questions?', *Agriculture, Ecosystems and Environment*, **104**(1), 113–34.
Verbeek, M. (2002), *A Guide to Modern Econometrics*, Chichester: John Wiley and Sons, Ltd.
Walker, B. and W. Steffen (1997), 'An overview of the implications of global change for natural and managed terrestrial ecosystems', *Conservation Ecology*, **1**(2), Article 2, accessed 4 February 2014 at http://www.consecol.org/vol1/iss2/art2/.
Wing, I.S. (2011), 'Computable general equilibrium models for the analysis of economy–environment interactions', in P. Nijkamp and A.A. Batabyal (eds), *Research Tools in Natural Resource and Environmental Economics*, Singapore: World Scientific Publishing Company, pp. 255–306.

10. Quantifying and valuing ecosystem services: an application of ARIES to the San Pedro River Basin, USA

Kenneth J. Bagstad, Darius Semmens, Ferdinando Villa and Gary W. Johnson

10.1 INTRODUCTION

A large body of research exists that identifies and values ecosystem services – the benefits that ecosystems provide to humans (MA, 2005) – and their underlying ecological processes. However, the development of software decision support tools that integrate ecology, economics and geography that can be independently used within the public, private, academic and NGO sectors is a more recent phenomenon (Ruhl et al., 2007; Daily et al., 2009). Spurred by growing demand for more sophisticated analysis of the social and economic consequences of land management decisions, the US Department of Interior – Bureau of Land Management (BLM) launched a pilot project with the US Geological Survey (USGS) to assess the usefulness and feasibility of ecosystem service assessment and valuation tools to provide inputs to decision-making. The project analysed ecosystem services in the US portion of the San Pedro River watershed, which includes the BLM-managed San Pedro Riparian National Conservation Area (SPRNCA), to improve the understanding of complex social and ecological relationships that transcend administrative divisions.

The BLM manages some 99 million hectares, primarily in the western United States, and 283 million hectares of sub-surface mineral estate. BLM's multiple-use mission requires that it appropriately balance non-extractive uses such as habitat conservation, recreation and archaeological heritage protection and the extractive use of resources such as timber, oil and gas, coal, uranium, and other minerals. Decisions on land and resource allocation across these uses are made through the development of resource management plans (RMPs) and the analysis of proposed projects through environmental impact statements (EISs) and environmental assessments (EAs), as part of the National Environmental Policy Act (NEPA) process (BLM, 2005). RMPs set the overall land use stipulations for a given management area, normally a Field Office, which usually covers up to several million hectares. Within that framework, EISs and EAs identify the benefits and harms, both environmental and social, for proposed projects, when considering several alternative actions.

Although ecosystem service assessment and valuation is appropriate for inclusion in NEPA and other planning documents, it has rarely been used in this way to date, with the exception of historically well-quantified non-market values such as recreation (Ruhl et al., 2007). Without tools and standards for measuring, quantifying and valuing ecosystem services, government agencies and the general public are unlikely to support their incorporation into agency decision-making processes. The recent emergence of ecosystem service tools offers initial insight into how service quantification can become part of such decision-making processes.

A full review of 17 ecosystem service modelling and valuation tools and a detailed analysis of modelling results for two spatially explicit ecosystem service modelling tools – Artificial Intelligence for Ecosystem Services (ARIES) and Integrated Valuation of Ecosystem Services and Tradeoffs (InVEST, Tallis et al., 2013) – in the San Pedro have been described elsewhere (Bagstad et al., 2013a, 2013b). A parallel project conducted by BSR (formerly Business for Social Responsibility) explored the application of ecosystem service tools for private-sector decision-making, also focused on the San Pedro (BSR, 2011). In this chapter, we detail the application of ARIES to the San Pedro.

ARIES is an open-source modelling framework developed by scientists at the Basque Centre for Climate Change (BC3), University of Vermont's Gund Institute for Ecological Economics, and USGS Geosciences and Environmental Change Science Center, in partnership with other conservation and ecological economics NGO partners. ARIES seeks to address several scientific and modelling challenges in ecosystem service research and application, including: (1) the need to purchase and be proficient in commercial modelling and/or GIS software, (2) the difficulty of using existing ecological process models to generate outputs that are relevant for ecosystem service quantification, (3) the difficulty in addressing trade-offs between generalized and locally specific models, (4) the inability of past models to capture spatial dynamics of ecosystem services (i.e., between points of provision and use and spatial flows between these), and (5) a lack of clear expression of uncertainty in past modelling efforts (Villa et al., 2014).

ARIES is accessible via a web browser, allowing users to access a library of ecosystem service models and spatial datasets that spans multiple levels of detail from local to global. ARIES plans integration with existing ecological process models in future releases of the system, but in its late 2012 release uses a combination of Bayesian network models, deterministic models and spatial data to quantify and map *source* locations that provide ecosystem services, *use* locations of human beneficiaries, and *sink* locations – the landscape features that deplete ecosystem service flows. These models account for both the ecological processes that generate ecosystem services and the socioeconomic features influencing their demand. A family of agent-based algorithms, collectively termed Service Path Attribution Networks (SPANs), accounts for the spatial dynamics of ecosystem service *flows* (Johnson et al., 2012). ARIES encodes a set of artificial intelligence-based decision rules that enable particular model components to be added or subtracted under appropriate circumstances (e.g., to include different components under certain climatic regimes, above specified population or income thresholds, or for specific biomes). Once generalized global models have been developed, ARIES will be capable of providing both low-cost first estimates of ecosystem service values in parts of the world where detailed local models have not yet been developed and tested and local models that are sensitive to regionally specific factors and make use of high-quality data. ARIES handles uncertainty via Bayesian network modelling and Monte Carlo simulation (Bagstad et al., 2011; Villa et al., 2014) and produces uncertainty maps to accompany the result maps, so that areas where the predictions are more or less reliable can be easily visualized. ARIES model outputs are thus spatially explicit biophysical or relative values for ecosystem services and their uncertainty. Economic valuation estimates can then be applied to these model outputs to derive monetary values for ecosystem service changes under scenarios, using directly estimated economic values or benefit transfer.

ARIES models have been developed for ten case-study sites around the world, and customization of existing and new models is a relatively straightforward process (Bagstad et al., 2011). However, as of late 2012, global models are not yet available for ARIES, limiting its use to the current case study regions (southern Arizona, southern California, Colorado, western Washington, Vermont, Veracruz [Mexico], the Dominican Republic, Ontario, Tanzania and Madagascar). Global models accessible via a web browser and using preloaded spatial data, when available, will increase the potential for future ARIES releases to have widespread use in decision-making.

In this chapter, we describe an ARIES case study application to the San Pedro River watershed. We developed and tested ARIES models for five locally important services: carbon sequestration and storage, surface water supply, aesthetic viewsheds, open space proximity and recreational values. We quantified changes in biophysical, abstract and monetary values of services for urban growth, mesquite management and water augmentation scenarios, and discuss future ecosystem service modelling needs for the San Pedro.

10.2 STUDY AREA

The San Pedro River flows north from its headwaters near Cananea, Sonora, into the United States (southeast Arizona), where it eventually flows into the Gila River, a major tributary of the Colorado River. The San Pedro watershed is divided into two major sub-watersheds: the Upper San Pedro, which extends from the river's headwaters to a geological constriction north of Benson, Arizona known as 'The Narrows', and the Lower San Pedro, which flows from the Narrows north to the Gila River (Figure 10.1). The Upper San Pedro has been the subject of considerable scientific study, which has characterized its geomorphology, hydrology, riparian vegetation and avifauna (Moran et al., 2008; Stromberg and Tellman, 2009). Additional studies have focused on the economic value generated by riparian habitat, particularly for recreational value (Orr and Colby, 2002; Colby and Orr, 2005; Brookshire et al., 2010). The combination of intense conservation interest in the biologically significant Upper San Pedro and the threat of groundwater decline due to pumping from urban growth, particularly near Sierra Vista and Benson, has led to the integration of past research into a decision support system designed for use by local watershed groups (e.g., the Upper San Pedro Partnership and Benson Community Watershed Alliance, Serrat-Capdevila et al., 2009).

The Lower San Pedro watershed, by contrast, is less populated and has received less scientific attention than the Upper San Pedro. However, the Lower San Pedro also has high conservation significance, particularly given the threat of groundwater pumping upstream. In recent years, conservation groups including The Nature Conservancy have been working to secure water rights for instream flows on the Lower San Pedro, and some accompanying hydrologic, ecological and economic studies have been conducted there.

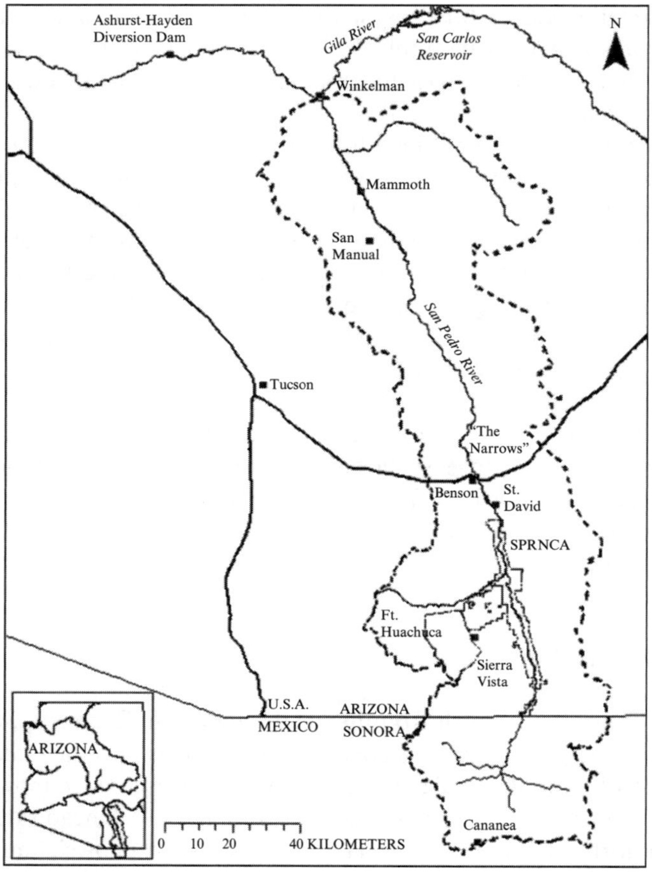

Figure 10.1 Study area map

10.3 DATA, SCENARIOS AND ECOSYSTEM SERVICES FOR ANALYSIS

We obtained many of the spatial data needed to populate the ARIES models from the US Environmental Protection Agency (EPA) San Pedro Data Browser (Kepner et al., 2003). Due to limitations in spatial data and the modelling language that were available when modelling was conducted in 2010 and 2011 we were unable to extend the modelling efforts to the Mexican portion of the watershed. Many of the ecosystem service models used Bayesian networks to quantify ecosystem service provision, use and depletion by landscape features. We constructed these models following a literature review and input from a group of over 20 scientists and managers at a model-review workshop held in September 2010. Bagstad et al. (2011) describe the full model structure and data sources for the ARIES application to the San Pedro.

10.3.1 Scenarios

Baseline conditions for our analysis are for the year 2000, for which many spatial datasets are available, including the Southwest Regional Gap Analysis Project land-cover dataset and urban growth scenarios comparing the year 2000 to alternative 2020 scenarios (Steinitz et al., 2003). We ran ARIES models for baseline conditions and three different scenario categories: urban growth, mesquite management and water augmentation. These three scenario groups vary widely in the spatial extent, type and degree of impacts on ecosystem services. While some climate change modelling work has taken place on the San Pedro (Serrat-Capdevila et al., 2007; Dixon et al., 2009), this work is either not spatially explicit or not up to date with results from the Intergovernmental Panel on Climate Change (IPCC) Fourth Assessment Report. We thus considered but did not apply climate change scenarios for this project.

We compared urban growth scenarios using the year 2000 baseline plus 'open' and 'constrained' growth scenarios for 2020, based on assumptions made by Steinitz et al. (2003) about population growth and alternative development regulations. The open development scenario assumes 50 per cent greater-than-expected population growth by the year 2020 with relaxed restrictions on the location of new development, while the constrained scenario combines 50 per cent less-than-expected population growth with concentration of new development near already developed areas. We expect widespread urban growth to reduce biotic carbon sequestration and storage, to increase runoff, and to simultaneously reduce the potential provision of viewshed, open space and recreational values but to increase the use of these services as new residents move into the watershed.

Widespread mesquite (*Prosopis velutina*) encroachment has occurred in desert scrub and grassland ecosystems in the US southwest in recent decades (Kepner et al., 2000), causing an interest in using fire and/or mechanical removal to control mesquite and restore grasslands. We identified 922 hectares within the San Pedro Riparian National Conservation Area (SPRNCA) with potential for conversion of mesquite shrubland into native grasslands to evaluate the ecosystem service effects of proposed mesquite management activities. We expect mesquite management to reduce carbon storage and increase water yield, based on greater woody biomass and evapotranspiration in mesquite shrublands than grasslands (Nie et al., 2011).

The Bureau of Reclamation's (BOR) proposed extension of the Central Arizona Project from Tucson to Sierra Vista provided the basis for a third category of scenarios (BOR, 2007). This action would bring water from the Colorado River to the San Pedro in order to increase groundwater levels and surface flow within the SPRNCA, with the goal of maintaining or improving aquatic and riparian ecosystem quality. We used alternative scenarios for riparian conditions (Brookshire et al., 2010) to evaluate potential ecosystem service changes resulting from a uniform 0.5 m rise in groundwater across the SPRNCA or achievement of year-round surface flow within the SPRNCA (surface flow is currently intermittent to ephemeral in parts of the SPRNCA, particularly near its downstream end). We expect water augmentation to improve the quantity of services provided within the riparian corridor, though quantification of these changes is difficult for many services, as data that precisely link surface flow frequency, vegetation change and changes in service provision are sparse.

10.3.2 Ecosystem Services for Analysis

Through stakeholder discussions, we identified four groups of ecosystem services of interest for the San Pedro: carbon, water, biodiversity and cultural services. Carbon that is sequestered or stored by ecosystems has value for mitigating the impacts of anthropogenic greenhouse gas emissions on climate change. Water on the San Pedro is used for domestic supply, irrigation and mining, and supports upland and riparian ecosystems that themselves provide a variety of services (Stromberg and Tellman, 2009). We do not explicitly consider the provision of riparian ecosystem water needs as an ecosystem service, since it lacks direct human beneficiaries, though it underpins other values. We similarly do not consider biodiversity to be an ecosystem service, as direct anthropocentric value must be derived from it in order to qualify as such (e.g., the MA [2005] considers biodiversity to be a supporting service). In the San Pedro, biodiversity supports recreational activities including wildlife viewing, hunting, and most importantly, birding (Colby and Orr, 2005). The economic value of several cultural services has been estimated for the San Pedro or nearby regions, which informed valuation, benefit transfer and modelling. Aesthetic values are often revealed by increased property values for real estate near open space or with access to high-quality views. This is relevant for the San Pedro given forecasted population growth, which could dramatically alter viewsheds and open space depending on the patterns of future development (Steinitz et al., 2003).

10.4 RESULTS

10.4.1 Carbon Sequestration and Storage

The ARIES carbon sequestration and storage models quantify and map: (1) carbon sequestration and its uncertainty, based on a probabilistic model and sequestration values from the literature and spatial datasets; (2) potential stored carbon release due to fire, and its uncertainty, based on stored carbon values from the literature and spatial datasets. Overlaying a polygon of fire locations in a particular year can provide a more precise estimate of carbon release from wildfire, such as for representative average or high-intensity wildfire years; and (3) anthropogenic greenhouse gas emissions, based on spatial data. These three final maps let a user quantify the regional carbon balance by subtracting stored carbon release and greenhouse gas emissions from sequestration to estimate whether a region is a net carbon sink or source.

We mapped carbon sequestration and potential stored carbon release and their uncertainty for the urban growth and mesquite management scenarios, along with baseline greenhouse gas emissions (Figure 10.2). We present tabular results for all values below. Results suggest loss in carbon sequestration under both urban growth (110–115 thousand tons of sequestration per year) and mesquite management scenarios (148 tons of sequestration per year). All scenarios also result in reduced potential for stored carbon release, with greater reduced potential in the open than constrained development scenario. This results from fire suppression, which typically occurs on and around newly developed land, lowering the risk for loss of stored carbon from fire. While Bayesian network training is available in ARIES (Villa et al., 2014), the quality of training data was such that the trained networks produced more spatially homogeneous results and had poor sensitivity to changing scenario conditions, with a very minimal improvement in fit to the training data relative to untrained results. We thus used untrained networks in obtaining these results, using prior conditional probabilities carefully selected to produce results in agreement with hypothesized correlations and validated on a subset of the available observed values. Improved national- and global-scale training data, under development in late 2012, may improve the quality of ARIES outputs generated with trained Bayesian networks.

10.4.2 Surface Water Supply

The ARIES surface water supply models quantify: (1) annual precipitation as the source of surface water, (2) evapotranspiration and infiltration and their uncertainty as sinks that deplete the quantity of surface water for downstream flows, (3) use areas where surface water is extracted by human beneficiaries, and (4) flows representing water movement across the landscape. These processes are currently modelled at an annual time step. ARIES does not currently include a groundwater flow model. Groundwater use data are not available in Arizona, severely limiting any opportunities for spatially explicit modelling of groundwater use and flows. This limits our analysis to only surface water flows and users – two small agricultural surface diversions near St. David, Arizona. The current ARIES surface water flow algorithm attempts to satisfy each water user's needs by capturing water near to each user before looking further upstream. Therefore, if enough water is available in the neighbourhood of all users, far-upstream impacts will not typically affect the model results.

We mapped theoretical and actual sources, sinks and surface water users for the urban growth and mesquite management scenarios (Figure 10.3). Theoretical values show potential evapotranspiration and infiltration while actual values show results when connected with a flow model (i.e., accounting for what happens to water as it flows across the landscape). The most interesting results are for the theoretical sink values – potential landscape-scale infiltration and evapotranspiration. There was a 2.7 per cent decrease in watershed-wide theoretical sink strength under the open development scenario and a 2.3 per cent decrease under the constrained development scenario. Mesquite management reduced theoretical sink strength within the SPRNCA by 0.3 per cent. These results mean that less infiltration and evapotranspiration are likely to reduce surface water flows under these scenarios, largely due to reduced infiltration from increased urbanization and lower evapotranspiration for the mesquite management scenario.

10.4.3 Aesthetic Values: Viewsheds and Open Space Proximity

The ARIES viewshed and open space proximity models quantify the contribution of nature toward amenity values, typically measured using hedonic pricing for real estate. These models produce maps of: (1) sources of high-quality open space or visually desirable views; (2) sinks that degrade these features, including visual blight such as transmission lines or mines for viewsheds or highways that reduce privacy, increase noise and block open space access at neighbourhood scales; and (3) users of open space or viewsheds, in this case homeowners. Sources, sinks and users are linked by the appropriate flow

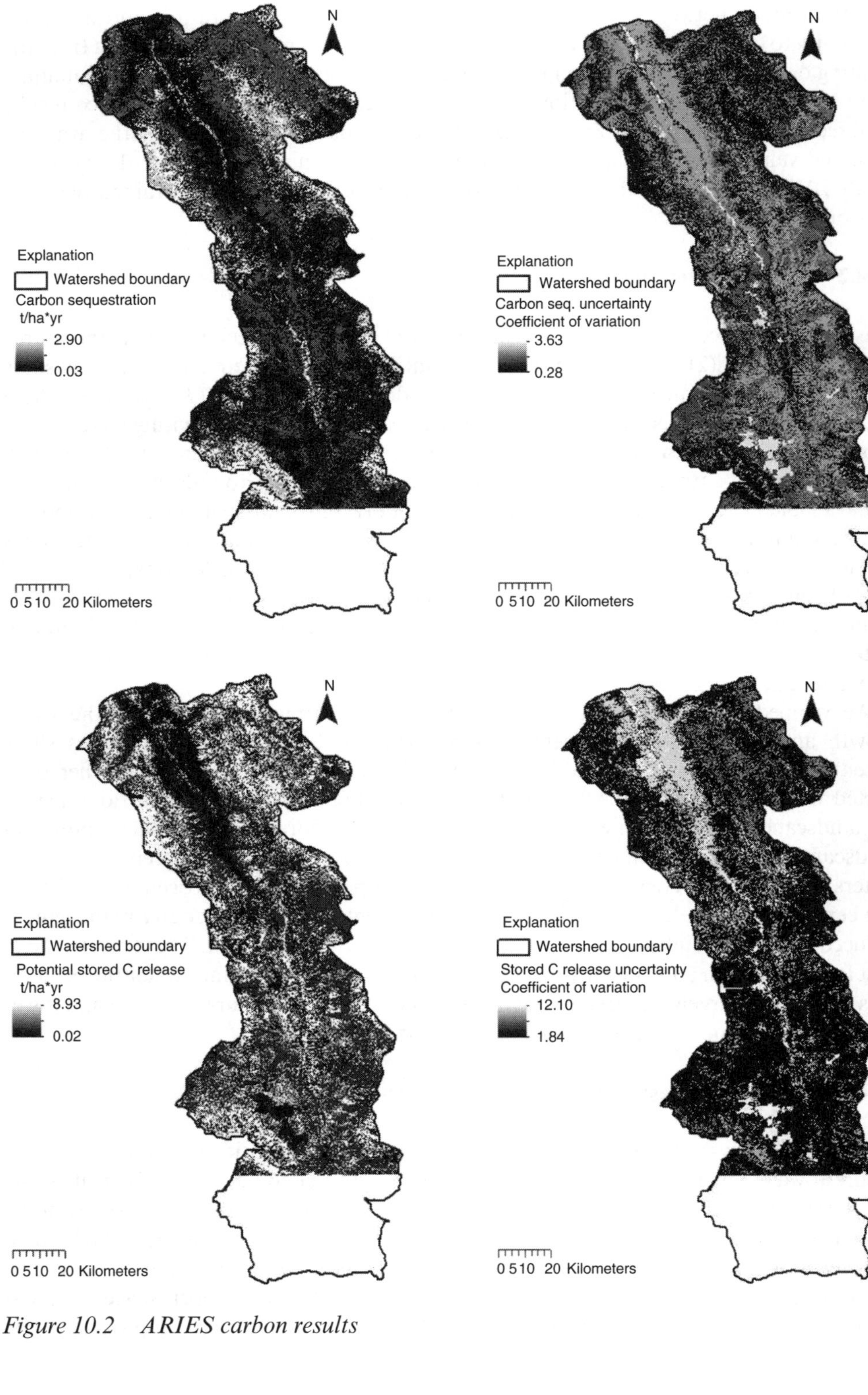

Figure 10.2 ARIES carbon results

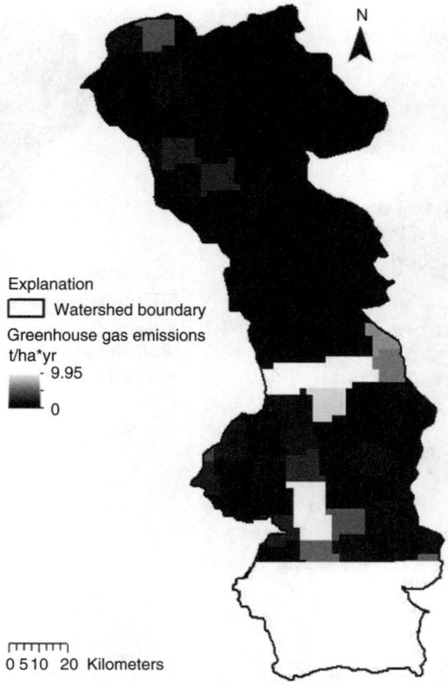

Figure 10.2 (continued)

model – a line-of-sight model for viewsheds and a walking access simulation model for open space proximity. Decay functions and parameterization are described in Bagstad et al. (2013c). Uncertainty estimates, produced as outputs of the Bayesian network models and propagated through the flow models, are provided for all outputs. These model outputs are quantified using relative rankings (i.e., values of 0–100 for each cell), which could be linked to hedonic values for real estate.

We mapped the theoretical source (viewshed or open space quality independent of the location of users and sinks) and actual use values (dependent on user presence and flow characteristics) for open space proximity and viewsheds (Figures 10.4–10.5). We then modelled changes in these values under urban growth scenarios. Lacking data on how changes in mesquite cover or surface flow translate into higher or lower aesthetic value, we did not run the models for these scenarios, though we did calculate baseline values for the SPRNCA. We found a decrease in theoretical viewshed quality of 0.04–0.1 per cent and a decrease in theoretical open space quality of 1.1–2.9 per cent. We found actual use to increase 15.7–40.2 per cent for open space proximity and 240–555 per cent for viewsheds, with greater changes occurring in the open than the constrained development scenario.

10.4.4 Recreation

The ARIES recreation models quantify, using relative rankings (i.e., values of 0–100 for each cell), the potential contribution of nature toward different recreational activities.

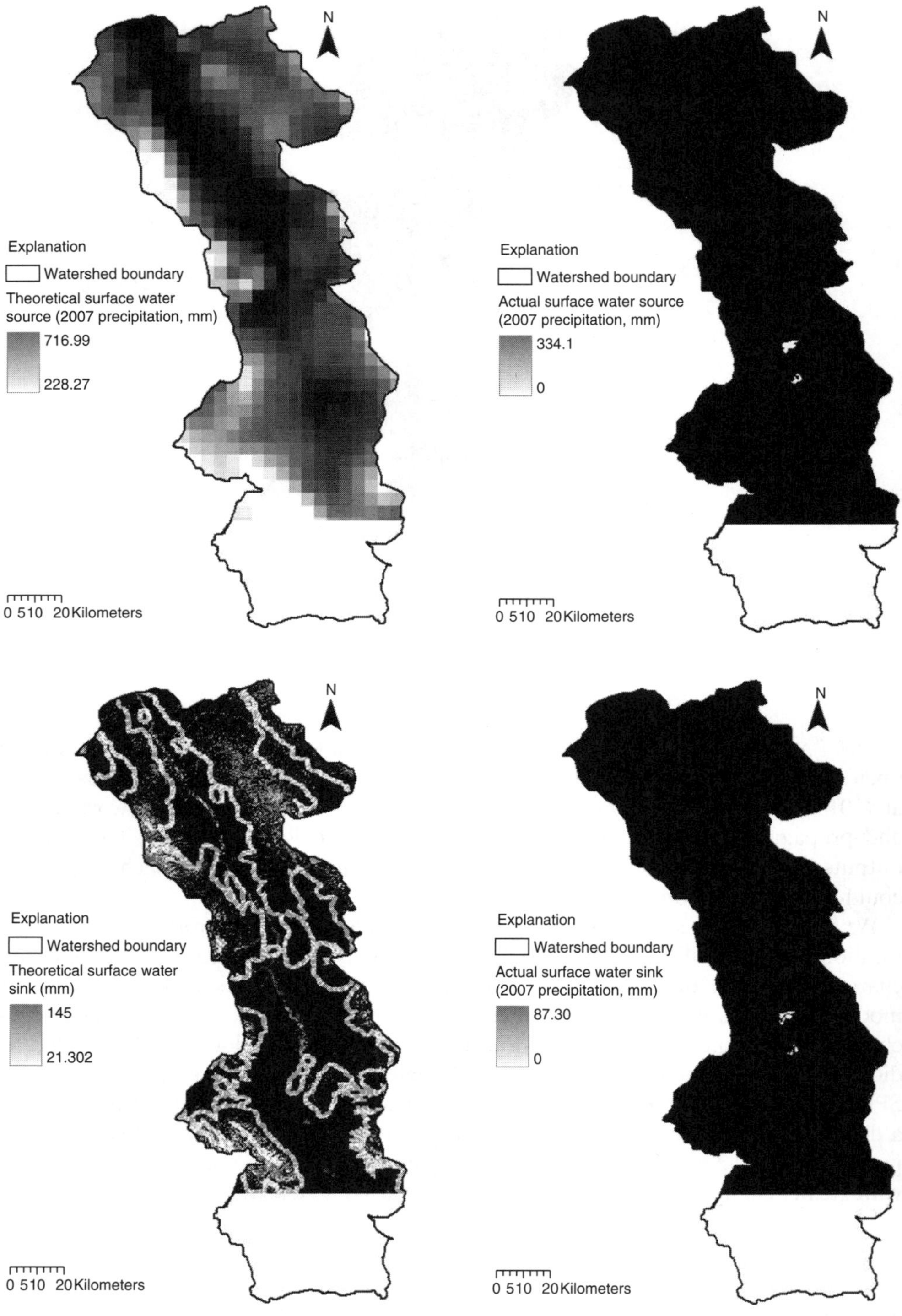

Figure 10.3 ARIES water results

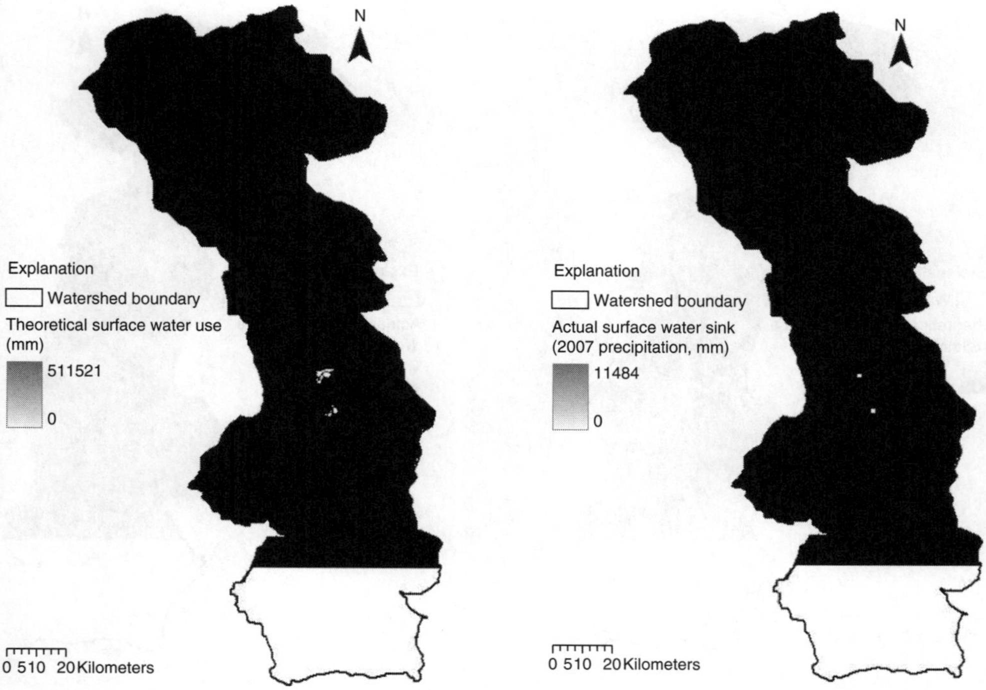

Figure 10.3 (continued)

Other influences on recreational value include site accessibility, infrastructure and visitor preferences, so the ecosystem service component of recreation is simply the relative contribution of underlying natural features toward the quality of the recreational experience. We modelled the relative value and uncertainty of potential birding, wildlife viewing and hunting sites for important game species (Figure 10.6). Recreational use models would show the locations of potential users of recreation sites, while flow models would link users to potential sites through a transportation network model. Fully parameterized recreational use and flow models require higher-quality data on visitor use, preferences and points of origin than are available for most recreation sites, including those on the San Pedro, however. We modelled changes in potential recreation value for birding, wildlife viewing and hunting under urban growth and water augmentation scenarios, as we lacked information on how mesquite management would influence recreational value.

We found a 3.3–4.9 per cent increase in recreation potential for the SPRNCA associated with groundwater augmentation, corresponding to improvements in wildlife habitat quality. Potential recreation values declined 0.6–4.7 per cent in the constrained growth scenario and 1.1–5.4 per cent in the open growth scenario. Both growth scenarios produce new development spread across the landscape, which has negative effects on habitat quality and connectivity. Urban growth, however, also leads to an increase in new residents, who may be recreational users, which could *increase* the value of recreation despite declines in habitat quality. We can only address this point through modelling of recreational use, which was not conducted.

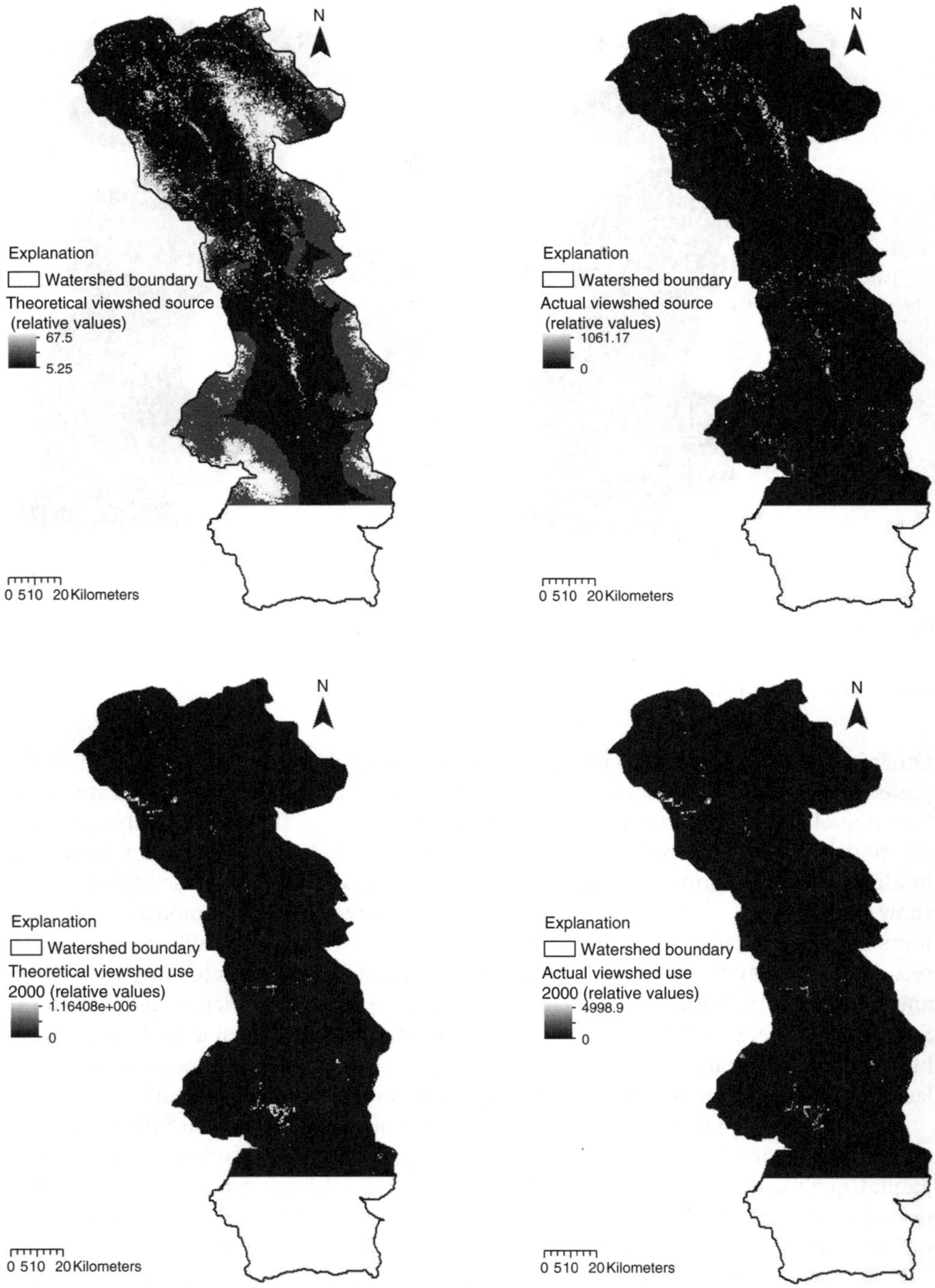

Figure 10.4 ARIES viewshed results

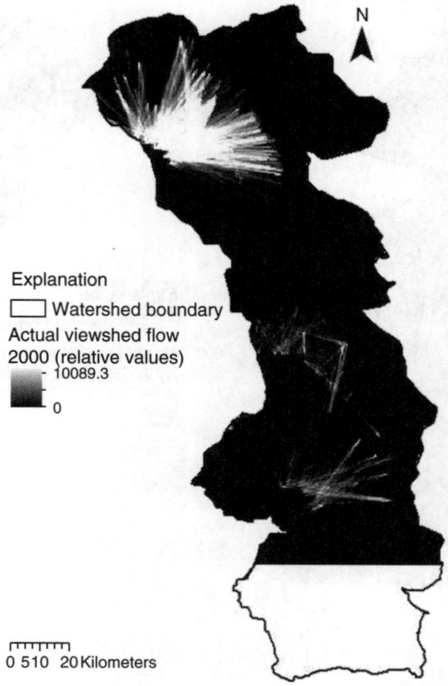

Figure 10.4 (continued)

10.4.5 Valuation

To value carbon sequestration, we used a set of conservative valuation assumptions (US$22/ton social cost of carbon [SCC], 7 per cent discount rate, 0 per cent rate of annual change in the SCC, median values from Tol's [2005, 2008] meta-analyses) and non-conservative valuation assumptions (US$89/ton SCC, 1 per cent discount rate, 6 per cent annual change in the SCC, based on Stern, 2007). We express all monetary values in 2011 US dollars. There is substantial debate in the literature about the SCC and the assumptions that go into those estimates, and even high-end values may underestimate the SCC (Ackerman and Stanton, 2012). However, these estimates provide defensible 'bookend' values. We did not use market prices for carbon due to the extreme price fluctuation that has occurred on both European markets and the now-defunct Chicago Climate Exchange, due largely to the artificial constraints imposed on these markets.

Water can be valued using market price, replacement cost, or willingness to pay (WTP, Table 10.1), though these methods rarely account for the full social, ecological, or commodity value of water. Market prices for water in the San Pedro watershed range from US$0.34 to US$0.90 per cubic metre, based on water rates for the Arizona American Water Company's service in the region (WIFA, 2009). Replacement costs for water range from US$1.07 to US$2.42 per m^3 (BOR, 2007). Piper and Martin (1997) found values for household WTP for improved water quality and water system reliability to range from US$4.63 to US18.06 per month in the rural western United States. Using the regression coefficients from Piper and Martin (1997, Table 4) and values for the independent

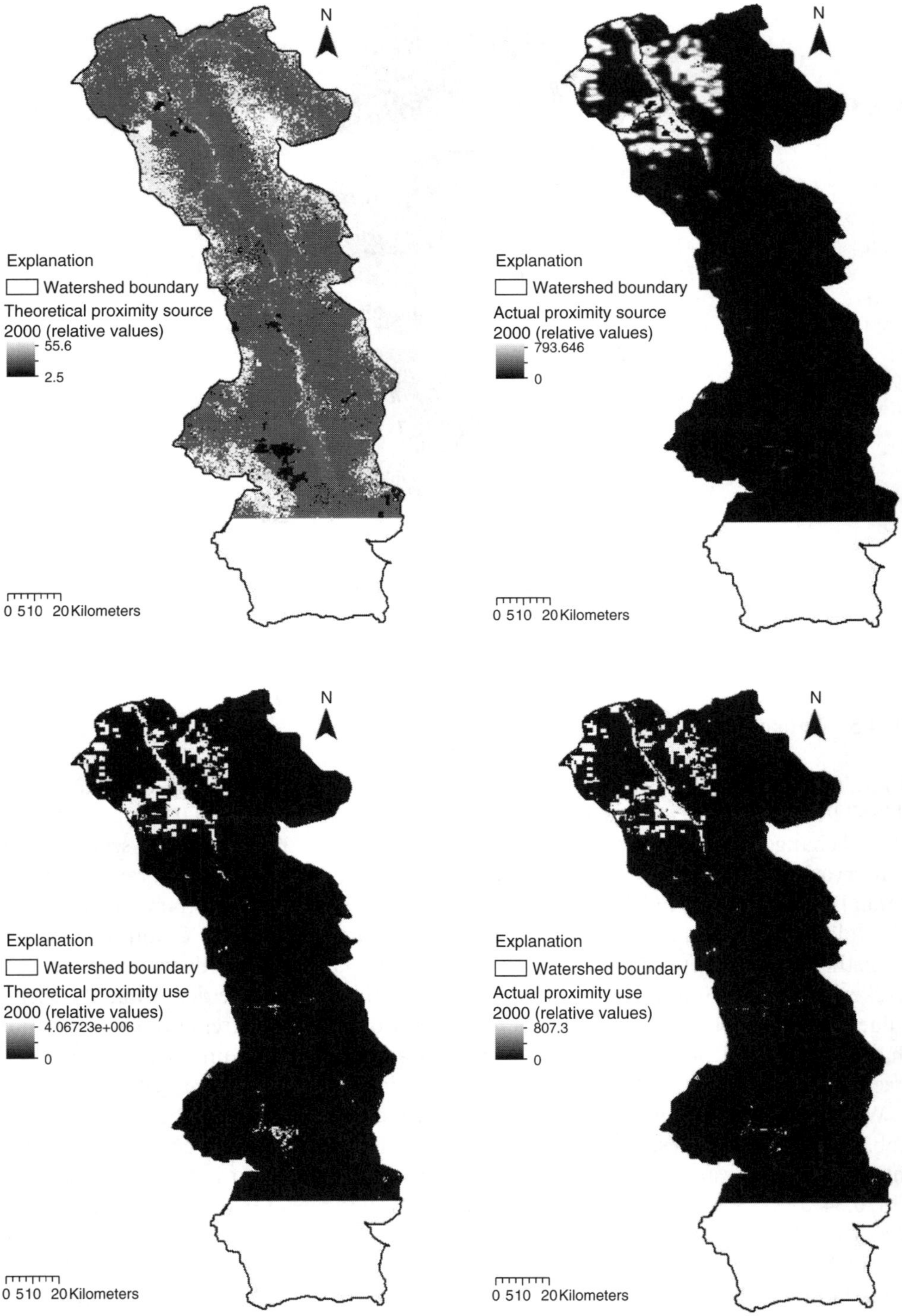

Figure 10.5 ARIES open space proximity results

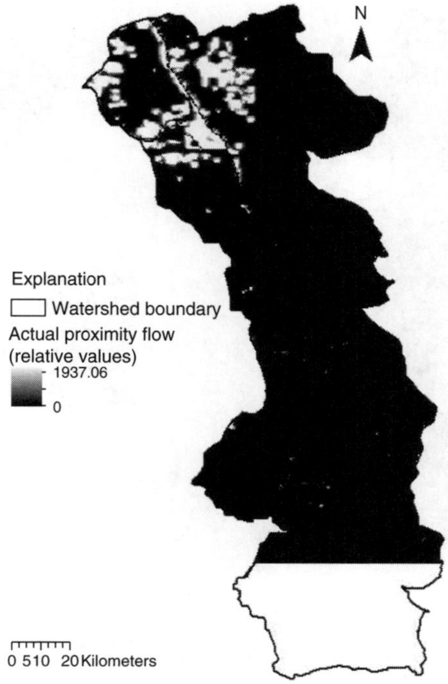

Figure 10.5 (continued)

variables for Cochise County, Arizona, we estimate household WTP to range from US$4.51 to US$14.99 per month. This is in addition to an average water bill of US$53.70 per month (WIFA, 2009) and raises the sum of current water expenditures plus WTP to US$0.40–0.93 per m^3, somewhat closer to the lower bound replacement cost.

Raw water volumes are difficult to interpret socially and ecologically. For the urban growth scenarios, increased runoff from impervious surfaces leads to a variety of water quality and quantity problems, the most important for the San Pedro and other arid regions being reduced infiltration that can recharge groundwater. To more rigorously value water yield changes, we would need to apply more precise biophysical models linking changes in water yield to other groundwater and aquatic system impacts while valuing changes to water supply for people in terms of increased groundwater pumping or municipal water costs. Since the models lacked such precision on either ecological effects or other economic costs, our monetization of water yield change should be interpreted as a relatively non-conservative estimate.

Regrettably, the underlying equations from the most locally applicable open space proximity hedonic studies (Sengupta and Osgood, 2003; Bark et al., 2009) do not readily lend themselves to value transfer to the San Pedro. Lacking studies from the region on the value of scenic views, we were similarly unable to obtain economic values to apply toward the viewshed results. Bourassa et al. (2006) found that view value premiums were greater when high-quality views were scarcer. In other words, when many new housing units have high-quality views – a likely outcome for the San Pedro urban growth scenarios – we might expect the value premium for each unit to be relatively small. Although monetary values could also

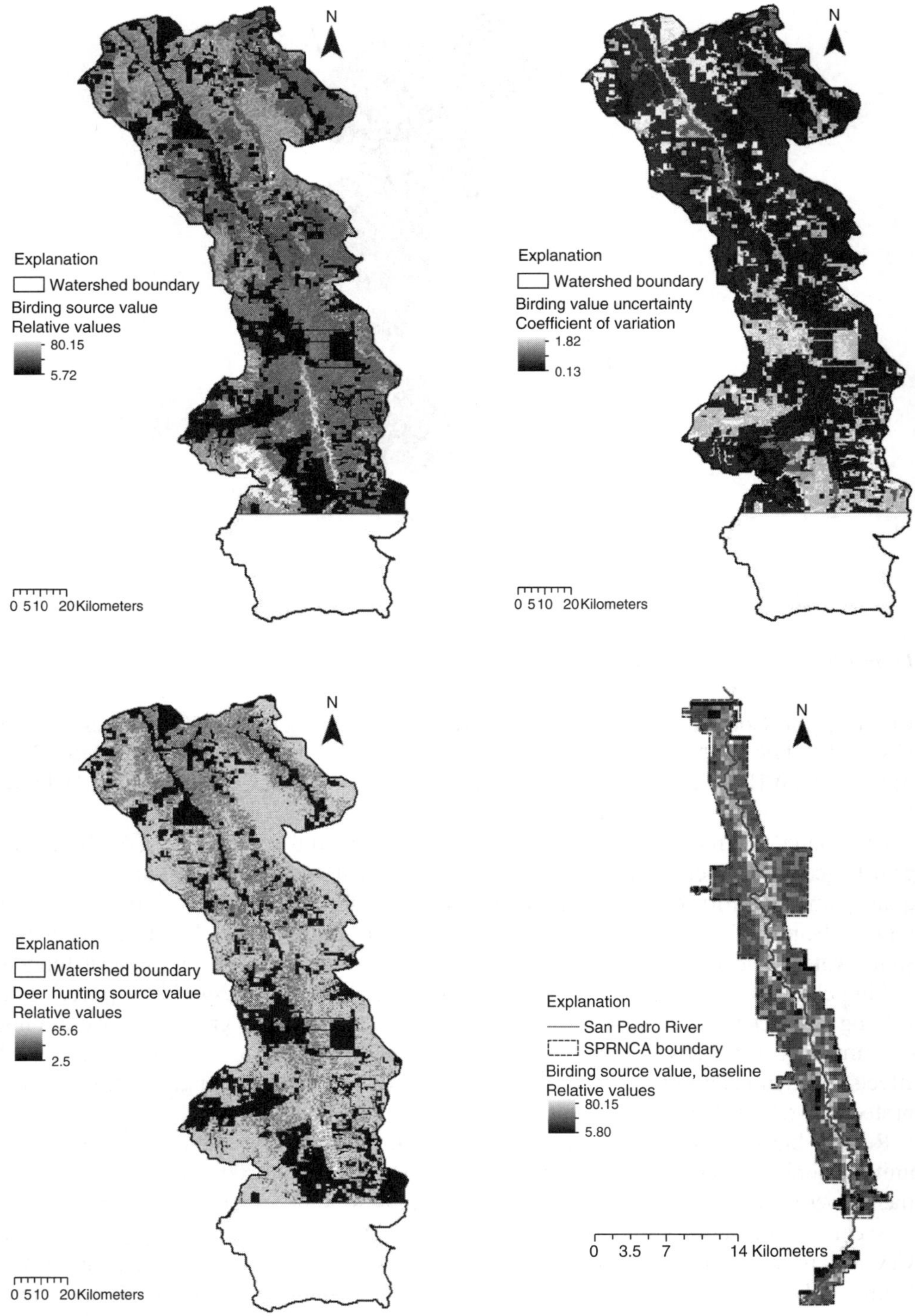

Figure 10.6 ARIES recreation results

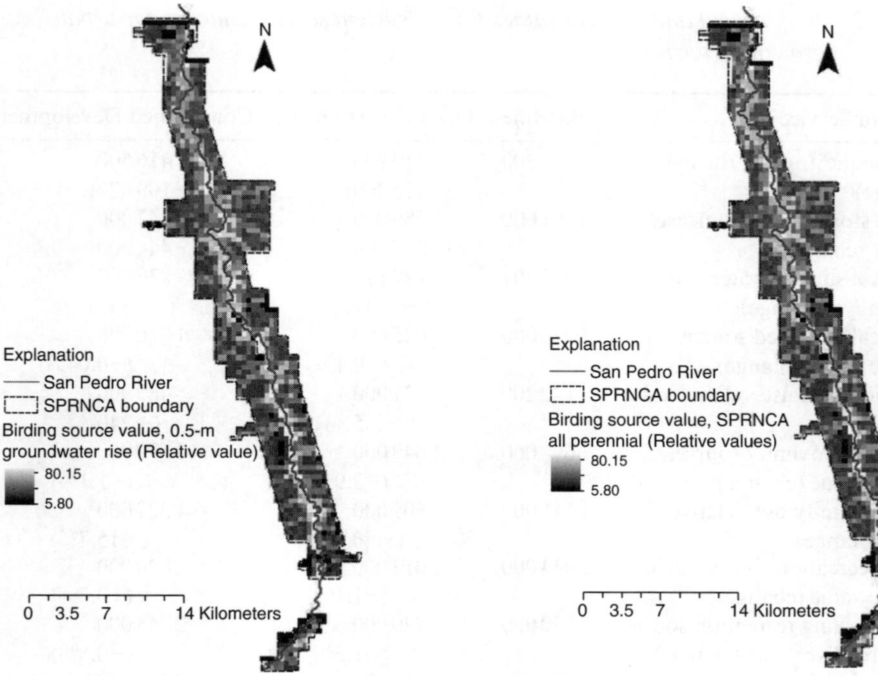

Figure 10.6 (continued)

Table 10.1 *Commodity values for water in the San Pedro River watershed (2011 US dollars)*

Method	Value/m^3
Market price – lower bound	0.34
Market price – upper bound	0.90
Replacement cost – lower bound	1.07
Replacement cost – upper bound	2.42
WTP – lower bound	0.40
WTP – upper bound	0.93

be applied to the change in recreational site quality, proper quantification of recreational value change would require development of a more extensive visitation model than was possible for this study (Loomis, 1995). We thus apply dollar values only for carbon and water supply but not for viewsheds, open space proximity, or recreation in the following section.

10.4.6 Scenarios

Urban growth scenarios

Traditionally quantified benefits such as increased employment and municipal tax base are often used to justify the negative environmental and social impacts that can

Table 10.2 Biophysical and relative values for ecosystem service changes for urban growth scenarios

Ecosystem Service	Baseline	Open Development	Constrained Development
Carbon sequestration, tons/yr (change)	526 200	410 900 (−115 300)	416 600 (−109 600)
Potential stored carbon release, tons/yr (change)	3 291 000	2 789 000 (−502 000)	2 847 000 (−444 000)
Theoretical surface water sink, 1000 m³/yr (change)	235 407	229 176 (−6 231)	230 067 (−5 340)
Theoretical viewshed source, relative value (change)	1 026 000	1 025 000 (−0.1%)	1 026 000 (−0.04%)
Actual viewshed use, relative value (change)	142 200	931 000 (+555%)	483 900 (+240%)
Theoretical proximity source, relative value (change)	1 698 000	1 649 000 (−2.9%)	1 679 000 (−1.1%)
Actual proximity use, relative value (change)	1 146 000	1 607 000 (+40.2%)	1 327 000 (+15.7%)
Birding, recreation source value, relative value (change)	2 034 000	2 012 000 (−1.1%)	2 020 000 (−0.7%)
Wildlife viewing recreation source value, relative value (change)	2 770 000	2 729 000 (−1.5%)	2 748 000 (−0.8%)
Javelina hunting recreation source value, relative value (change)	2 489 000	2 459 000 (−1.2%)	2 475 000 (−0.6%)
Dove hunting recreation source value, relative value (change)	2 583 000	2 513 000 (−2.7%)	2 529 000 (−2.1%)
Deer hunting recreation source value, relative value (change)	2 483 000	2 441 000 (−1.7%)	2 457 000 (−1.1%)
Quail hunting recreation source value, relative value (change)	2 601 000	2 461 000 (−5.4%)	2 478 000 (−4.7%)

accompany urban growth. Consideration of ecosystem services as an opportunity cost of urban growth can more fully show the costs and benefits of urban expansion. Since the aesthetic and recreation models did not produce results amenable to monetization, multicriteria analysis is the most feasible way to incorporate these results into decision-making. Multicriteria analysis is useful for comparing results using non-monetary units (Villa et al., 2002), and allows stakeholders or decision-makers to express preferences for total 'baskets' of ecosystem services produced under alternative scenarios.

Our results show a reduction in carbon sequestration, potential stored carbon release, theoretical surface water sinks (i.e., evapotranspiration and infiltration), viewshed, open space and recreational value quality, with greater changes occurring in the open development scenario. They also show large gains in actual viewshed and open space proximity use, based on the increase in new housing units in the watershed (Table 10.2). Applying the range of monetary values described in the valuation section above for carbon and water to their modelled biophysical quantities, annual social costs in the constrained development scenario ranged from US$2.7 to US$12.1 million and costs for the open development scenario were US$3.3 to US$16.0 million (Table 10.3). Although ARIES

Table 10.3 Ecosystem service value changes for urban growth scenarios – value ranges under alternative valuation assumptions (US$)

	Open Development	Constrained Development
Annual value of sequestered carbon	(−2436000 −10258000)	(−2412000 −9756000)
Annual cost of reduced infiltration	(−822000 −5777000)	(−327000 −2301000)
Total annual cost	(−3258000 −16035000)	(−2739000 −12057000)

Table 10.4 Biophysical values for ecosystem service changes for mesquite management scenario

Ecosystem Service	Pre-mesquite Management	Post-mesquite Management
Carbon sequestration, tons/yr (change)	14152	14004 (−148)
Potential stored carbon release, tons/yr (change)	96900	94200 (−2700)
Theoretical surface water sink, m^3/yr reduced evapotranspiration (change)	5561000	5545000 (−16000)

provides uncertainty estimates for each cell, it is difficult to meaningfully aggregate these uncertainty estimates across the landscape into a single uncertainty measure. We thus present only the summed values below, without associated uncertainty information.

Mesquite management scenario

Restoration projects such as mesquite management fall under the SPRNCA's mandate 'to protect, enhance, and maintain the riparian area and the aquatic, wildlife, archaeological, paleontological, scientific, cultural, educational, and recreational resources of the public lands surrounding the San Pedro River in Cochise County, Arizona' (16 USC. 460xx – Establishment, para. (a)). In real-world terms, however, restoration decisions hinge on prioritization of scarce resources and the cost of and public response to alternative options. Ecosystem services provide a quantifiable metric for comparison with other costs and benefits. We quantified a reduction in carbon sequestration and evapotranspiration, effectively increasing water yield (Table 10.4).

Applying monetary values to the changes in carbon sequestration and water yield, we found the summed annual social cost of lost carbon storage and benefit of increased water yield to be between –US$7750 and +US$35135 (Table 10.5). Unlike the urbanization scenario, for which increased water yield equated to reduced groundwater recharge and a flashier hydrograph, increased water yield in the mesquite management scenario occurs largely through reduced evapotranspiration as mesquite is replaced by grassland (Nie et al., 2011). We thus considered it a positive value.

Table 10.5 Ecosystem service value changes for mesquite management scenario – value ranges under alternative valuation assumptions (US$)

	Post-mesquite Management
Annual value of sequestered carbon	(−3 250 to −13 210)
Annual reduced evapotranspiration	+5 460 to 38 400
Total annual value	(−7 750) to +35 150

Table 10.6 Relative rankings and change for recreation source values for water augmentation scenarios

	SPRNCA Baseline	Uniform 0.5 m Groundwater Rise Across SPRNCA	Entire SPRNCA Perennial
Birding	65 500	68 000 (+3.9%)	68 700 (+4.9%)
Wildlife viewing	86 100	89 200 (+3.7%)	90 300 (+4.9%)
Javelina hunting	78 800	82 000 (+4.1%)	82 400 (+4.6%)
Dove hunting	83 100	85 900 (+3.4%)	86 300 (+3.9%)
Deer hunting	75 300	78 300 (+4%)	78 700 (+4.6%)
Quail hunting	84 000	86 800 (+3.3%)	87 200 (+3.8%)

Water augmentation scenarios

Changes in relative rankings for recreation site quality for the SPRNCA are presented in Table 10.6. Without more detailed information on the vegetative changes associated with water augmentation scenarios it was not possible to apply the other ecosystem service models. Improved scenarios that better described the biophysical change associated with water augmentation, combined with better underlying datasets and models, could support better ecosystem service quantification for these scenarios. Valuing the effects of water augmentation scenarios on recreational use would have required primary data collection that was beyond the scope of this study, so only non-monetary site quality results are presented.

10.5 DISCUSSION AND CONCLUSIONS

10.5.1 Study Implications

The promise of using ecosystem service assessment to support decision-making within government agencies rests on being able to systematically measure impacts in a credible, quantifiable and replicable manner (Daily et al., 2009). Ecosystem service impacts can often be described qualitatively without the need for models. However, modelling approaches can quantify results spatially in biophysical units, relative rankings, weighted preferences, or dollars for baseline conditions and a range of potential management scenarios. Maps of impacts, trade-offs and values can facilitate clearer communication to decision makers and the public, and can be an important addition to the decision process. This project was intended as a proof-of-concept for ecosystem service modelling tools and was not expressly intended to guide specific management decisions, yet it does offer insight into the use of ecosystem service modelling tools in decision-making.

Before undertaking this project, stakeholders on the San Pedro held preconceived ideas that urban growth, particularly at low densities, would entail some loss of ecosystem services, that mesquite management would involve trade-offs between carbon, water and habitat quality, and that water augmentation for the San Pedro could provide some ecosystem service benefits. Our results map and quantify ecosystem service trade-offs and provide monetized values for some services, and generally conformed to our initial expectations of the types and degree of ecosystem service changes under the scenarios. All scenarios illustrate some degree of uncertainty in the results, based on the models themselves and the economic values and discount rate applied. For the mesquite management scenario, the total value ranges from a loss to a gain, depending on how water and carbon are valued. Because of this uncertainty and the inability to monetize habitat quality, incorporation of economic values into a strict cost–benefit framework for decision-making is likely inappropriate for this type of management decision.

Rising demand for ecosystem services can lead to increases in their value, even as ecosystems are being degraded. The ARIES viewshed and proximity results are good examples of this – in both urbanization scenarios, landscape quality declines while at the same time becoming more valuable as ecosystem service use increases with more users present on the landscape. It is thus important that rising ecosystem service values should not always be equated to improvements in ecosystem quality.

Since ARIES produces ecosystem services and uncertainty maps for up to 16 source, sink, use and flow characteristics, further discussions with decision-makers would be valuable to better understand which flow and uncertainty information is most helpful for the decision process. It would also be useful to know how uncertainty is or could be used in decision-making. Are outcomes with a highly uncertain but large gain in ecosystem service provision preferable over small gains with greater certainty? A better understanding of how to communicate and use uncertainty and spatial flow information would thus be valuable for the ecosystem service science and policy community. Additionally, some of our results, particularly for water yield, translated poorly into metrics that were amenable to valuation. In each case, a better understanding of stakeholder needs in decision--making would help modellers. This is a broad issue faced by researchers across the field of ecosystem services (Boyd and Krupnick, 2009). Preliminary consensus is emerging around the concept of using ecological endpoints or 'final ecosystem goods and services' (FEGS, Boyd and Banzhaf, 2007; Nahlik et al., 2012) as metrics that can be measured or modelled to better support valuation. FEGS link directly to green accounting frameworks and avoid the problem of double counting noted in some past valuation efforts. This approach meshes well with the beneficiary-based approach taken by ARIES (Villa et al., 2014).

10.5.2 Next Steps for ARIES Modelling in the San Pedro

Despite the strengths of the ARIES modelling platform – its flexibility, intelligent modelling approach, and ability to account for uncertainty and spatial flows of services – it had several key limitations as of late 2012 (many of these limitations having been addressed as this chapter went to press in early 2014, see Villa et al., 2014). Constructing custom ARIES models for a new area can be a time-consuming process, though it is substantially less time consuming upon completion of this project than at its start. It also requires use of an open source but at the time minimally documented modelling language. While this made construction of new models infeasible for third party users, future improvements including a fully documented, graphical user interface (GUI)-based modelling language were underway in late 2012, and should make independent modelling much more feasible. ARIES' late 2012 release ran in a command-line interface that is not accessible for external users. However, access to the ARIES web tool through an internet browser was planned for a future release. While ARIES models account for local influences on service provision and use, a globally applicable, generalized model for each ecosystem service had not yet been developed as of late 2012, though such models are planned for future releases. This means that ARIES could only be run in its case study regions. Even for these regions, further model testing and refinement could improve the models' quality and relevance for decision-making. ARIES' late 2012 release had difficulty in handling cross-boundary datasets, and despite its intent to be able to run under conditions of data scarcity, we were unable to run the models for the Mexican side of the watershed. Better handling of cross-boundary datasets and linkages to external models, planned for future ARIES releases, will improve model accuracy and credibility.

Additionally, comparative analysis of ARIES and other spatially explicit ecosystem service modelling systems, such as InVEST, to diverse but common settings will aid modellers and decision-makers in understanding the strengths, limitations and areas for improvement of alternative ecosystem service modelling approaches across diverse contexts. Comparative results for InVEST and ARIES are presented elsewhere (Bagstad et al., 2013b), and showed similar types, though different magnitudes, of ecosystem service changes under scenarios. Both tools would thus lead to similar conclusions about ecosystem services trade-offs in decision-making. However, these tools do use different ecosystem service metrics and modelling approaches or philosophies – both with scientific validity – so further comparative applications, perhaps in concert with more robust biophysical models, will be useful in understanding when different approaches more accurately quantify ecosystem services.

For the San Pedro, planned enhancements to the ARIES system have implications for model flexibility, accuracy and access by managers interested in applying ecosystem services to decision-making. Improvements to the modelling language and online interface should allow more people to use and customize ARIES ecosystem service models for the San Pedro and other sites around the world. Improvements to ARIES' model library, including global models and intelligent model selection, should facilitate modelling of other ecosystem services for the San Pedro beyond the five modelled as part of this study. Planned linkages to external models would improve model trust and credibility by linking to well-accepted external biophysical models for the San Pedro. Further model testing and refinement, including Bayesian network training using new training datasets, should improve model accuracy. Finally, improved cross-boundary data handling will support ecosystem service mapping and valuation on both sides of the US–Mexico border, improving our understanding of cross-boundary resource management challenges and impacts.

ACKNOWLEDGEMENTS

Rob Winthrop and Joel Larson coordinated project work with the BLM. Numerous other individuals assisted with project scoping and expert model review as acknowledged in Bagstad et al. (2013a). Partial funding for this work was provided by the US Geological Survey's Mendenhall Research Fellowship Program. We thank Ron Raunikar for providing constructive editorial comments on an earlier draft of this chapter. Any use of trade, product, or firm names in this chapter is for descriptive purposes only and does not imply endorsement by the US Government.

REFERENCES

Ackerman, F. and E.A. Stanton (2012), 'Climate risks and carbon prices: revising the social cost of carbon', *Economics: The Open-Access, Open-Assessment E-Journal*, **6**(10), 1–25.

Bagstad, K.J., F. Villa, G. Johnson and B. Voigt (2011), *ARIES – Artificial Intelligence for Ecosystem Services: A Guide to Models and Data, Version 1.0 Beta*, Bilbao: The ARIES Consortium.

Bagstad, K.J., D. Semmens, S. Waage and R. Winthrop (2013a), 'A comparative assessment of tools for ecosystem services quantification and valuation', *Ecosystem Services*, **5**, 27–39.

Bagstad, K.J., D. Semmens and R. Winthrop (2013b), 'Comparing approaches to spatially explicit ecosystem service modeling: a case study from the San Pedro River, Arizona', *Ecosystem Services*, **5**, 40–50.

Bagstad, K.J., G.W. Johnson, B. Voigt and F. Villa (2013c), 'Spatial dynamics of ecosystem service flows: a comprehensive approach to quantifying actual services', *Ecosystem Services*, **4**, 117–25.

Bark, R.H., D.E. Osgood, B.G. Colby, G. Katz and J. Stromberg (2009), 'Habitat preservation and restoration: do homebuyers have preferences for quality habitat?', *Ecological Economics*, **68**(5), 1465–75.

Bourassa, S.C., M. Hoesli and J. Sun (2006), 'The price of aesthetic externalities', *Appraisal Journal*, **74**(1), 14–29.

Boyd, J. and S. Banzhaf (2007), 'What are ecosystem services? The need for standardized environmental accounting units', *Ecological Economics*, **63**(2–3), 616–26.

Boyd, J. and A. Krupnick (2009), 'The definition and choice of environmental commodities for nonmarket valuation', Resources for the Future Discussion Paper No. 09–35, Washington, DC: Resources for the Future.

Brookshire, D.S., D. Goodrich, M.D. Dixon, L.A. Brand, K. Benedict and K. Lansey et al. (2010), 'After restoration: a framework for preserving semi-arid regions in the Southwest', *Journal of Contemporary Water Research and Education*, **144**, 60–74.

BSR Environmental Services, Tools, and Markets Working Group (2011), *New Business Decision-making Aids in an Era of Complexity, Scrutiny, and Uncertainty: Tools for Identifying, Assessing, and Valuing Ecosystem Services*, San Francisco, CA: BSR.

Bureau of Land Management (BLM) (2005), *BLM Land Use Planning Handbook*, H-1601–1, Washington, DC: US Department of the Interior, Bureau of Land Management.

Bureau of Reclamation (BOR) (2007), *Appraisal Report: Augmentation Alternatives for the Sierra Vista Sub-watershed, Arizona: Lower Colorado Region*, Denver, CO: US Department of the Interior Bureau of Reclamation.

Colby, B.G. and P. Orr (2005), 'Economic tradeoffs in preserving riparian habitat', *Natural Resources Journal*, **45**(1), 15–32.

Daily, G.C., S. Polasky, J. Goldstein, P.M. Kareiva, H.A. Mooney and L. Pejchar et al. (2009), 'Ecosystem services in decision-making: time to deliver', *Frontiers in Ecology and the Environment*, **7**(1), 21–8.

Dixon, M.D., J.C. Stromberg, J.T. Price, H. Galbraith, A.K. Fremier and E.W. Larsen (2009), 'Potential effects of climate change on the Upper San Pedro riparian ecosystem', in J.C. Stromberg and B. Tellman (eds), *Ecology and Conservation of the San Pedro River*, Tucson, AZ: University of Arizona Press, pp. 57–72.

Johnson, G.W., K. Bagstad, R. Snapp and F. Villa (2012), 'Service Path Attribution Networks (SPANs): a network flow approach to ecosystem service assessment', *International Journal of Agricultural and Environmental Information Systems*, **3**(2), 54–71.

Kepner, W.G., C.J. Watts, C.M. Edmonds, J.K. Maingi, S.E. Marsh and G. Luna (2000), 'Landscape approach for detecting and evaluating change in a semiarid environment', *Environmental Monitoring and Assessment*, **64**(1), 179–95.

Kepner, W.G., D.J. Semmens, D.T. Heggem, E.J. Evanson, C.M. Edmonds and S.N. Scott (2003), 'The San

Pedro River Geo-data Browser and Assessment Tools', EPA/600/C-03/008, accessed 5 February 2014 at http://www.epa.gov/nerlesd1/land-sci/san_pedro/index.html.
Loomis, J.B. (1995), 'Four models for determining environmental quality effects on recreational demand and regional economics', *Ecological Economics*, **12**(1), 55–65.
Millennium Ecosystem Assessment (MA) (2005), *Millennium Ecosystem Assessment: Living Beyond Our Means – Natural Assets and Human Well-Being*, Washington, DC: World Resources Institute.
Moran, M.S., W.E. Emmerich, D.C. Goodrich, P. Heilman and C.D. Holifield Collins et al. (2008), 'Preface to special section on Fifty Years of Research and Data Collection: US Department of Agriculture Walnut Gulch Experimental Watershed', *Water Resources Research*, **44**(5), W05S01.
Nahlik, A.M., M.E. Kentula, M.S. Fennessy and D.H. Landers (2012), 'Where is the consensus? A proposed foundation for moving ecosystem service concepts into practice', *Ecological Economics*, **77**(C), 27–35.
Nie, W., Y. Yuan, W. Kepner, M.S. Nash, M. Jackson and C. Erickson (2011), 'Assessing impacts of landuse and landcover changes on hydrology for the Upper San Pedro watershed', *Journal of Hydrology*, **407**(1–4), 105–14.
Orr, P. and B.G. Colby (2002), *Nature-oriented Visitors and Their Expenditures: Upper San Pedro River Basin*, Department of Agricultural and Resource Economics, University of Arizona, accessed 5 February 2014 at http://ag.arizona.edu/AREC/pubs/san_pedro_report.pdf.
Piper, S. and W.E. Martin (1997), 'Household willingness to pay for improved rural water supplies: a comparison of four sites', *Water Resources Research*, **33**(9), 2153–63.
Ruhl, J.B., S.E. Kraft and C.L. Lant (2007), *The Law and Policy of Ecosystem Services*, Washington, DC: Island Press.
Sengupta, S. and D.E. Osgood (2003), 'The value of remoteness: a hedonic estimation of ranchette prices', *Ecological Economics*, **44**(1), 91–103.
Serrat-Capdevila, A., J.B. Valdes, J. Gonzalez Perez, K. Baird, L.J. Mata and T. Maddock III (2007), 'Modeling climate change impacts – and uncertainty – on the hydrology of a riparian system: the San Pedro Basin (Arizona/Sonora)', *Journal of Hydrology*, **347**(3–4), 48–66.
Serrat-Capdevila, A., A. Browning-Aiken, K. Lansey, T. Finan and J.B. Valdes (2009), 'Increasing socio-ecological resilience by placing science at the decision table: the role of the San Pedro Basin (Arizona) Decision Support System Model', *Ecology and Society*, **14**(1), Article 37.
Steinitz, C., H. Arias, S. Bassett, M. Flaxman, T. Goode and T. Maddock III et al. (2003), *Alternative Futures for Changing Landscapes: The Upper San Pedro River Basin in Arizona and Sonora*, Washington, DC: Island Press.
Stern, N. (2007), *The Economics of Climate Change: The Stern Review*, Cambridge, UK: Cambridge University Press.
Stromberg, J.C. and B. Tellman (eds) (2009), *Ecology and Conservation of the San Pedro River*, Tucson, AZ: University of Arizona Press.
Tallis, H.T., T. Ricketts, A.D. Guerry, S.A. Wood, R. Sharp and E. Nelson et al. (2013), *InVEST 2.6 User's Guide*, Stanford, CA: The Natural Capital Project.
Tol, R.S.J. (2005), 'The marginal damage costs of carbon dioxide emissions: an assessment of the uncertainties', *Energy Policy*, **33**(16), 2064–74.
Tol, R.S.J. (2008), 'The social cost of carbon: trends, outliers, and catastrophes', *Economics: The Open-Access, Open-Assessment E-Journal*, **2**(25), 1–22.
Villa, F., L. Tunesi and T. Agardy (2002), 'Zoning marine protected areas through spatial multiple-criteria analysis: the case of the Asinara Island National Marine Reserve of Italy', *Conservation Biology*, **16**(2), 515–26.
Villa, F., K.J. Bagstad, B. Voigt, G.W. Johnson, R. Portela, M. Honzak and D. Batker (2014), 'A methodology for adaptable and robust ecosystem services assessment', *PLOS ONE*, **9**(3), e91001.
Water Infrastructure Finance Authority of Arizona (WIFA) (2009), *2008 Water and Wastewater Residential Rate Survey for the State of Arizona*, Phoenix, AZ: Water Infrastructure Finance Authority of Arizona.

11. Optimal selection of clustered conservation lands using integer programming: the case of Fort Stewart in Georgia, USA

Sahan T.M. Dissanayake, Hayri Önal, James D. Westervelt and Harold E. Balbach

C61 Q57 Q58
Q24 Q28

11.1 INTRODUCTION

Suitable habitat areas for many rare, threatened, or endangered species in North America may be located on or near military installations in the USA (Stein et al., 2008). Figure 11.1 shows that Department of Defense (DoD) lands have the highest density of and second highest distribution of endangered and imperiled species amongst all the Federal land management agencies.

While military training may cause some habitat deterioration, military control of the lands actually prevents destructive urban and agricultural development. This has the potential to create a concentration of valuable habitat in and around military installations (Orth and Warren, 2006). Besides protection of the lands from alternative economic uses, the DoD also allocates a significant amount of human capital and land for conservation efforts toward protecting and managing wildlife habitat in and around military installations. In 2006, the DoD spent US$4.1 billion on environment-related expenses, of which US$1.4 billion was for environment restoration and US$204.1 million was for conservation (Benton et al., 2008). At the same time, new and conventional training requirements are increasing, making it difficult to set aside land solely for conservation purposes within military installations and elevating the importance of managing military lands for multiple competing objectives. As an alternative to more costly arrangements, such as purchasing land and sharing land with other agencies, effective utilization of the existing lands for conservation and military purposes can be accomplished by designing an optimum landscape that places conservation and military training areas in a desirable spatial configuration. This issue is the main motivation of this chapter where we present alternative optimum conservation reserve designs that incorporate multiple dependent species.

Fort Stewart, GA is an example of a military base supporting biological conservation. Fort Stewart currently has a substantial population of the gopher tortoise (*Gopherus polyphemus*), hereafter referred to as GT. The GT is a keystone species and is a candidate to be listed as Threatened under the Endangered Species Act. Fort Stewart also has a population of the gopher frog (*Rana capito*), hereafter referred to as GF, which is classified as a Species at Risk (Cash et al., 2008). The GF depends to a degree on GT burrows for survival and access to ephemeral ponds. In an effort to best co-manage the GT and GF populations, Fort Stewart is looking into the optimal selection of habitat areas that can be made available for the protection of these two species (among others, such as

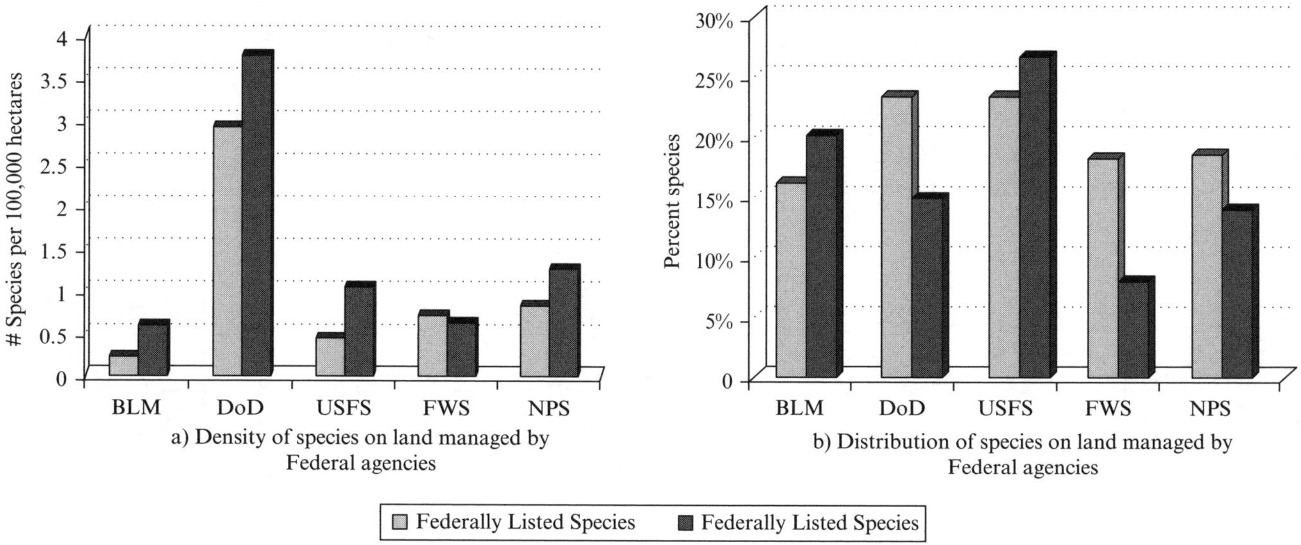

Note: BLM – Bureau of Land Management; DoD – Department of Defense; USFS – United States Forest Service; FWS – Fish and Wildlife Service; NPS – National Park Service.

Source: a. Figures modified from Stein et al. (2008).

Figure 11.1 Density and distribution of federally listed and imperiled species[a]

indigo snake and striped newt). This study aims to develop optimum land use strategies for the installation by incorporating various ecologically important considerations when determining the best possible management areas for both GTs and GFs without unduly hampering the military training activities.

11.2 LITERATURE REVIEW

We approach this problem in a manner similar to that involved in the design of 'reserves' for protection of certain sensitive species. The use of mathematical models in this area goes back to the 1980s (Kirkpatrick, 1983). The term 'reserves' in the standard sense is not entirely appropriate, however, in the case of conservation efforts on military installations where protection of targeted species and management of their habitats are always subject to training and mission requirements and Congressional authority. Therefore, we use the term 'Conservation Management Area (CMA)' with regard to the application. We do not present an extensive literature review of the reserve design model in this chapter. We summarize the literature briefly here and guide the interested reader to the extensive list of references.

Mathematical programming methods, in particular linear integer programming, have been used widely in the literature of biological conservation and reserve design. Initial studies used mostly heuristic methods for this purpose (Vane-Wright et al., 1991; Nicholls and Margules, 1993; Pressey et al., 1993, 1997; Margules and Pressey, 2000). Later, formal optimization models were introduced to either determine a least-cost site selection that provides suitable habitat to each and every target species (Underhill, 1994; Possingham et al., 2000; Rodrigues and Gaston, 2002) or maximize the number of species covered subject to a budget or area limitations (Camm, 1996; Church et al., 1996; Ando et al., 1998). Typically, the above optimum site selection model results in highly sparse and dispersed configurations (Figure 11.2).[1] Recognizing this deficiency, several integer programming models have been developed particularly in the past ten years to incorporate various forms of spatial

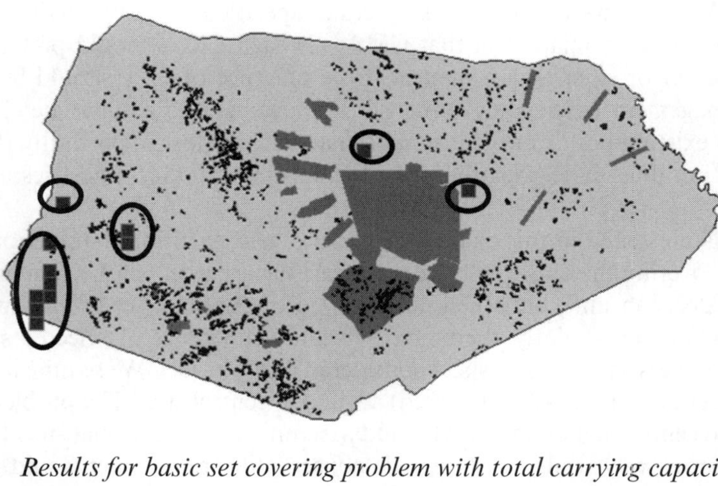

Figure 11.2 Results for basic set covering problem with total carrying capacity index of 5000

considerations, such as reserve connectivity, compactness, fragmentation, relocation, and so on (Williams and ReVelle, 1998; Cova and Church, 2000; Nalle et al., 2002; Önal and Briers, 2002, 2003, 2005, 2006; Cabeza, 2003; Cerdeira et al., 2005; Cerdeira and Pinto, 2005; Williams et al., 2005; Önal and Wang, 2008; Tóth et al., 2009; Dissanayake et al., 2012).

Prior work on this current problem of designing an optimum landscape that places conservation and military training areas in a desirable spatial configuration has focused on relocation of species (Dissanayake et al., 2012), and joint uses of military lands for training and conservation purposes (Dissanayake et al., 2011). This chapter builds upon and extends those studies by developing a model that incorporates multiple species that have different conservation requirements yet share the same areas. The model is applied to data representing GF and GT at Fort Stewart, Georgia.

11.3 GENERAL METHODS

The mathematical programming models developed here identify the conservation management areas (CMAs) to achieve/maintain desired levels of GT and GF populations while giving special emphasis to the location and size of the CMAs. Since GTs are a ground-bound species, the selected areas should be as 'compact' as possible, and preferably 'contiguous', in order to allow movement of individuals in the selected areas and facilitate interaction within and among multiple populations in those areas. A compact CMA would also be easier to fence, if needed. Furthermore, since GTs are a keystone species and GFs rely on GT burrows to survive, incorporating GF management areas into the model would further increase the efficiency of CMA selection because joint management of two species is always more efficient than independent management of individual species. Since GFs depend on access to water for a portion of their life cycle, the distances of GF sites to both ponds and the nearest GT habitat sites need to be considered when determining the best GT sites.

In light of the above, specifying the most suitable CMAs for GTs must involve various important ecological and spatial considerations including the following: (1) each designated CMA must have a minimum size, either specified in terms of the land area or in terms of the GT population in that CMA; (2) each CMA should preferably have a compact (circular or square-like) shape; (3) the presence of GFs should be considered for joint management efficiency, (4) the GF management areas must be close to both GT sites and existing ponds in the installation area; and most importantly (5) land uses for conservation must be compatible with the existing military land uses and training activities.

The models presented in this chapter consider a grid partition that comprises square land parcels covering the entire installation area.[2] Each parcel (site) is assumed to be an independent decision unit, but when determining the optimal set of sites we take into account their functions jointly, therefore the spatial coherence of selected sites, in particular compactness, is a crucial aspect considered in the model. We require a CMA to be formed by a set of sites packed (clustered) around a 'central site'. The problem is then to determine the central site of each CMA and assignment of individual sites to the CMA in an endogenous way while satisfying the conservation requirements and considering alternative spatial and species criteria in cluster formation.[3]

11.3.1 Base Model

In the application described in this chapter we start with a 'base model' that achieves clustered conservation areas. The base model is identical to the base model introduced in Dissanayake et al. (2011, 2012) and addresses the problem of constructing *n* compact CMAs, each covering a minimum sustainable GT population and collectively covering a desired GT population within the available budget. We present the base model below but refer the reader to Dissanayake et al. (2011, 2012) for a detailed description of it.

The notation used in the text is as follows. We denote the set of all sites by L and denote individual sites by $k, l, j \in L$. Site selection and assignment to a CMA is represented by a binary variable X_{lk}, where $X_{lk} = 1$ if site k is selected and belongs to the CMA centered at site l and $X_{lk} = 0$ otherwise. Note that by construct $X_{ll} = 1$ for all central sites l, that is, the central site of each CMA must belong to that CMA. We also note that sites in the most heavily used military training areas are not considered for inclusion in any CMA, therefore we set $X_{lk} = 0$ if site k is part of a training area. The symbol d_{lk} denotes the distance between site l and site k, and e_k denotes the existing population of GT in site k. The number of CMAs to configure is denoted by n, which is specified exogenously but varied when designing alternative optimal configurations. Each CMA is required to sustain a minimum GT population, denoted by p. Finally, the total GT population in all the selected areas is represented by tp. Using this notation the base model is given below:

Minimize $\sum_l \sum_k X_{lk} * d_{lk}$	Objective function: Sum of distances to centers
s.t.:	Constraints:
i) $\sum_l X_{ll} = n$	(i) Total number of CMAs = n
ii) $\sum_l X_{lk} \leq 1 \, \forall k$	(ii) Each site can belong to at most one CMA
iii) $\sum_k X_{lk} * e_k \geq p \, \forall l$	(iii) Minimum population for each CMA
iv) $\sum_l \sum_k X_{lk} * e_k \geq tp$	(iv) Minimum total population
v) $\sum_k X_{lk} \leq m X_{ll} \, \forall l$	(v) Site assignment can be done only if site l is a center ($X_{ll} = 1$)
vi) $\sum_l \sum_k X_{lk} * c_k \leq b$	(vi) Total cost must not exceed the available budget
$X_{lk} = 0, 1 \, \forall \, l, k$	Binary restrictions for site selection variables (1 if site k belongs to a reserve centered at site l, 0 otherwise)

The base model identifies the most suitable clusters to be considered as CMAs for GTs. However, it does not incorporate GF considerations. We next present a modification to the base model that determines GT and GF management areas simultaneously.

11.3.2 Simultaneous Selection of CMAs for GTs and GFs

The best CMAs for both GTs and GFs must also contain a minimum number of GF sites that are within 2 km of an existing pond. This is necessary since the GF life cycle requires access to a reliable water source and the maximum distance from a water source is known as 2 km, although shorter distances are preferable.[4]

In addition to the notation used earlier we define a new binary variable Y_k for site k, where $Y_k = 1$ if site k is selected as a designated GF habitat area and $Y_k = 0$ otherwise.[5] We also define the following symbols: f denotes the desired minimum number of sites assigned as GF parcels; dp_k denotes the distance between site k and the nearest pond, and $\bar{d}$ denotes the maximum allowed distance between a designated GF site and the nearest pond. Adding the two constraints below to the base model incorporates the GF management area requirements:

(vii) $Y_k \leq \sum_l X_{lk} \text{ for all } k$

(viii) $\sum_{k:dp_k \leq \bar{d}} Y_k \geq f$

(ix) $Y_k = 0, 1 \text{ for all } k$

Constraint (vii) ensures that only sites selected as GT sites can be considered as a GF site. In other words, if site k is designated as a GF site (i.e., $Y_k = 1$) then it must be assigned to some GT CMA centered at site l ($X_{lk} = 1$). Constraint (viii) ensures that the model selects at least f sites for GF protection. Note that a GF site can be considered as a designated site only if its distance from a pond is at most $\bar{d}$, as implied by the condition underlying the summation in (viii).

11.4 DATA

The data manipulation and model implementation was conducted using readily available software as presented in Figure 11.3.

Habitat suitability and military training information was obtained from personnel at Fort Stewart with the assistance from the Army Corps' Engineer Research and Development Center (ERDC) research scientists. The data were extracted into grid shape files using GeoDa and ArcGIS and imported into GAMS using Excel. The results were computed in GAMS and mapped using ArcGIS.

The current military training areas are shown in Figure 11.4a, the GT suitability is depicted in Figure 11.4b, and the locations of the ponds are shown in Figure 11.4c. A 55 × 30 grid file, where each grid cell is a 1000 m × 1000 m square, was created using GeoDa and the grid shape file was spatially joined with the above shape files using the spatial join tool in ArcGIS. The spatial join gives the grid file the attributes of the shape file. To ensure that each grid cell represents a density of the original data, the 'sum' option was used when joining the habitat suitability data. The grid cell values for Figure 11.4b are given as the sum of suitable points (the GT suitability raster map[6] was converted to point shape file) within the grid cell. The suitability index ranges from 0 to 600.[7]

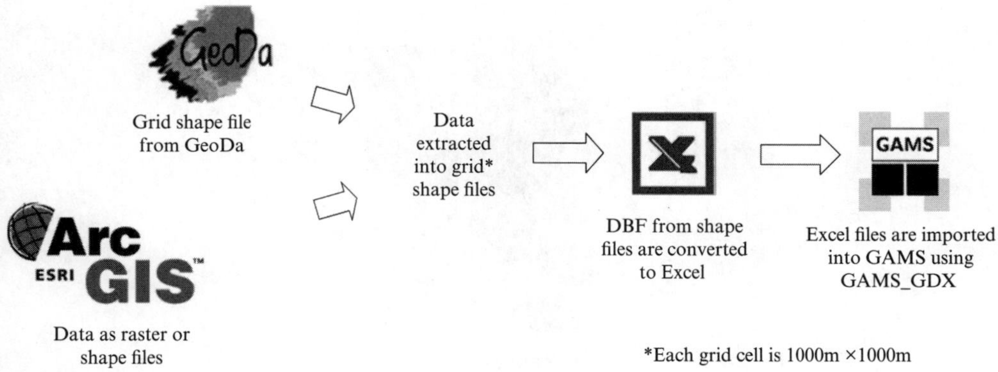

Figure 11.3 Data processing and software

11.5 RESULTS AND DISCUSSION

The models described above were solved using GAMS/CPLEX version 21.6 on a PC with an Intel Core 2 Duo processor and 2 GB of RAM running Windows XP with most of the presented results being generated after a few hours. In practice the GT populations can be moved either to a single large CMA or multiple smaller CMAs (all located outside the high-intensity military training areas). Therefore, we solve the model for a range of values for the number of CMAs. There are two reasons for specifying more than one CMA. First, in an effort to safeguard the population from total destruction (e.g., disease risk) we may seek to separate the populations into smaller disconnected subpopulations located in different parts of the installation. Second, designing multiple and relatively small conservation areas increases the flexibility for the military when further expansion of training areas is needed in future.

In all of the runs described below the minimum population for each CMA was specified as 5000.[8] The base model was solved with one, two, three and four CMAs. The joint species model was first solved for a minimum of ten GF parcels and then for 20 GF parcels.[9] A wide range of potential parameter values were tested after discussions with the Fort Stewart natural resource managers. We only present some of those results here to highlight the models' ability to: (1) optimally select the CMAs, (2) illustrate the workings of the models, and (3) demonstrate the trade-offs between incorporating different spatial and species criteria in site selection.

11.5.1 Base Model Results

The base model results are shown in Figures 11.5a, 11.6a, 11.7a and 11.8a for one, two, three and four CMAs respectively. Each map shows the installation with its range and pond locations. The results of the model runs are shown as square boxes representing GT sites that are clustered into the requested number of CMAs. The required number of GF are contained within the GT sites, and for Figure 11.5.b are denoted with an "F". Comparing the base model results in panel a in Figures 11.5–11.8 with the suitability

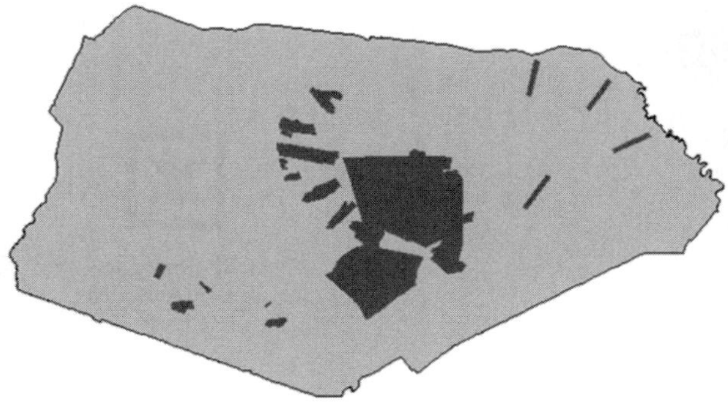

a. Ranges

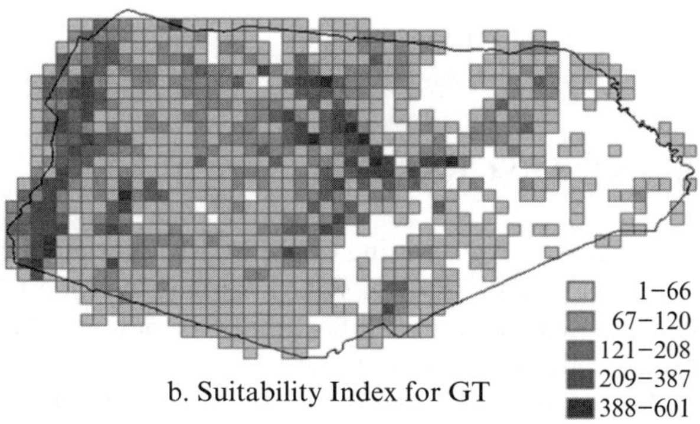

b. Suitability Index for GT

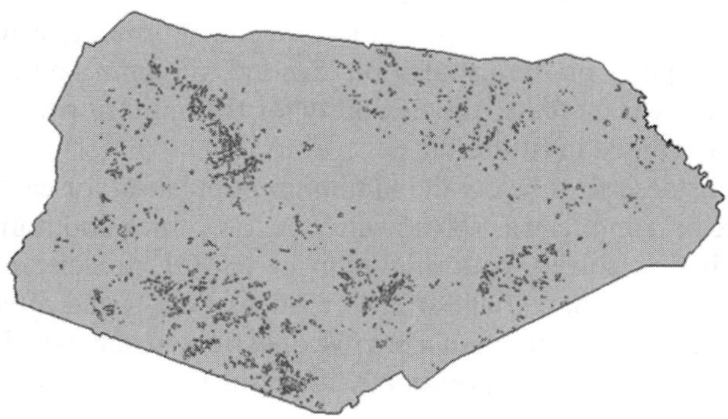

c. Location of Ponds

Figure 11.4a–c Summary of data

map given in Figure 11.4b illustrates that the base model simply selects from amongst the most densely packed and best available sites to form contiguous and compact CMAs. The optimal solution with one large conservation area (Figure 11.5a) shows that this area would be located at the southwest corner of the installation. The CMA is contiguous but the compactness of the CMA is poor and the selected sites are meandering in shape. Also, the solution has 16 sites as opposed to the 12 sites in the basic set covering problem

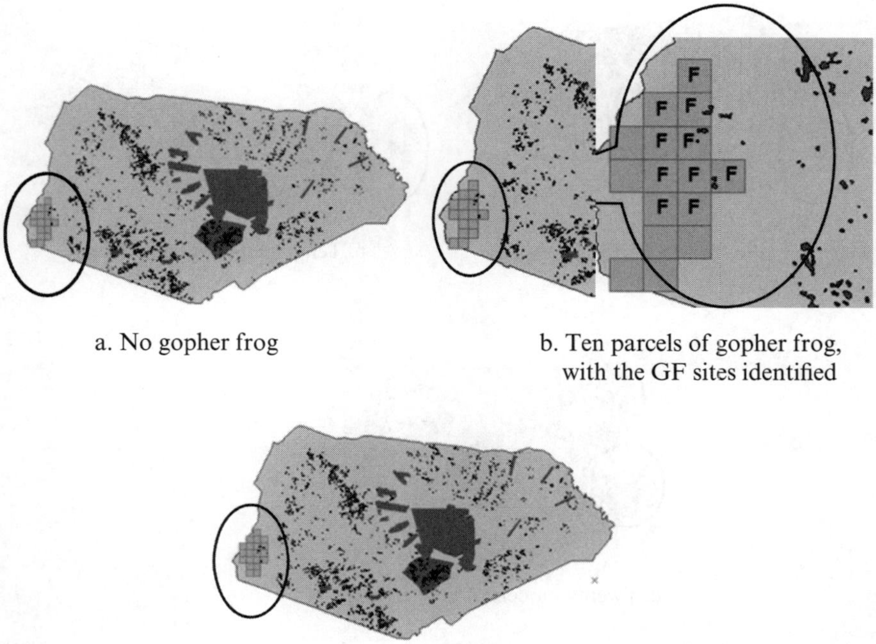

a. No gopher frog

b. Ten parcels of gopher frog, with the GF sites identified

c. Twenty parcels of gopher frog

Figure 11.5 Results for one cluster of GT with total carrying capacity index of 5000

(see Figure 11.2). The lack of compactness and the increase in the number of selected sites are both driven primarily by the fact that the model is forced to choose one cluster of habitat sites that meet the population criteria and the only available large quantity of good-quality sites are in that part of the installation. The good-quality sites in other parts of the installation are not in the solution for two reasons: (1) those sites are under military use, or (2) those sites are located far apart from each other.

For the two-CMA case the model chooses two clusters with seven and eight sites, respectively (Figure 11.6a) for a total of 15 sites. Although the two clusters are again selected in the southwest corner of the installation, allowing for two clusters enables the model to achieve the population goal with one less site than the previous (one-cluster) case. The three-CMA case selects a total of 14 sites (Figure 11.7a), with two clusters in the southwest part of the installation and one cluster in the north-central part of the installation. Finally, the four-CMA case selects 13 sites from three separate areas as shown in Figure 11.8a. This clearly demonstrates that as more CMAs are considered the model is able to choose fewer and better sites in different parts of the installation, decreasing the total area needed for the same level of conservation. Unlike the one big CMA scenario, the two-, three – and four-CMA configurations are comprised of compact clusters of sites as opposed to the meandering configuration in Figure 11.5a. Based on these results, we may conclude that if the size of the total area of all CMAs is a concern, forming four CMAs, two located in the southwest, one located in the west-central area and one located

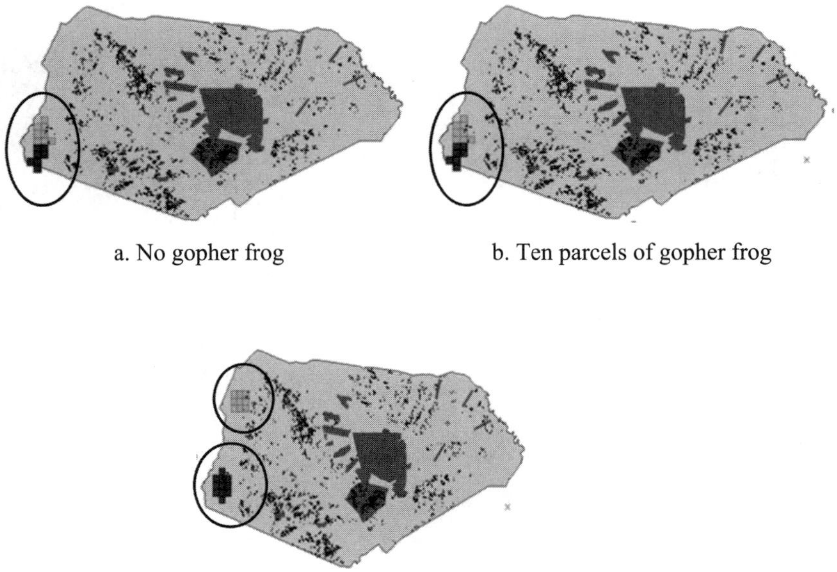

a. No gopher frog　　b. Ten parcels of gopher frog

c. Twenty parcels of gopher frog

Figure 11.6　Results for two clusters of GT with total carrying capacity index of 5000

in the north-central areas, is the best strategy as it selects 13 sites only. It is noteworthy to state that this alternative includes just one more site than the scattered configuration given in the set covering solution (Figure 11.2).

11.5.2 Joint Management Results

The results of the joint management of both GTs and GFs are shown in Figures 11.5–11.8 for one, two, three and four CMAs respectively. In each of the Figures 11.5–11.8, Figure b displays the results for at least ten GF sites ($f = 10$) and Figure c displays the results for at least 20 GF sites ($f = 20$). The optimal solution with one large conservation area and ten GF sites (Figure 11.5b) shows that this area would again be located at the southeast corner of the installation and is identical to the solution without GF considerations. This is because as depicted in Figure 11.5b, there are ten sites in that area that are within 2 km distance from a pond in that solution. When the number of GF sites is increased to 20 sites the selected sites are still in the southwest corner of the installation, but the locations change since the model now has to add more sites that are located within 2 km from a pond.

The results for two CMAs are shown in Figure 11.6. For ten GF sites the optimal configuration is similar to the base model solution. When 20 GF sites are required the results change dramatically, however, as the model selects one compact CMA with nine sites and another one with 11 sites, that are located away from each other and close to the locations of the ponds. The results for three CMAs are shown in Figure 11.7. In the base model solution one CMA was located in the north-central region away from

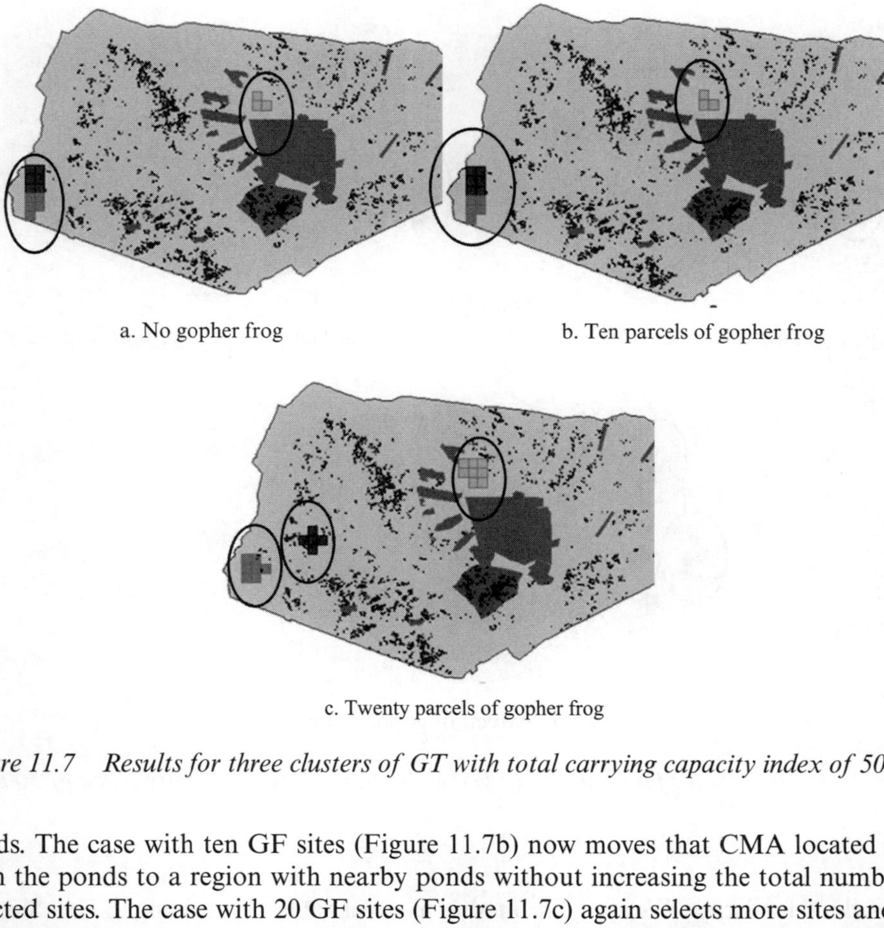

a. No gopher frog

b. Ten parcels of gopher frog

c. Twenty parcels of gopher frog

Figure 11.7 Results for three clusters of GT with total carrying capacity index of 5000

ponds. The case with ten GF sites (Figure 11.7b) now moves that CMA located away from the ponds to a region with nearby ponds without increasing the total number of selected sites. The case with 20 GF sites (Figure 11.7c) again selects more sites and has three CMAs that are located in different regions of the base. The solution including four CMAs (Figure 11.8) shows that it is possible to meet the 5000 GT population and the ten GF site goals with only 13 sites, just one more site than the set covering solution and same as the GT-only solution. When requiring 20 GF sites the optimal selection includes more sites that are located in the west side of the installation and part of the nicely grouped compact GT clusters. Clearly this is a much more preferred configuration, as opposed to the spatially unrestricted (and thus scattered) configuration shown in Figure 11.1b. In general, allowing for four CMAs results in more compact CMAs since the model is able to place the smaller CMAs in the most suitable areas, yet the individual CMAs are large enough to support a minimum viable population of GTs assumed in the analysis.

11.6 CONCLUSIONS

This chapter presents an example of how linear integer programming formulations can be used to design conservation management areas (CMAs) for inter-dependent species

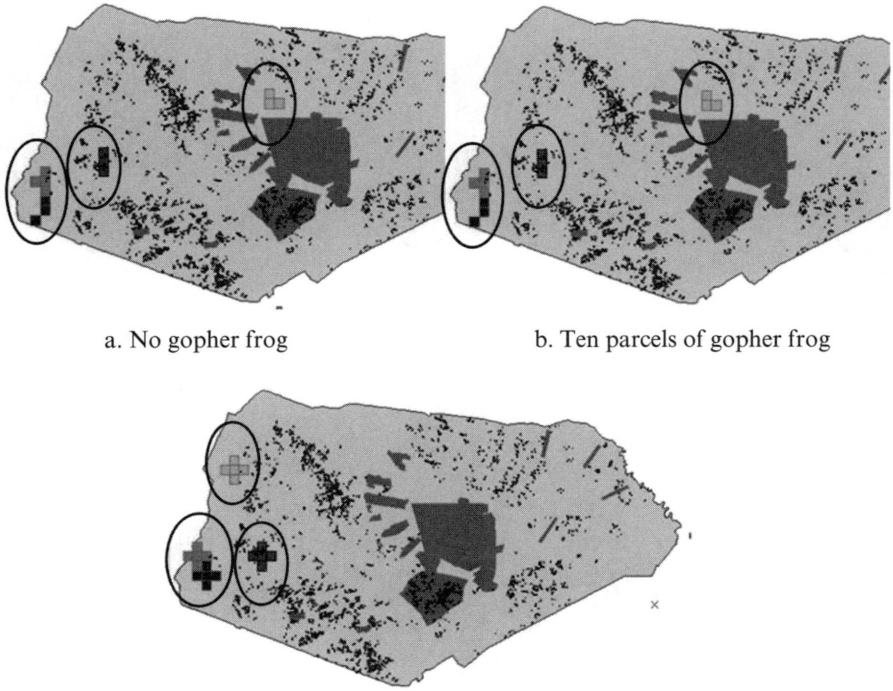

a. No gopher frog

b. Ten parcels of gopher frog

c. Twenty parcels of gopher frog

Figure 11.8 Results for four clusters of GT with total carrying capacity index of 5000

in a working landscape. We apply the models to a real dataset pertaining to Fort Stewart, a military installation where protection of gopher tortoise, a keystone species at risk, is of concern. The modeling approach presented here is more general than protection of GTs in a military installation, however, and is applicable to selecting conservation areas for any terrestrial species with a dependent secondary species. Though the models are complex, the empirical evidence demonstrates that they are computationally convenient (can be solved within a reasonable computation time, at least for the dataset used here).

The results of the models are consistent with intuition and reflected the desired outcomes:

- The models selected compact GT clusters.
- Considering multiple CMAs reduces the total amount of managed areas by selecting fewer and better sites.
- Incorporating GF requirements into the GT analysis does not change the results for a small number of GF sites but the results change considerably when a large number of GF sites are required.

We note that both the single and joint species conservation management models are solvable in a short computation time, which suggests that the formulations presented here can be applied to much larger datasets. In all cases, the optimum solutions were obtained only in a few minutes of processing time. It should also be noted that adding extra requirements to the model, such as the additional spatial considerations or additional dependent species conservation requirements, may force the model to select from among less suitable parcels when the best parcels do not meet the specified criteria. This can lead to the selection of larger CMAs, or poorer compactness of some CMAs. Therefore, there is a trade-off between incorporating additional requirements and the economic efficiency in optimal selection of conservation CMAs. These results provide general guidelines and will be useful for on-the-ground decision-makers.

According to the results it is possible to identify the optimum selection of compact sites that form up to four centrally placed CMAs within the boundaries of the particular military installation. The CMAs become smaller and more compact, and comprise higher-quality sites as the allowed number of CMAs is increased. However, they may be dispersed throughout the installation area. When GF considerations are included, the model identifies CMAs that simultaneously serve as good GT habitats and also GF habitats, indicating that ecological considerations for multiple species can be incorporated jointly in a unified framework.

ACKNOWLEDGMENTS

This research was partially supported by the US Army Engineer Research and Development Center-Construction Engineering Research Laboratory (ERDC-CERL) Project No. W81EWF-7204–6330 and Cooperative State Research, Education, and Extension Service (CREES) Project No. ILLU 05–0361.

NOTES

1. The results for a basic set covering problem (SCP) with total carrying capacity index of 5000 for the dataset used in this chapter are presented in Figure 11.2, which clearly shows the scattered sites (the selected sites are the dark square cells, identified by the surrounding circles, the lighted polygons show the training areas and the darker points identify the location of ephemeral ponds as detailed in Section 11.4).
2. The square-cell assumption is not restrictive. The approach developed here can be applied to other geometric forms, such as triangles, rectangles, polygons, or even irregular forms.
3. This model is an extension of classic p-median problem (Garfinkel et al., 1974). Similar models for clustering have been used previously in the literature of reserve design (Williams, 2008).
4. Having the ponds located less than 1 km is better in practice; in this particular example given the resolution of the data we use 2 km to illustrate how the model functions.
5. As formulated we require that only sites selected as GT sites can be considered as GF sites.
6. GT suitability values were calculated by Dr. James Westervelt and Dr. Tracey Tuberville.
7. The carrying capacity values in the suitability map are GT/ha. The number of tortoises in each grid cell = (suitability value of grid cell/121)*100. A one-hectare land parcel can support between two to five GTs. This is equivalent to supporting between 200–500 GTs per site at the 1000 m × 1000 m resolution.
8. The minimum sustainable population size for gopher tortoises varies considerably; see Styrsky et al. (2010) for an analysis of GT population threshold levels.
9. The only GF criteria we required are that a GF site has to be a GT site and also be within 2 km of a pond. These criteria can be refined based on the available data: for example, instead of considering all ponds, only ponds that are larger than a certain size or have water during the GF breeding season.

REFERENCES

Ando, A., J. Camm, S. Polasky and A. Solow (1998), 'Species distributions, land values, and efficient conservation', *Science*, **279**(5359), 2126–8.

Benton, N., J.D. Ripley and F. Powledge (eds) (2008), *Conserving Biodiversity on Military Lands: A Guide for Natural Resources Managers*, Arlington, VA: NatureServe.

Cabeza, M. (2003), 'Habitat loss and connectivity of reserve networks in probability approaches to reserve design', *Ecological Letters*, **6**(7), 665–72.

Camm, J.D. (1996), 'A note on optimal algorithms for reserve site selection', *Biological Conservation*, **78**(3), 353–5.

Cash, W.B., J.B. Jensen and D.J. Stevenson (2008), 'Gopher frog *Rana (Lithobates) capito*', in J.B. Jensen, C.D. Camp, J.W. Gibbons and M.J. Elliott (eds), *Amphibians and Reptiles of Georgia*, Athens, GA: University of Georgia Press, pp. 485–7.

Cerdeira, J.O. and L.S. Pinto (2005), 'Requiring connectivity in the set covering problem', *Journal of Combinatorial Optimization*, **9**(1), 35–47.

Cerdeira, J.O., K.J. Gaston and L.S. Pinto (2005), 'Connectivity in priority area selection for conservation', *Environmental Modeling and Assessment*, **10**(3), 183–92.

Church, R.L., D.M. Stoms and F.W. Davis (1996), 'Reserve selection as a maximal covering location problem', *Biological Conservation*, **76**(2), 105–12.

Cova, T.J. and R.L. Church (2000), 'Contiguity constraints for single-region site search problems', *Geographical Analysis*, **32**(4), 306–29.

Dissanayake, S., H. Önal and J.D. Westervelt (2011), 'Optimum selection of conservation reserves: extensions to multiple land use', *Military Operations Research*, **16**(1), 65–76.

Dissanayake, S., H. Önal, J.D. Westervelt and H.E. Balbach (2012), 'Incorporating species relocation in reserve design models: an example from Fort Benning GA', *Ecological Modeling*, **224**(1), 65–75.

Garfinkel, R.S., A.W. Neebe and M.R. Rao (1974), 'An algorithm for the m-median plant location problem', *Transportation Science*, **8**(3), 217–36.

Kirkpatrick, J. (1983), 'An iterative method for establishing priorities for the selection of nature reserves: an example from Tasmania', *Biological Conservation*, **25**(2), 127–34.

Margules, C.R. and R.L. Pressey (2000), 'Systematic conservation planning', *Nature*, **405**(6783), 243–53.

Nalle, D.J., J.L. Arthur, C.A. Montgomery and J. Sessions (2002), 'Economic and spatial impacts of an existing reserve network on future augmentation', *Environmental Modeling and Assessment*, **7**(2), 99–105.

Nicholls, A. and C. Margules (1993), 'An upgraded reserve selection algorithm', *Biological Conservation*, **64**(2), 165–9.

Önal, H. and R.A. Briers (2002), 'Incorporating spatial criteria in optimum reserve network selection', *Proceedings of the Royal Society of London. Series B: Biological Sciences*, **269**(1508), 2437–41.

Önal, H. and R.A. Briers (2003), 'Selection of a minimum–boundary reserve network using integer programming', *Proceedings of the Royal Society of London. Series B: Biological Sciences*, **270**(1523), 1487–91.

Önal, H. and R.A. Briers (2005), 'Designing a conservation reserve network with minimal fragmentation: a linear integer programming approach', *Environmental Modeling and Assessment*, **10**(3), 193–202.

Önal, H. and R.A. Briers (2006), 'Optimal selection of a connected reserve network', *Operations Research*, **54**(2), 379–88.

Önal, H. and Y. Wang (2008), 'A graph theory approach for designing conservation reserve networks with minimal fragmentation', *Networks*, **51**(2), 142–52.

Orth, P.B. and S.D. Warren (2006), *Disturbance-dependency of Threatened and Endangered Species Found on US Army Lands*, CEMML TPS, 06–17, Center for Environmental Management of Military Lands, Colorado State University.

Possingham, H., I. Ball and S. Andelman (2000), 'Mathematical methods for identifying representative reserve networks', in S. Ferson and M. Burgman (eds), *Quantitative Methods for Conservation Biology*, Dusseldorf: Springer-Verlag, pp. 291–306.

Pressey, R., C. Humphries, C. Margules, R. Vane-Wright and P. Williams (1993), 'Beyond opportunism: key principles for systematic reserve selection', *Trends in Ecology & Evolution*, **8**(4), 124–8.

Pressey, R., H. Possingham and J. Day (1997), 'Effectiveness of alternative heuristic algorithms for identifying indicative minimum requirements for conservation reserves', *Biological Conservation*, **80**(2), 207–19.

Rodrigues, A.S.L. and K.J. Gaston (2002), 'Optimization in reserve selection procedures – why not?' *Biological Conservation*, **107**(1), 123–9.

Stein, B.A., C. Scott and N. Benton (2008), 'Federal lands and endangered species: the role of military and other federal lands in sustaining biodiversity', *Bioscience*, **58**(4), 339–47.

Styrsky, J.N., C. Guyer, H. Balbach and A. Turkmen (2010), 'The relationship between burrow abundance and area as a predictor of gopher tortoise population size', *Herpetologica*, **66**(44), 403–10.

Tóth, S.F., R.G. Haight, S.A. Snyder, S. George, J.R. Miller, M.S. Gregory and A.M. Skibbe (2009), 'Reserve selection with minimum contiguous area restrictions: an application to open space protection planning in suburban Chicago', *Biological Conservation*, **142**(8), 1617–27.

Underhill, L. (1994), 'Optimal and suboptimal reserve selection algorithms', *Biological Conservation*, **70**(1), 85–7.

Vane-Wright, R.I., C. Humphries and P. Williams (1991), 'What to protect? Systematics and the agony of choice', *Biological Conservation*, **55**(3), 235–54.

Williams, J.C. (2008), 'Optimal reserve site selection with distance requirements', *Computational Operations Research*, **35**(2), 488–98.

Williams, J.C. and C.S. ReVelle (1998), 'Reserve assemblage of critical areas: a zero-one programming approach', *European Journal of Operations Research*, **104**(3), 497–509.

Williams, J.C., C.S. ReVelle and S.A. Levin (2005), 'Spatial attributes and reserve design models: a review', *Environmental Modeling and Assessment*, **10**(3), 163–81.

12. QUICKScan: a pragmatic approach for decision support in ecosystem services assessment and management

Manuel Winograd, Marta Pérez-Soba and Peter Verweij

12.1 INTRODUCTION

Despite the increasing availability of environmental and socio-economic data and the last generation of process-driven models for ecosystem and ecosystem services (ES) assessment and management, their exploitation and use to support decision – and policy-making is often limited due to issues of scale, knowledge gaps, lack of user-friendly tools and capacities to translate and understand users' demands and needs (Grant et al., 2008; de Groot et al., 2010; TEEB, 2010). More particularly, there is a lack of a decision support toolbox (DST) that specifically considers changes and dynamics in the capacity of ecosystems, from a multi-functional perspective, or that allows decision-makers to explore impacts on ES and alternatives from ecosystem management multi-scale and multi-level perspectives (Winograd, 2010). Understanding and using the concept of ES by decision – and policy-makers is a fundamental step towards their operationalization (Maes et al., 2011). However, this understanding has proved to be rather challenging since it involves connecting and integrating the environmental and economic sciences with the decision-making process (TEEB, 2010). Many potential conflicts/trade-offs or agreements/synergies between ES in multiple sectors and on multiple scales make it difficult to gain a comprehensive view of the impacts of a measure (TEEB, 2010; Maes et al., 2011). In addition, decisions need to be based on facts and sound evidence and the multifaceted questions and the cross-cutting nature of the issues involved need to be assessed, explored and answered in a short period of time to fit the policy development time horizon. This complexity demands the support of tools to build a stakeholder-based process, using appropriate data and producing usable information and strengthening user capacities to assess ecosystems and ES (Pérez-Soba et al., 2012; Verweij et al., 2012; Winograd et al., 2012).

In consequence, the gap between the scientific offer and decision-/policy-making demand creates asymmetries in the identification of, definition of and responses to users' needs. Thus, what are furthermore needed are approaches and a toolbox that provide a policy–science interface and interaction with 'translators' (Winograd et al., 2012). This means providing the policy-making process with an appropriate toolbox, functionalities and interfaces to explore options, thus allowing decision-makers to play with choices and alternative impacts, and facilitating the marriage of quantitative analysis and qualitative judgements in support of flexible decision-making (Verweij et al., 2012). The DST, and the associated processes we build, are not independent of the means, contexts, approaches and methods designed for proper development and applications that respond to user demands. Thus, supporting decision – and policy-making processes implies, in a changing context, identifying the relevant levels of decision, the suitable spatial scales of actions and the appropriate time schedule to support the process (ibid.). Actually, there is a need to anchor DST use and value around answering and analysing the challenge of achieving policy coherence through dedicated management of our natural capital and maintenance of ecosystem services (EEA and ALTERRA, 2011). Thus, in the case of Europe we have at least three main contexts where the DST could be used. For thematic assessments (e.g., land changes, ecosystems dynamics), for policy evaluation (e.g., CAP Reform, Habitats Directive and Water Directive application and coherence) and for exploring cross-cutting issues (e.g., resources efficiency, adaptation assessment) (Winograd, 2010).

The guiding principles of QUICKScan (QS) could be described in the following way (Winograd, 2010; Pérez-Soba et al., 2012):

- The DST is not aimed at high-tech development, but more on exchanging and exploring results and translating them into understandable/usable/appropriate outputs for decision support.
- The DST avoids complicated and sophisticated toolbox development, and substitution of techniques (e.g., models, GIS) for the problem at hand (e.g., it explores options and support decisions).
- The DST implies that what is important is the *process*, so transparency in the development of the DST enables improvement in its capacity and uses.
- The DST should ensure its appropriate use to improve the communication of results and exchanges between scientists, technicians, and decision-makers to guarantee user ownership and close the gap between all stakeholders.

12.2 POLICY REQUIREMENTS AND EFFECTS

The DST seeks to explore and analyse the range of potential effects of new policies either to assess costs, impacts and trade-offs; increase transparency of decision-making; integrate cross-cutting issues; or to engage all involved stakeholders (Verweij et al., 2012). Many policy assessment tools exist, ranging from complex computer models simulating real-world processes; cost–benefit and cost-effectiveness tools; multi-criteria analysis tools; scenario analysis tools; checklists and decision trees; and methods to structure group processes in which policy-makers, tool developers and experts play an active role (Lipsett et al, 2011).

The DST and processes we build are not independent of the means, approaches and methods designed for proper development that respond to users' needs and demands. More particularly, in the case of QUICKScan (QS), as an open DST, it can only be developed, tested and used if the process includes, in the beginning, the use of an iterative evolutionary software and use development method as well as knowledge-building (Verweij et al., 2011). Each iteration adds functionality to the tool via the function of uses test. This method gives maximum steering flexibility in the outcome of the product and provides training and insight into the tool while it is being developed. It implies strong learning-by-doing team-building and cascade work that is only possible when all involved actors work together in an interdisciplinary manner (e.g., developers, managers, users, experts and scientists should understand each other and not just add a piece to the

Table 12.1 The gap between the offer of and the demand for the DST

	Scientific Offer	Policy Demands
Approach	Powerful modelling	Flexible tool
Toolbox	Closed	Open
Time	Anticipation (10–100 years)	Policy cycle (2–10 years)
Risks	Accepted	Avoided
Analysis	Expert support	Decision support
Uses	Implement solutions	Explore options
Runs	Complex validation	Easy iteration

Sources: Bradshaw and Borchers (2000); Winograd (2007).

puzzle) (Verweij et al., 2011; Winograd et al., 2012). The approach was explicitly to avoid DST 'black boxes', and rather to produce an open and transparent set of functionalities and user/developer networks (Verweij et al., 2012). This means, based on step-by-step lessons learned, providing, within the current institutional, budgetary and capacity contexts, appropriate opportunities to apply, remodel and reshape the development and use of the DST, thus increasing the flow, pooling insight, capacities and resources.

At the beginning of the process the DST development was carried out in a step-by-step precautionary way to ensure a suitable and usable toolbox. Thus, a first exploratory and test phase allowed the identification of needs and demands, available products and information, as well institutional capacities, limitations and resources. As a result, the second phase was oriented to build and test the first interface with a minimum set of targeted functionalities to be used in a different policy context. Finally, the third phase was defined to end the interface and test the needed spatial and statistics functionalities in different policy contexts (Verweij et al., 2012). The main goal is to close the gap between scientific offer and decision-/policy-making demand that creates asymmetries in the identification, definition and responses to users' needs. For instance, as depicted in Table 12.1, a big gap between producers and users of DSTs exists in relation, for example, to time frame, risk assessment, analysis and metric needs.

Thus, what is furthermore needed are pragmatic approaches that provide a policy–science interface with co-development of the DST framework, toolbox and applications between users (e.g., policy advisors, decision-makers, researchers) and producers (e.g., software engineers, modellers, researchers) and within scientific (e.g., ES assessment, environmental impact assessment [EIA]) and policy assessment (e.g., resources efficiency assessment, CAP Reform assessment) (Verweij et al., 2011, 2012). This means providing policy-making processes with appropriate functionalities to explore options, allow decision-makers to play with choices and alternatives impacts; and facilitate the marriage of quantitative analysis and qualitative judgements in support of taking flexible decisions (ibid.).

12.3 USER DEMANDS

The DST should help, improve and integrate the identification of relevant levels of decision, the appropriate data and information, the suitable spatial scales of actions and the appropriate time schedule to support the process. In the case of Europe, there are at least three main contexts where QS could be used. For thematic assessments (e.g., land changes, ecosystem dynamics), for policy evaluation (e.g., CAP Reform, Habitats Directive assessment) and for the exploration of emerging issues (e.g., resources efficiency, adaptation assessment) (EEA and ALTERRA, 2011).

In fact, supporting decision – and policy-making processes implies responding to urgent issues in a short time frame (e.g., 6–12 months), for instance, land use changes related to intensification or abandonment of agriculture given the EU Common Agriculture Policy (CAP) or carbon dynamics changes related to restoration or rehabilitation of degraded lands given the targets of EU Habitats Directive. But there also is a need to assess convenient or inconvenient responses and effects in a medium time frame (e.g., 1–5 years), for instance, food security and agriculture sustainability related to the goals and targets of the CAP Reform or the land cover and land use changes as consequence of the Green Infrastructure's and Habitats Directive's new targets to restore around 15 per cent of the European degraded ecosystems (EC, 2010a, 2010b). Finally, there is also a need to explore the implications and identify emerging issues in a long-term time frame (4–12 years). For instance, land cover and land use changes as the result of resources efficiency policy frameworks or adaptation responses and actions as consequence of climate variability and climate changes (EC, 2010, 2010b, 2010c, 2010d).

However, context also refers to the process involved in the development and uses of the DST. For instance, there is a gap between developers, decision-makers and scientists in the identification of functionalities, capacities to apply the toolbox and uses of the outputs and results. In fact, the different contexts increase the gap between toolbox functionalities and applications and uses. For instance, scientific contexts relate to development of new methods, tools functionalities and metrics, while decision-makers' contexts relate to availability of user-friendly interfaces, open functionalities and appropriate metrics (Winograd, 2010).

For DST development it would be advisable to tune functionality tests to users' needs to avoid a 'black box' that will not be used. Given that the toolbox, and the process, works with available data and information needed for policy-making it is clear that an effort should be made to produce and create the appropriate functionalities that fit within the policy context and decision requirements. The toolbox itself must be introduced as an environment within which to work in dialogue and exchange more than a toolbox that will solve all of the possible questions of decision-makers (Verweij et al., 2012). In fact, policy – and decision-makers need easy-to-use, credible and appropriate data and information to explore options and actions in a changing policy context, while data and information that feed the toolbox are complex to use, and not appropriate to a changing policy context (Winograd et al., 2012). At the end, the development of highly interactive, open and transparent products (and applications) like QS, implies a balance between tool development, process-building and pooling resources and capacities (e.g., too much tool development becomes controlled by developers, too much process-building becomes controlled by facilitators) (Verweij et al., 2011).

12.4 THE QUICKSCAN SUPPORT TOOL

The QS toolbox is a pragmatic approach that basically constitutes a working environment to explore in a given time (WHEN) and for a given area (WHERE) the impacts and effects of alternatives and options for a given policy context (WHO) (Winograd et al., 2012). QS is a framework, a toolbox and applications: framework, because it facilitates process organization and content integration; toolbox, because the interface with functionalities enables the building and execution of workflows and models; application, because the participatory process and the participatory use allows the creation and analysis of knowledge and understanding in a policy context.

The main five characteristics of the approach have been previously described (Pérez-Soba et al., 2012; Verweij et al., 2012; Winograd et al., 2012) and are summarized below:

- It is user-friendly (exploring/describing/anticipating policy changes/effects/implications).
- It is flexible, scalable and open (important for user/use needs).
- It is transparent (back-tracing of results, underlying principles/assumptions and uncertainty analysis).
- It deals with complexity, integration and iteration (which increase with any additional function/component/scale to be integrated).
- It incorporates qualitative and quantitative approaches and information (using the advantages/knowledge/insight of each).

12.4.1 The Framework

To use a medical metaphor, the QS approach focuses on symptom pictures of the socio-ecological system, including diagnosis, 'now-casting', exploring therapies and evaluating the effects and implications. In medicine, diagnosis of any clinical symptoms is not the end, but only a means for exploring and specifying an appropriate therapy (Winograd, 2007). In the analogy, QS is not restricted to the analysis of current conditions and trends, but ultimately aims at the exploration of options and alternatives, the evaluation of impacts and trade-offs and the identification of hotspots (Winograd et al., 2012). The framework therefore addresses five aspects (Figure 12.1):

1. What aspects, in a policy context, are relevant with respect to human and ecosystem well-being?
2. What typical 'pictures' of the past and actual conditions and trends exist?
3. What elements and interactions are relevant for the persistence of these patterns, trends, impacts?
4. Which strategies and options can be devised to preserve, restore, use, improve, mitigate or adapt?
5. Which hotspot areas, services or land covers/uses could be identified as targets for policy actions?

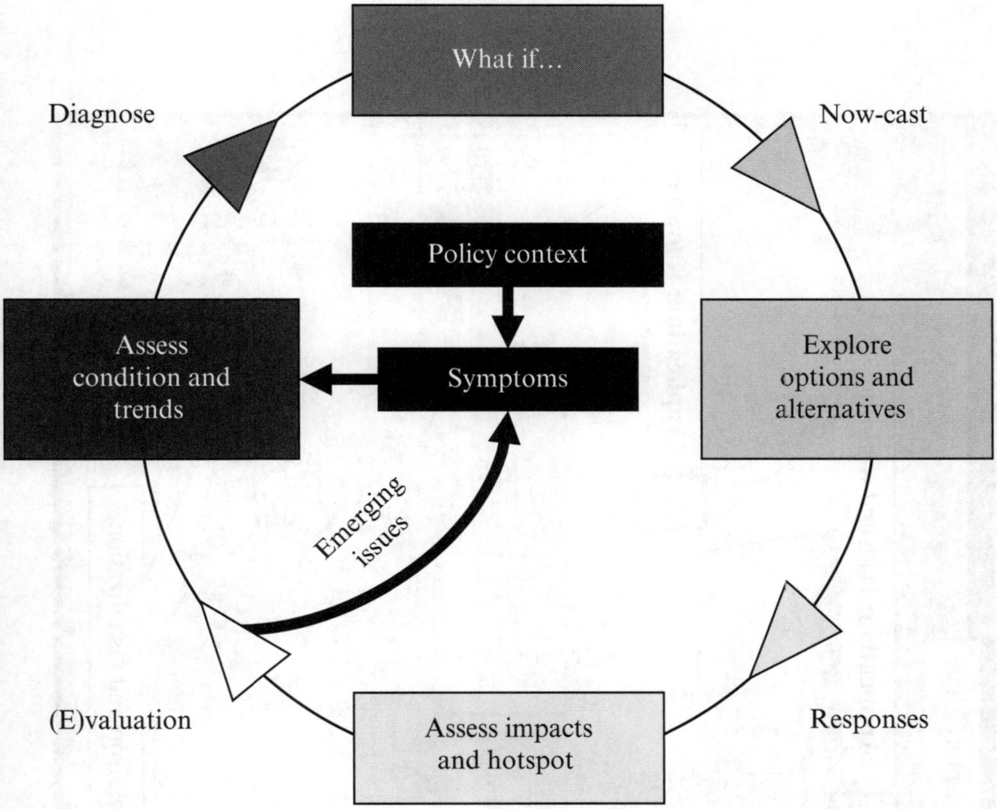

Sources: Winograd (2007); EEA and ALTERRA (2011).

Figure 12.1 The QUICKScan framework

12.4.2 The Toolbox

The QS toolbox (Figure 12.2) encompasses a modelling environment to assess societal and environmental conditions, diagnose patterns and interactions, explore alternative responses and evaluate the impacts of those responses and the linked trade-offs (Winograd et al., 2012). To do so, the application environment must be filled with spatial and statistical data for a given policy context or different alternatives. It enables the definition of 'what if', 'what then', and 'what else' rules and links those to available data to create derived information (Verweij et al., 2011, 2012). Typically, the rules use quantitative classifications or qualitative typologies to help formulate the objective (Verweij et al., 2011). Rules may also be linked together to form a chain of rules. Alternative (chains of) rules are used to capture different options. Derived data from alternatives can be aggregated (e.g., by administrative units, or biophysical units such as catchments, or climatic zones) to be displayed in tables and charts for overviews or in a radar chart to explore and visualize trade-offs (Pérez-Soba et al., 2012; Winograd et al., 2012).

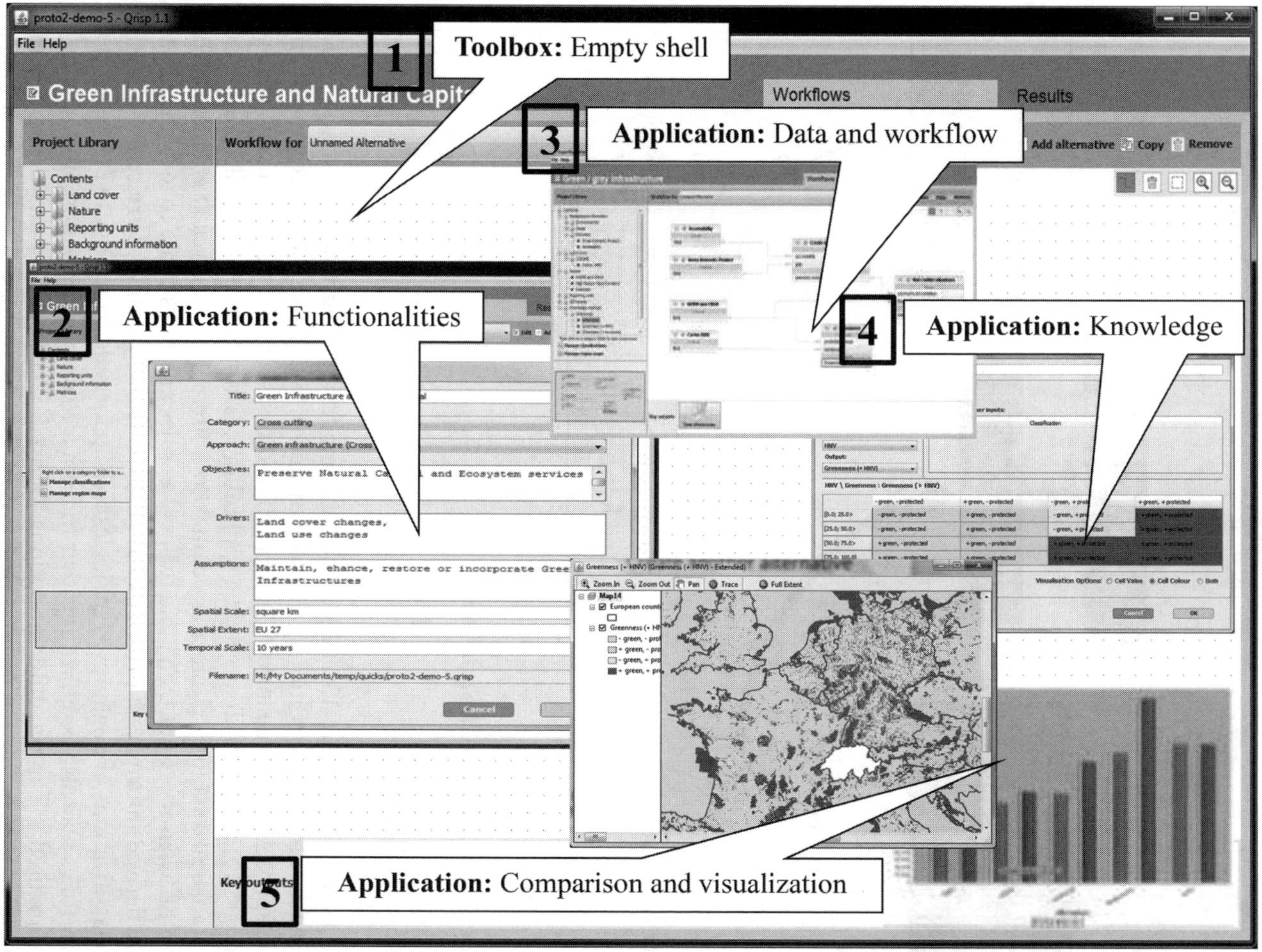

Figure 12.2 The QUICKScan toolbox

12.4.3 The Metrics

Measures and indicators for socio-ecological systems and dynamics are underdeveloped (TEEB, 2010; Maes et al., 2011). Ecosystem dynamics and ecosystem services need metrics to be measured and monitored. Given the pragmatic approach of QS, metrics should be defined based on the function of the data available and the information needed for helping and supporting the definition and identification of current states and trends, component thresholds, boundaries and targets.

Ecosystem services embrace many components, including the stock of natural resources from which goods can be extracted and the flows of ecosystems that provide and maintain services (Haines-Young and Potschin, 2011). The stocks and flows rely on ecosystem structures and functions such as land cover and land use, yields and primary productivity or ecotones and carbon storage in soil. Figure 12.3 shows a summary of a metrics proposed, covering structures and functions (Winograd, 2010; Verweij et al., 2011; Pérez-Soba et al., 2012).

12.5 APPLICATION OF THE QUICKSCAN TOOLBOX FOR EU POLICY SUPPORT

Until now, European or national biodiversity policies that aimed to reduce or stop the loss of biodiversity, essentially focused on the protection and conservation of endangered habitats, ecosystems and species (EC, 2010a, 2010b; EEA, 2010a, 2010b; Maes et al., 2011). In other words, policy targets are linked more with structural aspects related to land cover changes and dynamics (e.g., stopping fragmentation, improving connectivity) than with multi-functional aspects related to land use changes and dynamics (e.g., integration of 'high nature value' [HNV], use of ecotones) that characterize ecosystem services. For instance, on a European scale, an example of a structural conservation approach is the development of the Natura 2000 network established under the Habitats Directive (EC, 2010a).

But the inclusion of multi-functional aspects suggests a change in policy and scientific assumptions. In fact, it implies at least two main sides of the same coin: from one side assessing whether by protecting ecosystems and the services they provide, we maintain, increase or lose benefits to our well-being and biodiversity resources; from the other side mainstreaming the evaluation of natural capital and ecosystem services in other policies (e.g., CAP, Cohesion Policy) in order to assess trade-offs and synergies between policies and different land cover/use. Such assessments and evaluation necessitate the development of spatially explicit ecosystem service information and DST in order to estimate where ecosystem services are produced, to quantify the changes in service provision, regulation and maintenance over time, to describe the production of ecosystem services as a function of patterns of land cover/use and to explore alternative policies and actions (de Groot et al., 2010; EEA, 2011; Maes et al., 2011).

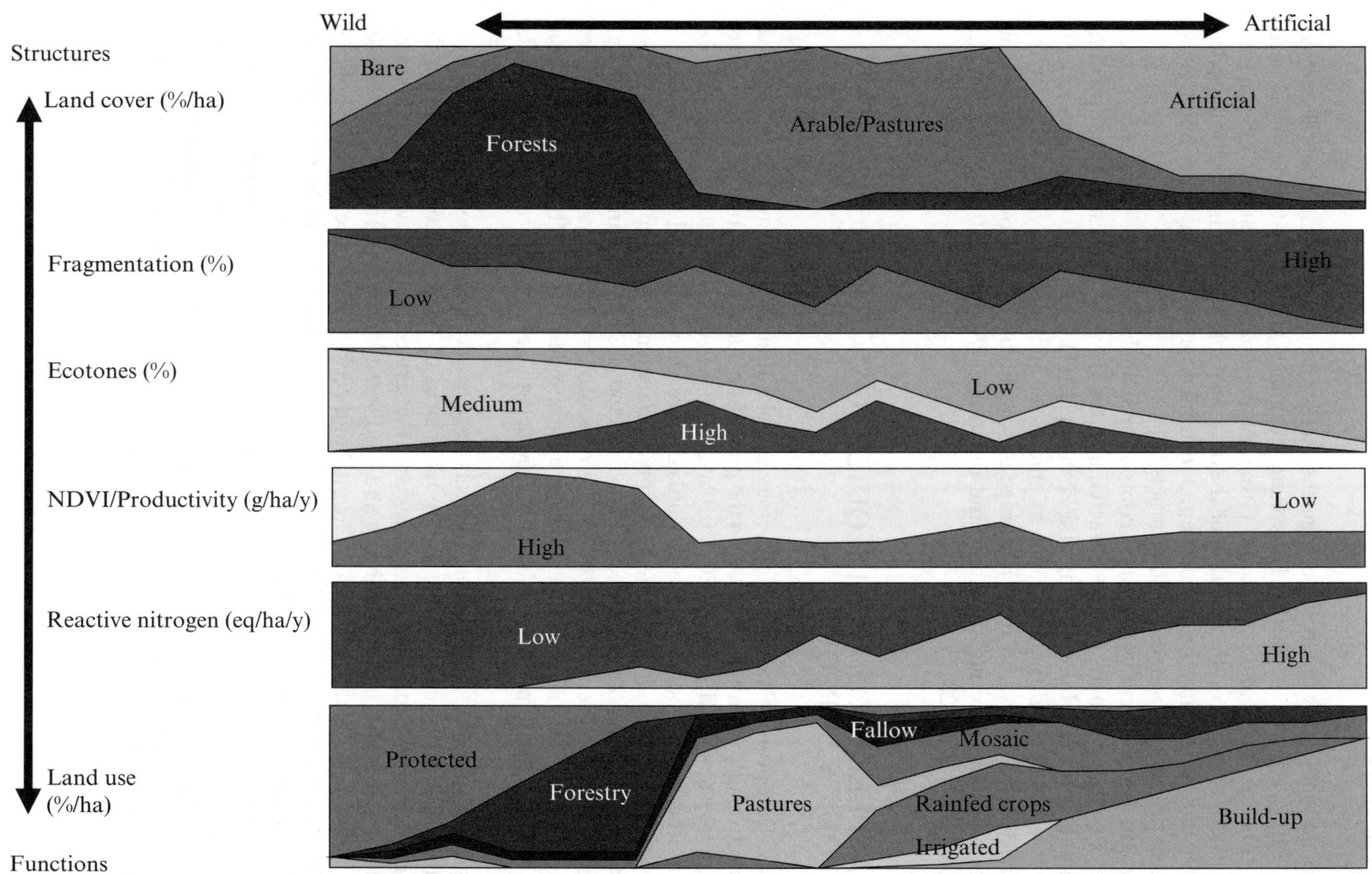

Sources: EEA and ALTERRA (2011); Winograd et al. (2012).

Figure 12.3 The QUICKScan metrics

12.5.1 Policy Context

Given the policy context and users' demands, the applications depicted in this chapter refer to the assessment of land changes and ecosystem dynamics in the context of the Habitats Directive and CAP Reform to explore and evaluate changes in Green (e.g., Natura 2000 [see EC, 2007], Common Database on Designated Areas [CDDA], HNV) and Grey (e.g., green urban areas, urban sprawl, infrastructure developments) Infrastructure and ecosystem services. The assumptions are based on actual land change conditions and trends as well as policy target definition by the European Commission (e.g., EU Biodiversity Strategy to 2020 [EC, 2011]; The CAP Towards 2020; Regional Policy Contributing to Sustainable Growth in Europe 2020 [EC, 2010c, 2010d]). To use real targets and figures, we applied those related to protecting and restoring biodiversity and ecosystem services (e.g., restoring at least 15 per cent of degraded ecosystems; maintaining/enhancing Green Infrastructure) as entry points to exploring actions that imply the main transitions and changes (e.g., restoration, protection, afforestation, abandonment, intensification) (Figure 12.4) as well to connect and explore trade-offs with other policy targets (e.g., CAP Reform and Cohesion Policy) (EC, 2010a, 2010b, 2010c).

12.5.2 Definition of Policy Alternatives

To build the application from a user perspective, a day-and-a-half workshop was held, divided into several parts (Verweij et al., 2011):

- *Day 1 morning session*: to delineate the policy context and brainstorm on alternatives and how to measure the success of the alternatives, as well as define the key output to compare alternatives. The morning session was attended by policy advisors, project managers, technical experts and a process facilitator.
- *Day 1 afternoon session*: implement alternatives in workflows in the QS toolbox by linking available GIS and statistical data with knowledge rules created by the experts. Several alternatives were created: (1) protected nature areas in a business-as-usual base; (2) exclude non-nature land use from protected nature areas and include nature areas outside of the protected zones (e.g., include city parks and forest and exclude roads and buildings); (3) additionally include all European areas that have farmland with 'high nature value' (HNV); (4) additionally include natural ecotones, transition areas between two adjacent but different plant communities; (5) additionally include agriculture areas prone to intensification inside Natura 2000 and CDDA (e.g., pastures and mosaics, wetlands, forests and semi-natural vegetation as well HNV) as intensification hotspots; (6) additionally intensify peri-urban sprawl (e.g., by 3, 5 and 10 per cent) to identify affected hotspots.
- *Day 1 afternoon session*: the QS team produced, based on defined workflows and alternatives, the different 'What if' information, analysis and outputs.
- *Day 2 morning session*: present the results to the policy advisors, project managers and technical experts using the QS toolbox and define the next steps needed (Verweij et al., 2011).

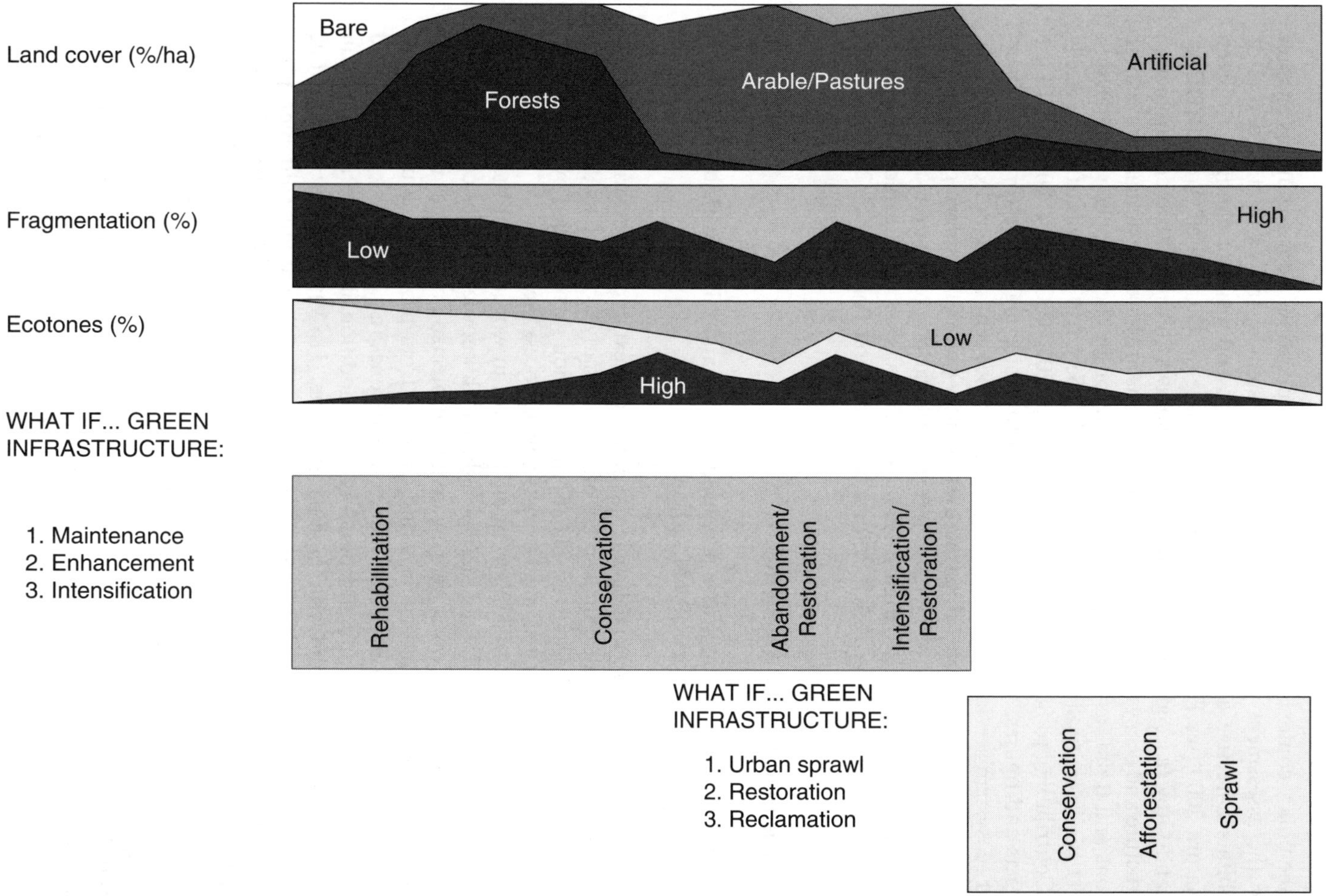

Source: Winograd et al. (2012).

Figure 12.4 Examples of structural indicators and 'What if'

12.5.3 Results

Green Infrastructure is an ideal test bed for development of the QS environment because of its multi-thematic and cross-cutting nature. QS can be used to explore and analyse questions such as 'What is the current and potential state and condition of Green Infrastructure in the EU-27, if considering different Green Infrastructure components?'; 'How can Green Infrastructure be measured and improved?' 'Which are the main ecosystems services that Green Infrastructure can provide, maintain or support?' and 'Which are the trade-offs between different regulation and policies related to Green Infrastructure?'

In all cases Green Infrastructure can be explored or assessed either as a purely structural theme, by looking at different land cover types and administrative declarations (e.g., protected areas, national designated areas), or it can be explored with a more functional approach, which seeks to identify different land use and land functions that might not be measured using purely structural means, but in all cases, the same steps and indicators were used.

For structural and functional changes and dynamics a selected set of ecosystem services can be assessed. For instance, service functions related to maintenance of lifecycle (e.g., pollination), biological control (e.g., pest, disease) and gene/populations (e.g., gene banks, habitats, ecosystems); sink functions related to regulation of flows (e.g., erosion, runoff, storage), waste (e.g., remediation, recycling, dilution) and quality (e.g., water quality, soil fertility, climate) and resources functions related to provision of food (e.g., cropping, animal production, harvesting), materials (e.g., fibres, minerals, genetic resources) and energy (e.g., biofuels, wind, hydro, solar) (Haines-Young et al., 2011).

Figure 12.5 shows the main average results (in per cent for EU-27 and by country) of a 'What if' for different Green Infrastructure alternatives, for instance, maintaining actual Green Infrastructure (e.g., Natura 2000 and CDDA) or different enhancement of Green Infrastructure (e.g., incorporating natural and semi-natural forest, non-protected woodlands and pastures, HNV and ecotones) (Verweij et al., 2011).

Figure 12.6 shows the results of different alternatives in a spatially explicit way to assess where changes happen. As depicted on the different maps, the structural changes and impacts can be easily visualized for the different alternatives as well assessing the potential effects of increasing the functional integrity of the wider ecological network of Europe. To assess functional impacts and trade-offs the radar chart in Figure 12.6 shows the differences between the alternatives regarding the loss or improvement of ecosystem services, including maintenance, regulation and provision. Even if no single definition of Green Infrastructure exists, the results underline the effects of main features including connectivity, multi-functionality, conservation and rehabilitation (EEA, 2011; Verweij et al., 2011). Obviously the link between Green Infrastructure and ecosystem services shows high synergy. Indeed, the purpose of Green Infrastructure maintenance and/or enhancement can be defined as maintaining, strengthening and restoring ecosystems and the services they provide. From the analysis of the results it is clear that ecosystem services, and the potential benefits of Green Infrastructure, are linked across the categories of maintenance (in particular pollination, pest and disease control, gene banks, habitats and ecosystems) and regulation (in particular erosion, runoff, storage and water quality and soil fertility).

It is important to note that the approach and the application are exploratory and

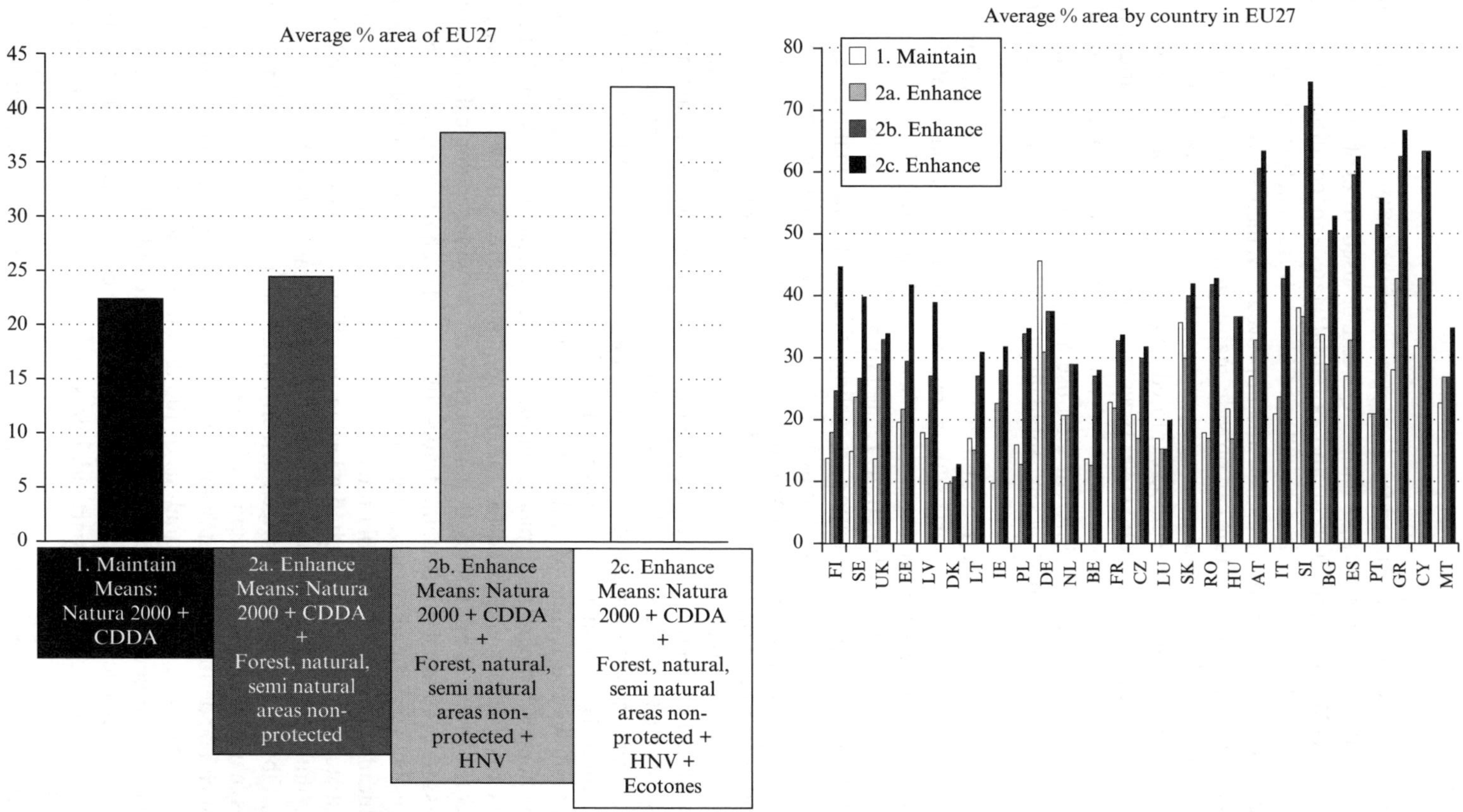

Source: Verweij et al. (2011).

Figure 12.5 Results of 'What if': Green Infrastructure maintenance and/or enhancement

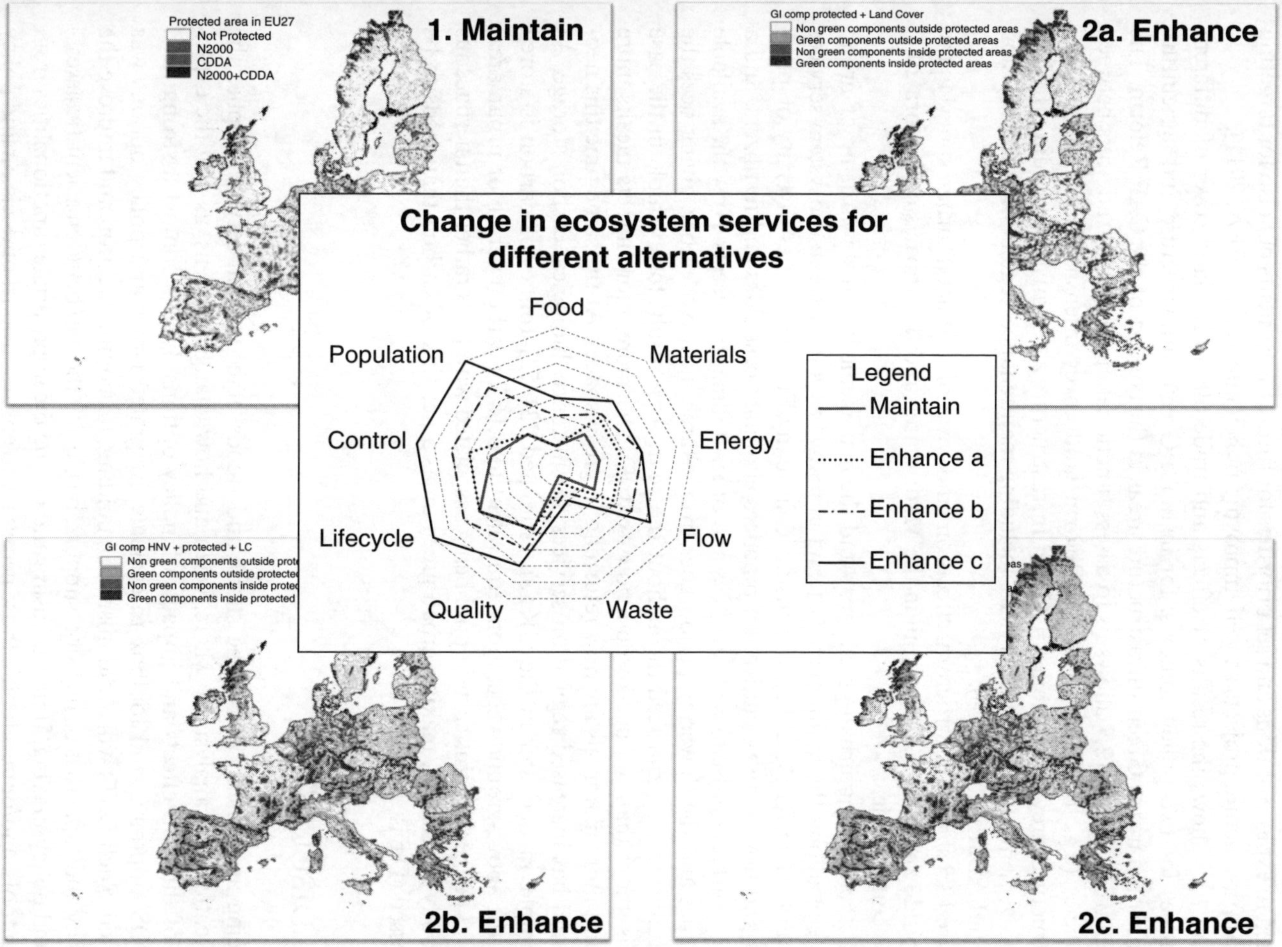

Source: Verweij et al. (2011).

Figure 12.6 Results on effects and trade-offs of different alternatives for a given 'What if': Green Infrastructure

not tailored as exact methods for measuring Green Infrastructure; rather, they should be considered as valuable inputs to explore (in)convenient actions with (in)convenient implications. It important to recognize that mapping and assessing Green Infrastructure is needed to explore options and set priorities for future investments but also for targeting Green Infrastructure projects as well improving policy integration (EEA, 2011).

Figure 12.7 shows the results of effects and trade-offs in the land cover of different alternatives for two policy contexts, for instance Green Infrastructure and agriculture intensification inside Green Infrastructure areas in the context of the CAP Reform. The radar chart in Figure 12.7 allows us to assess functional impacts and trade-offs between the policies (e.g., black line = CAP Reform land cover intensification inside Green Infrastructure, grey dotted tones = Green Infrastructure alternatives) regarding the loss or improvement of ecosystem services, including maintenance, regulation and provision (EEA and ALTERRA, 2011).

Figure 12.8 shows the structural effects on land cover of a 'What if' urban sprawl in the countryside and with a zoom detail in the Madrid area and the impacts on Natura 2000 and the countryside.

In fact, enhancing urban green areas and the management and stewardship of urban ecosystems allows the maintenance of and in many cases increase in ecosystem services (EEA and ALTERRA, 2011). For instance, in many European cities diversity of plantations and flowers rises, cultivation practices eliminate pesticides and fertilizers, and a variety of urban landscapes attract bees and other pollination insects. Now the available landscapes and the flowering period goes from March until November, giving bees the opportunity to stay and feed throughout the year. As a result, for example in the case of Paris, more than ten bee-keepers now exist. Furthermore, more insects means more birds and more mammals are now returning to green spaces. At the same time this neo-ecosystem has increased regulation services, for instance, biological carbon storage. As an example, in Leicester in the UK, an estimated 231 521 tonnes of carbon is stored within the above-ground vegetation, equivalent to 3.16 kg carbon per m^2 of urban area, with 97.3 per cent of this carbon pool being associated with trees rather than herbaceous and woody vegetation, making urban green areas a net sink of carbon that needs to be re-assessed (Davis et al., 2011).

12.6 DISCUSSION

Producing assessments about the status and trends of ecosystem services implies the identification and analysis of where, when and how change happens, as well the exploration of different effects and impacts of policy options. In this context a strong need for a 'QS' approach and toolbox to evaluate integrated issues and policy options was identified. Such DSTs will avoid misunderstanding, miscommunication and improve the efficiency and capacities of decision support, by closing the gap between scientific assessment and policy-making. It implies using and testing the appropriate functionalities from existing tools, and/or developing those necessary to ensure support of decision and policy processes and to integrate, improve and use the different data and information (e.g., land, water, carbon and biodiversity accounting). It will allow the best use and 'value-added' of the information and data produced and ensure integration and transversal analysis

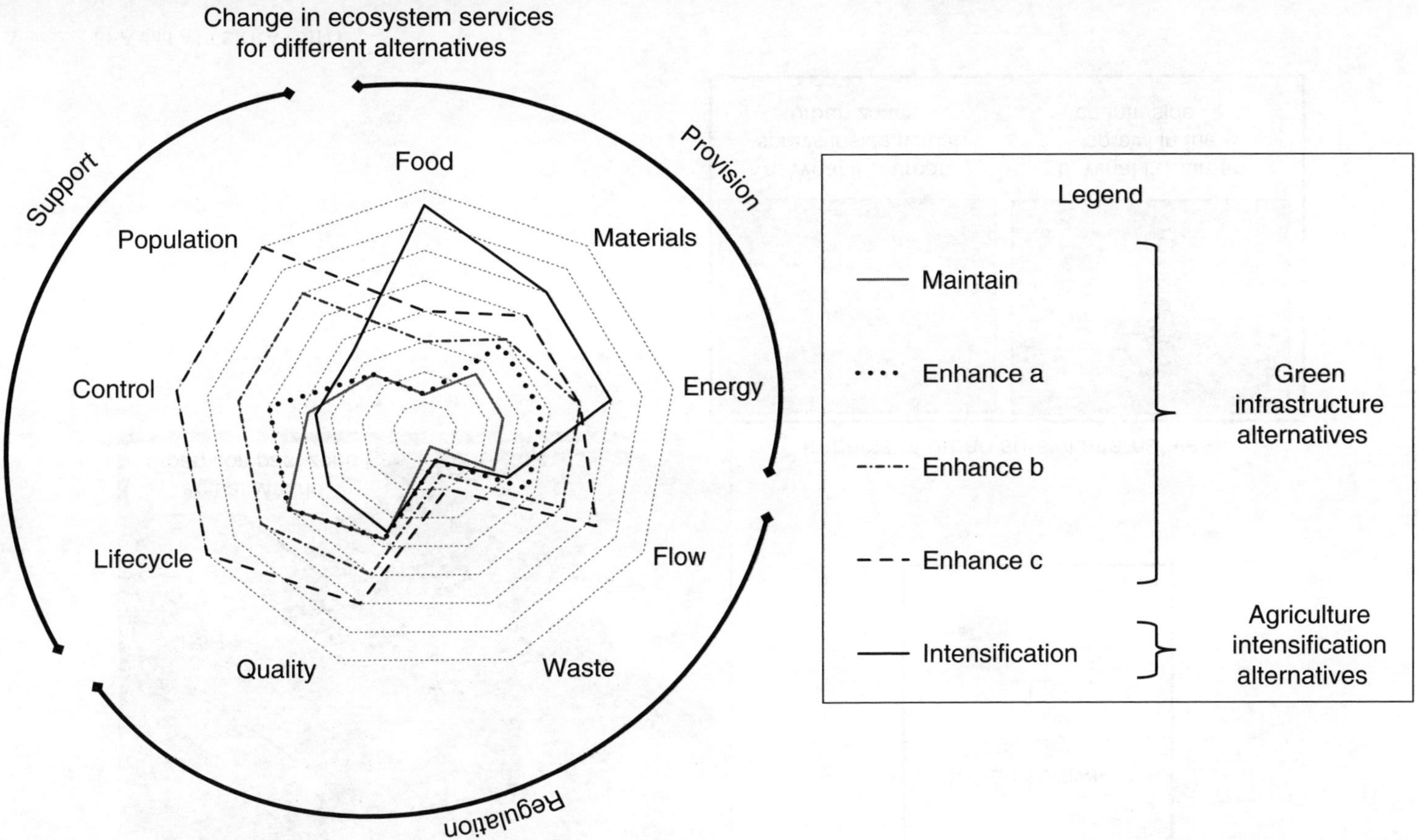

Source: EEA and ALTERRA (2011).

Figure 12.7 Results on effects and trade-offs of different alternatives for two 'What if': Green Infrastructure and CAP Reform

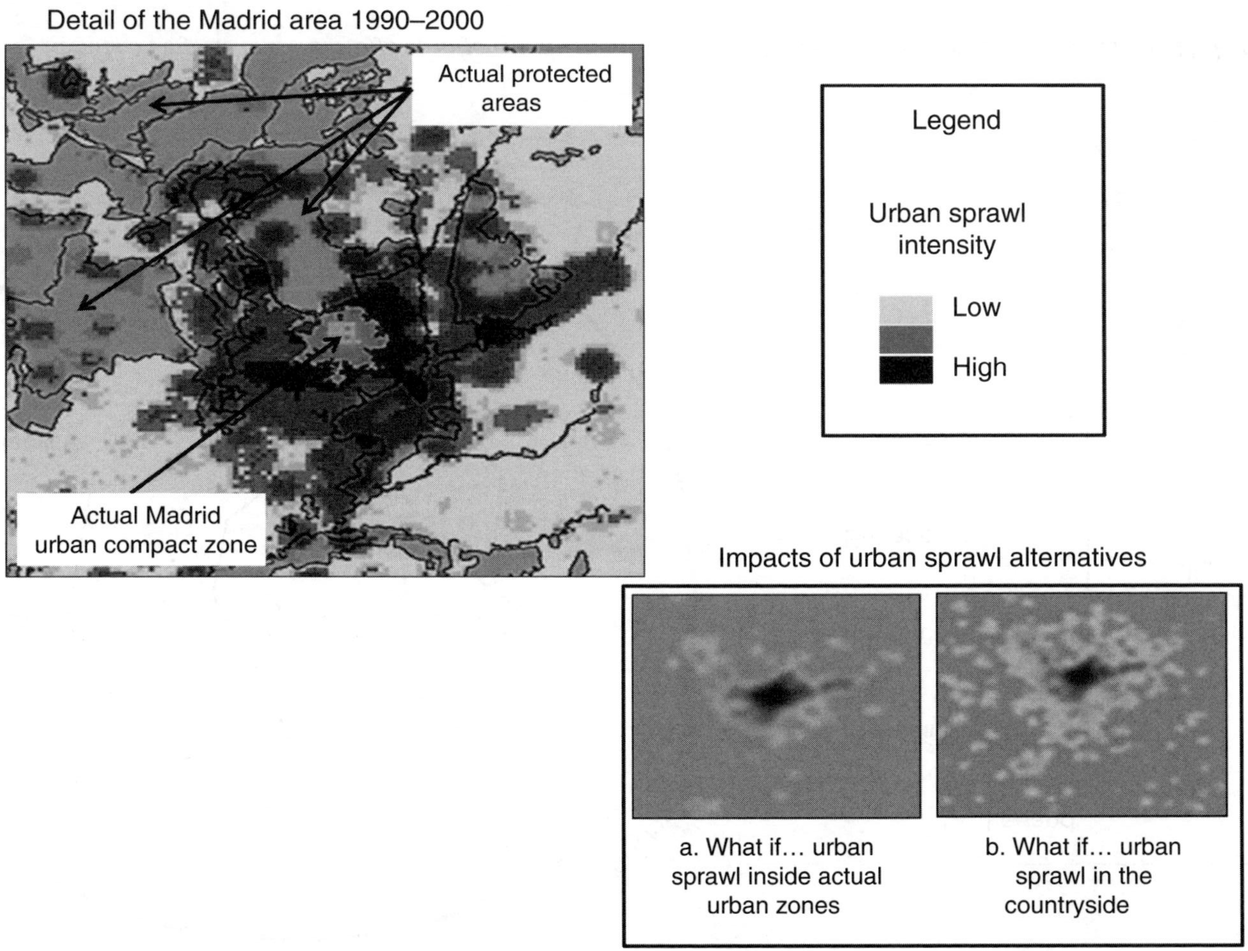

Source: EEA and ALTERRA (2011).

Figure 12.8 Results of effects of 'What if': urban sprawl in the countryside

of the main policy and cross-cutting issues (e.g., adaptation to climate change, resource efficiency, land planning). What is furthermore needed are approaches that provide support to decision-makers, not answers; that allow decision-makers to play with choices and alternatives, as they do in their day-to-day business; and that allow the marriage of quantitative analysis and qualitative judgements (precautionary principle, managing uncertainties, proactive and flexible responses) in support of decisions to be taken. Finally, the means to communicate results and insights and the capacities to use the data and the toolbox, are equally important, The toolbox and the information by themselves will not make better decisions.

So, participatory modelling, development and uses are key aspects of the QS approach. Despite political pressure, complex models are hardly used. Recommendations to improve their usage include the provision of training, communication materials and advancement of usability, which requires additional resources (Verweij et al., 2012). Still, complex computer models are not found to be transparent enough by the decision-makers who seek to understand the modelling and workflow behind the results to be able to cope with uncertainty (scientific as well political). In addition, multifaceted policy questions need to be answered in a short period of time. The complex tools and models are not ready to deal with this urgency as often new policy questions require careful model adaptation, expansion, linkage to other models (Knapen et al., 2012) and calibration. Modelling results are often sent to policy-makers in a report or policy brief and might be exemplified during a short presentation. Incorporation of feedback to the modelling results necessitates another time-consuming iteration of the complex modelling and model validation at the modeller's office, after which another workshop may be used to present the new findings. A gap between workshops leads to a loss of engagement and interest (Kok et al., 2011), but more importantly it may take longer than the time horizon of the policy development.

Visualization and interpretation tools are essential to support the discussion and interaction between users and producers and are capable of speeding up the decision-making process (Verweij et al., 2012). QS is capable of developing storylines, selecting indicators for measuring the objective achievement, gaining and processing stakeholder knowledge and jointly creating new workflows, knowledge as is gained by participatory modelling as well as comparing alternatives and visualizing and zooming hotspot areas (ibid.). QS can do this within the time frame of a two-day workshop. However, such a workshop needs careful preparation. Experts must be found and data gathered and made available to the QS tool. Preparation also means running through likely alternatives and scenarios and thinking of proxies to use for unavailable, or non-existing data. However, sometimes a more in-depth study is required, which may be solved by using a (set of) complex model(s). QS can be also used to identify those cases. Assessments tend to be performed at a late stage in the policy process. Because of the late timing such assessments tend to have little or no effect on the policy shaping.

The use of QS as an integrated approach helps in knowledge organization given that it allows users to ensure that components and information are linked and organized in an appropriate manner. Also it constitutes a useful working environment with empty shell and graphical user interfaces to build, edit and execute modular application and use functionalities. Finally, the application allows the use of the working environment to test, improve and develop easy iteration of particular projects for different users' needs.

The current knowledge about structures and functions of Green Infrastructure is not sufficient to provide a complete assessment of the contribution of ecosystem services and support of socio-ecological systems (EEA, 2010a, 2010b, 2010c). However, data and knowledge available provides valuable information on their status and trends, which is sufficient to target and explore future policy priorities and actions (EC, 2010a, 2010b).

For example, the trend in the case of Green Infrastructure maintenance is an increase in provision of crops from agro-ecosystems, timber and climate regulation from forests, water flow regulation from rivers and wetlands. However, there will be a decrease in livestock production, freshwater capture fisheries and wild foods in semi-natural grasslands, wetland and rivers. Finally, there will be an increase in timber provision, freshwater provision, water/erosion/natural hazard regulation and recreation/ecotourism from forests and mountains. In the case of enhancement of Green Infrastructure, with some effective actions (e.g., restoring green urban areas, connecting non-protected areas, changing land use practices, reducing disturbance of substrate and over-grazing, controlling fragmentation and hydrological disruption, controlling invasion of alien species or damaging species, nutrient removal, planting of herbaceous plants or grasses, planting of trees, soil amendments) the protected areas could increase from 22.5 per cent of the EU-27 territory to 24.5 per cent (option a), 37.5 per cent (option b) and 42.5 per cent (option c) (Verweij et al., 2011). However, what is important with the different options is not the percentage of the EU-27 territory under protection but the use, restoration, preservation or improvement of ecosystems and services for our well-being. For instance, the enhancement option (c) allows an increase in all regulation and maintenance services compared with the business-as-usual option. In the case of structural actions, for example in the case of ecotones, they may exist along a broad belt or in a small pocket, such as a forest clearing, where two local communities blend together. Conservation, restoration, creation and use of these areas as another building block of Green Infrastructure are key to ensure connectivity, decrease fragmentation and improve ecosystem services (in particular, regulation and maintenance). To decrease the effects of stress factors like land changes, habitat fragmentation, pollution and natural disaster impacts, actions that improve connectivity and promote 'ecological coherence' of a green network should strengthen the adaptive capacity of Europe's ecosystems. Thus, when planning green network coherence and adaptation to climate change there is a need to know, for example, actual and future condition of sites, location and distribution of ecosystem, eco-region and habitat types.

12.7 CONCLUSIONS

As a main conclusion we can highlight that what is important with the DST, more than the toolbox itself, is the approach to ensure co-construction between stakeholders (including developers, decision-makers and experts). Special attention should be given to the process, in order to ensure transparency in the tool development, guarantee capacity strengthening to use the toolbox and communicate results and improve exchange between users and developers (e.g., scientists, technicians and decision-makers). In this way, it could be possible to ensure user ownership. The development of a highly interactive, open and transparent product like the QS toolbox implies a balance between tool development, process-building and pooling resources and capacities to avoid the situation where too much tool development becomes controlled by developers or that too much process-building becomes controlled by facilitators.

From the toolbox development perspective, we can highlight as a barrier, related to decision – and policy-makers, the lack of interest in DST tools and methods. Among other reasons, one of the main causes of this lack of interest is the absence of suitability, reliability, usability and credibility in the results produced, given the gap between the developments of toolbox functionalities, the needs in a dynamic policy context and the communication and use of results. In fact, different contexts increase the gap between toolbox functionalities and applications and uses. For instance, scientific contexts relate to development of new methods, tool functionalities and metrics, while decision-makers' contexts relate to availability of user-friendly interfaces, open functionalities and appropriate metrics. Also, we need to take into account the demands and attitudes of users to toolboxes and information. For instance, there exists an increasing demand for more easy-to-use and transparent toolboxes, but if a toolbox is too easy to use the results are not credible, and if the toolbox use is complicated the results are not used.

From a toolbox development perspective, we can highlight as advantages that the DST allows us to bridge the gap between developers, decision-makers and scientists through defining real exchanges, application-building and capacity strengthening. Nevertheless given that software development has its own dynamics; it therefore would be advisable to better tune tests of functionalities and users' needs meeting with the progress of the software development. Given that DST, and the process, works with data available and needed for policy-making, as well as data that might result from complex models, it is clear that an effort should be made to produce and create the appropriate functionalities and data that fit with the policy context and decision needs. QS itself must be introduced to decision-makers before asking the users to formulate questions to be solved with it. There is a need to explain that QS in particular is a toolbox that enables us to build models and constitute a work environment, and emphasize that the tool itself is not a model.

From the toolbox use perspective, we can highlight as barriers the difficulties in communicating results for different audiences, users and uses, as visualization in different formats for different policy context is needed. For instance, communication and form became more important than analysis capacities and content. Policy-makers need easy-to-use, credible and appropriate data and information to explore options and actions in a changing policy context, while data and information that feed the toolbox are complex to use, and not appropriate to a changing policy context.

Finally, from the use of the toolbox perspective, we can highlight as advantages, that given the QS approach, a parallel between development of the DST and use of the DST helps in the testing and use of the toolbox, for instance using the QS in workshops with users (policy-/decision-makers). However, as mentioned, special attention needs to be given to workshops, given the heterogeneity of the participants, interests and policy needs/contexts. To ensure and improve the use of QUICKScan in decision-making, effort needs to be concentrated in using the toolbox to create and strengthen users' capacities instead of using it as a mapping and data machine. This is why the software tool should be used in the discussions, ensuring contact time with policy-makers, experts and policy analysts. It will help to explain what different parts of the workflow do, and the role of the knowledge matrix in the whole workflow. Given that workshop participants and users do not have the same mind set, goals and context, it might be wise to make some starting applications in a given policy context for a given issue. Finally, applications for capacity strengthening need to be designed and carried out to ensure decision-makers' and other users' toolbox ownership and sustainability of the process.

ACKNOWLEDGEMENTS

The authors would like to thank their many colleagues from the EEA (Oscar Gomez, Jock Martin, Alex Oulton and Ronan Uhel) and ALTERRA (Janneke Roos Klein-Lankhorst and Michiel van Eupen) for their constructive contribution, criticism and support.

REFERENCES

Bradshaw, G.A. and J.G. Borchers (2000), 'Uncertainty as information: narrowing the science-policy gap', *Conservation Ecology*, **4**(1), Article 7.
Davis, Z.G., J. Edmondson, A. Heinemeyer, J. Leake and K. Gaston (2011), 'Mapping an urban ecosystem service: quantifying above-ground carbon storage at a city-wide scale', *Journal of Applied Ecology*, **48**(5), 1125–34.
de Groot, R.S., R. Alkemade, L. Braat, L. Hein and L. Willemen (2010), 'Challenges in integrating the concept of ecosystem services and values in landscape planning, management and decision-making', *Ecological Complexity*, **7**(3), 260–72.
EC (2007), *Natura 2000 Newsletter*, No. 22, June 2007, accessed 7 February 2014 at http://ec.europa.eu/environment/nature/info/pubs/docs/nat2000newsl/nat22_en.pdf.
EC (2010a), *Green Infrastructure*, June 2010, accessed 6 February 2014 at http://ec.europa.eu/environment/nature/info/pubs/docs/greeninfrastructure.pdf, Brussels: DG Environment, EC.
EC (2010b), *EU Biodiversity Strategy to 2020*, accessed 6 February 2014 at http://ec.europa.eu/environment/nature/biodiversity/comm2006/pdf/2020/1_EN_ACT_part1_v7%5B1%5D.pdf.
EC (2010c), *The CAP Towards 2020*, accessed 6 February 2014 at http://eur-lex.europa.eu/LexUriServ/LexUriServ.do?uri=COM:2010:0672:FIN:en:PDF, Brussels: DG Agriculture, EC.
EC (2010d), *Regional Policy Contributing to Sustainable Growth in Europe 2020*, accessed 6 February at http://ec.europa.eu/regional_policy/sources/docoffic/official/communic/sustainable/swd_sec2011_92.pdf, Brussels: DG Regio, EC.
EC (2011), *Our Life Insurance, Our Natural Capital: An EU Biodiversity Strategy to 2020*, COM, Brussels: EC.
EEA (2010a), *The European Environment State and Outlook 2010, Land Use*, accessed 7 February 2014 at http://www.eea.europa.eu/soer/europe/land-use, Copenhagen: EEA.
EEA (2010b), *The European Environment State and Outlook 2010, Biodiversity*, accessed 7 February 2014 at http://www.eea.europa.eu/soer/europe/biodiversity, Copenhagen EEA.
EEA (2010c), *EU 2010 Biodiversity Baseline*, Technical Report No. 12, accessed 7 February 2014 at http://www.eea.europa.eu/publications/eu-2010-biodiversity-baseline, Copenhagen: EEA.
EEA (2011), *Green Infrastructure and Territorial Cohesion: The Concept of Green Infrastructure and its Integration into Policies Using Monitoring Systems*, EEA Technical Report No. 18, Copenhagen: EEA.
EEA and ALTERRA (2011), *The Quick Scan Approach: Toolbox to Support Decision and Policy-making, Guidelines for Use and Application, Guidelines # 1 to 10, Capacity Building Materials*, Copenhagen: EEA and ALTERRA.
Grant, F., J. Young, P. Harrison, M. Sykes, M. Skourtos and M. Rounsevell et al. (2008), *Ecosystem Services and Drivers of Biodiversity Change*, Report of the RUBICODE e-conference, April 2008, EC.
Haines-Young, R. and M. Potschin (2011), 'Common International Classification of Ecosystem Services (CICES): 2011 update', paper prepared for discussion at the expert meeting on ecosystem accounts organized by the UNSD, the EEA and the World Bank, London, EEA, Contract No. EEA/BSS/07/007, Centre for Environmental Management, University of Nottingham, UK.
Haines-Young, R., M.B. Potschin and F. Kienast (2011), *Quick Scan Modelling Approaches: The Use of Multi--criteria Approaches for Mapping Changes in Ecosystem Service Potential at European Scales*, Final Report, EEA/BSS/07/007, Centre for Environmental Management, University of Nottingham, UK.
Knapen, M., S. Janssen, O. Roosenschoon, P. Verweij, W. De Winter and M. Uiterwijk et al. (2012) 'Evaluating

OpenMI as a model integration platform across disciplines', *Environmental Modelling & Software*, **39**, 274–82.

Kok, K., I. Bärlund, A. Dubel, M. Flörke, P. Magnuszewski, J. Sendzimir and M. van Vliet (2011), 'Lessons learnt: summary of scenarios: multi-scale stories, conceptual models and policy actions', Deliverable 2.12, FP6 SCENES: Water Scenarios for Europe and Neighbouring States.

Maes, J., L. Braat, K. Jax, M. Hutchins, E. Furman and M. Termansen et al. (2011), *A Spatial Assessment of Ecosystem Services in Europe: Methods, Case Studies and Policy Analysis – Phase 1*, PEER Report No. 3, Ispra: Partnership for European Environmental Research.

Pérez-Soba, M., M. Winograd and P. Verweij (2012), 'QUICKScan: a decision-making support toolbox to improve assessments of ecosystems and ecosystem services', TEEB Conference 2012, Leipzig, Germany.

Poveda, C. and M. Lipsett (2011), 'A review of sustainability assessment and sustainability/environmental rating systems and credit weighting tools', *Journal of Sustainable Development*, **4**(6), 36–55.

TEEB (2010), *The Economics of Ecosystems and Biodiversity: Mainstreaming the Economics of Nature: A Synthesis of the Approach, Conclusions and Recommendations of TEEB*, accessed 6 February 2014 at http://www.teebweb.org/wp-content/uploads/Study%20and%20Reports/Reports/Synthesis%20report/TEEB%20Synthesis%20Report%202010.pdf.

Verweij, P.J.F.M., J. Roos-Klein Lankhorst, M. Pérez-Soba, M.J.R. Knapen, W.P. Winter and M. van Eupen (2011), *Develop a QUICKScan Application for Decision Support in the Context of Strategic Environmental Assessment*, Final Report Phase 1 to the restricted call for tender EEA/IEA/10/001, ALTERRA Wageningen UR.

Verweij, P., M. Winograd, M. Pérez-Soba, R. Knapen and Y. van Randen (2012), 'QUICKScan: a pragmatic approach to decision support', in R. Seppelt, A.A. Voinov, S. Lange and D. Bankamp (eds), *2012 International Congress on Environmental Modelling and Software Managing Resources of a Limited Planet, Leipzig, Germany*.

Winograd, M. (2007), 'Sustainability and vulnerability indicators for decision-making: lessons learned from Honduras', *International Journal for Sustainable Development*, **10**(1/2), 93–105.

Winograd, M. (2010), 'QUICKScan: an emerging process to support EEA/EU ecosystem assessments and decisions', EEA Scientific Committee Seminar 'The European Ecosystem Assessment (EURECA) process', Copenhagen: EEA.

Winograd, M., M. Pérez-Soba, P. Verweij and R. Knappen (2012), 'QUICKScan: a pragmatic approach for bridging gaps in the science–policy interface; LIAISE OPEN DAY', Bilbao, Spain.

PART III

ECOSYSTEM SERVICES AND CONSERVATION POLICY

13. Ecosystem service valuation and the allocation of land

*R. David Simpson**

13.1 INTRODUCTION

Readers of this volume will probably not need to be informed of current interest in 'ecosystem services'. Natural ecosystems provide services such as pollination, flood protection and water purification. Conservationists hope that such systems will prove to be of greater value providing these services than if converted to other uses. Peter Kareiva and Susan Ruffo (2009, p. 3) write that: 'The idea of "ecosystem services" . . . gives us a framework to measure nature's contribution to human well-being, and to understand the cost of its loss . . . This is why, now more than ever, we need to embrace ecosystem services as a basis for conservation' (see also Tallis and Kareiva, 2005; Turner et al., 2007; Daily and Matson, 2008; Turner and Daily, 2008).

Arguments for conservation grounded in ecosystem services are problematic on several levels, however. First, some of the services provided by natural ecosystems are global public goods, such as carbon sequestration and the protection of biodiversity. It is certainly desirable to maintain such services, but the conservation community has been trying for decades to motivate international transfers to conserve natural ecosystems (see, for example, Pearce and Moran, 1994). It is not clear that an appeal to the value of ecosystem services for providing global public goods is adding much of anything new – and such appeals have often fallen on deaf ears in the past (Pearce, 2007).

If the ecosystem services conservation advocates hope to promote more local public goods, one immediate question is: why have local communities not assured the provision of such services for their own good? One answer to that question could be that local communities do not always solve the problems of collective action that prevent them from implementing mutually beneficial measures. There is certainly something to this: how did some communities get to be so much poorer than others, if not by failing to overcome such problems?

There is an alternative explanation, however: not everyone is convinced that ecosystem services are as valuable as some others claim (see, for example, Sagoff, 2011). Naidoo and Ricketts (2006) (see also Kremen et al., 2000) find that local ecosystem service values in the area they studied are not sufficient to offset conservation costs. Other authors note that some ecosystem services demonstrate sharply diminishing returns (Tallis, 2006), ready substitutes exist for some ecosystem services (Turpie et al., 2003; Odling-Smee, 2005), and that some ecosystem services are likely to prove of little value in protection against extreme events (Kareiva and Ruffo, 2009).

Somewhat curiously, many writings extolling the conservation potential of ecosystem services also reveal some uncertainty regarding the value of such services. Ricketts et al. (2004, p. 12 579) state the paradox succinctly: 'Although the societal benefits of native ecosystems are *clearly immense*, they remain *largely unquantified* for all but a few services' (emphasis added). If something is 'largely unquantified' the reader is left to wonder on what basis the authors have concluded that it is 'clearly immense'. The juxtaposition of similar sentiments appears in other contributions to the literature on ecosystem services. Shortly after the passage I quoted above in which Kareiva and Ruffo (2009, p. 3) write that 'we need to embrace ecosystem services as a basis for conservation' they add that:

> getting beyond the platitude of nature's value has proven to be a challenge for both science and policy. Why? . . . Because we do not have enough science to back up our hypotheses of how and when services are delivered . . . In short, because we have not proven, on the ground, that these ideas work.

Daily et al. (2009, p. 21) write in the text of their article that 'Natural capital, and the ecosystem services that flow from it, are usually undervalued – by governments, businesses, and the public – if indeed they are considered at all', yet state in the same article's abstract that '[w]e have not yet developed the scientific basis . . . for incorporating natural capital into resource- and land-use decisions on a large scale' and then go on to say that the authors '[p]ropose a conceptual framework and sketch out a strategic plan for delivering on the promise of ecosystem services'.

Again, the reader is left to wonder how it is that he or she can conclude that ecosystem services are usually undervalued by a range of actors when an article claiming that they are, lists among its objectives the development of a research agenda to substantiate the point.[1] We may read such work simply as suggesting that more research is required to see if the case can be made that natural ecosystems provide more valuable services than would the systems that replace them, but it is important to understand that that point has not yet been firmly and generally established.

As noted above, some authors find that local ecosystem service values do not offset the opportunity costs of conservation (Kremen et al., 2000; Naidoo and Ricketts, 2006). Other carefully constructed estimates of ecosystem service values beg the question of whether the benefits estimated would justify forgoing alternative land uses. Costanza et al. (2008) find that the marginal value of a hectare of coastal wetland could be as high as US$51 000 per hectare per year for its storm protection services. Annualized at a 5 percent discount rate this yearly figure would yield a net present value in excess of a million dollars. This is an impressive sum. However, the opportunity cost of land in urban areas can be a substantial multiple of even that amount. Another very meticulously conducted study of the value of natural forest fragments for the provision of pollinating services to coffee plantations again found appreciable values (Ricketts et al., 2004), but it is not clear that the value of the forest would have been greater than it would have been if it were cleared and converted to plantation, even if by doing so the productivity of the existing plantation land declined (Simpson, 2010). Moreover, the pollination value of the adjoining forest areas was rendered moot when the plantation was converted to pineapple, a crop that does not require insect pollination (Macaulay, 2006).

The value of ecosystem services has not been firmly and generally established. Regrettably, I do not propose to fill in this gap here. Estimating the value of non-market goods remains a very difficult undertaking.[2] I pursue a more modest objective. I ask, under what conditions might we expect the services provided by preserved or restored natural ecosystems to be most valuable? I also pose a related question: when might preserving or restoring the services of natural ecosystems motivate the most conservation?

The values of ecosystem services depend crucially on context. Many of the services generate value only because the natural ecosystems that provide them adjoin decidedly unnatural systems. Water purification, for example, is a valuable service only because there is both a source of impurities upstream of the system providing it and a population that benefits from having clean water downstream. Pollination services are most valuable when crops requiring pollination are planted at an unnatural density in the vicinity of the natural areas sheltering pollinators. Protection against storms and flooding is most valuable when a sizable cluster of built structures is at risk. It is apparent on a moment's reflection, then, that the conservation incentives afforded by ecosystem services must be limited in one important sense: the greater the area conserved, the less will be area that stands to benefit from conservation.

Of course, different types of areas benefit from different types of ecosystem services, and it would be difficult to develop anything like a 'one-size-fits all' model of the relationship between habitat conservation and the value of ecosystem services. In this chapter I consider illustrative examples. In the next section I review some general principles and consider their illustration in some examples from the literature. Following that, I develop in some detail a schematic model that might be of particular relevance in many developing country scenarios of conservation interest. It describes an economy with one sector whose product is essential (call it 'agriculture'), which depends on land, and in which production can be enhanced by ecosystem services, and another whose output is less critical and is less dependent on both land and ecosystem services.

Some common themes emerge both from the examples in the literature and the new model I present. One, as I have already noted, is simply that the value of the services of natural systems depends on the proximity and extent of *un*natural systems that benefit from them. Another is that effects may seem counterintuitive. If conserved natural habitats are very effective in providing valuable ecosystem services, it may mean that we would optimally conserve *fewer* such habitats than we would if they were less effective. A recurring theme is that if 'a little goes a long way, there's little need for a lot'.

Another observation, along similar lines to the first, is that societal circumstances matter. It should not be surprising that a community would be more likely to rely on ecosystem services if its population is low and substitutable inputs are less abundant, but the consequence of this observation may be that an appeal to ecosystem services as a motivation for conservation may be least effective in those communities in which the pressure on natural systems is greatest. Conversely, however, it is not necessarily the case that an economy reduced to subsistence – one in which all resources are the devoted to the production of food – would devote any land to the provision of ecosystem services. Even when only limited other resources are available it could still make sense to devote all land directly to production.

I develop these ideas in the remainder of this chapter. In the next section I discuss the ways in which natural ecosystems might provide local public goods to benefit nearby human communities, and emphasize the limits this places on the conservation potential of an ecosystem services approach. I then consider a few examples from the literature underscoring the point. Following that, I introduce and explore a model of ecosystem services in a highly stylized model. I introduce the model in the third section, develop its basic results in the fourth, and then consider some special cases of interest in the first. I draw out some policy implications in the sixth section, and review and conclude in the seventh.

A Note on Context and Philosophy

It will be obvious to the reader that I have some serious doubts about what we might, following Turner and Daily (2008), classify as the 'ecosystem services framework' for conservation policy. This may raise some questions in the reader's mind that I might do well to anticipate.

Is the intended message of this chapter that conservation is not important? No! Disagreements concerning the means of achieving certain objectives should not be confused with disputes over the objectives themselves. Having said that, however, promotion of the ecosystem services framework does raise two important issues in conservation policy. The first is that not all conservationists agree that conservation and land uses that might most benefit from ecosystem service provision will prove compatible. Terborgh (1999), for example, writes that the goal of conservation should be the preservation of large, pristine systems, a scenario that is hard to square with the provision of ecosystem services to adjoining farms and cities. This is in marked contrast to recent work from Kareiva and colleagues (2011), who argue that very altered landscapes may still support a considerable diversity of life. Others who agree more closely with Terborgh have published strident rejoinders to Kareiva et al. (for example, Suckling, 2011).

The second concern is that emphasis on a not as yet widely substantiated ecosystem services value paradigm could undercut conservation efforts by devoting resources to an approach that could yet be discredited. I think it reasonable to take an agnostic view as to the general applicability of ecosystem service values to motivate conservation. There are, however, at present many reasons to doubt that many ecosystem service values will prove significant, particularly those that might most easily be appropriated by the communities generating them. More research may well be in order, but until the results are more definitive we should be on guard against excessive claims. There are, however, a great many impediments to conducting reliable, replicable work (see Simpson, 2010).

So, I fear we are not yet close to an empirical resolution of empirical questions. In this predicament, perhaps turning to some conceptual modeling might help guide us toward more fruitful empirical inquiries. As the statistician George E.P. Box famously remarked, 'All models are wrong, but some are useful'.

My model certainly demonstrates the first part of this adage, but I hope that it might also prove useful in certain respects. I suspect that economists will find the 'counterintuitive' results I refer to above considerably less paradoxical than might the proverbial person in the street. They amount to little more than illustrations of the diamonds and water paradox: things are valuable to the degree that they are scarce, and hence, once they become abundant, they lose (marginal) value.

I also hope, however, that the modeling effort I describe below might prove useful precisely *because* the model is wrong. I am not aware of any logical flaws in the model I develop below: its conclusions follow from its premises. A more important question concerns whether the premises are complete. There are several reasons to suppose that they are not. I will discuss the model's limitations in greater detail in the final section, but for now let me note its largest omission: an appreciation of ecosystem services beyond those whose benefits are enjoyed locally. My own view is that these services – particularly, the preservation and stewardship of the rest of creation – comprise the most compelling reasons for conserving biodiversity and the ecosystems that support it. This is not to say that more local ecosystem services cannot or do not motivate conservation in some areas, or that such local services do not constitute contributory reasons for conservation. However, my sense is also that the conservation community has been down the road before of trying to justify conservation on the basis of local benefits, and that that experience was not successful[3] (Terborgh et al., 2002). I fear that the emphasis on ecosystem services may represent another attempt to convince local communities that their own interests will be served by conservation, when attempts to do so in the past have not been convincing. Conservation efforts, particularly in those parts of the world in which biodiversity tends to be greatest and incomes tend to be lowest, will still likely require a transfer of funds from wealthier to less-developed countries. I would not want enthusiasm for ecosystem services to distract us from consideration of this possibility.

13.2 ECOSYSTEM SERVICES IN THE PROVISION OF LOCAL PUBLIC GOODS: GENERAL CONSIDERATIONS AND SPECIFIC EXAMPLES

Ecosystem services are valuable because they provide local public goods such as pollination, pest control, or water regulation services to agriculture, and flood and storm protection and water purification to cities. That this is true to some extent is beyond dispute.[4] The more important and economically relevant questions concern the degree to which such services are underprovided because they cannot be fully appropriated by those who provide them, that is, because they are local public goods.

This is an interesting question in many cases. Sagoff (2011) presents examples of instances in which people do provide themselves with such local public goods. There are a number of both historical and recent examples of water suppliers paying to protect upstream catchments (Heal, 2000; Perrot-Maître, 2006). These cases testify to some appreciation of the water protection services afforded by natural systems. Moreover, some examples offered in the literature as to the local benefits afforded by ecosystem services beg the question of why private parties would not 'internalize the externality' by acquiring the natural areas providing the services. For example, Ricketts et al. (2004), found that the Finca de Santa Fe coffee plantation in Costa Rica benefited from the pollination services afforded by the relatively small areas of remnant forests surrounding it. If this were the case, why didn't its owners simply acquire the forest to conserve it?[5]

Even if we concede the point that ecosystem services may be underprovided with respect to the interests of local producers, the question remains as to how much they are underprovided. Recall the example given above of the flood protection benefits studied by Costanza et al. (2008), who find that coastal wetlands provide valuable protection against flooding and storm damage. Their estimates of value varied considerably, but could be as high as US$51 000 per hectare per year – annualized at a 5 percent interest rate in perpetuity, a million dollars per hectare. Does this constitute a compelling argument for restoring larger expanses of coastal wetlands? The answer, not surprisingly, is 'It depends'. As an economic proposition, coastal wetlands should be preserved or restored until the value of the service they provide is equal to the opportunity cost of forgoing alternative uses of the land. In heavily developed areas of the United States land often commands prices in the millions of dollars per hectare. It is not clear that Costanza et al. were identifying a market failure.[6]

Let us follow the logic of this example a little further. Suppose that the study had found a much larger value for the flood protection service afforded by coastal wetlands. In such circumstances we might more confidently state that more land should be devoted to coastal wetlands, but how much more? If the marginal value of coastal wetlands is high, it must be because the marginal hectare of land is significantly reducing expected damages to the properties it protects. The more protection is afforded by the first hectare of land set aside for this purpose, however, the less value at risk remains to be protected by others. It may be, then, that finding high values at current margins implies that values decline rapidly as more area is conserved, and the net quantitative effect of accounting for land preserved would be modest.

If, on the other hand, coastal wetlands did not provide high marginal values in protecting otherwise vulnerable properties it could make sense to set aside more land for coastal wetlands. This would also have its limitations, however. Structural reinforcement might be an imperfect substitute for tempering the severity of storms, but if coastal wetlands did not provide such services very effectively, it would be more cost-effective to rely on the structural measures and forgo the wetlands.

Another example underscores these points. Interception, storage and neutralization of reactive nitrogen are an important ecosystem service. Natural areas such as wetlands and riparian buffer strips of grass, trees and other vegetation can be used to treat runoff water from agriculture, sewage treatment and other sources. In a recent paper (Simpson, 2011) I sketched a model of measures that might be taken to meet nutrient reduction targets from agricultural land in the Chesapeake Bay watershed in the eastern United States. Principles analogous to those I outlined above for storm protection arise in the context of nitrogen retention as well. The scientific literature suggests that preserved areas of natural vegetation can be quite effective in reducing the nitrogen loading reaching downstream waters (Mayer et al., 2007). My results suggest that *more* land would optimally be devoted to nitrogen retention if each hectare devoted to this purpose were *less* effective. There is a limit to this phenomenon, however. There is a threshold level of effectiveness below which farmers constrained by regulation would choose to apply less fertilizer to their fields than to withhold land from production so as to treat the runoff from the extra fertilizer they have applied.

In the next section I develop another model of the allocation of land between the provision of ecosystem services and an activity that benefits from the services generated. In contrast to the examples I have just cited I try to generalize somewhat in the model below, embedding the ecosystem-service-using sector in a slightly broader notion of an economy.

13.3 THE MODEL

I want to develop a simple and illustrative model of how ecosystem services might be employed in the agricultural sector of a simple economy. Let me begin with a list of features we might want such a model to illustrate, and then proceed to demonstrate that the model I propose incorporates these features. It will quickly be apparent that the model I propose is not realistic in many other respects. If the reader will temporarily indulge me now, I will defer defense of these assumptions until the final section.

Consider, then, the following stylized features of a simple economy:

1. Available resources will be divided between the production of food and other uses.
2. Land may be important in agriculture, but is not a constraining factor in other activities.
3. Food is essential, but the diamonds-and-water paradox applies to it: it will be very dear when it is in short supply, but very cheap when it is abundant relative to other goods.
4. While it is not often the case in fact, as a social ideal we should allocate production equitably among the population.
5. Land that is not devoted to agriculture may provide a variety of ecosystem services that will enhance agricultural production.
6. The ecosystem services provided by 'idle' land substitute for resources that might be employed elsewhere in the economy if they were not required in agriculture.

Let me now begin to construct a model that will illustrate these features. Beginning with the third and fourth attributes, suppose that the utility enjoyed by each of N individuals is:

$$U = \alpha \ln(Q/N) + \tilde{X}/N \tag{13.1}$$

where Q is agricultural consumption and $\tilde{X}$ is the amount of resources available for consumption or use in some other sector of the economy. I will not be explicit about what $\tilde{X}$ consists of; it could be some combination of capital equipment, other productive inputs, or, for that matter, available labor.

I assume that both agricultural production and residual resources are distributed equally to all citizens. Overall social welfare will be defined as N times individual utility:

$$W = NU = \alpha N \ln Q - \alpha N \ln N + \tilde{X} \tag{13.2}$$

To reduce notational clutter I will suppress α in what follows, on the argument that I am free to choose the units in which population is measured; conventionally, we consider the number of individuals, but there is no reason we could not speak of dozens, scores, and so on. The middle term on the right-hand side of (13.2) will not feature in the optimization problem that follows. While the choice of the units in which population is measured will affect the numerical magnitude of the welfare objective, it will not affect the results that follow.

I will now embody the first and second desired features in the model: that non-land resources are divided between agriculture and other uses, while land is not important in the provision of non-food products. Let:

$$\tilde{X} = \bar{X} - X \tag{13.3}$$

where $\widetilde{X}$ is the total endowment of resources and X is the amount of resources employed in agriculture.

Suppose that agricultural production may be described as a quadratic function:

$$q = y - \gamma y^2 \tag{13.4}$$

where q is output *per hectare of land* devoted to agriculture, and y is a composite input, also measured per hectare of land. Expression (13.4) may be regarded a canonical form, a second-order approximation to any agricultural production function in which some inputs are required if there is to be any production.

Next I will embed the fifth and sixth features enumerated above, that land not employed directly in agriculture provides ecosystem services that enhance productivity, and that these ecosystem services substitute for resources that could be employed elsewhere. So, let us assume that land used in production is homogeneous, and that the composite input is given by:

$$y = \frac{X + \phi(1 - A)}{A}, \tag{13.5}$$

where X is the aggregate quantity of the resources employed in the agricultural sector, A is the total amount of land used in production, and ϕ is a constant. Normalize the total area of land to one, and thus suppose that $1 - A$ represents the land preserved to provide ecosystem services. The assumptions embodied in (13.5) are extreme. In reality, land is not perfectly homogeneous, nor are ecosystem services and other resources perfect substitutes. It is, however, a staple of the literature on ecosystem services that the services of nature can be substituted for alternative resources that are provided in markets. By making the assumption of substitutability, I am doing no more than taking the existing literature on its own terms. If it were *not* possible to easily substitute between ecosystem services and purchased inputs, there would be little point in advocating such measures.

The constant ϕ measures the rate at which ecosystem services provided by natural systems can be traded off against purchased inputs. I will say that ϕ measures the *effectiveness* of preserved land in providing ecosystem services: the higher is ϕ, the greater is the contribution of any given area of preserved land.

Total production when an area $A \leq 1$ is cultivated is the product of output per hectare of land and the amount of land used directly in production:

$$Q = qA = X + \phi(1 - A) - \gamma\frac{[X + \phi(1 - A)]^2}{A} \tag{13.6}$$

Using (13.3) and (13.6) in the social welfare objective (13.2), and recalling that the utility-weighting constant, α, has been subsumed, the welfare objective may be written as:

$$W = N\ln\left(X + \phi(1 - A) - \gamma\frac{[X + \phi(1 - A)]^2}{A}\right) - N\ln N + (\overline{X} - X) \tag{13.7}$$

Differentiating with respect to X, the first-order condition for welfare maximization is that:

$$\frac{\partial W}{\partial X} = \frac{N}{Q}\left[1 - 2\gamma\frac{X + \phi(1 - A)}{A}\right] - 1 \begin{array}{c} > \\ = \\ < \end{array} 0 \quad \text{as} \quad \begin{array}{c} X = X \\ 0 \leq X \leq \bar{X} \\ X = 0 \end{array} \tag{13.8}$$

The first-order condition for welfare maximization with respect to A is that:

$$\frac{\partial W}{\partial A} = -\frac{N}{Q}\phi + \frac{N}{Q}\gamma\frac{(X + \phi)^2 - \phi^2 A^2}{A^2} \begin{array}{c} > \\ = \end{array} 0 \quad \text{as} \quad \begin{array}{c} A = 1 \\ 0 < A \leq 1 \end{array} \tag{13.9}$$

Both first-order conditions are stated as potential inequalities, with strict inequality obtaining if the amount of resources devoted to agriculture is constrained either by their total availability, $\bar{X}$, or the requirement that resources devoted to agriculture be non-negative. Similarly, the total amount of land applied directly in agriculture cannot exceed the land available, and this constraint may sometimes bind. Note that there would be no point in devoting all available land to the provision of ecosystem services, as there would then be none on which to produce food, and the marginal product of the consumption of food is unbounded as production approaches zero.

13.4 AN INTERIOR SOLUTION

Let us begin by characterizing an interior solution, in which some resources are employed both in the agricultural and the other sector of the economy, and some land is withheld from production to generate ecosystem services. Solving (13.8) for X when (13.8) is an equality, we have:

$$X = \frac{N - Q}{2N\gamma}A - \phi(1 - A) \tag{13.10}$$

Using (13.10) to substitute for X in (13.9), rearranging, and simplifying, we find that:

$$\frac{(N - Q)^2}{4NQ} = \phi\gamma \tag{13.11}$$

Adding 1 to each side of (13.11):

$$\frac{(N + Q)^2}{4NQ} = \phi\gamma + 1 \tag{13.12}$$

Thus, taking the square root of the quotient of (13.11) divided by (13.12):

$$\frac{N - Q}{N + Q} = \sqrt{\frac{\phi\gamma}{\phi\gamma + 1}} \tag{13.13}$$

from which we find that:

$$Q^* = N\left(\frac{\sqrt{\phi\gamma + 1} - \sqrt{\phi\gamma}}{\sqrt{\phi\gamma + 1} + \sqrt{\phi\gamma}}\right). \tag{13.14}$$

So, as long as an interior solution obtains in which some of the resource is employed in both sectors of the economy and some land is withheld from production, agricultural output per person remains constant.

Rearranging (13.8), the first-order condition with respect to X, we have:

$$\frac{N - Q}{2\gamma N}A = X + \phi(1 - A). \tag{13.15}$$

Using (13.15) in the expression for output (13.6), we have:

$$Q = \frac{N - Q}{2\gamma N}A - \gamma A\left[\frac{N - Q}{2\gamma N}\right]^2, \tag{13.16}$$

or:

$$Q = \frac{N^2 - Q^2}{4\gamma N^2}A. \tag{13.17}$$

Recalling (13.11) and (13.12):

$$N - Q = 2\sqrt{NQ\phi\gamma}, \tag{13.18}$$

and:

$$N + Q = 2\sqrt{NQ(\phi\gamma + 1)}. \tag{13.19}$$

So, using (13.18) and (13.19) in (13.17):

$$A^* = \frac{\gamma N}{\sqrt{\phi\gamma(\phi\gamma + 1)}} \tag{13.20}$$

When it is optimal to devote some of the scarce resource to agricultural production and when it is optimal to preserve some of the landscape for the provision of ecosystem services, the fraction of land employed directly in agricultural production will be proportional to the population.

Having now derived expressions for output, Q^*, and land use, A^*, it remains to determine X^*, the use of the scarce resource in agriculture, when neither land nor resource constraints bind. Rearranging the first-order condition with respect to A (13.9), when it holds as an equality and $X > 0$:

$$X^* = \sqrt{\phi\left(\frac{1 + \phi\gamma}{\gamma}\right)}A^* - \phi \tag{13.21}$$

Substituting for A^* from (13.20), we arrive at a compact result:

$$X^* = N - \phi \tag{13.22}$$

The amount of resources devoted to agriculture depends entirely on the comparison between the number of people who must be provided with food and the efficiency with which land withheld from agriculture enhances the productivity of agriculture. Of course if the effectiveness parameter ϕ is large enough with respect to the size of population all non-land resources would be put to other uses and, conversely, if $\overline{X}$, the total quantity of resource available, were small relative to the difference $N - \phi$ all resources would be devoted to agriculture. Heuristically, when resources are scarce food is essential, and agriculture will have first claim to scarce resources.

13.5 CONSTRAINED OPTIMA

Summing up what we have just derived, when not all land is devoted (directly) to farming and some other resources are devoted to farming, expressions (13.20) and (13.22) describe how much land and the other resource, respectively, are absorbed in agricultural production. It is also possible to derive corner solutions. There are three more possibilities:

- All other resources might be devoted to food production.
- All land might be devoted to food production.
- Some land might be devoted to providing ecosystem services, but all resources would be directed elsewhere.

These possibilities might be characterized as a subsistence outcome, one with no reliance on ecosystem services, and a third with complete reliance on ecosystem services, respectively. Let me briefly characterize each.

Subsistence

The case in which all resources would optimally be devoted to food production would describe an economy in desperate straits, where subsistence food production would be pursued to the exclusion of any other activity. An economy would rise out of this subsistence state when resources rose to the level relative to population that the marginal utility of consumption of resources in the 'other' sector of the economy exceeded the value of the marginal product (that is, marginal utility × marginal product) of their employment in agriculture. If all resources are devoted to agriculture, then from the first-order condition with respect to X (13.8):

$$\frac{N - Q}{N} \geq 2\gamma \frac{\overline{X} + \phi(1 - A)}{A} \tag{13.23}$$

and from the first-order condition with respect to A (13.9):

$$\gamma \frac{(\overline{X} + \phi)^2 - \phi^2 A^2}{A^2} \geq \phi \tag{13.24}$$

A question of some interest is whether a subsistence economy – one in which the entire stock of the resource $\overline{X}$ is devoted to food production – would ever choose to cultivate the entire area of land available to it, which, recall, has been normalized to one. If it would, expression (13.24) would be:

$$\bar{X}^2\gamma + 2\bar{X}\phi\gamma \geq \phi \tag{13.25}$$

Setting $A = 1$ in (13.23):

$$\frac{N - Q}{N} \geq 2\gamma\bar{X}, \tag{13.26}$$

so (13.25) could be written as:

$$\bar{X}^2\gamma N \geq \phi Q. \tag{13.27}$$

When all land is devoted directly to food production, and all resources are as well, output is:

$$Q = \bar{X}(1 - \gamma\bar{X}). \tag{13.28}$$

Thus (13.27) becomes:

$$N \geq \frac{\phi(1 - \gamma\bar{X})}{\gamma\bar{X}}. \tag{13.29}$$

Heuristically, no land would be set aside to provide ecosystem services in a subsistence economy if such preserved land were not sufficiently effective in enhancing agricultural productivity – that is, if ϕ were small. No land would be preserved to provide ecosystem services also if plentiful resources were available, that is, if $\bar{X}$ were large enough. It does not necessarily follow, then, that an economy reduced to subsistence would necessarily set aside land for the provision of ecosystem services.

All Land Devoted to Food Production

From expressions (13.20) and (13.22), it is readily seen that if N is large enough relative to the effectiveness parameter, f, all land will be devoted directly to food production. Land is too scarce and, consequently, valuable to devote it to the provision of even productivity--enhancing ecosystem services.

Complete Reliance on Ecosystem Services

We saw in expression (13.20) that when other factors are available in sufficient quantities, population is not too large, and conserved land is relatively effective it is optimal to set aside some land for the provision of ecosystem services. It seems intuitive that the more effective such conserved land is in the provision of ecosystem services, the more land would optimally be conserved.

This is only true up to a point, however. From (13.22), for values of ϕ in excess of population, N, no other resources would be devoted to agriculture. This might seem to be a conservationists' dream: ecosystem services are so important that agriculture should be completely reliant upon them. Under such circumstances, wouldn't we expect more land to be devoted to conservation than if there were some reliance on other resources as inputs as well? The answer is, in this case (these results are illustrative of reasonable possibilities, not general propositions), 'No'. When only some of the available land and no resources should optimally be employed in agriculture it is easily seen by setting $X = 0$ in (13.9) that:

$$A^0 = \sqrt{\frac{\phi\gamma}{1 + \phi\gamma}}. \tag{13.30}$$

Now the amount of land devoted directly to production is *increasing* in the effectiveness parameter. It follows by comparison of (13.20), (13.22) and (13.30), then, that the maximum amount of land conserved for given values of the other parameter, γ, would arise when $\phi = N$ and:

$$A^{\max} = \frac{N\gamma}{\sqrt{N\gamma(N\gamma + 1)}}. \tag{13.31}$$

An interesting implication of (13.31) is that *regardless* of the effectiveness with which conserved land provides productivity-enhancing ecosystem services, the optimal amount of land to conserve declines to nothing for large populations, if sufficient other resources are available.

13.6 SOME POLICY-RELEVANT IMPLICATIONS

Summarizing what we have just seen, other things being equal, a community is less likely to rely on conserved land to provide ecosystem services to augment the productivity of agriculture if its population is larger. The more effective conserved land is in providing ecosystem services, the more likely it is that a community will wish to conserve land for this purpose. Beyond a certain critical level of effectiveness, however, the more effective is the conserved land in providing ecosystem services, the *less* land will be conserved for this purpose.

The objective of many papers in the ecosystem services literature is to demonstrate that land on the margin will generate higher values if it is preserved in a more or less natural state so as to provide some ecosystem services than if it is devoted directly to production. In the spirit of such inquiries, let us consider a situation in which all land is devoted to production, and then suppose that one hectare is restored to provide ecosystem services. What would the marginal value of this first hectare be?

Evaluate (13.8), the first-order condition with respect to X at $A = 1$ to find:

$$\frac{\partial W}{\partial X} = \frac{N}{Q}[1 - 2\gamma X] - 1 = 0 \tag{13.32}$$

When $A = 1$, the left-hand side of (13.9) becomes:

$$\frac{\partial W}{\partial A} = \frac{N}{Q}[\gamma[(X + \phi)^2 - \phi^2] - \phi] < 0 \qquad (13.33)$$

Using (13.32) to eliminate *N*/*Q*:

$$\frac{\partial W}{\partial A} = \frac{\gamma X^2}{1 - 2\gamma X} - \phi < 0. \qquad (13.34)$$

Because *X* is independent of ϕ the choice of other input does not depend on the effectiveness of land in providing ecosystem services when, by assumption, land is not providing any ecosystem services, the value of *A* is decreasing in the effectiveness of *A* at the margin when no land is preserved. Put in a more natural way, when the effectiveness of conserved land is higher, the value of conserving some land is higher.

This is a statement about the value of conserving *some* land, however. Comparing (13.34) and (13.30), the value of the first hectare of land devoted to the provision of ecosystem services would be highest when it would be optimal to devote very little land to this purpose.

Alternative Policy Interventions

The point of many of the results I have illustrated above is that there may be circumstances in which greater reliance on ecosystem services will *not* be attractive to a country at a certain stage in its development. What approaches might, then, be effective in inducing greater conservation? Much of the appeal of ecosystem services as a conservation strategy likely lies in the perception that they will serve to motivate conservation without the need for outside funding (except, perhaps, to inform developing countries of the opportunities they are missing). If one were to abandon the hope that this would prove possible, what other alternatives present themselves?

The most obvious would be to offer compensation for conserving more land. If the first-order conditions hold as equalities, the country should be indifferent between conserving a little more land and using a little more capital in agriculture. Payment in the form of other resources ought to be accepted in exchange for a promise to conserve more land.

Such arrangements might prove problematic, however, unless performance can be enforced. If the country is not constrained in its use of other resources in agriculture, it would prefer to employ the additional resources in the other sector without changing its allocation of land between production and conservation: neither expression (13.20) nor (13.22) depends on $\bar{X}$, its total stock of resources. Worse yet, if the country were in subsistence mode and constrained in its use of other resources in agriculture, it can be shown by rearranging (13.9) that:

$$A = \frac{\bar{X} + \phi}{\sqrt{\phi(1/\gamma + \phi)}} \qquad (13.35)$$

the more other resources are available, the *greater* will be the area of land cultivated: other resources will be substituted for the ecosystem services provided by conserved land.

The remaining alternative would be to provide food aid in an effort to reduce the area of land that would be devoted to agriculture. We can illustrate this possibility by redefining expression (13.6), describing agricultural consumption as:

$$Q = X + \phi(1 - A) - \gamma\frac{[X + \phi(1 - A)]^2}{A} + q \tag{13.36}$$

where q is food-aid-in-kind. Making this amendment, the derivations recorded in expressions (13.7)–(13.15) remain valid, and we can write an implicit expression analogous to (13.17) as:

$$Q = \frac{N^2 - Q^2}{4\gamma N^2}A + q \tag{13.37}$$

As the first term on the right-hand side is domestic agricultural production, it cannot be negative. Since expression (13.14) remains valid, however, Q, total food consumption from both domestic and donor sources must be independent of the amount of food aid, q. Thus we can implicitly differentiate (13.37) to find that:

$$\frac{dA}{dq} = -\frac{4\gamma N^2}{N^2 - Q^2} \tag{13.38}$$

In addition to its obvious humanitarian merits, providing food aid might also divert land from direct agricultural production.[7]

13.7 DISCUSSION AND CONCLUSION

In this section I first discuss the intuition underlying the results derived above and then turn to the questions of whether a model this simple and schematic provides any useful insights. I believe that it does, although we may learn more from considering the implications of the model's omissions than we do from its results per se. I consider those omissions and their implications at the end of the section.

The most interesting results of the model concern what happens as population increases. Other things being equal, the higher is population, the less land should be preserved to provide ecosystem services to enhance agricultural productivity. It is worth emphasizing that this result does not come about because there is greater demand for land for other uses. It is an unrealistic assumption of the model that there are no other uses of land. This assumption is conservative, however, in the sense that no ad hoc characterization of demand for land in the other sector can be driving results.

The result is driven by the fact that food is a necessity. When population is high, food consumption per person will be low, other things being equal. The marginal rate of substitution between food and other consumption must then be low. Hence it is important not only to assure high yields, but also to assure that such yields are achieved on as many acres as possible.

These intuitions drive the chapter's basic results. Even a subsistence economy should not always preserve some land to provide ecosystem services. Land should be preserved to provide ecosystem services in a populous economy only if that land produces ecosystem services very effectively. Moreover, if preserved land does provide ecosystem services effectively enough to justify forgoing its direct use in production, the preservation of only a very small area of land can be justified in a populous economy; if land is worth preserving, it must be because it is very effective in providing ecosystem services, and if it is very effective, a little must go a long way.

Let me briefly discuss next some of the many things this model omits. It is admittedly curious to propose a model in which *population* matters but in which *labor* does not enter explicitly. As mentioned above, labor might be a component of the aggregate 'resources' I have been reluctant to define more specifically. The best excuses I can offer for not dealing with the composition of resources more explicitly are simply that I have sacrificed detail for tractability and, I hope, insight, and that similar lumping together of inconveniently heterogeneous resources has been done for the same reason in other work that seeks to derive illustrative results in schematic models (consider, for example, Dasgupta and Heal, 1974).

Another curious feature for a chapter whose title purports to offer insights on a developing economy is the lack of any dynamics. It would be interesting to consider what changes in land and resource use might be motivated by the desire to accumulate a larger stock of resources (we might, in this context, call it 'capital') for latter employment. Again, I have resisted the temptation to further complicate a model by introducing dynamic optimization when, despite its apparent simplicity, analysis of the model already requires more equations, inequalities, limits, and so on, than might appear to be justified by its simple insights.

It is unrealistic to suppose that the ecosystem services provided by preserved land can be perfectly substituted by other resources. This assumption does, however, seem broadly consistent with the observation that developing agriculture may be composed of varying combinations of traditional farms, in which land is allowed to lie fallow to recover productivity and fertilizer and pesticides are not applied, with modern farms that rely on purchased inputs for fertilization and pest control. Moreover, much of the literature on ecosystem services is predicated on the premise that the services of nature are good substitutes for other inputs. If they were *not* good substitutes, there would be little point or need to argue for their greater employment.

It is also obvious that I have attempted no sophisticated spatial analysis – or, for that matter, introduced any spatial dimension at all. In the real world, of course, different elements of the landscape are differentiated by their topography, hydrology, soil fertility, proximity to population centers, and a host of other factors. Again, I will simply say that I have tried to avoid complications.

It is not clear to me, however, that any of these avoided complications would change the basic conclusions of the model. These conclusions are that, other things being equal, wealthier and more populous economies would preserve less land for the provision of ecosystem services. Embellishing the model to incorporate some of the omitted features noted above might introduce some peripheral differences. In a model where land quality and location were considered, for example, some land would be 'preserved', but not so much to provide ecosystem services as simply because it is not sufficiently productive to be worth cultivating. The larger question is whether it makes sense to set aside land that has a high opportunity cost in terms of foregone production in order to provide more ecosystem services. I would submit that, in a relatively wealthy and populous economy, it does not.

This point, however, underscores some omissions of the model that do matter. In a wealthy and populous economy people likely care far more about another type of ecosystem service that does not enter into my model at all: the moral and aesthetic satisfactions afforded by preserving wild areas and the biodiversity they shelter. This *is* a crucial omission, though one that I would argue is not so much a failing of the model per se as one of its most important implications: different arguments are likely to be more compelling in different societies. It could well be that, in poor countries in which arable land is relatively abundant, an appeal to ecosystem services to motivate the preservation of natural areas would be compelling. One has to wonder, however, if such an appeal would really be necessary in such circumstances. Do people in such countries not realize this? In wealthier and more populous societies, however, it likely makes more sense to appeal to less tangible motives.[8]

Before concluding I should circle back to another aspect of dynamics that I may have dismissed too quickly above. While it is not clear that expanding the model to consider the accumulation of manufactured capital would greatly affect results, I have assumed that the direct employment of land in food production does not detract from the natural capital available should a different allocation be desired in the future. This could be problematic. While much of the current concern with ecosystem services is motivated by a sense that natural systems could unravel if overstressed, I am aware of little work that quantifies such claims. My model suggests that documenting the mechanism by which such unraveling might occur could also provide more compelling motivation for conservation than appealing to the productivity-enhancing effects of ecosystem services.

NOTES

* Any opinions expressed are those of the author only, and do not necessarily reflect those of the US Environmental Protection Agency.

1. It is also worth noting the title of this article: 'Ecosystem services in decision-making: time to deliver'. The clear implication is that the concept has not yet 'delivered'.
2. See an earlier paper, Simpson (2010), for some of the reasons.
3. I have in mind particularly international experience with 'integrated conservation and development projects' over the last three decades. For a critical review, see Terborgh et al. (2002), or Simpson (2004).
4. Although some have suggested that natural ecosystems can also provide disamenities, such as cross-pollination of crops (Sagoff, 2011), or reduction in water available for agriculture (Lele et al., 2011).
5. While various answers might be proposed to this question, the best may be that they knew it would not be useful to them for long. Douglas Macauley (2006) reports that the plantation was subsequently converted to pineapple cultivation, and no longer benefited from natural pollination. See also Simpson (2010), in which I suggested that the benefits of increased pollination, while appreciable, may not have been as high as the opportunity costs of forgoing cultivation of the land.
6. It is fair to note, though, that land providing flood protection services could well be providing other services, such as recreational use, wildlife habitat, or scenic vistas. These public values should be added when comparing them with the private benefits forgone.
7. I am mindful that here particularly we ought not to take the results of a very schematic model too seriously. Another consequence of food imports is to displace production in the recipient country, which may have adverse consequences well beyond the scope of this simple depiction.
8. This raises the difficult but pivotal question of what it is that motivates people to contribute to the provision of public goods. A number of contributions to economics and related literatures have delved into this question. While they document that the received theory of self-interested economic agents 'free-riding' on the

contributions of others is inadequate to explain observed behavior, a general theory of the private provision of public goods remains elusive (see, for example, Fehr and Schmidt, 1999; Meier, 2006; and Reeson, 2008). I am grateful to an anonymous reviewer for suggesting the relevance of this literature to my subject).

REFERENCES

Costanza, R., O. Pérez-Maqueo, M. Luisa Martinez, P. Sutton, S.J. Anderson and K. Mulder (2008), 'The value of coastal wetlands for hurricane protection', *Ambio*, **37**(4), 241–8.
Daily, G.C. and P.A. Matson (2008), 'Ecosystem services: from theory to implementation', *Proceedings of the National Academy of Sciences of the United States of America*, **105**(28), 9455–6.
Daily, G.C., S. Polasky, J. Goldstein, P.M. Kareiva, H.A. Mooney and L. Pejchar et al. (2009), 'Ecosystem services in decision-making: time to deliver', *Frontiers in Ecology and Environment*, **7**(1) 21–8.
Dasgupta, P. and G. Heal (1974), 'The optimal depletion of exhaustible resources', *The Review of Economic Studies, Vol. 41*, Symposium on the Economics of Exhaustible Resources, 3–28.
Fehr, E. and K.M. Schmidt (1999), 'A theory of fairness, competition and cooperation', *Quarterly Journal of Economics*, **114**(3), 817–68.
Heal, G. (2000), *Nature and the Marketplace: Capturing the Value of Ecosystem Services*, Washington, DC: Island Press.
Kareiva, P. and S. Ruffo (2009), 'Using science to assign values to nature', *Frontiers in Ecology*, **7**(3), 29–37.
Kareiva, P., R. Lalasz and M. Marview (2011), 'Conservation in the Anthropocene', accessed 8 February 2014 at http://thebreakthrough.org/index.php/journal/past-issues/issue-2/conservation-in-the-anthropocene/.
Kremen, C., J.O. Niles, M.G. Dalton, G.C. Daily, P.R. Ehrlich and J.P. Fay et al. (2000), 'Economic incentives for rain forest conservation across scales', *Science*, **288**(5472), 1828–32.
Lele, S., J. Vaidyanathan and S. Hegde (2011), 'Influence of forest cover change on watershed functions in the Western Ghats: a coarse-scale analysis', accessed 11 February 2014 at http://atree.org/sites/default/files/pubs/slele/recently-completed-studies/abstract-coarse-scale-analysis.pdf.
Mayer, P.M., S.K. Reynolds Jr., M.D. McCutcheon and T. Canfield (2007), 'Meta-analysis of nitrogen removal in riparian buffers', *Journal of Environmental Quality*, **36**(4), 1172–80.
McCauley, D. (2006), 'Selling out nature', *Nature*, **443**(7107), 27–8.
Meier, S. (2006), *The Economics of Non-Selfish Behaviour*, Cheltenham, UK and Northampton, MA, USA: Edward Elgar Publishing.
Naidoo, R. and T.H. Ricketts (2006), 'Mapping the economic costs and benefits of conservation', *PLoS Biology*, **4**(11) 2153–64.
Odling-Smee, L. (2005), 'Conservation: dollars and sense', *Nature*, **437**(7059), 614–16.
Pearce, D. (2007), 'Do we really care about biodiversity?' *Environmental and Resource Economics*, **37**(1), 313–33.
Pearce, D. and D. Moran (1994), *The Economic Value of Biodiversity*, London: Earthscan.
Perrot-Maître, D. (2006), *The Vittel Payments for Ecosystem Services: A 'Perfect' PES Case?*, London: International Institute for Environment and Development.
Reeson, A. (2008), 'Institutions motivations and public goods, socio-economics and the environment in discussion', *CSIRO Working Chapter Series* No. 2008–01.
Ricketts, T.H., G.C. Daily, P.R. Ehrlich and C. Michener (2004), 'Economic value of tropical forest to coffee production', *Proceedings of the National Academy of Sciences of the United States of America*, **101**(34) 12579–82.
Sagoff, M. (2011), 'The quantification and valuation of ecosystem services', *Ecological Economics*, 70(3), 497–502.
Simpson, R.D. (2004), 'Conserving biodiversity through markets: a better approach', *PERC Policy Series* No. 32.
Simpson, R.D. (2010), 'The "ecosystem service framework": a critical assessment', in P. Kumar and M.E. Wood (eds), *Valuation of Regulating Services of Ecosystems: Methodology and Applications*, London: Routledge.
Simpson, R.D. (2011), 'Allocating land for an ecosystem service: a simple model of nutrient retention with an application to the Chesapeake Bay watershed', *Canadian Journal of Agricultural Economics*, **59**(2), 259–80.
Suckling, K. (2011), 'Conservation for the real world', accessed 8 February 2014 at http://thebreakthrough.org/journal/debates/conservation-in-the-anthropocene-a-breakthrough-debate/conservation-for-the-real-world.
Tallis, H. (2006), 'Introduction to ecosystem services', *Science Chronicles*, **4**(3).
Tallis, H. and P. Kareiva (2005), 'Ecosystem services', *Current Biology*, **15**(18), 746–8.
Terborgh, J. (1999), *Requiem for Nature*, Washington, DC: Island Press.
Terborgh, J., C. van Shaik, L. Davenport and M. Rao (2002), *Making Parks Work: Strategies for Preserving Tropical Nature*, Washington, DC: Island Press.

Turner, R.K. and G.D. Daily (2008), 'The ecosystem services framework and natural capital conservation', *Environmental and Resource Economics*, **39**(1), 25–35.
Turner, W.R., K. Brandon, T.M. Brooks, R. Costanza, G.A.B. da Fonseca and R. Portela (2007), 'Global conservation of biodiversity and ecosystem services', *BioScience*, **57**(10), 868–75.
Turpie, J.K., B.J. Heydenrych and S.J. Lamberth (2003), 'Economic value of terrestrial and marine biodiversity in the Cape Floristic Region: implications for defining effective and socially optimal conservation strategies', *Biological Conservation*, **112**(1–2), 233–51.

14. Biodiversity prospecting over time and under uncertainty: a theory of sorts

Amitrajeet A. Batabyal and Peter Nijkamp*

14.1 INTRODUCTION

The notion of biological diversity or biodiversity has now become a fashionable concept. This concept refers to the variability among living organisms from all terrestrial and marine sources and from the ecological complexes of which these organisms are a part (Nunes et al., 2003). Biodiversity itself can be of various types such as genetic, species, ecosystem and functional. The loss of biodiversity is generally considered to be very costly from a societal standpoint and hence many studies have now attempted to assess the economic value of biodiversity (Nunes and Nijkamp, 2011). In this regard, a comparative, meta-analytic review of the economic valuation of biodiversity can be found in Nijkamp et al. (2008).

Over the past couple of years, several studies have been devoted to the economic analysis of genetic diversity in the context of the commercial search among genetic codes contained in living organisms in order to develop chemical compounds of industrial and pharmaceutical value in agricultural, industrial and medical applications (see Simpson et al., 1996; Swanson, 1996; Grifo et al., 1997). This state of affairs has given rise to intriguing questions about the willingness to pay by biotechnological companies for genetic diversity as inputs into commercial products such as anti-cancer drugs (see Macilwain, 1998; Sonner, 1998; Neto and Dickson, 1999; ten Kate and Laird, 1999).

An interesting recent survey article on the value of conserving genetic resources for research and development (R&D) is contained in Sarr et al. (2008). These authors assess the extent to which society is able to invest now in order to prepare for future risks and uncertainties in the arrival of various biological and medical contingencies. Such issues have given rise to a new strand of literature on what is now known as 'biodiversity prospecting' or bioprospecting.

Specifically, biodiversity prospecting refers to 'the exploration of biodiversity for commercially valuable genetic and biochemical resources' (Reid et al., 1993a, p. 1). As noted by Eisner (1989, 1992), Reid et al. (1993b), and others, ecologists, environmentalists and taxonomists have been saying for some time that it should be possible to justify the conservation of biodiversity on the basis of its many pharmaceutical and other commercial applications. Even so, interest among pharmaceutical firms in particular in biodiversity prospecting has been muted until the completion of the now prominent agreement between Costa Rica's National Biodiversity Institute (INBio) and the United States--based pharmaceutical giant Merck and Company.

The September 1991 agreement between INBio and Merck contained two key provisions. First, INBio would provide Merck with a whole host of chemical extracts from wild plants, insects and micro-organisms from Costa Rica's conserved wildlands for Merck's drug screening program. In turn, Merck would provide INBio with a research and sampling budget of US$1 135 000 and royalties on commercial products arising from the INBio-provided chemical extracts. Laird (1993) and Sittenfeld and Gamez (1993) rightly note that this agreement has proved to be a watershed event in the history of biodiversity prospecting. In addition, this agreement has given rise to significant interest in designing 'win-win' biodiversity prospecting contracts that, inter alia, provide an explicit economic rationale for the conservation of biodiversity in many different parts of the world.

As our thinking on the subject of biodiversity prospecting has progressed, our understanding of the merits and demerits of this activity has become nuanced. Therefore, it is fair to say that in the context of the conservation of biodiversity, broadly speaking, the contemporary literature in economics consists of a group of researchers who are positively inclined towards biodiversity prospecting, a second group that is largely neutral about the utility of biodiversity prospecting, and a third group that sees little of value in biodiversity prospecting. We now summarize the findings of representative contributions from each of these three groups.

14.2 REVIEW OF THE LITERATURE

14.2.1 The Positive Perspective

The Uruguay Round of the General Agreement on Tariffs and Trade (GATT) proposed that, inter alia, trade-related intellectual property rights or TRIPS be conferred on international firms on a whole host of life forms and on biotechnology. Given this background, Bhat (1996) points out that establishing intellectual property rights to products derived from genetic and biochemical resources is necessary but not sufficient for biodiversity prospecting and the survival of three kinds of biological resources. He urges developing nations to create institutions and policies that will enable local communities to receive the 'benefits of biodiversity conservation and prospecting' (Bhat, 1996, p. 205).

What lessons can one learn from the successes and the eventual failure of Shaman Pharmaceuticals, a once promising player in the biodiversity prospecting market? Clapp and Crook (2002) contend that even though Shaman Pharmaceuticals eventually failed, the key lesson to be learned from this failure is that because of rapid technological change, new models and institutions are needed for drug development from natural products. In addition, it is important to comprehend that 'bioprospecting' and 'ethnobotanical searches' will continue to be salient activities because 'natural products will remain important to drug development' (Clapp and Crook, 2002, p. 79).

Some of the world's most biologically diverse resources are to be found in the tropics. This much is well known and agreed upon. Kala (2006) focuses on the Himalayan region in northern India and notes that it has often not been possible to meet the increasing demand for medicinal plants from both pharmaceutical firms and herbal healers because of several constraints. One such constraint stems from the 'specific ecological requirements of many Himalayan medicinal plant species' (Kala, 2006, p. 370). This notwithstanding, this researcher sees value in biodiversity prospecting and, as such, he discusses ways in which the medicinal plants sector might be developed and managed.

14.2.2 The Neutral Perspective

Mulholland and Wilman (2003) conduct an interesting intertemporal analysis of the properties of a biodiversity prospecting contract between a host nation and a pharmaceutical firm. As expected, in this analysis, the host nation's stock of biodiversity and genetic information are the key inputs in the production of high-quality samples. These authors demonstrate that even with complete property rights, contracts are second best because it is impossible to perfectly monitor the host nation's inputs in the process of drug discovery. More generally, it is shown that contracts 'vary due to the different degrees of observability of host-country inputs, and incomplete or ineffective property rights' (Mulholland and Wilman, 2003, p. 417).

Clearly, biodiversity prospecting can create incentives for the private conservation of what are often known as 'biodiversity hotspots'. Given this situation, will a market for biological resources give rise to sufficient incentives for private conservation? This pertinent question is analyzed by Di Corato (2007). This researcher develops and studies a market framework and shows that different market equilibria are possible and that these different equilibria have different implications for the extent of conservation. In particular, the 'industry structure on the supply side' (Di Corato, 2007, p. 44) is shown to have a fundamental bearing on the private incentive to conserve biodiversity.

Ozturk and Ozturk (2008) begin their analysis of the biological screening of medicinal plants in developing countries with two observations. First, they note that the importance of medicinal plants has been increasing over time for both pharmaceutical firms and traditional users. Second, they contend that environmental changes stemming from global warming are likely to have a non-trivial impact on the natural resources of developing countries. In light of these two observations, these authors rightly note that there are threats stemming from the biological screening of medicinal plants. This notwithstanding, Ozturk and Ozturk (2008) clearly state that if the stages and the techniques of pharmacological screening and the differences between natural and synthetic products are well understood then there also exist many opportunities from biodiversity prospecting for pharmaceutical firms in developing nations.

14.2.3 The Negative Perspective

In a prominent paper, Simpson et al. (1996) analyze the valuation of biodiversity for use in pharmaceutical research. The central contention of the authors of this paper is that even though biodiversity prospecting has been much lauded for being an effective mechanism for discovering novel pharmaceutical products and for conserving biodiversity, theoretical analysis warrants a much more cautious approach. Specifically, these authors *value* the marginal species on the basis of its incremental contribution to the likelihood of making a commercial discovery. It is shown that even under favorable assumptions, the upper bound on the value of the marginal species is modest. This finding is then extended to valuing the marginal hectare of habitat. This exercise leads these authors to conclude that 'the incentives for habitat conservation generated by private pharmaceutical research are . . . at best, very modest' (Simpson et al., 1996, p. 163).

The success of arguments promoting biodiversity prospecting as a way of conserving biodiversity ultimately depends on the *value* of biodiversity for use in new pharmaceutical research. Given this state of affairs, Craft and Simpson (2001) attempt to estimate the above-mentioned value using two models of competition among differentiated products. Analysis of both models confirms a key finding in the Simpson et al. (1996) paper discussed in the previous paragraph. Specifically, it is shown that the 'value to private researchers of the "marginal species" is likely to be small' (Craft and Simpson, 2001, p. 1). This negative finding notwithstanding, these researchers stress that the models being analyzed have very different implications for the social value of biodiversity. Therefore, these researchers conclude their analysis with a plea for a better understanding of the true meaning of biodiversity.

Extending the arguments made in the two papers discussed in the preceding two paragraphs, Sedjo and Simpson (2005) focus on the nexuses between investments in biodiversity prospecting and incentives for biodiversity conservation. Their analysis leads these researchers to contend that investments in biodiversity prospecting are *unlikely* to increase incentives for conservation by much. This contention is explained by noting that if the value of the marginal species were noteworthy then investments already ought to have been made to exploit this species. On the other hand, if this value is not noteworthy then it is unlikely that additional investments in biodiversity prospecting will lead to any substantial increase in the incentives for biodiversity conservation. This line of reasoning leads these authors to conclude that if we believe that biodiversity is salient then strategies more effective than biodiversity prospecting need to be found to ensure its conservation.

14.3 IMPLICATIONS AND CONTRIBUTIONS OF THIS CHAPTER

Our review of the literature in Section 14.2 above leads to three conclusions. First, the various studies we have discussed have certainly advanced our understanding of the many nexuses between biodiversity prospecting on the one hand and biodiversity conservation on the other. Second, we see that the picture concerning the desirability of biodiversity prospecting as a tool for promoting conservation is mixed. Finally, even though biodiversity prospecting can, at least in some circumstances, be a desirable tool for promoting biodiversity conservation, there are virtually no theoretical studies of biodiversity prospecting that explicitly account for the facts that this process takes place over time and under uncertainty.

Therefore, to fix ideas and to provide answers to previously unstudied questions about biodiversity prospecting, in this chapter we model and analyze the activities of a drug-producing pharmaceutical firm (PF) that searches for potentially useful chemicals made by wild organisms in a specific conservation area. This PF is able to assign quality levels to the wild organisms in the conservation area. From a drug production standpoint, organism quality is a proxy for the possible usefulness of the underlying chemicals in an organism. At each date, our PF must decide whether to search for a new wild organism with a certain quality or to produce the drug in question with an extant wild organism with its own quality.

Our theoretical analysis sheds light on the following four hitherto unstudied questions concerning biodiversity prospecting. (1) If our PF discards a wild organism with a certain quality at a specific point in time then is it ever optimal for it to use this same organism at a future date? (2) If our PF uses a particular organism with its quality at a specific point in time then ought it to continue to produce the drug in question with the chemicals from this organism at all subsequent points in time? (3) Is there a threshold level of organism quality such that an optimal course of action requires our PF to use (discard) all organisms with qualities that are above (below) this threshold? (4) What are the impacts of increases in an exogenous income source and the discount factor on our PF's threshold quality?

The rest of this chapter is organized as follows. Section 14.4 adapts the analysis in McCall (1970)[1] and in Batabyal and Beladi (2010) and delineates a dynamic and stochastic model of biodiversity prospecting by a PF that involves a choice between the search for new wild organisms with distinct qualities and drug production using an extant wild organism with its own quality. Section 14.5 provides an analysis of the first question mentioned in the preceding paragraph. Section 14.6 studies the second question stated in the previous paragraph. Section 14.7 sheds light on the third question from the previous paragraph. Section 14.8 discusses the preceding paragraph's final question. Section 14.9 concludes and then discusses potential extensions of the research described in this chapter.

14.4 A MODEL OF BIODIVERSITY PROSPECTING

Consider the decision problem faced by a PF that operates in a dynamic and stochastic environment. In our model, time is discrete, the discount factor at time t is ρ^t,our PF's profit at time t is $\pi(t)$, and its risk-neutral objective function is $\sum_{t=0}^{\infty}\rho^t\pi(t)$. To keep the following analysis straightforward, we assume that our PF is unable to either borrow or to lend and hence its profit in any time period is equal to the income it generates in this same time period.

At time t, if our PF uses the chemicals in a wild organism[2] of quality $q(t)$ then we suppose that it can generate income given by $i(t) = q(t)$. Also, at time $t = 0$, without loss of generality, we suppose that our PF begins drug production with the chemicals from a wild organism of quality $q(0) = 0$. From then on, at each date, this PF can either produce the drug in question using the chemicals from any one of the wild organisms it has already found or it can spend this time period searching for a new wild organism. In each time period in which our PF searches for the chemicals from a new wild organism, it obtains an independent realization from a time-invariant cumulative distribution function $G(q)$, which is defined over the bounded and closed interval $[0, \hat{q}]$.

The decision problem confronting our PF at each date is to determine whether to search for a new wild organism or to produce the drug in question with the chemicals from one of the wild organisms it has already found. Because there is no borrowing or lending in our model, at every date, profit equals current income, which, in turn, equals the value/quality of the drug. In symbols, we have $\pi(t) = i(t)$. The reader will note that at every date, our PF has a significant choice before it that affects the quality of the input (chemicals from a wild organism) that is actually available to it. In particular, by searching for additional time periods, which is costly in terms of forgone drug production, this PF can potentially ameliorate the quality levels of the chemicals from the various wild organisms that are available to it.

Let us assume that our PF can use the chemicals from any one of the wild organisms that it has found in the past to produce the pertinent drug at any point in time. In addition, in principle, it can also stop producing the drug at any date and go back to searching for new wild organisms. Given this state of affairs, our first task now is to formulate our PF's maximization problem recursively. To do so, we shall use the techniques of stochastic dynamic programming.[3]

Let $q^m(t) = \max\{t' \in [0, t]\}\, q(t')$ denote the wild organism with the highest quality available to our PF at time t. Note that this PF will not use the chemicals from any wild organism with quality in the set $\{q(0), \ldots, q(t)\}$ that is not equal to $q^m(t)$ and hence we can think of $q^m(t)$ as the payoff relevant state variable at time t. Suppose our PF searches at time t. Then it produces the drug in question with the chemicals from the wild organism of quality $q^m(t)$ and receives the continuation *value* that we shall represent with $V\{q^m(t + 1) = q^m(t)\}$ given that it does *not* find a new wild organism. Once again, suppose that our PF searches at time t. Now, in an alternate scenario, our PF does not produce the drug at time t but it receives the continuation value $V\{q^m(t + 1) = max(q^m(t), \tilde{q})\}$ where $\tilde{q}$ is the quality of the chemicals in the wild organism found at time t.

Combining the two observations from the preceding paragraph, our PF's maximization problem can be formulated in a recursive manner. This formulation gives us:

$$V\{q^m\} = max\{q^m + \rho\, V(q^m), \rho E[V\{max(q^m, \tilde{q})\}]\} \tag{14.1}$$

where $E[\cdot]$ is the expectation operator. Using integrals, the expectation on the RHS of (14.1) can be expressed in a more suitable manner. We get:

$$V\{q^m\} = max\{q^m + \rho V(q^m), \rho \int_0^{\hat{q}} V\{max(q^m, \tilde{q})\} dG(\tilde{q})\} \tag{14.2}$$

This completes the task of formulating our PF's maximization problem. We now proceed to study the main properties of this PF's optimal course of action by concentrating on the first question posed in the penultimate paragraph of Section 14.3. In particular, if our PF discards a wild organism with a certain quality at a specific point in time then we want to know whether it is ever optimal for it to use this same organism to produce the drug in question at a later point in time.

14.5 TO USE OR NOT TO USE A PREVIOUSLY DISCARDED ORGANISM

In order to answer the above query, we must first express this query in mathematical terms. To this end, suppose that our PF has decided that it will not produce the drug in question using the wild organism with quality q' at time t. Then, what we want to demonstrate is that it will *never* use this organism with quality q' at time $t + s$ where $s > 0$. Put differently, if our PF discards a wild organism with a particular quality at a certain date, then it will *never* use this same organism for any possible realization of events between dates t and $t + s$.

To demonstrate this 'no use' result, let us assume that our PF decides to use a wild organism with a certain quality when it is indifferent between using this organism and searching for additional periods of time. Also, we assume that the set of wild organisms with distinct qualities found through search are strictly ordered at the top.[4] We are now ready to provide a proof by contradiction. To this end, suppose that when our PF has access to the set of wild organisms with qualities $q^t = \{q(0), \ldots, q', \ldots, q(t)\}$, it chooses to search for new wild organisms at time t and that it uses the wild organism with quality q' at time $t + s$ for some $s > 0$. Now, since our PF has decided to search at time t, we know that the following strict inequality:

$$q^m(t) + \rho V\{q^m\{(t)\} < \rho \int_0^{\hat{q}} V\{max(q^m(t), \tilde{q})\} dG(\tilde{q})\} \tag{14.3}$$

must hold. Also, because our PF uses the wild organism with quality q' at time $t + s$, we have $q^m(t + s) = q'$. These last two results tell us that the following weak inequality also holds:

$$q^m(t + s) + \rho V\{q^m(t + s)\} \geq \rho \int_0^{q} V\{max(q^m(t + s), \tilde{q}\} dG(\hat{\tilde{q}}\}. \tag{14.4}$$

We know that the wild organism with the highest quality at time $t + s$ is at least as good as the wild organism with the highest quality at time t for any realization of events between times t and $t + s$. In symbols, this means that $q^m(t + s) \geq q^m(t)$. Also, we know that the function $V\{\cdot\}$ is weakly increasing. Given this finding, the inequalities in (14.3) and (14.4) together tell us that:

$$q^m(t + s) + \rho V\{q^m(t + s)\} > q^m(t) + \rho V\{q^m(t)\}. \tag{14.5}$$

Since the function $V\{.\}$ is weakly increasing, we can use the inequality in (14.5) to determine that $q^m(t + s) > q^m(t)$. On the other hand, we know that $q' = q^m(t + s)$ and $q^t = \{q(0), \ldots, q', \ldots, q(t)\}$, which implies that $q^m(t + s) = q^m(t)$, which is plainly *not* possible. This gives us the contradiction we seek and, as such, we have also demonstrated that in an optimal course of action, our PF *never* uses a wild organism it has discarded in the past. We now proceed to study a second basic property of our PF's optimal course of action. Recall from the discussion in the penultimate paragraph of Section 14.3 that this involves answering the following query. If our PF produces the drug in question with a particular quality wild organism at a specific point in time then ought it to continue to produce the drug with this organism at all succeeding points in time?

14.6 DRUG PRODUCTION CONTINUITY WITH AN ORGANISM IN USE

Suppose our PF uses the wild organism with quality q' at time t. Then, some thought ought to convince the reader that the query at the end of the preceding paragraph involves ascertaining whether, for all dates $s \geq t$, it makes sense for our PF to stop drug production with the wild organism with quality q' and to go back to searching for new wild organisms.

Given that our PF has agreed to produce the drug in question with the wild organism with quality q' at time t, it is obvious that $q^m(t) = q'$. Also, given the decision to use this organism, the weak inequality:

$$q' + \rho V\{q'\} \geq \rho \int_0^{\hat{q}} V\{max(q', \tilde{q})\} dG(\tilde{q}) \tag{14.6}$$

must hold. Since our PF does not search for wild organisms at time $t + 1$, we get $q^m(t + 1) = q^m(t) = q'$. Hence, from the inequality in (14.6) it follows that our PF also uses the wild organism with quality q' at time $t + 1$. Now, it follows by mathematical induction that our PF will continue to produce the pertinent drug with the wild organism with quality q' for all dates $s \geq t$ and will *never* go back to searching for new wild organisms. The next topic on the agenda is to study a third basic property of our PF's optimal course of action. Recall that this involves ascertaining whether there is a threshold level of organism quality such that an optimal course of action requires our PF to use (discard) all organisms that are above (below) this threshold.

14.7 THE THRESHOLD LEVEL OF ORGANISM QUALITY

Before moving on, let us briefly summarize the implications of our analysis thus far in Sections 14.4 through 14.6. First, in Section 14.5 we demonstrated that our PF *never* uses a discarded wild organism from the past and hence there is no loss of generality in assuming that the payoff relevant state variable is the most recent wild organism that is found. Second, in Section 14.6 we established that our PF *never* goes back to searching for wild organisms once it has decided to produce the drug in question with the chemicals from a particular wild organism. This means that the value to our PF from deciding to use a wild organism with quality q – for drug production – can be expressed as $V^{use}(q) = q/(1 - \rho)$. Third, given these two points, our PF's maximization problem can be formulated as a stationary or time-independent problem in which the function $V\{\cdot\}$ has a piecewise linear form.[5]

With this summary in place, let us analyze the case in which our PF is not producing the drug in question and it receives an exogenous amount of income denoted by i_e.[6] Now, using an approach similar to that employed in Section 14.4, we want to state our PF's maximization problem and to examine the potential existence of a threshold level of wild organism quality. To begin our analysis, we suppose that $i_e < \hat{q}$. If this were not the case then it would *never* be optimal for our PF to use any wild organism with its distinct quality and the trivial optimal solution to our PF's maximization problem would involve always searching for wild organisms at all points in time.

Given the summary in the first paragraph of this section, when $i_e < \hat{q}$, without any loss of generality, the maximization problem for our PF can be recursively written as:

$$V\{q\} = max\left[\frac{q}{1 - \rho}, i_e + \rho \int_0^{\hat{q}} V\{\tilde{q}\} dG(\tilde{q})\right]. \tag{14.7}$$

Since $V\{\cdot\}$ is the maximum of a constant function and a linear function, $V\{\cdot\}$ is itself a piecewise linear function. This tells us that our PF's optimal policy does involve a *threshold rule*. In other words, there exists a threshold level of organism quality $\hat{T}$ such that our PF decides to use all organisms with qualities that are above $\hat{T}$ and it discards all organisms with qualities $q < \hat{T}$ and continues to search for new wild organisms. Mathematically, the pertinent threshold satisfies:

$$\frac{\hat{T}}{1-\rho} = i_e + \rho \int_0^{\hat{q}} V\{\tilde{q}\} dG(\tilde{q}). \tag{14.8}$$

For a wild organism with quality $q < \hat{T}$, we have $V\{q\} = V\{\hat{T}\}$ and for a wild organism with quality $q > \hat{T}$, we have $V\{q\} = q/(1-\rho)$. With these two pieces of information, (14.8) tells us that:

$$\frac{\hat{T}}{1-\rho} = i_e + \left\{\frac{\rho\hat{T}}{1-\rho} G(\hat{T}) + \rho \int_{\hat{T}}^{\hat{q}} \frac{q}{1-\rho} dG(q).\right. \tag{14.9}$$

Now, if we subtract the identity:

$$\frac{\rho\hat{T}}{1-\rho} = \frac{\rho\hat{T}}{1-\rho} G(\hat{T}) + \rho \int_{\hat{T}}^{\hat{q}} \frac{\hat{T}}{1-\rho} dG(q). \tag{14.10}$$

from both sides of (14.9), then the threshold quality level $\hat{T}$ solves:

$$\hat{T} = i_e + \frac{\rho}{1-\rho} \int_{\hat{T}}^{\hat{q}} (q - \hat{T}) dG(q). \tag{14.11}$$

We can think of the LHS of (14.11) as the cost of forgoing drug production with the wild organism with quality $\hat{T}$. In contrast, the RHS of (14.11) is the expected benefit to our PF from one more round of searching for wild organisms. Clearly, for the organism with threshold quality level $\hat{T}$, these two values have to equal each other. In other words, this simply means that for the wild organism with threshold quality level $\hat{T}$, our PF is indifferent between the two actions of producing the drug in question and continuing the search for wild organisms.

We have now demonstrated that in the modeling setup of this chapter, there exists a threshold level of organism quality $\hat{T}$ in the sense that an optimal course of action requires our PF to use (discard) all organisms with qualities that are above (below) this threshold. Also, we have solved for this threshold quality level in (14.11). We now proceed to our last task in this chapter and this entails an analysis of the effects of increases in the exogenous income i_e and the discount factor ρ on our PF's threshold organism quality level $\hat{T}$.

14.8 CHANGES IN THE THRESHOLD LEVEL OF ORGANISM QUALITY

We first concentrate on the exogenous income i_e. Let us represent the RHS of (14.11) with the function $\zeta(i_e, \hat{T})$. Observe that $\xi(\cdot,\cdot)$ is decreasing in the threshold organism quality level $\hat{T}$. Also, we know that $\zeta(i_e, 0) > 0$ and, by assumption, that $\zeta(i_e, \hat{q}) = i_e < \hat{q}$. This tells us that the function ζ(ie,) crosses the 45 degree line. Since this is a decreasing function, it crosses the 45 degree line exactly once and hence the equation $= \zeta(i_e, \hat{T})$, has a unique solution $\hat{T} \in (0, \hat{q})$ for any value of exogenous income i_e. In addition, because the function $\zeta(i_e, \hat{T})$ is increasing in i_e, the unique solution $\hat{T}$ is also increasing in i_e. This tells us that the threshold organism quality level $\hat{T}$ *rises* as the exogenous income i_e *rises.* From an intuitive standpoint, this result is telling us the following. When our PF receives additional benefits from searching for wild organisms, it has a greater incentive to continue to search and hence it requires a *higher* threshold to use an organism and produce the drug in question.

Given the discussion in the preceding paragraph, the effect of an increase in the discount factor ρ is quite simple to determine and therefore we provide only an intuitive delineation of the relevant effect. The reader should note that an increase in ρ makes our PF more patient or more concerned about the future. Hence, it should be clear to the reader that this *increased* patience or concern for the future *increases* the threshold level of organism quality $\hat{T}$. In their analysis of the value of the marginal species, Simpson et al. (1996) do not consider the effects of discounting. However, they claim that the introduction of discounting is likely to strengthen their finding that the upper bound on the value of the marginal species is modest. The implication is that this, most likely, will further discourage biodiversity prospecting. Our analysis of this chapter's model of biodiversity prospecting with discounting leads to a somewhat more specific result. We find that an increase in the discount factor raises the wild organism use threshold and hence this is likely to delay drug production. This concludes our discussion of the fourth and last question of this chapter.

14.9 CONCLUSIONS

In this chapter, we examined the activities of a PF that, at each date, had to decide between producing the drug in question with an existing organism with its distinct quality and searching for new wild organisms. Our examination shed light on four hitherto unstudied questions in the existing literature on biodiversity prospecting. First, we demonstrated that if our PF discards a wild organism with a certain quality at a specific point in time then it will never use this same organism at a subsequent point in time to produce the drug in question. Second, we determined that if our PF decides to engage in drug production with an organism of a certain quality at a specific point in time then it will continue to produce the drug in question with the chemicals from this organism at all subsequent points in time. Third, we showed that there exists a threshold organism quality level and that our PF's optimal policy involves using (discarding) all organisms with qualities above (below) this threshold. Finally, we analyzed the effects of increases in an exogenous income source and the discount factor on our PF's threshold level of organism quality.

The analysis in this chapter can be extended in a number of directions. Here are two suggestions for extending the research described here. First, Simpson et al. (1996, p. 165) have rightly noted that all theoretical models are built on a number of simplifying assumptions and our model certainly fits this description. Specifically, in the model of this chapter, we have significantly compressed the amount of time it takes a PF to make a determination of the quality of a wild organism. In reality, this quality determination exercise is an elaborate process involving the active participation of several trained groups of individuals. Therefore, it would be useful to analyze the wild organism *quality determination* aspect of biodiversity prospecting in greater detail.

Second, we have focused exclusively on the private incentives facing a PF. Clearly, the *social* incentives for biodiversity conservation are almost certainly higher than the private incentives. Therefore, it would be instructive to analyze a model that explicitly accounts for the fact that consumer surplus from new drug development may well exceed the profits of a PF by a large margin. Studies of biodiversity prospecting that incorporate these features of the problem into the analysis will provide additional insights into an activity that many believe can play a positive role in conserving some of our most valuable natural resources.

NOTES

* Batabyal acknowledges financial support from the Gosnell endowment at RIT. The usual disclaimer applies.

1. See Ljungqvist and Sargent (2005) for a lucid exposition of the McCall (1970) model.
2. For concreteness, the reader may want to think of this wild organism as a wild plant species.
3. See Ross (1983), Puterman (2005) and Acemoglu (2009) for textbook treatments of stochastic dynamic programming.
4. With this assumption, we are ruling out instances in which there are two maxima so that our PF moves back and forth between using these two wild organisms and a wild organism that is not used at time t but used at time $t + s$.
5. See equation (16.28) in Acemoglu (2009, p. 558) for additional details on this point.
6. We shall not concern ourselves with the source of this income but it could arise, for instance, from licensing activities.

REFERENCES

Acemoglu, D. (2009), *Introduction to Modern Economic Growth*, Princeton, NJ: Princeton University Press.
Batabyal, A.A. and H. Beladi (2010), 'A model of entrepreneurial activity with two actions over time and under uncertainty', *Letters in Spatial and Resource Sciences*, **3**(2), 45–54.
Bhat, M.G. (1996), 'Trade-related intellectual property rights to biological resources: socioeconomic implications for developing countries', *Ecological Economics*, **19**(3), 205–17.
Clapp, R.A. and C. Crook (2002), 'Drowning in the magic well: Shaman Pharmaceuticals and the elusive value of traditional knowledge', *Journal of Environment and Development*, **11**(1), 79–102.
Craft, A.B. and R.D. Simpson (2001), 'The value of biodiversity in pharmaceutical research with differentiated products', *Environmental and Resource Economics*, **18**(1), 1–17.
Di Corato, L. (2007), 'Bioprospecting and market for biodiversity conservation', *International Journal of Ecological Economics and Statistics*, **7**(W07), 44–67.
Eisner, T. (1989), 'Prospecting for nature's chemical riches', *Issues in Science and Technology*, **6**(2), 31–4.
Eisner, T. (1992), 'Chemical prospecting', in F.H. Bormann and S.R. Kellert (eds), *Ecology, Economics, and Ethics*, New Haven, CT: Yale University Press, pp. 196–202.
Grifo, F., D. Newman, A.S. Fairfield, B. Bhattacharya and J.T. Grupenhoff (1997), 'The origin of prescription drugs', in F. Grifo, J. Rosenthal and T. Lovejoy (eds), *Biodiversity and Human Health*, Washington, DC: Island Press, pp. 131–63.
Kala, C.P. (2006), 'Problems and prospects in the conservation and development of the Himalayan medicinal plants sector', *International Journal of Sustainable Development*, **9**(4), 370–89.
Laird, S.A. (1993), 'Contracts for biodiversity prospecting', in W.V. Reid, S.A. Laird, C.A. Meyer, R. Gamez, A. Sittenfeld and D.H. Janzen et al. (eds), *Biodiversity Prospecting*, Washington, DC: World Resources Institute, pp. 99–130.
Ljungqvist, L. and T.J. Sargent (2005), *Recursive Macroeconomic Theory*, Cambridge, MA: MIT Press.
Macilwain, C. (1998), 'Bid to block Yellowstone enzymes deal', *Nature*, **392**(6672), 117.
McCall, J. (1970), 'Economics of information and job search', *Quarterly Journal of Economics*, **84**(1), 113–26.
Mulholland, D.M. and E.A. Wilman (2003), 'Bioprospecting and biodiversity contracts', *Environment and Development Economics*, **8**(3), 417–35.

Neto, R.B. and D. Dickson (1999), '$3m deal launches major hunt for drug leads in Brazil', *Nature*, **400**(6742), 302.

Nijkamp, P., G. Vindigni and P.A.L.D. Nunes (2008), 'Economic valuation of biodiversity: a comparative study', *Ecological Economics*, **67**(2), 217–31.

Nunes, P.A.L.D. and P. Nijkamp (2011), 'Biodiversity: economic perspectives', unpublished manuscript, Amsterdam: Free University.

Nunes, P.A.L.D., J.C.J.M. van den Bergh and P. Nijkamp (2003), *The Ecological Economics of Biodiversity*, Cheltenham, UK and Northampton, MA, USA: Edward Elgar Publishing.

Ozturk, N. and Y. Ozturk (2008), 'Opportunities and threats for biological screening of medicinal plants: importance for the pharmaceutical industries in developing countries', *World Review of Science, Technology, and Sustainable Development*, **5**(2), 124–39.

Puterman, M.L. (2005), *Markov Decision Processes*, Hoboken: Wiley.

Reid, W.V., S.A. Laird, R. Gamez, A. Sittenfeld, D.H. Janzen, M.A. Gollin and C. Juma (1993a), 'A new lease on life', in W.V. Reid, S.A. Laird, C.A. Meyer, R. Gamez, A. Sittenfeld, D.H. Janzen, M.A. Gollin and C. Juma (eds), *Biodiversity Prospecting*, Washington, DC: World Resources Institute, pp. 1–52.

Reid, W.V., S.A. Laird, C.A. Meyer, R. Gamez, A. Sittenfeld, D.H. Janzen, M.A. Gollin and C. Juma (eds) (1993b), *Biodiversity Prospecting*, Washington, DC: World Resources Institute.

Ross, S.M. (1983), *Introduction to Stochastic Dynamic Programming*, San Diego, CA: Academic Press.

Sarr, M., T. Goeschl and T. Swanson (2008), 'The value of conserving genetic resources for R&D: a survey', *Ecological Economics*, **67**(2), 184–93.

Sedjo, R.A. and R.D. Simpson (2005), 'Investments in biodiversity prospecting and incentives for conservation', Discussion Paper, Washington, DC: Resources for the Future.

Simpson, R.D., R.A. Sedjo and J.W. Reid (1996), 'Valuing biodiversity for use in pharmaceutical research', *Journal of Political Economy*, **104**(1), 163–85.

Sittenfeld, A. and R. Gamez (1993), 'Biodiversity prospecting by INBio', in W.V. Reid, S.A. Laird, C.A. Meyer, R. Gamez, A. Sittenfeld and D.H. Janzen et al. (eds), *Biodiversity Prospecting*, Washington, DC: World Resources Institute, pp. 69–98.

Sonner, S. (1998), 'Suit tries to block "bioprospecting" in Yellowstone', *Idaho News* (6 March).

Swanson, T. (1996), 'The reliance of northern economies on southern biodiversity', *Ecological Economics*, **17**(1), 1–8.

ten Kate, K. and S.A. Laird (eds) (1999), *The Commercial Use of Biodiversity*, London: Earthscan.

15. Game theory and marine protected areas: the effects of conservation autarky in a multiple-use environment

Maarten J. Punt, Hans-Peter Weikard and Ekko C. van Ierland

Q24 Q28 Q57
Q25 Q22 Q58

15.1 INTRODUCTION

Marine Protected Areas (MPAs) are gaining momentum as a possible tool for the protection and management of the ecosystem services provided by our seas and oceans. Their general approval is demonstrated by the fact that the parties to the Convention on Biological Diversity have set the target that 10 per cent of the world's oceans should be protected (Convention on Biological Diversity, 2010), even though currently no more than approximately 1 per cent are protected (Spalding et al., 2010).

In the literature the effectiveness and the possibilities of MPAs are debated. Some authors seem to consider MPAs as a panacea (e.g., Bohnsack, 1993); others are more sceptical about the possibilities (e.g., Hannesson, 1998; Anderson, 2002). This divergence is partly caused by the different meanings that people attach to 'MPA', especially to the term 'protected' (Jones, 2001). If 'protected' only means 'protected from fishing' and the goal is to protect or improve fisheries, then MPAs are no-take zones. If they are to be protected from all human uses, then the term 'marine reserve' might be more appropriate. Sometimes 'protected' is interpreted to mean 'protected from some uses but not from others'; in that case MPAs are essentially a zoning tool. In this chapter we will consider MPAs as areas that are protected from extractive uses and affect multiple ecosystem services, in our case food production (fisheries) and cultural and option use services (species protection).

Setting aside all terminological vagueness, there is also no consensus on MPAs' effects on fish stocks and fisheries in particular. Although generally positive effects on fish stocks have been found within MPAs (see Lester et al., 2009 for a global synthesis), it remains unclear how much of the increase in stocks spills over to the fishery. Economic models have also shown the importance of the regime outside the MPA, that is, whether open access prevails (Hannesson, 1998; Sanchirico and Wilen, 2001; Anderson, 2002) or whether a limited entry system exists (Sanchirico and Wilen, 2002; Costello and Kaffine, 2010). Moreover, an important aspect is the scale of governance of MPAs, and related, whether or not they are located within an Exclusive Economic Zone (EEZ), because the property rights regime matters. If one focuses on the local scale, i.e., within the EEZ, the issues of interest are mostly the behavioural responses of the fishermen and other excluded users. On a regional scale, however, strategic aspects of MPA designation arise because of their transboundary characteristics, if their effects are not limited to an EEZ and spill over to other countries. Finally, MPAs in the High Seas, beyond individual countries' EEZs, where the common heritage of mankind applies, are essentially a weakest-link public good. This means that all countries have to agree on the implementation of MPAs installed in the High Seas, suggesting that the maximum size that can be obtained reflects the preferences of the country whose support for the MPA is the weakest (Punt et al., 2013).

In this chapter we focus on MPAs on the regional scale, that is, the effects of MPA designation in an EEZ of one country on MPA designation in the EEZs of other countries. It has been pointed out by several authors that location of both reserves and fishing activities matters for the effectiveness of MPAs as fisheries management tools (Smith and Wilen, 2003; Sanchirico, 2004; Sanchirico et al., 2006). The same has been found for MPAs with conservation as their main goal (Chan et al., 2006; Edgar et al., 2008; Weeks et al., 2010). On a regional scale, questions of location become even more important because MPAs have public good characteristics: their benefits are generally non-excludable and non-rivalrous. The fishery effects of MPAs may spill over to other countries, and when a species is protected in one location the incentive to protect it in another location may become weaker. The spillover effect for fisheries has been investigated by Ruijs and Janmaat (2007). They found that migration patterns are the most important factor determining incentives to assign MPAs. The effect of location for species protection has been investigated in a terrestrial setting by, for example, Rodrigues and Gaston (2002) and Jantke and Schneider (2010) and in a coastal setting by Kark et al. (2009). These authors, however, compare international cooperation with independent action where impacts of others' actions are ignored. Following the trade literature we call the latter situation 'conservation autarky'.

From an economist's point of view, conservation autarky might seem unlikely because countries could choose to act strategically and free-ride on each other's contributions. There are, however, valid arguments for conservation autarky: countries may not know about each other's conservation efforts or about the presence of certain species in other countries. Another possible reason is that they do not trust or value the conservation in another country and hence do not consider it a substitute for protection in their own country. Moreover, conservation autarky and cooperation may be interpreted as looking at the same problem at different scale levels where cooperation represents the larger regional scale and conservation autarky the smaller local scale. Both strategic behaviour and conservation autarky can be observed in practice: Denmark's refusal to assign protected area status to the Dogger Bank (a shallow sand bank in the North Sea) can be interpreted as strategic behaviour – Denmark is free-riding on conservation efforts by Germany, the Netherlands and the UK, who have already declared their parts a protected area. Conservation autarky often occurs as a result of parochialism: for example, in the Western world much money is spent on protecting species that are relatively safe globally, such as wolves, bald eagles and grizzly bears (Hunter and Hutchinson, 1994). Obviously, viewed from a regional or global scale, conservation autarky is generally inefficient because species tend to be over-protected or at least not protected at the best locations. Still, conservation autarky may be preferred over strategic behaviour where species are under-protected due to free-riding.

Because MPAs affect multiple ecosystem services, they can produce multiple benefits at the same time. The increased biomass within MPAs implies conservation benefits and in the case of, for example, tropical reefs, increased tourism benefits, but may also imply larger fish with increased fertility. Thus, MPAs can increase the growth rate of the fish stocks, allowing for larger harvest when some of this growth spills over into neighbouring areas. The multiple benefits of MPAs have been addressed by, for example, Beattie et al. (2002), Brown et al. (2001), Boncoeur et al. (2002) and Wood and Dragicevic (2007). Most authors, however, focus either on the local scale, that is, within one EEZ, or simply ignore the public good effects. Punt et al. (2010) have focused on the multiple uses of MPAs and how public good aspects affect MPA assignment by individual countries. Their finding was that if countries focus exclusively on fishing or conservation benefits of the MPA, cooperation may actually produce worse results than non-cooperation, that is, under strategic behaviour, when all uses are considered. Thus, the effects of conservation autarky are complicated further by the fact that MPAs have multiple uses. In this chapter we investigate the effects of conservation autarky in a multiple-use environment in more detail using Punt et al.'s model (2010). A similar analysis was carried out by Bode et al. (2011). They focus, however, on an environment where multiple agents benefit from different uses, whereas in our case the planners in the countries benefit from the same uses.

The important aspect for policy-makers of the multiple-use environment is that the optimal MPA size is different for different ecosystem services. As such, there is a trade-off between the ecosystem services. This is especially true if some of the ecosystem services are extractive such as fisheries, whereas others are not, such as the benefits of species conservation. Generally, the extractive use requires a smaller size than the conservation use. In the Netherlands, for example, the designation of marine reserves has led to fierce discussions between fishers, NGOs with a conservation focus such as Greenpeace, scientists and policy-makers; although, finally, a compromise has been reached (Rijksoverheid, 2011).

If, in addition to multiple-use considerations, some of the benefits spill over to other countries and possibilities for free-riding or conservation autarky exist, the decision becomes even more complicated. In this chapter we demonstrate how the effects of the multiple ecosystem services and the policy options of strategic behaviour and conservation autarky interact.

Our contribution is to investigate the effects of conservation autarky with a game--theoretic model in a multiple-use environment. In our setting we find that conservation autarky actually improves welfare over strategic behaviour, although payoffs are not as high as under full cooperation, but only if *all uses* are considered. If not, then strategic behaviour may be better than conservation autarky.

15.2 THE MODEL

15.2.1 General Characteristics

We use the model of Punt et al. (2010) to explore the effects of conservation autarky in a multiple-use environment. This model combines two separate models of MPA designation, one for MPAs for fisheries and one for MPAs for conservation of species. We assume conservation autarky (i.e., not accounting for conservation benefits due to protection by others) is only relevant in a conservation context. In the fisheries model, spillovers affect the harvest directly because we consider a model where countries are fishing a common stock, and therefore conservation autarky does not occur. We first consider the outcomes of conservation autarky in the separate conservation model. We then introduce the fisheries model and combine it with the conservation model to investigate the effects of conservation autarky in a multiple-use environment.

Consider a regional sea that is shared by a set N of n symmetric countries. The regional sea is completely claimed by EEZs and consists of a single ecosystem. As an example one could think of the North Sea or the Black Sea. Each country has the possibility to assign an MPA within their respective EEZs. For convenience we will normalize the total sea area to one, and assume that each of the n countries has the jurisdiction over an equal share of the sea implying that each EEZ has size $1/n$. Finally, we assume that the effects of MPAs are additive, that is, the final effect of MPAs is equal to the final effect of an MPA size M, where:

$$M = \sum_{i \in N} M_i, \tag{15.1}$$

and $0 \leq M_i \leq 1/n$ are the individual contributions.

15.2.2 Conservation Autarky in the Conservation MPA Model

In the conservation model the main goal of MPAs is the protection of species. Therefore, we measure the effect of the MPA as the number of species that can be expected within the MPA. To this end we use the ecological concept of the species–area curve. The species–area curve describes the number of species that can be expected to be found given the size of the area. It is explained by either the passive sampling effect (MacArthur and Wilson, 2001) or the habitat diversity hypothesis (Williams, 1943) and is considered to be one of the fundamentals of ecology (Rosenzweig, 1995). Its general form is $S = kA^z$ or, when log-transformed $\ln S = \ln k + z \ln A$, with S the number of species, A the area and k and z positive parameters. Although originally a concept from terrestrial ecology, species–area curves can be applied equally well in marine conservation (Neigel, 2003; Levin et al., 2009). For convenience we use the log-transformed version of the species–area curve.

We assume that the benefits of protecting $\ln S$ species are b_P and that these benefits accrue in equal shares to the different countries. The costs of installing an MPA are assumed to be linear in MPA size for each country, that is, $c_p M_i$. The total net benefits of conservation D (i.e., diversity benefits), when all countries cooperate are then:

$$D^{FC}(M) = b_p(\ln k + z \ln M) - c_p M, \tag{15.2}$$

with superscript FC indicating full cooperation. If countries behave strategically (indicated by superscript S), they only consider their own net benefits D_i^S and take the contribution by others as given:

$$D_i^S(M_i) = \frac{1}{n} b_p(\ln k + \ln(M_i + (n-1)M_j)) - c_p M_i \;\; \forall i \in N. \tag{15.3}$$

In contrast, when countries behave independently, as under conservation autarky (CA), they still consider only their own benefits, but now they ignore contributions by others:

$$D_i^{CA}(M_i) = \frac{1}{n} b_p(\ln k + \ln M_i) - c_p M_i \ \forall i \in N. \tag{15.4}$$

Taking first-order conditions of (15.2), (15.3) and (15.4), and solving gives the following optimal MPA sizes:

$$\begin{aligned} M^{FC,*} &= \frac{b_p z}{c_p} \\ M_i^{S,*} &= \frac{b_p z}{n^2 c_p} \quad \forall i \in N \\ M_i^{CA,*} &= \frac{b_p z}{n c_p} \quad \forall i \in N. \end{aligned} \tag{15.5}$$

The effects of parameters on MPA sizes can readily be seen from these solutions. MPA sizes increase if benefits go up (b_p) or the slope of the species area curve increases (z), but go down if the costs increase (c_p). Moreover, the total MPA designated under strategic behaviour is smaller than under conservation autarky ($nM_i^{*,S} = \frac{b_p z}{nc_p} < \frac{b_p z}{c_p} = nM_i^{*,CA}$). The first-order condition of (15.2) produces an overall MPA size M^* but is silent on the contributions by individual countries, whereas (15.3) and (15.4) produce individual solutions. If we assume that all countries contribute equally under full cooperation we find individual contributions of $M_i^{FC,*} = \frac{b_p z}{nc_p} \ \forall i \in N$, that is, conservation autarky produces the same solution as full cooperation. This surprising result is a feature of the symmetry in this conservation game: if countries behave strategically in this game each contribution by an individual country is immediately offset by an equal reduction by another country. If countries ignore the contribution of others, strategic conservation leakage cannot occur, resulting in the same total MPA as under full cooperation.

15.2.3 The Fisheries MPA Model

The full details of the fisheries MPA model are spelled out in Punt et al. (2010). Here we will only summarize the model. The fisheries model is a static Gordon-Schaefer model, altered such that MPAs can be accommodated. The MPA alters the growth of the fish stock:

$$F(X, M) = \left(r_o + r_M \sum_{i \in N} M_i\right) X(1 - X), \tag{15.6}$$

with $F(X,M)$ the growth of the fish stock, r_o the original growth rate, r_M the growth bonus due to the MPA because habitats are restored, and X the size of the fish stock. The carrying capacity is normalized to 1 and assumed to be proportional to the area under consideration.

In a standard Gordon-Schaefer model the catchability of the stock is described by a single parameter. In our model we assume that the individual players can only fish parts of the stock because they have EEZs and that the catchability in this area is further reduced by the introduction of an MPA. Therefore, we use a function for catchability that collapses to the original catchability parameter q_o if the MPA is zero and the number of players is one:

$$Q_i(M_i) = q_o - q_M\left(\frac{n-1}{n} + M_i\right) \quad \forall i \in N. \tag{15.7}$$

Thus, q_M is the marginal catchability reduction due to area that cannot be fished because it is owned by other players ($\frac{n-1}{n}$) or protected (M_i). Under full cooperation, $Q_i(M_i)$ collapses to $Q(M) = q_o - q_M M$. In equilibrium the total catches equal the growth of the stock, which means that the equilibrium stock is:

$$\begin{aligned} R(M)X(1-X) &= \sum_{i\in N}(Q_i(M_i)E_i)X \Leftrightarrow \\ X &= 1 - \frac{\sum_{i\in N}(Q_i(M_i)E_i)}{R(M)}. \end{aligned} \tag{15.8}$$

Given the price of fish p, and the cost per unit of effort c_E, the respective objective functions for full cooperation and strategic behaviour are then:

$$\pi^{FC}(E, M) = pQ(M)EX - c_E E \tag{15.9}$$

$$\pi^S(E_i, M_i) = pQ_i(M_i)E_iX - c_E E_i \quad \forall i \in N. \tag{15.10}$$

As explained before we assume that conservation autarky is irrelevant in the fisheries game because countries fish a common stock. Substituting the equilibrium stock from (15.8) in (15.9) and (15.10), solving first-order conditions gives the following equilibrium MPAs (Punt et al., 2010):

$$M^{FC,*} = \frac{q_o}{q_M} + \frac{c_E}{2pq_M} - \sqrt{\frac{c_E}{pq_M}\left(\frac{c_E}{4pq_M} + 2\left(\frac{q_o}{q_M} + \frac{r_o}{r_M}\right)\right)} \tag{15.11}$$

$$M_i^{S,*} = \frac{q_o}{q_M} - \left(1 - \frac{1}{n}\right) + \frac{nc_E}{2pq_M} - \sqrt{\frac{c_E}{pq_M} + \left(\frac{n^2c_E}{4pq_M} + (n+1)\frac{q_o}{q_M} + \left(1 + \frac{1}{n}\right)\frac{r_o}{r_M} - n + \frac{1}{n}\right)}$$

From (15.11) we see that MPA sizes depend on relative parameter values (q_o/q_M, $c_E/(pq_M)$ and r_o/r_M). As conservation autarky concerns conservation and not fishery, we assume that the solution for conservation autarky in the fisheries model coincides with strategic behaviour because under conservation autarky countries account only for their own benefits.

15.2.4 The Combined MPA Model

General setup

In the combined model, countries consider both the benefits from conservation and the benefits from the fisheries. We assume that under conservation autarky countries use the strategic welfare function for the fisheries profits. The combined welfare (W) functions for full cooperation, strategic behaviour and conservation autarky are respectively:

$$W^{FC}(E,M) = pQ(M)E\left(1 - \frac{Q(M)E}{R(M)}\right) - c_E E + b_p(\log k + z\log(M)) - c_p M, \quad (15.12)$$

$$W_i^S(E_i, M_i) = pQ_i(M_i)\left(1 - \frac{Q_i(M_i)E_i + (n-1)Q_j(M_j)E_j}{R(M_i, M_{-i})}\right) - c_E E_i$$

$$+ \frac{1}{n}b_p(\log k + z\log(M_i + (n-1)M_j)) - c_p M_i \ \forall i \in N, \quad (15.13)$$

$$W_i^{CA}(E_i,M_i) = pQ_i(M_i)\left(1 - \frac{Q_i(M_i)E_i + (n-1)Q_j(M_j)E_j}{R(M_i,M_{-i})}\right) - c_E E_i$$

$$+ \frac{1}{n}b_p(\log k + z\log(M_i)) - c_p M_i \ \forall i \in N, \quad (15.14)$$

where $M_{-i} = M_j \forall j \in N\backslash\{i\}$. The first-order conditions with respect to M or M_i are:

$$\frac{\partial W^{FC}}{\partial M} = \frac{pQ(M)E}{R(M)}\left(\frac{r_M Q(M)E}{R(M)} + q_M E\right) - pq_M E\left(1 - \frac{Q(M)E}{R(M)}\right) + \frac{b_p z}{M} - c_P = 0 \quad (15.15)$$

$$\frac{\partial W_i^S}{\partial M_i} = \frac{pQ_i(M_i)E_i}{R(M_i, M_{-i})}\left(\frac{r_M(Q_i(M_i)E_i + (n-1)Q_j(M_j)E_j)}{R(M_i, M_{-i})} + q_M E_i\right)$$

$$- pq_M E_i\left(1 - \frac{Qi(M_i)E_i + (n-1)Q_j(M_j)E_j}{R(M_i, M_{-i})}\right) + \frac{b_p z}{M_i + (n-1)M_j} - c_p = 0 \quad (15.16)$$

$$\frac{\partial W_i^{CA}}{\partial M_i} = \frac{pQ_i(M_i)E_i}{R(M_i, M_{-i})}\left(\frac{r_M(Q_i(M_i)E_i + (n-1)Q_j(M_j)E_j)}{R(M_i, M_{-i})} + q_M E_i\right)$$

$$- pq_M E_i\left(1 - \frac{Q_i(M_i)E_i + (n-1)Q_j(M_j)E_j}{R(M_i, M_{-i})}\right) + \frac{b_p z}{M_i} - c_p = 0 \quad (15.\ 17)$$

Table 15.1 Parameters used in the simulations

Game	Value	Unit
Parameters fisheries game		
P	25	Value per unit of harvest
c_E	5	Cost per unit of effort
r_o	0.2	–
r_M	0.8	–
q_o	1	(Unit of effort)$^{-1}$
q_M	1	(Unit of effort)$^{-1}$
Parameters conservation game		
b_p	4	Value per log no. of species
c_P	1	Cost per unit of MPA
k	2	–
z	0.2	–

Unfortunately, because effort and MPA size are jointly determined, the solution to (15.15), (15.16) and (15.17) consists of fourth-degree polynomials that cannot be solved analytically. Therefore, we explore the effects of conservation autarky in simulations.

Simulations

In the simulations we investigate a symmetric two-player setting. We selected a set of parameters such that an interior solution is guaranteed for full cooperation, strategic behaviour and conservation autarky. The parameters are shown in Table 15.1. In a later section we explore the effects of parameters with a sensitivity analysis. Using these parameters we calculate the MPA sizes, payoffs and other variables in the separate games and the combined game for full cooperation, strategic behaviour and conservation autarky. The results are shown in Table 15.2.

To calculate the individual results for full cooperation in Table 15.2 we divided the results of full cooperation by two. To calculate the individual results for conservation autarky and make them comparable to the other two, we used the MPA and effort values found by solving the conservation autarky setting in the respective profit and welfare functions of strategic behaviour.

Table 15.2 shows that conservation autarky pushes the game in the direction of full cooperation. General welfare (W_i), however, does not improve in the case of the conservation game. This is because in the conservation game the fisheries benefits and costs are ignored, when the decision of an MPA is taken. Therefore, the total welfare is reduced by moving from strategic behaviour to conservation autarky or full cooperation.

In Figure 15.1 we show the welfare of single countries as a function of MPA size assuming that the other country cooperates (for full cooperation) or plays the strategic equilibrium (for strategic behaviour) or the conservation autarky equilibrium (for conservation autarky). It can be seen from Figure 15.1 that conservation autarky dominates strategic behaviour over the full range, implying an improvement in terms of fisheries profits, conservation and total welfare.

Table 15.2 Results of the separate games for individual values

Variable	Fisheries Game		Conservation Game			Combined Game		
	FC	SB	FC	SB	CA	FC	SB	CA
M_i	0.19	0.06	0.4	0.2	0.4	0.22	0.11	0.15
π_i	0.73	0.25	0	0.16	0	0.71	0.25	0.22
H_i	0.06	0.04	0	0.05	0	0.06	0.04	0.04
E_i	0.14	0.12	0	0.2	0	0.15	0.15	0.18
X	0.66	0.63	1	0.78	1	0.68	0.67	0.71
D_i	0.81	0.45	0.90	0.81	0.90	0.83	0.66	0.75
ln S	0.5	0.26	0.65	0.51	0.65	0.53	0.38	0.45
W_i	1.54	0.7	0.90	0.97	0.90	1.55	0.91	0.97

Notes: FC = full cooperation; SB = strategic behaviour; CA = conservation autarky. In the fisheries and conservation game the effort level is optimal given the chosen MPA size. The individual values for FC were calculated by dividing the total FC results by two, the results for conservation autarky were calculated with the solutions found by solving the games and using these solutions in the individual payoff functions under strategic behaviour.

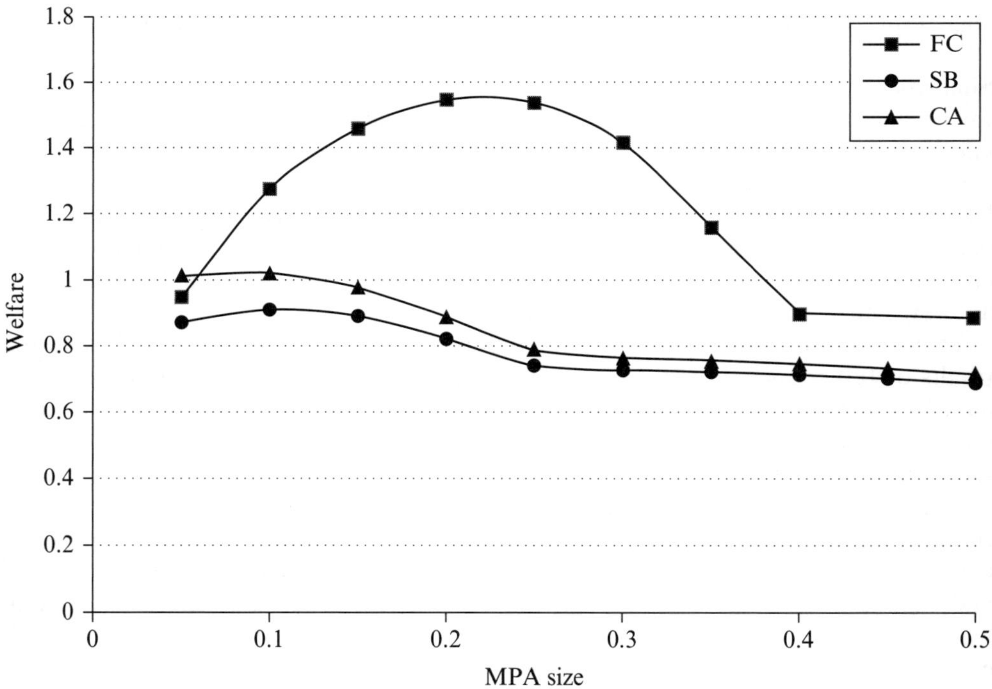

Note: FC = full cooperation; SB = strategic behaviour; CA = conservation autarky. It is assumed that under full cooperation the other country cooperates, and behaves according to the equilibrium values for strategic behaviour and conservation autarky. The welfare for conservation autarky was calculated with the welfare function of strategic behaviour, using the MPA sizes given by conservation autarky solutions.

Figure 15.1 Welfare of a single country as a function of the chosen MPA level

Even though the calculated optimal MPA size for conservation autarky is 0.15 in Table 15.2, this is not where the payoff of conservation autarky is maximized in Figure 15.1. We obtain this result because the payoff for conservation autarky in Figure 15.1 is calculated with the payoff function of strategic behaviour, given the MPA sizes of conservation autarky, and not with the payoff function for conservation autarky. We use the payoff function of strategic behaviour in order to make the results comparable. Using the strategic behaviour payoff function allows us to calculate welfare considering substitution possibilities, that is, welfare to individual countries from a global perspective.

The welfare level in Figure 15.1 under conservation autarky and strategic behaviour at the MPA level chosen in the conservation game (MPA size = 0.4 for conservation autarky and MPA size = 0.2 for strategic behaviour) is lower than the welfare level in Table 15.2 for the conservation game for these outcomes. This effect is reversed if we compare the outcomes for the fisheries game. The reason is the different assumption on behaviour of the other player. In Table 15.2, if countries play either the fisheries or conservation game they ignore the benefits from the other use, that is, both countries choose the MPA size specified in Table 15.2. In Figure 15.1 we assume that the other country chooses the respective equilibria for strategic behaviour and conservation autarky for the combined game.

We now explore how the differences between full cooperation, strategic behaviour and conservation autarky are influenced by the different parameters. We do not test the influence of the parameters k, z and c_P because k does not influence the equilibrium, only payoffs, and z and c_P clearly have, respectively, the same and the opposite effect of b_P. The results of the simulations are shown in Table 15.3. We have restricted ourselves to interior solutions. This means that certain parameter values cannot become too large or small relative to others. This applies to p and q_o that cannot become too small relative to c_E and q_M, respectively. Obviously the reverse holds for c_E and q_M relative to p and q_o.

Table 15.3 shows that the MPA size chosen under conservation autarky is affected differently by parameters than the size chosen under full cooperation or strategic behaviour. The reason is that parameters in such cases have two opposing effects: they affect the MPA size that is optimal from a fisheries point of view and they affect the value of the fisheries relative to the value of conservation.

A case in point is the price of fish p. As it becomes smaller the optimal MPA size from a fisheries perspective shrinks. However, the value of conservation becomes more important relative to the fisheries. As the optimal MPA size from a conservation point of view is in our setting larger than that of a fisheries point of view and this use becomes more important, the final result is ambiguous. This is shown in Table 15.3 where the optimal MPA size grows slightly under conservation autarky as p decreases, whereas for full cooperation and strategic behaviour the optimal MPA size decreases.

As the importance of conservation rises relative to the importance of the fisheries, the gap between strategic behaviour and conservation autarky becomes larger. For example, in simulations 26 to 29, where the benefits of conservation rise, the gap between strategic behaviour and conservation autarky becomes larger both for MPA size and realized welfare. The gap in welfare between conservation autarky and full cooperation also becomes larger, but this gap is still smaller than the gap between strategic behaviour and full cooperation. In contrast, if the fisheries part becomes more important, the gap

Table 15.3 Individual MPA size and welfare for different parameter sets under full cooperation, strategic behaviour and conservation autarky

No	Parameter Set	Full Cooperation		Strategic Behaviour		Conservation Autarky	
	p; c_E; r_o; r_M; q_o; q_M; b_p; c_p; k; z	M_i	W_i	M_i	W_i	M_i	W_i
1	20; 5; 0.2; 0.8; 1; 1; 4; 1; 2; 0.2	0.21	1.26	0.10	0.76	0.16	0.84
2	22.5; 5; 0.2; 0.8; 1; 1; 4; 1; 2; 0.2	0.21	1.40	0.10	0.83	0.15	0.90
3	27.5; 5; 0.2; 0.8; 1; 1; 4; 1; 2; 0.2	0.22	1.70	0.11	1.00	0.15	1.06
4	30; 5; 0.2; 0.8; 1; 1; 4; 1; 2; 0.2	0.24	1.86	0.12	1.10	0.15	1.15
5	32.5; 5; 0.2; 0.8; 1; 1; 4; 1; 2; 0.2	0.24	2.01	0.12	1.20	0.15	1.25
6	25; 2; 0.2; 0.8; 1; 1; 4; 1; 2; 0.2	0.30	2.24	0.19	1.57	0.21	1.61
7	25; 3; 0.2; 0.8; 1; 1; 4; 1; 2; 0.2	0.28	1.95	0.15	1.28	0.18	1.32
8	25; 4; 0.2; 0.8; 1; 1; 4; 1; 2; 0.2	0.24	1.72	0.12	1.06	0.16	1.12
9	25; 6; 0.2; 0.8; 1; 1; 4; 1; 2; 0.2	0.21	1.41	0.10	0.80	0.15	0.87
10	25; 5; 0.1; 0.8; 1; 1; 4; 1; 2; 0.2	0.24	1.43	0.12	0.88	0.16	0.94
11	25; 5; 0.15; 0.8; 1; 1; 4; 1; 2; 0.2	0.23	1.49	0.11	0.89	0.16	0.96
12	25; 5; 0.3; 0.8; 1; 1; 4; 1; 2; 0.2	0.21	1.68	0.09	0.94	0.13	1.02
13	25; 5; 0.4; 0.8; 1; 1; 4; 1; 2; 0.2	0.20	1.82	0.08	0.98	0.12	1.06
14	25; 5; 0.2; 0.4; 1; 1; 4; 1; 2; 0.2	0.22	1.32	0.10	0.84	0.15	0.92
15	25; 5; 0.2; 0.6; 1; 1; 4; 1; 2; 0.2	0.22	1.43	0.10	0.88	0.15	0.95
16	25; 5; 0.2; 1; 1; 1; 4; 1; 2; 0.2	0.22	1.66	0.11	0.94	0.14	1.01
17	25; 5; 0.2; 1.6; 1; 1; 4; 1; 2; 0.2	0.22	2.01	0.11	1.03	0.14	1.09
18	25; 5; 0.2; 0.8; 0.9; 1; 4; 1; 2; 0.2	0.19	1.45	0.08	0.70	0.12	0.79
19	25; 5; 0.2; 0.8; 0.95; 1; 4; 1; 2; 0.2	0.21	1.48	0.09	0.81	0.13	0.88
20	25; 5; 0.2; 0.8; 1.2; 1; 4; 1; 2; 0.2	0.28	1.83	0.18	1.30	0.22	1.36
21	25; 5; 0.2; 0.8; 1.4; 1; 4; 1; 2; 0.2	0.34	2.11	0.27	1.69	0.31	1.73
22	25; 5; 0.2; 0.8; 1; 0.6; 4; 1; 2; 0.2	0.35	1.91	0.28	1.58	0.33	1.62
23	25; 5; 0.2; 0.8; 1; 0.8; 4; 1; 2; 0.2	0.27	1.70	0.17	1.21	0.22	1.26
24	25; 5; 0.2; 0.8; 1; 1.1; 4; 1; 2; 0.2	0.20	1.49	0.09	0.78	0.12	0.85
25	25; 5; 0.2; 0.8; 1; 1.2; 4; 1; 2; 0.2	0.19	1.44	0.07	0.65	0.11	0.75
26	25; 5; 0.2; 0.8; 1; 1; 2; 1; 2; 0.2	0.19	1.03	0.07	0.49	0.10	0.53
27	25; 5; 0.2; 0.8; 1; 1; 3; 1; 2; 0.2	0.21	1.29	0.09	0.69	0.13	0.74
28	25; 5; 0.2; 0.8; 1; 1; 6; 1; 2; 0.2	0.24	2.09	0.13	1.38	0.18	1.48
29	25; 5; 0.2; 0.8; 1; 1; 8; 1; 2; 0.2	0.26	2.64	0.15	1.88	0.21	2.01

Note: As before the individual values for FC were calculated by dividing the total FC results by two, the results for conservation autarky were calculated with the solutions found by solving the games and using these solutions in the individual payoff functions under strategic behaviour.

between conservation autarky and full cooperation becomes larger, for example, in simulations 1 to 5 where the price of fish rises.

Overall, in our setting conservation autarky constitutes an improvement over strategic behaviour, but as can be seen from Table 15.2, that only holds if countries take all uses into account. Incidentally, the same holds for full cooperation.

15.3 DISCUSSION AND CONCLUSIONS

In this chapter we investigated the effect of conservation autarky on MPA assignment in a multiple-use environment. We investigated and compared the results with strategic behaviour and full cooperation, both in the single-purpose conservation game and in a multiple-use game where the benefits of conservation and fisheries are both considered.

We found that for the symmetric conservation game conservation autarky coincides with full cooperation. However, in the conservation game we only consider size and not location, apart from that MPAs are located in countries. With a more explicit specification of species locations, conservation autarky may result in more overprotection and no longer coincide with a fully cooperative solution (e.g., Bode et al., 2011; Punt et al., 2012).

The fact that conservation autarky coincides with full cooperation in the conservation game is also driving the result that if all uses are considered, playing the strategic equilibrium from the conservation game or the combined game makes a country better off than behaving either fully cooperative or ignoring the contributions of others (i.e., conservation autarky) and lets the MPA size be dictated by species conservation alone.

In contrast, if all uses are considered in the MPA decision, conservation autarky is generally better than strategic behaviour, producing higher welfare levels, although not as high as full cooperation. Conservation autarky produces even better results when the gains in the conservation game rise. Obviously, this is because the conservation autarky equilibrium in the separate conservation game is equal to the fully cooperative one, and hence increasing the benefits of conservation improves the performance of conservation autarky.

As usual in game theory, the fully cooperative outcome is the best in global terms for welfare, but the strategic outcome is the best for an individual country, if the other one also behaves strategically. Conservation autarky can constitute an improvement, but only if the other country acts independently as well. If the other country behaves strategically the final outcome can be similar to the outcome where all countries act strategically. It would seem that it should be easier to reach the fully cooperative outcome from a situation of conservation autarky than from strategic behaviour. To investigate this, however, we need a dynamic model rather than a static model such as this one.

A limitation of our game and therefore of the results, is the symmetric setting of the game. It has been shown by, for example, Pintassilgo et al. (2010) that results in fisheries games with a Gordon-Schaefer model can change dramatically by introducing asymmetry. Moreover, asymmetry in location of species, fishers and stock may play important roles in the effectiveness of MPAs (Sanchirico and Wilen, 2001; Smith and Wilen, 2003; Sanchirico et al., 2006; Ruijs and Janmaat, 2007). That does not reduce the validity of our model results for the symmetric case, but does call for a further exploration of asymmetric games in the future.

Although written in the context of our seas and oceans, the main results of this model and its implications can be easily extended to other ecosystems and their services. Most ecosystems provide multiple services and some of them are extractive in nature whereas others are not and many of them are transboundary. Protected areas are also frequently used in other ecosystems. Cases in point are forests where we have on the one hand multiple extractive uses such as timber, firewood and game, and on the other hand we have multiple non-extractive uses such as carbon credits and species conservation. Moreover, a number of these forests are transboundary, and have transboundary parks such as the Great Limpopo Transfrontier Park in South Africa, Zimbabwe and Mozambique or the adjoining Bavarian forest and Šumava National Park on the border between Germany and the Czech Republic.

We have shown in our setting, that conservation autarky constitutes an improvement over strategic behaviour. As such, if countries act independently this constitutes an improvement, compared to a situation where both countries free-ride. However, this only applies if all uses are considered. This chapter therefore stresses the importance that policy-makers consider all possible uses of protected areas, and the importance of clarity of concepts, that is, the meaning and goals of these areas. If multiple goals are to be served we have to find a compromise in the protected area size that is optimal from the different points of view, as well as the uses that are allowed, either through valuation or through economic impact assessment or other methods such as (spatial) multi-criteria analysis.

REFERENCES

Anderson, L.G. (2002), 'A bioeconomic analysis of marine reserves', *Natural Resource Modeling*, **15**(3), 311–34.

Beattie, A., U.R. Sumaila, V. Christensen and D. Pauly (2002), 'A model for the bioeconomic evaluation of marine protected area size and placement in the North Sea', *Natural Resource Modeling*, **15**(4), 413–37.

Bode, M., W. Probert, W.R. Turner, K.A. Wilson and O. Venter (2011), 'Conservation planning with multiple organizations and objectives', *Conservation Biology*, **25**(2), 295–304.

Bohnsack, J. (1993), 'Marine reserves: they enhance fisheries, reduce conflicts, and protect resources', *Oceanus*, **36**(3), 63–71.

Boncoeur, J., F. Alban, O. Guyader and O. Thébaud (2002), 'Fish, fishers, seals and tourists: economic consequences of creating a marine reserve in a multi-species, multi-activity context', *Natural Resource Modeling*, **15**(4), 387–411.

Brown, K., W. Adger, E. Tompkins, P. Bacon, D. Shim and K. Young (2001), 'Trade-off analysis for marine protected area management', *Ecological Economics*, **37**(3), 417–34.

Chan, K.M.A., M.R. Shaw, D.R. Cameron, E.C. Underwood and G.C. Daily (2006), 'Conservation planning for ecosystem services', *Plos Biology*, **4**(11), e379.

Convention on Biological Diversity (2010), *COP 10 Decision X/29 Marine and Coastal Biodiversity*, accessed 8 February 2014 at http://www.cbd.int/decision/cop/?id=12295.

Costello, C. and D.T. Kaffine (2010), 'Marine protected areas in spatial property-rights fisheries', *Australian Journal of Agricultural and Resource Economics*, **54**(3), 321–41.

Edgar, G.J., P.F. Langhammer, G. Allen, T.M. Brooks, J. Brodie and W. Crosse (2008), 'Key biodiversity areas as globally significant target sites for the conservation of marine biological diversity', *Aquatic Conservation: Marine and Freshwater Ecosystems*, **18**(6), 969–83.

Hannesson, R. (1998), 'Marine reserves: what would they accomplish?', *Marine Resource Economics*, **13**(3), 159–70, accessed 8 February 2014 at http://ideas.repec.org/a/ags/mareec/28105.html.

Hunter, M.L. and A. Hutchinson (1994), 'The virtues and shortcomings of parochialism: conserving species that are locally rare, but globally common', *Conservation Biology*, **8**(4), 1163–5.

Jantke, K. and U.A. Schneider (2010), 'Multiple-species conservation planning for European wetlands with different degrees of coordination', *Biological Conservation*, **143**(7), 1812–21, accessed 8 February 2014 at http://www.sciencedirect.com/science/article/pii/S0006320710001801.

Jones, P.J.S. (2001), 'Marine protected area strategies: issues, divergences and the search for middle ground', *Reviews in Fish Biology and Fisheries*, **11**(3), 197–216, accessed 8 February 2014 at http://www.ingentaconnect.com/content/klu/rfbf/2001/00000011/00000003/05093376.

Kark, S., N. Levin, H.S. Grantham and H.P. Possingham (2009), 'From the cover: between-country collaboration and consideration of costs increase conservation planning efficiency in the Mediterranean Basin', *Proceedings of the National Academy of Sciences*, **106**(36), 15368–73.

Lester, S.E., B.S. Halpern, K. Grorud-Colvert, J. Lubchenco, B.I. Ruttenberg and S.D. Gaines et al. (2009), 'Biological effects within no-take marine reserves: a global synthesis', *Marine Ecology Progress Series*, **384**, 33–46.

Levin, P.S., I. Kaplan, R. Grober-Dunsmore, P.M. Chittaro, S. Oyamada, K. Andrews and M. Mangel (2009), 'A framework for assessing the biodiversity and fishery aspects of marine reserves', *Journal of Applied Ecology*, **46**(4), 735–42.

MacArthur, R.H. and E.O. Wilson (2001), 'The theory of island biogeography', Princeton, NJ: Princeton University Press.

Neigel, J.E. (2003), 'Species–area relationships and marine conservation, *Ecological Applications*, **13**(sp1), 138–45.
Pintassilgo, P., M. Finus, M. Lindroos and G. Munro (2010), 'Stability and success of regional fisheries management organizations', *Environmental and Resource Economics*, **46**(3), 377–402.
Punt, M.J., H.-P. Weikard, R.A. Groeneveld, E.C. van Ierland and J.H. Stel (2010), 'Planning marine protected areas: a multiple use game', *Natural Resource Modeling*, **23**(4), 610–46.
Punt, M.J., H.-P. Weikard, E.C. van Ierland and J.H. Stel (2012), 'Large-scale marine protected areas for biodiversity conservation along a linear gradient: cooperation, strategic behavior or conservation autarky?', *Environmental and Resource Economics*, **53**(2), 203–28.
Punt, M.J., H.-P. Weikard and E.C. van Ierland (2013), 'Marine protected areas in the high seas and their impact on international fishing agreements', *Natural Resource Modeling*, **26**(2), 164–93
Rijksoverheid (Government of the Netherlands) (2011), 'Delen Noordzee verboden voor visserij door akkoord natuurbeweging, vissers en rijksoverheid' [Fisheries prohibited in parts of the North Sea in consequence of agreement between nature conservation organizations, fishers and the Dutch government], Ministry of Economic Affairs press release, 13 December 2011, accessed 11 February 2014 at http://www.rijksoverheid.nl/nieuws/2011/12/13/delen-van-noordzee-verboden-voor-visserij-door-akkoord-natuurbeweging-vissers-en--rijksoverheid.html.
Rodrigues, A.S.L. and K.J. Gaston (2002), 'Rarity and conservation planning across geopolitical units', *Conservation Biology*, **16**(3), 674–82.
Rosenzweig, M.L. (1995), *Species Diversity in Space and Time*, Cambridge, UK: Cambridge University Press.
Ruijs, A. and J.A. Janmaat (2007), 'Chasing the spillovers: locating protected areas in a trans-boundary fishery', *Land Economics*, **83**(1), 6–22.
Sanchirico, J.N. (2004), 'Designing a cost-effective marine reserve network: a bioeconomic metapopulation analysis', *Marine Resource Economics*, **19**(1), 41–65, accessed 8 February 2014 at http://ideas.repec.org/a/ags/mareec/28099.html.
Sanchirico, J.N. and J.E. Wilen (2001), 'A bioeconomic model of marine reserve creation', *Journal of Environmental Economics and Management*, **42**(3), 257–76.
Sanchirico, J.N. and J.E. Wilen (2002), 'The impacts of marine reserves on limited-entry fisheries', *Natural Resource Modeling*, **15**(3), 291–310.
Sanchirico, J.N., U. Malvadkar, A. Hastings and J.E. Wilen (2006), 'When are no-take zones an economically optimal fishery management strategy?', *Ecological Applications*, **16**(5), 1643–59.
Smith, M.D. and J.E. Wilen (2003), 'Economic impacts of marine reserves: the importance of spatial behavior', *Journal of Environmental Economics and Management*, **46**(2), 183–206.
Spalding, M., L. Wood, C. Fitzgerald and K. Gjerde (2010), The 10% target: where do we stand?' in C. Toropova, I. Meliane, D. Laffoley, E. Matthews and M. Spalding (eds), *Global Ocean Protection. Present Status and Future Possibilities*, Agence des Aires Marines Protégées/IUCN/WCPA/UNEP-WCMC/TNC/UNU/WCS, pp. 25–40.
Weeks, R., G.R. Russ, A.A. Bucol and A.C. Alcala (2010), 'Incorporating local tenure in the systematic design of marine protected area networks', *Conservation Letters*, **3**(6), 445–53.
Williams, C.B. (1943), 'Area and number of species', *Nature*, **152**(3853), 264–7.
Wood, L.J. and S. Dragicevic (2007), 'GIS-based multicriteria evaluation and fuzzy sets to identify priority sites for marine protection', *Biodiversity Conservation*, **16**(9), 2539–58.

16. The valuation of ecosystem services and their role in decision-making: constraints and ways forward

Anil Markandya and Marta Pascual

16.1 INTRODUCTION

There has been a great deal of interest in, and work on, the valuation of ecosystem services (ESS) over the last three decades. Since then, a wide range of benefits that nature provides to human beings has been identified and, to various degrees, values have been estimated for them. Yet, there remain some important concerns about what has been achieved so far. In this chapter we examine these concerns in some detail. The chapter begins with a summary of the state-of-the-art in the study of ecosystem services and, in particular, the role of economic valuation in this area (Sections 16.2 and 16.3). Section 16.4 looks at the actual use of economic valuation of ESS in the decision-making process, noting that its applicability is still relatively scarce. We believe this is unfortunate as it limits the mainstreaming of ESS into public decision-making, where money values are a key unit of account and an important contribution to determining relative priorities. Section 16.5 discusses the constraints of the ESS approach, which lead to it being less deployed as a tool of analysis, and suggests how these can be addressed. The main issues examined are the following:

- How satisfactory is it to have a single value total economic value (TEV) (in euros or dollars) for the complex range of services provided by ecosystems?
- To what extent do the methods of valuation capture the wide range of socio--cultural perceptions that exist about the different ESS?
- How can the equity concerns (winners and losers) be addressed in ESS valuation?
- How far can the spatial and temporal dimensions of ESS be fully captured and rectified?
- How satisfactory is the benefit transfer approach and is there any way round it?

Finally, Section 16.6 suggests how we can go forward from here and Section 16.7 shows what conclusions one can draw in order to improve the role of the valuation of ESS in decision-making.

16.2 ECOSYSTEM SERVICES STATE-OF-THE-ART

16.2.1 The Idea of ESS Has Really Caught On

The term ESS was first coined in 1981 by Ehrlich and Ehrlich as an attempt to build a common language for discussing linked ecological and economic systems (Liu et al.,

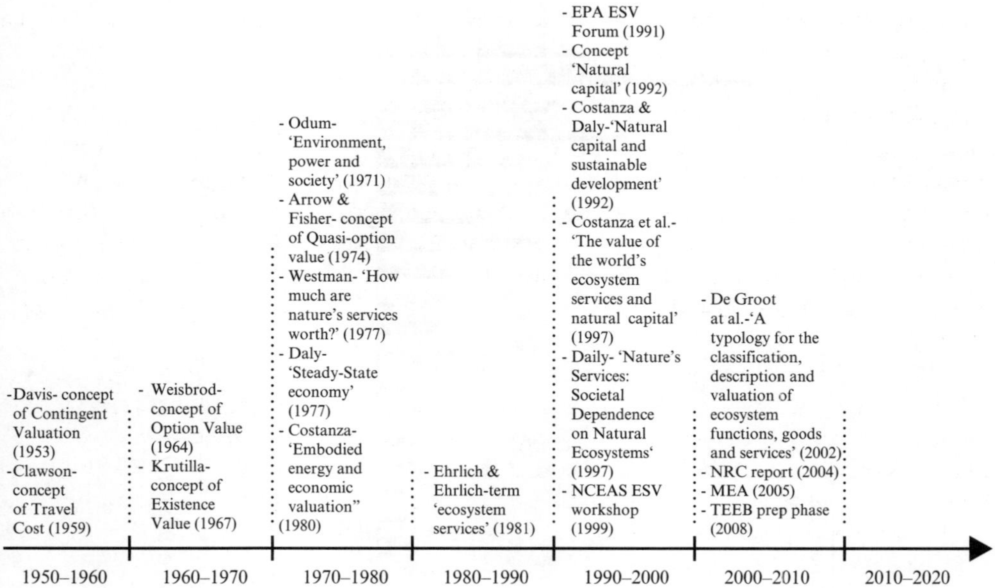

Source: Adapted from TEEB (2010) and Liu et al. (2010).

Figure 16.1 Milestones in the history of ecosystem services and the ecosystem services valuation (1950–2020)

2010). Before that, some work had been carried out addressing how much nature's services were worth (Westman, 1977), and even valuing certain nature services through single valuation methods (Figure 16.1). In the early 1990s, the interdisciplinary field of ecological economics developed the concept of natural capital (Costanza and Daly, 1992; Jansson and Jansson, 1994); however, it was not until the late 1990s that the concept got widespread attention with the publications by Costanza et al. (1997) and Daily (1997) (TEEB, 2010) (Figure 16.1).

Since the 1990s this area of research has grown exponentially as can be seen from the number of papers addressing ESS (Fisher et al., 2009; Figures 16.2a–b).

Furthermore, the term has started to be widely used and accepted by scientists and academic works as shown by the increasing citing trend over the ESS term (Figure 16.3).

The term ESS has also been widely used in various subject areas (Table 16.1), showing that this issue has moved into the realm of multidisciplinary research, building trans-boundary bridges and merging many disciplines.

16.2.2 Ecosystem Services and Human Well-being

Humanity's reliance upon nature for welfare and survival is complete. While, as said, the term itself may be relatively new, an understanding that nature provides services for human welfare goes back to the story of the Garden of Eden (Fisher et al., 2009). In this way, humans have always recognized the importance of what we now call ESS. Both Fisher et al. (2009) and TEEB (2010) note that the concept of ESS has become an

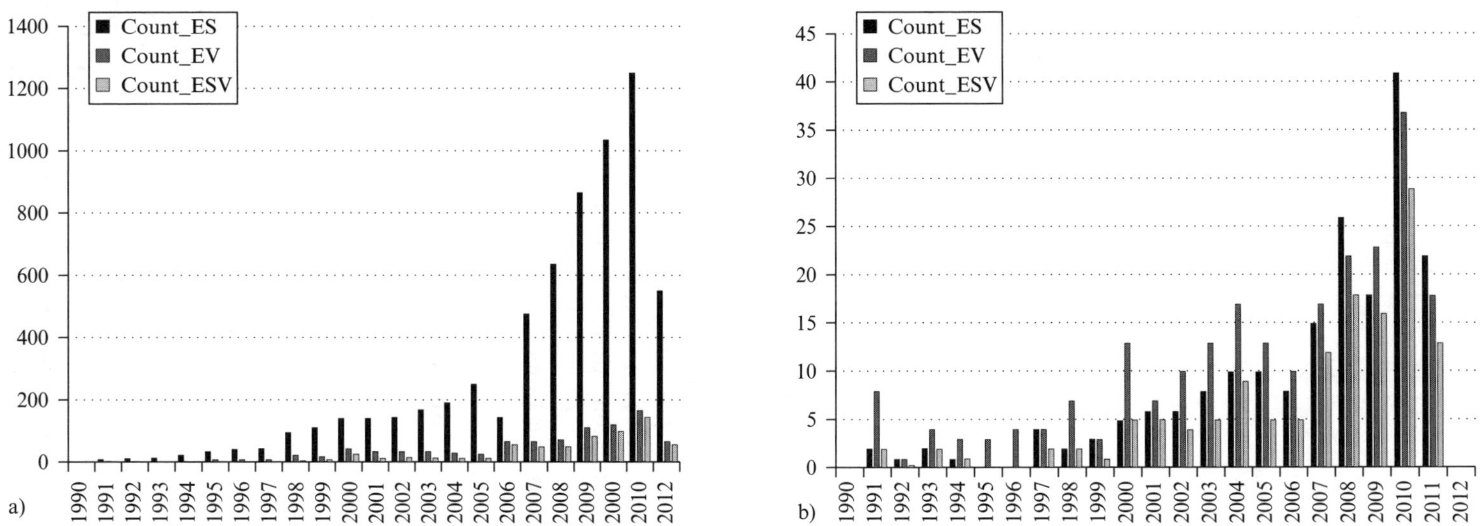

Sources: Figure (a) is constructed by examining ISI Web of Science (2012), an index of scientific publications, accessed June 2012. Figure (b) is constructed by examining EVRI, Environmental Valuation Reference Inventory, accessed June 2012.

Figures 16.2a–b *Evolution of peer-reviewed papers with the 'ecosystem services (ES)', 'environmental valuation (EV)' or 'ecosystem services valuation (ESV)' sentences either in the title, abstract or keywords*

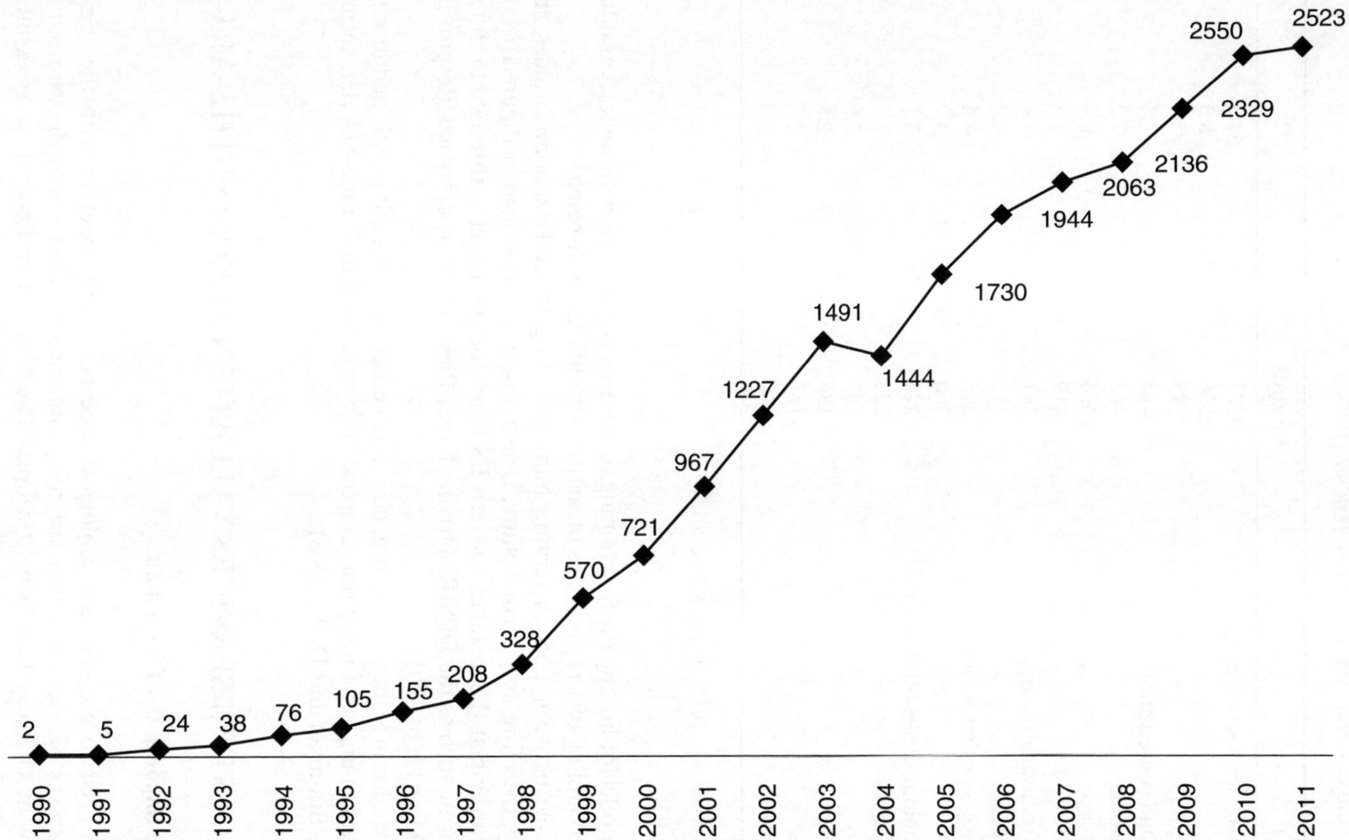

Source: Figure constructed by examining ISI Web of Science (2012), an index of scientific publications, accessed June 2012.

Figure 16.3 Evolution of cites in peer-reviewed papers with 'ecosystem services' as a keyword either in the title, abstract or keywords

Table 16.1 Number of records and percentages of the term 'ecosystem services' per subject area defined in ISI Web of Science

Subject Areas	Records	%
Environmental sciences, ecology	3627	54.5
Engineering	556	8.3
Agriculture	515	7.7
Biodiversity and conservation	484	7.3
Business economics	482	7.2
Computer science	424	6.4
Forestry	419	6.3
Marine and freshwater biology	361	5.4
Water resources	315	4.7
Science technology other topics	296	4.4
Geology	260	3.9
Life sciences, biomedicine other topics	244	3.7
Physical geography	220	3.3
Oceanography	193	2.9
Geography	185	2.8
Plant sciences	167	2.5
Other subject areas	1870	< 2.0

Source: Web of Science (2012), accessed June 2012.

important model for linking the functioning of the ecosystems to human welfare, as it has allowed the coupling of the functions of nature with human well-being.

This linkage between ESS as serving human, ecological and economic values and contributing, therefore, to human welfare, is also observable as shown in Figure 16.4; or even when looking at the linkages between ESS and societal benefits throughout different mechanisms such as the DPSIR (Driving forces/Pressures/States/Impacts/Responses) approach (Atkins et al., 2011).

What is for certain is that recognizing how much ecosystems benefit people and in what ways is critical to understanding the complex environment-related trade-offs that society faces today (Brauman and Daily, 2008).

16.3 ECOYSTEM SERVICES VALUATION STATE-OF-THE-ART

16.3.1 The Importance of Valuation

Why do we need to consider the ecological aspects of ESS together with the socio-economic aspects of the assets that are being obtained by them? Why do we need to measure how much ecosystems benefit people? The Millennium Ecosystem Assessment (MA) (MA, 2005), found that globally 15 of the 24 ESS investigated are in a state of decline and that this is likely to have a large and negative impact on future human welfare. Liu et al. (2010) also note that ecosystems are becoming increasingly scarce and that there

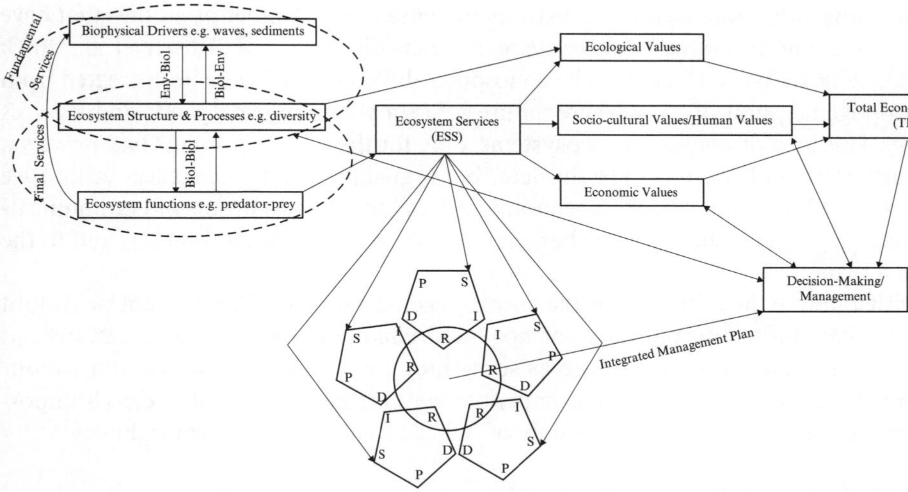

Sources: Adapted from de Groot et al. (2002); Liu et al. (2010) and Atkins et al. (2011).

Figure 16.4 The ecosystem services valuation and DPSIR framework

is a need to get society to acknowledge the value of natural capital. The MA also states that there is an urgent need for research on measuring, modelling and mapping of ESS, and in assessing changes in their delivery with respect to human welfare (MA, 2005; Carpenter et al., 2006; Sachs and Reid, 2006).

On the supply side, assets embodied in ecosystems, though poorly understood (Brauman and Daily, 2008), are experiencing rapid degradation and depletion in their capability of providing services. At the same time, the demand for ESS is rapidly increasing as populations and standards of living increase (MA, 2005). The two factors taken together allow us to conclude that ESS are becoming increasingly scarce.

It could be argued that this trend is partly due to the lack of proper valuation of the different ESS, leading to a lack of appreciation of their worth when making management decisions: it is impossible to manage what we do not value in some form in quantitative terms (TEEB, 2008). Sadly the importance of ESS is often widely appreciated only upon their loss (Daily et al., 2000).

As Liu et al. (2010) state, we use the terms 'priceless' and 'invaluable' when discussing the environment and yet this has proven woefully insufficient in terms of reducing or halting ecosystem degradation. The challenge then is to provoke society to acknowledge the value of ESS, and 'ecosystem services valuation' (ESV) is the tool that can tackle such a challenge.

ESV appeared as the way of linking ESS with economic values, acting as a vehicle to integrate ecological understanding and economic considerations (Chee, 2004), in order to be able to show the linkages that would allow scientists to redress the traditional neglect of ESS in policy decisions and show society and managers the need to account for ESS when making management decisions (Liu et al., 2010).

Right after the ESS term was developed, the term ESV was coined and started receiving ample attention, leading to a strong increase in the number of studies that have analysed the economic value of ecosystems in scientific literature (Turner et al., 2003; Hein et al., 2006) (Figure 16.2a–b). The concepts of ESS and ESV were both created from an anthropogenic point of view where nature offers a wide range of goods and services to society. This idea of conserving ecosystems, only for the sake of humans has, however, been widely criticized. We return to this later but arguably the criticism is less valid if we use ESS and ESV not only to derive economic values but also as a means of communicating the interdependency between the benefits of conservation both to humans and to the Earth as a planet.

Notwithstanding the criticisms of the anthropogenic nature of ESS concept we cannot deny that it has shifted our paradigm of how nature matters to human societies. Instead of viewing the preservation of nature as something for which we have to sacrifice our well-being, we now perceive the environment as natural capital, one of society's important assets (Liu et al., 2010). That is surely of benefit to nature in its own right.

16.3.2 Methodologies for Ecosystem Service Valuation

Innovative and promising methods are now available to quantify ESS and align economic incentives with their protection (Brauman and Daily, 2008). These include values based on 'market' data, that is, what is bought and sold for money. An example would be soil retention as an ESS providing the service of agricultural yields that can be valued in terms of the market price of the crops grown on the land (Ansink, 2008). In addition to market values a number of methods have now been developed that capture what are broadly termed 'non-market values'. These would include recreational use of land and water bodies, aesthetic values associated with some sceneries and landscapes, and some values associated with the pleasure of knowing that some forms of nature are simply there (Lazo, 2002). Table 16.2 summarizes the range of different ESS; some of these can be valued by using market values and some by non-market valuation methods.

16.3.3 Choosing Methods for Valuing

The appropriateness of different methods for valuing different services is presented in Table 16.3. In that table market-price-based methods include: (1) direct use of market prices, (2) methods based on measures of the link between ESS and productivity in terms of yields and so on, (3) methods based hedonic methods, where analysts estimate the impact of ecosystems of land and property values, and (4) travel cost methods, where the analysis consists of looking at what people spend to visit a site of some environmental or cultural interest. Non-market methods include: (1) estimation of the costs of avoiding damaging impacts such as soil erosion, (2) costs of replacing lost natural functions of ESS with artificial inputs, and (3) methods of eliciting information about values by asking people directly how much they are willing to pay for the services in question.

The table also provides some other important information that is critical to the valuation of ESS. First is the amenability of the service to valuation in monetary terms. Second is the transferability of values estimated in one site to another site (so-called benefit transfer). We rate that as low, medium or high. Third are the beneficiaries of the services and whether they are local, regional or global. Fourth is the spatial scale of the

Table 16.2 Functions, ecosystem services (ESS) (according to MA, 2005) and their most common context of valuation

Functions	ESS	Context of Valuation
Regulation	Gas regulation	Anthropogenic disturbance; climate change; land-use changes; regulation of CO_2/O_2 balance; maintenance of the ozone-layer (O_3); regulation of the SO_x levels; maintenance of clean, breathable air; prevention of diseases
	Climate regulation	Deforestation; reforestation; REDD+ (Reduced Emissions from Deforestation and Forest Degradation)
	Disturbance regulation	Flooding; extreme events; sea level rise; storms; droughts; human constructions
	Water regulation	Natural irrigation patterns; discharge and drainage changes; river regulation; medium for transportation
	Water supply	Filtering of vegetation cover; water storage; water supply for households, agriculture and industry
	Soil retention	Vegetation cover changes; root system changes; soil stabilization; erosion and sedimentation
	Soil formation	Flooding; sea level rise; restoration
	Nutrient regulation	Limiting nutrients; healthy ecosystems; gas/climate and water regulation
	Waste regulation	Water purification; organic/inorganic recycling
	Pollination	
	Biological regulation	Communities regulation; pest and diseases control
Provisioning	Food	Food scarcity; subsistence farming; human diet; wild plants and animals; food production mechanisms; land-use changes
	Raw material	Renewable biotic resources; energy resources; animal feed
	Genetic resources	Wild sources; cultivated plants and domesticated animals; genetic manipulation; biotechnological research
	Medicinal resources	New chemicals, pharmaceutical drugs; new medical tools
	Ornamental resources	Precious minerals; fashion; clothing; ceremonies; souvenirs; collectors' items
Cultural	Recreation	Place for relaxing, resting, refreshment; activities: e.g., walking, hiking, camping, surfing etc.
	Aesthetics	Scenery and landscape; views (real estate)
	Science and education	Environmental education; research; excursions; field laboratories; publications
	Spiritual and historic	Ethical and heritage values
Supporting	Refugium function	MPAs, geologically important areas; sanctuaries; no-take zones, protected sites; Natura 2000 sites; Habitats Directive; biodiversity
	Nursery function	Fisheries; overfishing; marine protected areas; biodiversity; food webs dynamics

Table 16.3 Analyses results

Functions	ESS	Amenability to ESV	Studied Ecosystems	Most Appropriate Method for Valuation (X)/Other Methods (O)								Transfer-ability to Sites	Bene-ficiaries	Spatial Scales	Spatial Scope km^2
				Market prices				Evidence		Surveys					
				M	P	H	TC	AC	RC	CV	CC				
Regulation functions	Gas reg.	Medium	Forest	O	O			X	X	X		High	Global	Global	$>10^6$
	Climate reg.	Low	Forest/Ocean/ Wetlands/Marshes	O	O	O		O	O	X		High	Global	Global	$>10^6$
	Disturbance reg.	High	Coasts/Wetlands/ Corals/Estuaries/ Marshes	O	O	O		X	O	O		Medium	Regional	Regional	10^4–10^6
	Water reg.	High	Rivers/Estuaries/ Marshes	X	X	X		X	X	X		Medium	Regional	Regional	10^4–10^6
	Water supply	High	Rivers/Lakes/ Aquifers	X	O		X	X	X	O		Medium	Regional	Ecosystem	<1
	Soil retention	Medium	Rivers/Estuaries/ Marshes	O	O	X	O	X	X	O		Medium	Regional	Landscape	1–10^4
	Soil formation	Medium	Coasts/Estuaries/ Marshes/Forests	X	O	X	O	X	X	O		Medium	Regional	Landscape	1–10^4
	Nutrient reg.	Medium	Oceans/Rivers/ Estuaries/Marshes	O	O			X	O	X		Medium	Global	Global	$>10^6$
	Waste reg.	High	Rivers/Estuaries/ Marshes/Forest	O	O			X	X	X		Medium to High	Local	Ecosystem	<1
	Pollination	Medium	Forests	X	X			O	O	O		Medium	Local	Regional	10^4–10^6
	Biological reg.	Medium	All	O	X			X	O			High	Global	Global	$>10^6$

Provisioning functions	Food	High	All	X	X				O	O		High	Regional / Global	Regional	10^4–10^6
	Raw material	High	All	X	X				O	O		High	Regional	Regional	10^4–10^6
	Genetic res.	High	All	X	O			X	O			Low	Global	Global	$>10^6$
	Medicinal res.	High	All	O	X			X	X	O		High	Global	Global	$>10^6$
	Ornamental res.	High	All	O		X		X	X	O		Medium	Global	Global	$>10^6$
Cultural functions	Recreation	High	All	O	O	O	X		O	X	X	Low	Regional	Regional	10^4–10^6
	Aesthetics	High	All	O		X	X		O	X	X	Low	Regional	Landscape	1–10^4
	Science & education	Low	All	O						X	X	High	Regional	Regional	10^4–10^6
	Spiritual & historic	Low	All	O	O		O		O	X	X	Low	Regional	Regional	10^4–10^6
Supporting functions	Refugium function	Medium	All	O			X		O	X	X	Low	Regional	Regional	10^4–10^6
	Nursery function	Medium	All	O	X		X			X	X	Low	Regional	Global	$>10^6$

Note: ESS = ecosystem services; reg.= regulation; res. = resources; ESV= ecosystem service valuation; M = market price; P = productivity; H = hedonic pricing; TC = travel cost; AC = avoidance cost; RC = replacement cost; CV = contingent valuation; CC = contingent choice.

Sources: From de Groot et al. (2002); MA (2005); Angulo-Valdés and Hatcher (2009); Liu et al. (2010); Fu et al. (2011).

service, which is at the following levels: ecosystem, landscape, regional or global. This is complemented by the spatial scope, which is measured in square kilometres.

The table reveals some interesting findings:

- The amenability to valuation is mostly medium to high, the exceptions being some of the cultural functions, particularly those relating to 'science and education' and 'spiritual and historic'. While the gaps are important the fact that most services have a reasonable chance of being valued is encouraging for the use of ESV.
- In terms of valuation methods an 'X' indicates the main method used and an 'O' as an alternative or other method. The table shows that, of the 22 ESS that are valued, only four have market prices and market-based methods and the main approach. At the same time only six have non-market methods as the main approach. That means that 12 ESS can be valued using either market or non-market methods. It is clear that the different methods have their advantages and disadvantages (Angulo-Valdés and Hatcher, 2009) and, in general, policy-makers prefer valuations based on market-based approaches. Indeed, since valuation methods are 'value articulating institutions' (TEEB, 2010), the choice of a method can strongly influence the outcome of a valuation exercise (ibid.); although there is only limited evidence to suggest that one approach is systematically biased in one direction.[1]
- The scope of transferring values from one site to another are evaluated as medium to high for 16 of the 22 ecosystem services, which is also encouraging for ESV, as it is not possible to value all sites in primary studies. Having said that, benefit transfer is a problem and we return to it in Section 16.5.
- The beneficiaries of different services can be local, regional or global. The level of beneficiary is of key importance when using the data on ESV is ecosystem management, as we show later.
- The spatial scale of the service is important, along with the level of beneficiary in determining how valuation studies are carried out and used in estimating benefits. The wider the spatial scope, the more heterogeneous will be the beneficiaries and the more likely that the average value will not be representative for all groups. We return to this issue in Section 16.5.

16.4 ARE ESV OUTCOMES BEING APPLIED IN THE REAL WORLD

Many authors (Smith, 2000; Silva and Pagiola, 2003; Adamowicz, 2004) have looked at the use of valuation research in policy. They find that some actions have indeed resulted from the ESV research, such as damage assessment cases in the USA; controls on some pollutants based on evaluations of human health; cost/benefit analysis of water resource planning; forest resource use planning; tax revenues from the improvement of the environmental quality and so on. However, there have not been as many applications as one would hope for, and the majority of the ESV studies have been of an academic nature and have not been intended to influence decisions.

Furthermore, when looking at the types of valuation methods used in those studies that link valuation to policy and management approaches, there is a clear bias in favour of market-based methods in most of them (especially avoided costs and changes in productivity methods, basically). This is unfortunate as it means that some important services are excluded, such as passive use values of natural resource valuation, leading to the possibility of placing zero values over such services (Adamowicz, 2004). As we noted above, there is no reason to discount non-market valuations, and when a particular ecosystem service is valued using both methods there is no evidence that the non-market estimates are more unreliable or biased in one direction.

Over time there has been movement towards a more trans-disciplinary approach to ESV research and this has helped facilitate valuation and make the numbers obtained more credible. Nevertheless, the contributions of ESV to ecosystem management have not been as significant as hoped, or as clearly defined. ESV researchers will therefore have to transcend disciplinary boundaries and synthesize tools, skills and methodologies from various disciplines; ESV research has to become more problem driven rather than tool driven because ultimately the success of ESV will be judged on how well it facilitates real-world decision-making and the conservation of natural capital (Liu et al., 2010).

In addition to the factors mentioned above, there are several shortcomings in ESV that limit its use in policy-making. In the next section we consider these and ask how they may be addressed.

16.5 SHORTCOMINGS OF ESV AND HOW THEY MAY BE ADDRESSED

16.5.1 Multiple Values Gathering into a Single TEV

The use of a single economic indicator of total economic value (TEV) has been the subject of much of the criticism of ESV. It has been argued that recognizing the existence of multiple values and encouraging an open and pluralistic discussion of values will lead to new solutions for conservation practice (Norton and Noonan, 2007).

We would argue that there are two issues here. One is the need to recognize that all aspects of the value of ESS cannot be captured in TEV and the second that TEV cannot even capture all the economic values in a narrow sense. On the first we would agree completely. Values from a given set of ESS include not only the monetary values of the services, but also the support they can provide to communities and the underpinning they serve in cultural values. Thus, a loss of ESS is more than simply the loss of the values of the services as measured by existing methods described in Section 16.3. But that is not to say that the methods described in that section are irrelevant: they are simply part of a bigger picture of the impacts of an ecosystem. To exclude them would be to exclude an important component of the overall picture.

The second is the criticism that even some economic values are not captured by the currently available toolkits. It is true to say that some environmental assets, such as biodiversity, are not adequately valued in economic terms, although a lot of progress has been made in capturing these values (TEEB, 2010). Where methods of valuation are insufficient it is important to recognize the plurality of information and to use physical indicators of the services provided where available. This information complements that supplied by ESV and the whole package is then assessed by policy-makers, either using their judgement or using tools such as multi-criteria analysis.

16.5.2 Socio-cultural Perceptions of ESV

ESV functions as a system of cultural projection of the forms of relationship between humans and the environment, reflecting the perceived realities, worldviews, mindsets and belief systems (TEEB, 2010). Since different stakeholders (or even individuals) perceive different benefits from the same ecosystem processes, their valuations can, at times, be conflicting (Turner et al., 2003; Hein et al., 2006). Indeed, there are ecosystem services that are valuable even though stakeholders do not perceive them: for example, climate regulation, which is of vital importance for human well-being but which is probably not perceived by a large portion of the earth's population (Fisher et al., 2009).

The ecological identity of individuals is revealed at various levels of the decision--making hierarchy, that is, from local to regional and further on to a global level (Kumar and Kumar, 2008), which makes it even harder to constrain the valuation. This is why, using the tools of psychoanalytic psychology and environmental psychology, Kumar and Kumar (2008) outline recent research findings that suggest redefining concepts such as ecological identity, self–other dichotomy, and the fostering of identification with nature, as issues that must be embraced in the valuation of ecosystem services.

These dimensions are important and exciting but, as yet, they are not formulated in a way that allows them to be operationalized in an ESV exercise. Yet it is important to take them into account where possible. One way of doing so is to make sure that non-market valuation methods are firmly rooted in local perceptions of the services that ESS provide. Where such perceptions fail to capture important services such as climate regulation, other methods need to be used, such as those based on scientific links between ecosystem functions and markets.

16.5.3 ESS Winners and Losers

One reason why ESV has been limited in its application is that policies and measures to conserve or enhance ecosystems involve winners and losers and the methods of valuation focus too much on the overall gains without looking closely enough at how these gains are distributed. An example would be a conservation project that includes among its benefits carbon capture and storage. This is a global benefit, which is of little importance to the local communities that will be affected by the project. These communities may gain from other benefits that the measure provides (such as reduced soil erosion) but they may also lose out if their traditional rights of access to the conservation area are restricted. In overall terms benefits may well be higher than costs but that is not all that matters. Unless some of these wider benefits can be transferred to those who stand to lose, the project has little chance of success. All project outcomes are obtained within a particular institutional setup, in which environmental policy and governance are currently embedded. It has been argued that ESV and its outcomes, if badly managed, could pave the way for the commoditization of ESS with potentially counterproductive effects in the long term for equity of access to ESS benefits (Gomez-Baggethun and Ruiz-Perez, 2011).

These aspects of any valuation exercise are of much greater importance than has been appreciated. The valuation exercise has to be conducted as an integral part of the assessment of gainers and losers and the measures have to be designed in such a way that: (1) the benefits are shared equitably, and (2) those who stand to lose are not able to prevent the effective functions of the project.

16.5.4 Creating Markets for ESS

Because most ESS are public goods, markets are not available to provide clear units of account. This point can be made most forcibly if we consider the challenge of creating markets for ESS, which, in practice, tend to stumble over the issue of trading units. When regulators attempt to compensate for ecological losses, they inevitably rely on rough units for trade, such as 'acres of wetland', 'pounds of nitrogen', or 'equivalent habitats' (Boyd and Banzhaf, 2007). The problem is that such units do not capture the full range of variation that an acre of wetland may represent in terms of services provided.

This issue comes up not only in regulation of ESS but also in the control of the environment more generally. The trade-off is between direct control methods, which can discriminate between locations in terms of what is and is not permitted and fiscal control methods such as taxes and charges or commodity markets, which use a unit of account that is unable to discriminate in this way. In practice some degree of compromise is possible, so that a market is defined in terms of, for example, acres of wetland, but hotspots where the wetland is of particularly high value are proscribed in terms of their tradability in a habitat banking system (see Barbier and Markandya, 2013 for a discussion of these markets).

16.5.5 Spatial/Temporal Issues of ESS (Local, National, Regional, Global Level)

ESV is a complex, spatial and institutional cross-scale problem, as many efforts focusing on particular parts of ecosystems or species, while effective at one level, fail to capture the linkages created by functional interdependencies within large ecosystems (TEEB, 2010).

The production and use of services from ecosystems vary spatially, along with the economic benefits and costs they generate. A spatially explicit assessment of the impacts of action and quantification of benefits and costs is therefore important. It is also helpful in showing the possible mismatch between the ecological and socio-economic scales of decision-making, service provision and use, and between winners and losers in different scenarios (ibid.). We contend that such an assessment is highly important when valuation of services is applied to support the formulation or implementation of ecosystem management plans (Hein et al., 2006; Luisetti et al., 2011).

Konarska et al. (2002) show how the spatial scale of measurement influences ESV, where a finer spatial resolution leads to an increase in the number of determined ecosystems, as well as to an increase in their final economic valuation. However, to date, there has been relatively little elaboration of the various spatial and temporal scales at which ecosystem services are supplied (Hein et al., 2006).

The fundamental challenge, then, is to understand the dynamics of ESS and human well-being as they interact from local to global scales in the context of multiple changing drivers (Carpenter et al., 2009). In order to show the importance of addressing these spatial issues of ESS, an example is presented here:

The case of rice as a food provision service

Almost half of the world's population depends on rice as the staple food provision resource; more than 3 billion people eat it every day and depend on it for 20 per cent of their daily calories (IRRI, 2012). It is the most important food crop of the developing world and, thus, its supply and market prices are directly related to poverty issues. An example of this relation could be seen in 2008 when rice prices tripled and the World Bank estimated that an additional 100 million people were pushed into poverty (ibid.).

Rice also plays an important role in many cultures as, for thousands of years different parts of the rice plant have been used in religious and ceremonial occasions, as medicine, and as inspiration and medium for a great number of artwork. However, when analysing it as an ESS that is being provided to humans as a food provision source, we realize that there are significant spatial differences between the supply (Figure 16.5a) and demand (Figure 16.5c) in rice areas worldwide. Furthermore, the economic benefits also appear to shift between the supply (Figure 16.5b) and the demand (Figure 16.5d) areas together with the exports/imports economic values. This example allows us to see that, if the valuation of rice as an ESS is done without considering its regional and even global implications, rice shortages may affect society far beyond its economic price and food security terms as any significant disruptions of rice supplies can and do have far-reaching social and political ramifications (ibid.).

16.5.6 Double Counting

Fu et al. (2011) note that the basic causes of double counting include ambiguous definitions and inconsistent classifications of ecosystem services, poor understanding of ecosystem complexity, inadequate recognition of exclusiveness and complementarities of individual ecosystem services, spatial-temporal scale dependence of ecosystem services, and overlap and lack of cross-referencing between ecosystem service valuation methods.

Measures for reducing double counting in ecosystem service valuation should, therefore, tackle these issues through: identifying the spatial-temporal scales of ecosystem services; valuing the final benefits obtained from ecosystem services; establishing consistent classification systems for ecosystem services; and selecting valuation methods appropriate for the study context (Fu et al., 2011). Furthermore, as Ansink et al. (2008) state, valuations should be undertaken either based on functions, or based on services, but not both, so as to avoid the overlooking or overlapping of values that occurs when valuing both fundamental and final services.

16.5.7 Benefit Transfer

Applying economic value estimates obtained from one location to a similar site in another location is referred to as benefit transfer. Among the potential pitfalls of such an approach, the correspondence (or lack thereof) between the locations is probably the most important for evaluating the probable validity of the benefit transfer. A common type of benefit transfer in ecosystem service valuation applies an estimate of value per hectare to all areas having the same land cover or habitat type, and is particularly susceptible to errors resulting from lack of correspondence.

Estimates of errors from using such methods can be quite large. Brander et al. (2012)

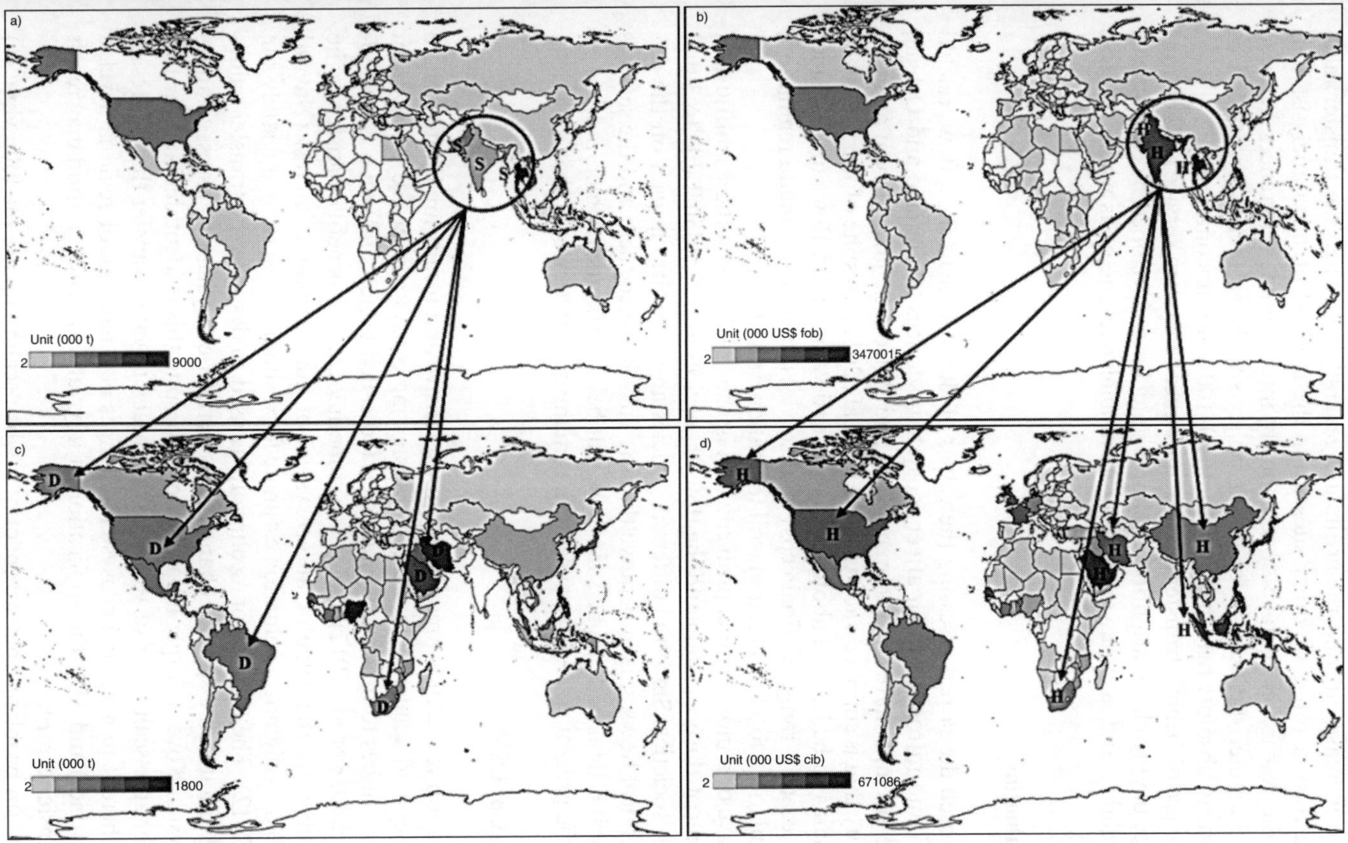

Note: S = supply areas; H = high economic value; D = demand areas; arrows represent the possible flow pathways from supply to demand areas.

Sources: Data and maps obtained from: IRRI (International Rice Research Institute), accessed June 2012 at http://ricestat.irri.org:8080/wrs. Data according to FAO (Food and Agriculture Organization) for 2008 (imports/exports quantities) and 2007 (economic values).

Figure 16.5 Rice as food provision ESS. (a) Tons of rice exporting countries; (b) rice exporting economically profitable areas (US$); (c) tons of rice importing countries; (d) rice importing economically profitable areas (US$)

provide a survey of potential errors from such transfers for a range of environmental valuations. The results show that as study and policy sites become more different, transfer errors tend to increase but that the errors are generally less when the transfer takes account of data from a range of other sites using the method of meta-analysis. This method allows the researcher to take systematically into account differences between the study sites' characteristics and those of the policy site. Nevertheless, errors can be expected and a mean absolute percentage error of 50–100 is not uncommon.

Enhancing the use of benefit transfers in this and other ecosystem service applications therefore requires paying closer attention to simple guidelines, developed by economists, for improving validity and accuracy and using meta-analytical methods where possible (Plummer, 2009).

16.5.8 Flow Valuation

Some work has been developed assessing the values of the biogeochemical flows related to water, carbon and nitrogen, which affect ecosystem services (Watanabe and Ortega, 2011), and some attempts have tried to map (Troy and Wilson, 2006) and even quantify the flows' quantity that is supplied each year (Maes et al., 2011). In spite of this, however, a great deal remains to be learned about the nature and value of the flows of ESS since they are connected to a range of final ecosystem services including climate regulation, hydrological regulation, food production, soil formation and others.

This is indicative of a more general problem, that is, the link between the scientific data on the functioning of the ecosystem and the flow of services as defined in Table 16.3 that are of interest to societies. Since the different policies and measures impact on the different biogeochemical flows, it is imperative that the links between these and the services of interest are well established. To be sure there will be uncertainties, which need to be quantified, but the present state of knowledge has many gaps that need to be filled.

16.5.9 'Scaling Up' ESV

As primary valuation research is time and money intensive and results are limited, there is a growing policy and academic interest in transferring ecosystem service values from existing valuation studies to other ecosystem sites (so-called policy sites) on a large geographic scale (Brander et al., 2012). Value (or sometimes called benefit) transfer is the procedure of estimating the value of an ecosystem of current policy interest (Navrud and Ready, 2007) by assigning a valuation estimate for a similar ecosystem at a study site (Brander et al., 2012). The concept of 'scaling up' is used to describe the transfer values that have been estimated for specific changes in a certain ecosystem to assess changes in multiple ecosystems over a larger area. However, despite this evident demand for this combined transfer and 'scaling up' of values, an approach to value transfer that addresses the challenges inherent in assessing ecosystem changes at a national or regional level is not yet available (ibid.) and, still, the application of any value transfer method may result in significant transfer errors.

Amongst the main problems that we encounter when scaling up are the following: (1) the lack of information (often scarce, fragmented, incomplete and of varying quality); (2) the aggregation amongst ecosystem services (as non-marginal changes in ecosystem service provision will affect the value of services from the remaining stock of ecosystems); and (3) the non-constancy of marginal values and critical thresholds (as valuing changes in the provision of ecosystem services) (EEA, 2010).

Notwithstanding these limitations we would argue that scaling up from a reasonable database of primary studies is useful, as long as the error bounds are recognized. Over time, as that database improves, we can expect the quality of the scaling up to improve as well but we cannot wait for enough primary studies to be in place to satisfy all sceptics (and we doubt if they will ever be satisfied) before we start making scaling up estimates.

16.6 FUTURE DIRECTIONS

Based on the above, and following Bao et al. (2007), we would argue that there are three major areas for future work on the valuation of ecosystem services: (1) improving accuracy of valuation; (2) association of results of valuation with policy and management decisions; and (3) transfer of payments for ecosystem services.

16.6.1 Improving the Accuracy of Valuation

We have noted that estimates of ESV are uncertain and subject to error. Improving the valuation of ESS will come from three sources. First are improved methods of benefit transfer or, ideally, the conduct of primary studies for individual ecosystems rather than the use of benefit transfer. Second is research on the links between the technical parameters of an ecosystem and its anthropogenic values. This requires taking account of the spatial scale of the services in a more comprehensive fashion than has been the case hitherto. Third is improvement in the methods themselves – the list of tools described in Table 16.3.

Of these the most important is arguably the question of benefit transfer and/or primary studies. When undertaking benefit transfer, researchers rely on databases of existing studies and various such databases of ESV studies exist. The Environmental Valuation Reference Inventory (EVRI), for instance, is a Canadian-run comprehensive storehouse of over 2000 international studies, providing values, methodologies, techniques and theories on environmental valuation. Now there are a number of others as well. The Environmental & Recreational (Non-Market) Values portal, from the National Ocean Economics Program, contains peer-reviewed and technical reports on the non-market value of a broad range of coastal and marine resources, while portals exist that are even specific to certain countries such as the New Zealand Non-Market Valuation Database, which enables easy identification of non-market valuation studies that have been undertaken in that country. However, no unique database portal exists where all valuation studies could be accessed in order to make better benefit transfer valuations analysis of ESS. We therefore suggest that an effort should be made in order to gather all information into a single database from whose insights we would all benefit.

The second way to improve the valuation of ecosystem services is through a better understanding of the linkages between the technical and economic parameters of the systems. This can be significantly improved through better use of spatial data. Indeed, digital spatial data has dramatically increased in quality and availability recently, particularly at the state level, which has great benefits for the mapping of ecosystem serv-

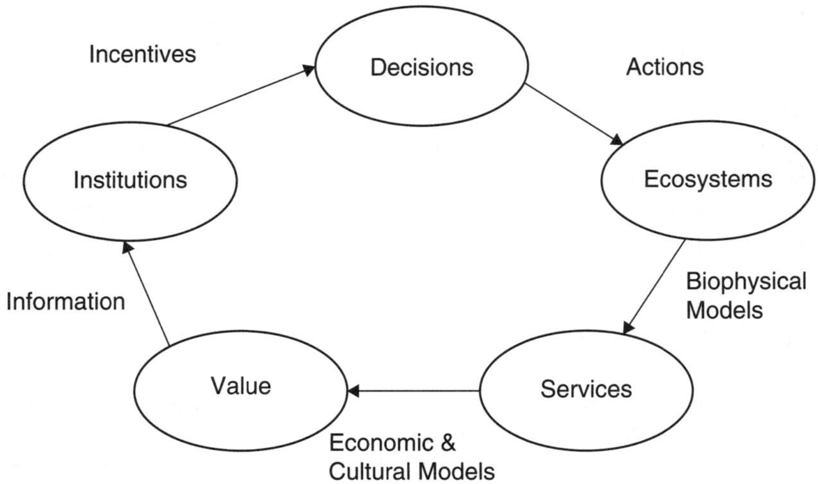

Source: Daily et al. (2009).

Figure 16.6 A framework showing how ecosystem services can be integrated into decision-making

ices. The mapping of ESS and their linking to service flow estimates allows us to identify critical areas in the delivery of ecosystem services. However, currently there are limited opportunities to access these spatial mappings, so we also propose the creation of a unique GIS ESV server where all this spatial information would be gathered. The lack of sufficient contextual variation in valuation studies makes it hard to transfer values from an ESS case study to another, but as more valuation studies are conducted across a range of socio-economic, demographic and regulatory conditions, such data will prove to be highly useful, making ESV exercises much more usable and spatially specific.

The third route to improving valuation lies in research on the methods themselves. While this is important and can yield policy-relevant results, it is, in our view, the least important of the three. Certainly there is no dearth of talented researchers working on these problems and one can be confident that the techniques will only get more accurate over time.

16.6.2 Enhancing Associations Between Valuation and Management

Valuation work often seeks to inform institutions about the best decision-making pathways when dealing with certain environmental issues. However, as we stated in Section 16.4, there is still a lack of application of ESV in management decisions. The links between the two are shown in Figure 16.6. By understanding and valuing natural capital and ESS we are able to make better decisions, resulting in better actions towards the ecosystems, which provide us with certain services. These ESS values are then measured, through economic and cultural models, and these values provide useful information that can help design the institutions that will guide resource management and policy (Daily et al., 2009).

We have already stressed one important way in which valuation can be made more useful and that is through a more comprehensive estimation of the gainers and losers; of where the benefits are going from the ESS and who would be the beneficiary of any action that would enhance or damage these services. Methods of valuation are available that take account of equity issues. These should be used more in ESV work than has been the case hitherto.

A second way of enhancing associations between valuation and management could include the building of 'bundles' around a particular topic of concern. For instance, a 'water security' bundle would consist of those services that provide clean water, including the availability of fresh water as well as regulating services protecting against floods or supplying purified water (Maes et al., 2011). The UNEP-WCMC (2011) technical report notes that bundling services in such a way could clarify the messages different services provide and hence enhance communication. Some possible bundles modified from Maes et al. (2011) are shown in Table 16.4.

16.6.3 Payments for Ecosystem Services (PES)

PES is an important instrument for promoting a more efficient use of the environment, so that beneficiaries pay for the services they receive, and those who have to take measures to ensure that the ecosystem functions well are compensated for the costs incurred. Unfortunately, not all PES schemes respect this arrangement and often the government is the one making the payment rather than the beneficiary. Nevertheless, there are some areas where the arrangements do work to ensure a greater degree of protection of ESS that provide key benefits to identifiable social groups (Barbier and Markandya, 2013).

ESV has a key role in ensuring that such schemes do in fact work for the broader social benefit. First, by identifying and measuring the ESS in money terms it informs the beneficiaries of what they gain from the ecosystem. Second, it can play a key role in ensuring that the providers of these services are correctly compensated. In order to do this it also has to estimate the cost side of the equation: what it costs to preserve the ESS in their present state or what it would cost to improve them.

If ESV is to fill this role, the valuations made have to be credible and understood by all parties. This means that a programme of information and confidence building is required. There are some success stories in this regard (e.g., investment in sustainable watershed management, avoiding expensive infrastructure investment while also enhancing local biodiversity and other local services). But many PES schemes do not work as effectively and there is a greater role that ESV can play to strengthen them.

16.7 CONCLUSIONS

Valuation of ESS in monetary terms is not to everyone's liking; some ecologists consider that it misses important contributions of ecosystems and others consider it downright misleading. In this chapter we have reviewed the work on such valuation and our assessment is more positive. ESV has contributed to showing that ecosystems are worth conserving and that doing so has a benefit to the economy rather than a

Table 16.4 Ecosystem service bundles contributing to the following benefits: timber production, food security, water security and health and well-being

Bundle Name	Capacity Indicators	Units	Flow Indicators	Units
Timber provision & production	Stock	Timber stock (m^3)	Stock potential	Timber increment (m^3 $year^{-1}$)
Food security	Crop	Share of cropland (%)	Crop production statistics	
	Livestock	Livestock density (no. km^{-2})	Livestock production statistics	
	Erosion control	Weighed area of protective ecosystems (ha ha^{-1})	Share in the yield of crop productions resulting from erosion	
	Pollination potential index		Share in the yield of crop production resulting from pollination	
	Soil quality and fertility		Share in the yield of crop productions resulting from soil quality	
	Pest control		Share in the yield of crop production resulting from pest control	
	Genetic diversity		Share in the yield of crop production resulting from genetic diversity	
Water security	Water provision	Share of wetland and water bodies (%)	Water flow available from aquatic ecosystems	Surface water flow (m^3 $year^{-1}$)
	Water regulation	Infiltration (mm)	Water flow regulated by terrestrial ecosystems	Sub-surface water flow (m^3 $year^{-1}$)
	Water purification	Nitrogen retention (%)	Removal of pollutants	In-stream nitrogen removal (ton $year^{-1}$)
Human health & well-being	Climate regulation	Carbon stock (ton C)	Climate regulation	Carbon sequestration (ton C $year^{-1}$)
	Air quality regulation	Dry deposition velocity m $year^{-1}$)		Atmospheric cleansing by vegetation removal of NO_x (ton $year^{-1}$)

Table 16.4 (continued)

Bundle Name	Capacity Indicators	Units	Flow Indicators	Units
	Coastal protection against storms	Share of coastal habitats (%)		Measures of the contribution of ecosystems in protection against natural disasters
	Recreation potential index			Visitor statistics measuring the flow of recreational and cultural services
	Other protection against other natural disasters (landslides, avalanches, floods etc.)			

Source: Modified from Maes et al. (2011).

cost. That is an important contribution. There are, however, still a number of areas where the estimates are highly uncertain and others where it is not possible to provide credible values. We have noted the main shortcomings of the current toolkit for ESV and suggested ways in which it may be strengthened. The major efforts, in our view, lie in better linking the economic valuation to the technical properties of the biogeochemical functioning of ESS and paying greater attention to spatial variability. We would also recommend that constructing of better and more coherent databases of the results of different studies will make the work in this area more accurate. Finally, we need to link the valuation work more closely to the assessment of who gains and loses and to develop protocols for showing combinations of monetary and non-monetary values of ESS. In this way the different valuation exercises will be of greater use to policy-makers.

NOTE

1. Some of the valuation work has shown that some market-based approaches such as hedonic pricing may overestimate the true value of an environmental service (Pearce and Markandya, 1989) but this is in a rather limited context and, in general, the evidence for an upward or downward bias in one method is limited. Indeed, more recent work on ESS from wetlands has found that market-based hedonic methods come up with statistically significant lower values (Brander et al., 2012).

REFERENCES

Adamowicz, W.L. (2004), 'What's it worth? An examination of historical trends and future directions in environmental valuation', *The Australian Journal of Agricultural and Resource Economics*, **48**(3), 419–43.

Angulo-Valdés, J.A. and B.G. Hatcher (2009), 'A new typology of benefits derived from marine protected areas', *Marine Policy*, **34**(3), 635–44.

Ansink, E., L. Hein and P. Hasund (2008), 'To value functions or services? An analysis of ecosystem valuation approaches', *Environmental Values*, **17**(4), 489–503.

Atkins, J.P., D. Burdon, M. Elliott and A.J. Gregory (2011), 'Management of the marine environment: integrating ecosystem services and societal benefits with the DPSIR framework in a systems approach', *Marine Pollution Bulletin*, **62**(2), 215–26.

Bao, Y., W. Wu, M. Wang and W. Liu (2007), 'Disadvantages and future research directions in valuation of ecosystem services in China', *International Journal of Sustainable Development and World Ecology*, **14**(4), 372–81.

Barbier, E.B. and A. Markandya (2013), *A New Blueprint for a Green Economy*, Abingdon, UK and New York: Routledge.

Boyd, J. and S. Banzhaf (2007), 'What are ecosystem services? The need for standardized environmental accounting units', *Ecological Economics*, **63**(2–3), 616–26.

Brander, L., I. Brauer, H. Gerdes, A. Ghermandi, O. Kuik and S. Navrud et al. (2012), 'Using meta-analysis and GIS for value transfer and scaling up: valuing climate change induced losses of European wetlands', *Environment and Resource Economics*, **52**(3), 395–413.

Brauman, K.A. and G.C. Daily (2008), 'Ecosystem services', *Encyclopaedia of Ecology*, Oxford: Academic Press, 1148–54.

Carpenter, S.R., R. DeFries, T. Dietz, H.A. Mooney, S. Polasky, W.V. Reid and R.J. Scholes (2006), 'Millennium Ecosystem Assessment: research needs', *Science*, **314**(5797), 257–8.

Carpenter, S.R., H.A. Mooney, J. Agard, D. Capistrano, R.S. DeFries and S. Díaz et al. (2009), 'Science for managing ecosystem services: beyond the Millennium Ecosystem Assessment', *PNAS*, **106**(5), 1305–12.

Chee, Y.E. (2004), 'An ecological perspective on the valuation of ecosystem services', *Biological Conservation*, **120**(4), 549–65.

Costanza, R. and H.E. Daly (1992), 'Natural capital and sustainable development', *Conservation Biology*, **6**(1) 37–46.

Costanza, R., R. d'Arge, R. de Groot, S. Farber, M. Grasso and B. Hannon et al. (1997), 'The value of the world's ecosystem services and natural capital', *Nature*, **387**, 253–60.

Daily, G.C. (ed.) (1997), *Nature's Services. Societal Dependence on Natural Ecosystems*, Washington, DC: Island Press.

Daily, G.C., T. Soderqvist, K. Aniyar, P. Arrow, P.R. Dasgupta and P.R. Ehrlich (2000), 'The value of nature and the nature of value', *Science*, **289**(5478), 395–6.

Daily, G.C., S. Polasky, J. Goldstein, P.M. Kareiva, H.A. Mooney and L. Pejchar et al. (2009), 'Ecosystem services in decision making: time to deliver', *Frontiers in Ecology and the Environment*, **7**(1), 21–8.

de Groot, R.S., M.A. Wilson and R.M.J. Boumans (2002), 'A typology for the classification, description and valuation of ecosystem functions, goods and services', *Ecological Economics*, **41**(3), 393–408.

EEA (2010), *Scaling Up Ecosystem Benefits: A Contribution to The Economics of Ecosystems and Biodiversity (TEEB) Study*, Copenhagen: European Environment Agency, p. 44.

Ehrlich, P. and A. Ehrlich (1981), *Extinction: The Causes and Consequences of the Disappearances of Species*, New York: Random House.

Environmental & Recreational (Non-Market) Values – National Ocean Economics Program, accessed 10 February 2014 at: http://oceaneconomics.org/nonmarket/NMsearch2.asp.

EVRI – Environmental Valuation Reference Inventory, accessed 10 February 2014 at https://www.evri.ca/Global/Splash.aspx.

Fisher, B., R.K. Turner and P. Morling (2009), 'Defining and classifying ecosystem services for decision making', *Ecological Economics*, **68**(3), 643–53.

Fu, B.-J., C.H. Su, Y.-P. Wei, I.R. Willett, Y.-H. Lü and G.-H. Liu (2011), 'Double counting in ecosystem services valuation: causes and countermeasures', *Ecological Research*, **26**(1), 1–14.

Gomez-Baggethun, E. and M. Ruiz-Perez (2011), 'Economic valuation and the commodification of ecosystem services', *Progress in Physical Geography*, **35**(5), 613–28.

Hein, L., K. van Koppen, R.S. de Groot and E.C. van Ierland (2006), 'Spatial scales, stakeholders and the valuation of ecosystem services', *Ecological Economics*, **57**(2), 209–28.

IRRI (International Rice Research Institute) (2012), accessed June 2012 at http://ricestat.irri.org/vis/wrsMap_main.html.

ISI Web of Science (2012), accessed June 2012 at http://www.accesowok.fecyt.es/.

Jannson, A. and B. Jansson (1994), 'Ecosystem properties as a basis for sustainability', in A. Jansson et al. (eds), *Investing in Natural Capital*, Washington, DC: Island Press.

Konarska, K.M., P.C. Sutton and M. Castellon (2002), 'Evaluating scale dependence of ecosystem service valuation: a comparison of NOAA-AVHRR and Landsat TM datasets', *Ecological Economics*, **41**(3), 491–507.

Kumar, M. and P. Kumar (2008), 'Valuation of the ecosystem services: a psycho-cultural perspective', *Ecological Economics*, **64**(4), 808–19.

Lazo, J.K. (2002), 'Economic valuation of ecosystem services: discussion and application', *Drug and Chemical Toxicology*, **25**(4), 349–74.

Liu, S., R. Costanza, S. Farber and A. Troy (2010), 'Valuing ecosystem services: theory, practice, and the need for a transdisciplinary synthesis', *Annals of the New York Academy of Sciences*, **1185**, 54–78.

Luisetti, T., R.K. Turner, I.J. Bateman, S. Morse-Jones, C. Adams and L. Fonseca (2011), 'Coastal and marine ecosystem services valuation for policy and management: managed realignment case studies in England', *Ocean & Coastal Management*, **54**(3), 212–24.

Maes, J., M.L. Paracchini and G. Zulian (2011), *A European Assessment of the Provision of Ecosystem Services – Towards an Atlas of Ecosystem Services*, JRC Scientific and Technical Reports.

MA (2005), *Ecosystems and Human Well-Being*, Washington, DC: Island Press.

Navrud, S. and R. Ready (2007), *Environmental Value Transfer: Issues and Methods*, Berlin: Springer.

Norton, B.G. and D. Noonan (2007), 'Ecology and valuation: big changes needed', *Ecological Economics*, **63**(4), 664–75.

Pearce, D.W. and A. Markandya (1989), *The Benefits of Environmental Policy*, Paris: OECD.

Plummer, M.L. (2009), 'Assessing benefit transfer for the valuation of ecosystem services', *Frontiers in Ecology and the Environment*, **7**(1), 38–45.

Sachs, J.D. and W.V. Reid (2006), 'Environment – investments toward sustainable development', *Science*, **312**(5776), 1002.

Silva, P. and S. Pagiola (2003), 'A review of the environmental costs and benefits in world bank projects', World Bank Environmental Economic Series, Paper No. 94, Washington, DC: World Bank.

Smith, V.K. (2000), 'JEEM and non-market valuation: 1974–98', *Journal of Environmental Economics and Management*, **39**(2), 351–74.

TEEB (2008), *The Economics of Ecosystems & Biodiversity. An Interim Report*, Brussels: European Communities.

TEEB (2010), *The Economics of Ecosystems and Biodiversity: Ecological and Economic Foundations*, edited by Pushpam Kumar, London: Earthscan.

The New Zealand Non-Market Valuation Database, accessed 14 February 2013 at http://www2.lincoln.ac.nz/nonmarketvaluation/.

Troy, A. and M.A. Wilson (2006), 'Mapping ecosystem services: practical challenges and opportunities in linking GIS and value transfer', *Ecological Economics*, **60**(2), 435–49.

Turner, R.K., J. Paavola, P. Cooper, S. Farber, V. Jessamy and S. Georgious (2003), 'Valuing nature: lessons learned and future research directions', *Ecological Economics*, **46**(1), 493–510.

UNEP-WCMC (2011), *Developing Ecosystem Service Indicators: Experiences and Lessons Learned from Sub-- global Assessments and Other Initiatives*, Technical Series No. 58.

Watanabe, M.D.B. and E. Ortega (2011), 'Ecosystem services and biogeochemical cycles on a global scale: valuation of water, carbon and nitrogen processes', *Environmental Science & Policy*, **14**(6), 594–604.

Westman, W.E. (1977), 'How much are nature's services worth?' *Science*, **197**(4307), 960–64.

17. Optimal species preservation policy in a symbiotic relationship between species

Shiri Zemah-Shamir, Benyamin Shitovitz and Mordechai Shechter

17.1 INTRODUCTION

Biodiversity is defined as the variety within the living world, and describes the relationships between the species, within ecosystems and beyond. The relationships between the species are often very important to the construction of the habitat's biodiversity. According to the predominant opinion in both economics and ecology, damage to a certain species in the habitat could negatively affect the harmony of the habitat and of the species that depend on it. As Berlow et al. (1999) state, 'understanding how the strengths of species interactions are distributed among species is critical for developing predictive models of natural food webs as well as for developing management and conservation strategies' (p. 2206).

Financial resources for biodiversity conservation programs are not sufficient to protect all habitats and species. This situation requires choosing conservation priorities in order to support the most species at the least cost. Moran et al. (1996, 1997) note that no single correct method exists for establishing biodiversity conservation priorities at any level of organization. Nevertheless, the question of how to determine priorities for maintaining or increasing biodiversity under a limited budget constraint has concerned environmental economists since the seminal work by Weitzman (1998). He developed 'a more-or-less consistent conceptual framework and a more-or-less usable measure on the value of diversity that can tell us how to trade off one form of diversity against the other' (Weitzman, 1995, p. 21). In addressing the Noah's ark problem – whereby Noah had to decide which species he should take aboard the ark to survive, and which were to die out – Weitzman applied diversity theory (1992) to provide quantitative indicators of which species to preserve. Furthermore, in his seminal work on the Noah's ark problem (1998), Weitzman assumed that the survival probabilities are independent and the costs function is linear. One of Weitzman's conclusions is that the optimal policy is an extreme policy. Therefore, in the Noah's ark model, almost all species go aboard either in full or not at all.

When central planners aim to maximize the social welfare that is obtained from biodiversity, they must take into account the interactions between the species, since one species often cannot exist without the other. The extinction of a species due to the absence of the species it depends on damages species' diversity directly and human society's benefit indirectly. Indeed, decisions made concerning the preservation of certain species that ignore their effects on other species might harm the stability of the ecosystem and its recovery capacity. Such action would also cause a decline in the benefits to human society. For example, cutting down pine trees would harm the pine mushrooms (that live in commensalism with pine trees), which would in turn cause the habitat to change its species composition and thus damage other values such as recreation, landscape, soil quality, among others. In numerous other cases as well, when the quantity of one species (that delineates the habitat's nature and landscape) is reduced, the ecosystem is damaged by the invasion of other species.

Therefore, the goal of this chapter is to determine the optimal preservation policy in cases where many species depend on others for their survival, and to find the ranking criteria by which to act. Species preservation requires a budget, which is limited (by human society), and consequently the central planner (for instance, the Israel Parks Authority) must determine a policy under budget constraints. This policy dictates the biodiversity for years to come, as well as the anticipated social benefits from species preservation.

17.2 BIODIVERSITY PRESERVATION

In the modern world, we observe a multitude of development pressures that originate in demography and economic growth, and the consequent damage to the habitat is inevitable. Moreover, intensification of agriculture activities affects biodiversity and creates biodiversity-related conflicts (Henle et al., 2008). Some argue for sustainable development, claiming that each generation must leave 'enough and as good for the next'. Others yet follow the natural inheritance approach, positing that the present generation does not have the right to deplete basic resources from economic and ecological aspects. Since natural resources cannot be passed on intact, that is to say, unused, this approach proposes that future generations be financially compensated for every devaluation of natural resources that was caused by the present generation.

Although in recent years, a deceleration was marked in the rise of polluting emissions, pollutants are still emitted. Indeed, while conservation efforts are increasing, biodiversity continues to decrease (Rands et al., 2010). The destruction of natural habitats and other factors, which are believed to be the causes of species' extinction, can be very expensive to compensate for. Therefore, it is only natural to ask what the value of biodiversity preservation is (Craft and Simpson, 2001).

We assume that an optimal level of environmental protection exists, which changes under various conditions. Hence, we can conclude that there is also an optimal level of biodiversity, linked to different uses of the soil. For example, agricultural land requires a different level of environmental protection than tropical forests do, and as a result, the optimal level of biodiversity required for agriculture is different from the optimal level required for the preservation and continuity of tropical forests.

The need for environmental protection does not necessarily mean that we must preserve all species in all places (which would indeed be impossible); rather, it is essential to preserve the species that are used as products or services that people depend on (Perrings, 1999). Decisions concerning protected species reflect species' value, perception and the uncertainty the society that makes these decisions derives from them. Choosing areas for preservation or maintenance based on economic criteria is rare in the literature. The uncertainty and the changes to the value of biodiversity require dynamic evaluation, which can facilitate the relearning of long-term preservation decisions.

Assessing the value of biodiversity provides the quantitative basis for making decisions about biodiversity preservation. As noted above, biodiversity faces many global and local threats. Thus, it is important that all sectors (government, business and individuals) take into consideration that biodiversity contributes to our well-being and invest in preservation actions (Rands et al., 2010). Yet under current conditions, the financial resources of various biodiversity programs are not sufficient to protect all of the habitats and species. Indeed, even under optimal collaboration, the human race is incapable of protecting everything. We must therefore decide upon priorities, in order to support and preserve most species at minimal costs. As Moran et al. (1996, 1997) note, allocation of resources to maximize the classic goal of the economic problem is not possible.

17.3 THEORIES TO DETERMINE CRITERIA OF BIODIVERSITY PRESERVATION

Various conservation program tools have developed over the last 20 years, which relate differently to the determination of preservation priorities for genetic diversity, species diversity and ecosystems diversity. In this section we present the criteria from the global level to the genetic level.

Moran et al. (1996, 1997) suggest an index to rate global investments. Any investment in biodiversity is influenced by damage (threat) and sustainability (success). Successful intervention by means of investment would depend on the probability or likelihood of success and the level of threat common in a certain country. This index is applied to determine regional ratings for various diverse areas in general, and for areas rich in endemic species in particular (such as Central and South America and the Caribbean, among others). The index binds a number of socio-economic variables and uses the cost-effectiveness approach. This method provides available information about costs, but effectiveness is defined in non-financial terms. Moran et al.'s (1996) Cost Effectiveness Priority Investment Index (CEPII) is derived by subtracting the result of the rating from information about threat and the probability of successful human intervention. The index is composed of a characteristic divided into threat or success (ρ), multiplied by change to this characteristic in case of limited government intervention or no intervention at all. This value is normalized to chosen values of species abundance and endemism per square kilometer ($\Delta\hat{B}$) and divided by the average cost per unit that derives from the intervention to preserve biodiversity between countries (C). Since this information is unavailable, the closest estimate is used: the international level of investment.

This CEPII specifies the rating of countries that should be invested in, but it provides relative rather than absolute values. In addition, there is a lack of data about local priorities, demands and needs, so that the relative values the formula yields are also lacking. In their 1997 paper, Moran et al. defined $\hat{B}$ as the sum of tall vegetation, mammals, birds, reptiles and amphibians. The authors concentrated on tropical forests (as opposed to the 1996 paper) and expanded the spatial level and location in addition to the habitat they examined. This criterion fits the global discussion and we will not focus on it.

Another proposed criterion involves the genetic level, which Weitzman discusses. Weitzman relies on a biodiversity function he created, which primarily describes the distance and difference between the species. One of his seminal papers, 'On Diversity' (1992), illustrates how logical assessment of the 'diversity value' of a collection of objects can be created from basic information about the difference-distance between any two objects in the group. Weitzman assumes the existence of one big universe that includes a large number of potential species, and suggests a distance between every pair of species that is not negative and symmetrical. Namely, the basic problem is to create appropriate assessment of the diversity of a group (Q), indicated by $V(Q)$, when values of distance-difference are given. Weitzman portrays the group by means of a tree diagram, in which the length of each branch describes a family of existing species that have a historical-evolutionary connection. If two species are close and one of them is extinct, this is not a catastrophe. However, when both species are extinct, the evolutionary branch is compromised, and that might be a catastrophe. He presents various approaches to the discussion on biodiversity – the ecological approach and the economic approach, combined with a number of functions. The economic approach prefers species that are distant from each other – that is to say, if two species in a group are to be preserved, then it is preferable to preserve distant species. Another result Weitzman obtains is that if one species is equal to other species in policy considerations, then its existence or extinction are irrelevant to the decision-making process regarding its preservation or extinction, relative to subgroups of other species. Weitzman also proves that according to the biodiversity function, the species with the biggest distance from other species has the highest value.

Weitzman (1995) defines the 'diversity function' as a function that evaluates the diversity of a collection of given elements. The expectation of the biodiversity function is the sum of the diversity functions of a variety of collections of species weighted by the survival probabilities of all the collections. The function's goal is to maximize the present value of diversity expectation (in addition to the direct values of the elements) subject to the preservation budget constraints. In this paper, for the first time,[1] Weitzman compares a species to a library, noting two advantages to the metaphor: first, a library is emotionally neutral, lacking the emotional value of animals, and second, concentrating on libraries and the books they hold makes it much easier to focus on what specifically is preserved, when speaking of the diversity function. Weitzman proposes the 'bead model', which describes a tree diagram in which the branches represent the evolutionary process over time, and the species are located at the tips of the branches. Each species is composed of a large number (M) of beads bound together. In his metaphor, the species is a library and the beads are books. At the beginning of the process, one species exists. Over time, more species are added that are slightly different from one another (in their genetic set or, in the library model, the book collection expands). All the species in the bead model are composed of equally long strings of beads. The difference or distance between each pair of species can be defined as the number of beads between the species. This is, in fact, the diversity function – a function that measures the distance or difference between species.

The ranking criterion, which Weitzman developed in his paper 'The Noah's Ark Problem' (1998), is based on his former papers. Weitzman tries to present a practical way to think about the economics of biodiversity, mediated by an abstract mathematical model. He compares biodiversity economics to the Noah's ark problem, which should be expressed as an analytical problem and which characterizes biodiversity preservation problems under budget constraint. One of the constraints in the allegory is the space constraint, determined by the budget and Noah's labor. The central question Weitzman asks is how to determine the basic priority of preservation or biodiversity enhancement. Specifically, what is the cost–benefit formula or criterion that can serve to rate priorities between various biodiversity preservation projects? In what follows below, we present an extension to Weitzman's ranking criterion.

17.4 EXTERNALITIES, FUNCTIONS AND ECOSYSTEM PROCESSES

Biodiversity yields a direct and indirect impact on human society, but has an external impact on nature itself: on ecosystems, habitats and other species (Rands et al., 2010). An ecosystem can add innumerable interactions between species, in which each species represents a certain niche and contributes to the resilience and stability of the ecosystem. In this section, we describe the links between the species and the ecosystem, and their interactions with society.

In ecosystems, the 'players' are organisms, which cumulatively form the species. A species can have direct and/or indirect benefit for the human race, as raw materials to produce goods and commodities or to support other species; but species have incomplete markets. Crocker and Tschirhart (1992) describe the stationary equilibrium of an ecosystem, in which the demand of species for other species is equal to the supply of other species, and energy is preserved. The ecosystem, consequently, is integrated into the economy and the impact of human intervention can be monitored. Human intervention creates externalities that affect the ecosystem and shift the system's equilibrium, which in turn influences the benefit and/or human production processes. Ecosystems exhibit complex and dynamic connections, as do economic systems. Economic systems were built successfully, so if similarity exists between the complexity of economic systems and that of ecological systems, a model can be likewise constructed. The similarity is manifested in the demand, supply, competition and marginal substitution rates of ecology.

Crocker and Tschirhart's approach enables incorporation of an ecological model in a simple economic model, for example to compare decision-makers in ecosystems to organisms that unite at the species level. They present a model of the influences of human intervention on the ecosystem, as if only one consumer and one producer existed in the world. The consumer prefers a single end product (bread) and spare time. The end product is produced by labor and an interim product (wheat). The interim product is produced by labor, and the consumer enjoys no direct benefit from it. The ecosystem, which also plays a role in the process, is composed of a predator that eats a mouse that eats wheat kernels. The predators provide the consumer with negative benefit. Therefore, the consumer's utility function is composed of bread, spare time units in which the predators can be killed, and actual spare time. The production function is composed of labor to produce wheat kernels and dependence on predators. The dependence is the impact of the externalities on the ecosystem.

In the next section, we describe the links between the species, and their interactions.

17.5 RELATIONSHIPS BETWEEN SPECIES THROUGH BIODIVERSITY

A connection exists between species' diversity and the ecosystem; the variety of species largely determines the functional characteristics of the ecosystem (Woodward, 1993). In an ecosystem, populations of various microorganisms live side by side, mixed with each other in some ways, and we can therefore expect some relations between the individuals of various species populations[2] (Perevolotsky and Polak, 2001). Species are categorized as structural or interstitial (Huston, 1994), with interrelations existing between the two kinds, in addition to other interrelations. Structural species, such as trees or corals, create the physical structure of the ecosystem (or habitat). Interstitial species live in the environmental system created by the structural species, and are directly influenced by structural species – for instance, microorganisms or fungi. The influence is one way – interstitial species have almost no effect on structural species.

Population pairs and their mutual influences are often used to classify other interactions. The following are six possible results:

0,0 Neutral: both partners to the interaction are indifferent to each other.
0,+ Commensalism:[3] one partner benefits from the presence of the other, and the other neither benefits nor loses.
+,+ Mutuality: both partners in the interaction benefit from each other.
+,− Predation[4] (including herbivorism[5] and parasitism[6]): the fitness of one organism increases at the expense of the other organism.
−,0 Amensalism: disturbed competition – one partner is indifferent in its fitness to the presence of the other, which is damaged.
−,− Competition: the fitness of both partners decreases because of their interaction.

Additional types of relationships have been classified as well. The first involves a dominant host species called a 'keystone species' (Kotliar et al., 1999) – defined as 'a species the removal of which would have a disproportionately big influence on its environment' (Shkedy, 2009, personal communication) – and its dependent. The dependent species is known as 'keystone species dependent' (Berlow et al., 1999). Not every system has keystone species, namely, the presence of a species that improves the habitat and increases the chances of other species to exist, without damaging the survival chances of the (beneficial) facilitating species. Another relationship involves landscape modulator species, which determine the composition of the species in the habitat, for instance the common oak or mastic tree (Shkedy, 2009, personal communication).

17.6 CHARACTERIZING THE EXPECTATION OF THE BIODIVERSITY VALUE FUNCTION

17.6.1 Introduction

Weitzman (1995, 1998) regarded species as if they were books in a library. His focal question was, what the best way to preserve libraries? The parallel question being, what is the best way to preserve specific species, that is, the problem of optimal preservation of biodiversity under budget constraints? The basic unit in the libraries model is the library, and in the abstract, each library unit represents a certain species. An important result was that assuming independent survival probabilities between species and linear costs function, the optimal policy should be extreme (either preserve or die out). The only species that does not support this conclusion has interior survival probability determined by budget equality.

In the symbiotic Noah's ark problem, a central planner allocates a given budget to maximize the expectation of the biodiversity value function. One of the species is the keystone species, and the others are keystone-dependent species; the latter may be beneficiaries and/or predators, and they have a symbiotic relationship with the keystone species and vanishing biodiversity whenever the keystone species becomes extinct. In this case, the optimal policies in different costs function regimes may be interior policies – some of the species may survive. Specifically, generalization of the biodiversity value function yields: $W(\pi_0; \pi_1, \ldots, \pi_K) = \pi_0 M_0 + \pi_0 \pi_1 E_1 + \ldots + \pi_0 \pi_K E_K$, where $M_0, E_1, \ldots, E_K$ are positive constants. For this special case, with a single keystone-dependent species, inserting a linear budget constraint, we obtain a concave function in π_0, which might lead to an interior optimal policy. Moreover, under such an optimal policy, at least half of the budget is invested in the keystone species' preservation. Finally, the optimal preservation policy π_0^* of the keystone species is also the optimal policy of the general case where there are several ($K \geq 1$) similar keystone-dependent species, and is not dependent on K.

17.6.2 The Expectation of the Biodiversity Value Function

We denote by π_0 the survival probability of the keystone species and by π_k for $k = 1, \ldots, K$, the conditional survival probability of the kth keystone dependent species, conditioned on the event that the keystone species survives. Analogously, let A_0 be the set of books in the central library, and let us denote by M_0 the number of books in A_0. For $k = 1, \ldots, K$, A_k denotes the set of books in the kth professional library.
We assume:

(*Assumption 17.1*) For $1 \leq k' \neq k'' \leq K$ $A_{k'} \cap A_{k''} \subset A_0$ (thus, for example, books in the collection of the Physics library, which are not included in A_0, are disjointed from books in the Chemistry collection, which are not included in A_0). We denote the number of books in the kth professional library, which are not included in the Central library, by $|A_k \backslash A_0| = E_k > 0$

(*Assumption 17.2*) The total number of the books in a set S of libraries, where $S \subset \{0,1, \ldots, K\}$ (which is the diversity of books) is:

(a) $$V(S) = \begin{cases} M_0 + \sum\limits_{k \in S \backslash \{0\}} E_k & 0 \in S \\ 0 & \textit{otherwise} \end{cases}$$

(*Assumption 17.3*) The probability of survival of library 0 and the kth library together is $\pi_0 \cdot \pi_k$ (based on the multiplicative formula that: $Prob(A \cap B) = P(A) \cdot P(B|A)$).

Theorem 1
The expectation of the biodiversity value function is satisfied in this model:

(b) $$W(\pi_0; \pi_1, \ldots, \pi_K) = \pi_0 M_0 + \pi_0 \pi_1 E_1 + \ldots + \pi_0 \pi_K E_K.$$

In particular $\frac{\partial W}{\partial \pi_0} > 0$ and $\frac{\partial W}{\partial \pi_k} > 0$ for all $k = 1, \ldots, K$ and $\pi = (\pi_0; \pi_1, \ldots, \pi_K) \in (0, 1]^{K+1}$.

Proof
Denote the *random biodiversity value function* by $V(\cdot) = X_0(\cdot) + \sum_{k=1}^{K} X_k(\cdot)$ where $X_0(\cdot)$ is the random variable $X_0(S) = \begin{cases} M_0 & 0 \in S \\ 0 & \textit{otherwise} \end{cases}$ and $X_k(S) = \begin{cases} E_k & k \in S \\ & 0 \in S \\ 0 & \textit{otherwise} \end{cases}$ for $k = 1, \ldots, K$. Since $EX_0(\cdot) = M_0 \pi_0$ and $EX_k(\cdot) = E_k \pi_0 \pi_k$, we obtain:

$$W(\pi_0; \pi_1, \ldots, \pi_K) = EX_0(\cdot) + EX_1(\cdot) + \cdots + EX_K(\cdot) = \pi_0 M_0 + \pi_0 \pi_1 E_1 + \cdots + \pi_0 \pi_K E_K.$$

We get from this formula for the *biodiversity value function* that by differentiation, the partial derivatives of W are strictly positive in $(0, 1]^{K+1}$.

Corollary 1
Assume $E_1 = \ldots = E_K$, then $W(\pi_0; \pi_1, \ldots, \pi_K) = \pi_0 M_0 + \pi_0 E_1 (\pi_1 + \cdots + \pi_K)$.

17.7 THE CENTRAL PLANNER PROBLEM

The central planner maximizes the expectation of the biodiversity value function under budget constraint $B > 0$. In our model, the range of π is I^{K+1}, where $\pi = (\pi_0; \pi_1, \ldots, \pi_K)$ and $I = [0, 1]$ is the unit interval. The central planner solves the problem of:

$$\text{MAX } W(\pi_0; \pi_1, \ldots, \pi_K) = \pi_0 M_0 + \pi_0 \pi_1 E_1 + \cdots + \pi_0 \pi_K E_K$$
$$\text{s.t.} \quad C(\pi_0; \pi_1, \ldots, \pi_K) \leq B$$
$$\pi \in I^{K+1}$$

Proposition 1
Assume that $C(\pi_0; \pi_1, \ldots, \pi_K)$ is defined on I^{K+1} and is a continuous and monotonic increasing function, and that $C(0; 0, \ldots, 0) = 0$ and $C(1; 1, \ldots, 1) \geq B$. Then the central planner problem has a non-empty set of maximizers and each optimal policy satisfies the budget equality.

Proof
The set $\{(\pi_0; \pi_1, \ldots, \pi_K) \in I^{K+1}: C(\pi) \leq B\}$ is non-empty, closed and bounded set in I^{K+1}. By the continuity of $W(\pi_0; \pi_1, \ldots, \pi_K)$, the result follows. Next, to prove the budget equality, we assume that $C(1; 1, \ldots, 1) \geq B$ or in the linear case: $C_0 + KC_1 \geq B$ then there is budget equality for $\pi \in I^{K+1}$.

17.8 THE EXISTENCE OF AN OPTIMAL POLICY

Theorem 2
In our symbiotic model with keystone species and K keystone-dependent species, under Assumption (17.1) and if the budget $B > 0$ and the costs function $C(\pi_0; \pi_1, \ldots, \pi_K)$ is defined on I^{K+1} and is a continuous and monotonic increasing function, and $C(0; 0, \ldots, 0) = 0$ and $C(1; 1, \ldots, 1) \geq B$, then $\hat{\pi}_0 > 0$.

Proof
As we mentioned in Proposition 1, the continuity of $W(\pi_0; \pi_1, \ldots, \pi_K)$, creates an optimal policy, because continuous function over non-empty closed and bounded set has maximum and minimum. Since the biodiversity value function is increasing over $R_+^{K+1}$and $C(1; 1, \ldots, 1) \geq B$, there is a budget constraint. Particularly, there is budget equality for any optimal policy. Obviously, if $B> 0$ then $\hat{\pi}_0 > 0$ for any optimal policy $\hat{\pi}$. The assumption about increasing monotony is derived from the partial derivation of the biodiversity value function (i.e., the marginal utilities):

(a) $$\frac{\partial W}{\partial \pi_0} = M_0 + \pi_1 E_1 + \cdots + \pi_K E_K > 0$$

(b) $$\frac{\partial W}{\partial \pi_k} = \pi_0 E_k > 0 \qquad k = 1, \ldots, K$$

17.9 THE INTERIORITY AND UNIQUENESS OF THE OPTIMAL POLICY

Assume now that $C(\pi_0; \pi_1, \ldots, \pi_K)$ is a separable costs function $C(\pi_0; \pi_1, \ldots, \pi_K) = C_0(\pi_0) + C_1(\pi_1) + \cdots + C_K(\pi_K)$ and assume that:

(*Assumption 17.4*) $$C_0(\pi_0): [\underline{\pi_0}, \overline{\pi_0}] \rightarrow \Re_+$$

(*Assumption 17.5*) $$C_0(\underline{\pi_0}) = 0$$

(*Assumption 17.6*) $$C_0' > 0,\ C_0'' \geq 0$$

(*Assumption 17.7*) $$C_k(\pi_k): [\underline{\pi_k}, \overline{\pi_k}] \rightarrow \Re_+\ \forall k = 1, \ldots, K$$

(*Assumption 17.8*) $$C_k(\underline{\pi_k}) = 0$$

(*Assumption 17.9*) $$C_k' > 0,\ C_k'' \geq 0$$

(*Assumption 17.10*) $$C(\overline{\pi}) = C_0(\overline{\pi_0}) + \sum_{k=1}^{K} C_k(\overline{\pi_k}) > B > 0$$

(*Assumption 17.11*) $$C_0(\overline{\pi_0}) \leq B$$

Theorem 3
Under the above assumptions and a continuous convex costs function $C(\pi)$ and positive budget $B> 0$, if $\hat{\pi}$ is an optimal policy in the symbiotic model, then $\hat{\pi}_0 = \tilde{\pi}_0$. That is, the survival probability of the keystone species is unique. In particular, if

$C_0(\pi_0)$ is convex in π_0 and $C(0; \pi_1, \ldots, \pi_K)$ strictly convex in $(\pi_1, \ldots, \pi_K)$, that is, $C(\pi) = C_0(\pi_0) + C_1(\pi_1) + \ldots + C_K(\pi_K)$, a unique optimal policy exists: $\hat{\pi} = \tilde{\pi}$.

The convex cost function implies that the marginal cost is increasing or constant; in this case, its implication is that the expectation of the biodiversity value function will be concave.

Proof
See Zemah Shamir and Shitovitz (2014).

Now, we extend our assumption to be:

(*Assumption 17.12*) $$C_0'(\overline{\pi_0}) = \infty$$

This assumption seeks to describe the cases in which the marginal costs of total prevention or preservation costs are extremely high, even to the extent of infinity (for example: the case of full fire prevention).

Theorem 4
Under Assumptions (17.4)–(17.12), for any optimal policy $(\hat{\pi}_1, \ldots, \hat{\pi}_K)$, the unique survival probability of the keystone species $\hat{\pi}_0$ is interior, that is, $0 < \hat{\pi}_0 < \overline{\pi}_0$ and $C_0(\hat{\pi}_0) < B$.

The optimal survival probability is interior, in contrast to Weitzman's extreme optimal policy, and this implies that the money to be spent on the keystone species won't exceed the budget.

Proof
Since $C(\hat{\pi}) = B < C(\overline{\pi})$, assume by negation that $\hat{\pi}_0 = \overline{\pi}_0$, therefore there is a species $1 \leq k \leq K$ in which $\hat{\pi}_k < \overline{\pi}_k$. In particular, for species 0 and k, since $\hat{\pi}$ is an optimal corner solution at $\hat{\pi}_0 = \overline{\pi_0}$ and $\hat{\pi}_k < \overline{\pi_k}$. We obtain that first-order condition for the corner solution is:

$$\frac{MW_{\pi_0}}{MW_{\pi_k}} = \frac{M_0 + \sum_{k=1}^{K} \pi_k E_k}{\pi_0 E_k} \geq \frac{MC_{\pi_0}}{MC_{\pi_k}} = \frac{C_0'(\overline{\pi_0})}{C_k'(\hat{\pi}_k)} = \frac{\infty}{C_k'(\hat{\pi}_k)} = \infty$$

A contradiction.

17.10 THE SPECIES PRESERVATION RANKING CRITERION OF NOAH'S ARK

Weitzman (1998) set a ranking criterion based on the distance between the species and their expected utilities weighted with their survival probabilities. In our symbiotic model, we derive a ranking criterion based on the marginal contribution of the species to its marginal preservation costs.

17.10.1 The General Case

Applying Theorem 4, the central planner problem is now:

$$\text{MAX} \quad M_0 + (\pi_1 E_1 + \ldots + \pi_K E_K)$$

$$\text{s.t.} \quad \underline{\pi_1} \leq \pi_1 \leq \overline{\pi_1}$$

$$\underline{\pi_K} \leq \pi_K \leq \overline{\pi_K}$$

$$C_1(\pi_1) + \ldots + C_K(\pi_K) \leq B - C_0(\hat{\pi}_0) \equiv \tilde{B}$$

Note that based on Theorem 4, $\tilde{B} = B - C_0(\hat{\pi}_0) > 0$.

The central planner problem equivalent (since M_0 is constant) to this maximization problem:

$$\text{MAX} \quad \pi_1 E_1 + \ldots + \pi_K E_K$$

$$\text{s.t.} \quad \tilde{B} - C_1(\pi_1) - \ldots - C_K(\pi_K) \geq 0$$

$$\underline{\pi_1} \leq \pi_1 \leq \overline{\pi_1}$$

$$\underline{\pi_K} \leq \pi_K \leq \overline{\pi_K}$$

Using Kuhn-Tucker's theorem and its conditions, we obtain a parameter $\alpha^* > 0$, such that for any $k = 1, \ldots, K$:

If $\dfrac{E_k}{C_k'(\hat{\pi}_k)} < \alpha^*$, then $\hat{\pi}_k = \underline{\pi_k}$, that is, the species dies out.

If $\dfrac{E_k}{C_k'(\hat{\pi}_k)} > \alpha^*$, then $\hat{\pi}_k = \overline{\pi_k}$, that is, the species survives.

For $\dfrac{E_k}{C_k'(\hat{\pi}_k)} = \alpha^*$, no specific conclusion.

The α^* parameter is the shadow value (price) of the species' optimal policy.

17.10.2 The Linear Case

Theorem 5

Assume that $\underline{\pi_1} = \underline{\pi_2} = \ldots = \underline{\pi_K} = 0$, $\overline{\pi_1} = \overline{\pi_2} = \ldots = \overline{\pi_K} = 1$ and the marginal costs function for the keystone species is $C_0'(\overline{\pi_0}) = \infty$, while the rest of the costs function is the form of $C_k(\pi_k) = C_k \cdot \pi_k$, then for any optimal policy $\hat{\pi}$, there is $\alpha^* > 0$, such that for any $k = 1, \ldots, K$:

If $\dfrac{E_k}{C_k} < \alpha^*$, then $\hat{\pi}_k = 0$, that is, the species dies out.

If $\frac{E_k}{C_k} > \alpha^*$, then $\hat{\pi}_k = 1$, that is, the species survives fully.

17.11 THE GENERAL CASE OF $K \geq 2$ KEYSTONE-DEPENDENT SPECIES

In this section, we introduce the central planner problem in diverse scenarios. We examine the symmetric case in different costs functions.

17.11.1 The Case with K Identical Keystone-dependent Species – The General Symmetric Case

Assume now that:

(*Assumption 17.13*) $\qquad C(\pi_0; \pi_1, \ldots, \pi_K)$

is a convex and symmetric function in all of its K last coordinates.

Conclusion 1
There is an optimal policy $\hat{\pi} \in D$, where D is a non-empty closed subset of I^{K+1}, with the Equal Treatment Property, which means that for identical species (with symmetric costs and for all $1 \leq k' \neq k'' \leq K$ $E_{k'} = E_{k''}$) the probabilities in this optimal policy satisfies $\hat{\pi}_{k'} = \hat{\pi}_{k''}$.

Remark 1
A simple but not trivial example of symmetric convex cost function at its last K variables is:

$$C(\pi_0; \pi_1, \ldots, \pi_K) = C(\pi_0; 0, \ldots, 0) + \pi_1^2 + \ldots + \pi_K^2.$$

Theorem 6
Assume now that $E_1 = E_2 = \ldots = E_K \equiv E_1$, under Assumptions (17.1) and (17.13) and that $C(\pi_0; \pi_1, \ldots, \pi_K)$is convex and symmetric in $\pi_1, \ldots, \pi_K$, there is an optimal policy $\hat{\pi}$ where $\hat{\pi}_0$ for the keystone species and symmetric optimal policy for the keystone-dependent species $\tilde{\pi}_1 = \frac{\pi_1 + \ldots + \pi_K}{K}$. Particularly, our $(K + 1)$ species' problem is equivalent to a single keystone species and a single keystone dependent species (see Zemah Shamir, 2011).

Proof
For $\hat{\pi} \in D$ derived by Conclusion 1, the expectation of the biodiversity value function is:

$$W(\pi_0, \pi_1, \ldots, \pi_K) = \pi_0 M_0 + \pi_0 \pi_1 E_1 + \ldots + \pi_0 \pi_K E_K =$$

$$= \pi_0 M_0 + \pi_0 \tilde{\pi}_1 E_1 K$$

While the costs function is:

$$C(\hat{\pi}_0, \tilde{\pi}) \equiv C(\hat{\pi}_0; 0, \ldots, 0) + KC(0, \tilde{\pi}_1) \leq C(\hat{\pi}_0, 0, \ldots, 0) + C(0, \pi_1, \ldots, \pi_K).$$

This is derived from the symmetry and convexity, therefore the average of $(\hat{\pi}_1, \ldots, \hat{\pi}_K)$ is a symmetric optimal policy within $C(0, \pi_1, \ldots, \pi_K) \geq KC(0, \tilde{\pi}_1)$.

17.11.2 The Case with *K* Identical Keystone-dependent Species – The Linear Symmetric Case

In this subsection, we applied Weitzman's linear costs function in order to observe the differences between both models.

Theorem 7

Assume now that $E_1 = E_2 = \ldots = E_K \equiv E_1$, under Assumptions (17.1) and (17.13), and that the assumption $C(\pi_0; \pi_1, \ldots, \pi_K)$ is linear and symmetric in $\pi_1, \ldots, \pi_K$, that is, $C_1 = C_2 = \ldots = C_K \equiv C_1$. In this case, there is an optimal policy $\hat{\pi}$ where $\hat{\pi}_0$ is for the keystone species and symmetric optimal policy for the keystone-dependent species. In particular, this $(K + 1)$ species' problem is equivalent to a single keystone species and a single keystone dependent species with linear costs function (see Zemah Shamir, 2011).

The central planner problem in this case is:

$$\text{MAX} \qquad \pi_0(M_0 + \sum \pi_k E_1)$$

$$\text{s.t.} \qquad \pi_0 C_0 + \pi_1 C_1 + \ldots + \pi_K C_1$$

This central planner problem is equivalent to the two-species model, where $\tilde{\pi}_1 = \frac{\pi_1 + \ldots + \pi_K}{K} \in [0, 1]$ and $\tilde{C}(\pi_0, \tilde{\pi}_1) = C(\pi_0; \frac{\tilde{\pi}_1}{K}, \ldots, \frac{\tilde{\pi}_1}{K})$.

By solving the maximization problem above, and assuming linear cost function, an interior optimal policy might occur. In this case,

$$\text{MAX} \qquad \pi_0 M_0 + \pi_0(\underbrace{KE_1}_{\tilde{E}_1}\tilde{\pi}_1)$$

$$\text{s.t.} \qquad \pi_0 C_0 + \tilde{\pi}_1 \underbrace{KC_1}_{\bar{C}_1} 1 = B$$

in this equivalent problem in which $(\hat{\pi}_0, \hat{\tilde{\pi}}_1)$ is an interior optimal policy and $\hat{\pi}_0 C_0 > \frac{B}{2}$. That is, more than 50 percent of the budget is spent on the keystone species and the rest of the budget is spent on the other species, that is, the keystone-dependent species divided by the number of identical species.

17.12 DISCUSSION AND CONCLUSION

In his paper 'The Noah's Ark's Problem' (1998), Weitzman presented the classic problem of species preservation under a budget constraint. Many studies were conducted in his footsteps (Weikard, 2002; van der Heide et al., 2005, and others) that expanded the problem and introduced criticism of it. A major criticism was the lack of reference to the relationships between the species. Although this topic has been addressed on the level of ranking criteria (van der Heide et al., 2005) and probabilities (Baumgartner, 2002), the literature has not dealt with the wider issue of solving the central planner's problem – specifically, solving the Noah's ark problem when there is a keystone species and the species boarding the ark depend on the keystone species' survival. In addition, various studies failed to address the linear cost function presented by Weitzman.

In this chapter, we expanded the Noah's ark problem to an uncertain environment, which includes a keystone species, the survival of which other species are dependent on. We used a variety of cost functions to obtain comprehensive results for various applications. We examined a linear cost function and an increasing convex monotonous function, that is, where the marginal cost tends to infinity for the upper bound.

Our obtained results suggest an optimal policy different from the one Weitzman presented (1998). In the case of $K > 1$ keystone-dependent species, under various assumptions about the cost function, we find that the keystone species has a unique interior optimal survival probability. When we determine the optimal policy for the keystone species, we can establish ranking criteria for species preservation in Noah's ark. A keystone-dependent species whose marginal cost ratio is lower than the shadow price will not board Noah's ark – it will not survive. If the shadow price is lower than the marginal cost ratio, the species will be preserved. When the ratio is equal, we cannot know the fate of that species with certainty.

An optimal policy always exists for the Noah's ark scenario with keystone-dependent species. We examined various types of Noah's arks. One type involves a symmetrical ark, namely, the cost of the dependent species is equal for each k species, as is their biodiversity value. In such a case, the optimal symmetric policy is single (and an infinite number of asymmetric optimal policies are possible, all with a single optimal survival probability $\hat{\pi}_0$ for the keystone species). In this case, we found that the keystone species' (i.e., Noah's) survival probability to be single, interior and not dependent on the number of keystone-dependent species. We therefore conclude that the cost of the keystone species is not dependent on the number of K species.

In the case of a Noah's ark with an asymmetrical effective cost–benefit between the species, we can determine preservation-ranking criteria. The criteria is determined by the rank of cost-effective analysis according to the inequalities, and in case of a strong inequality, all the optimal survival probabilities to the left of the strong inequality get 0 probability, and all the optimal survival probabilities to the right get 1 probability. In the center, we get one single interior symmetric survival probability, which is determined by budget equality.

The applications to Noah's ark are manifold, starting from Weitzman's library model, in which the library collections (the books) could be destroyed by fire. In this study, we have posited one central library and $K \geq 2$ professional libraries (for instance, one national library and any number of professional libraries such as Physics, Chemistry, Math, etc.), and the scientific diversity of the library group is the number of the various books in all the professional libraries that belong to that group. In this case, the lending process is performed through the central library (inter-library lending).

One important application of this line of thought is in the field of banking. If the central branch of a bank is out of action for some reason, then data transfer to other branches is incomplete. In this case, payments and other fiscal operations could be damaged. Yet beyond this, these applications are suitable to other organizations that have a central 'head office' and branches in various dispersed locations. The Israel Nature and Parks Authority provides one example of such an organization: its Head Scientist and management are located in Jerusalem, and the country is divided into districts. In this case, the expenditure for each district is lower than the cost of the head office.

NOTES

1. He expands on this in his 1998 paper.
2. In each type of interaction and mutual influence, the individual influences and is influenced, but the result is evident on the population level – the population grows, decreases, moves, and so on.
3. Commensalism is a type of non-competitive interaction between organisms, in which one species benefits from the proximity and the other species is not affected either way.
4. An interaction in which one organism feeds on the other organism; the predator's fitness increases while the prey's fitness decreases.
5. For example, an animal species eats a plant species, partially or fully.
6. Two species in an obligatory relationship; the parasite is metabolically dependent on the host.

REFERENCES

Baumgartner, S. (2002), 'Optimal investment in multi-species protection when species interact in an ecosystem', paper presented at the Conference 'Healthy Ecosystems, Healthy People – Linkages Between Biodiversity, Ecosystem Health and Human Health' 6–11 June 2002, Washington, DC, and at the 2nd World Congress of Environmental and Resource Economists, 24–27 June 2002, Monterey, CA.

Berlow, E.L., S.A. Navarrete, C.J. Briggs, M.E. Power and B.A. Menge (1999), 'Quantifying variation in the strengths of species interactions', *Ecology*, **80**(7), 2206–24.

Craft, A.B. and R. Simpson (2001), 'The value of biodiversity in pharmaceutical research with differentiated products', *Environmental and Resource Economics*, **18**(1), 1–17.

Crocker, T.D. and J. Tschirhart (1992), 'Ecosystems, externalities and economies', *Environmental and Resource Economics*, **2**(6), 551–67.

Henle, K., D. Alard, J. Clitherowe, P. Cobb, L. Firbank and T. Kull et al. (2008), 'Identifying and managing the conflicts between agriculture and biodiversity conservation in Europe – a review', *Agriculture, Ecosystems and Environment*, **124**(1–2), 60–71.

Huston, M. (1994), *Biological Diversity – the Coexistence of Species on Changing Landscape*, New York: Cambridge University Press.

Kotliar, N.B., B.W. Baker, A.D. Whicker and G. Plumb (1999), 'A critical review of assumptions about the prairie dog as a keystone species', *Environmental Management*, **24**(2), 177–92.

Moran, D., D. Pearce and A. Wendelaar (1996), 'Global biodiversity priorities; a cost-effectiveness index for investments', *Global Environmental Change*, **6**(2), 103–19.

Moran, D., D. Pearce and A. Wendelaar (1997), 'Investing in biodiversity. An economic perspective on global priority setting', *Biodiversity and Conservation*, **6**(9), 1219–43.

Perevolotsky, A. and G. Polak (2001), *Ecology: Theory and the Israeli Experience*, Jerusalem: Carta (in Hebrew).

Perrings, C. (1999), 'Lecture notes', FEEM Seminar on Biodiversity: An Economic and Naturalistic Integrated Approach, Sicily, November 1999.

Rands, M.R.W., W.A. Adams, L. Bennun, S.H.M. Butchart, A. Clements and D. Coomes et al. (2010), 'Biodiversity conservation: challenges beyond 2010', *Science*, **329**(5997), 1298–303.

van der Heide, C.M., J.C.J.M. van den Bergh and E.C. van Ierland (2005), 'Extending Weitzman's economic ranking of biodiversity protection: combining ecological and genetic considerations', *Ecological Economics*, **55**(2), 218–23.

Weikard, H.-P. (2002), 'Diversity functions and the value of biodiversity', *Land Economics*, **78**(1), 20–27.

Weitzman, M.L. (1992), 'On diversity', *Quarterly Journal of Economics*, **107**(2), 363–405.

Weitzman, M.L. (1993), 'What to preserve? An application of diversity theory to crane conservation', *Quarterly Journal of Economics*, **108**(1), 157–83.
Weitzman, M.L. (1995), 'Diversity functions', in C. Perrings, K.-G. Mäler, C. Folke, C.S. Holling and B.-O. Jansson (eds), *Biodiversity Loss; Economic and Ecological Issues*, Cambridge, UK: Cambridge University Press, pp. 21–43.
Weitzman, M.L. (1998), 'The Noah's ark problem', *Econometrica*, **66**(6) 1279–98.
Woodward, F.I. (1993), 'How many species are required for a functional ecosystem', in E.D. Schulze and A.H. Mooney (eds), *Biodiversity and Ecosystem Function*, Berlin: Springer, pp. 271–92.
Zemah-Shamir, S. (2011), 'The symbiotic Noah's ark problem with two types of species', working paper published at The Foerder Institute for Economic Research.
Zemah-Shamir, S. and B. Shitovitz (2014), 'Optimizing the expectation of a coalitional game form in symbiotic uncertain environment under budget constraint', working paper.

18. Biodiversity, poverty and development*

Charles Palmer and Salvatore Di Falco

18.1 INTRODUCTION

The Millennium Ecosystem Assessment (MA, 2005) was a landmark attempt to assess the state of the world's ecosystems and the consequences of ecosystem change for human well-being. It found that the structure and functioning of global ecosystems have changed more rapidly between 1950 and 2000 than at any comparable period in human history. During this 50-year period, the world's population doubled while the global economy grew six-fold, leading to rapid increases in demand for ecosystem goods and services. For example, food production more than doubled while wood harvests for pulp and paper production tripled. These increases in demand were met both by increased consumption of available supply and by raising production through the application of new technologies and increasing the areas under production.

Biodiversity is crucial for the production of a range of marketed and non-marketed ecosystem goods and services, from consumptive goods such as timber, meat and medicines to hydrological services, soil management and biosphere resilience. A growing body of evidence shows how it supports system productivity and how its loss can have adverse effects on ecosystem functioning (e.g., Naeem et al., 1994; Tilman and Downing, 1994; Zhu et al., 2000; Loreau and Hector, 2001; Cork et al., 2002; Hooper et al., 2005; Landis et al., 2008). Entering into force in 1993, the Convention on Biological Diversity (CBD) declared the conservation of biological diversity 'a common concern of humankind' and an integral part of economic development.

In this chapter, we examine the evidence for the role of terrestrial biodiversity and biodiversity conservation in economic development and poverty, at both the macro (e.g., country) and micro scale (e.g., farm). This is very relevant in order to appreciate the role of biodiversity in socio-ecological systems. The micro scale, more specifically, enables a more precise exploration of *how* biodiversity influences ecosystems services that are ultimately instrumental for development, for example, in supporting food security. Our focus is on those areas and countries with high endowments of biodiversity, which also tend to be poor (Fisher and Christopher, 2007; Barrett et al., 2011).

We begin in Section 18.2 with background to the themes covered in the chapter along with the definitions used. In Section 18.3 we present evidence for general relationships between biodiversity and economic development before providing a closer examination of the links between biodiversity and ecosystem services. Key for food production and supply, agricultural biodiversity is used to illustrate these links. Given projected, future population increases and continued economic growth and consumption, Section 18.4 discusses research on future scenarios for biodiversity and development. Future threats to biodiversity need to be met via effective policies to protect biodiversity. Section 18.5 first examines evidence for a relationship between biodiversity protection and economic growth at the country scale before focusing on the very low incomes of rural people in biodiverse developing countries. We then present evidence for two types of policy, protected areas and bioprospecting, which aim to capture some of the public good benefits from biodiversity conservation, and their impacts on the welfare of the resource-dependent poor. Section 18.6 concludes.

18.2 BACKGROUND AND DEFINITIONS

For many countries, ecosystem goods and services have long contributed to human well-being and economic development; the latter defined in terms of the growth of gross domestic product. Agriculture, fisheries and forestry have long been crucial to countries' development strategies, providing capital for investments in other sectors and for the alleviation of poverty (MA, 2005). Consumption goods such as timber, fuel, meat, medicines and fruits have obvious use values. These more easily excludable and rivalrous, that is, private, benefits are ones that are typically included in countries' national accounts. Agriculture in 2000, for example, provided work and income-earning opportunities for half the world's total labour force, and accounted for almost a quarter of GDP in countries with per capita incomes of less than US$765.

The production of ecosystem goods and services that exploit the consumptive benefits of biodiversity has the potential to be sustainable but instead they are often over-harvested (Albers and Ferraro, 2007). Over-harvesting, along with land-use change, climate change, invasive species and pollution have all contributed to biodiversity loss (MA, 2005). Indeed, anthropogenic interference in the natural environment has been detrimental to biodiversity to the extent that the Earth could be in the midst of its sixth 'mass extinction' (Barnosky et al., 2011).[1] More specifically, the MA (2005) documents the following:

- the conversion of substantial proportions of the land area of a number of key terrestrial 'biomes' to agriculture;
- the decline in the population size and range of a number of taxonomic groups;
- the homogenization of the distribution of species across the world;
- a decline in the number of known species;
- a decline in genetic diversity among cultivated species;
- a continued rise in the global extinction rate by as much as 1000 times over background rates typical during the Earth's history.

As this list suggests, 'biodiversity' encompasses a range of levels, scales and attributes, and is thus impossible to capture in a single metric. In general, it defines the variety of living things in terms of genes within species, species themselves, and ecosystems. The latter refers to the typically complex interrelationships between species and their habitats.[2] As a consequence, biodiversity policies tend to focus on indicators such as the amount of land under protection or the International Union for Conservation of Nature (IUCN) Red List indicators (lists of vulnerable species) instead of, for example, numbers of species (Albers and Ferraro, 2006). Regarding the latter, only a proportion of all species have been taxonomically classified, and no one knows the true number of species, even to the nearest order of magnitude (Wilson, 1986).

However biodiversity is defined, it plays an important role in the production of a wide range of ecosystem services, which are unlikely to be priced in markets or included in GDP data (see Atkinson et al., 2012). These include soil management, pollination services and ecosystem resilience. Moreover, intrinsic, non-use values are often attached to biodiversity, for example, for the existence of certain species. There may also be value in retaining the option of future potential benefits of biodiversity, for example, new pharmaceutical products derived from the discovery of a yet to be discovered species of plant. Such benefits tend to be non-excludable and non-rivalrous, which implies that the investment in and protection of biodiversity is a public good. In recent decades, a range of policy initiatives and instruments, implemented at different scales (international, national, subnational) by different actors (international agencies, national governments, NGOs, etc.) and embodying different mechanisms (command-and-control, market-based), have been implemented with the aim of protecting biodiversity. These include 'bioprospecting' and protected area networks, for example, National Parks. Bioprospecting involves searching for, collecting and deriving genetic material from organisms collected in the wild that can be used in commercialized pharmaceutical, agricultural, industrial or chemical processing end products.

Such policies have to contend with the fact that biodiversity is unevenly distributed across the world. In particular, much biodiversity is concentrated in less-developed countries.[3] For example, around half of all terrestrial species are located in one-tenth of the Earth's land surface, with many found in areas of tropical forests (Wilson, 1986). Indeed, biodiversity 'hotspots' tend to be concentrated in rural areas where livelihoods depend disproportionately on natural capital embodied in forests, rangelands, soils, water and wildlife (Myers et al., 2000; Barrett et al., 2011). The challenge of policy is thus not only to provide closer to socially optimal levels of biodiversity but also to ensure that resource-dependent people are not made worse off as a result of policy implementation.

In summary, biodiversity clearly contributes to economic growth and development, with evidence of a trade-off between these. The next section, 18.3, first examines the research undertaken on a possible (simple) relationship between income growth and biodiversity before looking at the more complex relationships between biodiversity and ecosystem services. Given recorded losses in biodiversity in recent decades, and the multiple drivers of these losses, for example, land-use change, invasive species, Section 18.4 reviews research that generates projections of future losses and the drivers in these scenario analyses. In general, these have widely varying predictions. Of course, such analyses are based on numerous assumptions. One relates to the prevailing policy environment in which the public good benefits of biodiversity are addressed. Section 18.5 reviews two important policies for biodiversity conservation, one driven by the public sector (protected areas) and one by the private sector (bioprospecting). While bioprospecting has not been adopted as widely as its proponents would have predicted, protected area networks continue to expand and are the commonest means of biodiversity conservation around the world. However, there are concerns regarding their relative effectiveness, and their impacts on the poor, that is, those who are most dependent on the consumptive outputs of ecosystems for their incomes and livelihoods.

18.3 BIODIVERSITY, ECOSYSTEM SERVICES AND DEVELOPMENT

In this section, we begin with an examination of the evidence for a simple, direct relationship between biodiversity and conventional economic growth and development. However, understanding the relationship between biodiversity and incomes requires knowledge of *how* biodiversity influences the production of ecosystem services since the latter are ultimately instrumental for development. On their own, relationships between measures of biodiversity and incomes may be insufficient for drawing meaningful policy guidance. Due to complexity, the relationship between biodiversity and many different types of ecosystem service remains, however, relatively little understood (Albers and Ferraro, 2006). One crucial exception is the relationship between biodiversity and agricultural production. We review an emerging body of evidence that sheds light on this relationship.

Biodiversity and Economic Growth and Incomes

The relationship between per-capita income and measures of environmental degradation has come to be framed according to the Environmental Kuznets Curve (EKC) hypothesis (Dasgupta et al., 2002). This follows an inverted U-shaped relationship in which environmental degradation initially rises with increasing incomes but then, at some level of income, subsequently declines.[4] At higher income levels, a shift to less environmentally degrading economic activities, more effective environmental regulation and societal preferences for less degradation begins to drive improvements in environmental quality. Thus, a trade-off between economic growth and environmental quality is hypothesized only to exist when countries are relatively poor. There is, however, only limited empirical evidence in support of the EKC. This evidence tends to focus on various measures of air pollution (e.g., Selden and Song, 1994; Grossman and Krueger, 1995). Yet, a limited number of studies have explored the relationship between economic growth and biodiversity loss.

Naidoo and Adamowicz (2001) examine the link between numbers of threatened species, as classified by the IUCN, and per-capita gross national product (GNP), using cross-sectional data from over 100 countries. Their main finding is that increasing GNP appears to be associated with reductions in threatened species numbers for some taxa, particularly birds. Yet, for most of the other taxa surveyed, including plants, mammals, amphibians, reptiles, fishes and invertebrates, they find no evidence for a relationship between numbers of threatened species and GNP. A number of factors may explain the result for bird species, including the idea of birds as a charismatic taxonomic group with high societal demand for its continued preservation.

Note, however, that Naidoo and Adamowicz (2001) focused on the relationship between per-capita income and the relative degree of extinction threat. Hence, they did not measure biodiversity losses per se. Dietz and Adger (2003) investigate the relationship between economic growth, biodiversity loss and efforts to conserve biodiversity using a combination of panel and cross-sectional data for a sample of countries containing biodiverse tropical forest. More in keeping with the EKC hypothesis, they suggest that if economic growth drives biodiversity losses, for example, through destruction of habitats, then an inverse relationship between the two might be expected. But where increasing incomes are associated with an increase in demand for biodiversity conservation then investments in biodiversity protection should rise, leading to a corresponding decline in the rate of biodiversity loss.

Utilizing a species–area relationship to proxy for predicted species richness and a comparison of fixed and random effects models, Dietz and Adger fail to find an EKC between income and rates of species loss. Moreover, they do not find an EKC between income and rates of habitat loss. Mills and Waite (2009) reanalysed the data used by Dietz and Adger and attempted to address some of the econometric issues associated with the dataset. Despite finding some initial support for the EKC, primarily due to differences between countries, the overall conclusion is that there is relatively little empirical evidence for an EKC between income and biodiversity.

More recently, Perrings and Halkos (2010) examine the relationship between gross national income (GNI) per capita and threats to biodiversity. They look at four taxonomic groups: mammals, birds, plants and reptiles. In their econometric analysis, they control for important potential confounding factors such as climate, population density, land area and protected area status. Using the number of species in each taxonomic group under threat (according to the 2004 IUCN Red List) as the response variable, they model the relation between the level of threat and GNI per capita in a sample of 73 countries. They found that the relation between income and species under threat is quadratic for all terrestrial species. The turning points are, however, different for the different taxonomic groups. Income growth is found to be strongly correlated with increasing levels of threat to biodiversity. This may be due to the dependence of poorer countries' economies on agriculture. Thus, income growth depends on the expansion of agricultural lands into more 'marginal' areas, which are often habitats for wild species as well. This effect can be compounded by agricultural intensification. Managed agro-ecosystems are progressively simplified. It should be stressed, however, that the cross-sectional nature of this analysis raises some concerns about the statistical robustness of these results. Both unobserved heterogeneity and endogeneity can impact the estimates. Moreover, the large variation in the set of estimated turning points also generates some doubts on the usefulness of this sort of exercise for understanding the relationship between income and biodiversity.

Biodiversity and Ecosystem Services

Agricultural biodiversity refers to all diversity within and among species found in crop and domesticated livestock systems, including wild relatives, interacting species of pollinators, pests, parasites and other organisms (Qualset et al., 1995; Wood and Lenné, 1999). Similar to species hotspots, 'centres of origin/diversity', that is, hotspots of wild genetic diversity, have been identified for major crop plants, which also tend to be concentrated in tropical and sub-tropical regions. Domesticated biodiversity (i.e., crops) is located in agricultural landscapes (in situ). It is complemented by wild relatives stored in gene banks and breeders' collections (Smale, 2006). Biodiversity is utilized by farmers and breeders to adapt crops to different and changing production environments. Crop biodiversity is also important for both the functioning of ecosystems and the generation of many other ecosystem services (e.g., Naeem et al., 1994; Tilman and Downing, 1994; Tilman et al., 1996; Wood and Lenné, 1999; Loreau and Hector, 2001). We focus on crop biodiversity and agricultural production. Crop biodiversity is shown to be critical in attempts to achieve food security. This view has been widely documented in the applied and agricultural economics literature. Here a growing body of research focuses on the same research question, but using different methods reveals similar findings.

Evenson and Gollin (1997) provide evidence of the role of genetic diversity on agricultural yields. The role of biodiversity on productivity is also found to be positive and not negligible by Di Falco et al. (2007) and Smale et al. (1998). These findings are based on two different empirical approaches: aggregate panel data and farm-level cross-section analyses. Aggregate panel data analysis makes use of regional or district-level datasets to estimate aggregate production functions in which biodiversity is typically modelled as an input to the production process (e.g., Smale et al., 1998; Widawsky and Rozelle, 1998; Omer et al., 2007). The benefit of this approach is the implementation of a fixed-effects estimator that removes time-invariant unobserved heterogeneity. However, the scale of these analyses does not allow one to control for farm agro-ecological characteristics, and such analyses implicitly assume that the underlying theoretical model can be scaled up to the macro level. The second approach of using farm-level cross-section analysis, although overcoming the aggregation problem, has the obvious shortcoming of neglecting dynamics (Di Falco and Chavas, 2009).

In using farm-level panel data from the Central Highlands of Ethiopia, a more recent paper by Di Falco et al. (2010) attempts to circumvent these shortcomings. The dataset, collected in 2002 and 2005, was from a survey of 1500 farm households. The adoption of farm-level panel data, besides helping to manage the problem of endogeneity, enabled the researchers to address the issue of time-invariant heterogeneity at the household level (e.g., farmers' ability or farm-specific unobserved characteristics). Again, the study is conducted in a setting where environmental conditions are difficult due to poor soil quality and challenging weather conditions: the drought-prone and moisture-stressed production environment of Ethiopia. The empirical strategy is very comprehensive. It assesses the relationship between productivity, diversity and rainfall and addresses the possible endogeneity of diversity in productivity. Di Falco et al. (2010) jointly estimate two separate equations representing farm productivity and the determinants of biodiversity, respectively. This analysis provides useful information on the determinants of crop biodiversity at the farm level and sheds light on the way farmers use in situ diversity (see also Benin et al., 2004; van Dusen and Taylor, 2005) in food production.

Omer et al. (2007) adopted a stochastic production frontier approach to empirically test the hypothesized positive relationship between biodiversity stock and optimal levels of crop output. This analysis is based on data from a panel of UK specialized cereal farms for the period 1989–2000. Their results support the theoretical hypothesis. Increases in biodiversity can thus lead to a continual outward shift in the output frontier. Agricultural transition toward biodiversity conservation may be consistent with an increase in crop output in already biodiversity-poor modern agricultural landscapes. The relationship between risk exposure and crop biodiversity has also attracted empirical attention. Smale et al. (1998) studied the relationships between crop biodiversity and wheat production in the Punjab of Pakistan. They find that genealogical distance and number of varieties are associated with higher mean yield. Widawsky and Rozelle (1998), using data from regions of China, find, on the other hand, that the number of planted varieties reduces both the mean and the variance of rice yield. This finding is consistent with empirical research undertaken by Di Falco and Perrings (2005) and Di Falco et al. (2007). These three studies use a mean and variance framework. Risk exposure is therefore captured by the variance of crop yields or revenues.

Evidence of the narrowing of the genetic base of crops can thus indicate that farmers in the developing world are in fact becoming more and more vulnerable to environmental risk – such as the vagaries of weather.

18.4 LOOKING AHEAD: BIODIVERSITY AND DEVELOPMENT

Having established the scale and nature of biodiversity losses in Section 18.2 and examined the critical role of biodiversity in the production of ecosystem services – by way of agricultural production – in Section 18.3, this section reviews research that generates scenarios for future trends in biodiversity loss, with implications for biodiversity policy (see Section 18.5) and development. For the most part, the proximate threats remain as documented by the MA (2005) for the period 1950–2000: land use and habitat change; overexploitation; invasive alien species, pollution and climate change. We begin by examining projections for forces that ultimately influence the proximate threats.

Underlying Drivers

Underlying all the proximate threats outlined above are continued increases in the human population, rising incomes and changing consumption patterns across the world. Between 1800 and 2011, the global, human population increased from 1 to 7 billion, although growth rates have been falling in recent decades. This expansion is expected to continue for several more decades before reaching close to 10 billion by 2050 (UN, 2011). A large conservation science literature emphasizes the threat to biodiversity and ecosystem functioning as a consequence of population growth (e.g., Cincotta and Engelman, 2000; Balmford et al., 2001; McKinney, 2001; Ceballos and Ehrlich, 2002; Harcourt and Parks, 2003).

McKee et al. (2003), for example, develop an empirical cross-country model of the relationship between human population density and the number of threatened mammal and bird species. Their multiple regression analysis shows that two predictor variables, human population density and species richness, account for the majority of the variability in log-transformed densities of threatened species. They then use the regression model with projected population sizes of each country to find that the number of threatened species is, on average, expected to increase 7 per cent by 2020, and 14 per cent by 2050. While this might seem relatively low, the authors conclude: 'If other taxa follow the same pattern as mammals and birds . . . then we are facing a serious threat to global biodiversity' (p. 163). Note also that this is the projected rise in threatened species due to projected human population growth alone and hence does not consider other factors. Similar to the EKC studies, some caution is, however, in order when assessing empirical results based on small sample sizes and country-scale data. The cross-sectional nature of the analysis is highly prone to statistical problems (e.g., unobserved heterogeneity and endogeneity) that can bias the econometric estimates.

With the expansion of the world's population, human societies have undergone a remarkable transition in which the majority of people now live in urban areas rather than rural ones. Much future population growth is thus expected to occur in urban areas; almost 2 billion new urban residents are expected by 2030, mainly in relatively small cities in developing countries (UN, 2012). Urbanization is expected to have significant effects on ecosystem services (e.g., Forman, 2008; Martine et al., 2008). Effects are expected both directly through the expansion of urban areas and indirectly through changes in consumption and pollution as people migrate into cities (McKinney, 2002; Liu et al., 2003; McGranahan and Satterthwaite, 2002).[5]

Grimm et al. (2008) discuss the direct effects of urbanization on biodiversity. First, within cities, urbanization reduces both species richness and evenness for most biotic communities, although there are exceptions. For example, plant species richness and evenness can be higher than in 'wildlands' due to the highly heterogeneous patchwork of habitats and human introductions of exotic species. Second, urbanization alters the species composition of communities due to human-induced changes such as altered temperatures, light, noise and air pollution. This posits urbanization as a strong evolutionary force. Indirect effects are much more difficult to isolate empirically, although tentative evidence has begun to emerge. For example, Defries et al. (2010) use satellite-based estimates of forest loss from 2000 to 2005 to assess economic, agricultural and demographic correlates across 41 tropical countries. Utilizing two methods of analysis, linear regression and regression tree, they show that forest loss is positively correlated with urban population growth and exports of agricultural products. Their results highlight the importance of urban-based and international demands for agricultural products as potential drivers of deforestation, although yet again note the dependence on small numbers of observations and coarse country-scale data. Thus, while the results are suggestive more work is needed to understand the causal chains between the demand for certain products in one area and forest conversion in another.

Proximate Drivers

Much research focuses on the issue of habitat conversion (driven by population and economic growth) as the key driver of biodiversity loss. This has been incorporated by the MA (2005) indicators of the proximate drivers of biodiversity loss, which have intensified in recent years (see Butchart et al., 2010). The most important driver is in fact habitat change (e.g., land-use change). The list of drivers is extended by crucial physical modifications determined by the process of climate change. Although research typically focuses on single drivers, these are often synergistic. For example, land-use change can result in greater nutrient loading if land is converted to high-intensity agriculture, increased emissions of greenhouse gases (if forest is cleared), and increased numbers of invasive species (due to the disturbed habitat) (MA, 2005). The process of economic integration among countries can also play an important role in future biodiversity change. The number and distribution of species can indeed be affected by globalization of economic systems.

Looking ahead, the MA developed four plausible scenarios for ecosystems and human well-being. These explored two global development paths, one in which the world becomes increasingly globalized and the other in which it becomes increasingly regionalized. In addition, they explored two different approaches to ecosystem management, one in which actions are reactive and most problems are addressed only after they become obvious and the other in which ecosystem management is proactive and policies deliberately seek to maintain ecosystem services over the long term (MA, 2005). Under all four MA scenarios, the projected changes in drivers to 2050 result in significant growth in consumption of ecosystem services, continued loss of biodiversity, and further degradation of some ecosystem services. More specifically:

- Demand for food crops is projected to grow by 70–85 per cent, and demand for water by between 30 per cent and 85 per cent. Water withdrawals in developing countries are projected to increase significantly, although these are projected to decline in industrial countries (*medium certainty*).
- Food security is not achieved and child malnutrition is not eradicated (and is projected to increase in some regions in some MA scenarios) despite increasing food supply and more diversified diets (*medium certainty*).
- A deterioration of the services provided by freshwater resources (such as aquatic habitat, fish production and water supply for households, industry and agriculture), particularly in the scenarios that are reactive to environmental problems (*medium certainty*).
- Habitat loss and other ecosystem changes are projected to lead to a decline in local diversity of native species (*high certainty*). Globally, the number of plant species is projected to be reduced by around 10–15 per cent as the result of habitat loss alone (*low certainty*), and over-harvesting, invasive species, pollution and climate change will further increase the rate of extinction.

Predicting the response of biodiversity to climate change has recently become an active field of research, again to be found mainly in the conservation science literature (e.g., Dillon et al., 2010; Gilman et al., 2010; Beaumont et al., 2011; Dawson et al., 2011; McMahon et al., 2011). Although there is relatively limited evidence of current extinctions caused by climate change, research suggests that climate change could surpass habitat destruction as the biggest threat to biodiversity over the next few decades (Leadley et al., 2010). However, the multiplicity of approaches and the resulting variability in projections make it difficult to obtain clarity with respect to the future of biodiversity under different climate change scenarios (Pereira et al., 2010). Bellard et al. (2012) reviewed both the ranges of possible impacts of climate change that operate at individual, population, species, community, ecosystem and biome scales and the different responses that could occur at individual, population or species levels. They show that species can respond to climate change challenges by shifting their climatic niche along three non-exclusive axes: time (e.g., phenology), space (e.g., range) and self (e.g., physiology). While current estimates are highly variable, the majority of models indicate serious consequences for biodiversity. The worst-case scenarios suggest extinction rates that would qualify as the sixth mass extinction in the Earth's history.

18.5 BIODIVERSITY PROTECTIONS, POLICY AND WELFARE IMAPACTS

Biodiversity plays a crucial role in the production of a wide range of ecosystem services. In turn, these services often generate positive externality effects, which are rarely – if ever – internalized by actors who derive benefits from the consumptive outputs of ecosystems. Hence, policies may be implemented in order to protect biodiversity and overcome market failures. Both underlying and proximate drivers of biodiversity losses need to be addressed. Such policies, if effective, could play an important role in minimizing the probability of extinction rates commensurate with a mass extinction event. In this section, we first examine the limited evidence for a relationship between biodiversity protection and economic growth at the country scale. We then investigate possible interlinkages at a finer scale, focusing on the very low incomes of rural people in biodiverse developing countries. In the second part of this section, we review two different yet important policy instruments utilized to protect biodiversity: protected areas and bioprospecting.

Biodiversity Protection, Economic Growth and Poverty

Given the possibility that more effective environmental regulation could help drive the increase in environmental quality at higher levels of income, Dietz and Adger (2003) examined the relationship between economic growth and biodiversity conservation using their cross-country dataset. The latter was proxied, first by the proportion of land area defined as a national park or protected area and second, the degree of reporting to the Convention on International Trade in Endangered Species (CITES). The results suggest that the extent of government policy on biodiversity protection increases with economic development. More specifically, Dietz and Adger suggest that 'low levels of income in a country may be correlated with restrictions on government enforcement of CITES and other environmental legislation' (2003, p. 30).

Increases in incomes may only lead to conservation if the extracted products are 'inferior' goods that are replaced by other, more-preferred and less-degrading goods as incomes rise (Albers and Ferraro, 2006). Within countries, however, there is little evidence that rising incomes have led to more biodiversity protection. Instead, where investment opportunities are limited to agriculture, increased incomes have been shown to result in greater conversion of habitat and thus biodiversity loss. For example, Zwane (2007) examined the relationship between income and land clearing for households living in tropical forest in Peru. Econometric analysis of panel data found lagged income to be positively correlated with clearing, though at a decreasing rate. Due to labour market constraints, clearing is found to be positively correlated with household labour availability. In conclusion, small increases in the incomes of the poorest were unlikely to reduce deforestation in the Peruvian context.

There is little doubt regarding the important role of ecosystem services in supporting the incomes and livelihoods of the rural poor in developing countries (MA, 2005; WRI, 2005; TEEB, 2010; World Bank, 2011). Turner et al. (2012) demonstrate how biodiverse areas supply important – and valuable – ecosystem services to the poor, even where the range of services considered is restricted to consumptive outputs alone, for example, food, fuel, clean water. Yet these services are often insufficient to lift people out of poverty. In a recent special issue of the *Proceedings of the National Academy of Sciences USA* 'On Biodiversity Conservation and Poverty Traps', the introductory paper by Barrett et al. (2011) identifies four classes of mechanisms that define the links between biodiversity conservation and poverty. More specifically, they focus on any self-reinforcing mechanism that causes poverty, however measured, to persist, that is, poverty traps:

1. Dependence of the poor on inherently limited natural resources in order to meet consumption needs.
2. Vulnerabilities shared between the poor and biodiverse ecosystems, that is, due to poverty, population growth and environmental degradation.
3. Failure of sociopolitical and economic institutions, including missing and imperfect markets, for example, as in the study by Zwane (2007) summarized above.
4. Unintended consequences and a lack of adaptive management as a consequence of decisions made over the use and extraction of natural resources, for example, the development of mineral resources leading to changes in watershed ecosystems that impact negatively on downstream communities.

Since one or more of these four mechanisms might apply in any given setting, the policy implications that follow may vary widely (Barrett et al., 2011). Thus, any starting point for policy design requires that 'careful site-specific diagnostics' are undertaken first (ibid., p. 13910; see also Pfaff and Robalino, 2012).

Biodiversity Policy and Welfare Impacts

Recent decades have witnessed widespread experimentation with different policies to conserve biodiversity around the world, from regulatory and command-and-control instruments to newer generations of voluntary and market-based instruments (see Albers and Ferraro, 2006; Miteva et al., 2012). Regarding the latter, incentives for supplying closer to socially optimal levels of biodiversity could be delivered through payments for ecosystem services (PES). Alternatively, incentives could be supplied more indirectly. Either capital and labour could be redirected away from activities that underlie habitat loss such as certain types of agriculture, or commercial activities that supply ecosystem services could be encouraged via 'joint production' (Ferraro and Kiss, 2002; Bulte and Engel, 2006). Examples of the former include agricultural intensification and the development of off-farm labour opportunities; the latter include sustainable forestry, some non-timber forest products and eco-tourism. In this sub-section, we discuss the evidence for welfare impacts of two types of policy only: one command-and-control instrument, protected areas, and one market-based instrument, bioprospecting.

Protected areas

Dietz and Adger (2003) highlight the important role of protected areas in conserving biodiversity. It is the most common policy implemented to protect biodiversity in both developed and developing countries (MA, 2005). For example, in the most recent Global Forest Resources Assessment (FAO, 2010), National Parks, game reserves, wilderness areas and other legally established protected areas cover more than 10 per cent of the total forest area in most countries and regions. Yet, Myers (2003) reports that just over a third of the world's 25 biodiversity hotspots are found in protected areas. Typically implemented by government agencies, these areas have varying degrees of restriction with regard to their use, from 'strict nature reserves' through to areas in which some extractive uses may be permitted (IUCN, 2003). In principle, legal restrictions prevent anthropogenic disturbance, thus contributing to the maintenance or recovery of ecosystem services (Ferraro et al., 2012).

In many developing countries, conserving biodiversity through the establishment of protected areas may impose an uncomfortable trade-off for policy-makers: people often rely on resources within these areas and human uses may not be compatible with biodiversity protection (Albers and Ferraro, 2006). However, *effective* protected area management requires that governments have the ability to enforce their property rights, which is often not the case even where they might be reasonably well-defined.[6] Local people dependent on natural resources for incomes and livelihoods often have property rights claims in protected areas, which may conflict with those of governments (Engel et al., 2010). One consequence is that many protected areas in developing countries have, in effect, a de facto open access regime.[7] There may be incentives for local people to over--harvest resources without regard to the costs imposed on others, including both locals and those beyond the local level.

Where property rights to protect habitats and ecosystems are relatively clearly defined and enforced, people's access to consumptive outputs from resources might be constrained. Patrols and fines for illegal extraction, for example, can impose high costs on local people (Bulte and Engel, 2006). Much research on the socioeconomic impacts of protected areas in developing countries tends, however, to show little more than 'that protected areas are established near poor people and provide both opportunities and constraints to economic development' (see also Wilkie et al., 2006; Coad et al., 2008; Ferraro et al., 2012, p. 35).[8] Barrett et al. (2011) highlight a number of challenges for studies attempting to draw inference from associations between the physical and social aspects of biodiversity protection and conservation policy: the absence of landscape-level controls, the absence of sufficient baseline or historical data, and the general absence of credible counterfactual analysis (see also Miteva et al., 2012).

A recent, growing body of empirical research appears to support the economic intuition underlying the trade-off between biodiversity protection and welfare (Andam et al., 2010; Sims, 2010; Barrett et al., 2011). Specifically, this research applies programme evaluation techniques, including randomized field experiments, matching methods and econometric analyses, to assess the environmental effectiveness and welfare impacts of policies to conserve biodiversity. Such techniques are applied in order to overcome selection bias due to the fact that biodiversity policy interventions are commonly implemented in a non-random manner, both over time and space. Protected areas, for both political and economic reasons, are often situated in areas with few profitable, alternative uses (Albers and Ferraro, 2006).

A number of examples of recent research on the welfare impacts of protected areas can be found in the special issue 'On Biodiversity Conservation and Poverty Traps', introduced earlier. First, Ferraro et al. (2011) use geospatial data and econometric analyses to estimate the impacts of protected areas on poverty and deforestation across diverse sites in Costa Rica and Thailand. Little evidence is found that creating protected areas traps historically poorer areas in poverty. Yet, the spatial characteristics associated with the most poverty alleviation are not always the ones associated with the slowing down of deforestation. Second, in Uganda, Naughton-Treves et al. (2011) use satellite data along with primate censuses, forest transects, household surveys, and econometric analyses and found that Kibale National Park protected both forest and primates. Through matching methods they found no evidence that the Park constituted a poverty trap. Third, McNally et al. (2011) analyse the impacts of Saadani National Park on local households in

Tanzania, using satellite data, survey data and econometric analyses. Restricting access to mangrove timber, the Park was found to increase people's income dependence on fishing and shrimping as well as increase people's indirect benefits from mangrove protection. This study clearly highlights the possibility of overspill effects from policy implementation (see also Miteva et al., 2012).

Bioprospecting

Although protecting biodiversity is a global public good, there can be private benefits from actions that lead to biodiversity protection (Albers and Ferraro, 2006). For example, non-governmental organizations such as The Nature Conservancy (TNC) depend on voluntary contributions in order to privately supply biodiversity through the acquisition of habitat, both in developed and developing countries. One highly debated source of private incentives for biodiversity protection is bioprospecting.

As noted earlier, bioprospecting is the search for valuable compounds from wild organisms. This involves searching for, collecting and deriving genetic material from samples of biodiversity that can be used in commercialized end products. It has been touted as a mechanism for both discovering new pharmaceutical products and saving endangered ecosystems via the financing of conservation. For example, Rausser and Small (2000) claimed that the value of protecting certain ecosystems for bioprospecting can be quite high. Given that the annual market size for products based on genetic resources has been estimated to lie within the range of US$220–300 billion (Deke, 2008), there would appear to be strong enough private incentives for protecting biodiversity through bioprospecting.[9]

This view, however, has been challenged in a seminal paper by Simpson et al. (1996). In this paper, both the marginal benefits and costs of conservation are addressed from a biosprospecting angle. The increase in the total value of an additional species is therefore considered. The marginal value is, however, decreasing due to redundancy. There are many species that may perform the same function. The total economic cost of losing some species is also included, which is found to be negligible for both high and low levels of species. This, of course, creates a problem for conservation.

Contracts have been negotiated between pharmaceutical firms and the government or individuals who control biodiverse ecosystems, although the numbers of private partnerships remains small. For example, the National Biodiversity Institute (INBio) of Costa Rica negotiated a contract with Merck in 1989 (Sedjo, 1992). In this, US$1 million was paid to INBio for rights to screen plants for useful chemicals over two years and INBio was to receive royalties in the event of any successful commercial applications. In general, however, there is relatively little empirical information regarding these transactions (Deke, 2008).

While doubts have been raised about bioprospecting's potential to both effectively conserve biodiversity and alleviate poverty, very few studies have assessed both the socioeconomic and environmental impacts in an empirically robust manner (Barrett et al., 2011). Yet, the socioeconomic impacts in developing countries are likely to be afflicted by the same kinds of problems as discussed for protected areas. In particular, property rights to genetic resources may be difficult to specify and even more difficult to enforce. Furthermore, there may be missing legal frameworks for regulating benefits sharing from genetic resources (Dhillion et al., 2002) and excessive bureaucracy may divert finance research and development away from the conservation of in situ genetic material (ten Kate and Laird, 1999).

18.6 CONCLUSIONS

According to Barnosky et al. (2011), the recent loss of species while dramatic does not yet qualify as a 'mass extinction', at least from a palaeontological perspective. Only a few per cent of assessed, that is, known species, have been lost in recent decades, although species may have been lost that had never been described in the first place. Yet, there are clear indications that losing species currently in the 'critically endangered' category could lead to a mass extinction on a scale that has only occurred five times in the previous 540 million years. If so, another mass extinction could occur within 'a few centuries' (Barnosky et al., 2011, p. 56). Thus, it would seem that biodiversity faces a bleak future at least in the long run given the projections of proximate and underlying drivers of change presented in Section 18.4. Although many recent reviews and projections point in this direction, we note that projections of biodiversity loss do vary widely according to the assumptions and data used in the models. Requiring further research is biodiversity's role in ecosystem resilience and in providing a kind of buffer against reaching a 'tipping point' in system dynamics (see Helm and Hepburn, 2012). Ecosystem collapse could potentially speed up rates of biodiversity loss.

Recent years have witnessed intensifying attention from policy-makers in response to the documented decline of biodiversity, and a realization that the undersupply of biodiversity conservation results from biodiversity's transboundary public good characteristics. Such characteristics imply a need both for international institutions that can govern biodiversity effectively and a continued focus on improving policy design for effective biodiversity protection. Indeed, Butchart et al. (2010) note recent upticks in indicators of attempts to counter biodiversity losses, including extent and biodiversity coverage of protected areas, sustainable forest management, and policies to deal with invasive species. However, even where such policy responses are shown to be effective – still a relatively rare process as discussed for protected areas and bioprospecting in Section 18.5 (see also Miteva et al., 2012) – they still often fail in enabling local, resource-dependent people to capture some of the social value of biodiversity. And where policies constrain people's access to and use of natural resources, they could even be made worse off as a consequence of policy implementation.

Policy design and implementation is further complicated by the high degree of uncertainty regarding our knowledge and understanding of biodiversity and its interlinkages with ecosystem services. As illustrated with the case of agricultural biodiversity in Section 18.3, however, progress on research is being made and sensible policy responses have been forthcoming. Maintaining diverse plant varieties on farmers' fields, in-situ conservation, vis-à-vis storing germplasms in gene banks, for example, is increasingly regarded as an effective way of conservation of plant genetic resources (Benin et al., 2004; Bezabih, 2008). At the heart of whether in-situ conservation could be pursued as a fruitful strategy of keeping important germplasms alive is whether it generates farm-level benefits that are internalized by farmers. Here, policies such as PES could provide the necessary incentives for local conservation activities that yield wider social benefits.

An emerging body of empirical evidence appears to suggest, however, that individual policies or strategies, for example, protected areas or PES, may be unable to reconcile both biodiversity conservation and poverty alleviation objectives in multiple settings across the world (Barrett et al., 2011). For instance, protecting biodiversity from invasive species may require both monitoring and eradication activities in existing protected areas (Albers and Ferraro, 2006). Siting of areas and the management of buffer zones among areas and their surroundings could reduce the opportunities for invasive species to take hold. Similarly, climate change impacts may be mitigated to some degree if the siting of protected areas and the management of nearby land and wildlife corridors allows species to move and adapt gradually to changes in climate. While some of the activities connected with these policy strategies such as eradication or management efforts could potentially generate local employment and other, associated benefits, they are unlikely to have a broader impact on poverty. To achieve this requires policies, separate from biodiversity conservation, that actually identify and tackle the root causes of poverty. If, on the other hand, these are in some way related to resource use and dependence then there may be little option but to carefully design policies that attempt to address both the loss of biodiversity and poverty.

Assuming that policies can be designed to effectively conserve biodiversity and at the minimum, 'do no harm' to the poor, there remains the question of how such policies might be financed. Given ever-tightening constraints on public expenditures in many of the world's developed economies, attention is shifting to alternative sources of finance for biodiversity protection. Recently, there has been much speculation about the role of Reducing Emissions from Deforestation and Degradation (REDD+) in financing the preservation of tropical forests rich in both carbon and biodiversity (e.g., Gardner et al., 2012). Although REDD+ was originally posited as a strategy primarily to mitigate the effects of climate change, a series of 'safeguards' relating to biodiversity protection and the livelihoods of the poor were adopted by parties to the United Nations Framework Convention on Climate Change, at Cancun in 2010. In principle, REDD+ could enable a number of changes to the governance and use of forests leading to increased levels of biodiversity protection, for example, in situ conservation through the establishment of new protected areas and associated corridors for connecting landscapes. Yet, mechanisms for long-term financing of REDD+ remain uncertain and policies for operationalizing REDD+ will need to overcome the same challenges that have bedevilled the management of natural ecosystems in recent decades. As the world's population grows and becomes increasingly urbanized, our dependence upon biodiversity for our incomes and livelihoods is unlikely to diminish. Stemming biodiversity loss is thus critical for ensuring our future prosperity. However, this requires not only a willingness to pay for biodiversity protection but also that we actually pay for biodiversity, in the process implementing effective policies on the basis of learning from past policy experiences – for good or ill.

NOTES

* With permission from Oxford University Press (© OUP), this chapter is reprinted from C. Palmer and S. Di Falco (2012), 'Biodiversity, poverty, and development', *Oxford Review of Economic Policy*, **28**(2), 48–68.

1. Of the 4 billion species estimated to have evolved on the Earth over the last 3.5 billion years, some 99 per

cent no longer exist. Thus, extinction is relatively common, although it is balanced by speciation (Barnosky et al., 2011). Mass extinction occurs when extinction rates accelerate relative to origination rates such that over 75 per cent of species disappear within a geologically short interval – typically less than 2 million years, in some cases much less (ibid.).

2. Ecosystem complexity can be characterized by, for example, the possibility of sudden, unpredictable non-linear changes, feedback effects and vulnerability to sudden shocks such as from fire or disease (Barrett et al., 2011).
3. In 1998, the NGO Conservation International identified 17 countries with exceptional endowments of biodiversity. Excluding Australia and the United States, the political group, Like-Minded Megadiverse Countries (LMMC) brought together the interests and concerns of 15 of these countries plus other less-developed countries, in 2002. It claims to represent about 80 per cent of the world's biodiversity and 45 per cent of the world's population. Members include: Bolivia, Brazil, China, Colombia, Costa Rica, Democratic Republic of the Congo, India, Ecuador, Indonesia, Kenya, Madagascar, Malaysia, Mexico, Papua New Guinea, Peru, Philippines, South Africa and Venezuela (Deke, 2008). While acknowledging differences in per-capita incomes among these countries, these are important examples of biodiverse 'less-developed' or 'developing' countries as defined in this chapter.
4. The EKC is based on analogy with the observations of Kuznets (1955), who explored the U-shaped relationship between income inequality and changes in per-capita income. Hepburn and Bowen (2013) provide a recent conceptual and synthetic analysis of the relationship between economic growth and environmental limits, including more discussion on the conceptual and empirical relevance of the EKC.
5. We note, however, that urban living can be associated with more 'environmentally friendly' modes of living due to people living in closer proximity, leading, for example, to less intensive per-capita energy usage and lower greenhouse gas emissions (see Glaeser, 2011).
6. Where protected areas are effective in conserving biodiversity, their impacts may, however, be diminished by the displacement of extractive activities to unprotected areas nearby (e.g., see Andam et al., 2008).
7. In addition to the breakdown of a management and authority system, the purpose of which is to introduce and enforce a set of norms of behaviour among individuals with respect to a given resource, note that open access also results from the mere absence of such a system (see Bromley, 1991).
8. Although outside the scope of this review, Ferraro et al. (2012) also note that there are no studies that have examined the cost-effectiveness of protected areas.
9. This figure is based on sales on world markets for products sold in the healthcare (e.g., pharmaceuticals, cosmetics), agriculture (e.g., seeds, crop protection), and 'other biotechnology' (e.g., bioenergy) sectors (see Deke, 2008).

REFERENCES

Albers, H.J. and P. Ferraro (2006), 'The economics of terrestrial biodiversity conservation in developing nations', in R. López and M. Toman (eds), *Economic Development and Environmental Sustainability*, Oxford: Oxford University Press.

Andam, K.S., P.J. Ferraro, A. Pfaff, G. Arturo Sanchez-Azofeifa and J.A. Robalino (2008), 'Measuring the effectiveness of protected area networks in reducing deforestation', *Proceedings of the National Academy of Sciences*, **105**(42), 16089–94.

Andam, K.S., P.J. Ferraro, K.R.E. Sims, A. Healy and M.B. Holland (2010), 'Protected areas reduced poverty in Costa Rica and Thailand', *Proceedings of the National Academy of Sciences*, **107**(22), 9996–10001.

Atkinson, G., I. Bateman and S. Mourato (2012), 'Recent advances in the valuation of ecosystem services and biodiversity', *Oxford Review of Economic Policy*, **28**(2), 22–47.

Balmford, A., J.L. Moore, T. Brooks, N. Burgess, L.A. Hansen, P. Williams and C. Rahbek (2001), 'Conservation conflicts across Africa', *Science*, **291**(5513), 2616–19.

Barnosky, A.D., N. Matzke, S. Tomiya, G.O.U. Wogan, B. Swartz and T.B. Quental et al. (2011), 'Has the Earth's sixth mass extinction already arrived?', *Nature*, **471**(7336), 51–7.

Barrett, C.B., A.J. Travis and P. Dasgupta (2011), 'On biodiversity conservation and poverty traps', *Proceedings of the National Academy of Sciences*, **108**(34), 13907–12.

Beaumont, L.J., A. Pitman, S. Perkins, N.E. Zimmermann, N.G. Yoccoz and W. Thuiller (2011), 'Impacts of climate change on the world's most exceptional ecoregions', *Proceedings of the National Academy of Sciences, USA*, **108**(6), 2306–11.

Bellard, C., C. Bertelsmeier and P. Leadley (2012), 'Impacts of climate change on the future of biodiversity', *Ecology Letters*, **15**(4), 365–77.

Benin, S., M. Smale, J. Pender, B. Gebremedhin and S. Ehui (2004), 'The economic determinants of cereal crop diversity on farms in the Ethiopian Highlands', *Agricultural Economics*, **31**(2–3), 197–208.
Bezabih, M. (2008), 'Agrobiodiversity conservation under an imperfect seed system: the role of community seed banking schemes', *Agricultural Economics*, **38**(1), 77–87.
Bromley, D.W. (1991, 'Property rights and property regimes in natural resource policy', in *Environment and Economy: Property Rights and Public Policy*, Oxford: Blackwell, pp. 14–40.
Bulte, E. and S. Engel (2006), 'Conservation of tropical forests: addressing market failure', in R. López and M. Toman (eds), *Economic Development and Environmental Sustainability*, Oxford: Oxford University Press, pp. 412–45.
Butchart, S.H.M., M. Walpole, B. Collen, A. van Strien, J.P.W. Scharlemann and R.E.A. Almond et al. (2010), 'Global biodiversity: indicators of recent declines', *Science*, **328**(5982), 1164–8.
Ceballos, G. and P.R. Ehrlich (2002), 'Mammal population losses and the extinction crisis', *Science*, **296**(5569), 904–7.
Cincotta, R.P. and R. Engelman (2000), *Nature's Place – Human Population and the Future of Biological Diversity*, Washington, DC: Population Action International.
Coad, L., A.L. Campbell, L. Miles and K. Humphries (2008), 'The costs and benefits of protected areas for local livelihoods: a review of the current literature', Cambridge, UK: United Nations Environment Programme, World Conservation Monitoring Centre.
Cork, S.J., W. Proctor, D. Shelton, N. Abel and C. Binning (2002), 'The ecosystem services project: exploring the importance of ecosystems to people', *Ecological Management and Restoration*, **3**, 143–6.
Dasgupta, S., B. Laplante, H. Wang and D. Wheeler (2002), 'Confronting the Environmental Kuznets Curve', *The Journal of Economic Perspectives*, **16**(1), 147–68.
Dawson, T.P., S.T. Jackson, J.I. House, I.C. Prentice and G.M. Mace (2011), 'Beyond predictions: biodiversity conservation in a changing climate', *Science*, **332**(6025), 53–8.
DeFries, R.S., T. Rudel, M. Uriarte and M. Hansen (2010), 'Deforestation driven by urban population growth and agricultural trade in the twenty-first century', *Nature Geoscience*, **3**(3), 178–81.
Deke, O. (2008), *Environmental Policy Instruments for Conserving Global Biodiversity*, Berlin: Springer-Verlag.
Dhillion, S.S., H. Svarstad, C. Amundsen and H.C. Bugge (2002), 'Bioprospecting: effects on environment and development', *Ambio*, **31**(6), 491–3.
Dietz, S. and W.N. Adger (2003), 'Economic growth, biodiversity loss and conservation effort', *Journal of Environmental Management*, **68**(1), 23–35.
Di Falco, S. and J.-P. Chavas (2009), 'On crop biodiversity, risk exposure and food security in the Highlands of Ethiopia', *American Journal of Agricultural Economics*, **91**(3), 599–611.
Di Falco, S. and C. Perrings (2005), 'Crop biodiversity, risk management and the implications of agricultural assistance', *Ecological Economics*, **55**(4), 549–66.
Di Falco, S., J.-P. Chavas and M. Smale (2007), 'Farmer management of production risk on degraded lands: the role of wheat variety diversity in the Tigray region, Ethiopia', *Agricultural Economics*, **36**(2), 147–56.
Di Falco, S., M. Bezabih and M. Yesuf (2010), 'Seeds for livelihood. Crop biodiversity in Ethiopia', *Ecological Economics*, **69**(80), 1695–1702.
Dillon, M.E., G. Wang and R.B. Huey (2010), 'Global metabolic impacts of recent climate warming', *Nature*, **467**(7316), 704–6.
Engel, S., C. Palmer and A. Pfaff (2010), 'Participatory protection in theory and application: paper tigers, fences & fines, or negotiated co-management?', Duke Environmental Economics Working Paper Series No. 10–02, Nicholas Institute for Environmental Policy Solutions, Duke University.
Evenson, R.E. and D. Gollin (1997), 'Genetic resources, international organizations, and improvement in rice varieties', *Economic Development and Cultural Change*, **45**(3), 471–500.
FAO (2010), *Global Forest Resources Assessment 2010*, Rome: Food and Agriculture Organization of the United Nations.
Ferraro, P. and A. Kiss (2002), 'Direct payments to conserve biodiversity', *Science*, **298**(5599), 1718–19.
Ferraro, P.J., M.M. Hanauer and K. Sims (2011), 'Conditions associated with protected area success in conservation and poverty reduction', *Proceedings of the National Academy of Sciences, USA*, **108**(34), 13913–18.
Ferraro, P.J., K. Lawlor, K.L. Mullan and S.K. Pattanayak (2011), 'Forest figures: ecosystem services and policy evaluation in developing countries', *Review of Environmental Economics and Policy*, **6**(1), 20–44.
Fisher, B. and T. Christopher (2007), 'Poverty and biodiversity: measuring the overlap of human poverty and the biodiversity hotspots', *Ecological Economics*, **62**(1), 93–101.
Forman, R. (2008), *Urban Regions: Ecology and Planning Beyond the City*, New York: Cambridge University Press.
Gardner, T.A., N.D. Burgess, N. Aguilar-Amuchastegui, J. Barlow, E. Berenguer and T. Clements et al. (2012), 'A framework for integrating biodiversity concerns into national REDD+ programmes', *Biological Conservation*, **154**, 61–71.

Gilman, S.E., M.C. Urban, J. Tewksbury, G.W. Gilchrist and R.D. Holt (2010), 'A framework for community interactions under climate change', *Trends in Ecology and Evolution*, **25**(6), 325–31.

Glaeser, E. (2011), *Triumph of the City*, New York: Penguin.

Grimm, N.B., S.H. Faeth, N.E. Golubiewski, C.L. Redman, J. Wu, X. Bai and J.M. Briggs (2008), 'Global change and the ecology of cities', *Science*, **319**(5864), 756–60.

Grossman, G. and A. Krueger (1995), 'Economic growth and the environment', *Quarterly Journal of Economics*, **110**(2), 352–77.

Harcourt, A.H. and S.A. Parks (2003), 'Threatened primates experience high human densities adding an index of threat to the IUCN Red List criteria', *Biological Conservation*, **109**, 137–49.

Helm, D.R. and C. Hepburn (2012), 'Introduction to special issue on biodiversity', *Oxford Review of Economic Policy*, **28**(2).

Hepburn, C. and A. Bowen (2013), 'Prosperity with growth: economic growth, climate change and environmental limits', in R. Fouquet (ed.), *Handbook on Energy and Climate Change*, Cheltenham, UK and Northampton, MA, USA: Edward Elgar Publishing.

Hooper, D.U., F.S. Chapin, J.J. Ewel, A. Hector, P. Inchausti and S. Lavorel et al. (2005), 'Effects of biodiversity on ecosystem functioning: a consensus of current knowledge', *Ecological Monographs*, **75**(1), 3–35.

IUCN (2003), *United Nations List of Protected Areas*, Cambridge, UK: IUCN and UNEP-WCMC.

Kuznets, S. (1955), 'Economic growth and income inequality', *American Economic Review*, **45**(1), 1–28.

Landis, D.A., M.M. Gardiner, W. van der Werf and S.M. Swinton (2008), 'Increasing corn for biofuel production reduces biocontrol services in agricultural landscapes', *Proceedings of the National Academy of Sciences, USA*, **105**(51), 20552–7.

Leadley, P., H.M. Pereira, R. Alkemade, J.F. Fernandez-Manjarres, V. Proenca and J.P.W. Scharlemann et al. (2010), 'Biodiversity scenarios: projections of 21st century change in biodiversity and associated ecosystem services', in Secretariat of the Convention on Biological Diversity (ed.), Montreal, Technical Series No. 50.

Liu, J.G., G.C. Daily, P.R. Ehrlich and G.W. Luck (2003), 'Effects of household dynamics on resource consumption and biodiversity', *Nature*, **421**(6922), 530–33.

Loreau, M. and A. Hector (2001), 'Partitioning selection and complementarity in biodiversity experiments', *Nature*, **412**(6842), 72–6.

Martine, G., G. McGranahan, M. Montgomery and R. Fernandez-Castilla (eds) (2008), *The New Global Frontier: Urbanization, Poverty and Environment in the 21st Century*, London: Earthscan.

McGranahan, G. and D. Satterthwaite (2002), 'The environmental dimensions of sustainable development for cities', *Geography*, **87**(3), 213–26.

McKee, J.K., P.W. Sciulli, C.D. Fooce and T.A. Waite (2004), 'Forecasting global biodiversity threats associated with human population growth', *Biological Conservation*, **115**(1), 161–4.

McKinney, M.L. (2001), 'Role of human population size in raising bird and mammal threat among nations', *Animal Conservation*, **4**(1), 45–57.

McKinney, M.L. (2002), 'Urbanization, biodiversity, and conservation', *Bioscience*, **52**(10), 883–90.

McMahon, S.M., S.P. Harrison, W.S. Armbruster, P.J. Bartlein, C.M. Beale and M.E. Edwards et al. (2011), 'Improving assessment and modelling of climate change impacts on global terrestrial biodiversity', *Trends in Ecology & Evolution*, **26**(5), 249–59.

McNally, C., E. Uchida and A. Gold (2011), 'The effect of a protected area on the tradeoffs between short-run and long-run benefits from mangrove ecosystems', *Proceedings of the National Academy of Sciences, USA*, **108**(34), 13945–50.

Millennium Ecosystem Assessment (MA) (2005), *Ecosystems and Human Well-being: Synthesis. Summary for Decision-makers*, Washington, DC: Island Press.

Mills, J.H. and T.A. Waite (2009), 'Economic prosperity, biodiversity conservation, and the Environmental Kuznets Curve', *Ecological Economics*, **68**(7), 2087–95.

Miteva, D.A., S.K. Pattanayak and P.J. Ferraro (2012), 'Evaluation of biodiversity policy instruments: what works and what doesn't?', *Oxford Review of Economic Policy*, **28**(1), 69–92.

Myers, N. (2003), 'Conservation of biodiversity: how are we doing?' *The Environmentalist*, **23**(1), 9–15.

Myers, N., R.A. Mittermeier, C.G. Mittermeier, G.A.B. da Fonseca and J. Kent (2000), 'Biodiversity hotspots for conservation priorities', *Nature*, **403**(6772), 853–8.

Naeem, S., L.J. Thompson, S.P. Lawler, J.H. Lawton and R.M. Woodfin (1994), 'Declining biodiversity can affect the functioning of ecosystems', *Nature*, **368**(6473), 734–7.

Naidoo, R. and W. L. Adamowicz (2000), 'Effects of economic prosperity on numbers of threatened species', *Conservation Biology*, **15**(4), 1021–9.

Naughton-Treves, L., J. Alix-Garcia and C.A. Chapman (2011), 'Lessons about parks and poverty from a decade of forest loss and economic growth around Kibale National Park, Uganda', *Proceedings of the National Academy of Sciences, USA*, **108**(34), 13919–24.

Omer, O., U. Pascual and N.P. Russell (2007), 'Biodiversity conservation and productivity in intensive agricultural systems', *Journal of Agricultural Economics*, **58**(2), 308–29.

Pereira, H.M., P.W. Leadley, V. Proenca, R. Alkemade, J.P.W. Scharlemann and J.F. Fernandez-Manjarres et al. (2010), 'Scenarios for global biodiversity in the 21st century', *Science*, **330**(6010), 1496–501.
Perrings, C. and G. Halkos (2010), 'Biodiversity loss and income growth in poor countries: the evidence', ecoSERVICES Working Paper, Tempe, Arizona: ASU.
Pfaff, A. and J. Robalino (2012), 'Protecting forests, biodiversity and the climate: predicting policy impact to improve policy choice', *Oxford Review of Economic Policy*, **28**(2), 164–79.
Qualset, C.O., P.E. McGuire and M.L. Warburton (1995), 'Agrobiodiversity: key to agricultural productivity', *California Agriculture*, **49**(6), 45–9.
Rausser, G.C. and A.A. Small (2000), 'Valuing research leads: bioprospecting and the conservation of genetic resources', *Journal of Political Economy*, **108**(1), 1–33.
Sedjo, R. (1992), 'Property rights, genetic resources, and biotechnological change', *Journal of Law and Economics*, **35**(2), 199–213.
Selden, T. and D. Song (1994), 'Environmental quality and development: is there a Kuznets Curve for air pollution emissions?' *Journal of Environmental Economics and Management*, **27**(2), 147–62.
Simpson, R.D., R.A. Sedjo and J.W. Reid (1996), 'Valuing biodiversity for use in pharmaceutical research', *Journal of Political Economy*, **104**(1), 163–85.
Sims, K. (2010), 'Conservation and development: evidence from Thai protected areas', *Journal of Environmental Economics and Management*, **60**(2), 94–114.
Smale M. (2006), *Valuing Crop Biodiversity*, Wallingford: CABI.
Smale, M., J. Hartell, P.W. Heisey and B. Senauer (1998), 'The contribution of genetic resources and diversity to wheat production in the Punjab of Pakistan', *American Journal of Agricultural Economics*, **80**(3), 482–93.
TEEB (The Economics of Ecosystems and Biodiversity) (2010), *Mainstreaming the Economics of Nature: A Synthesis of the Approach, Conclusions and Recommendations of TEEB*, United Nations Environment Programme.
ten Kate, K. and S.A. Laird (2000), *The Commercial Use of Biodiversity: Access to Genetic Resources and Benefit Sharing*, London: Earthscan.
Tilman, D. and J.A. Downing (1994), 'Biodiversity and stability in grasslands', *Nature*, **367**(6461), 363–5.
Tilman, D., D. Wedin and J. Knops (1996), 'Productivity and sustainability influenced by biodiversity in grassland ecosystems', *Nature*, **379**(6567), 718–20.
Turner, W.R., K. Brandon, T.M. Brooks, C. Gascon, H.K. Gibbs and K. Lawrence (2012), 'Global biodiversity conservation and the alleviation of poverty', *BioScience*, **62**(1), 85–92.
UN (2011), *World Population Prospects: The 2010 Revision, Highlights and Advance Tables*, Department of Economic and Social Affairs, Population Division Working Paper No. ESA/P/WP.220, New York: United Nations.
UN (2012), *World Urbanization Prospects, The 2011 Revision*, online data, accessed 4 April 2012 at http://esa.un.org/unpd/wup/index.htm.
van Dusen, M.E. and J.E. Taylor (2005), 'Missing markets and crop diversity: evidence from Mexico', *Environment and Development Economics*, **10**(4), 513–31.
Widawsky, D. and S. Rozelle (1998), 'Varietal diversity and yield variability in Chinese rice production', in M. Smale (ed.), *Farmers, Gene Banks, and Crop Breeding*, Boston, MA: Kluwer.
Wilkie, D.S., G.A. Morelli, J. Demmer, M. Starkey, P. Telfer and M. Steil (2006), 'Parks and people: assessing the human welfare effects of establishing protected areas for biodiversity conservation', *Conservation Biology*, **20**(1), 247–9.
Wilson, E.O. (1986), 'The current state of biodiversity', in E.O. Wilson (ed.), *Biodiversity*, Washington, DC: National Academy Press, pp. 3–18.
Wood, D. and J.M. Lenné (eds) (1999), *Agrobiodiversity: Characterization, Utilization and Management*, Wallingford: CABI.
World Bank (2011), *The Changing Wealth of Nations: Measuring Sustainable Development in the New Millennium*, Washington, DC: World Bank.
WRI (World Resources Institute) in collaboration with United Nations Development Programme, and World Bank (2005), *World Resources 2005: The Wealth of the Poor – Managing Ecosystems to Fight Poverty*, Washington, DC: WRI.
Zhu, Y., H. Chen, J. Fan, Y. Wang, Y. Li and J. Chen et al. (2000), 'Genetic diversity and disease control in rice', *Nature*, **406**(6797), 718–22.
Zwane, A.P. (2007), 'Does poverty constrain deforestation? Econometric evidence from Peru', *Journal of Development Economics*, **84**(1), 330–49.

19. Biodiversity conservation and ecosystem services provision: a tale of confused objectives, multiple market failures and policy challenges

Jessica Coria, Elizabeth Robinson, Henrik G. Smith and Thomas Sterner

19.1 INTRODUCTION

Human impacts on the environment are intensifying, raising seriously the fundamental questions of how to protect nature itself and how to best allocate the limited resources available for biodiversity conservation. Protecting 'Nature' is, however, a complex task that begs the question of what nature to protect and how. We believe there has been a tendency to use the word 'biodiversity' as a proxy for what we want to protect since it sounds more precise. This is, however, somewhat of a misunderstanding: biodiversity can be measured in a number of different ways and some measures would not give the answers we are intuitively seeking. This brings us to the question of whether we should put the emphasis on biodiversity itself or on the ecosystem services that may flow from a well-functioning ecosystem? The first approach focuses on biodiversity for its own sake, independent of human needs or preferences. The second focuses on preserving biodiversity for its role in promoting ecosystem services for the provision of goods, basic-life support services and human enjoyment of nature (Balvanera et al., 2001; Mace et al., 2012). So far there has been a relative imbalance in the attention devoted to the former as most research and funding in conservation has been oriented toward biodiversity per se, with until recently, little concrete effort towards conserving biodiversity for its role in ecosystem service production.

Conservation efforts often focus on maintaining biodiversity as measured by the number of species present at a particular place. Since threats to biodiversity are distributed unevenly, the spatial limits of biodiversity are recognized, spawning a group of terms – α, β and γ diversity – differentiating between local species richness (α diversity, the number of species at a location), the regional species pool (γ diversity, the number of different species that could be at a location) and variability between localities (β diversity). In the long term, however, persistence of species requires not only maximizing their representation in places where they are currently present, but crucially also minimizing the probability of their being lost (Rodrigues et al., 2006). The International Union for the Conservation of Nature (IUCN) Red List of Threatened Species (www.iucnredlist.org) is the accepted standard for species global extinction risk. Both governmental and non-governmental organizations increasingly rely on the IUCN Red List to inform priorities, influence legislation and guide conservation investment (Rodrigues et al., 2006).

Concentrating on the number of species and taking threat rankings at face value to define conservation priorities reduces biodiversity to a simple metric that is easy to comprehend; however, when it comes to the link between biodiversity and ecosystem functioning, much evidence points to a strong role for species identity (Bengtsson, 1998; Díaz et al., 2005). That is, the composition of ecological communities and even the presence of individual dominant species (e.g., keystone species or ecosystem engineers) can play a key role in controlling ecosystem function, rather than the number of species per se (Chapin et al., 2000).

What is the implication for policy-making? Emphasizing the conservation of biodiversity instead of biodiversity-related ecosystem services may not lead to very different policy conclusions if biodiversity would just have direct, indirect and option use values (for instance, values assigned to products harvested by people, resilience and potential future uses, respectively). However, biodiversity conservation is also justified by the existence of underlying positive values that are independent of utilitarian viewpoints. These views stem in part from a recognition that biodiversity represents a form of human heritage and that the attitudes to nature are driven by feelings of responsibility or stewardship over inherited natural systems. This has been codified in the Convention on Biological Diversity, stating that biodiversity should be protected also for its own sake.

On the other hand, economic valuation of biodiversity-related ecosystem services is highly dependent upon our understanding of the relationship between biodiversity conservation and ecosystem services; ascribing direct, indirect and option use values to ecosystem services requires an understanding of the interlinked production of services and the integration of multiple services temporally and at regional and global scales since (like ecological functioning) economic values are context, space and time dependent.

In terms of ecological functioning, the extent to which the choices of protecting biodiversity versus biodiversity-related ecosystem services are likely to coincide depends on complex and yet little understood interactions between biodiversity and resulting ecosystem services (Mace et al., 2012). For instance:

- There is theoretical and empirical work that suggests a positive correlation between changes in biodiversity and the way ecosystems function (Hooper et al., 2005; Worm et al., 2006; Rey Benayas et al., 2009). For example, conservation on larger scales may result in significant reduction of carbon emissions from deforestation (Venter et al., 2009). But there is also evidence in the other direction, that is, many ecosystem services are provided not by whole ecosystems but by functional groups of species that are either resilient to change, or easily substitutable. Consequently these services will continue to be provided despite rare species fluctuation or loss (Ridder, 2008; Gaston and Fuller, 2008; Gaston, 2010). The importance of species richness, however, may increase with the number of ecosystem functions considered (Gamfeldt et al., 2008).
- Many ecosystem services may be little affected by small losses of biodiversity, but changes beyond certain thresholds or 'tipping points' might result in drastic ecosystem changes (Balvanera et al., 2001). High species diversity may result in response diversity, increasing the adaptive capacity of ecosystems (Elmqvist et al., 2003).
- Finally, the effect of particular species on ecosystem functioning may vary across space and time, such that a species redundant at a particular time or space may still be important for the long-term stability of ecosystem processes (Hooper et al., 2005; Lyons et al., 2005). A more even distribution in abundance of species

may enhance ecosystem functioning and increase the flow of ecosystem services (Crowder et al., 2010).[1]

Thus, the two choices of protecting biodiversity or protecting ecosystem services are more likely to coincide when (1) the number of ecosystem services considered increases (Isbell et al., 2011); (2) functional redundancy is important as a buffer against random natural events (Hooper et al., 2005); and (3) the focus is on stability rather than flow (Srivastava and Vellend, 2005). The two approaches are less likely to coincide if species conservation only focuses on species richness, with no regard for their frequency distribution.

In practice, for most biodiversity-related ecosystem services the devil will be in the details; uncertainty regarding the relationship between biodiversity and ecosystem service flows, the role of biodiversity as a buffer to change and the occurrence of irreversible tipping points is central to the design of successful conservation policies. In any case, focusing on conserving biodiversity at local scales for the benefit of ecosystem services may result in ignorance of conserving species of conservation concern, and it is unlikely to be enough to fulfil the ambition in the Convention on Biological Diversity to preserve species for future generations. Rather, synergies between conservation efforts for species per se and conservation for the benefit of biodiversity ecosystem services need to be considered to make conservation cost-efficient, while many management actions still need to focus on the separate goals for conservation.[2]

A wide variety of policy approaches can be used to enhance biodiversity conservation and increased production of biodiversity-related ecosystem services. These comprise legal and ethical tools such as liability laws, property rights and moral suasion; institutional innovations; command-and-control approaches such as product, input, or technology standards; and economic incentive approaches such as subsidies or tax reductions for adopting desired technology and production practices, or the levying of taxes and fees on sanctioned engagement in otherwise prohibited behaviours. All of these approaches currently are employed to varying degrees. However, the need to explore innovative financial mechanisms for the implementation of the Convention on Biological Diversity and its 2010 biodiversity targets was explicitly established in the Millennium Ecosystem Assessment (2005). There exists a belief that the expansion or creation of new market-based approaches to providing ecosystem services will better achieve conservation goals by making conservation financially feasible, despite the many market failures affecting the provision of some ecosystem services (i.e., the public good nature of the services, externalities affecting their provision, incomplete information, and so on). Payments for environmental services (PES) are increasingly becoming the subject of national and international development strategies as means to finance biodiversity conservation around the world (Landell-Mills and Porras, 2002; Pagiola et al., 2002; Farley and Constanza, 2010). Indeed, in barely three decades a rapidly growing number of ecosystem functions have been characterized as services, valued in monetary terms and, to a lesser extent, incorporated into markets and payment for ecosystem services (Gómez-Baggethun et al., 2010).

In this chapter we analyse the many complexities involved in the design of 'optimal policies' to protect (1) biodiversity versus (2) biodiversity-related ecosystem services and analyse to what extent the criteria in (1) and (2) set against each other or create synergies. We also analyse how payments for ecosystem services affect the relationship between biodiversity and ecosystem services conservation.

This chapter is organized as follows. Section 19.2 describes the different categories of ecosystem services defined as discrete and identifiable end products, which is in turn a prerequisite for the establishment of ecosystem markets. Section 19.3 analyses the economic importance of space and time and how they affect the relationship between biodiversity conservation and biodiversity-related ecosystem services. Section 19.4 analyses the role of payments for ecosystem services as a means to internalize use values derived from ecosystems and to finance biodiversity conservation. Finally, Section 19.5 concludes the chapter.

19.2 WHAT ECOSYSTEM SERVICES SHOULD WE PRESERVE AND HOW SHOULD THEY BE MEASURED?

Ecosystem processes and raw materials are essential in food, fibre and biochemical production; they play a major role regulating air quality, climate and water, and they also provide cultural and aesthetic benefits. If ecosystem services are to provide an effective framework for natural resource decisions, they must be classified in a way that allows comparisons and trade-offs amongst the relevant set of potential benefits. According to the Millennium Ecosystem Assessment (2005), the full range of benefits from ecosystems can be organized as in Table 19.1.

A related question regards whether ecosystem services arising from natural resource management should be described primarily in terms of delivering a particular structure and composition of services, or in terms of maintaining a certain range, redundancy and intensity of ecosystem processes. Defining ecosystem services as discrete and identifiable end products is necessary for quantification, which in turn is a prerequisite for the establishment of ecosystem service markets (Kroeger and Casey, 2007). Indeed, the 'commodification process' involves three main stages (Kosoy and Corbera, 2010): (1) framing an ecological function as a service, (2) assigning it a single exchange value, and (3) linking providers and users of the services in a market exchange. Some, however (including Kosoy and Corbera, 2010), have argued that this commodification might have negative ethical consequences on the way nature is perceived and the way human nature interactions are constructed. Further, they argue that it has implications for unequal social relations because ecosystems complexity may be disregarded in order to facilitate market transactions based on a single exchange value, which imposes a trend towards monetary, market-driven conservation. This market-driven approach is perceived to fail to account for value in a broader sense (beyond monetary value) and so obliterate other social and ecological qualities embedded in these services, which are perceived at different scales by those who benefit from the ecosystem services. Furthermore, power asymmetries across those involved in price formation mechanisms would arise due to the unequal access to these services, for example, market-driven conservation obscures the existence of structural poverty conditions and the fact that the poor sell cheap.

The above suggests that caring for nature is incompatible with valuing nature in terms of its contributions to human well-being. However, arguably there are bigger threats to both nature and to the poor than the commodification of ecosystem services. In fact, many of the most destructive practices of large-scale clear cutting of forests, overharvesting or monoculture cultivations are undertaken precisely to provide market-

Table 19.1 Categories of ecosystem service and examples of related services

Type of Service	Service
Provisioning services	Food
	Fibre
	Genetic resources
	Biochemicals, natural medicines, etc.
	Ornamental resources
	Fresh water
Regulating services	Air quality regulation
	Climate regulation
	Water regulation
	Erosion regulation
	Disease regulation
	Pest regulation
	Pollination
Cultural services	Cultural diversity
	Spiritual and religious values
	Recreation and ecotourism
	Aesthetic values
	Knowledge systems
	Educational values
Supporting services	Soil formation
	Photosynthesis
	Primary production
	Nutrient cycling
	Water cycling

Source: Millennium Ecosystem Assessment (2005).

able 'ecosystem services' – that is, food and fibre – that are sold to the benefit of some private entrepreneur. One can argue that, for example, it is the lack of commodification of the standing forest – the lack of institutions, laws and policies that enable the realization of the value of the forest in situ – that facilitates its destruction. We thus see that there is also a private/public dimension to the debate on ecosystem services and when we speak of 'PES' – payments for ecosystem services – we refer typically to benefits and sometimes even payments that benefit communities and individual landowners and that encourage these communities and individuals to conserve rather than convert the ecosystem.

Valuation of biodiversity-related ecosystem services is one piece of helpful information in the complex task of sustainably managing our natural assets. If biodiversity can be valued inherently, incorporating ecosystem services into conservation agendas becomes less of a paradigm switch and more of a shift in emphasis or an expansion. The real test, however, of whether an ecosystem service will facilitate conservation is not whether academics can evaluate the service but whether someone – or some organization – is able and willing to do what is necessary to secure it. As pointed out by Chan et al., (2007), western conservationists are clearly more willing and able to pay for biodiversity conservation than, say, African peasants (who may be willing but unable) and multinational corporations (which may be able but unwilling).

19.3 PROTECTING BIODIVERSITY VS BIODIVERSITY-- RELATED ECOSYSTEM SERVICES: TRADE-OFFS ACROSS SPACE, TIME, ECOSYSTEM SERVICES AND IRREVERSIBILITY

Various factors affect the capacity of stocks of biological natural capital to supply ecosystem services. The biodiversity/service production function is determined by the complex interplay of biological and physical processes that vary across scales and ecosystems. For instance, in the case of regulating services (e.g., climate regulation) and supporting services (e.g., nutrient cycling), aggregate stocks of natural capital are more important than in the case of provisioning services, where the value of flows of services is most dependent on local stocks and the technology available to convert stocks into valuable supplies of goods (Vira and Adams, 2009).[3] The relationship between cultural services and stocks of biodiversity is a little more complex and can vary over time; declining stocks can increase the marginal cultural values attached to surviving individuals of a particular species, suggesting an inverse relationship between the value of flows of services and the stock size, and, paradoxically for conservationists, that stock recovery may actually reduce the value of associated cultural service flows (Gault et al., 2008).

On the other hand, ecosystem management choices made by humans change the type, magnitude and mix of services provided by ecosystems (ES). Typically, trade-offs occur as the provision of some ES are reduced as a consequence of increased use of another (Foley et al., 2005; Rodriguez et al., 2006). For instance, enhancing the provision of services such as food and timber has led to declines in many other ecosystem services, including regulating services as nutrient cycling and flood regulation. These trade-offs may be an explicit choice, but they might also arise due to the lack of knowledge about the interactions among ecosystem services that occur in space and time. When it comes to time, management decisions are often said to be too focused on the immediate provision of an ES at expenses of this same ES or other services in the future (Rodriguez et al., 2006). An economic analysis would suggest that this is particularly the case when ownership of resources is not secure, thereby reducing the incentives to plan for future benefits. A particular aspect of the temporal dimension is that there may be a degree of irreversibility that refers to the likelihood that a perturbed ES may not return to its original state if its provision actually destroys or irreversibly damages important stocks of natural capital (Scheffer, 2009). When it comes to space, the analysis of spatial patterns of ecosystem services helps us to understand how the distributions of different services compare, how the value of an ecosystem service varies across an ecological and human landscape (mediated, for example, by patterns of human settlement, population densities and transport costs), and where trade-offs and synergies might occur (Raudsepp-Hearne et al., 2010). However, simple mapping of ecosystem services in relation to local management practices may be an oversimplification, since the flow of many services is not only affected by the local conditions, but also by the management of the surrounding landscape (Tscharntke et al., 2005).

There is a growing literature on the economic importance of space and time for biodiversity conservation and biodiversity-related ecosystem services that we review in the following sections.

19.3.1 Optimizing Resource Use Over Time and Irreversibility

Since the early 1900s authors like Hotelling and Ramsey have raised concerns on the external effects that resource depletion could have on future generations, and elaborated on dynamic optimization models to determine the allocation of resource use over time that maximize firms' profits and/or social welfare. When it comes to non-renewable resources – which unlike renewable resources exhibit no growth or regenerative processes – the fundamental result of the seminal work by Hotelling (1931) is that the price of a non-renewable resource in a competitive market would raise at the interest rate and that the quantity extracted should continuously decline until the resource is exhausted. The simple rule was derived under very restrictive assumptions, and resource economists have developed more complex versions of the Hotelling rule, attempting to improve its empirical validity by adding more realistic assumptions.[4] However, the fundamental meaning of the rule remains the same: optimality requires that resource prices reflect the scarcity of non-renewable resources. A similar story applies to renewable natural resources. Ever since Clark's famous paper on the discount rate and the extinction of animal species (Clark, 1973), it has been recognized that a higher discount rate increases the optimal rate of exploitation of a renewable resource, and so increases the likelihood of extinction, yet latter studies have shown that the discount rate might have an uncertain effect in some circumstances (see Farzin, 1984).[5]

What is the effect of discounting on the relationship between the focus on biodiversity versus biodiversity-related ecosystem services' conservation? It might seem obvious that focusing on flows would benefit the present and focusing on stocks would benefit the future. In the simple world of economic textbooks, this is, however, not so. With correct discounting and with correct measurement of nature's stocks and flows, it would not make any difference. Optimization of flows would take into account the need – also in the future – of intact stocks.

In reality, with poor measurement and understanding of nature, with incomplete property rights and excessively high discount rates, the choice between nature's stocks and ecosystem services will pit the interests of current and future people against one another. On the one hand, conservation efforts intended to dissipate development or resource extraction pressures in order to protect biodiversity necessarily entail costs to those local people who currently rely on such activities, in the form of a reduced flow of provisioning services. On the other hand, different categories of biodiversity-related ecosystem services tend to change over different time scales, making it difficult for policy-makers to evaluate fully the trade-offs among them.

If uncertain property rights are the cause of current rates of overexploitation of many natural resources, we would need some policy interventions. The solution is easier said than done: establish and enforce well-defined property rights. However, the very lack of enforced property rights suggests that the government itself is either unwilling or unable to solve the problem. As discussed by Ostrom (1990), one fundamental cause may be that common pool resources have a productivity that is too low or too variable to motivate the cost of creating and defending private property rights (for instance, the cost of fencing).

Inertia strongly influences the time frame for solving different ecosystem-related problems. For example, supporting services such as soil formation and primary production and regulating services such as water and disease regulation tend to change over much longer time scales than provisioning services. Agricultural intensification results in a slow loss of soil biodiversity with negative consequences for productivity, but may be compensated by increased used of agro-chemicals (Giller et al., 1997). As a consequence, impacts on more slowly changing supporting and regulating services are likely to be overlooked in pursuit of increased use of provisioning services (Millennium Ecosystem Assessment, 2005).

A related aspect of the temporal dimension of the management of natural resources is irreversibility. Despite most of the time, change in ecosystems and their services is gradual and incremental (and detectable and predictable, at least in principle), there are many examples of non-linear and sometimes abrupt changes in ecosystems. In these cases, the ecosystem may change gradually until a particular pressure on it reaches a threshold, at which point changes occur relatively rapidly as the system shifts to a new state. For instance, some coral reef ecosystems have undergone sudden shifts from coral-dominated to algae-dominated reefs. The trigger for such phase shifts – which are essentially irreversible (at least on the time scales traditionally used in economic planning) – is increased nutrient input, leading to eutrophic conditions, and removal of herbivorous fishes that maintain the balance between corals and algae. Once a threshold is reached, the change in the ecosystem takes place within months and the resulting ecosystem, although stable, is less productive and less diverse.

Tipping points concerning biodiversity-related ecosystem services mediated by biodiversity change may be caused by multiple drivers such as land-use change, climate change, overexploitation and pollution. Tipping points occur when effects of global change are augmented by positive feedbacks, when there are thresholds beyond which shifts to alternative stable states occur, when changes induced by drivers are long-lasting and hard to reverse and there are significant time lags between dynamics of drivers and the expression of impact on the system (Leadley et al., 2010). For example, a linear loss of biodiversity may, through loss of keystone species or loss of key functional groups, result in abrupt non-linear changes in ecosystem functioning and therefore the flow of ecosystem services (Chapin et al., 2000). Habitat loss may result in sudden extinction of species when critical thresholds are reached (Hanski, 1999). Pollinator loss may result in parallel loss of pollinated plants (Biesmeijer et al., 2006), which may feedback on the population dynamics of the pollinators. Empirical studies have confirmed the existence of tipping points in different ecological systems (Scheffer et al., 2001; Leadley et al., 2010). When there are alternative stable states, the process of change may be difficult to reverse (Sheffer, 2009; Leadley et al., 2010). For example, when change results in species extinctions, this may cause change to become irreversible unless the species can recolonize from elsewhere. In spite of much evidence, it should be pointed out that experimental evidence demonstrates that alternative stable states are but one of several ways in which ecosystems can respond to change (Schröder et al., 2005), and that it is a major empirical question to determine the extent of alternative stable states in real ecosystems.

Regime shifts affect the services that people derive from ecosystems. Capabilities for predicting some non-linear changes are improving, but for most ecosystems and for most potential non-linear changes, while science can often warn of increased risks of change, it cannot predict the thresholds where the change will be encountered. Marginal changes in management methods could thus have huge effects on some ecosystems and on future management opportunities. Due to multiple steady states, several strategy paths may fulfil marginal conditions derived using optimal control theory or dynamic programming, but some of them may be suboptimal (Crepin, 2007).

The presence of thresholds, combined with uncertainty motivates precautionary approaches because social planners must account for the option value associated with the system's future potential production of goods and services. The economics literature (Arrow and Fisher, 1974; Henry, 1974; Epstein, 1980; Freixas and Laffont, 1984) shows substantial evidence of an irreversibility effect – risk-neutral and risk-averse agents should act more carefully to prevent future irreversible risks if more information is expected to be available in the future. Thus, the existence of biodiversity-related alternative stable states, suggest that we should determine boundaries for the degradation of biodiversity that should not be passed (Rockström et al., 2009).[6]

Finally, it has been argued that renewable natural capital, such as biodiversity, cannot be treated in the same way as manufactured and financial capital (Blignaut and Aronson, 2008). The reason is that the value of natural capital will not erode over time. Furthermore, the cost of restoring biodiversity may be larger in the future than today. Therefore, the value of remaining natural capital (biodiversity) must increase by at least an equivalent degree as the rate of depletion. Hence, biodiversity as a natural capital may need to be valued using a negative discount rate. When conserving biodiversity for its own sake, this is not related to an economic argument, but to an ethical one, as, for example, codified by the Convention on Biological Diversity. What is the value of a conserved species in the future, when it comes to ethics? It is hard to envision that the value should be less; if anything it will be higher due to fewer species existing in the world (with all likelihood). It has been argued that approaches other than economic arguments may not be applicable for the conservation of species for ethical reasons, but that other approaches such as a safe minimum standard approach should be preferred (Bulte and van Kooten, 2001).

19.3.2 Scale and Spatial Models of Biodiversity-related Ecosystem Services

Scale may influence the management of biodiversity for ecosystem services for two major reasons. First, ecological processes may be scale-dependent, necessitating management at scales larger than, for example, an individual property. If relevant biodiversity only responds to management at large scales then actions of property owners need to be coordinated (Zang et al., 2007). Second, if the effect of biodiversity conservation occurs at a larger spatial scale than the individual property or is displaced from the individual property, there may be little incentive for the landowner to benefit biodiversity to promote ecosystem services (Lant et al., 2008; Fisher et al., 2009). For example, promoting crop pollination by enhancing the abundance and diversity of pollinators may require incentives in the form of agri-environment schemes and coordination of the action across several farms.

Moreover, since the internalization of environmental externalities generally relies on political institutions, and since these tend to be most responsive to local concerns, externalities that involve immediate local costs will tend to be addressed before externalities generating costs that occur in the future at distant locations (Barbier, 1997). Furthermore, the value of local conservation may well differ among countries due to different income levels. Indeed, evidence has shown that biodiversity conservation has been especially challenging in the poorest countries, partly because conservation through the designation of protected areas has often adversely affected the people displaced in the process. Even though Integrated Conservation and Development Projects (ICDPs) have been implemented to address this problem by including compensation for those who have lost access to resources due to the biodiversity conservation, they have led to mixed results (Tallis et al., 2008; Coria and Calfucura, 2012).

On the other hand, the local benefits yielded by global biodiversity conservation are closely connected to the productivity of managed agricultural or forest land. Since the proportion of the population that directly depends on agriculture is generally highest in poor countries, the localized consequences of global biodiversity conservation (both positive and negative) are generally expected to be greatest in those countries (Perrings, 2007). In some sense, the main international mechanism for making payments for biodiversity-related development assistance, the Global Environment Facility, was established precisely to cover the global interest in local conservation. The concept of incremental cost, which in principle determines payments made by the GEF, is a measure of the difference between the cost a country would be prepared to bear in the provision of an environmental public good and the cost of meeting global demand for the same public good. As pointed out by Perrings and Halkos (2012), the fact that the GEF is widely regarded by scientists as underfunded is an indication that donor countries have a lower estimate of the incremental cost of local conservation than the scientific community.

Policy-makers often implement explicitly spatial policies that recognize the spatial nature of biodiversity and ecosystem services, taking into account, for example, sizing and siting of protected areas, edge effects and corridors. As an example, the SLOSS debate (single large or several small) addressed the impact on biodiversity conservation of protecting a single large reserve as against several smaller reserves. Larger reserves have the benefit of supporting more species (species richness) than a number of small reserves of the same total area and so were thought to be better for biodiversity conservation. But more recently some have argued that, depending on the extent of and nature of spatially correlated risks, having several smaller reserves can spread the risk of not losing a species entirely if fire, disease, or pests wipe out one protected area (Busby et al., 2012), and many small reserves may better capture spatial variation in ecological conditions (Tscharnkte et al., 2002; Moilanen et al., 2009).

Even where biodiversity-related ecosystem services and biodiversity conservation are scale-neutral, edge effects and fragmentation may well be an issue. Borders of protected areas can become sinks and so the ratio of perimeter length to area of the reserve becomes important, particularly for fauna with wide-ranging areas (Woodroffe and Ginsberg, 1998). Fragmentation increases considerably the ratio of de facto perimeter to area and is believed to be a key driver of biodiversity loss (Busby et al., 2012). (On the other hand the 'perimeter' may also be positive from the viewpoint of human access to ecosystem services or the flow of ecosystem services from the reserves into other the surrounding landscapes [Smith et al., 2010].) Fragmented patches of habitat may exhibit lower resilience and therefore reduced potential to provide ecosystem services. Corridors are a spatial policy that enable the creation of an integrated network of protected areas and therefore reduce inter-reserve fragmentation (Leal et al., 2005). Wildlife corridors are increasingly seen as an important approach to conserving biodiversity (Lindenmayer and Nix, 1993; Gilbert-Norton et al., 2010; Hodgson et al., 2011), or to reintroduce biodiversity such as into large-scale agro-ecosystems that have lost their natural vegetation (Altieri, 1999).

Much of the ecological-oriented literature has in the past implicitly ignored how people react to policy and the feasibility of regulations over access to and use of reserves. For example, when a protected area or reserve is planned, and indeed most of the SLOSS debate, there is an implicit assumption that planners can achieve what they want to achieve – that is, that there is going to be full compliance with choices over boundaries and access rules. Further, only recently has the issue of displacement of anthropogenic activities from a protected area to less protected areas been included in analyses of the effectiveness of efforts to protect biodiversity and ecosystem services.

These changes in people's behaviour as a response to environmental policy are reflected in terms of labour allocation, what resources and what quantities households collect, and which forests and which areas of forests are used, implying that the spatial pattern of resource extraction is likely to be changed, not just within the confines of the particular protected area, and therefore the spatial pattern of biodiversity and ecosystem services from the landscape as a whole. Changes in spatial patterns of biodiversity measures and ecosystem services are therefore often the result of human activities that are responses to policies that may not themselves be explicitly spatial. And because human actions change landscapes, and therefore implicitly patterns and qualities of biodiversity and the ability of those landscapes to provide ecosystem services, it is important for policy-makers to understand how their policies influence human decisions to alter those landscapes and so predict spatial patterns of human responses (Bockstael, 1996).

Rural households, particularly in low-income countries, tend to be highly dependent on the local resource base, both for extractive – and therefore ecosystem degrading – and non-extractive ecosystem services. These households might collect fuel wood from a nearby forest; earn a livelihood from fishing; or hunt bush meat for home consumption and for sale. Typically, policies designed to protect biodiversity and ecosystem services in such areas place a burden on these resource-dependent local communities. Indeed, if such policies did not affect local people, and did not affect how these people interact with the resource base, then arguably the policy either would not be needed or would not be effective. Amani Nature Reserve in Tanzania is an example of an area that according to national regulations must be protected from any extraction within its boundaries (Burgess and Rodgers, 2004). However, the Amani management committee recognized that villagers did not have access to alternative forests from which to collect the non--timber forest products such as fuel wood that were essential to their livelihoods, or space and resources to plant trees on their own land. The management therefore introduced a de facto buffer zone within the reserve boundaries, reducing the effective size of the reserve, but recognizing the reality of the socio-economic setting and landscape within which the reserve is situated (Robinson et al., 2011).

Spatial analyses of natural resource use often take as a starting point the impact of distance costs on human activities that affect biodiversity and ecosystem services. Von

Thunen's theoretical model of optimal land use across a landscape has often been taken as the natural starting point for many of the explicitly spatial analyses of deforestation (Chomitz and Gray, 1986; Nelson and Hellerstein, 1997; Pfaff, 1999). The concept comes from economic geography and develops a simple one-dimensional linear landscape in which the relative costs of transporting different agricultural commodities to a central market determine agricultural land use around a city. Economic rents to particular land uses vary with location even with homogeneous resources because market access costs are a function of distance from markets. This model has been used to explore the forest–agriculture margin under different policies. Although a relatively simple concept, much of the empirical deforestation literature finds that better road access and shorter distance to towns correlates to lower levels of forest cover, which implies that patterns of forest cover correspond to patterns of roads and cities. Ahrends et al. (2010) apply such a framework to forest degradation in Tanzania. They find timber-induced degradation in a ring closer to Dar es Salaam with a wider concentric ring of NTFP-induced degradation (that includes charcoal) further away from the city.

Key to research that builds off the von Thunen framework is that transport costs are central to spatial economics, and that location matters, whether that involves getting produce from farm to market, bringing fuelwood from the forest to the hearth, or transporting charcoal to urban centres of demand. These approaches may not be explicitly spatial, but they typically have a spatial implication/consequence. Importantly, this framework predicts that distance matters for how people interact with ecosystems. In empirical papers, distance from village to resource often emerges as a significant determinant of how much forest biomass is collected (MacDonald et al., 2001; Sills and Abt, 2003). Distance to markets is also often significant, reflecting in the literature a recognition that forest product markets are far from perfect, often modelled with the inclusion of some discrete fixed cost to market access (Omamo, 1998; Key et al., 2000; Robinson et al., 2002, 2008).

There have been far fewer attempts to link spatial policies over natural resources to the impact on biodiversity and the resulting ecosystem services provided by the particular resource that takes into account the spatial reaction of people to such policies. A number of papers recognize explicitly that rural people's response to a policy can change the pattern, and therefore value, of ecosystem services, though this link is not always made explicit. For example, several papers highlight the impact of a spatial exclusionary policy (such as the reclassifying of a forest as a REDD [Reducing Emissions from Deforestation and Forest Degradation] forest) on the spatial pattern of forest degradation through 'leakage', the neighbourhood displacement of activities from that newly designated REDD forest to less protected forests. The protection of one particular forest patch is increasingly recognized to have the potential to result in the displacement of some extraction from the now-protected area to less-protected forest areas. For example, there is empirical evidence of an 'intensification' of resource use in a peripheral zone outside of a park in Madagascar (Ferraro, 2002). Lokina et al. (2008) find evidence of such forest deterioration in unprotected forests following the re-enforcement of Tanzania's restrictions on access to government-owned 'preservation' forests. Lewis (2002) suggests that depending on the type of ecosystem, 'concentrating previously dispersed activities into certain parts of the forest may actually increase the negative ecological impact'. That is, the policy of protecting a forest from human degradation and deforestation may at a landscape level have a negative impact on the overall provision of ecosystem services.

How this displacement of activity affects the ecosystem services provided by a particular forest or landscape of forests depends not only on how people change their spatial interaction with the landscape, but also how that change translates into the ecosystem services provided by the landscape. Robinson et al. (2011) introduce the explicitly spatial concept of an 'ecological damage function' (EDF) that maps the spatial degradation of the resource into loss of ecosystem services. In their paper a very simple model is employed, which assumes that there is one resource providing one ecosystem service. The authors identify four EDFs that reflect stylized representations of ecosystems in practice. A pristine-only EDF values only fully protected ecosystem and so represents the most extreme of EDFs in which any degradation renders the particular area where the degradation occurs without ecological value: this is therefore the most extreme threshold. In contrast, the biomass-proportional EDF values the total amount of biomass and not its spatial location, and so is perhaps most relevant to carbon sequestration. The ecosystem services EDF provides less value per unit of biomass if it is degraded but does not lose its value entirely. The eco-threshold EDF recognizes that limited degradation may cause limited ecological damage, but there is some threshold of degradation below which the ecosystem services provided by the protected area fall off rapidly. Many ecosystem services display this kind of non-convexity. As examples, hydrological benefits have been found to be insensitive to slight degradation but there is a point at which a highly degraded forest makes little contribution to water flow (Wu and Boggess, 1999; Robinson et al., 2011). It is likely that for a particular forest or protected area there are multiple ecosystem services and each maps through a different EDF into a different level of ecological damage.

Protected areas are one of the key policy tools for biodiversity conservation. And indeed many studies of protected areas have found that they are effective at reducing degradation – and therefore by implication protecting biodiversity and ecosystem services – within their boundaries. But these studies have typically simply considered whether resources within a protected area are more or less degraded than those outside. Yet unless those protected areas have been located randomly, the results are open to bias (Andam et al., 2008). Policy-makers that want to maximize the impact on deforestation might choose areas most at threat to locate new reserves; whereas policy-makers that want to ensure that protected land remains forested might choose those locations where there is least anthropogenic threat (Pfaff et al., 2009). Similarly, the choice of biodiversity measure is likely to influence reserve location. For example, if valued species are a key indicator then lands that face a lower pressure might be chosen. Globally, less-threatened land has tended to be where reserves are located (Joppa and Pfaff, 2009). There might also be a choice between pure species numbers (which might be highest in some inhospitable swamp) or protecting 'charismatic' species with beauty and viewing value such as found in tiger or panda reserves.

Sizing and siting decisions tend to have different drivers in high-income and low-income countries. In high-income countries costs of purchasing land can change the optimal pattern of conservation (Andam et al., 1998). But in low-income countries how well markets for labour and forest products function influences how people react to a protected area and therefore the anthropogenic pressure on a protected area and the relative costs of protecting a particularly located and sized reserve. Robinson et al. (2011) show conceptually that where markets function well people can more easily switch from reliance on resources inside a protected area to wage and product markets. But when markets do not function so well people are likely either to substitute resources found outside the protected area or simply to continue to use it, albeit now illegally. Biodiversity-rich areas located in remote areas may be subject to fewer pressures from markets and human population. But if rural communities are highly dependent on resources found in these areas, it may be hard to site a reserve because these communities have few alternatives to using the resources found in a proposed reserve. In general, how people react to a protected area, and therefore its effectiveness, is influenced by markets – well-functioning markets allow relatively easy substitution from the protected area to the market.

Buffer zones are an example of an explicitly spatial policy designed to meet the dual goals of protecting park resources and possibly the biodiversity within the park while also providing benefits to nearby households (Wells et al., 1992; Naughton-Treves et al., 2005; Dudley, 2008). The UNESCO Biosphere Reserves are a specific example of spatial policy to achieve multiple aims. The reserves comprise three adjacent areas, a core zone, buffer zone and transition area, which each have different roles with respect to conservation and livelihoods. Yet there has been little use of spatial models to inform the optimal size and access/management rules for buffer zones that take into account how people respond to the spatial pattern of reserve and buffer areas (Robinson et al., 2011).

The protection of biodiversity requires enforcement of access rules, and in low-income countries that enforcement is often limited by scarce resources or a lack of political will and so rarely perfect (Albers, 2010). Yet to date, economists have made very little contribution to addressing the role of enforcement in driving spatial patterns of degradation and fragmentation of habitats. Albers (2010) and Robinson et al. (2011) demonstrate that when enforcement is imperfect, policy-makers may be better off concentrating that enforcement effort into a smaller reserve surrounded by a buffer zone where extraction of resources is permitted, rather than trying and failing to fully protect a larger reserve. These papers are based on the idea that there is a strategic interaction between those with a limited budget responsible for protecting a reserve and those who have traditionally (albeit illegally) extracted from that reserve.

Sims (2013) provides a rare analysis of the interaction between protected areas and spatial patterns of protecting those reserves and the spatial patterns of fragmentation that result from the interaction of illegal extraction and imperfect enforcement. Her research shows how the spatial pattern of enforcement affects the level of fragmentation within a protected area. For example, enforcement that is concentrated along the boundaries of a park may push illegal clearing further into the interior of a protected area, thereby increasing fragmentation. Fragmentation in reserves therefore results from the inability of policy-makers to enforce access restrictions perfectly and a lack of understanding as to how spatial patterns of enforcement affect fragmentation.

19.3.3 On the Need for Multiple Policy Instruments

Biodiversity and ecosystems provide goods and services with a mix of values, some of which are tangible and marketable, whereas others are of a public or common good nature. It is not surprising that the use and non-use values associated with biodiversity and ecosystems will require different policy instruments. Direct-use values, especially when reflecting provisioning services, are privately appropriable, and can often be addressed with economic instruments. Indirect-use values reflecting regulating services such as water regulation and purification, and non use values, including existence values and many cultural services provided by ecosystems, are more difficult to address, since due to their public good nature, markets would lead to insufficient provision. Uncertainty, irreversibility and biodiversity hotspots complicate the problem even more as they require the design of more sophisticated instruments that ensure that the boundaries for the degradation of biodiversity as a whole and in vulnerable areas are not passed. Moreover, even if we leave the question of market failures to one side, we would expect to find differences in the socially optimal level of biodiversity conservation in different countries due to differences in national income, species richness, threats to biodiversity and conservation infrastructure (Perrings and Halkos, 2012), and hence in the stringency of policy instruments for biodiversity conservation across countries.

Due to the mentioned features of the biodiversity conservation problem, it is clear that a single policy instrument is unlikely to achieve optimality. We need a combination of policies to ensure both the protection of the flows and the stock. Furthermore, due to multiple market failures, coordination of policies to address the interaction of these failures will often be more efficient than policies that address these failures separately (Bennear and Stavins, 2007).

The ecological, socio-economic and political constraints of conservation are so difficult to address that there is no generic 'optimal combination of instruments', as they will vary depending on the type of ecosystem service, time and context. Nevertheless, if we build on traditional evaluation criteria, the mixes should promote conservation effectiveness and efficiency to the largest extent possible. In order to promote political feasibility, they should minimize social impacts and be compatible with the institutional frameworks in place. Clearly, this is easier said than done. However, the focus on ecosystem services might help in the task of strengthening the conservation of biodiversity; the current Conventions on Biodiversity are framed in terms of avoiding species loss instead of bringing biodiversity back to reasonable levels. Protecting the flow as well as the stock could be both more relevant to human needs and also more effective as it implicitly means that you would not allow the stock to go under a threshold.

On the other hand, it is important to keep in mind that biodiversity conservation and biodiversity-related ecosystem service management are already subject to different policy mixes that originate from the interaction of environmental policy and different sectorial policies aiming at increasing the provision of certain types of ecosystem services. In most instances these sectorial policies have negative impacts on biodiversity conservation through, for example, infrastructure development, clear cutting of forest to provide land for agriculture, and subsidies for the extraction of raw materials or for energy or fuel that consequently lead to higher impacts on the environment. It is ironic that in reality one of the most relevant issues for biodiversity conservation is not subsidies for conservation but the prevalence of perverse subsidies for overexploitation. Perverse subsidies are so common that subsidy removal is actually classified as an environmental policy instrument (Coria and Sterner, 2011).

19.4 PAYMENTS FOR BIODIVERSITY-RELATED ECOSYSTEM SERVICES AND BIODIVERSITY CONSERVATION

This section starts with a brief description of the concept, and it continues with a discussion of the use of PES in agricultural lands and forests, before dealing with current practical design issues. Regarding the different types of payments for ecosystem services, Scherr et al. (2004) classify them in four categories:

1. Public payment schemes, where the local authorities or the government decide which ES are priorities for conservation and implement payment schemes targeted to preserve these services. Examples include conservation easements (guarantees that such land will not be logged or farmed); programmes to co-finance investments in conservation efforts; and the PES system in Costa Rica.
2. Open trading under regulatory cap or floor, where a government defines a mandatory level of a specific ecosystem service to be provided, but to achieve this level the regulated party can decide whether to directly comply or to compensate by paying others who are in the position to supply the service more cheaply. The most developed programme is for wetland mitigation under the Clean Water Act of 1972, which led to the development of numerous wetland mitigation banks in the USA (Bayon, 2006).
3. Self-organized private deals involving closed transactions between off-site beneficiaries and service providers. Examples include the deal between the Costa Rican National Institute of Biodiversity (INBio) and Merck, Sharpe and Dohme, Inc. The agreement called for INBio to provide Merck with plant, insect and microbiological samples that could be tested for biological activity in exchange for up-front payments to INBio, training opportunities for Costa Rican scientists at Merck facilities, collaborative projects with Costa Rican universities, transfer of processing technologies to INBio, and royalty arrangements developed in the event that marketable pharmaceuticals resulted from the collaboration.
4. Eco-labelling, whereby the ES component is embedded in a traded product; producers sell products produced under a management system certified to enhance environmental service provision. Examples include the Forest Stewardship Council wood and non-wood certification. However, some consider eco-labelling as not being a kind of PES since it is based on a different mechanism to induce sustainable use of biodiversity (Wunder, 2005).

Assessments of PES schemes find that most current schemes tend to be local-level arrangements. In contrast, large-scale PES schemes are mainly government driven, working at the state, provincial, national level and international level, with some of them being part of the carbon sequestration markets created by the Kyoto Protocol on Climate Change, as for instance, REDD and CDM. A major challenge to implement large-scale PES schemes is that the ecosystem services that they target are non--excludable.[7] For instance – as discussed previously – some of the ecosystem services provided by biodiversity, such as climate regulation, are pure public goods, and it is impossible to exclude anyone from benefiting from them. In addition, the climate regulation both spans local and global scales and is omni-directional. For addressing non--

excludability we require instead collective institutions based on cooperation – instead of competition – that either create the conditions required for private sector payments, or accept the public good nature of the services and pay for them directly. Cooperation should be at the same spatial scale over which the service is provided (Farley and Constanza, 2010).

Today, most PES projects have focused on one or more of the following services: biodiversity, carbon and water regulation, or landscape beauty. Furthermore, there is only a limited number of PES projects in the developing world, but the number is rapidly increasing through national programmes (that tend to be government driven) to local PES projects (which are smaller and tend to be financed by the private sector) (Wendland et al., 2010). In what follows, we describe the use of PES for preserving biodiversity-related ecosystem services in agricultural lands and forests.

19.4.1 Payments for Ecosystems Services in Agricultural Lands and Forests

Agri-environmental policies (AEPs) in the United States and the European Union are examples of payments for environmental services. The programmes sponsor environmental services targeted at reducing negative externalities, such as nutrient run-off and soil erosion and an increase in positive externalities, such as scenic vistas, and preserving of farming heritage (Baylis et al., 2008). the USA began providing payments to farmers to protect soil and reduce the production of certain crops in excess of supply in the 1930s, while in Europe AEPs were not developed until the 1980s (ibid.). Even though the motivations for AEPs in the two regions are similar, there are some important differences regarding the services targeted by the programmes. For instance, the bulk of US conservation expenditures target the reduction of the negative externalities produced by agriculture. Thus, until the mid-1980s the focus was preventing the loss of topsoil, but it has been widened since then to incorporate the reduction of agricultural water pollution, as well as ensuring that farming does not result in the draining of wetlands and the loss of wildlife habitat (ibid.). Instead, the EU has taken a wider view, including not only negative externalities but also many aspects of traditional farming that are perceived as desirable outcomes. For instance, the Environmentally Sensitive Areas (ESA) programme, launched in the United Kingdom in 1986, was the first agri-environmental programme in the European Union (EU). The ESA programme was intended to protect valued landscapes and habitats and to improve public enjoyment of the countryside through long-term (usually ten-year) voluntary agreements between the government and farmers. Environmental services purchased under the ESA programme included improved habitat for birds, biodiversity (such as in species rich grassland), landscape beauty and historic preservation (Dobbs and Pretty, 2008).

Furthermore, there are differences in the way that negative externalities are addressed in these two regions: whereas US policy focuses mainly on the by-products of extensification – that is, the use of excessive amounts of environmentally sensitive land – EU AEPs focus mainly on by-products of intensification of farming – that is, the use of too many non-land inputs per unit of land.

Regarding PES aimed to avoid deforestation and/or encourage afforestation, Costa Rica pioneered the use of formal PES mechanisms by establishing a country-wide programme called Pago por Servicios Ambientales (PSA) in 1997, which aimed to reverse the severe deforestation rates existing at that time (Pagiola, 2008). To date, a growing number of PES-like mechanisms are spread thought the world and international cap and trade programmes, such as Reduced Emission from Deforestation and Degradation (REDD), are in place to articulate international PES schemes.

19.4.2 Key Design Issues

An important challenge related to the implementation of PES is the existence of important information asymmetries (Ferraro, 2008). These asymmetries arise since landowners have better information than the conservation agents about the opportunity costs of supplying environmental services when negotiating the contract (adverse selection). Furthermore, since it is costly to monitor contract compliance, the landowners have fewer incentives to fully fulfil the contractual responsibilities (moral hazard). Indeed, in the European Union, great efforts have been spent to monitor if farmers fulfil the contracts for receiving agri-environment schemes. Unfortunately, less effort has been spent on evaluating their efficiency in protecting biodiversity and ecosystem services.

Furthermore, as long as participation is voluntary, payments are flat and there is a big deal of variation in the opportunity costs of supplying environmental services, PES would lead to a concentration of contracts on unproductive lands run by farmers with the lowest land-use intensities (i.e., lowest opportunity cost), a problem known in the literature as 'adverse selection'. Indirect evidence of adverse selection is found in most AEPs. For instance, several major evaluations of the ESA programme conclude that the programme was effective in enrolling many farmers in the entry-level contract tiers, but the scheme did not generally offer sufficient economic incentives to attract high levels of enrolment in the intensive farming areas; incentives offered induced enrolment of a substantial fraction (40–90 per cent) of farmers in eligible areas characterized by grazing and less intensive agriculture, but enrolment was lower in areas characterized by more intensive arable production (Dobbs and Pretty, 2008). This pattern of participation suggests that environmental benefits coming from AEPs are primarily in the form of preserving environmental features already in place: monetary incentives were adequate to arrest intensification in more marginal areas where it might not have been profitable to increase the level of intensification anyway, but insufficient to reverse intensification in more productive arable areas (Kleijn et al., 2001, 2006; Kleijn and Sutherland, 2003, Rundlöf and Smith, 2006). There is also evidence that some of the schemes affect biodiversity at local scales, but few studies have linked local conservation effects to national trends in biodiversity (Kleijn et al., 2011).

The previous evidence does not mean that AEPs do not provide net benefits, but the schemes were limited in their success in enrolling farmers in higher payment tiers, tiers that required more substantial changes in farming practices. Furthermore, an important reason for limited success of AESs is that the landscape context of agri-environment schemes is not considered (ibid.). As a consequence, the uptake of schemes, such as organic farming, is largest in regions where the opportunity cost is lowest and not where the effect on biodiversity is largest (Rundlöf and Smith, 2006). It is also not always clear if the goal of agri-environment schemes is to promote local biodiversity for the benefit of biodiversity-related ecosystem services or national biodiversity for conservation per se. For example, organic farming may promote local biodiversity more in intensively farmed regions, but be more important for rare species in less intensively farmed areas (Smith et al., 2010).

Similar evidence has been found in the case of Costa Rica. For instance, Ortiz et al. (2003) find that 76.8 per cent of the forest area under PSA contracts would likely have been conserved or managed with limited forest interventions in the absence of PSA. Hartshorn et al. (2005) also find more than 70 per cent of PSA forest protection contracts to be on land with production capacities that only allow forest management/protection (51 per cent), or severely limited agriculture (20 per cent). Finally, Sills et al. (2006) detected statistically significant but very small effects on land-use patterns between PSA and non-PSA land.

Paying for forest protection on land that requires no protective measures is an inefficient use of scarce conservation funds (Wünscher et al., 2008). This means that the success of PES promoting conservation largely depends on the programmes' ability to encourage clear additionality. However, additionality is not a selection criterion in most PES programmes, though they could clearly benefit from improved spatial targeting. Babcock et al. (1997) classify targeting approaches for conservation programmes into those that target (1) benefits, (2) costs, or (3) benefit-to-cost ratios. Wünscher et al. (2008) empirically test the potential of the targeting tool to boost the financial efficiency of PES programmes in Costa Rica in terms of expected additional environmental services (benefits) per dollar spent. They address three kinds of possible inefficiencies: (1) the desired land-use activity would be adopted even in the absence of payments; (2) the payment is not high enough to lead to the socially desirable land use; (3) the payment encourages a land use with positive externalities that are worth less than the cost. They address (1) by applying deforestation probabilities: a forest at low risk of deforestation would provide less additionally if enrolled, so it should have a lower likelihood of being selected. They address (2) through flexible payments that are aligned to landowner participation costs. Regarding (3), they reduce the likelihood of the benefit exceeding the cost by selecting first the sites with the largest service–cost ratio. They conclude that targeting to reduce inefficiency (2) yields the most significant increase in environmental service delivery. Nevertheless, estimating participation cost might not be an easy task since non-monetary attitudes of the landowners, for example, risk behaviour, mistrust, or conservation preferences could influence the level of the landowner's real willingness to accept. They suggest inverse auctions as an alternative that could take such subjective attitudes into account. Technically, it appears that an auction system could easily be integrated into most PES programmes in place. If landowners have to apply formally for programme participation, their bid could be part of the application process. They also argue that auction systems might also be a powerful way of making payment differentiation politically acceptable because service sellers suggest the price themselves. However, it could increase the service buyer's transaction costs, at least in the short run, as it would require changes in administrative processes.

Unfortunately, the existence of subsidies that promote excessive use of ecosystem services has had a negative impact on the uptake of AEPs. The high crop and livestock-related payments received by farmers under the EU's Common Agricultural Policy (CAP) contributed to the disincentives to participate, especially in higher tiers (Baylis et al., 2008). Furthermore, they also increase land values, adding to landowners' resistance to subsidy reductions.

Similar problems are created by subsidies to forestry or fishing. Although removal of production subsidies would produce net benefits, it would not occur without costs. The farmers and fishers benefiting directly from the subsidies would suffer the most immediate losses, but there would also be indirect effects on ecosystems both locally and globally. In some cases it may be possible to transfer production subsidies to other activities that promote ecosystem stewardship, such as payment for the provision or enhancement of regulatory or supporting services. Compensatory mechanisms may be needed for the poor who are adversely affected by the immediate removal of subsidies. On the other hand, reduced subsidies within the OECD may lessen pressures on some ecosystems in those countries, but they could lead to more rapid conversion and intensification of land for agriculture in developing countries and would thus need to be accompanied by policies to minimize the adverse impacts on ecosystems there.

As discussed before, interactions among ecosystem services occur when multiple services respond to the same driver of change or when interactions among services themselves cause changes in one service to alter the provision of another. The linkages among ecosystem services are not fixed, but can shift through time due to policies that address ecosystem services. For example, afforestation for carbon sequestration is a potential strategy for the mitigation of global climate change in regions of the world where the net carbon storage of a tree plantation would be greater than that of the ecosystem it replaces. The Kyoto Protocol's Clean Development Mechanism (CDM) provides economic incentives for afforestation by allowing countries to offset a portion of their greenhouse gas emissions through carbon sequestration projects. However, afforestation may have negative impacts on other ecosystem services, including water supply, soil fertility and biodiversity values (Chisholm, 2010). Furthermore, in some cases, we are not limited to simply responding to existing synergies and trade-offs, but can actually manage ecosystem strength and even their existence. For instance, creation of riparian buffers alters the trade-off between agricultural production and water quality by limiting the effect of the driver (fertilizer use) on water quality, but has hardly any effect on the impact of the driver on agricultural yields (Bennett et al., 2009).

'Bundling' ecosystem services refers to merging multiple values from a piece of property under a single credit type. For example, if a landowner restores an area of riparian forest, it results in improvements to more than one ecosystem service, including reducing stream temperature, improving wildlife habitat, sequestering carbon and mitigating potential floods. Bundling allows these services to be sold under a single credit type – for example, an ecosystem credit – and it might provide a way for landowners to get paid for the broader benefits they are providing (Deal et al., 2012). Understanding the interactions among ecosystem services is, however, a necessary precondition to bundling, yet a fundamental problem is the lack of understanding of where trade-offs and synergies among ecosystem services might occur. This suggests that a critical area of research in ecosystem services includes studies that identify common sets of correlated ecosystem services and the situations (landscapes and management regimes) in which they typically occur. Once such bundles have been identified, research to understand the mechanisms behind their grouping (e.g., are the services responding to the same driver or are they interacting?) can help us better manage the relationships among ecosystem services, designing bundles that reduce the trade-offs and create synergies in addition to simply avoiding or taking advantage of them where they already exist (Bennett et al., 2009). For example, Raudsepp-Hearne et al. (2010) identified patterns of interactions among 12 ecosystem services (including provisioning, cultural and regulating services) in Quebec, Canada. They perform a principal component analysis of the variations in the set of ecosystem services, concluding that more than half of the ecosystem service variance could be explained through two components. The first component corresponded to an ecological gradient that varies from fully forested land to fully agricultural land and explained 34 per cent of variance. The second corresponded to social gradient – ranging from tourism and recreation on one end to pork production and deer hunting on the other – explained an additional 17 per cent of ecosystem variance. Furthermore, they find strong trade-offs between provisioning and both regulating and cultural ecosystem services in a peri-urban agricultural landscape. In their landscape, food production in areas with low regulating ecosystem services was not affected by these trade-offs, but the loss of soil-regulating services was costly to farmers that have to replace these services, tourism operators that have to suspend water recreation, and governments that have to pay for water quality treatment and improvement. Compared to previous literature, they found a much larger negative correlation among ecosystem services, which seems to be linked to the scale of the study and the set of ecosystem services analysed.

Bundling of services could increase the benefits to conservation and the political support for a payment scheme by increasing the beneficiaries of the programme.[8] It also eliminates the potential for the provision of one ecosystem service to crowd out another and may provide a more administratively efficient process for integrating different ecosystem services that are managed by different regulatory agencies (Deal et al., 2012). Moreover, if the services bundled have benefits across spatial scales, this is also a way to expand the potential market and increase payments to a particular area, yet it might increase transaction costs. However, bundling of services under a single credit type may also require buyers to purchase services that buyers do not want (Chan et al., 2007; Kroeger and Casey, 2007; Townsend et al., 2011) and it is important to be able to 'unbundle' services from a broader suite of services to meet regulatory requirements for a specific service. Ideally, an integrative accounting system would 'bundle' ecosystem services at a landscape scale and accommodate current jurisdictional limitations by separating out regulated credits and 'stacking' them alongside other parts of the bundled services (Deal et al., 2012). For instance, Costa Rica's National Forestry Environmental Service's programme bundles the provisioning of carbon sequestration, watershed protection, biodiversity conservation and scenic beauty services, and markets them to different buyers (Kemkes et al., 2010).

19.5 CONCLUSIONS

Biodiversity and ecosystem services are in serious jeopardy and the best hope to protect them is to create and align diverse incentives for conservation wherever possible and to integrate these into the larger policy-making arena. It is a major challenge to reverse the degradation of ecosystems while meeting increasing demands for their services. The evidence gathered over the last decades indicates that this challenge can be met if a series of major interventions are promptly developed. In fact, growth in the use of ecosystem services over the past five decades was generally much less than the growth in GDP

(Millennium Ecosystem Assessment, 2005). This change reflects structural changes in economies, but it also results from new technologies and new management practices and policies that have increased the efficiency with which ecosystem services are used and provided substitutes for some services. Even with this progress, though, the absolute level of consumption of ecosystem services continues to grow, which indicates that much more is needed to keep pace with growing pressures and demands.

Historically, conservation has largely relied on considerations of intrinsic value. This has been manifestly insufficient as a response to the increasing threats to biodiversity, particularly in the world's poorest regions, where considerations of intrinsic and spiritual values are often trumped by the needs for survival or used to exclude significant segments of the population from the benefits from their ecosystem resources. The reality, of course, is that our planet is a mosaic of systems providing people with different bundles of ecosystem services and disservices. We cannot manage these systems effectively if we do not actively seek to measure the flows of these services, examine who is benefiting from them, and consider a range of policies, incentives, technologies and regulations that could encourage better management and sharing of the benefits.

Global commons are increasingly a part of the policy debate as there is increasing recognition and willingness to act to reduce and reverse the effects of the long-run de facto open access of the atmosphere, oceans and overall stock of biodiversity. These global commons are threatened with degradation as a result of collective human activity but are much trickier to deal with than local natural resource commons. Not only are global commons degraded on a large scale by many diverse actors, but also often the outcomes are spatially dispersed from the causes. In the case of climate change, those countries most negatively affected by climate change are often those low-income countries that have least caused the harm.

Local conservation depends on the existence of mechanisms to translate global willingness to pay for conservation into local incentives. In the absence of such incentives biodiversity conservation is unlikely to have high priority in the poorest countries. There has been a 'relative' success in supporting local conservation throughout the creation of markets and the emergence of ecosystem services pricing. This is largely due to the fact that ownership of these services can be easily established through property rights. However, it must be recognized that many of the prices for these services are still far from perfect and that establishing prices for regulating, cultural and supporting ecosystem services raises a serious challenge since many of these services are public goods by definition. Moreover, the inter-linkages between ecosystem services make the assignment of property rights slightly more complex. The use of many provisioning services has a direct impact on the flow of regulating, supporting and cultural services. In most instances, these impacts are negative. Therefore, extraction of timber from forest at rates above a certain threshold will cause the decline in water regulation, flood regulation and erosion control.

Even confining our interest to preservation of ecosystem services, however, is complex as there are likely to be many different goal conflicts. One might be between production of classical provisioning services and regulating, cultural or supportive services to use the terminology of the MEA. Put simply, we might find that narrowly managed monocultures give us the most yields in terms of tons of food or fibre but relatively less in terms of aesthetics, wildlife, resilience and or services such as pollination. When we have a multitude of goals we typically find that we need to employ also a number of policy instruments and we have in this chapter explored some ways in which such instruments might be combined and(or) diversified in time and space.

Finally, the current focus on ecosystem services may sometimes be bundled with conservation of biodiversity for ethical reasons. In particular, measures can be developed that benefit both regulating and supporting ecosystem services on the one hand and nature conservation on the other. However, this will not always be the case and the focus on ecosystem services should not overshadow the need for traditional nature conservation. Similarly, we cannot merely focus on preservation of species or conservation of protected areas; we need also to think of sensible, balanced, management of the big majority of areas that are in fact exploited commercially.

NOTES

1. Management for biodiversity conservation in delimited areas may, because of the mobility of organisms, affect ecosystem services also in surrounding landscapes (Öckinger and Smith, 2007; Kohler et al., 2008).
2. For example, organic farming may result in increased biodiversity at local and farm scales, thus benefiting ecosystem services such as pollination (Andersson et al., 2012), but may not be cost-efficient in preserving Red List species (Hodgson et al., 2010; Smith et al., 2010). More generally, it has been argued that measures to preserve biodiversity in production landscapes ('land-sharing') may be inefficient as a conservation measure compared to assigning separate areas for conservation ('land-sparing') (Green et al., 2005; Phalan et al., 2012). However, it has also been suggested that the land-sharing/land-sparing concept is an oversimplification of a complex reality (Fisher et al., 2011).
3. For example, differences in harvesting technologies can lead to different flows of outputs from similar stocks. Furthermore, though a high carbon in soils results in high biodiversity, nutrient retention and therefore a high crop production, in the short run farmers can partly compensate for reduced carbon using inorganic fertilizers and other technology.
4. For instance, Pindyck (1978) shows that exploratory activity has the effect of reducing the rate of increase of price. Moreover, extraction costs may increase inversely with the remaining stock of reserves, which also leads the shadow price to rise less than the rate of interest over time (Farzin, 1992). Instead, André and Smulders (2004) consider the effect of endogenous technological progress in extraction showing that the path of the price follows a different trajectory (technological progress produces a U-shaped price path, yet in the long run the exhaustion effect should overcome the cost reduction effect due to technological progress, and resource prices should still increase).
5. We shall distinguish between two cases. First, the unit harvesting cost is independent of the size of the stock being harvested. Here, the optimal rate of rate of exploitation is determined by the requirement that the marginal growth rate of the resource be equal to the discount rate; a higher discount rate raises the optimal rate of exploitation and lowers the optimal standing stock for a resource with a concave growth function. Second, there is a stock-dependent unit cost of harvesting (a positive 'stock effect'). Here the discount rate plays a dual role. It represents the required rate of return on a growing asset, leading to smaller standing stock when it increases, but also the opportunity cost of capital to be invested in harvesting equipment (a higher discount rate thus means more costly harvesting, which in turn implies a less intensive optimal harvesting and a larger standing stock). Hence, the discount rate is active in both of its roles simultaneously, and its effects on optimal harvesting and standing stock would depend on the capital intensity of the harvesting process and the cost of capital (Hannesson, 1987).
6. It has also been pointed out that proposing a boundary may increase the risk of prolonged degradation if, in fact, conditions continue to degrade also before the proposed threshold is reached (Schleisinger, 2009).
7. Non-rivalry can be also a challenge, as many local, regional and global services are open access. The implementation of market-based mechanisms requires 'propertization'. That is, the creation of property rights where none currently exist. Therefore, the first implicit step for implementing market-based mechanism is for the 'collective' institution to declare property rights for itself so that it can then determine the supply of the services available. The next step is to distribute the property rights: who has the right to the resource and for how long? Unfortunately, propertization will only take account of the non-rival character of many ecosystem services, since unlike non-rivalry, non-excludability is inherently a physical characteristic and not a policy variable.

8. For instance, Townsend et al. (2011) analyse the value of bundling payments for environmental services (PES) from watershed restoration, including water quality improvement and carbon sequestration coupled with wood production. They compare the net returns of reforestation and bundling to the value of existing agricultural land use of 408 000 ha of the Warren–Tone watershed (WT) in south-western Australia. A hydrological model (LUCICAT) was used to define the relationships between reforestation/deforestation and water yield and quality, thus providing a basis for valuing the hydrological benefits of reforestation. Various land-use change scenarios were examined, with modelling indicating more than 70 per cent reforestation is required to restore stream salinity to a potable threshold. They conclude that reforestation was unprofitable when only wood revenues were considered with a discount rate of 9.5 per cent, but was profitable at carbon prices of at least AU$22 t CO_2e. For the latter, reforestation activities driven by the economic benefits from selling wood and carbon generate an externality benefit in the form of improved water quality without the need for any additional payment to landholders. However, payments for activities that lead to improvements in water quality could represent a new, additional source of income for landholders on the proviso that there is sufficient reforestation to reach the potable threshold. Alternatively, costs could be imposed on those whose land-use practices cause the release of salt into waterways.

REFERENCES

Ahrends, A., N.D. Burgess, S.A.H. Milledge, M.T. Bulling, B. Fisher and J.C.R. Smarti et al. (2010), 'Predictable waves of sequential forest degradation and biodiversity loss spreading from an African city', *Proceedings of the National Academy of Sciences of the United States of America*, **107**(33), 14556–61.

Albers, H.J. (2010), 'Spatial extraction and enforcement in developing country protected areas', *Resource and Energy Economics*, **32**(2), 165–79.

Altieri, M.A. (1999), 'The ecological role of biodiversity in agroecosystem', *Agriculture, Ecosystems and Environment*, **74**(1–3), 19–31.

Andam, K.S., P.J. Ferraro, A. Pfaff, G.A. Sanchez-Azofeifa and J.A. Robalino (2008), 'Measuring the effectiveness of protected area networks in reducing deforestation', *PNAS*, **105**(42), 16089–94.

Andersson, G., M. Rundlöf and H.G. Smith (2012), 'Organic farming improves pollination success in strawberries', *PLoS One*, **7**(2), e31599.

André, F.J. and S. Smulders (2004), 'Energy use, endogenous technical change and economic growth', Universidad Pablo de Olavide, Department of Economics and Business Administration, manuscript.

Arrow, K.J. and A.C. Fisher (1974), 'Environmental preservation uncertainty, and irreversibility', *Quarterly Journal of Economics*, **88**(1), 312–19.

Babcock, B.A., P.G. Lakshiminarayan, J.J. Wu and D. Zilberman (1997), 'Targeting tools for the purchase of environmental amenities', *Land Economics*, **73**(3), 325–9.

Balvanera, P., G.C. Daily, P.R. Ehrlich, T.H. Ricketts, S.A. Bailey and S.A. Kark et al. (2001), 'Conserving biodiversity and ecosystem services', *Science*, **291**(5511), 2047.

Barbier, E.B. (1997), 'Introduction to the Environmental Kuznets Curve special issue', *Environment and Development Economics*, **2**(4), 369–82.

Baylis, K., S. Peplow, G. Rausser and K. Simon (2008), 'Agri-environmental policies in the EU and United States: a comparison', *Ecological Economics*, **65**(4), 753–64.

Bayon, R. (2006), 'Biodiversity offsets – from laws to theory', presentation at the seminar 'Les mécanismes de compensation une opportunité pour les secteurs économiques et financiers et les gestionnaires de la diversité biologique', French Ministry of Ecology and Sustainable Development, Paris, 6 July 2006.

Bengtsson, J. (1998), 'Which species? What kind of diversity? Which ecosystem function? Some problems in studies of relations between diversity and ecosystem function', *Applied Soil Ecology*, **10**(3), 191–9.

Bennear, L. and R. Stavins (2007), 'Second-best theory and the use of multiple policy instruments', *Environmental and Resource Economics*, **37**(1), 111–29.

Bennett, E.M., G.D. Petterson and L.J. Gordon (2009), 'Understanding relationships among multiple ecosystem services', *Ecological Letters*, **12**(12), 1394–404.

Biesmeijer, J.C., S.P. Roberts, M. Reemer, R. Ohlemuller, M. Edwards and T. Peeters et al. (2006), 'Parallel declines in pollinators and insect-pollinated plants in Britain and the Netherlands', *Science*, **313**(5785), 351–4.

Blignaut, J. and J. Aronson (2008), 'Getting serious about maintaining biodiversity', *Conservation Letters*, **1**(1), 12–17.

Bockstael, N.E. (1996), 'Modeling economics and ecology: the importance of a spatial perspective', *American Journal of Agricultural Economics*, **78**(5), 1168–80.

Bulte, E. and G.C. van Kooten (2000), 'Economic science, endangered species, and biodiversity loss', *Conservation Biology*, **14**(1), 113–19.

Burgess, N. and A. Rodgers (2004), 'Protected area categories: why they matter for the Eastern Arc and Coastal Forests in Tanzania – briefing note', Project Report No. 05, Conservation and Management of the Eastern Arc Mountain Forests Project (CMEAMF), September.

Busby, G.M., H.J. Albers and C.A. Montgomery (2012), 'Wildfire risk management in a landscape with fragmented ownership and spatial interactions', *Land Economics*, **88**(3), 496–517.

Chan, K.M.A., R.M. Pringle, J. Ranganathan, C.L. Boggs, Y.L. Chan and P.R. Ehrlich et al. (2007), 'When agendas collide: human welfare and biological conservation', *Conservation Biology*, **21**(1), 59–68.

Chapin, F.S., E.S. Zavaleta, V.T. Eviner, R.L. Naylor, P.M. Vitousek and H.L. Reynolds et al. (2000), 'Consequences of changing biodiversity', *Nature*, **405**(6783), 234–42.

Chisholm, R.A. (2010), 'Trade-offs between ecosystem services: water and carbon in a biodiversity hotspot', *Ecological Economics*, **69**(10), 1973–87.

Chomitz, K.M. and D.A. Gray (1996), 'Roads, land use, and deforestation: a spatial model applied to Belize', *The World Bank Economic Review*, **10**(3), 487–512.

Clark, C.W. (1973), 'Profit maximization and the extinction of animal species', *Journal of Political Economy*, **81**(4), 950–61.

Coria, J. and E. Calfucura (2012), 'Ecotourism and the development of indigenous communities: the good, the bad, and the ugly', *Ecological Economics*, **73**(C), 47–55.

Coria, J. and T. Sterner (2011), 'Natural resource management: challenges and policy options', *Annual Review of Resource Economics*, **3**(1), 203–30.

Crepin, S. (2007), 'Using fast and slow processes to manage resources with thresholds', *Environmental and Resource Economics*, **36**(2), 191–213.

Crowder, D.W., T.D. Northfield, M.R. Strand and W.E. Snyder (2010), 'Organic agriculture promotes evenness and natural pest control', *Nature*, **466**(7302), 109–12.

Deal, R.L., B. Cochran and G. LaRocco (2012), 'Bundling of ecosystem services to increase forestland value and enhance sustainable forest management', *Forest Policy and Economics*, **17**, 69–76.

Díaz, S., D. Tilman, J. Fargione, F.S. Chapin, R. Dirzo and T. Kitzberger et al. (2005), 'Biodiversity regulation of ecosystem services', in *Ecosystems and Human Well-being*, Washington, DC: Island Press.

Dobbs, T.L. and J. Pretty (2008), 'Case study of agri-environmental payments: the United Kingdom', *Ecological Economics*, **65**(4), 765–75.

Dudley, N. (2008), *Guidelines for Applying Protected Area Management Categories*, Gland, Switzerland: International Union for Conservation of Nature.

Elmqvist, T., C. Folke, M. Nyström, G. Peterson, J. Bengtsson, B. Walker and J. Norberg (2003), 'Response diversity, ecosystem change, and resilience', *Frontiers in Ecology and the Environment*, **1**(9), 488–94.

Epstein, L.S. (1980), 'Decision-making and the temporality of uncertainty', *International Economic Review*, **21**(2), 269–84.

Farley, J. and R. Constanza (2010), 'Payments for ecosystem services: from local to global', *Ecological Economics*, **69**(11), 2060–68.

Farzin, Y.H. (1984), 'The effect of the discount rate on depletion of exhaustible resources', *Journal of Political Economy*, **92**(51), 841–51.

Farzin, Y.H. (1992), 'The time path of scarcity rent in the theory of exhaustible resources', *Economic Journal*, **102**(413), 841–51.

Ferraro, P.J. (2002), 'The local costs of establishing protected areas in low-income nations: Ranomafana National Park, Madagascar', *Ecological Economics*, **43**(2–3), 261–75.

Ferraro, P.J. (2008), 'Asymmetric information and contract design for payments for environmental services', *Ecological Economics*, **65**(4), 810–21.

Fisher, B., R. Kerry Turner and P. Morling (2009), 'Defining and classifying ecosystem services for decision making', *Ecological Economics*, **68**(3), 643–53.

Foley, J.A., R. DeFries, G.P. Asner, C. Barford, G. Bonan and S.R. Carpenter et al. (2005), 'Global consequences of land-use', *Science*, **309**(5734), 570–74.

Freixas, X. and J.J. Laffont (1984), 'On the irreversibility effect', in M. Boyer and R. Kihlstrom (eds), *Bayesian Models in Economic Theory*, Dordrecht: Elsevier.

Gamfeldt, L., H. Hillebrand and P.R. Jonsson (2008), 'Multiple functions increase the importance of biodiversity for overall ecosystem functioning', *Ecology*, **89**(5), 1223–31.

Gaston, K.J. (2010), 'Valuing common species', *Science*, **327**(5962), 154–5.

Gaston, K.J. and R.A. Fuller (2008), 'Commonness, population depletion and conservation biology', *Trends in Ecology and Evolution*, **23**(1), 14–19.

Gault, A., Y. Meinard and F. Courchamp (2008), 'Consumers' taste for rarity drives sturgeons to extinction', *Conservation Letters*, **1**(5), 199–207.

Gilbert-Norton, L., R. Wilson, J.R. Stevens and K.H. Beard (2010), 'A meta-analytic review of corridor effectiveness', *Conservation Biology*, **24**(3), 660–68.

Giller, K.E., M.H. Beare, P. Lavelle, A.M.N. Izac and M.J. Swift (1997), 'Agricultural intensification, soil biodiversity and agroecosystem function', *Applied Soil Ecology*, **6**(1), 3–6.
Gómez-Baggethun, E., R. de Groot, P.L. Lomas and C. Montes (2010), 'The history of ecosystem services in economic theory and practice: from early notions to markets and payment schemes', *Ecological Economics*, **69**(6), 1209–18.
Green, R.E., S.J. Cornell, J.P.W. Scharlemann and A. Balmford (2005), 'Farming and the fate of wild nature', *Science*, **307**(5709), 550–55.
Hannesson, R. (1987), 'The effect of the discount rate on the optimal exploitation of renewable resources', *Marine Resource Economics*, **3**(4), 319–29.
Hanski, I. (1999), *Metapopulation Ecology*, Oxford: Oxford University Press.
Hartshorn, G., P.J. Ferraro and B. Spergel (2005), *Evaluation of the World Bank–GEF Ecomarkets Project in Costa Rica*, Washington, DC: World Bank.
Henry, C. (1974), 'Investment decisions under uncertainty: the "irreversibility effect"', *American Economic Review*, **64**(6), 1006–12.
Hodgson, J.A., W.E. Kunin, C.D. Thomas, T. Benton and D. Gabriel (2010), 'Comparing organic farming and land sparing: optimizing yield and butterfly populations at a landscape scale', *Ecology Letters*, **13**(11), 1358–67.
Hodgson, J.A., A. Moilainen, B.A. Wintle and C.D. Thomas (2011), 'Habitat area, quality and connectivity: striking the balance for efficient conservation', *Journal of Applied Ecology*, **48**(1), 148–52.
Hooper, D.U., F.S. Chapin, J.J. Ewel, A. Hector, P. Inchausti and S. Lavorel et al. (2005), 'Effects of biodiversity on ecosystem functioning: a consensus of current knowledge', *Ecological Monographs*, **75**(1), 3–35.
Hotelling, H. (1931), 'The economics of exhaustible resources', *The Journal of Political Economy*, **39**(2), 137–75.
Isbell, F., V. Calcagno, A. Hector, J. Connolly, W. Stanley Harpole and P.B. Reich et al. (2011), 'High plant species diversity is needed to maintain ecosystem services', *Nature*, **477**(7363), 199–202.
Joppa, L.N. and A. Pfaff (2009), 'High & far: how the world's protected areas have avoided threat', Duke University Nicholas School of the Environment Working Paper.
Kemkes, R.J., J. Farley and C.J. Koliba (2010), 'Determining when payments are an effective policy approach to ecosystem service provision', *Ecological Economics*, **69**(11), 2075–84.
Key, N., E. Sadoulet and A. de Janvry (2000), 'Transactions costs and agricultural supply response', *American Journal of Agricultural Economics*, **82**(2), 245–59.
Kleijn, D. and W.J. Sutherland (2003), 'How effective are European agri-environment schemes in conserving and promoting biodiversity?' *Journal of Applied Ecology*, **40**(6), 947–69.
Kleijn, D., F. Berendse, R. Smit and N. Gilissen (2001), 'Agri-environment schemes do not effectively protect biodiversity in Dutch agricultural landscapes', *Nature*, **413**(6857), 723–5.
Kleijn, D., R.A. Baquero, Y. Clough, M. Diaz, J. De Esteban and F. Fernandez et al. (2006), 'Mixed biodiversity benefits of agri-environment schemes in five European countries', *Ecology Letters*, **9**(3), 243–54.
Kleijn, D., M. Rundlöf, J. Scheper, H.G. Smith and T. Tscharntke (2011), 'Does conservation on farmland contribute to halting the biodiversity decline?' *Trends in Ecology & Evolution*, **26**(9), 474–81.
Kohler, F., J. Verhulst, R. van Klink and D. Kleijn (2008), 'At what spatial scale do high-quality habitats enhance the diversity of forbs and pollinators in intensively farmed landscapes?' *Journal of Applied Ecology*, **45**(3), 753–62.
Kosoy, N. and E. Corbera (2010), 'Payment for ecosystem services as commodity fetishism', *Ecological Economics*, **69**(6), 1228–36.
Kroeger, T. and F. Casey (2007), 'An assessment of market-based approaches to providing ecosystem services on agricultural lands', *Ecological Economics*, **64**(2), 321–32.
Landell-Mills, N. and I.T. Porras (2002), *Silver Bullet or Fool's Gold? A Global Review of Markets for Forest Environmental Services and their Impact on the Poor, Instruments for Sustainable Private Sector Forestry Series*, London: International Institute for Environment and Development.
Lant, C.L., J.B. Ruhl and S.E. Kraft (2008), 'The tragedy of ecosystem services', *BioScience*, **58**(10), 969–74.
Leadley, P., H.M. Pereira, R. Alkemade, J.F. Fernandez-Manjarrés, V. Proença, J.P.W. Scharlemann and M.J. Walpole et al. (2010), *Biodiversity Scenarios: Projections of the 21st Century Change in Biodiversity and Associated Ecosystem Services*, Report of the Convention on Biological Diversity, Montreal, Canada.
Leal, I.R., J.M.C. Da Silva, M. Tabarelli and T.E. Lacher (2005), 'Changing the course of biodiversity conservation in the Caatinga of Northeastern Brazil', *Conservation Biology*, **19**(3), 701–6.
Lewis, J. (2002), 'Scarcity and abundance. Contrasting conceptions of the forest in Northern Congo-Brazzaville, and issues for conservation', Ninth International Conference on Hunting and Gathering Societies, Edinburgh Conference Centre, Heriot-Watt University, 9–13 September.
Lindenmayer, D.B. and H.A. Nix (1993), 'Ecological principles for the design of wildlife corridors', *Conservation Biology*, **7**(3), 627–30.
Lokina, R.B., E.J.Z. Robinson and H. Eggert (2008), 'Determinant of the effectiveness of participatory

forest management in Tanzania', presented at the European Association of Environmental and Resource Economists 16th Annual Conference, Gothenburg, Sweden, 25–28 June.

Lyons, K.G, C.A. Bringham, B.H. Traut and M.W. Schwartz (2005), 'Rare species and ecosystem functioning', *Conservation Biology*, **19**(4), 1019–24.

MacDonald, D.H., W.L. Adamowicz and M.K. Luckert (2001), 'Fuelwood collection in North-Eastern Zimbabwe: valuation and caloric expenditures', *Journal of Forestry Economics*, **7**(1), 29–52.

Mace, G.M., K. Norris and A.H. Fitter (2012), 'Biodiversity and ecosystem services: a multi-layered relationship', *Trends in Ecology and Evolution*, **27**(1), 19–26.

Millennium Ecosystem Assessment (2005), *Ecosystems and Human Well-being: Current State and Trends: Findings of the Condition and Trends Working Group*, Washington, DC: Island Press.

Moilanen, A., K.A. Wilson and H. Possingham (2009), *Spatial Conservation Prioritization*, Oxford: Oxford University Press.

Naughton-Treves, L., M. Buck and K. Brandon (2005), 'The role of protected areas in conserving biodiversity and sustaining local livelihoods', *Annual Review of Environment and Resources*, **30**, 219–52.

Nelson, G.C. and D. Hellerstein (1997), 'Do roads cause deforestation? Using satellite images in econometric analysis of land use', *American Journal of Agricultural Economics*, **79**(1), 80–88.

Öckinger, E. and H.G. Smith (2007), 'Semi-natural grasslands as population sources for pollinating insects in agricultural landscapes', *Journal of Applied Ecology*, **44**(1), 50–59.

Omamo, S. Were (1998), 'Transport costs and smallholder cropping choices: an application to Siaya District, Kenya', *American Journal of Agricultural Economics*, **80**(1), 116–23.

Ortiz, E., L.F. Sage Mora and C. Borge Carvajal (2003), *Impacto del Programa de Pago por Servicios Ambientales en Costa Rica como Medio de Reducción de la Pobreza en los Medios Rurales* [*Impact of the Payment for Ecosystem Services Program in Costa Rica as a Means to Reduce Rural Poverty*], San José, Costa Rica: RUTA.

Ostrom, E. (1990), *Governing the Commons: The Evolution of Institutions for Collective Action*, Cambridge, UK and New York: Cambridge University Press.

Pagiola, S. (2008), 'Payments for environmental services in Costa Rica', *Ecological Economics*, **65**(4), 712–24.

Pagiola, S., J. Bishop and N. Landell-Mills (2002), 'Making market-based mechanisms work for forests and people', in S. Pagiola, J. Bishop and N. Landell-Mills (eds), *Selling Forest Environmental Services – Market--based Mechanisms for Conservation and Development*, London and Sterling: Earthscan.

Perrings, C. (2007), 'Pests, pathogens and poverty: biological invasions and agricultural dependence', in A. Kontoleon, U. Pascual and T. Swanson (eds), *Biodiversity Economics: Principles, Methods, and Applications*, Cambridge, UK: Cambridge University Press, pp. 133–65.

Perrings, C. and G. Halkos (2012), 'Who cares about biodiversity? Optimal conservation and transboundary biodiversity externalities', *Environmental and Resource Economics*, **52**(4), 585–608.

Pfaff, A.S.P. (1999), 'What drives deforestation in the Brazilian Amazon? Evidence from satellite and socioeconomic data', *Journal of Environmental Economics and Management*, **37**(1), 26–43.

Pfaff, A., J. Robalino, G.A. Sanchez-Azofeifa, K. Andam and P. Ferraro (2009), 'Park location affects forest protection: land characteristics cause differences in park impacts across Costa Rica', *The B.E. Journal of Economic Analysis & Policy*, **9**(2) (Contributions), Article 5.

Phalan, B., M. Onial, A. Balmford and R.E. Green (2012), 'Reconciling food production and biodiversity conservation: land sharing and land sparing compared', *Science*, **333**(6047), 1289–91.

Pindyck, R. (1978), 'The optimal exploration and production of nonrenewable resources', *Journal of Political Economy*, **86**(51), 841–61.

Raudsepp-Hearne, C., G.D. Peterson and E.M. Bennett (2010), 'Ecosystem service bundles for analyzing tradeoffs in diverse landscapes', *PNAS*, **107**(11), 5242–7.

Rey Benayas, J.M., A.C. Newton, A. Díaz and J.M. Bullock (2009), 'Enhancement of biodiversity and ecosystem services by ecological restoration: a meta-analysis', *Science*, **325**(5944), 1121–4.

Ridder, B. (2008), 'Questioning the ecosystem services argument for biodiversity conservation', *Biodiversity and Conservation*, **17**(4), 781–90.

Robinson, E.J.Z., J.C. Williams and H.J. Albers (2002), 'The influence of markets and policy on spatial patterns of non-timber forest product extraction', *Land Economics*, **78**(2), 260–71.

Robinson, E.J.Z., H.J. Albers and J.C. Williams (2008), 'Spatial and temporal modelling of community non-timber forest extraction', *Journal of Environmental Economics and Management*, **56**(3), 234–45.

Robinson, E.J.Z., H.J. Albers and J.C. Williams (2011), 'Sizing reserves within a landscape: the roles of villagers' reactions and the ecological-socioeconomic setting', *Land Economics*, **87**(1), 233–49.

Rockström, J., W. Steffen, K. Noone, Å. Persson, F.S. Chapin III and E.F. Lambin et al. (2009), 'A safe operating space for humanity', *Nature*, **461**(7263), 472–5.

Rodrigues, A.S.L., J.P. Pilgrim, J.L. Lamoreux, M. Hoffmann and T.M. Brooks (2006), 'The value of the Red List for conservation', *Trends in Ecological Evolution*, **21**(2), 71–6.

Rundlöf, M. and H.G. Smith (2006), 'The effect of organic farming on butterfly diversity depends on landscape context', *Journal of Applied Ecology*, **43**(6), 1121–7.

Scheffer, M. (2009), *Critical Transitions in Nature and Society*, Princeton, NJ: Princeton University Press.
Scheffer, M., S. Carpenter, J.A. Foley, C. Folke and B. Walker (2001), 'Catastrophic shifts in ecosystems', *Nature*, **413**(6856), 591–6.
Scherr, S., A. Khare and A. White (2004), *For Services Rendered. Current Status and Future Potential of Markets for Ecosystem Services of Tropical Forests: An Overview*, Technical Series 21, ITTO.
Schleisinger, W.H. (2009), 'Planetary boundaries: threshold risk prolonged degradation', *Nature Reports: Climate Change*, DOI:10.1038/climate.2009.93, 112–3.
Schröder, A., L. Persson and A.M. De Roos (2005), 'Direct experimental evidence for alternative stable states: a review', *Oikos*, **110**(1), 3–19.
Sills, E.O. and K.L. Abt (2003), *Forests in a Market Economy – Forestry Sciences, Vol. 72*, Dordrecht/Boston: Kluwer Academic Publishers.
Sills, E., R. Arriagada, S. Pattanayak, F. Ferraro, L. Carrasco, E. Ortiz and S. Cordero (2006), 'Impact of the PSA program on land use', paper presented at the Workshop on Costa Rica's Experience with Payments for Environmental Services, San José, Costa Rica, 25–26 September 2006.
Sims, K. (2013), 'Do protected areas reduce forest fragmentation? A microlandscapes approach', *Environmental Resource Economics*, DOI 10.1007/s10640–013–9707–2.
Smith, H.G., E. Öckinger and M. Rundlöf (2010), 'Biodiversity and the landscape ecology of agri-environment schemes', *Aspects of Applied Biology*, **100**, 225–32.
Srivastava, D.S. and M. Vellend (2005), 'Biodiversity-ecosystem function research: is it relevant to conservation?' *Annual Review of Ecological Systems*, **36**, 267–94.
Tallis, H., P. Kareiva, M. Marvier and A. Chang (2008), 'An ecosystem services framework to support both practical conservation and economic development', *PNAS*, **105**(28), 9457–64.
Townsend, P.V., R.J. Harper, P.D. Brennan, C. Dean, S. Wu, K.R.J. Smettem and S.E. Cook (2011), 'Multiple environmental services as an opportunity for watershed restoration', *Forest Policy and Economics*, **17**(C), 45–8.
Tscharntke, T., I. Steffan-Dewenter, A, Kruess and C. Thies (2002), 'Contribution of small habitat fragments to conservation of insect communities of grassland-cropland landscapes', *Ecological Applications*, **12**(2), 354–63.
Tscharntke, T., A.M. Klein, A. Kruess, I. Steffan-Dewenter and C. Thies (2005), 'Landscape perspectives on agricultural intensification and biodiversity – ecosystem service management', *Ecological Letters*, **8**(8), 857–74.
Venter, O., W.F. Laurance, T. Iwamura, K.A. Wilson, R.A. Fuller and H.P. Possingham (2009), 'Harnessing carbon payments to protect biodiversity', *Science*, **326**(5958), 1368.
Vira, B. and W.M. Adams (2009), 'Ecosystem services and conservation strategy: beware the silver bullet', *Conservation Letters*, **2**(4), 158–62.
Wells, M., K. Brandon and L. Hannah (1992), 'People and parks: linking protected area management with local communities', Washington, DC: World Bank, World Wildlife Fund and US Agency for International Development.
Wendland, K.J., M. Honzák, R. Portela, B. Vitale, S. Rubinoff and J. Randrianarisoa (2010), 'Targeting and implementing payments for ecosystem services: opportunities for bundling biodiversity conservation with carbon and water services in Madagascar', *Ecological Economics*, **69**(11), 2093–107.
Worm, B., E.B. Barbier, N. Beaumont, J. Emmett Duffy, C. Folke and B.S. Halpern et al. (2006), 'Impacts of biodiversity loss on ocean ecosystem services', *Science*, **314**(5800), 787–90.
Woodroffe, R. and J.R. Ginsberg (1998), 'Edge effects and the extinction of populations inside protected areas', *Science*, **280**(5372), 2126–8.
Wu, J. and W.G. Boggess (1999), 'The optimal allocation of conservation funds', *Journal of Environmental Economics and Management*, **38**(3), 302–21.
Wunder, S. (2005), 'Payments for environmental services: some nuts and bolts', Occasional Paper No. 42, Bogor: CIFOR.
Wünscher, T., S. Engel and S. Wunder (2008), 'Spatial targeting of payments for environmental services: a tool for boosting conservation benefits', *Ecological Economics*, **65**(4), 822–33.
Zang, W., H. Taylor, C. Ricketts, K. Kremen, K. Carney and S.M. Swinton (2007), 'Ecosystem services and dis-services to agriculture', *Ecological Economics*, **64**(2), 253–60.

PART IV

SHEDDING LIGHT ON NON-MARKET VALUES OF ECOSYSTEM SERVICES

20. A choice experiment to value the recreational benefits of coral reefs: a case study of Ras Mohammed National Park, Egypt

Rady T. Tawfik and R. Kerry Turner

20.1 INTRODUCTION

Ras Mohammed National Park was declared in 1983 and covers an area of 460 km^2. The area includes the islands of Tiran and Sanafir and all shorelines fronting the Sharm El-Sheikh tourism development area. The park is home to some of the most spectacular coral reefs and best-known SCUBA diving areas in the world. This recognition is based on the diversity of flora and fauna, clear warm water devoid of pollutants, its proximity to shorelines and its breathtaking beauty. This combination plus its accessibility in most weather conditions and relative proximity for European tourists form the basis of Ras Mohammed's popularity as a tourist destination. The number of visitors to Ras Mohammed increased from hundreds in 1988 to more than 1 million in 2009. A total of 495 382 people visited the old boundaries of Ras Mohammed in the fiscal year 2008/09, of which 471 142 (95 per cent) were foreign tourists (EEAA, 2009). The volume of tourists and the intensive recreational use by snorkellers and SCUBA divers have degraded the reefs. The reef degradation and the loss of productivity and biodiversity could have serious consequences. It is threatening the ecological services provided to millions of other dependent species and the numerous benefits provided to people. Lack of awareness, insufficient enforcement of protective legislation, market failure and undervaluation of benefits are the root causes of several threats to coral reefs.

If decision-makers are aware of coral reefs' benefits, their values and the amount of money that reefs bring to their economy, then a more concerted and united effort can be effectively established. Moreover, this can empower the park management and justify the commitment to conserve and manage these valuable ecosystems. Economic valuation also helps to efficiently assess different alternatives, exhibit impacts on reef services, penalize environmentally degrading activities, incentivize sustainable uses, increase awareness, galvanize support for conservation, and establish solid partnerships among the different stakeholders. The challenge for reef-associated activities in Egypt is to generate considerable economic benefits while maintaining the reef ecosystem on which it depends. This study aims to assist in this debate through valuing the coral reef recreational benefits within Ras Mohammed National Park, identifying the associated measurement issues that make the valuation a unique challenge, determining the factors that affect the demand for reef sites in the park, and investigating visitor perceptions towards park attributes. It also uses the example of recreational benefits to show that improvements in choice experiments can correct for some of the shortcomings and measurement errors, and thus produce more accurate valuation estimates, and demonstrates the different methods for deriving WTP estimates and the affect of model specification and preference assumption on the results.

20.2 CHARACTERISTICS OF THE STUDY AREA

The unique geological and bio-geographic features of the Red Sea provide an appropriate environment for numerous species and habitats. It may be the most diverse coral reef area apart from the coral reefs in Southeast Asia (Spalding et al., 2001). The salinity varies from 36.5 parts per thousand (ppt) in the south to more than 41 ppt in the north in summer with minimal freshwater inflows and high rates of evaporation (Kotb et al., 2004). The water is clearer in the north (40–50 m) compared to the south (~5 m) (Hassan et al., 2002). The climate in the Red Sea region is extremely hot and arid, with rainfall averaging less than 5 mm per year (Head, 1987). The water temperatures range between 21–30°C (Hawkins and Reports, 1994). The clear, warm and saline water of the Red Sea offers an optimal environment for coral. In the north, the Sinai Peninsula divides the Red Sea into the Gulfs of Suez and Aqaba, which both have markedly different morphologies. The coral diversity is greater in the central, northern Red Sea and the Gulf of Aqaba (see Table 20.1 and Figure 20.1).

The coral reef ecosystems found in Ras Mohammed are recognized internationally as among the world's best. They vary from shallow slopes with sandy plateaus (e.g., Turtle Beach) to steep walls (e.g., Shark Reef and Shark Observatory) (Pearson and Shehata, 1998). The fringing reefs are the most common reef type in the park, with a reef flat ranging between 5–50 m along the coastline and reef slope depth ranges from 10–85 m (PERSGA, 2003). Genera diversity was found to peak at 10–15 m, and the coral community characterized by a shallow *Acropora* zone down to 5 m, followed by a *Millepora* zone at 5–10 m depth, a *Xeniid* zone at 10–40 m depth, and an overlapping *Montipora* zone at 20–25 m depth (Kotb et al., 1996). It is believed that the popularity of Ras Mohammed as a destination for tourists depends on the natural attractiveness, the aesthetic value and the diversity of coral reefs. In addition, the endemic species living in this area give it a global significance as a repository of biodiversity (Kotb et al., 2004). It is deemed the major centre for marine environmental education in Egypt.

Table 20.1 Number of genera and species of reef-building corals in the Egyptian Red Sea

Region	Genera	Species
Gulf of Aqaba	47	120
Gulf of Suez	25	47
North Red Sea	45	128
Central Red Sea	49	143
South Red Sea	31	74

Source: Abou Zaid (2000).

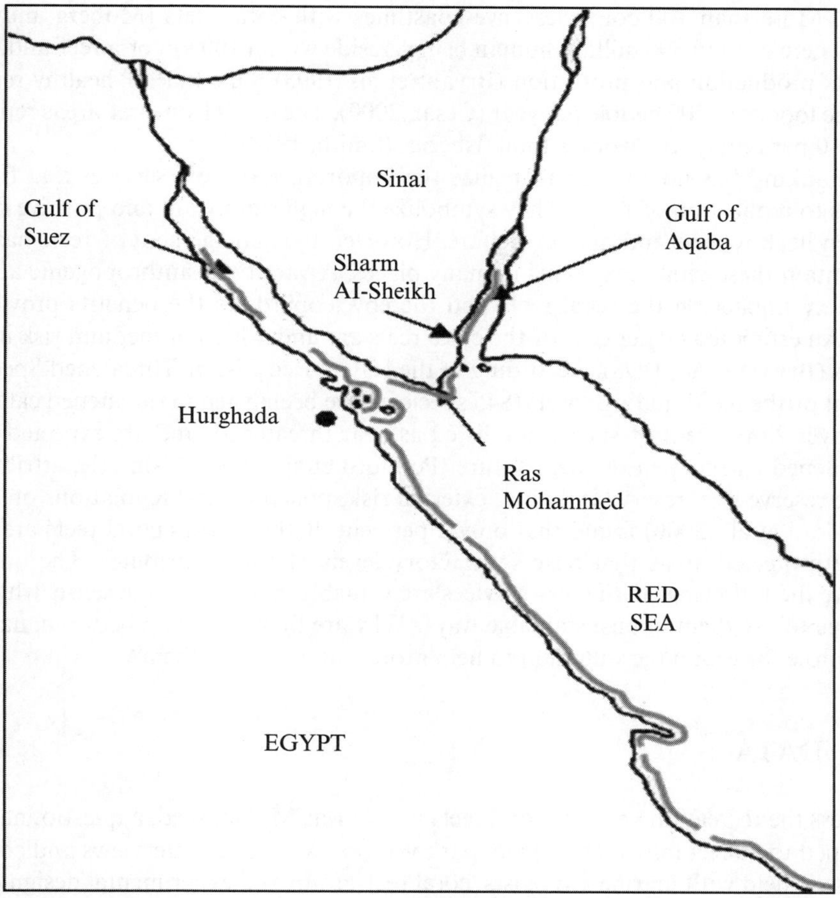

Figure 20.1 Coral reef coverage along the Egyptian coastline

20.3 CORAL REEF ECOSYSTEM

Coral reefs provide a wide range of services with valuable benefits to humanity, physical as well as moral, as means of life support and quality of life. They are the most biologically diverse marine ecosystems of the ocean, and probably on the Earth (Cesar, 2000), despite covering only 0.2 per cent of the marine environment area (Reaka-Kudla, 1997; Veron et al., 2009). They are the home to about 33 per cent of all described marine species (McAllister, 1991; Reaka-Kudla, 1997). They rival the tropical rainforests in terms of diversity and outstrip other mega-diversity ecosystems (Spalding et al., 2001). A single reef may accommodate 200 species of coral, 300 species of fish and 10000–100000 invertebrates (Cesar, 2000). Up to 60000 reef-living animals and plants have been described to date (Reaka-Kudla, 1994), but the total number of species inhabiting the world's reefs may be between 0.5 and 2 million (Spalding et al., 2001). Coral reefs are also among the most productive ecosystems, which provide protein and other benefits for millions of people. More than 100 countries have coastlines with coral reefs (Moberg and Folke, 1999) where around 500 million human beings reside within 100 km of a reef and benefit from its production and protection (Bryant et al., 1998). One km^2 of healthy reef may produce food for 2500 people per year (Cesar, 2000). The catch from reef areas represents about 10 per cent of the world's total fisheries (Smith, 1978).

Humankind has now begun to realize the importance of reef services and the constraints to replication of them. They symbolize the high significant and positive relation between healthy reefs and human welfare. However, the management of reefs has failed to maintain these vital ecosystems in many places. A variety of anthropogenic activities negatively impact on the coral reefs and thereby jeopardizes the benefits provided by them. An estimated 80 per cent of the coral reefs are under high or medium risk in many regions (Bryant et al., 1998). According to the IUCN Red List of Threatened Species, 27 per cent of the reef-building corals (845 species) have been listed in threatened categories, while over 20 per cent of species are listed as near threatened, and are expected to join a threatened category in the near future (Polidoro et al., 2008). Using the attributes of marine reserve size, reserve isolation, external risk, poaching and regulations on extraction, Mora et al. (2006) found that only 2 per cent of the world's coral reefs are within marine protected areas that have satisfactory levels of these attributes. The questions then are the following. If the reef services are valuable in an economic sense, why do we continue to use them in unsustainable ways? Why are they not being better managed? Is there a role for economic valuation to help protect and manage them?

20.4 DATA

To assess the recreation value of coral reefs within Ras Mohammed, a questionnaire was developed to collect information from park visitors. A series of interviews and consultations were held with marine biologists, coral reef group and experimental design experts from the University of East Anglia in the UK, an anthropologist from the University of London, and tour operators, park managers and staff from Ras Mohammed. The aim was to design the survey and to confirm questions' appropriateness and attributes' definitions. Focus groups were arranged to ensure respondent understanding of the questions and of their ability to complete the choice tasks. Then a pilot survey was carried out to test readability of the questionnaire and identify the potential problems with the survey and its administration. Suggestions were made on the most effective manner that problems could be solved. Consequently, many improvements were added to the main survey, which worked very well in the field. The questionnaire included a short introduction explaining the reason for it. The first section was designed to elicit respondents' experience visiting the reef sites in the park. The next section comprised background information, and opinion on the importance and the satisfaction with reef quality and other attributes. The third section covered the choice experiments. The final section was on socioeconomic characteristics of the respondents.

The visitor survey was conducted between March and August of 2008 to obtain data on the perception, socioeconomic characteristics of visitors to Ras Mohammed and their attitudes towards coral reefs. These months represent low and peak seasons for visitors. Because of the diverse nationalities of tourists, English, Italian, Russian and Arabic formats of the questionnaire were used. Samples were taken by handing the questionnaire to visitors on boat trips (Naama Jetty and Travco Jetty) and at the Ras Mohammed gate, then meeting them after their visit and during their lunch time at Main Beach, Marsa Ghozlani and Yolanda (the most popular beaches to visitors from land) and on the boat. These were considered to be ideal circumstances where visitors would have sufficient time to complete the survey with few distractions. Completed questionnaires were obtained from 1200 respondents.

The survey was designed under the assumption that there are two distinct populations: International Tourists (IT) and National Tourists (NT). For the IT sample, visitors from 19 countries participated in the survey. The main countries represented in the sample are Italy (42 per cent), Russia (16 per cent), United Kingdom (12 per cent), Poland (8 per cent), France (5 per cent), Germany (4 per cent), Austria (4 per cent), Netherlands (2 per cent), USA (2 per cent), and others (5 per cent). While, the NT sample included participants from 24 governorates mainly from Cairo (20 per cent), Alexandria (12 per cent), Giza (8 per cent), Dakahlia (7 per cent), Ismailia (6 per cent), Sharqia (6 per cent), South Sinai (5 per cent), Monufia (5 per cent) and others (31 per cent).

20.5 SPECIFIC ISSUES IN THE METHODOLOGY

'Reality often counts for little, perceptions count for everything' (Hensher et al., 2005, p. 94). In contrast to some valuation methods that are limited to particular benefits, stated preferences methods (e.g., contingent valuation [CV], choice experiments [CE]) have the potential to be used widely in valuing different benefits provided by coral reefs. The CE method is an application of random utility theory combined with the characteristics theory of value (Lancaster, 1966). Respondents derive utility from the characteristics or the attributes of goods rather than from the goods themselves (Alpizar et al., 2001). The method depends on the estimation of a response between choice probabilities and attribute levels. The probability of choosing an alternative increases as the levels of desirable attributes rise relative to the levels of the attributes in the other alternatives (Bennett, 1999). Thus, the respondent *i* will choose alternative *g* over alternative *h* if and only if:

$$\text{Prob}(U_{gi} > U_{hi}, \forall h \neq g) = \text{Prob}\{V_{gi} + \varepsilon_{gi} > V_{hi} + \varepsilon_{hi}\}.$$

There are many reasons for the growing interest in choice experiments. The technique has many advantages over CV, travel cost method (TCM) and other valuation techniques. For instance, it is superior to CV in terms of modelling substitution possibilities (Boxall et al., 1996) and to TCM in terms of avoiding co-linearity between attributes (Adamowicz et al., 1994) and being able to study attribute levels beyond the observed range. Morrison et al. (1996) argued that CE may be less prone to some potential biases that affect CV such as strategic, payment vehicle, start point, part-whole, and hypothetical bias. Many concerns affecting CV are alleviated by the repeated sampling method of CE (Carson, 1991 cited in Boxall et al., 1996). The individual attributes, as well as situational changes, can be valued and more information is brought out from each respondent using CE compared to CV. Compensating amounts of other goods

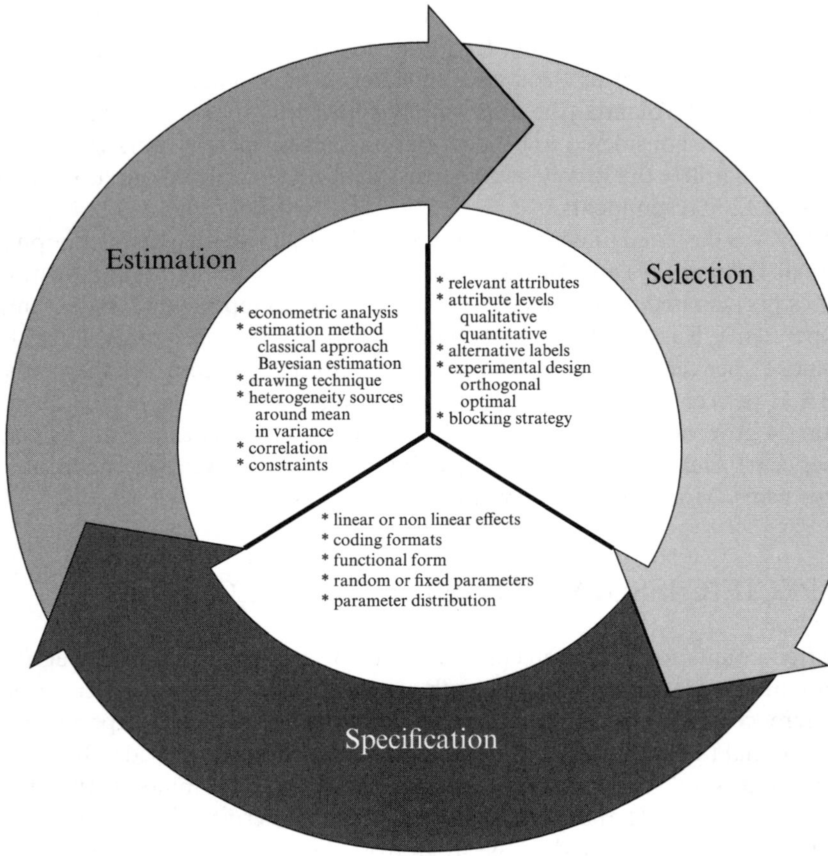

Figure 20.2 The complexity of choice experiments

(not based on money) can be estimated in the case of damage to a particular attribute (Adamowicz et al., 1995). Moreover, the estimates provided by CE (specifically the unlabelled experiments) are likely to be less site-specific and more suited to benefit transfer. The popularity of this method has resulted in numerous applications in different fields. However, CE takes a longer time, needs greater skills and higher costs and imposes a higher level of complexity on respondents than CV (Figure 20.2). Furthermore, the CE requires the use of experimental design and making many important decisions such as the relevant attributes, the number of attributes, the number of levels, how the attributes and levels should be described, the econometric analysis, coding format, correlation, interaction effects, estimation method, drawing techniques, distribution, restrictions and heterogeneity sources. Thus, every method has its own idiosyncrasies that bear different challenges.

In order to efficiently show the influences on choice behaviour, there are some factors that should be considered such as selection and measurement of the relevant attributes, functional form of the attributes and the parameters and socioeconomic characteristics as proxies for unobserved attributes (i.e., selection, measurement and specification).

20.5.1 Experimental Design

Blamey et al. (2002) stressed that the attributes should be measurable, demand-relevant and policy-relevant. The choice experiments involved several rounds of design and testing. Literature reviews and expert consultations were conducted to determine what attributes do generally matter to the diver/snorkeller. Focus groups and interviews with park managers and staff from SCUBA diving centres were made to select and revise which attributes are most important to visitors of Ras Mohammed. The clarity and simplicity of attribute meaning and measurement were considered and tested (e.g., reef quality was used instead of coral cover and percentages were used instead of absolute numbers for some attribute levels). Some attributes were excluded because of ambiguity over their meaning (e.g., recreational facilities) or difficulty with their measurement and forecasting their future levels for applications (e.g., reef area). The selected levels were established as feasible to respondents, relevant to park managers and in line with the shape of utility expression.

The design should balance between using all relevant attributes on the one hand and a reasonable level of complexity on the other (in other words, between validity and reliability). Blamey et al. (1997) noted that studies with fewer attributes produce more reliable results. However, results become less valid if highly relevant attributes were omitted. Mazotta and Opaluch (1995) found that using more than four to five attributes in a choice set may reduce the quality of the data collected because of the task complexity. Particular attention was paid in this study to ensure a simple choice set was presented that could be easily interpreted and answered.

DeShazo and Fermo (2002) found that the cognitive burden generating from introducing greater information is likely to outweigh the growth in consistency. In addition, it is difficult to administer the large experimental design and the respondents may have not enough time. Therefore, a fractional factorial design combined with a blocking strategy was used to reduce the number of choice sets presented to respondents. The results of the focus groups and the pilot study showed that each respondent could answer four choice sets while he or she demonstrated fatigue with eight choice sets.

Unlabelled experiments were used for choice treatment combinations. Hensher et al. (2005) illustrated that the independently and identically distributed terms (IID) assumption is likely to be achieved under unlabelled experiments than under labelled experiments. Moreover, unlabelled experiments are preferred if the aim is to estimate willingness to pay as respondents are forced to focus more on the outcome attributes, which is desirable from an economic valuation perspective, while labels may prompt respondents to make their choices on the basis of the label alone (Blamey et al., 1997).

20.5.2 Fixed Parameters vs Random Parameters

There are an increasing number of applications and growing popularity of using random parameter models to estimate willingness to pay and account for the preference heterogeneity. They have developed into becoming 'key innovation' (Train, 2003), 'the state of the art' (Sillano and Ortúzar, 2005) and 'the frontier' (Greene, 2007) in the analysis of discrete choice. Because they are more flexible and powerful, the random parameter models have overshadowed other models. They have the ability to treat correlated and heteroskedastic alternatives, increase the opportunity of identifying sources of preference heterogeneity, make the discrete choice model less restrictive in its behavioural assumptions, model a wide range of behaviour, allow for unrestricted substitution patterns, accommodate a variety of model specifications, and approximate any random utility model with total precision (McFadden and Train, 2000; Hensher and Greene, 2003). In addition, these methods are preferable if the sampled individuals are drawn from a larger population (Greene, 2007) and 'simply because people are different' (Eggert and Olsson, 2009). However, the various forms of their estimation and the practical issues associated with the different methods confuse and complicate the selection of the appropriate model in addition to the difficulty in interpreting the results (Sillano and Ortúzar, 2005). The random utility expression is restated and the structure of the random parameter vector β_i is presented as follows (Hensher and Greene, 2003):

$$U_{jti} = \beta_i X_{jti} + \varepsilon_{jti}$$

$$\beta_i = \beta + \Delta z_i + \eta_i = \beta + \Delta z_i + \Gamma v_i$$

where t is the choice situation, z_i is observed data, η_i is a random term whose distribution over individuals relies on underlying parameters, v_i represents a vector of uncorrelated random variables and Γ is a lower triangular matrix that allows the random parameters to be correlated. Each element of β_i has mean and standard deviation and specified as a random parameter as opposed to a fixed parameter that treats the standard deviation as zero. The unobserved heterogeneity is accommodated by the standard deviation of β_i and allows the individuals to have different β_i instead of a single β for the entire sample.

'Observed heterogeneity is captured by entering the relevant attributes of the individual while unobserved heterogeneity is captured by entering random terms' (Greene and Hensher, 2007, p. 610). Bhat and Castelar (2002) noted that the unobserved heterogeneity typifies the unobserved differences across individuals in the intrinsic preferences for a choice alternative (preference heterogeneity) and/or in the sensitivity to the choice alternative characteristics (response heterogeneity). Therefore, the error term is decomposed into an individual-specific (inter-respondent) effect and an observation-specific effect (intra-respondent) (Hess and Rose, 2007). Hensher et al. (2005) stressed that the selection of the random parameters and their distributions and the number and types of draws should be examined. McFadden and Train (2000) used the Lagrange multiplier tests to reject or maintain the fixed parameters in the model. A simpler alternative test was suggested by Hensher and Greene (2003) by assuming that all parameters are random and investigating their estimated standard deviations. Revelt and Train (2000) favoured fixing the cost coefficient to make the distribution of the marginal WTP for an attribute equal to the distribution of that attribute's coefficient. However, this overestimates the WTP by making the denominator in the equation smaller (Sillano and Ortúzar, 2005). Moreover, the marginal utility of income is expected to be different among respondents; simply because individuals with high income have different values of a monetary unit than lower-income people. Torres et al. (2009) noted that assuming the homogeneity for a coefficient that is likely to be heterogeneous could have negative effects on the welfare estimates.

Different distributions could be used on each attribute. Every distribution has advantages and disadvantages. For example, the log-normal distribution allows constraining the parameters to be strictly positive, however it carries undesirable effects, such as a biased mean value caused by its long right-hand tail and producing positive parameter values may be an incorrect sign for certain parameters. Also, the triangular, uniform and normal distributions may present implausible signs for some attributes because of the standard deviation, which can imply behaviourally inconsistent WTP values. Therefore, it may be better to impose constraints on the distribution, such as constructing the standard deviation of each random parameter as a function of the mean. Hensher and Greene (2003) argued that using triangular distribution with imposing constraints on the spread guarantees a behaviourally correct sign of WTP. Alternatively, many studies have solved this problem by removing information that is incorrectly signed (see, for example Sillano and Ortúzar, 2005 and Greene et al., 2006).

There is no consensus about the number of draws required to produce a stable set of parameter estimates, however, this number increases with more complex models. An efficient approximation to the actual function for likelihood-based estimation is obtained with a sufficiently large number of draws (Greene and Hensher, 2007). The computation procedure can be much faster without sacrificing precision by using intelligent draw techniques such as Halton sequence, shuffled uniform vectors, antithetic and quasi-random sampling. Bhat (2001) found that using Halton draws reduces the number of draws required for estimation by a factor of 90 per cent or more. Hensher (2001) examined Halton sequences and concluded that a small number of intelligent draws, as low as 25, can produce stability and model fits. However, problems with high correlation were observed between standard Halton sequences in high dimensions. Randomized scrambled and shuffled Halton sequences have been employed to overcome the preceding problem (Bhat, 2003; Hess and Polak, 2003).

20.5.3 Implementation

The sets of options were presented to the visitors to determine how they would like to see Ras Mohammed reef sites managed and which characteristics matter to them. These options were defined in terms of four attributes: reef quality (*REEF*); uncrowded conditions (*PEOPLE*); number of dive sites (*D_SITES*); and the possible increase in entrance fees (*FEES*). These attributes were considered the most appropriate for the study objectives and the policy implementation. Reducing the level of congestion and maintaining the reef quality are considered to be a mechanism to manage reef carrying capacity and allow certain tourists to enjoy less crowded reef sites. The need to assess the preferred number of dive sites as an important attribute was stressed by park managers. This attribute plus *PEOPLE* represent the horizontal and vertical extensions for the park. The willingness to pay for park entrance to access the reef sites, with some improvements to the park and reef quality, was expected to be greater than the current entrance fees. The range of increase in entrance fees was chosen according to the results of the focus groups and the pilot survey. However, any higher amount than US$25 was felt to be unrealistic for a daily entrance fees to Ras Mohammed. Four levels were used to secure sufficient variation in the alternative option. Table 20.2 lists the attributes and levels presented in the choice experiments. The attributes of the alternative option were expressed as

Table 20.2 Attributes and levels used in the choice experiments

Attribute	Short Name	Levels
Increase in reef quality	*REEF*	No change; 15%; 30%; 45%
Congestion level	*PEOPLE*	Usual number; 25% fewer people; 50% fewer people; 75% fewer people
Number of dive sites	*D_SITES*	15; 20; 25; 30
Increase in entrance fees	*FEES*	US$5; US$10; US$15; US$20

increments to the current situation. Thus, the values of interest are the additional benefits and costs resulting from the implementation of the alternative policy. The model framework was established in accordance with the concept of change at the margin and consistent with the principles of benefit–cost analysis.

20.6 RESULTS

The results for the base model (binary logit) were presented before progressing to the random parameter models. Hensher and Greene (2003) considered this sequence is essential to investigate the data, the attributes' functional form and the sensibility of the results. It refers to the increase in behavioural richness and the choice modelling maturity that result from this progression.

20.6.1 Binary Logit Models

The preferences of international and national tourists were expected to be distinct. This assumption was confirmed by conducting a likelihood ratio test (Swait and Louviere, 1993; Hearne and Salinas, 2002). The formula for this test is:

$$-2\,(LL_{\textit{pooled data}} - LL_{\textit{international tourists}} - LL_{\textit{national tourists}}) = 54.8 \sim \chi^2_5$$

Given that the corresponding critical chi-square value at the 95 per cent confidence level is 11.07, the equality of the combined parameters between the two sets was rejected. Since the two populations represent different preference orderings and have underlying models with different parameters, two models were presented (Table 20.3).

The two models are statistically significant (chi-square equal to 339.970 and 527.067 for the international tourists and national tourists respectively with four degrees of freedom and p-values equal to zero). In both models all the attributes are statistically significant and have the expected signs (i.e., higher reef quality, lower congestion, more dive sites and lower entrance fees will result in higher utility level and a higher probability of that alternative option being selected). It is worth noticing that the coefficient of congestion level in the NT model is significant at 90 per cent probability level (not 95 per cent). Therefore, whereas the international tourists prefer less people at reef sites, this attribute is not highly significant among national tourists.

The contingency table of the predicted choice outcomes as based on the model pro-

Table 20.3 Results from logit models[a]

Variable	International Tourists (IT)		National Tourists (NT)	
	Coefficient	*p*-value	Coefficient	*p*-value
CONSTANT	−0.993662	0.0000	−0.704863	0.0018
REEF	0.036770	0.0000	0.056975	0.0000
PEOPLE	0.013509	0.0000	0.003217	0.0540
D_SITES	0.031006	0.0001	0.027455	0.0010
FEES	−0.074814	0.0000	−0.088171	0.0000
Log-likelihood	−1492.605		−1392.649	
Chi-squared	339.9700	0.0000	527.0673	0.0000
Hosmer-Lemeshow chi-squared	46.48249	0.0000	48.22675	0.0000
Correct prediction	66.50%		72.25%	
Observations	2400		2400	

Note: a. The logit models were estimated using NLOGIT, version 4.0 (Greene, 2007).

duced versus the actual choice outcomes was examined to determine model performance. The choice model correctly predicted the actual outcome for 66.5 per cent and 72.3 per cent of the total number of cases for international and national tourists respectively. Figure 20.3 presents an ROC (receiver operating characteristics) curve, which produces a measure of fit and can be used to compare models (Greene, 2007). A greater area under the ROC curve means a greater model fit. An area of 0.5 implies a model with no fit. For example, the area under the ROC curve was 0.71 in the IT model and 0.75 in the NT model. The second chart depicts the cross-tabulation of predicted values versus observed values. Since the number of observations are predicted to be a '0' when the actual value is 0 is significant and almost equals the number of observations that are predicted to be a '1' when the actual value is 1, this shows that both models are stable, balanced and perform well.

20.6.2 Random Parameter Models

Five specifications of random parameters were estimated (Table 20.4) in which (1) the underlying attribute parameters were randomized (RPL1); (2) the heterogeneity around the mean was considered (RPL2); (3) the heteroskedasticity of the standard deviation was allowed (RPL3); (4) the correlated parameters were incorporated (RPL4); and (5) the distribution of random parameters was constrained (RPL5).

Base model with random parameters only (RPL1)

The parameter values for the attributes and their corresponding standard deviation are significant (Table 20.4). Statistically significant parameter estimates for derived standard deviations of the experiment attributes in the two models refer to the presence of heterogeneity over the sampled population around the mean parameter estimate. As such, different respondents have parameter estimates that may be different from the sample population mean parameter estimate. In comparison to the binary logit models (BNL), the estimation of RPL models results in a substantial improvement of fit and

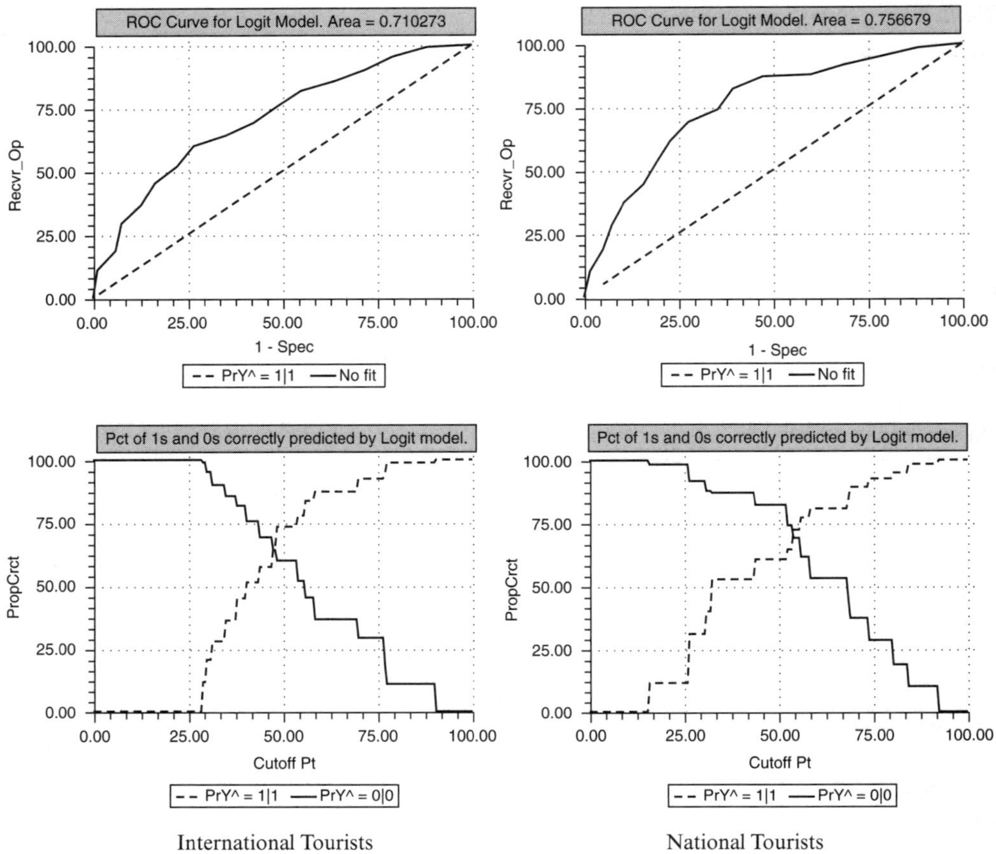

Figure 20.3 ROC curve and cross-tabulation of predicted values versus observed values

the hypothesis of homogeneity of the model's parameters is rejected (–2LL values were 304.33 and 227.25 for IT and NT respectively, which are greater than the corresponding critical chi-square value of 9.48).

Different distributional forms were assigned to test for better model fits and the model was re-estimated with a greater number of draws to ensure results stability. Hensher and Greene (2003) used the ratio of the mean to its standard deviation for the random parameters to examine the stability of the model. Following the same approach, 200 intelligent draws (shuffled Halton sequences) were found producing statistically similar results to higher draws (300, 500 and 1000). To illustrate the difference across respondents in the coefficients, centipede plots were produced for the respondent expected values of the coefficient on experiment attributes. For each of the 600 respondents, the range was given by the mean ± two standard deviations. This range captures at least 95 per cent of the distribution. The individual specific point estimates are presented by the dots in the centres of the bars. Each leg of the centipede plot presents the confidence interval for the coefficient for that visitor. Figure 20.4 shows that there is a considerable amount of variation across respondents in both means and standard deviation. In addition to individual-level

Table 20.4 Results from random parameters models

Variable	RPL1		RPL2		RPL3		RPL4		RPL5	
	Coeff.	*p*-value	Coeff.	*p*-value	Coeff.	*p*-value	Coeff.	*p*-value	Coeff.	*p*-value
International Tourists										
Non-random parameters										
CONSTANT	−0.932	0.00	−0.976	0.00	−0.705	0.00	−1.058	0.00	1.915	0.00
Means for random parameters										
REEF	0.056	0.00	0.088	0.00	0.056	0.00	0.057	0.00	0.063	0.00
PEOPLE	0.022	0.00	0.023	0.00	0.023	0.00	0.024	0.00	0.026	0.00
D_SITES	0.045	0.00	0.048	0.00	0.043	0.00	0.051	0.00	0.088	0.00
FEES	−0.156	0.00	−0.165	0.00	−0.171	0.00	−0.158	0.00	−0.181	0.00
Scale parameters for dists. of random parameters							Diagonal elements of Cholesky matrix		Derived std. dev. of parameter distributions	
REEF	0.138	0.00	0.132	0.00	0.084	0.00	0.113	0.00	0.063	0.00
PEOPLE	0.084	0.00	0.083	0.00	0.063	0.00	0.048	0.00	0.026	0.00
D_SITES	0.053	0.00	0.068	0.00	0.114	0.00	0.021	0.19	0.088	0.00
FEES	0.332	0.00	0.329	0.00	0.239	0.00	0.176	0.00	0.181	0.00
Heterogeneity in the means of random parameters										
REEF:SUB			−0.011	0.02						
REEF:MEMBER			0.025	0.01						
REEF:EDU			−0.002	0.03						
REEF:INCOME			0.000	0.06						
REEF:FAMILY			−0.004	0.03						
PEOPLE:SUB			0.011	0.00						
PEOPLE:CERT			−0.009	0.00						
PEOPLE:INFO			0.003	0.02						
PEOPLE:MEMBER			0.011	0.04						
PEOPLE :FAMILY			−0.004	0.00						
FEES:MALE			0.018	0.01						
FEES:AGE			−0.001	0.00						
FEES:MEMBER			−0.045	0.03						

Table 20.4 (continued)

Variable	RPL1		RPL2		RPL3		RPL4		RPL5	
	Coeff.	*p*-value	Coeff.	*p*-value	Coeff.	*p*-value	Coeff.	*p*-value	Coeff.	*p*-value
Heterogeneity in the means of random parameters										
FEES:EDU			0.003	0.05						
FEES:INCOME			0.000	0.07						
Heterogeneity in variances of random parameters										
REEF:EDU					0.022	0.12				
PEOPLE:EDU					−0.008	0.51				
D_SITES:EDU					0.027	0.04				
FEES:EDU					0.131	0.00				
Below diagonal elements of Cholesky matrix										
PEOPLE:REEF							−0.069	0.00		
D_SITES:REEF							0.017	0.34		
D_SITES:PEOPLE							−0.146	0.00		
FEES:REEF							−0.013	0.68		
FEES:PEOPLE							0.036	0.18		
FEES:D_SITES							−0.249	0.00		
Log-likelihood	−1340		−1310		−1346		−1334		−1380	
Chi-squared	304	0.00	365	0.00	293	0.00	317	0.00	566	0.00
Shuffled Halton draws	200		200		200		200		200	
Individuals	600		600		600		600		600	
Observations	2400		2400		2400		2400		2400	
National Tourists										
Non-random parameters										
CONSTANT	−0.655	0.00	−0.653	0.00	−0.513	0.01	−0.792	0.00	1.069	0.00
Means for random parameters										
REEF	0.073	0.00	0.075	0.00	0.069	0.00	0.084	0.00	0.091	0.00
PEOPLE	0.005	0.00	0.009	0.00	0.006	0.00	0.008	0.00	0.006	0.00
D_SITES	0.029	0.00	0.019	0.02	0.024	0.00	0.047	0.00	0.058	0.00
FEES	−0.112	0.00	−0.160	0.00	−0.109	0.00	−0.156	0.00	−0.155	0.00

Scale parameters for dists. of random parameters							Diagonal elements of Cholesky matrix		Derived std. dev. of parameter distributions	
REEF	0.170	0.00	0.172	0.00	0.235	0.00	0.190	0.00	0.091	0.00
PEOPLE	0.007	0.04	0.018	0.00	0.005	0.82	0.005	0.41	0.006	0.00
D_SITES	0.040	0.00	0.031	0.00	0.073	0.09	0.131	0.00	0.058	0.00
FEES	0.214	0.00	0.216	0.00	0.201	0.01	0.032	0.00	0.155	0.00
Heterogeneity in the means of random parameters										
REEF:REPEAT			−0.017	0.00						
REEF:SUB			0.020	0.00						
PEOPLE:MALE			−0.004	0.07						
D_SITES:REPEAT			0.021	0.00						
D_SITES:MEMBER			0.014	0.00						
FEES:EDU			0.003	0.02						
Heterogeneity in the variances of random parameters										
REEF:EDU					−0.017	0.21				
PEOPLE:EDU					0.031	0.91				
D_SITES:EDU					0.086	0.02				
FEES:EDU					0.280	0.00				
Below diagonal elements of Cholesky matrix										
PEOPLE:REEF							0.021	0.00		
D_SITES:REEF							0.078	0.00		
D_SITES:PEOPLE							0.070	0.00		
FEES:REEF							−0.221	0.00		
FEES:PEOPLE							0.053	0.06		
FEES:D_SITES							−0.419	0.00		
Log-likelihood	−1279		−1264		−1283		−1272		−1300	
Chi-squared	227	0.00	257	0.00	220	0.00	241	0.00	727	0.00
Shuffled Halton draws	200		200		200		200		200	
Individuals	600		600		600		600		600	
Observations	2400		2400		2400		2400		2400	

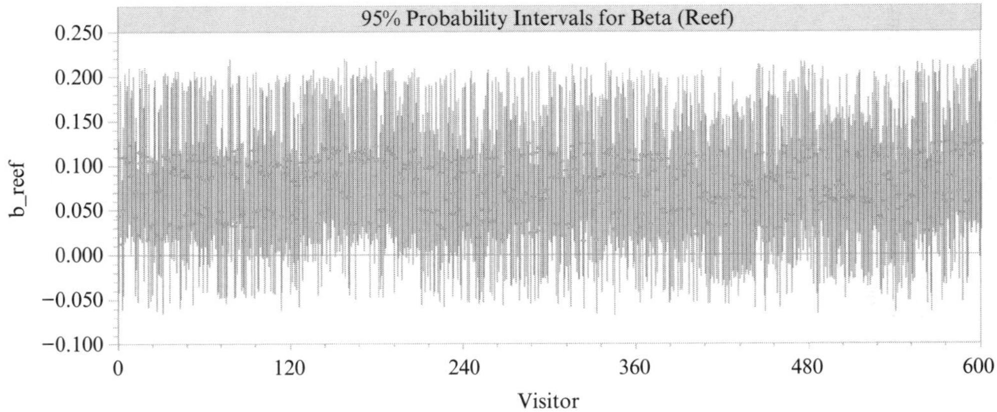

Figure 20.4 Confidence intervals for conditional means (IT model)

heterogeneity, these graphical summaries of estimates present general conclusions about relationships among variables.

Revealing preference heterogeneity around the mean of random parameters (RPL2)
The preference heterogeneity around the mean and its sources can be revealed by using the random parameters instead of creating the interaction effects through the data. The hierarchical formulation of the random parameter models enables the analyst to incorporate both observed and unobserved heterogeneity into the models (Greene et al., 2006). The parameters are randomly distributed over individuals with means and variances that may rely on demographic and attitudinal variables (Greene and Hensher, 2007). The significant interaction terms explain why preference heterogeneity may exist. For example, foreign tourists who visit other reef sites who have higher education or have larger families are more sensitive to reef quality, while those with higher incomes or are members of environmental organizations tend to be less reef-quality-sensitive (Table 20.4).

Heterogeneity in variances (heteroskedasticity) (RPL3)
The assumption of equal variances may distort the prediction of the model and come at substantial cost (Greene, 2007). The random parameter models relax this assumption and allow the unequal variances to be dependent on individual characteristics. As can be seen from Table 20.4, education has a statistically significant influence on number of dive sites and entrance fees. The positive sign on both *D_SITES* and *FEES* suggests that more educated visitors are much more heterogeneous in terms of the marginal utility associated with these attributes.

Correlated parameters (RPL4)
The random parameter models allow the error components in the choice sets to be correlated. Having done that, the standard deviations are no longer independent and the Cholesky decomposition matrix parameters should be used instead. Significant diagonal elements in the Cholesky decomposition matrix suggest significant variance directly attributable to the underlying random parameters while significant below-diagonal (off-

diagonal) elements refer to significant cross-product correlations among the random parameters previously confounded with the standard deviation parameter estimates. For example, the amount of variance attributable for reef quality and people are 0.113 and 0.048 respectively while –0.069 is attributable to the cross-correlations between them (Table 20.4).

Restricting the distribution (RPL5)

Following Hensher and Greene (2003) the sign and the range of a triangular parameter were restricted by constraining the spread to that of the mean of the random parameter in order to derive behaviourally plausible WTP values. The symmetry of this appealing distribution around the mean makes the results interpretation easier and bypasses the biased mean value caused by the long tail of the log-normal distribution. The two models are statistically significant (chi-square value of 566.12 for IT and 726.79 for NT with five degrees of freedom and *p*-values equal to 0). The means of random parameters are statistically significant and of the expected signs. However, the comparison of log-likelihood functions of these models with those of the base models suggests worsening in the model's fit.

20.6.3 Comparison of Models

Five RPL models were estimated for each set, starting with the base model and moving to more complicated and general models. These models were presented in order to explore the gains in the behavioural outputs of the choice models. Using a likelihood ratio test, the null hypothesis of parameter consistency across the models was rejected in the comparison between the binary logit model and the random parameter models and between RPL2 and RPL1 (model fit enhancement). The log-likelihood is flat at RPL3 while RPL4 is not any better than RPL1. Significant changes in chi-square and deterioration of model fit occurred when the distribution is constrained (RPL5).

The flat profile of values across most RPL models indicates to little if any behavioural improvement when proceeding from the base model to the more complex models.

Table 20.5 Model comparison and log-likelihood ratio test

Model	Chi-square Statistic	Degrees of Freedom	Critical Value at 5%
International tourists			
RPL1 vs BNL	304	4	9.48
RPL2 vs RPL1	60.69	15	24.99
RPL3 vs RPL1	−10.93	4	9.48
RPL4 vs RPL1	12.18	6	12.59
RPL5 vs RPL1	−80.11	4	9.48
National tourists			
RPL1 vs BNL	227	4	9.48
RPL2 vs RPL1	29.85	6	12.59
RPL3 vs RPL1	−7.01	4	9.48
RPL4 vs RPL1	13.48	6	12.59
RPL5 vs RPL1	−42.26	4	9.48

Allowing for correlation or interaction leads to improvements in model fit, which are, however, not significant when taking into account the additional parameters, or smaller than those obtained with the recognition of the repeated choice nature in expression of preference heterogeneity. Hensher et al. (2005) noted that the inclusion of the separate attributes along with their interaction is likely to induce multicollinearity. Furthermore, the constrained triangular distribution may be problematic with this inclusion (Greene, 2007). For instance, Greene et al. (2006) elucidated that the constraint condition may be not satisfied and the parameter distribution cannot be guaranteed to be limited to one side of zero when heterogeneity around the mean or in the variance of the random parameter are incorporated. Finally, using a heteroskedastic model with correlated parameters may make the model inestimable (Greene, 2007). On the basis of the above discussion, the recommended model structure is the base model (RPL1).

20.6.4 Marginal Willingness to Pay

Comparison between BNL and RPL models

There are different methods to derive WTP estimates. They could be calculated by the ratios of population means. However, the resultant values are derived from the coefficients of the average individual (which perceives the mean marginal utility of the population) for each parameter and are not the mean values of WTP and should not be used in cost–benefit analysis (Sillano and Ortúzar, 2005). Furthermore, if the underlying parameters are estimated as random parameters, then the WTP calculations should consider this specification. Using the ratios of population means to derive WTP values ignores the sampling variance, which makes the extra estimation effort ineffectual. In addition to such point estimates, the WTP could be derived using all the information in the distribution. Simulation is used in this way, drawing from the estimated covariance matrix for the parameters (Hensher and Greene, 2003). The mean WTP is calculated for each draw and this process is repeated for many draws. That provides the estimated mean WTP (the means of the ratios). For selecting a final WTP value, Sillano and Ortúzar (2005) referred to the superior explanatory power of the RPL models and the extra variance explained by them. WTP values can be estimated using either the unconditional parameter estimates or the conditional parameter estimates. In the former, the population must be stimulated and large random draws are taken for each parameter, allowing frequencies to be calculated by sampling WTP distribution where the individual-level parameters are calculated using the simulated maximum likelihood estimates and conditioning them with the respondent choices. The unconditional parameter estimates yield some negative and behaviourally implausible WTP values. Sillano and Ortúzar (2005) argued that removing parts from the distribution seems to be rational when the WTP values are derived for the sampled population. Therefore, small and equal percentages (3 per cent) were cut off from each tail of the WTP distribution for the experiment attributes (truncated distribution). In addition, it may be desirable to impose constraints on the random parameter distributions (e.g., Hensher and Greene, 2003 suggested using constrained distribution in order to guarantee non-negative WTP measures). Although the constrained distribution may outperform the truncated distribution because of the concern associated with arbitrarily removing part of the distribution, a behavioural rationale should exist for imposing such constraints.

Table 20.6 WTP values derived from BNL and RPL models

	BNL	Unconditional Parameters	Unconditional (Truncated) Distributions	Conditional (Unconstrained) Distributions	Conditional (Constrained) Distributions
International tourists					
Reef quality					
Mean	*0.49*	*0.0038*	*0.2399*	*0.49597*	*0.3884*
Std. dev.		14.5112	1.4936	0.40973	0.1591
Minimum		−353.4960	−6.1625	−0.0884	0.1643
Maximum		195.4700	7.6008	1.85042	0.7632
Uncrowded conditions					
Mean	*0.18*	*−0.0392*	*0.0804*	*0.20527*	*0.1586*
Std. dev.		7.3565	0.7572	0.20878	0.0605
Minimum		−179.2470	−3.1653	−0.2416	0.0755
Maximum		99.0530	3.8121	0.87082	0.3128
No. of dive sites					
Mean	*0.41*	*0.0833*	*0.2204*	*0.36624*	*0.5472*
Std. Dev.		8.4278	0.8674	0.28577	0.2501
Minimum		−205.2220	−3.4980	−3.3815	0.2294
Maximum		113.6060	4.4955	1.41715	1.1959
National tourists					
Reef quality					
Mean	*0.65*	*−0.1959*	*0.5667*	*0.79351*	*0.63657*
Std. dev.		28.6163	2.5174	0.58493	0.26538
Minimum		−766.9810	−10.4251	−0.0228	0.2046
Maximum		348.4510	11.4188	2.7263	1.31606
Uncrowded conditions					
Mean	*0.04*	*0.0023*	*0.0444*	*0.05591*	*0.04324*
Std. dev.		1.5784	0.1389	0.02676	0.01202
Minimum		−42.2908	−0.5619	0.03352	0.02876
Maximum		19.2324	0.6430	0.16951	0.0804
No. of dive sites					
Mean	*0.31*	*−0.0008*	*0.2373*	*0.30527*	*0.40663*
Std. Dev.		8.9319	0.7858	0.16276	0.15391
Minimum		−239.3360	−3.1936	0.1643	0.20353
Maximum		108.8210	3.6245	0.98148	0.86813

Different WTP estimates were obtained to investigate the effect of model specification and preference assumption on the results. Table 20.6 depicts the WTP for each attribute for the standard logit model together with corresponding figures of the RPL models. The results obtained by the conditional RPL models are consistent with those of the binary logit models. The foreign tourist is WTP an extra US$0.5 for each 1 per cent increase in the reef quality, US$0.2 for each 1 per cent decrease in the congestion level and US$0.4 for each additional dive site, while the national tourist is WTP an extra LE0.7[1] for each

1 per cent increase in the reef quality, LE0.05 for each 1 per cent decrease in the congestion level and LE0.3 for each additional dive site (the exchange rate in August 2008 was US$1 = LE5.5).

Based on the total number of visitors to Ras Mohammed of 495 382 (471 142 international tourists and 24 240 national tourists) in the year 2008/09 (EEAA, 2009), the annual WTP on top of the existing entrance fees was estimated to be US$238 656 for each 1 per cent increase in the reef quality, US$94 994 for each 1 per cent decrease in the congestion level and US$189 779 for each additional dive site.

WTP values derived from different specifications of RPL models

Progressing from the base model to more complicated models provides an analytical way of investigating the gains in the behavioural outputs of interest. The results suggest that accounting for heterogeneity in the variance tends to reduce the mean WTP values for the underlying attributes in the two sets, while accounting for heterogeneity around the mean or allowing for correlated parameters produce mean WTP values close to those of the base model.

WTP values for different respondent groups

In order to distinguish between respondent segments, we identify which attributes are perceived to be valuable for different visitor types and investigate the impacts of socio-economic characteristics on the model parameter values and welfare estimates, and the results were conditioned on these characteristics. In the IT model, the WTP for higher reef quality is greater when the respondent is male, old, a member of an environmental organization, has high income, has a small family, or visits the reef sites only in Ras Mohammed. Also, the highest WTP values are for the visitors from the UK and USA, where the visitors from Poland have the lowest WTP. For NT, the respondents holding a diving certificate, have snorkelling skills, are female, young, or a graduate are WTP more for improving reef quality. In addition, the respondents from Dakahlia and Ismailia have the greatest WTP, while the lowest WTP values are for respondents from South Sinai and Monufia. Interestingly, the respondents who have rich information about coral reefs or have small families are WTP more for uncrowded conditions in the IT model. In terms of nationality, Italians and Russians have the highest WTP values for this attribute.

20.6.5 Elasticities and Marginal Effects

The elasticity for fees attribute is calculated as –0.44 and –0.49 for IT and NT respectively. This suggests that a 1 per cent increase in entrance fees will decrease the probability of choosing the alternative option by 0.44 per cent in the IT model and 0.49 per cent in the NT model, ceteris paribus. This is consistent with the demand theory (raising the price is likely to decrease the demand). However, the entrance fees elasticity is relatively inelastic (<1). For the park management, this suggests that the revenue gained by any increase in the entrance fees will outweigh the negative impacts the fee increase will bring. Another noteworthy issue is the small percentage of elasticity for the congestion level attribute in the NT model (0.05), which implies that any changes in this attribute will slightly affect the choice outcomes. The inverse effect is evidence for reef quality attribute. It is also

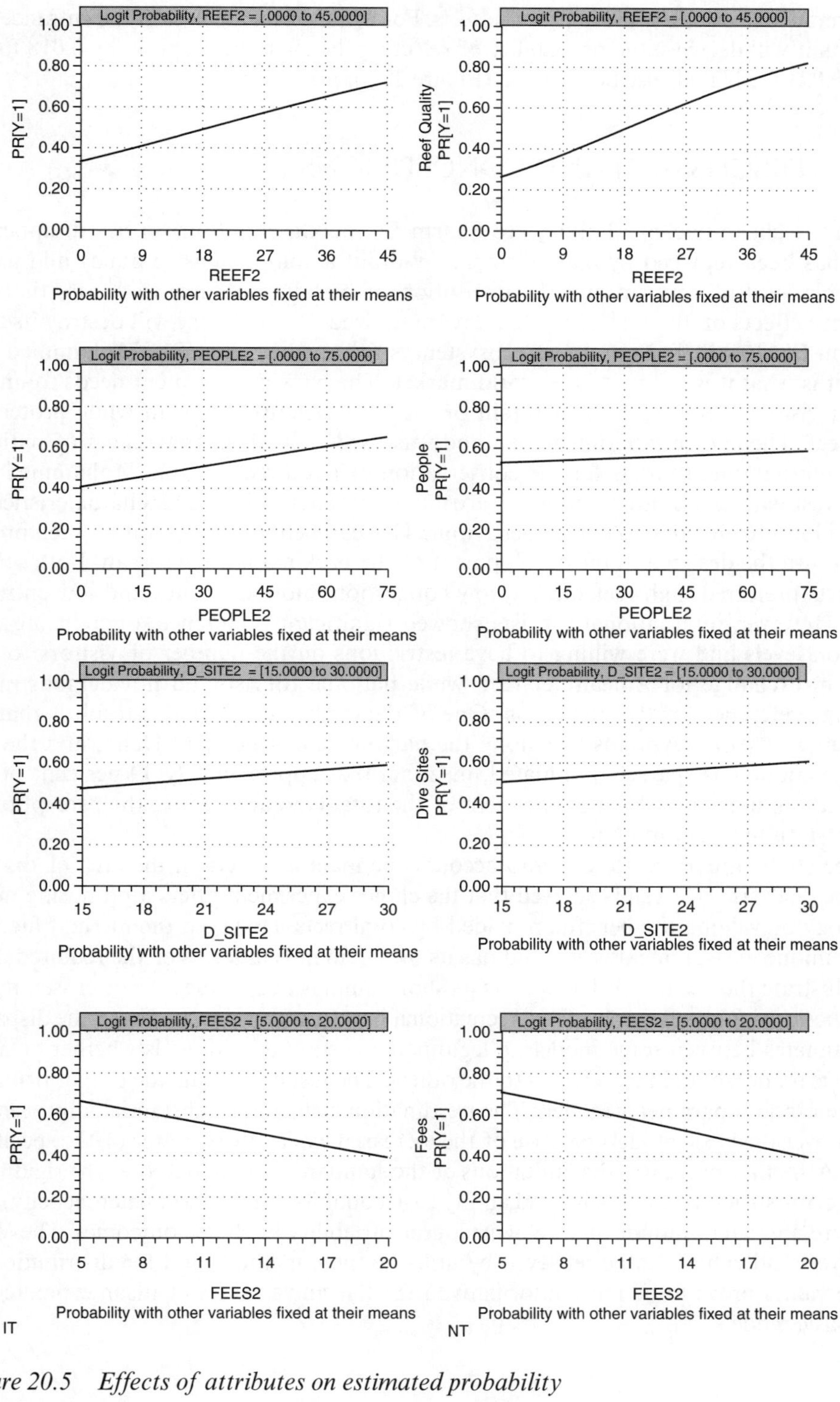

Figure 20.5 Effects of attributes on estimated probability

informative to calculate the marginal effects. For instance, an increase in the entrance fees of 1 unit will decrease the probability of selecting the alternative option by 0.018 for IT and 0.021 for NT, all else being equal (Figure 20.5).

20.7 DISCUSSION AND CONCLUSIONS

With the cheap package holidays to Sharm El-Sheikh, elite tourism has disappeared and has been replaced by mass tourism. Without a comprehensive policy and a sustainable level of tourism and the institution of certain measures to ensure that any adverse effects on the reef ecosystem are minimized, this industry will destroy itself in Sharm El-Sheikh. The coral reef ecosystem is the greatest asset Ras Mohammed has, and it is what it is selling to the world market. The park management needs to understand visitor preferences to maintain or increase benefits for them while protecting the reef. The choice experiments method was used to analyse preferences of national and international tourists for the conservation of coral reefs at Ras Mohammed and to investigate the contributions of attributes of alternatives and characteristics of individuals to elucidating choice behaviour. This can help in incorporating such preferences into the design and the development of the park management plan. Both sets of tourists preferred high reef quality, low congestion, more dive sites and low entrance fees. However, international tourists showed significant preference for reducing congestion levels and were willing to have restrictions on the number of visitors to reef sites in exchange for a healthier reef, while national tourists did not demonstrate a strong preference for this reduction. One of the explanations of this result is that the vast majority of Egyptians live along the narrow Nile Valley and Delta, and the rest of the country is sparsely populated, meaning that approximately 99 per cent of the population uses only about 5.5 per cent of the total land area. Thus, the perception of congestion may be different.

The study attempted to take into account the main advances in the area of discrete choice analysis. The results showed that the choice experiment offers a promising methodology of valuing the benefits provided by coral reefs. However, the method faces its own unique measurement issues and has its own demands in terms of the required data. To illustrate the CE method as well as its shortcomings and virtues, we discussed recent advances, using the example of the recreational benefit of coral reefs. Given the disparity in estimates between some models, a legitimate question to address is whether or not a specific model should be preferred to the others. The results indicate the outperformance of the random parameter models. Their estimation results in a substantial improvement of fit over the basic models because of the increased explanatory power of the specification. Also, they overcome the limitations of the standard logit models (i.e., the rigidity of their error structure and the limited ability to account for unobserved heterogeneity) and capture a greater amount of true behavioural variability in choice outcomes. The WTP for every individual can be retrieved by utilizing these methods and the distribution of these values prove to be more informative than the single values of mean estimated by the basic models.

NOTE

1. LE = livre égyptienne, or Egyptian pound.

REFERENCES

Abou Zaid, M. (2000), 'Overview of the status of Red Sea coral reefs in Egypt', unpublished report.

Adamowicz, W., J. Louviere and M. Williams (1994), 'Combining stated and revealed preference methods for valuing environmental amenities', *Journal of Environmental Economics and Management*, **26**(3), 271–92.

Adamowicz, W., P. Boxall, M. Williams and J. Louviere (1995), *Stated Preference Approaches for Measuring Passive Use Values: Choice Experiments Versus Contingent Valuation*, Report for University of Alberta, Department of Rural Economy.

Alpizar, F., F. Carlsson and P. Martinsson (2001), *Using Choice Experiments for Non-market Valuation*, report for Göteborg University, Department of Economics.

Bennett, J.W. (1999), *Some Fundamentals of Environmental Choice Modelling*, Choice Modelling Research Report No. 11, Canberra: University College, the University of New South Wales.

Bhat, C.R. (2001, 'Quasi-random maximum simulated likelihood estimation of the mixed multinomial logit model', *Transportation Research Part B: Methodological*, **35**(7), 677–93.

Bhat, C.R. (2003), 'Simulation estimation of mixed discrete choice models using randomized and scrambled Halton sequences', *Transportation Research Part B: Methodological*, **37**(9), 837–55.

Bhat, C.R. and S. Castelar (2002), 'A unified mixed logit framework for modelling revealed and stated preferences: formulation and application to congestion pricing analysis in the San Francisco Bay area', *Transportation Research Part B: Methodological*, **36**(7), 593–616.

Blamey, R.K., J.C. Rolfe, J.W. Bennett and M.D. Morrison (1997), *Environmental Choice Modelling: Issues and Qualitative Insights*, Choice Modelling Research Report No.4, Canberra: University College, University of New South Wales.

Blamey, R.K., J.W. Bennett, J.J. Louviere, M.D. Morrison and J.C. Rolfe (2002), 'Attribute causality in environmental choice modeling', *Environmental and Resource Economics*, **23**(2), 167–86.

Boxall, P.C., W.L. Adamowicz, J. Swait, M. Williams and J. Louviere (1996), 'A comparison of stated preference methods for environmental valuation', *Ecological Economics*, **18**(3), 243–53.

Bryant, D., L. Burke, J. McManus and M. Spalding (1998), *Reefs at Risk: A Map-based Indicator of Threats to the World's Coral Reefs*, Washington, DC: World Resources Institute.

Carson, R.T. (1991), 'Constructed markets', in J.B. Braden and C.D. Kolstad (eds), *Measuring the Demand for Environmental Quality*, Amsterdam: North-Holland.

Cesar, H. (2000) (ed.), *Collected Essays on the Economics of Coral Reefs*, Kalmar, Sweden: CORDIO, University of Kalmar.

DeShazo, J.R. and G. Fermo (2002), 'Designing choice sets for stated preference methods: the effects of complexity on choice consistency', *Journal of Environmental Economics and Management*, **44**(1), 123–43.

EEAA (2009), *The Annual Report of Income Department for 2008/09*, Sharm El-Sheikh: Ras Mohammed National Park, South Sinai Protectorates Library.

Eggert, H. and B. Olsson (2009), 'Valuing multi-attribute marine water quality', *Marine Policy*, **33**(2), 201–6.

Greene, W.H. (2007), *Nlogit Version 4.0 Reference Guide*, New York: Econometric Software Inc.

Greene, W.H. and D.A. Hensher (2007), 'Heteroscedastic control for random coefficients and error components in mixed logit', *Transportation Research Part E: Logistics and Transportation Review*, **43**(5), 610–23.

Greene, W.H., D.A. Hensher and J. Rose (2006), 'Accounting for heterogeneity in the variance of unobserved effects in mixed logit models', *Transportation Research Part B: Methodological*, **40**(1), 75–92.

Hassan, M., M.M.A. Kotb and A.A. Al-Sofyani (2002), 'Status of coral reefs in the Red Sea and Gulf of Aden', in C.R. Wilkinson (ed.), *Status of Coral Reefs of the World: 2002*, GCRMN Report, Townsville: Australian Institute of Marine Science, pp. 45–52.

Hawkins, J.P. and C.M. Roberts (1994), 'The growth of coastal tourism in the Red Sea: present and future effects on coral reefs', *Ambio*, **23**(8), 503–8.

Head, S.M. (1987), 'Corals and coral reefs of the Red Sea', in A.J. Edwards and S.M. Head (eds), *Key Environments: Red Sea*, Oxford: Pergamon Press, pp. 128–51.

Hearne, R.R. and Z.M. Salinas (2002), 'The use of choice experiments in the analysis of tourist preferences for ecotourism development in Costa Rica', *Journal of Environmental Management*, **65**(2), 153–63.

Hensher, D.A. (2001), 'The valuation of commuter travel time savings for car drivers: evaluating alternative model specifications', *Transportation*, **28**(2), 101–18.

Hensher, D.A. and W.H. Greene (2003), 'The mixed logit model: the state of practice', *Transportation*, **30**(2), 133–76.
Hensher, D.A., J.M. Rose and W.H. Greene (2005), *Applied Choice Analysis: A Primer*, Cambridge, UK: Cambridge University Press.
Hess, S. and J. Polak (2003), 'An alternative method to the scrambled Halton sequence for removing correlation between standard Halton sequences in high dimensions', paper presented at the European Regional Science Conference, Jyväskylä.
Hess, S. and J.M. Rose (2007), 'Some lessons for working with repeated choice data', paper presented at the 11th Triennial WCTR Conference, Berkeley, CA, 24–28 June.
Kotb, M.M.A., R.G. Hartnoll and A.F. Ghobashy (1996), 'Coral reef community structure at Ras Mohammed in the northern Red Sea', *Tropical Zoology*, **4**(2), 269–85.
Kotb, M.M.A. and M. Abdulaziz et al. (2004), 'Status of the coral reefs of the Red Sea and Gulf of Aden in 2004', in C.R. Wilkinson (ed.), *Status of Coral Reefs of the World: 2004*, GCRMN Report, Townsville: Australian Institute of Marine Science, pp. 137–54.
Lancaster, K. (1966), 'A new approach to consumer theory', *Journal of Political Economy*, **74**(2) 132–57.
Mazotta, M. and J. Opaluch (1995), 'Decision making when choices are complex: a test of Heiner's hypothesis', *Land Economics*, **71**(4), 500–515.
McAllister, D.E. (1991), 'What is a coral reef worth?' *Sea Wind*, **5**(1), 21–4.
McFadden, D. and K.E. Train (2000), 'Mixed MNL models for discrete response', *Journal of Applied Econometrics*, **15**(5), 447–70.
Moberg, F. and C. Folke (1999), 'Ecological goods and services of coral reef ecosystems', *Ecological Economics*, **29**(2), 215–33.
Mora, C., S. Andrefouet, M.J. Costello, C. Kranenburg, A. Rollo and J. Veron et al. (2006), 'Enhanced: coral reefs and the global network of marine protected areas', *Science*, **312**(5781), 1750–51.
Morrison, M.D., R.K. Blamey, J.W. Bennett and J.J. Louviere (1996), *A Comparison of Stated Preference Techniques for Estimating Environmental Values*, Choice Modelling Research Report No. 1, Canberra: University College, University of New South Wales.
Pearson, M.P. and A.E. Shehata (1998), 'Protectorates management for conservation and development in the Arab Republic of Egypt', *Parks*, **8**(2), 29–35.
PERSGA (2003), *Coral Reefs in the Red Sea and Gulf of Aden: Surveys 1990 to 2000 Summary and Recommendations*, Technical Series No. 7, Jeddah: The Regional Organization for the Conservation of the Environment of the Red Sea and Gulf of Aden.
Polidoro, B.A., S.R. Livingstone, K.E. Carpenter, B. Hutchinson, R.B. Mast and N. Pilcher et al. (2008), 'Status of the world's marine species', in J.-C. Vié, C. Hilton-Taylor and S.N. Stuart (eds), *The 2008 Review of The IUCN Red List of Threatened Species*, Gland, Switzerland: IUCN.
Reaka-Kudla, M.L. (1994), 'Biodiversity of coral reefs', in C. Heasley (ed.), *Science and a Changing World*, AAAS94, Program & Abstracts, San Francisco, p. 22.
Reaka-Kudla, M.L. (1997), 'Global biodiversity of coral reefs: a comparison with rainforests', in M.L. Reaka-Kudla and D.E. Wilson (eds), *Biodiversity II: Understanding and Protecting Our Biological Resources*, Washington, DC: Joseph Henry Press.
Revelt, D. and K.E. Train (2000), 'Customer-specific taste parameters and mixed logit', Working Paper No. E00–274, Department of Economics, University of California at Berkeley, CA.
Sillano, M. and J.D.D. Ortúzar (2005), 'Willingness-to-pay estimation with mixed logit models: some new evidence', *Environment and Planning A*, **37**(3), 525–50.
Smith, S.V. (1978), 'Coral-reef area and the contributions of reefs to processes and resources of the world's oceans', *Nature*, **273**(5659), 225–6.
Spalding, M.D., C. Ravilious and E.P. Green (2001), *World Atlas of Coral Reefs*, prepared at the UNEP World Conservation Monitoring Centre, Berkeley, CA: University of California Press.
Swait, J. and J.J. Louviere (1993), 'The role of the scale parameter in the estimation and comparison of multinomial logit models', *Journal of Marketing Research*, **30**(3), 305–14.
Torres, C., N. Hanley and A. Riera (2009), 'The implications of incorrect utility function specification for welfare measurement in choice experiments', Stirling Economics Discussion Papers, University of Stirling, Department of Economics.
Train, K.E. (2003), *Discrete Choice Methods with Simulation*, Cambridge, UK: Cambridge University Press.
Veron, J.E.N., O. Hoegh-Guldberg, T.M. Lenton, J.M. Lough, D.O. Obura and P. Pearce-Kelly et al. (2009), 'The coral reef crisis: the critical importance of <350 ppm CO_2', *Marine Pollution Bulletin*, **58**(10), 1428–36.

21. Using ecological information in choice experiments to value ecosystem services restoration programmes in East Asia

Yohei Mitani and Ståle Navrud

21.1 INTRODUCTION

Since the Law for the Promotion of Nature Restoration in Japan took effect in 2003, a nature restoration programme is being implemented nationwide. In the shallow lake ecosystem of Kushiro Mire wetlands, which was a pioneer project in this programme, an Ecosystem Restoration Strategy was formulated in 2005. As more restoration projects are considered not only in Japan, but also worldwide, it is important to understand public preferences for ecosystem restoration, in order to document the economic benefits of costly restoration projects, and determine efficient restoration goals. We use Lake Takkobu, which is part of the Kushiro Mire wetlands pilot project, as our case study.

Lakes and wetlands are productive ecosystems, providing various services ranging from flood and flow control to water quality maintenance, biodiversity and recreational benefits. As wetlands create not only use values but also substantial non-use values (Wilson and Carpenter, 1999; Lupi et al., 2002; Brander et al., 2006), we use the stated preference (SP) technique of choice experiments (CEs) to map public preferences and estimate their use and non-use benefits of restoring ecosystem services.

When conducting ecosystem restoration, it is crucial for the government and local residents to set the restoration goals. The restoration goal set would affect local industries, recreational use, restoration costs and so on, and thus conflicts between different stakeholder groups could occur. Therefore, it is important to document the social benefits of different restoration goals (Kiker, 2001), and how these benefits depend on the ecosystem services affected. As CEs enable multi-attribute and multi-level valuation, the method is very well adapted to value the social benefits of ecosystem restoration projects.

Previous CE studies of wetlands indicate that respondents' preferences for wetland restoration and management vary across individuals and groups (Carlsson et al., 2003; Milon and Scrogin, 2006; Birol et al., 2006). These studies show that if people's preferences for conserving and using natural environments are heterogeneous, this can cause conflict between people or groups with regard to nature restoration. Thus, having information about preference heterogeneity becomes critical with regard to consensus building.

Our study aims at adding to the scarce literature valuing restoration of wetland ecosystem services in East Asia, by conducting a CE and provide economic estimates for individual wetlands ecosystem services. This can assist policy-makers in implementing ecosystem restoration projects in Lake Takkobu in the Kushiro Mire wetland, and provide important lessons for similar type projects in other similar wetlands in Japan and elsewhere in East Asia.

The chapter is organized as follows: Section 21.2 provides a literature review of wetlands valuation studies, Section 21.3 presents the findings from ecological studies and the resulting restoration plans/goals we plan to value in our case study area. Section 21.4 describes our CE survey design and implementation of the survey instrument. Section 21.5 shows the econometric model applied in this CE, and Section 21.6 reports the results. Section 21.7 concludes, and outlines the policy implications of implementing ecosystem restoration in Lake Takkobu.

21.2 LITERATURE REVIEW

There have been a rapidly growing number of valuation studies of wetland ecosystem services over the last four decades, and several meta-analyses of these studies have been performed. Brouwer et al. (1999) conducted a meta-analysis of 30 contingent valuation (CV) studies of wetlands in developed countries. Results show that methodological design, study country, response rate, flood control and water quality all have a statistically significant effects on the estimated value. Woodward and Wui (2001) used 39 wetland valuation studies that employed not only CV but also travel cost (TC) and hedonic pricing (HP) methods in their meta-analysis, and examined the possibility of benefit transfer. They concluded that more empirical studies are needed in order to value wetland ecosystem services through benefit transfer. Brander et al. (2006) consider 191 wetland valuation studies in 25 countries from the 1970s to 2004, and identify 215 value estimates from 80 studies that contained the information necessary for inclusion in a meta-analysis. They found that socio-economic characteristics like population density and GDP per capita are important in explaining wetland values per km^2. Brander et al. (2012) have successfully extended this meta-analysis, and combined it with GIS data to perform benefit transfer and scaling up the observed unit values to cover larger geographical areas.

These meta-analyses confirm that the number of valuation studies of wetlands has increased rapidly over time. While the use of stated preference methods has also increased, the majority of these studies employ the CV method. However, as ecosystem restoration projects usually consist of measures that can address single attributes and/or ecosystem services of the wetlands, there is a need for multi-attribute valuation. This can be achieved by CE, while CV can only value a single attribute, or a package of attributes in combination.

In spite of the rapidly growing interest in using CE in environmental valuation over the last decade, there are still few CEs applied to restoration of wetlands or other ecosystem. Some of the earliest applications include the following. Carlsson et al. (2003) estimate economic values of each attribute of wetland management in Staffanstorp in Sweden. Hanley et al. (2006) evaluate environmental attributes of river ecosystems in England and Scotland in the United Kingdom. Christie et al. (2006) elicit public preferences for different attributes of biodiversity in Cambridgeshire and Northumberland in the United Kingdom. Birol et al. (2006) estimate the value of several wetland attributes in the Cheimaditida wetland in Greece.

In this chapter, we focus on valuing the benefits of restoring shallow lake ecosystems, of which there are few studies. Lupi et al. (2002) discuss the importance of valuing wetland restoration, and apply CV to wetland mitigation in Michigan. They find that people care about the several attributes of wetlands, from habitat quality to wetland acreage. Milon and Scrogin (2006) utilize pairwise rating conjoint analysis to evaluate wetland ecosystem restoration projects in the Everglades National Park in Florida. They demonstrate that people place more importance on ecological structural restoration such as the number of species than ecological functional restoration such as water levels.

Choice experiments designed to capture preference heterogeneity among individuals have shown that respondents' preferences for environmental goods often vary across individuals and groups (Train, 1998; Provencher and Bishop, 2004; Scarpa and Thiene, 2005).

Discrete choice models capturing heterogeneity have also been applied in valuation of wetland ecosystems. Using a random parameter logit (RPL) model, Carlsson et al. (2003) show that there exist heterogeneous preferences for wetland ecosystem management, and also indicate that the relative magnitude of the heterogeneity differs for each attribute. Milon and Scrogin (2006) use a latent segment logit (LSL) model to estimate the preferences for wetland restoration, and find that there are three homogeneous groups, and that the sign of their stated value can differ across groups. They also insist that LSL analysis is useful in understanding the causes of the preference heterogeneity. Birol et al. (2006) indicate that people's preferences for wetland ecosystem management vary in the RPL analysis, and that the LSL model is better specified.

This chapter contributes in four ways to the wetland valuation literature by: (1) adding a CE study to the scarce environmental valuation literature in Japan and East Asia, (2) applying the CE to value individual ecosystem services of ecosystem restoration projects, (3) uncovering preference heterogeneity for the different CE attributes/ecosystem services; and (4) basing the ecosystem services valuation on comprehensive ecological data from field surveys in the case area studied.

21.3 ECOYSTEM RESTORATION IN LAKE TAKKOBU

Ecosystem restoration in Lake Takkobu is one of five pilot project areas for ecosystem restoration projects in the Kushiro Mire wetlands in Hokkaido, Japan. The ecosystem services we evaluate are derived from a discussion with the conservation ecologists working in the area, and the restoration goals to be valued are derived from historical ecological data about the lake.

Lake Takkobu is a small and shallow lake with a surface area of 1.36 km^2, and with an average depth of 1.90 m. The lake is located in the eastern part of Kushiro Mire, which is the largest Ramsar Convention designated wetland in Japan. We chose this lake as our study site because: (1) the estimated values can be transferred (Navrud and Ready, 2007) to other similar sites, and provide important input to restoration decisions for other shallow lakes and wetlands in Japan and the rest of East Asia, (2) much ecological information including historical data exists for the lake, and (3) the ecosystem restoration project in the lake was on-going at the time of our survey.

Ecological survey results showed that the status of the lake ecosystem has been deteriorating over the last few decades, and also indicated that the main source of degradation of water quality is human activities since the 1970s, and especially livestock excreta (Takamura et al., 2003). According to a satellite image of the lake in 1992 created by Geospatial Information Authority of Japan, approximately 80 per cent of the surface area was covered by various aquatic plant species including submerged plants. The existence of submerged plants indicates high water quality. According to an ecological survey of the lake in 1991, there were 20 aquatic plant species including six endangered ones (Kadono et al., 1992). The water quality at that time was very good, and various aquatic plants including many submerged plants were found in the lake.

However, the results of an ecological survey conducted in 2000 showed that the number of aquatic plant species in the lake had started to decline. At the time of our CE survey some species seem to be extinct, and some other species were threatened by extinction. Rapid environmental deterioration of the lake, symbolized by the appearance of *Anabaena* bloom, has occurred over the last few decades (Takamura et al., 2003). An ecological survey conducted in 2003 showed that the percentage of surface area covered by aquatic plants had declined to approximately 20 per cent. The number of aquatic plant species had also declined to 14 species including two endangered ones, implying that some species have become extinct and other species have been moving towards extinction since 1991 (Takamura et al., 2003). Correlated with the species decline, water quality was also deteriorating. Previously unseen, *Anabaena* bloom started to appear more frequently.

In a shallow lake ecosystem such as Lake Takkobu two ecologically stable states exist: (1) a clear-water state where submerged plants are abundant and water is transparent, and (2) a turbid-water state where phytoplankton is abundant and no submerged plants exist. Drastic and rapid changes in ecosystems from one steady state to another, which are difficult or impossible to reverse, are called ecological regime shifts (Carpenter, 2003). Due to human activities in the upstream of or around lakes, shallow lakes often switch from a clear-water state to a turbid-water state.

Figure 21.1 shows a hypothetical example of time series for water quality (solid thin line), where there exist two ecosystem regimes, that is, clear-water and turbid-water regimes, with stochastic fluctuations due to internal and external causes.[1] The dotted lines show the average water quality for each regime. During a regime, the water quality tends to return to the regime average after stochastic disturbances. In this example, we have three large disturbances. The first disturbance (toward worse quality) does not result in a regime shift because the disturbance does not cross the threshold for the clear-water regime. The thresholds are related to the concept of resilience. Resilience is the amount of disturbance required to cross a threshold causing a regime shift (Carpenter, 2003). The second big disturbance is large enough to result in a shift to the turbid-water state. A regime shift happens when a disturbance crosses a threshold beyond the capacity of resilience. The third disturbance does not result in a shift back to the clear-water regime, even though the water quality increases above the bottom of the first big disturbance. Thus, the threshold for a regime shift is different for the clear-water and turbid-water states as shown in Figure 21.1. This is called hysteresis, where the conditions required to change a regime in one direction are different from the conditions required to change the regime back to the original state (Carpenter, 2003). Note that the thresholds, as well as resilience, can change over time.

The ecosystem condition in Lake Takkobu in 1991 can be defined as a clear-water state, while in 2003 a regime shift from the clear-water state to the turbid-water state seemed to be on the verge of occurring (Takamura et al., 2003). If no restoration measure was undertaken in 2003, the water quality would deteriorate further, the number of

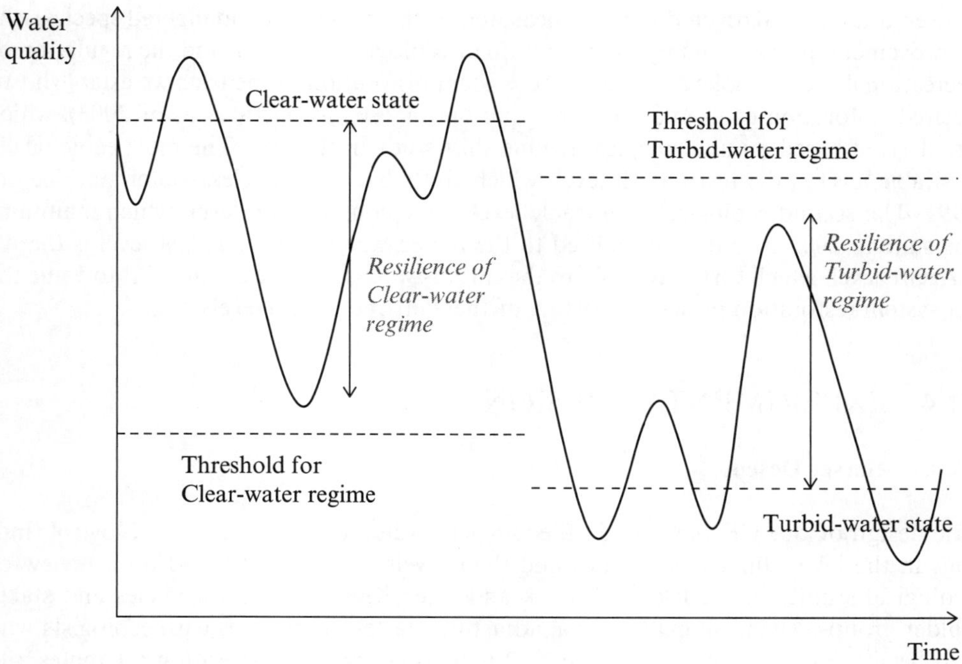

Source: Based on Carpenter (2003, Figure 2), modified by the authors.

Figure 21.1 Regime shift, thresholds and resilience in shallow lakes

Table 21.1 Choice experiment attributes/ecosystem service, and restoration goals (defined by reference year and ecosystem state)

Attributes/Ecosystem Service	Restoration	Conservation	No Measure
Biodiversity (# aquatic plant species)	20 species	14 species	7 species
Endangered species (# endangered species)	6 species	2 species	0 species
Water quality (i.e., ecosystem function)	Clear	Fair	Unclear
Recreational availability (i.e., potential demand)	Not available	Available	Available
Reference year	Around 1991	Around 2003	Years later
Ecosystem states	Clear water	Regime shifting	Turbid water

aquatic plants would decline further, and the lake would shift to the turbid-water state (Nakamura, 2007).

Table 21.1 presents ecosystem restoration attributes and targets used in this study. A review of ecological studies in the lake leads us to have four restoration attributes: the restoration of biodiversity (measured by the number of aquatic plant species in the lake), the conservation of endangered species (measured by the number of endangered species), the improvement of water quality as restoration of ecological function, and the regulation of recreational use. As policy targets for ecosystem restoration in the lake, we establish two desired restoration goals and the status quo option (i.e., *No measure* as of 2003), which are derived based on an ecological regime shift study in the lake. The most ecologically desirable level is the restoration level, which shifts back to the clear-water state before 1991. The second ecologically desirable level is the conservation level, which maintains the state in 2003 before it has shifted to the turbid-water state. The last level is the *No measure* level, which certainly shifts to the turbid-water state. We employ CE to value the ecosystem restoration project consisting of these attributes and levels.

21.4 EXPERIMENTAL DESIGN

21.4.1 Survey Design

The design of our CE was based on ecological evaluation criteria and ecological findings in the lake. Our CE attributes and their levels are selected based on a review of ecological studies in the lake, fieldwork and interviews with local agencies and stakeholder groups. Then, we made a final adjustment after discussions with ecologists who had been working at the lake. Table 21.2 reports ecosystem restoration attributes and levels used in our CE. We have four attributes, which can be said to represent four different ecosystem services: (1) the restoration of biodiversity measured by the number of aquatic plant species existing in the lake; (2) the conservation of endangered species measured by the number of registered endangered species in the lake; (3) the improvement of water quality; and (4) the regulation of recreational use. We refer to these four attributes as *Biodiversity*, *Endangered Species*, *Water Quality*, and *Recreational Availability*, respectively:

- *Biodiversity* is a measure that aims to restore or conserve biodiversity by restoring and conserving the number of aquatic plant species in the lake. This attribute asks the public's preferences for the restoration of species diversity. Although biodiversity is the totality of genes, species and ecosystems in a region, we deal only with the species diversity in our study, which refers to the variety of species within a region (i.e., the lake).
- *Endangered Species* is a measure that avoids the extinction of scarce aquatic plant species across the country and reduces the risk of the extinction by restoring and preserving the number of endangered species that inhabit the lake.
- *Water Quality* is a measure that improves water quality or prevents water quality degradation in the lake. A three-step water quality ladder is applied; from unclear to fair to clear, which represents the best water quality.
- *Recreational Availability* is a measure that restricts the recreational use of camping and canoeing in or around the lake, in relation to ecosystem restoration. Lake Takkobu provides a recreational environment for camping, canoeing and fishing. In implementation of this ecosystem restoration project, it is important to capture potential demand for recreational use of the lake.

As described in the previous section, we established two restoration endpoints and one status quo as policy targets for ecosystem restoration in the lake, based on the ecological regime shift studies in the lake. The restoration level (*Restoration*) restores the state back to the clear-water state present before 1991. The conservation level (*Conservation*) maintains the state in 2003 just before it shifted to the turbid-water state. In the status quo level (*No Measure*) no measures are taken, in which the lake shifts to the turbid-water state.

A large number of unique ecosystem restoration profiles can be constructed from the combinations of these attributes and levels. *Biodiversity*, *Endangered Species*, and *Water Quality* have three levels each. *Recreational Availability* has two levels. *Cost*, a one-time payment to the restoration fund, has five levels of 500, 1000, 2000, 4000, and 8000 JPY (100 JPY = 1.20 USD = 0.95 €). Therefore, the number of total ecosystem restoration profiles is $3^3 \times 2 \times 5 = 270$. We eliminated two unrealistic combinations of alternatives: the pair of the *Restoration* level of *Biodiversity* and the *No Measure* level of *Water Quality* as well as the pair of the *Restoration* level of *Endangered Species* and the *No Measure* level of *Water Quality*. Then, we employed an orthogonal design, which consists of only the main effects, resulting in 64 pair-wise comparisons of alternative ecosystem restoration profiles (Louviere et al., 2000). These were randomly blocked into eight different versions, each version consisting of eight choice sets. Each choice set contains two ecosystem restoration profiles and a *No Measure* status quo option as shown in Figure 21.2. The inclusion of the status quo baseline alternative in the choice sets is instrumental in achieving welfare measures that are consistent with demand theory (Bateman et al., 2002).

Our questionnaire contained five sections: (1) an introductory section that asks questions about respondents' knowledge, attitudes and motivations, (2) an information section that describes ecosystem restoration in Lake Takkobu, including ecological

Table 21.2 Attributes, units and levels of the choice experiment

Attributes	Levels	Descriptions
Biodiversity	20 species	Restoring to 20 species, the level of 1991
(# Aquatic plant	14 species	Conserving 14 species, the level of 2003
species)	7 species	With no measure, decreasing to 7 species
Endangered Species	6 species	Restoring to 6 endangered species, the level of 1991
(# Endangered	2 species	Conserving 2 endangered species, the level of 2003
species)	0 species	With no measure, all endangered species will extinct
Water Quality	Clear	Restoring to clear-water state. Drinkable, swimmable, fishable
	Fair	Maintaining fair state. Swimmable, fishable
	Unclear	With no measure, deteriorating to turbid-water state Fishable
Recreational	Available	No regulation. Free to camp, canoe
Availability	Not available	Regulation of camping and canoeing
Cost	500, 1000, 2000, 4000, 8000 JPY,[a] one time payment	

Note: a. 100 JPY = 1.20 USD = 0.95 € (2012).

	Plan 1	Plan 2	No Plan
Cost (JPY)	500 JPY	4000 JPY	0 JPY
# Aquatic Species	7 (conserve)	20 (restore)	7 (conserve)
# Endangered Species	2 (conserve)	2 (conserve)	0 (no project)
Water Quality	Unclear	Clear	Unclear
Recreational Use	Available	Not Available	Available

Click most preferable plan →	○1	○2	○3

Note: Amounts in Japanese yen (JPY), 2005 price level; 100 JPY = 1.20 USD = 0.95 € (2012 exchange rates).

Figure 21.2 An example of the choice tasks of the choice experiment valuation question. Each respondent was subjected to eight such choice tasks

information on biodiversity, endangered species and regime shifts with visual aids, (3) a review section asking questions about respondents' understanding and perceptions, (4) a valuation section containing the valuation scenario, eight value elicitation questions like the choice set shown in Figure 21.2, and a follow-up question that aims to identify protest bids; and (5) a socio-economic characteristics section.

21.4.2 Data Collection

The main survey was implemented in October 2006 as an internet survey, after a pre-test with 108 respondents in July of the same year had checked whether the instrument was working well. Regarding the internet survey as the survey mode, there were major concerns regarding coverage and sampling error because not everybody has access to the internet (Bateman et al., 2002). However, over the last decade internet coverage has increased rapidly, and now seems high enough in most developed countries to get representative samples of the population through careful sampling procedures and/or the use of internet panels. A review of comparative studies of internet surveys and other survey modes in countries with relatively high internet penetration show encouraging results for internet surveys (Lindhjem and Navrud, 2011b). Results from one of the few fully controlled comparative studies shows that the estimated mean WTP per household from internet surveys is somewhat lower, but not significantly different from WTP in in-person interviews (Lindhjem and Navrud, 2011a). One important advantage of internet surveys is that they allow for more extensive use of visual aids including photos and graphical presentations, which we utilized in our survey.

The survey was conducted by a professional survey firm. E-mail invitations and respondent passwords were sent to a random sample of 1400 respondents recruited from their nationwide, non-random, pre-recruited panel of respondents. The number of

Table 21.3 *Characteristics of respondents in the sample and the national population*

Characteristics	Sample Mean	Sample S.D.	Population Mean[a]
Gender (% male)	0.53	0.50	0.49
Mean age	36.26	11.63	43.3
Mean number of people in household	3.22	1.311	2.55
Mean gross income (×100 in 2005 JPY)	665.3	387.9	295.8

Source: Source data of the national population is http://www.stat.go.jp/data/kokusei/2005/kihon1/00/hyodai.htm; last accessed 21 February 2014.

valid responses was 425, giving a response rate of 30 per cent, calculated as the ratio of completed responses to invitations sent. Our response rate is higher or similar to other SP studies using internet panels (Berrens et al., 2004; Lindhjem and Navrud, 2011a), but higher than internet surveys sent to a random sample of e-mail addresses (Lindhjem and Navrud, 2011b). Our survey was conducted in order to be representative of the Japanese population in terms of gender and age. Table 21.3 reports socio-economic characteristics of our respondents. Compared to the characteristics of the Japanese population, gender is almost the same, average age is about seven years younger, the number of people in the household is a little higher, and the income is much higher. Regarding respondents' visits of Lake Takkobu and the Kushiro wetlands, 17 per cent of the respondents have visited the Kushiro wetlands. Ten per cent answered that they did not know the wetland, only 4 per cent of respondents have visited Lake Takkobu, and the majority (83 per cent) answered that they did not know the lake, implying that the lake would have a very low recreational use potential for our nationwide sample.

21.5 ECONOMETRIC MODEL

21.5.1 Modelling Background

Discrete choice models are applied in the analysis of choice experiment data. The conditional logit (CL) model, which has been widely and frequently used in CEs, assumes homogeneous preference where utility parameters of each attribute are fixed over individuals. However, recent applications have shown that respondents' preferences for environmental goods often vary across individuals (Provencher and Bishop, 2004). This may lead to biased estimates of welfare measures for changes in restoration projects and adversely affect policy and management decisions. In addition, the CL model requires the assumption of the independence of irrelevant alternatives (IIA).

In this chapter, we estimate a mixed logit (ML) model, which enables us to identify the presence or absence of the preference heterogeneity by allowing the utility parameters to have a continuous distribution. Compared with the CL model, the ML model is not only able to capture the preference heterogeneity, but is also a more general model, relaxing the assumption of IIA (Train, 2003).

21.5.2 Mixed Logit Model

We here describe the structure of ML model. Let U_{ij} denote respondent i's utility from alternative j in a choice set C as follows:

$$U_{ij} = \beta_i' x_{ij} + \varepsilon_{ij}. \tag{21.1}$$

The first term on the right-hand side is the observable component of U_{ij}. The second term is the unobservable component, that is, error term. The coefficient vector β_i denotes respondent i's utility parameter, and x_{ij} is the attribute vector associated with alternative j for respondent i. Now, we assume that each ε_{ij} is distributed with a type I extreme value distribution and β_i is distributed with a normal distribution $N(b, W)$ with mean b and covariance matrix W. Then the mixed logit choice probability that the respondent i chooses alternative j from her choice set C becomes as follows (McFadden, 1974; Train, 2003):[2]

$$P_{ij}(b, W) = \int \prod_{t=1}^{T} \left[\frac{\exp(\beta_i' x_{ijt})}{\sum_{k \in C} \exp(\beta_i' x_{ikt})} \right] . \phi(\beta | b, W) d\beta, \tag{21.2}$$

where $\phi\ (b \mid \beta, W)$ is the density function of normal distribution.

In general, the integral in Equation (21.2) does not have a closed form (Brownstone and Train, 1999). Thus, we approximate the integral through simulation. The procedure of the calculation of choice probability is as follows:

(1) Draw β from $\phi\ (\beta \mid b, W)$ R times.[3]
(2) Insert each drawn β into a conditional probability $L_{ij}(\beta)$ and calculate R choice probabilities:

$$L_{ij}(\beta) = \frac{\exp(\beta' x_{ij})}{\sum_{k \in C} \exp(\beta' x_{ik})}.$$

(3) Average the results:

$$SP_{ij} = \frac{1}{R} \sum_r L_{ij}(\beta^r),$$

where β^r is the rth drawn β.

Note that SP_{ij} is an unbiased estimator of P_{ij}, that is, $E(SP_{ij}) = P_{ij}$ (Train, 2003). In this way, using the simulated probability SP_{ij}, the simulated log-likelihood function (SLL) is constructed as follows:

$$SLL(b,W) = \sum_i \sum_j \delta_i^j \ln SP_{ij}(b,W), \tag{21.3}$$

where δ_i^j denotes the dummy variable equalling 1 when respondent i chooses alternative j, and 0 otherwise. The estimated parameters (b, W) are simulated maximum likelihood (SML) estimators that maximize SLL (McFadden and Train, 2000). Note that the ML model has two characteristics particularly useful for the purpose of our analysis. First, this model can deal with problems of heterogeneity across respondents' preferences, as it allows utility parameters to vary randomly across individuals according to a continuous probability distribution. Second, the ML model allows for a complete relaxation of IIA since the ratio of ML probabilities, P_{ij}/P_{ik}, depends on all other alternatives.

21.5.3 Welfare Measurement

Once the utility parameters have been estimated, a willingness to pay (WTP) welfare measure for an environmental change from x^0 to x^1 under the assumption of no income effects can be calculated as follows:

$$WTP = -\frac{1}{\beta_{Cost}}\left[\ln\left(\sum_{j\in C}\exp(\beta' x_j^1)\right) - \ln\left(\sum_{j\in C}\exp(\beta' x_j^0)\right)\right], \tag{21.4}$$

where β_{Cost} denotes the marginal utility of income. For example, the marginal willingness to pay (MWTP) for the water quality improvement from the status quo to the restoration level ($WaQ1$) can be calculated as follows (Hanemann, 1982):

$$MWTP_{WaQ1} = -\frac{\beta_{WaQ1}}{\beta_{Cost}}. \tag{21.5}$$

21.6 ESTIMATION RESULTS

21.6.1 Specification

Table 21.4 shows the list of variables used in our analysis. We assume a main effect model for the observable component V of the indirect utility function U in Equation (21.1) as follows:

$$V = \beta_0\, ASC_{SQ} + \beta_{Cost}\, COST + \beta_2\, BIO_{res} + \beta_3\, BIO_{con} + + \beta_4\, END_{res}$$
$$+ \beta_5\, END_{con} + \beta_6\, WAQ_{res} + \beta_7\, WAQ_{con} + \beta_8\, REC, \tag{21.6}$$

Table 21.4 Descriptions of variables

Variables	Descriptions
ASC_{SQ}	Equals 1 when *No Measure* is chosen (dummy)
$COST$	One time voluntary payment in JPY ($\times 10^3$, continuous)
BIO_{res}	Restoring to 20 aquatic plant species (dummy)
BIO_{con}	Conserving 14 aquatic plant species (dummy)
END_{res}	Restoring to 6 endangered species (dummy)
END_{con}	Conserving 2 endangered species (dummy)
WAQ_{res}	Improving water quality to clear-water state (dummy)
WAQ_{con}	Maintaining fair/clear-water state (dummy)
REC	Recreational use available (dummy)

Table 21.5 Results

Variables		Conditional Logit		Mixed Logit (ML)		Based on ML Est.	
		Est.	S.E.	Est.	S.E.	WTP	S.D./M.
ASC_{SQ}	M.	−1.051	0.086***	−1.492	0.118***		
$COST$	M.	−0.199	0.01***	−0.356	0.019***		
BIO_{res}	M.	0.347	0.068***	0.554	0.124***	1556.6	2.35
	S.D.			1.304	0.166		
BIO_{con}	M.	0.19	0.059***	0.235	0.104**	659.4	4.63
	S.D.			1.086	0.133***		
END_{res}	M.	0.352	0.07***	0.663	0.134***	1863.0	2.39
	S.D.			1.582	0.151***		
END_{con}	M.	0.129	0.061**	0.175	0.099*	492.3	5.29
	S.D.			0.928	0.136***		
WAQ_{res}	M.	0.888	0.079***	1.362	0.146***	3825.1	1.19
	S.D.			1.618	0.152***		
WAQ_{con}	M.	0.577	0.077***	0.842	0.124***	2365.6	1.06
	S.D.			0.893	0.164***		
REC	M.	−0.046	0.041	0.022	0.095	62.1	67.43
	S.D.			1.49	0.11***		
No. of obs		3400		3400			
Log-likelihood		−2847.8		−2555.2			
LRI		0.238		0.316			
AIC		5713.7		5142.5			
BIC		5768.8		5240.6			

Note: *, **, and *** = significant at the 10, 5 and 1 per cent level, respectively.

where ASC_{SQ} denotes an alternative specific constant (ASC), which is a dummy variable that equals 1 when *No Measure* (status quo) is chosen and 0 otherwise.

The valid sample used in the analysis, after removing missing values and protest answers, was 425 respondents × 8 choices = 3400 observations. We use GAUSS 8.0 with maxLik application to estimate the utility parameters in Equation (21.6).

21.6.2 Mixed Logit Estimation Result

All the parameters in the mixed logit model except ASC_{SQ} and *COST* are specified to be normally distributed[4] (Train, 1998; Revelt and Train, 1998; Carlsson et al., 2003), and distribution simulations are based on 200 Halton draws. The ML estimation result is reported in the second column of Table 21.5. We hold the *COST* parameter fixed for the purpose of calculating the marginal willingness to pay (MWTP).[5]

Our ML model performs quite well in the sense that all the estimates of utility parameters have the expected sign and are statistically significant at the 1 per cent level, except the mean estimates of BIO_{con} and END_{con}, which are statistically significant at the 5 and 10 per cent levels, respectively. Also, the mean estimate of *REC* is not statistically significant even at the 10 per cent level. All the estimates of standard deviation are statistically significant at the 1 per cent level, indicating that there is non-negligible heterogeneity in preferences for ecosystem restoration projects in the lake among our nationwide respondents.

For comparative purposes, we also report the estimation results of a conditional logit (CL) model in Table 21.5. Consistent with the result of the ML model, all the estimates have expected signs and are statistically significant at the 5 per cent level or less; except for *REC*. Table 21.5 also reports McFadden's likelihood ratio index (LRI) as well as Akaike information criteria (AIC) and the Bayesian information criteria (BIC). Comparing the ML with CL model, the substantial increase in the LRI from 0.238 to 0.316 indicates the advantage of applying the ML instead of the CL model.

The alternative specific constant ASC_{SQ} captures the average, systematic effect on utility of all factors that are not included in the utility function. The result shows that ASC_{SQ} has a negative sign and is significant at the 1 per cent level. Since ASC_{SQ} equals 1 when *No Measure* is chosen in our specification, the result indicates that ASC_{SQ} has a positive effect on the probability of choosing restoration programmes (i.e., Plan 1 or Plan 2), when compared with the probability of choosing the *No Measure* status quo option. The result also implies that the average impact of omitted variables may be influenced by a tendency towards ecosystem restoration that our model specification overlooks. In other words, environmental deterioration in the lake due to no restoration measure would have an additional negative effect on the nationwide citizens' utility, beyond what is captured by our attributes.

The third column of Table 21.5 reports the calculated marginal WTP (MWTP) based on the estimated utility parameters. Since all attributes are dummy variables in our model, MWTP in Equation (21.5) is simply WTP for a discrete change. The highest WTP, 3825 JPY, is found for improving water quality to the clear-water state (WAQ_{res}) while the lowest WTP, 492 JPY, is found for conserving two endangered species (END_{con}). The results indicate that the nationwide respondents have a positive WTP for the ecosystem restoration projects in the lake, and that the water quality improvement is the most highly valued ecosystem service among the ecosystem restoration attributes investigated in our choice experiment.

By comparing coefficients of variation among attributes, we examine the relative magnitude of the preference heterogeneity. The coefficient of variation is defined as the ratio of standard deviation to mean, that is S.D./M. Since this value is dimensionless, the S.D./M. can be used to compare variability for random variables with different means. The S.D./M. of each attribute is reported in the third column of Table 21.5, which shows that the magnitude of the heterogeneity in preferences for recreational availability (*REC*) is the highest, while the water quality improvement (WAQ_{res}) has the lowest ratio of standard deviation to mean. That is, the percentage of respondents who dislike the water quality improvement measure is relatively low, implying that it would be relatively easy to achieve consensus on this measure compared to measures addressing other ecosystem services.[6] For regulation of recreational use, however, the result implies that it can be difficult to achieve consensus.

21.6.3 Ecosystem Restoration Scenario Valuation

We estimate the economic benefits of two ecosystem restoration targets; the *Restoration* and the *Conservation* scenario, scenarios in terms of the respondents' mean WTP per person for the ecosystem restoration project over the *No Measure* status quo situation. Economic benefits are calculated as the sum of marginal WTP for three attributes/

ecosystem services: *Biodiversity*, *Endangered Species* and *Water Quality*. For the fourth ecosystem service, *Recreation Availability*, we found that whether the wetland was available for recreation or not did not have a significant effect on respondents' WTP. When the utility parameter is not statistically significant at the 10 per cent level, we assume zero WTP for this attribute. Note that we do not include the ASC in our calculation, implying that our scenario valuations can be underestimated by around 4000 JPY (48 USD). The mean WTP per person as a one-time payment for the *Restoration* scenario is 7244 JPY (87 USD) while the WTP for the *Conservation* scenario is 3517 JPY (43 USD). This implies that this nationwide sample of residents value the *Restoration* scenario approximately twice as highly as the *Conservation* scenario.

21.7 CONCLUSION AND POLICY IMPLICATIONS

Ecosystem restoration has over the past decade gained popularity as a preservation strategy and important addition to conservation due to the increased scarcity of undisturbed areas. However, there are few studies of households' preferences and willingness to pay (WTP) for restoration of ecosystem services. We use available ecological information in a choice experiment in order to estimate the economic value of conservation and restoration of different ecosystem services of the shallow lake ecosystem of Lake Takkobu in the RAMSAR site Kushiro Mire in Japan.

When the Ecosystem Restoration Strategy for Kushiro wetland was designed in 2005, it became a pilot project in Japan subjected to extensive ecological research in order to assess the impacts of restoration measures on ecosystem services. The restoration project valued here consists of four attributes derived from ecological information representing four different ecosystem services of this wetland area: (1) restoration of biodiversity, (2) protection of endangered species, (3) improvement of water quality, and (4) regulation of recreational use. Based on ecological survey data, we established two desired alternative restoration targets/endpoints: (1) the *Restoration* scenario, that is, restoration back to the level of stable state with clear-water quality in 1991, and (2) the *Conservation* scenario, which maintains the state in 2003 before it shifted to the stable state with turbid-water quality.

Several policy implications for conducting actual ecosystem restoration in Lake Takkobu can be drawn from our results. First, public preferences for the ecosystem restoration project in Lake Takkobu are heterogeneous, and thus it will be important to build consensus on the project. While people on average value the ecosystem restoration highly, not everybody thinks it is desirable to continue to conserve the ecosystem. Water quality improvements, which have both use and non-use values, seem to be the most feasible of the ecosystem restoration measures. Given the heterogeneity in preferences for ecosystem restoration in Lake Takkobu, there is a need for research exploring how to build consensus for the restoration measures.

Second, our result provides guidance to policy-makers on what restoration target they should choose from an economic valuation viewpoint. We find that the restoration level is valued at twice the conservation level. This indicates that the restoration level that is desirable ecologically is also most desirable based on public preferences and their expected welfare improvement from restoration. However, these social benefits aggregated over the national population need to be compared with the social costs in terms of investment and operating costs in a benefit–cost analysis in order to determine the economic efficient restoration target.

NOTES

1. The explanation in this paragraph is based on Carpenter (2003, Chapter 1).
2. Respondents make their decisions T times in choice experiments. We specify that utility parameters vary over individuals but are constant over choice situations for each individual. Then we denote the random utility function that respondent i chooses alternative j at choice situation t period as $U_{ijt} = \beta_i x_{ijt} + \varepsilon_{ijt}$.
3. We use a Halton sequence in our estimation. The necessary number of draws on Halton sequences to estimate parameters at a given confidence level is smaller than that on random sequences (Train, 2003).
4. The normal distribution was empirically preferred to the log-normal distribution for non-price coefficients.
5. There are two supporting reasons of this assumption: (a) the distribution of the MWTP for an attribute is simply the distribution of its coefficient, and (b) we expect to restrict the cost variable to be negative for all respondents (Carlsson et al., 2003).
6. The percentage of respondents who dislike attribute A is defined as $\Pr(WTP_A < 0)$.

REFERENCES

Bateman, I., B. Day, M. Hanemann, N. Hanley, T. Hett and M. Jones-Lee et al. (2002), *Economic Valuation with Stated Preference Techniques*, Cheltenham, UK and Northampton, MA, USA: Edward Elgar Publishing.

Berrens, R., A. Bohara, H. Jenkins-Smith, C. Silva and D. Weimer (2004), 'Information and effort in contingent valuation surveys: application to global climate change using national internet samples', *Journal of Environmental Economics and Management*, **47**(2), 331–63.

Birol, E., K. Karousakis and P. Koundouri (2006), 'Using a choice experiment to account for preference heterogeneity in wetland attributes: the case of Cheimaditida wetland in Greece', *Ecological Economics*, **60**(1), 145–56.

Brander, L.M., R.J.G.M. Florax and J.E. Vermaat (2006), 'The empirics of wetland valuation: a comprehensive summary and a meta-analysis of the literature', *Environmental and Resource Economics*, **33**(2), 223–50.

Brander, L.M., I. Bräuer, H. Gerdes, A. Ghermandi, O. Kuik and A. Markandya (2012), 'Using meta-analysis and GIS for value transfer and scaling up: valuing climate induced losses of European Wetlands', *Environmental and Resource Economics*, **52**(3), 395–413.

Brouwer, R., I.H. Langford, I.J. Bateman and R. Turner (1999), 'A meta-analysis of wetland contingent valuation studies', *Regional Environmental Change*, **1**(1), 47–57.

Brownstone, D. and K. Train (1999), 'Forecasting new product penetration with flexible substitution patterns', *Journal of Econometrics*, **89**(1–2), 109–29.

Carlsson, F., P. Frykblom and C. Liljenstolpe (2003), 'Valuing wetland attributes: an application of choice experiments', *Ecological Economics*, **47**(1), 95–103.

Carpenter, S.R. (2003), *Regime Shifts in Lake Ecosystems: Pattern and Variation, Vol. 15, Excellence in Ecology Series*, Oldendorf/Luhe, Germany: Ecology Institute.

Christie, M., N. Hanley, J. Warren, K. Murphy, R. Wright and T. Hyde (2006), 'Valuing the diversity of biodiversity', *Ecological Economics*, **58**(2), 304–17.

Hanemann, W. (1984), 'Welfare evaluation in contingent valuation experiments with discrete responses', *American Journal of Agriculture Economics*, **66**(3), 332–41.

Hanemann, W.M. (1982), 'Applied welfare analysis with qualitative response models', Working Paper No. 241, University of California, Berkeley.

Hanley, N., R.E. Wright and B. Alvarez-Farizo (2006), 'Estimating the economic value of improvements in river ecology using choice experiments: an application to the water framework directive', *Journal of Environmental Management*, **78**(2), 183–93.

Kadono, Y., T. Nakamura, K. Watanabe and K. Ueda (1992), 'Present state of aquatic macrophytes of three lakes in Kushiro Moor, Hokkaido, Japan', *Journal of Phytogeography and Taxonomy*, **40**, 41–6.

Kiker, C.F., J.W. Milon and A.W. Hodges (2001), 'Adaptive learning for science-based policy: the Everglades restoration', *Ecological Economics*, **37**(3), 403–16.

Lindhjem, H. and S. Navrud (2011a), 'Are internet surveys an alternative to face-to-face interviews in contingent valuation?' *Ecological Economics*, **70**(9), 1628–37.
Lindhjem, H. and S. Navrud (2011b), 'Using internet in stated preference surveys: a review and comparison of survey modes', *International Review of Environmental and Resource Economics*, **5**(4), 309–51.
Louviere, J., D. Hensher and J. Swait (2000), *Stated Choice Methods: Analysis and Application*, Cambridge, UK: Cambridge University Press.
Lupi, F., M.D. Kaplowitz and J.P. Hoehn (2002), 'The economic equivalency of drained and restored wetlands in Michigan', *American Journal of Agricultural Economics*, **84**(5), 1355–61.
McFadden, D. (1974), 'Conditional logit analysis of qualitative choice behaviour', in P. Zarembke (ed.), *Frontiers in Econometrics*, New York: Academic Press, pp. 105–42.
McFadden, D. and K. Train (2000), 'Mixed MNL models of discrete response', *Journal of Applied Econometrics*, **15**(5), 447–70.
Milon, J.W. and D. Scrogin (2006), 'Latent preferences and valuation of wetland ecosystem restoration', *Ecological Economics*, **56**(2), 162–75.
Nakamura, F. (2007), 'Restoration for Lake Takkobu in Kushiro Mire', *Japanese Journal of Limnology*, **68**(1), 61–3.
Navrud, S. and R. Ready (eds) (2007), *Environmental Value Transfer: Issues and Methods*, Dordrecht: Springer.
Provencher, B. and R. Bishop (2004), 'Does accounting for preference heterogeneity improve the forecasting of a random utility model? A case study', *Journal of Environmental Economics and Management*, **48**(1), 793–810.
Revelt, D. and K. Train (1998), 'Mixed logit with repeated choices', *Review of Economics and Statistics*, **80**(4), 647–57.
Scarpa, R. and M. Thiene (2005), 'Destination choice models for rock-climbing in the North-East Alps: a latent-class approach based on intensity of participation', *Land Economics*, **81**(3), 426–44.
Takamura, N., Y. Kadono, M. Fukushima, M. Nakagawa and B.O. Kim (2003), 'Effects of aquatic macrophytes on water quality and phytoplankton communities in shallow lakes', *Ecological Research*, **18**(4), 381–95.
Train, K. (1998), 'Recreation demand models with taste differences over people', *Land Economics*, **74**(2), 230–39.
Train, K. (2003), *Discrete Choice Methods with Simulation*, Cambridge, UK: Cambridge University Press.
Wilson, M.A. and S.R. Carpenter (1999), 'Economic valuation of freshwater ecosystem services in the United States: 1971–1997', *Ecological Applications*, **9**(3), 772–83.
Woodward, R.T. and Y. Wui (2001), 'The economic value of wetland services: a meta-analysis', *Ecological Economics*, **37**(2), 257–70.

22. A one-and-one-half bound contingent valuation survey to estimate the benefits of restoring a degraded coastal wetland ecosystem: the case study of Capo Feto, Italy

Giovanni Signorello, Joseph C. Cooper, Giuseppe Cucuzza and Maria De Salvo

22.1 INTRODUCTION

Over the past few decades many efforts have been made to monetize ecosystem services provided by wetlands. An examination of existing literature reviews of wetland studies (Boyer and Polasky, 2004; MEA, 2005; Turner et al., 2008; Wilson and Liu, 2008; Kumar, 2010; Barbier, 2011; Ghermandi et al., 2012) and meta-analyses that have been produced to synthesize empirical findings (Brouwer et al., 1999; Woodward and Wui, 2001; Brander et al., 2006; Liu and Stern, 2008; Ghermandi et al., 2010) as well as the recent non-market valuation literature in general, reveal that the pool of available valuation studies is strongly unbalanced in terms of its geographical coverage. Although in recent years a significant shift in the geographical distribution of wetland valuation studies has been observed, the overwhelming majority of existing primary published and unpublished studies and reports that document economic values of wetlands come principally from North America (Ghermandi et al., 2010). Few primary studies exist to date that cover the Mediterranean region, despite Mediterranean wetlands providing many valuable marketed and non-marketed benefits, including those associated with delivering services of global importance (Signorello, 1998; Alberini et al., 2004, 2005; Birol et al., 2006; EEA, 2010).

This chapter contributes to filling this empirical gap by reporting results of a contingent valuation study carried out to estimate benefits of restoring Capo Feto (Sicily), a degraded coastal wetland ecosystem, recently included in the Natura 2000 EU framework and in the Ramsar wetlands' list. The study was undertaken to provide policy-makers with information on the public benefits that the restoration of the wetland might generate. Individual's willingness to pay (WTP) was gathered by using the one-and-one-half bound (OOHB) elicitation format proposed by Cooper et al. (2002), hereafter denoted as CHS.

The OOHB approach consists of a survey design in which the respondent is given two prices up front and told that, while the exact cost of the item is not known for sure, it is known to lie within the range bounded by these two prices, a lower bound and an upper bound. One of the two prices is selected at random, and the respondent is asked whether she would be willing to pay this amount; she is then asked about the other price only if doing so would be consistent with the stated price range. For example, if the lower of the two prices was selected initially and she says 'yes' to this, she is then asked whether she would be willing to pay the higher price; but, if she says 'no' to the lower price, there is no follow-up question because that would go below the stated price range. With the OOHB format, the respondent is exposed to two bid amounts as in the double-bounded (DB) format suggested by Hanemann et al. (1991). However, unlike in DB, OOHB bids are disclosed in advance as lower and upper limits of cost interval, prior to any responses. The OOHB process explicitly incorporates cost uncertainty into the valuation scenario. In addition, this feature, by eliminating the element of surprise of the double-bounded (DB) format, has the potential to lessen or remove the inconsistency between the first and the second WTP questions in the responses to the two valuation questions (McFadden and Leonard, 1993; Cameron and Quiggin, 1994; Altaf and DeShazo, 1994; Kanninen, 1995; Herriges and Shogren, 1996; DeShazo, 2002). At the same time, as shown in CHS, the OOHB format, despite using less information than the DB format,[1] captures two-thirds of the gains in efficiency associated with the move from SB elicitation format to DB approach.

To the best of the authors' knowledge, only two unpublished studies have estimated the economic value of wetlands through OOHB CV surveys. The first study concerns the estimate of conservation value of coastal wetlands in Muthurajawela Marsh and Negombo Lagoon (Sri Lanka) (CEMARE, 2002). The second study relates to capturing public benefits from restoring and expanding culm[2] grassland in England (Burgess et al., 2008).

Our OOHB application diverges from other OOHB studies not only with respect to the type of wetland evaluated, but also in terms of econometric approaches used to analyse data on willingness to pay. We develop a spike-univariate parametric probit model, a semi-non-parametric normal model, and Turnbull non-parametric estimator. Moreover, to account for the possibility that the responses to the second bid might be biased, we also present similar econometric analysis for the single-bounded (SB) dichotomous choice format by ignoring the responses to the follow-up bid.

This chapter is organized as follows. The next section describes the Capo Feto site and the restoration project. Subsequent sections address the features of the OOHB format elicitation process, the econometric approaches utilized, the survey and data, and the empirical results. The last section draws conclusions.

22.2 THE POLICY SITE AND RESTORATION PROJECT

Capo Feto is a wetland located along the south-western coast of Sicily, 5 km west of Mazara del Vallo (37.60° N, 12.48° E) (Figure 22.1).

Capo Feto, which covers an area of 157 ha, shows the typical landscape of Mediterranean wetlands with pond and marsh zones, and migratory and stationary birds only in the winter season. In the summer, the site is almost dry. The ecosystem of Capo Feto is rich in biodiversity. For this reason Capo Feto is included (with the nearby Margi Spano wetland) in the EU Natura 2000 framework (code: ITA010006) as a Special Area of Conservation (SAC) under the EU Habitats Directive (92/43/EEC), and as a Special Protection Area (SPA) under the EU Birds Directive (79/401/EEC). Capo Feto supports six habitat types listed under Annex I of the EU Habitats Directive, one of which is a priority natural habitat under Article 1, namely coastal lagoons (habitat

Figure 22.1 Map of Capo Feto wetland

type 1150).[3] Of the numerous rare plant species, 15 are Sicilian endemic and included in the Red List of plants at risk of extinction. The adjacent marine ecosystem is dominated by the *Poseidonia oceanica*, which acts as a natural dam in creating the small lagoon. Capo Feto also supports a wide array of fauna diversity, including six mammal, two reptile, and two amphibian species, which are listed under Annex II and IV of the Habitats Directive and in Annex II and III of the Berne Convention as being at risk of extinction. Due to its particular geographical location near to Africa, Capo Feto is a key area along the migration routes of waterfowl coming from the Western Palearctic and from Eastern Europe. Thus, Capo Feto is an important site for wintering and resting in the migratory movements and breeding for many European bird populations. Approximately 45 bird species were identified in the site. Since July 2011, Capo Feto has been included in the list of wetlands recognized, under the Ramsar Convention, as being of international importance.

The natural ecosystems of Capo Feto remained largely intact until the 1950s when most subregions within it were drained to provide agricultural land, and other subregions were used for military purposes, urban development and tourism. In the 1980s the area was also crossed underground by the Italian–Algerian methane pipeline. In recent years, with the abandonment of agriculture and the decline in other economic activities in the area, the wetland has been illegally utilized as a dumping ground for debris from demolition of buildings. Fortunately these human pressures have not totally compromised the integrity of the wetland ecosystems.

To recover the original natural environment, a restoration project was proposed under the EU Life programme. In particular, the restoration project involved a variety of ecological activities, including the demolition of military buildings, electrical lines and other infrastructures, the removal of all debris, the conversion of agricultural lands to wetlands, the rehabilitation of habitats (ponds) and water channels, the covering of sand dunes with endemic grass, the restoration of sand banks, and the management of the restored wetland according the principles of conservation biology and wetland ecosystem services framework.

22.3 METHODOLOGY

Let C_i be the individual's true maximum WTP for the good that is the subject of the survey. While the individual knows her own WTP, to the observer it is a random variable with a given cumulative distribution function (CDF) denoted $G(C_i; \theta)$, where θ represents the parameters of this distribution, and that are to be estimated on the basis of the responses to the CV survey. The parameters will be functions of the variables in X_i, but this is left implicit in $G(C_i; \theta)$. The estimated CDF can be used to calculate the average maximum WTP or other summary welfare measures.

In the SB format, the *i*th respondent is asked if she would be willing to pay some given amount B_i^* (henceforth we refer to this as the 'bid') to obtain, say, a given improvement in environmental quality (Bishop and Heberlein, 1979). The probability of a 'Yes' response, or a 'No' response, $\pi_i^Y(B_i^*)$, can be cast in terms of a random utility maximizing choice by the respondent. As shown by Hanemann (1984), the response probabilities are related to the underlying WTP distribution by:

$$\pi_i^N \equiv \Pr\{\text{No to } B_i^*\} \equiv \Pr\{B_i^* > C_i\} = G(B_i^*;\theta) \tag{22.1a}$$

$$\pi_i^Y \equiv \Pr\{\text{Yes to } B_i^*\} \equiv \Pr\{B_i^* \leq C_i\} = 1 - G(B_i^*;\theta) \tag{22.1b}$$

The resulting log-likelihood function for the responses to a CV survey using the SB format is:

$$\ln L^{SB}(\theta) = \sum_{i=1}^{N} \{d_i^Y \ln[1 - G(B_i^*;\theta)] + d_i^N \ln G(B_i^*;\theta)\} \tag{22.2}$$

where $d_i^Y = 1$ if the ith response is Yes and 0 otherwise, while $d_i^N = 1$ if the ith response is No and 0 otherwise.

In the one-and-one-half bound format (OOHB) the respondent is presented with a range, $[B_i^-, B_i^+]$, where $B_i^- < B_i^+$. One of these two prices is selected at random and the respondent is asked whether she would be willing to pay that amount. She is asked about the second price only if that is compatible with her response to the first price. If the lower price, B_i^-, is randomly drawn as the starting bid, the three possible response outcomes are (No), (Yes, No) and (Yes, Yes); we denote the corresponding response probabilities π_i^N, π_i^{YN}, π_i^{YY}:

$$\pi_i^N = \Pr\{C_i \leq B_i^-\} = G(B_i^-;\theta) \tag{22.3a}$$

$$\pi_i^{YN} = \Pr\{B_i^- \leq C_i \leq B_i^+\} = G(B_i^+;\theta) - G(B_i^-;\theta) \tag{22.3b}$$

$$\pi_i^{YY} = \Pr\{C_i \geq B_i^+\} = 1 - G(B_i^+;\theta) \tag{22.3c}$$

If the higher price, B_i^+, is randomly drawn as the starting bid, the possible response outcomes are (Yes), (No, Yes) and (No, No). We denote the corresponding response probabilities π_i^Y, π_i^{NY}, π_i^{NN}:

$$\pi_i^Y = \Pr\{C_i \geq B_i^+\} = 1 - G(B_i^+;\theta) \tag{22.4a}$$

$$\pi_i^{NY} = \Pr\{B_i^- \leq C_i \leq B_i^+\} = G(B_i^+;\theta) - G(B_i^-;\theta) \tag{22.4b}$$

$$\pi_i^{NN} = \Pr\{C_i \leq B_i^-\} = G(B_i^-;\theta) \tag{22.4c}$$

We observe that in the OOHB format there is a symmetry between the ascending sequence and the descending sequence of valuation questions; then, all possible answers can be summarized in three groups: (No, No), (Yes, No), and (Yes, Yes). The probabilities associated with each possible answer are:

$$\pi_i^{N} = \pi_i^{NN} = \Pr\{C_i \leq B_i^-\} = G(B_i^-;\theta) \tag{22.5a}$$

$$\pi_i^{YN} = \pi_i^{NY} = \Pr\{B_i^- \leq C_i \leq B_i^+\} = G(B_i^+;\theta) - G(B_i^-;\theta) \tag{22.5b}$$

$$\pi_i^{YY} = \pi_i^{Y} = \Pr\{C_i \geq B_i^+\} = 1 - G(B_i^+;\theta) \tag{22.5c}$$

Let $d_i^N = 1$ if either the starting bid is B_i^- and the response is (No) or the starting bid is B_i^+ and the response is (No, No), and 0 otherwise; let $d_i^{YN} = 1$ if either the starting bid is B_i^- and the response is (Yes, No) or the starting bid is B_i^+ and the response is (No, Yes), and 0 otherwise; and let $d_i^{YY} = 1$ if either the starting bid is B_i^- and the response is (Yes, Yes) or the starting bid is B_i^+ and the response as (Yes), and 0 otherwise. Then, the log-likelihood function for the responses to a CV survey using the OOHB format is:

$$\ln L^{OOHB}(\theta) = \sum_{i=1}^{N} \{d_i^Y \ln[1 - G(B_i^+;\theta)] + d_i^{YN} \ln[G(B_i^+;\theta) - G(B_i^-;\theta)] d_i^N \ln G(B_i^-;\theta)\} \tag{22.6}$$

The specification above implicitly assumes that in the cases in which there is a follow-up response, the correlation between the two responses, call it ρ, is equal to 1. However, because the researcher will never be able to fully model the respondent's decision-making process (i.e., the research has insufficient information to consistently predict the respondent's response to the follow-up based on his response to the first bid), in practice this assumption may be too strong. Then, we modify the previous likelihood function into a hybrid likelihood function in which responses with a follow-up are distributed with a bivariate normal CDF (denoted by Φ), and those without a follow-up follow the univariate distribution, that is:

$$\ln L^{OOHB}(\theta) = \sum_{i=1}^{N} \{d_i^Y \ln[1 - G(B_i^+;\theta)] + d_i^{YN} \ln[\Phi(B_i^+;\theta;B_i^-;\theta;\rho)] + d_i^N \ln[G(B_i^-;\theta)]\} \tag{22.7}$$

where ρ is the correlation coefficient between the first and the follow-up response.

22.4 ECONOMETRIC ANALYSIS

To analyse the individual's OOHB survey responses to estimate the survivor function and calculate welfare measures, we use three different estimation approaches: (1) a univariate parametric model; (2) a semi-non-parametric distribution-free (SNPDF) OOHB estimator; and a (3) non-parametric OOHB estimator.

We implement a univariate parametric model based on the assumption of a normal distribution for the function $G(B_i^*; \theta)$ and on a linear model for the indirect utility function, that is:

$$G(B_i^*; \theta) = \Phi(-\alpha + \beta B_i^*) \tag{22.8}$$

where in this case, $\theta \equiv (\alpha, \beta)$ and median $\{C\} = \alpha/\beta$. This linear probit model is simple to estimate, and convenient to use especially when there is an interest in doing a comparison of parametric welfare estimates with analogue welfare estimates from a non-parametric model.

The semi-non-parametric distribution-free (SNPDF) approach was first applied to SB data by Creel and Loomis (1997) and is extended here to OOHB data. The reason for using the SNPDF approach is to reduce the sensitivity of our econometric analysis to specific parametric assumptions regarding the form of the distribution and serve as a specification check on the parametric model. A simple way to motivate the SNPDF approach is to observe that, with the normal distribution, the CV response probabilities corresponding to (22.5b) takes the form:

$$\pi_i^{YN} = G(B_i^+;\theta) - G(B_i^-;\theta) \equiv F[\Delta V(B_i^+)] - F[\Delta V(B_i^-)] \tag{22.5b$'$}$$

where $F(.)$ is the standard normal CDF and:

$$\Delta V(\beta) \equiv -\alpha + \beta B \tag{22.9}$$

is the utility difference function (Hanemann, 1984; Hanemann and Kanninen, 1999), which is increasing in the bid price, B. The SNPDF approach retains the normal CDF in the response probabilities, such as in (22.5b′), but replaces the linear utility difference with a Fourier flexible form, where (omitting quadratic term as in Creel and Loomis):

$$\Delta V(x, \theta_k) = x\beta + \sum_{\alpha=1}^{A} \sum_{j=1}^{J} \{v_{j\alpha} \cos[jk'_\alpha s(x)] - w_{j\alpha} \sin[jk'_\alpha s(x)]\} \tag{22.10}$$

where the vector x contains all arguments of the utility difference model, A and J are positive integers, and k_α are vectors of positive and negative integers that form indices in the conditioning variables. There exists a coefficient vector such that, as the sample size becomes large, $\Delta V(x)$ in (22.10) can be made arbitrarily close to a continuous unknown utility difference function for any value of x. If the bid price is the only explanatory variable, then k_α is a (1×1) unit vector and max(A) equals 1. We choose the same value for integer J as do Creel and Loomis, leading to:

$$\Delta V(B) = \gamma + \delta B + \delta_V \cos s(B) + \delta_w \sin s(B) \tag{22.11}$$

where $s(B)$ prevents periodicity in the model and is a function that shifts and scales the variable to lie in an interval less than 2π. Specifically, the variable is scaled by subtracting its minimum value, then dividing by the maximum value, and then multiply the resulting value by $2\pi - 0.00001$, which produces a final scaled variable in the interval $[0, 2\pi - 0.0001]$. When $\delta_v = \delta_w = 0$, (22.7) reduces to (22.5) with $\delta = \beta$ and $\gamma = -\alpha$: the normal WTP model is nested within the SNPDF model. The four coefficients of the utility difference function (22.11) are estimated by maximum likelihood, using the log-likelihood function in (22.6) for the OOHB data. While the SNPDF OOHB makes no assumptions regarding the distribution of the two marginals (for the first and follow-up questions), it does assume that the relationship between them follows the bivariate normal, with the correlation defined by the single parameter ρ.

Finally, we use a non-parametric approach, based on the Turnbull's estimator of Type II interval censored data (Klein and Moeschberger, 1997), to estimate the cumulative distribution function of WTP, and to calculate an unbiased lower-bound approximation to the expected WTP (Haab and McConnell, 1997). Responses to OOHB dichotomous choice questions inform in which interval might be the individual's WTP. If the lower price, B_i^-, is randomly drawn as the starting bid, the three possible intervals are the following: if the respondent rejects paying the lower price, B_i^-, her WTP is in the interval $[0, B_i^-]$. If she refuses to pay the second upper amount of the stated range, then her WTP is in the interval $[B_i^-, B_i^+]$. If the individual is willing to pay also the higher price, B_i^+, then her maximum WTP lies in the interval $[B_i^+, \infty]$. If the higher price, B_i^+, is randomly drawn as the starting bid, the three possible intervals are the following: if the individual accepts to pay that bid her WTP lies in the interval $[B_i^+, \infty]$; for a negative answer followed by as positive answer the WTP is in the interval $[B_i^-, B_i^+]$; for two consecutive negative answers the interval for the individual WTP is $[0, B_i^-]$.

In the OOHB application, the lower and the upper limits B_i^* of each interval can be ordered to form the following ascending sequence:

$$0 = B_0 < B_1 < B_2 < B_3 < B_4 < \ldots < B_M < B_{m+1} = \infty$$

where B_1 is the lowest bid presented in the survey, B_2 is the second lowest bid and B_M is the last lowest element of the bid vector. The corresponding order of the distribution-free function F is: $0 = F(B_0) < F(B_1) < F(B_2) < F(B_3) < F(B_4) < \ldots < F(B_M) < 1$ and the probabilities associated with each value are: $\pi_j = F(B_j) - F(B_{j-1})$ if and only if $F(B_j) \leq F(B_{j+1})$.

In the OOHB CV survey, the bounds are not presented to individuals in a sequential order. Instead, the range $[B_i^-, B_i^+]$ presented to respondents could be any combination of the above values. Then, several sub-intervals $[B_{j-1}, B_j]$ of the bid vector can be included in $[B_i^-, B_i^+]$. To account for this issue, Gentlemen and Geyer (1994) define an indicator variable d_{ij} that is equal to 1 if the sub-interval $[B_{j-1}, B_j]$ is a subset of $[B_{j-1}, B_j]$ and equal to 0 otherwise, and formulate the following log-likelihood function:

$$\ln L = \sum_{i=1}^{N} d_{ij} \left\{ \sum_{j=1}^{N} d_{ij} [F(B_j) - F(B_{j-1})] \right\} \tag{22.12}$$

which depends only on M values defined by probabilities $\pi_1, \pi_2 \ldots, \pi_M$ and it is subject to the conditions $\sum_{J=1}^{M+1} \pi_j = 1$ and $\pi_j \geq 0 \, \forall j$.

To estimate the above non-parametric maximum likelihood, several variants of the self-consistent (SC) algorithm suggested by Turnbull (1976) exist (Silverman, 1986; Wellner and Zhan, 1997; Day, 2007). In this application we use the so-called iterative convex minorant and self-consistent (ICM-SC) hybrid algorithm implemented and tested by Day (2007).

While the OOHB may yield more efficient results than the SB approach, the trade-off is the risk that responses to the first price may sometimes be inconsistent with the responses to the second, with the latter revealing a different benefit estimate. Several explanations have been proposed for this anomaly. Carson et al. (1992) suggest an explanation based on cost expectations: a respondent who said 'Yes' to the initial price sees the second price as a price increase, which he rejects; a respondent who said 'No' and is then offered a lower price may suspect that an inferior version of the item will be provided, which he also is disposed to reject. Altaf and DeShazo (1994) suggest that the second bid converts what had seemed to be a straightforward posted-price market into a situation involving bargaining; if this is bargaining, the respondent should say 'No' in order to drive the price down. Hence, to account for the possibility that the responses to the second bid may be biased, we also present results for the SB format by ignoring the responses to the follow-up bid, where the same models for OOHB are estimated. As the respondent did not know when answering the first WTP question that there was a possibility of a follow-up WTP question, the SB format is nested within the OOHB format.

22.5 THE SURVEY AND DATA

The Capo Feto survey design was pre-tested on a small group of individuals. A pilot-study survey implemented using a payment card elicitation format to get prior information on possible distribution of willingness to pay. After a few rounds of revisions, the final survey was administered through in-person interviews during the period June–July 2002 to individuals aged 18 years or older, selected at random, from the nine provinces of Sicily (Agrigento, Caltanissetta, Catania, Enna, Messina, Palermo, Ragusa, Siracusa and Trapani). The interviews were carried out by a professional survey company. The survey was administered to be representative of the Sicilian population in terms of gender and age.

The survey consisted of three sections. In the first section we posed general questions about wetlands, questions on attitudes toward general environmental issues, specific questions on Capo Feto, and questions on the use of related natural ecosystems and protected areas. The second section was devoted to the valuation scenario, and we posed questions on the willingness to pay followed up by other questions for identifying valid, protest and biased responses, for assessing the credibility and meaningfulness of the scenario presented in the survey. As payment vehicle for willingness to pay we used a lump-sum voluntary donation to a specific fund. In the last section we inserted questions on socio-economic characteristics of the individuals.

To facilitate communication, during the interviews a portfolio containing maps, cards and photographs on Capo Feto ecosystem that addressed the current (degraded) status of the wetland was shown to each respondent. The card covered the restoration project and the maps illustrated the new possible status of the wetland after the project. The

Table 22.1 Sample descriptive statistics (480 observations)

Variable	Mean	Std. Dev.
Heard of wetlands (%)	63.54	48.18
Heard of ecological importance of wetlands (%)	50.41	50.05
Perception of wetland as wastelands (%)	17.08	50.04
Visits in the last three years to protected areas (number)	3.81	6.35
Heard of Capo Feto (%)	8.35	27.66
Visits to Capo Feto (%)	5.83	23.46
Intention to visit Capo Feto (%)	84.37	36.35
Importance to restore Capo Feto (from 1 to 4)	1.94	79.07
Member of environmental groups (%)	3.12	17.41
Donation for social causes (%)	56.87	49.57
Gender (% male)	49.58	50.05
Age (number)	36.36	12.92
Education (from 1 to 5)	3.75	0.73
Distance from Capo Feto (km)	242.08	118.23

enumerators also reminded the respondents of their budget constraints and of alternative wetlands in Sicily.

The OOHB bid pairs (in euros) used in the valuation questions are (5; 10), (10; 15), (15; 20), (20; 30), (30; 50), and (50; 100). The final sample consists of 480 respondents, distributed between the nine provinces in approximate proportions to their populations, and managed according to a statistical design aimed at assuring the same size for each OOHB bid pair. Some descriptive statistics of the sample are presented in Table 22.1. We see that 63.54 per cent of the sample has in general heard of wetlands, 50.41 per cent believe this ecosystem to be ecologically important, and only 17.08 per cent perceived these wetlands to be wastelands. In the past three years the average number of visits to protected areas was 3.81. Only a small percentage (8.35) of the sample had heard of Capo Feto. A smaller percentage (5.83 per cent) of individuals visited the site, but the percentage of people interested in visiting it is high (84.37 per cent), although the average degree of importance assigned to preserve the Capo Feto wetland is lower (1.94, based on a 1–4 scale, with '1' being 'unimportant' to '4' being 'high importance'). In the sample, individuals are almost balanced in terms of gender, and their mean age is a bit lower to that of Sicilian population (39.1). Few respondents (3.12 per cent) belonged to environmental groups, and 56.87 per cent of them voluntarily donated money for social causes. On average, individuals in the sample lived 242 km from Capo Feto.

Tables 22.2 and 22.3 present the OOHB data format and the nested SB format data, respectively, showing the bid values and response summaries.

22.6 ECONOMETRIC RESULTS

Table 22.4 presents the maximum likelihood results for the SB and OOHB models, for both the parametric and SNPDF cases. For brevity and for the sake of comparison with results from the non-parametric estimator, we limit the scope of analysis to estimating the

Table 22.2 Summary of the responses to the OOH bound questions (480 observations)

Bid $[B^-, B^+]$ (€)	Lower Bound Bid (B^-) Offered First			Upper Bound Bid (B^+) Offered First			Sample Size
	No. of 'Yes' Responses	No. of 'No-Yes' Responses	No. of 'No-No' Responses	No. of 'No' Responses	No. of 'Yes-No' Responses	No. of 'Yes-Yes' Responses	
5; 10	13	8	19	23	5	12	80
10; 15	8	7	25	21	5	14	80
15; 20	22	3	15	29	0	11	80
20; 30	18	5	17	20	1	19	80
30; 50	21	7	12	15	6	19	80
50; 100	23	16	1	13	5	22	80

Table 22.3 Dataset for the first bound (SB) of the OOH bound questions (480 observations)

Bid $[B^-$ or $B^+]$ (€)	Sample Size	No. of 'Yes' Responses	% of 'Yes' Responses	No. of 'No' Responses	% of 'No' Responses
5	40	27	67.50	13	32.50
10	80	55	68.75	25	31.25
15	80	39	48.75	41	51.25
20	80	51	63.75	29	36.25
30	80	39	48.75	41	51.25
50	80	32	40.00	48	60.00
100	40	13	32.50	27	67.50

Table 22.4 Regression results

Variable	Coefficient (*t*-stat)			
	Single bound		OOH bound	
	Parametric	SNPDF	Parametric	SNPDF
Constant	0.3822	0.0221	0.4027	0.2215
	(4.261)	(1.458)	(4.999)	(1.995)
BID	−0.0101	0.1036	−0.0133	0.0579
	(−4.286)	(2.341)	(−6.413)	(1.874)
BIDu	–	−0.0935	–	−0.0817
		(−0.922)		(−1.096)
BIDv	–	−0.0262	–	−0.0366
		(−3.278)		(−5.600)
P	–	–	0.8144	0.8090
			(13.110)	(12.940)
Log-L.	−321.64	−320.75	−459.03	−459.37
Efron's R^2	0.0402	0.0448	–	–
Chi-sq.	19.336	21.796	–	–

Table 22.5 Non-parametric estimation using the hybrid self-consistent and iterative convex minorant algorithm (OOHB dataset)

Equivalence Class (€ Interval) [Lower Bound–Upper Bound]	Change in Density (PDF)	Cumulative Distribution (CDF)	Survivor Function (1 – CDF)	Standard Error of PDF
[0–5]	0.2200	0.2200	0.7800	0.0238
[5–10]	0.1144	0.3345	0.6655	0.0199
[10–15]	0.1032	0.4377	0.5623	0.0185
[15–20]	0.0273	0.4650	0.5350	0.0148
[20–30]	0.0560	0.5209	0.4791	0.0148
[30–50]	0.1240	0.6449	0.3551	0.0207
[50–100]	0.2130	0.8580	0.1420	0.0255
[100–∞][a]	0.1420	1.000	0.0000	0.0230

Note: a. For this left censored interval the upper bound bid is set to €999.

mean maximum WTP for the sample. Therefore, the bid (donation) is the only necessary explanatory variable. By design, the monetary variable (bid) presented to the respondent is uncorrelated with other possible explanatory variables. Hence, for the estimation of the mean welfare measure for the sample, the other explanatory variables are irrelevant (McFadden, 1994). Additional explanatory variables become useful, for example, when there is some interest in extrapolating, through benefit transfer techniques, the welfare measure according to these variables.

Coefficients are statistically significant and have the expected sign in both formats. That is, an increase in the bid reduces the probability of a positive answer. OOHB coefficients, as shown by the value of *t*-statistics, exhibit more efficiency than SB parameters. Note that the coefficient of variation for WTP is not much smaller for the OOHB approach than for the SB approach. The high estimated value of ρ and its high statistical significance indicate a strong correlation between the two responses in the OOHB format.

Table 22.5 presents the non-parametric results for the OOHB. The first column reports each equivalence class formed by the bid vector, that is, the interval within which the maximum likelihood estimate of the probability distribution ascribes the probability mass. The second column shows values of PDF calculated as step differences in the CDF values reported in the third column. The fourth column reports on the estimates survivor (demand) function, and the last column the estimated standard error of each step of PDF function.

Table 22.6 presents the non-parametric results for the SB format based on data shown in the previous Table 22.2. Since SB responses produce a 'type interval censoring data', we estimate the survivor function (1 – CDF), shown in the fourth column, through the classical Turnbull's estimator based on Ayer et al. (1955) and on the 'pool-adjacent-violators algorithm' (PAVA) (Haab and McConnell, 1997).

Table 22.7 presents WTP estimates, including those based on the coefficient estimates from Table 22.4. As it concerns the parametric and semi-parametric models, we calculate the E(WTP) values by integrating the density function between $B = 0$ and the maximum

Table 22.6 Turnbull non-parametric results[a] (SB dataset)

Bid [B^- or B^+] (€)	Sample Size	No. of 'Yes' Responses	% of 'Yes' Responses (Survivor Function) (1 – CDF)	No. of 'No' Responses	% of 'No' Responses (CDF)	Change in Density (PDF)
10	120	82	67.50	13	32.50	0.3167
20	160	90	56.25	70	43.75	0.1208
30	80	39	48.75	41	51.25	0.0750
50	80	32	40.00	48	60.00	0.0875
100	40	13	32.50	27	67.50	0.0750

Note: a. Pooled bid value is from upper end of boundary.

Table 22.7 Estimates of mean WTP (€) for restoring Capo Feto

Format	Approach	WTP Estimates (€)			
		Mean	Coefficient of Variation[a]	90% Confidence Intervals[b]	
SB	Parametric	45.32	0.0625	40.34	49.81
	SNPDF	43.75	0.1734	25.96	50.01
	Non-parametric Turnbull	41.48	–	35.09	48.07
OOHB	Parametric	40.37	0.0576	36.32	44.10
	SNPDF	39.63	0.2722	35.33	47.83
	Non-parametric HSCICMA[c]	31.71	–		

Notes:
a. The coefficient of variation is generated from the standard error of the empirical confidence interval.
b. The confidence intervals for parametric and semi-parametric approaches are estimated though a jack-knife analysis (1000 simulated datasets).
c. Hybrid self-consistent and iterative convex minorant algorithm.

amount (€100) in the bid vector. That means that we estimate a spike model that allows for indifference, with some positive probability, the individual has a zero WTP for the change in the environmental good. Indifference is equivalent to a probability mass, or spike, at Bid = 0. Regarding the non-parametric approach, we adopt a conservative approach as we take the lower bound of each interval to feature the whole interval for the unknown WTP. The parametric OOHB estimates are lower and more efficient than measures deduced from the parametric SB model. Semi-parametric estimates are close to parametric ones but they exhibit higher variation. The non-parametric lower bound estimate of mean WTP is, as expected, smaller than parametric E(WTP) values. The difference is more evident in OOHB format than in SB format, but the OOHB non-parametric model is different from the non-parametric model used in the SB case. Hence, the differences in the OOHB and SB non-parametric WTP estimates may be due largely to differences in model specification and not to elicitation differences.

22.7 POLICY IMPLICATIONS

The previous welfare estimates are used to determine the social profitability of the environmental project within the framework of cost–benefit analysis. The total benefits of changes are obtained by aggregating the mean WTP produced by the OOHB parametric approach over the entire sample frame population. Within Sicily in the year of the survey, there were 3 926 601 individuals aged 18 years or older (ISTAT, 2002). Therefore, the aggregate present value of the benefits amounts to €158 516 882,[4] corresponding to €30 290 per hectare per year.[5]

The total cost of implementing the project, including either one-off payments (e.g., investment in infrastructure and land purchase)[6] or discounted recurrent annual costs (e.g., on-going costs of managing and maintaining the restored ecosystem) amounts to €6 258 184, corresponding to €1197 per hectare per year.

Based on these estimates,[7] the comparison of benefits and costs demonstrate a substantial net benefit for implementing the environmental project, as the benefits of restoring and conserving the degraded wetland are 25.32 times larger than the investment needed. Thus, the project is still likely to be economically efficient even in the case that estimated benefits are biased upwards due to the hypothetical treatment and to the non-coercive payment vehicle used in the CV survey (Carson and Groves, 2007).

22.8 CONCLUSION

The main objective of the study was to examine whether or not devoting public financial resources to restore and protect Mediterranean wetlands can be economically viable investments in Southern Europe, with the specific application being the restoration and protection of the wetland ecosystem of Capo Feto in Sicily. To estimate the benefits of these activities to society, we adopted the total economic value (TEV) framework, and used the CVM survey approach.

The empirical exercise shows that the benefits of the examined environmental project are substantially larger than its costs. Given the large benefit–cost ratio even if the quantified economic benefits might be considered an overestimate due to possible response biases to the hypothetical scenario, the actual ratio is still likely to be greater than one.

This Capo Feto valuation study provides an additional observation to the empirical literature on wetland valuation in Europe. The estimated benefits can contribute to raising the awareness with decision-makers regarding the economic benefits of conserving and sustainable managing wetlands, especially when they play a significant role in delivering ecosystem services globally. Further, the estimated values can be used in benefit transfer and scaling up exercises to map values of similar ecosystems in the Mediterranean basin.

From a methodological perspective, the empirical results obtained in this CV study support the OOHB format as a useful mechanism for eliciting information on an individual's WTP. Within each elicitation format (SB and OOHB), the parametric and SNPDF mean WTP results are quite close. The Turnbull non-parametric estimator used in the SB case also produces a mean WTP estimate that is also not too different from the parametric and SNPDF estimates, although a statistical comparison across the estimation methods is beyond the scope of this chapter. The comparison of the parametric and

SNPDF methods across the two elicitation methods suggests that the OOHB mean WTP estimates are more conservative than the nested SB values. However, the non-parametric mean WTP estimate produced using the hybrid self-consistent and iterative convex minorant algorithm in the OOHB case is not convincing; not only it is significantly lower than the parametric and SNPDF OOHB estimates of WTP, it is also notably lower than all the SB WTP estimates. This result suggests that more work is necessary to produce reliable results using the hybrid self-consistent and iterative convex minorant algorithm.

ACKNOWLEDGEMENTS

The present study was conducted under the EC Life project LIFE99NAT/IT/006270 proposed for the environmental restoration of the Capo Feto wetland. The authors wish to express their sincere gratitude to Edoardo Politano, the team leader of the Life project. Moreover, they wish to thank Brett Day, CSERGE – University of East Anglia (UK), for providing codes to estimate non-parametric survivor function and lower bound mean WTP from OOHB data. The views presented herein are those of the authors, and do not necessarily represent the views of the Economic Research Service or the United States Department of Agriculture.

NOTES

1. By using the OOHB format we are not always able to ask the second valuation question: the second question will be appropriate half the time, on average, but not the rest of the time. This is the reason why CHS refer to this format as one-and-one-half bound format (OOHB).
2. A unique habitat in Devon and Cornwall, which developed above a geological formation known as the Culm Measures, laid down in the Carboniferous period. Culm is a word that comes from a local name for a soft, sooty coal.
3. Other habitat types in the Annex I are: annual vegetation of drift lines; *Salicornia* and other annuals colonizing mud and sand; Mediterranean salt meadows (*Juncetalia maritimi*); Mediterranean and thermo-Atlantic halophilous scrubs (*Sarcocornetea fruticosi*); embryonic shifting dunes.
4. These benefits are already expressed in present value terms as in the survey the WTP is stated as lump sum.
5. To convert the present value of total benefits in annual benefit flow we use a social interest rate of 3 per cent.
6. One-off management and investment costs are almost the 25 per cent of the total costs.
7. The annualized benefits per hectare are higher than mean estimates reported in Brander et al. (2006) and Ghermandi et al. (2010) for European wetlands. Also, annualized costs per hectare are higher than average total cost reported in Gantioler et al. (2010) for sites included in the Natura 2000 networks.

REFERENCES

Alberini, A., P. Rosato, A. Longo and V. Zanatta (2004), 'Information and willingness to pay in a contingent valuation study: the value of S. Erasmo in the Lagoon of Venice', Fondazione Eni Enrico Mattei Nota Di Lavoro No. 19.

Alberini, A., V. Zanatta and P. Rosato (2005), 'Combining actual and contingent behaviour to estimate the value of sports fishing in the Lagoon of Venice', Fondazione Eni Enrico Mattei Nota Di Lavoro No. 44.

Altaf, M. and J. DeShazo (1994), 'Bid elicitation in the contingent valuation method: the double referendum format and induced strategic behaviour', Working Paper, Harvard University, Arts and Sciences, Urban Planning Committee.

Ayer, M., H.D. Brunk, G.M. Ewing, W.T. Reid and E. Silverman (1955), 'An empirical distribution function for sampling with incomplete information', *Annals of Mathematical Statistics*, **26**(4), 641–7.

Barbier, E.B. (2011), *Capitalizing on Nature. Ecosystems as Natural Assets*, Cambridge, UK: Cambridge University Press.

Birol, E., K. Karousakis and P. Koundouri (2006), 'Using a choice experiment to account for preference heterogeneity in wetland attributes: the case of Cheimaditida wetland in Greece', *Ecological Economics*, **60**(1), 145–6.

Bishop, R. and T. Heberlein (1979), 'Measuring values of extra-market goods: are indirect measures biased?' *American Journal of Agricultural Economics*, **61**(4), 926–30.

Boyer, T. and S. Polasky (2004), 'Valuing urban wetlands: a review of nonmarket valuation studies', *Wetlands*, **24**(4), 744–55.

Brander, L.M., R.J.G.M. Florax and J.E. Vermaat (2006), 'The empirics of wetland valuation: a comprehensive summary and a meta-analysis of the literature', *Environmental and Resource Economics*, **33**(2), 223–50.

Brouwer, R., I.H. Langford, I.J. Bateman and R.K. Turner (1999), 'A meta-analysis of wetland contingent valuation studies', *Regional Environmental Change*, **1**(1), 47–57.

Burgess, D., N. Jackson, D. Hadley, K. Turner, S. Georgiou and H.B. Day (2008), 'Assessing the value of a scientifically important wetland ecosystem: the case of the culm grasslands', CSERGE Working Paper, Centre for Social and Economic Research on the Global Environment, University of East Anglia.

Cameron, T.A. and J. Quiggin (1994), 'Estimation using contingent valuation data from a dichotomous choice with follow-up questionnaire', *Journal of Environmental Economics and Management*, **27**(3), 218–34.

Carson, R. and T. Groves (2007), 'Incentive and informational properties of preferences questions', *Environmental and Resource Economics*, **37**(1), 181–210.

Carson, R., R. Mitchell, W.M. Hanemann, R. Kopp, S. Presser and P. Ruud (1992), *A Contingent Valuation Study of Lost Passive Use Values Resulting from the Exxon Valdez oil Spill*, report to the Attorney General of Alaska, NRDA Inc, La Jolla CA.

Centre for the Economics and Management of Aquatic Resources (CEMARE) (2002), *Effective Management for Biodiversity Conservation in Sri Lankan Coastal Wetlands, CVM Report II: An Estimation of Economic Value for Conservation of Wetlands*, University of Portsmouth.

Cooper, C.J., W.M. Hanemann and G. Signorello (2002), 'One-and-one-half bound dichotomous choice contingent valuation', *Review of Economics and Statistics*, **84**(4), 742–50.

Creel, M. and J. Loomis (1997), 'Semi-non-parametric distribution-free dichotomous choice contingent valuation', *Journal of Environmental Economics and Management*, **32**(3), 341–58.

Day, B.H. (2007), 'Distribution-free estimation with interval-censored contingent valuation data: troubles with Turnbull?' *Environmental and Resource Economics*, **37**(4), 777–95.

DeShazo, J.R. (2002), 'Designing transactions without framing effects in iterative question formats', *Journal of Environmental Economics and Management*, **43**(3), 360–85.

European Environment Agency (EEA) (2010), *Ecosystem Accounting and the Cost of Biodiversity Losses. The Case of Coastal Mediterranean Wetlands*, Technical Report No. 10, Copenhagen: EEA.

Gantioler, S., M. Rayment, S. Bassi, M. Kettunen, A. McConville, R. Landgrebe and H. Gerdes (2010), *Costs and Socio-economic Benefits Associated with the Natura 2000 Network*, final report to the European Commission, DG Environment on Contract ENV.B.2/SER/2008/0038, Brussels: Institute for European Environmental Policy/GHK/Ecologic.

Gentleman, R. and C.J. Geyer (1994), 'Maximum likelihood for interval censored data: consistency and computation', *Biometrika*, **81**(3), 618–62.

Ghermandi, A., J.C.J.M. van den Bergh, L.M. Brander, H.L.F. de Groot and P.A.L.D. Nunes (2010), 'The values of natural and human-made wetlands: a meta-analysis', *Water Resource Research*, **46**(12), W12516.

Ghermandi, A., P.A.L.D. Nunes, R. Portela, N. Rao and S.S. Teeluksingh (2012), 'Recreational, cultural and aesthetic services from estuarine and coastal ecosystems', in E. Wolansky and D. McLusky (eds), *Treatise on Estuarine and Coastal Science*, Amsterdam: Elsevier.

Haab, T.C. and K.E. McConnell (1997) 'Referendum models and negative willingness to pay: alternative solutions', *Journal of Environmental Economics and Management*, **32**(2), 251–70.

Hanemann, W.M. (1984), 'Welfare evaluations in contingent valuation experiments with discrete response data', *American Journal of Agricultural Economics*, **66**(3), 332–41.

Hanemann, W.M. and B. Kanninen (1999), *The Statistical Analysis of Discrete Response CV Data*, in I.J. Bateman and K.G. Willis (eds), *Valuing Environmental Preferences: Theory and Practice of the Contingent Valuation Method in the US, EU, and Developing Countries*, Oxford: Oxford University Press.

Hanemann, W.M., J. Loomis and B. Kanninen (1991), 'Statistical efficiency of double-bound dichotomous choice contingent valuation', *American Journal of Agricultural Economics*, **73**(4), 1255–63.

Herriges, J. and J. Shogren (1996), 'Starting point bias in dichotomous choice valuation with follow-up questioning', *Journal of Environmental Economics and Management*, **30**(1), 112–31.

ISTAT (2002), *Demografia in Cifre*, 2002, demo.istat.it.

Kanninen, B.J. (1995), 'Bias in discrete response contingent valuation', *Journal of Environmental Economics and Management*, **28**(1), 114–25.
Klein, J.P. and M.L. Moeschberger (1997), *Survival Analysis Techniques for Censored and Truncated Data*, New York: Springer.
Kumar, P. (ed.) (2010), *The Economics of Ecosystems and Biodiversity: Ecological and Economic Foundations*, London: Earthscan.
Liu, S. and D.I. Stern (2008), 'A meta-analysis of contingent valuation studies in coastal and near-shore marine ecosystems', Munich Personal RePEc Archive Paper No. 1170.
McFadden, D. (1994), 'Contingent valuation and social choice', *American Journal of Agricultural Economics*, **76**(4), 689–708.
McFadden, D. and G. Leonard (1993), 'Issues in contingent valuation of environmental goods: methodologies for data collection and analysis', in J.A. Hausman (ed.), *Contingent Valuation: A Critical Assessment*, Amsterdam: North-Holland.
Millennium Ecosystem Assessment (MEA) (2005), *Ecosystems and Human Well-Being: Wetlands and Water, Synthesis*, Washington, DC: World Resources Institute.
Signorello, G. (1998), 'Valuing birdwatching in a Mediterranean wetland', in R.C. Bishop and D. Romano (eds), *Environmental Resource Valuation: Applications of the Contingent Valuation in Italy*, Dordrecht: Kluwer Academic.
Silverman, B. (1986), *Density Estimation*, London: Chapman and Hall.
Turnbull, B. (1976), 'The empirical distribution function with arbitrary grouped, censored and truncated data', *Journal of Royal Statistical Society, Series B*, **38**, 290–95.
Turner, R.K., S. Georgiou and B. Fisher (2008), *Valuing Ecosystem Services. The Case of Multifunctional Wetlands*, London: Earthscan.
Wellner, J.A. and Y. Zhan (1997), 'A hybrid algorithm for computation of the non-parametric maximum likelihood estimator from censored data', *Journal of the American Statistical Association*, **92**(439), 945–59.
Wilson, M.A. and S. Liu (2008), 'Non-market value of ecosystem services provided by coastal and nearshore marine systems', in M. Patterson and B. Glavovic (eds), *Ecological Economics of the Oceans and Coasts*, Cheltenham, UK and Northampton, MA, USA: Edward Elgar Publishing.
Woodward, R.T. and Y.S. Wui (2001), 'The economic value of wetland services: a meta-analysis', *Ecological Economics*, **37**, 257–70.

23. A micro-econometric approach to deriving use and non-use values of in situ groundwater: the Vosvozis case study, Greece

Phoebe Koundouri, Vassilis Babalos, Mavra Stithou and Ioannis Anastasiou

23.1 INTRODUCTION

The objective of this work is to derive the in situ shadow price of unextracted groundwater in the Vosvozi aquifer, through modelling and empirically analysing the technology of vertically integrated agricultural firms that both extract and use groundwater as an input in their production. This shadow price, also referred to as the resource's scarcity rent or royalty, represents the marginal valuation of the individual agricultural producer for the resource left in situ and is not directly observable. In the model developed, the non-observability of the in situ shadow price of groundwater is caused by the fact that market transactions in vertically integrated agricultural firms occur only after groundwater has been extracted and used in the production of agricultural products; that is, there is no market for groundwater.

This research uses duality theory in order to derive information on the in situ shadow price of the resource and the effects of cumulative extraction on the marginal cost of extraction. First, we solve the 'restricted' version of the dual cost minimization problem of the vertically integrated agricultural firm. The solution of this problem establishes the relationship between the current (unobserved) in situ shadow price of groundwater in the unrestricted solution of the problem, with the derivatives of the observable and estimable restricted cost function. This exact same method has been employed in theoretical and applied work, for the derivation of the time path of in situ shadow prices of unextracted ore, to be used as a production input in the vertically integrated Canadian metal mining industry (Halvorsen and Smith, 1984, 1991).

Second, another method that allows derivation of the unobservable shadow price of in situ resources through the use of an input distance function is proposed. The relationship between the derivatives of the estimable input distance function with the unobserved shadow price of in situ groundwater is established. The derivation of this lemma is possible by the use of the duality between Shephard's input distance function and the cost function.

The key extension of our work on the existing literature is that it establishes that when cost, profit or revenue function representations are precluded (i.e., profit maximization or cost minimization are violated, resulting in distortions in the shadow prices of resources that are both produced and used as inputs in the production processes of vertically integrated firms), the restricted distance function provides an excellent analytical tool for estimating unobservable shadow prices of in situ natural resources produced and used as inputs in production processes of vertically integrated firms. We also review alternative methods of estimating distance function frontiers and argue for the superiority of the stochastic frontier model, adopted in the empirical analysis to follow. The stochastic frontier model exhibits two major advantages over alternative estimation methods: (1) it acknowledges that observed costs may deviate from an efficient cost frontier due to events that are both within and outside a firm's control, and (2) it allows firm-specific derivation of shadow prices, whereas other methods allow derivation of shadow prices for efficient firms only.

With regard to the empirical application of the Vosvozi case study it involves the use of micro (at farm level) dataset in order to estimate a restricted input distance function stochastic frontier and provide an estimate of the individual producer's valuation of the marginal unit of groundwater in the aquifer. This shadow price is central to the implementation of the EU Water Framework Directive (WFD) and EU Groundwater Directive, because it allows per farm estimation of the value of groundwater. It also allows the calculation of the difference between the current price charged for groundwater, that is, the current level of cost recovery. This in turn allows suggestions of policy instruments (economic and social tools) for the achievement of full cost recovery, as indicated by the WFD. Finally, a brief discussion on estimated farm-specific technical inefficiencies/efficiencies is provided, which indicates whether agricultural production can be made more efficient. If such potential exists then the relevant policy instruments can be identified.

The rest of the chapter is organized as follows. Section 23.2 provides a description of the case study area and the relevant dataset, while Section 23.3 outlines the empirical model. Results are presented in Section 23.4, while in Section 23.5 policy implications are commented upon. The chapter closes with conclusions offered in Section 23.6.

23.2 DESCRIPTION OF THE CASE STUDY AREA AND DATASET

23.2.1 Case Study[1]

Vosvozis catchment area covers an area of 340 km^2. The river's length is 40 km. Vosvozis River discharges into Ismarida Lake. In the coastal part of the study area a system of coastal lagoons is formed, where surface, groundwater and seawater interact. All the area of Ismarida Lake and the coastal lagoons form an extremely important ecosystem (Figure 23.1). Land uses in the Vosvozis River basin are mainly agricultural (cotton, corn, tobacco, sugar beets, barley and clover cultivations), cattle breeding, industrial (mainly in the form of cotton industry, dairy product industry and meat processing plants) and urban/residential. The area has 70000 inhabitants, while the main urban centre is Komotini town. Point sources of pollution are formed from industrial activities that discharge their wastewaters into Vosvozis River or into its tributaries in an uncontrolled manner and by private septic tanks (half of the population is served by such systems), which are point sources of pollution for groundwater. It should also be noted herein that Komotini's wastewater treatment plant discharges treated wastewaters in Vosvozis River. Special attention should be focused on Komotini's industrial area, which is not located within Vosvozis river basin but adjacent to it. This industrial area comprises plastic,

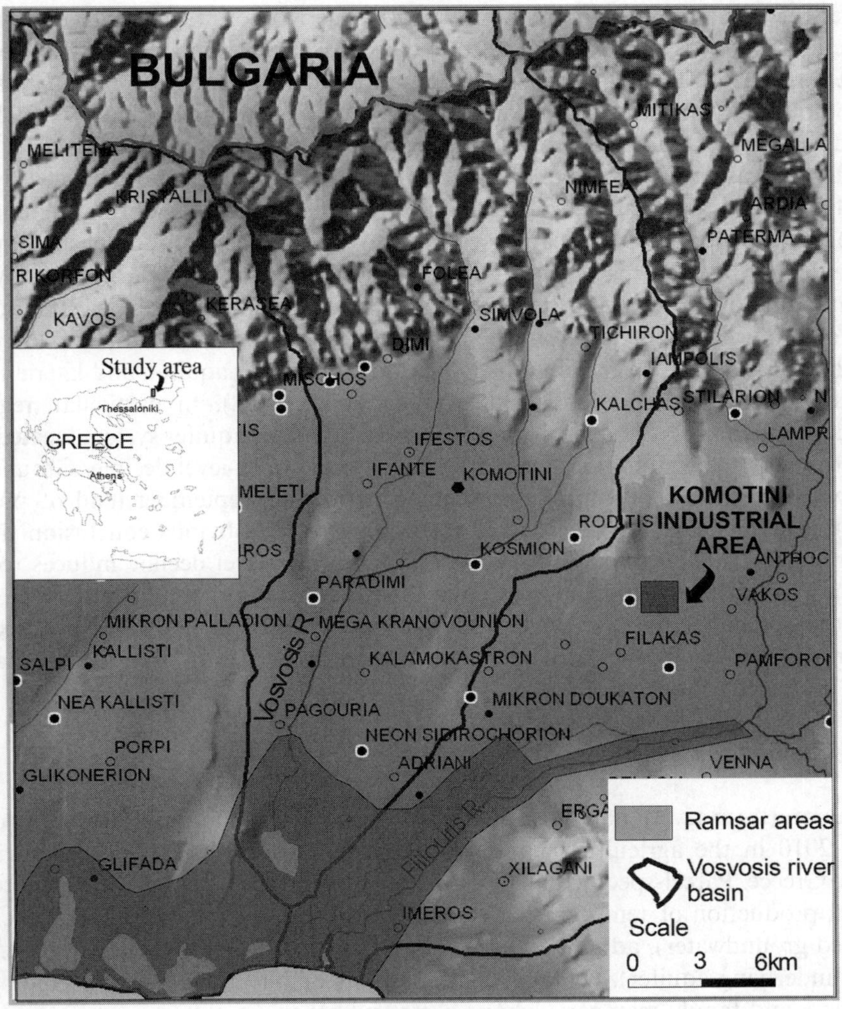

Figure 23.1 Location map of the study area

paper, wood, food processing plants, as well as a thermo-electric power-producing plant. Industrial wastewaters are disposed in Filiouris River (Figure 23.1), which discharges into the coastal lagoon ecosystem, thus forming a serious threat to it. In particular, the main threat to the wetland ecosystem is eutrophication, diminishing its aerial and seawater intrusion, which seriously affects the fragile wetland ecosystem. Agriculture is the dispersal source of pollution for the aquifer system studied, merely through the application of fertilizers and pesticides. The existing hydrochemical data from 25 irrigation boreholes within the study aquifer showed that groundwaters are seriously affected by nitrate pollution, with nitrates ranging from 30 to 100 mg/l. Besides the quality problems of the aquifer system, piezometric data for the last ten years indicate that there is a constant groundwater level drawdown, which ranges from 10 to 50 m in the examined boreholes (Moutsopoulos et al., 2008).

Water for human consumption is provided by the Komotini well field and by direct abstraction from Vosvozis River. The total daily discharge pumped from the Komotini well field reaches 23 000 m^3/d, providing domestic water to almost 70 000 inhabitants of the Komotini city and the surrounding settlements. The well field consists of 21 boreholes drilled in the study area, 15 of which are productive, while the remaining six are currently used as observational ones. The average discharge of the productive well ranges from 45 to 90 m^3/h. Groundwater pumping takes place mainly during summertime, whereas during the rest of the year Vosvozis River is used directly for domestic consumption, and when its water is of appropriate quality (because storm surges usually carry large amounts of sediment, thus making river water unsuitable for domestic use). The origin of the water extracted from the aquifer in the Komotini well field is the nearby river, that is, Vosvozis River, rain infiltrated directly into the aquifer, and lateral inflows from the northern mountains (Moutsopoulos et al., 2008). In particular, regarding groundwater dynamics Sidirohori aquifer, the second major aquifer system located in the southern part of the study area, shows serious groundwater-level decline. Groundwater drawdown from May (beginning of pumping period) to September (end of pumping period) in certain locations reaches 20 m, leading to the obvious conclusion that the aquifer system is overexploited. Moreover, groundwater level decline induces recharge from the Vosvozis River and Ismarida Lake, thus diminishing an important source for the life of the wetland ecosystem. Finally, another threat due to groundwater level decline is the intrusion of seawater in the wetland area, a serious alteration in the initial character of this protected ecosystem.

23.2.2 Dataset

The micro (farm-level) dataset was drawn from a Production Survey conducted during 2010 in the agricultural region of Vozvozi aquifer, located in the region of Thrace, Greece. Parcel-specific data include: area of holding, land use and tenure, area planted, production of temporary and permanent crops, production inputs (including extracted groundwater), administrative costs, hydrogeological characteristics (i.e., head of the underlying aquifer), personal characteristics of buyers and sellers, employment of holders and family members, labour costs and other investment and indirect costs. In particular, the dataset is an unbalanced panel of the same 100 cross-sections over the year 2010.

An important consideration in the estimation of production functions is the selection of the proper output and input variables. Following the relevant literature output is defined as the firm-specific total value from production of agricultural crops measured in euros and is denoted as y. It should be noted that the output variable has been deflated using the agricultural price index for Greece provided by Eurostat. Regarding model inputs as in Koundouri and Xepapadeas (2004) we have employed the following: farm-specific total area of non-irrigated land (variable x_1), farm-specific annual labour costs in euros (variable x_2), farm-specific total value of input costs (variable x_3) deflated using the agricultural price index, farm-specific yearly groundwater extraction (m^3) (variable x_4) and farm-specific water table head (dummy variable, variable x_5). With respect to the variable water table head we have constructed a dummy variable that differentiates the location of the farm in terms of water quality based on hydrogeological information. In particular, variable x_5 takes the values 1, 2 and 3 based on water quality (low, medium and good respectively).

23.3 METHODLOGY

The distance function representation of a production technology, proposed by Shephard ([1953] 1970), provides a multi-output primal alternative, which requires no aggregation, no prices and no behavioural assumption. A distance function may have either an input orientation or an output orientation. In empirical applications, distance functions have a number of advantages: (1) they do not necessarily require price data to compute the relevant parameters, only quantity data is needed; (2) they do not impose any behavioural hypothesis and (3) they allow the estimation of firm-specific inefficiencies.

In the context of the present study we opt for a translog stochastic input distance function (Aigner et al., 1977) for the case of K inputs and $M = 1$ output. To obtain the frontier surface (i.e., the transformation function) we set $D_i = 1$. Model estimation was performed employing STATA. Necessary restrictions for (1) homogeneity of inputs of degree +1, (2) symmetry and (3) separability between inputs and outputs have been imposed:

$$\ln(D_i/X_{Ki}) = \alpha_0 + \alpha_1 \ln y_i + \frac{1}{2}\alpha_2(\ln y_i)^2 + \sum_{k=1}^{K-1}\beta_k \ln x^*_{ki} + \frac{1}{2}\sum_{k=1}^{K-1}\sum_{l=1}^{K-1}\beta_k \ln x^*_{ki} \ln x^*_{li} + \sum_{k=1}^{K-1}\delta_k \ln x^*_{ki} \ln y_i$$

$$\text{for } i = 1, 2, \ldots N \text{ and } x_k = \frac{x_k}{x_K} \tag{23.1}$$

where i stands for the ith firm within the sample.

The frontier function has an error term with two components that are independent. The first component is a symmetric error term (V_i) that accounts for noise, which is assumed identically and independently distributed with zero mean and constant variance (N (0, σ^2_V)). The second component is an asymmetric error term (U_i) that accounts for technical inefficiency, which is assumed to follow an *iid* distribution truncated at zero (N (v, σ^2_u)). It should be noted that the two error components are independent.

Estimated values of $D_i = \exp(U_i)$ are obtained employing the conditional expectation $D_i = E(\exp(U_i/\Omega_i))$, where Ω_i equals $V_i - U_i$. If we alter notation $\ln(D_i)$ to U_i, Equation (23.1) is as follows:

$$-\ln(x_{Ki}) = TL\left(y_i, \frac{x_i}{x_{Ki}}, \alpha, \beta\right) + V_i - U_i \, i = 1, 2 \ldots N \tag{23.2}$$

23.4 RESULTS

The dependent variable of Equation (23.2) is irrigated land and the model was estimated by maximum likelihood. Results are presented in Table 23.1. Variable x_1 that stands for farm-specific total area of non-irrigated land was dropped from the estimation due to a

Table 23.1 Estimated parameters for the input distance function[a]

Variable	Parameter	ML Estimates	t-ratios[b]
Constant	α_0	−1.37	−0.73
Output	α_1	−0.18	−1.67
Labour	β_2	0.08	1.6
Costs	β_3	0.17	1.59
Water extraction	β_4	0.013	0.18
Head	β_5	0.68	5.72
	Log (likelihood)	−7.8002	
	γ	0.004	0.000
	σ^2	0.1	
	σ_u^2	0.0004	
	σ_v^2	0.104	

Notes:
a. The dependent variable is irrigated land. Number of cross-sections is 27.
b. Hypothesis tests are carried out at 95% confidence level.

large amount of missing values. Gross products and squared coefficients are not reported because they were excluded from the empirical model after a preliminary estimation that indicated that their estimated effects were not significantly different from zero.

Estimated coefficients have the anticipated signs (positive for inputs and negative for outputs). Coelli (1995) has derived a one-sided test for the presence of the inefficiency term and according to this the null hypothesis is about NO TECHNICAL INEFFICIENCY and we reject it. Moreover, the reported value of gamma (γ) that is close to zero indicates that the deviations from the frontier are entirely due to noise.

Firm-specific technical efficiencies are reported in Table 23.2. A firm is said to operate in an efficient manner if it is impossible to produce a larger amount of outputs with the given inputs or the same output with less of one or more inputs without increasing the amount of other inputs. Our results reveal a significant level of operational efficiency for the firms/farms in our sample. The mean efficiency level is 0.99. Technical inefficiency results from employing a larger amount of inputs than required in order to achieve a certain output level and is explicitly related to the lack of incentives faced by the owners of the firm. Technical inefficiency measures could help regulators to implement the designated policy regarding taxes and subsidies granted to each farm relying on the costs of a similar (in terms of input mix) but more efficient firm. This process is widely known as competitive benchmarking ('yardstick competition'). Such a regulatory framework can (1) raise the farm managers' incentives toward efficiency and (2) alleviate the informational asymmetry between the managers of the farms (agents) and the regulators or consumers of agricultural products (the principals).

In Table 23.3, we calculate the estimated in situ price, that is, value for farmers (use value) per cubic metre, of unextracted groundwater in the Vosvozi aquifer as in Koundouri and Xepapadeas (2004). The mean annual per farm minimum restricted cost function $\hat{C}_i^R$ is approximated by the mean annual per farm revenue. The change in

Table 23.2 Estimated firm efficiency levels

Firm	Efficiency
1	0.99014
2	0.99026
3	0.99009
4	0.99003
5	0.99000
6	0.98957
7	0.99048
8	0.99025
9	0.99006
10	0.99033
11	0.99001
12	0.99004
13	0.98981
14	0.99044
15	0.99045
16	0.98997
17	0.98973
18	0.98988
19	0.98987
20	0.99008
21	0.99017
22	0.99001
23	0.99014
24	0.98985
25	0.98971
26	0.99034
27	0.99014
Mean	0.99007

Table 23.3 Estimated in situ price of unextracted groundwater

Year	$\hat{C}_i^R$	$\frac{\theta \ln D_i^R}{\theta \ln W_i}$	W_i	μ_i
2010	€4083.61	€0.01/m³	18 686.33 m³	0.009 m³

the restricted distance function per unit change in groundwater extraction $\frac{\theta \ln D_i^R}{\theta \ln W_i}$ measured in € per cubic metre is the estimated parameter of the quantity of groundwater extraction from the stochastic distance function estimation, the results of which are presented in Table 23.3 and W_i is the mean groundwater extraction per farm, measured in m³.

Table 23.4 Taxonomy of groundwater valuation terminology

Physical State Terminology	Economic Terminology	Accounting Terminology	
		Stocks	Flows
A. Extractive values			
1. Municipal use values	Use values		*
2. Industrial use values	Use values		*
3. Agricultural use values	Use values		*
4. Other extractive use values	Use values		*
B. In situ values			
1. Ecological values	Use values	*	*
2. Buffer values	Use values	*	*
3. Subsidence avoidance values	Use values	*	*
4. Recreational values	Use values		*
5. Seawater intrusion values	Use values	*	*
6. Existence values	Non-use values		*
7. Bequest values	Non-use values		*

Source: Committee on Valuing Groundwater, National Research Council (1997).

23.5 POLICY IMPLICATIONS

The economic value of groundwater in a specific aquifer is derived from the use it can be put to, and therefore it originates from the benefits that it generates or the services that it provides. Local availability and quality compared to surface water are also determinants of its economic value. These are determined by factors such as population growth, economic development, pollution and climatic variability. Figure 23.4 offers an overview of the total economic value of groundwater according to which its services can be divided into two basic categories: extractive services and in situ services. The more familiar of these two components are the extractive values, while the in situ services include, for example, the capacity of groundwater to: (1) buffer against periodic shortages in surface water supplies; (2) prevent or minimize subsidence of the land surface from groundwater withdrawals; (3) protect against seawater intrusion; (4) protect water quality by maintaining the capacity to dilute and assimilate groundwater contaminants; (5) facilitate habitat and ecological diversity; and (6) provide discharge to support recreational activities (Committee on Valuing Groundwater, National Research Council, 1997). Discharge to ecosystems, rivers and lakes can be seen as a groundwater service of indirect (ecosystem) value (Kemper et al., 2002–06).

However, in many cases the human health focus ignores other functions of groundwater that humans might value such as the role of groundwater in ecological functions and in particular in providing an important contribution to unique terrestrial and aquatic ecosystems. As Kløve et al. (2011) note (p. 779):

> these systems are typically of high value as they support high biodiversity and provide the habitat for several endangered species. Some of these ecosystems and related water bodies have been protected to a certain extent by international conventions such as the Ramsar Convention and, in Europe, by several laws such as the Habitat and Water Framework Directive.

This is the case with our case study area in which groundwater dynamics interact with important ecosystems such as these of the area of Ismarida Lake and the coastal lagoons. The exclusion of these services and values may be due to the lack of knowledge regarding status of groundwater and impacts of land and water use, pollution and climate change.

Few studies have attempted to measure the value that people place on the ecological services that groundwater supplies, while a few do estimate non-use values related to quality (Jensen et al., 1995; Rozan et al., 1997; Press and Söderqvist, 1998; Hasler et al., 2005) or quantity (Koundouri et al., 2012) of groundwater. In particular, in Rozan et al. (1997) the estimated €52 per household/year in 1995 of non-user households to protect the Alsatian aquifer (France) is considered as a proxy of its existence value and is used to assess the economic non-use value of the aquifer. Similarly, Press and Söderqvist (1998) employed the contingent valuation (CV) method to estimate the benefits of groundwater protection in the Milan area (Italy) in order to also consider non-use values directly. The study elicited a high value of ITL640000 per household/year showing the broad values at stake in the preservation of groundwater. In addition, Jensen et al. (1995), using the CV method, estimated the willingness to pay (WTP) for groundwater protection from pollution at DKK1000 household/year elicited by an open-ended payment format, and at DKK2100 using the closed-ended format. Regarding the choice experiment (CE) method the applications are even fewer. Hasler et al.'s (2005) national CE study assessed the non-marketed benefits associated with increased protection of the groundwater resources and revealed an estimated WTP of €253/year for protected and naturally clean groundwater, not the need for purification, a WTP for good conditions for flora and fauna in waterways and lakes of €161/year, and a WTP for purified water of €122/year (2005 prices). Finally, in Koundouri et al. (2012) the case study of interest is Rokua in Northern Finland, a groundwater-dependent ecosystem very sensitive to climate change and natural variability that faces disturbance of the water dynamics and in particular of water quantity. Results of a CE survey indicate that an average household is willing to pay €22–23 (one-off payment) in order to ensure that water management does not allow the decline of total quantity of water available in groundwater aquifers, lakes and springs. As a result, the above prices in contrast with the in situ derived value from the Vosvozi case study reveal the important role of non-use values, which are of considerable magnitude when seen from the residents' perspective.

Furthermore, in Table 23.3 the reported in situ value of unextracted groundwater is much lower than the established in situ per cubic metre groundwater's total economic value. This total economic value is equal to the relevant backstop technology for water, which is, for example, the per cubic metre cost of desalination (at €0.05; see Koundouri, 2000). This divergence points to the significant non-use values of groundwater, such as option value and ecosystem resilience value, as well as alternative use values of economic sectors other than agriculture. Another point is raised after comparing our estimate of the individual farmers' valuation of the marginal unit of groundwater in the aquifer with the socially optimal shadow price of in situ groundwater derived for the Kiti aquifer in Cyprus in 1999 by Koundouri and Christou (2000). The in situ value (in Cyprus pounds) of the resource was determined to be £0.2017 per m^3 of water. As has also been noted in Koundouri and Xepapadeas (2004) where results were similar to this study, such a divergence can be rationalized in the presence of non-cooperative behaviour and common

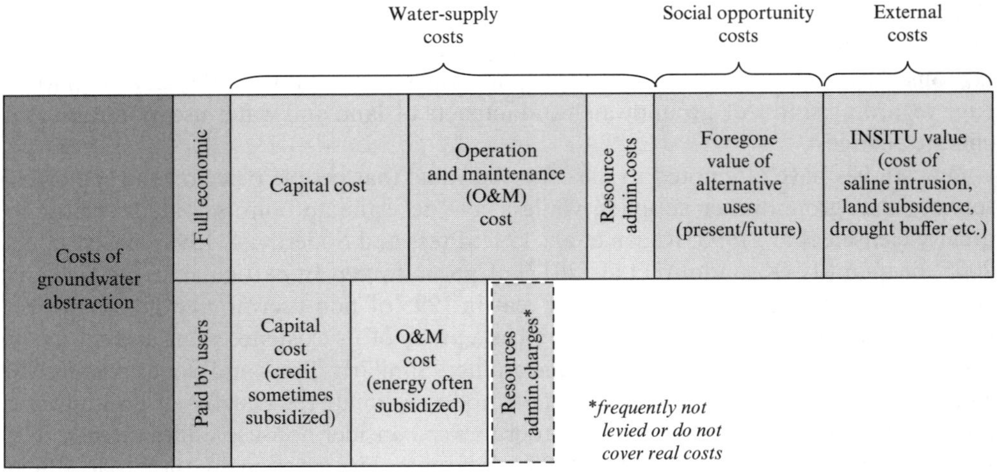

Source: Kemper et al. (2002–06).

Figure 23.2 The costs of groundwater use

pool externalities, as current users of the resource are willing to pay only the private cost and not the full social cost of their resource extraction.

In this context, it becomes apparent that the notion of total economic value can be used to inform decision-makers regarding the use of water resources, enabling the determination of the net benefits of policies and management actions, since what is commonly observed is that groundwater tends to be undervalued, especially where its exploitation is uncontrolled. In this situation the exploiter of the resource receives all the benefits of groundwater use but pays only part of the costs (Figure 23.2) – usually the recurrent cost of pumping and the capital cost of well construction, but rarely the external and opportunity costs (Kemper et al., 2002–06). The fact that groundwater is priced well below its value, has as a consequence its misallocation in two ways: (1) the groundwater resource is not efficiently allocated relative to alternative current and future uses; and (2) authorities responsible for resource management and protection devote inadequate attention and funding to maintaining groundwater quality (Committee on Valuing Groundwater, National Research Council, 1997). This is also the case of Vosvozi where no charge is imposed for water withdrawn, and the consumer, whether a public water supply entity, an individual, or a firm regards the cost as being confined to the energy used for pumping and the amortization of well construction and the costs of the treatment and distribution system. As a result, depletion and pollution continue as it is not recognized that groundwater has a high or long-term value. This is apparent by the difference between the estimated in situ shadow price of the stock of groundwater in Vosvozi (use value) and total economic value that explains the inefficiency of agriculture using water and paying only for its use value. That is, agriculture uses water efficiently as far as groundwater agricultural use value is concerned but seriously overexploits/overextracts groundwater as far as its total economic value is concerned.

In Greece, water use is charged by farm-specific total area and not by type of crop, to subsidize irrigation, and to have illegal private wells or when they are legal not to have metering to monitor the volumetric use of the resource. As a result, these practices have eroded the same farmers' resource availability in the longer term because of excessive groundwater abstraction.

Economic instruments can provide incentives to allocate and/or use groundwater more efficiently. There are two categories relevant to groundwater, namely those that focus upon (Kemper et al., 2002–06): (1) changing groundwater abstraction costs by (a) direct pricing through resource abstraction fees, (b) indirect pricing through increasing energy tariffs and (c) the introduction of water markets; (2) positive economic incentives for certain activities by (a) modifications to agriculture and food trade policies and (b) subsidies to encourage the use of more efficient irrigation technologies to achieve real water savings. Therefore, appropriate institutional foundations are required to provide farmers with the incentive to pay today for conserving in situ groundwater for future extraction and avoid myopic behaviour that resides from the absence of properly defined property rights for groundwater. Efficient pricing of the resource should incorporate marginal cost of extraction and scarcity rents. Regarding the latter the establishment of interactions between groundwater resources and ecosystem goods and services is of paramount importance in order to estimate resources' full total cost incorporating its scarcity value. Supplementary to this approach is the use of lump-sum payments to poor farmers at the beginning of the year to cover their estimated energy bill, in order to give them an incentive to use water more efficiently and consume less, maybe through a shift to higher-value crops (Kemper et al., 2002–06) and herbal, medicinal and aromatic plants. Hence, since they receive lump sum payments to offset their increased energy bills, they can actually gain twice by being more efficient. It is important, therefore, for our region under investigation to identify an avenue that combines promising production and efficient water use through the prism of sustainability.

Finally, the relatively new approach of payments for environmental services has often focused on supporting watershed protection and water quality enhancements that target the provision of surface water and groundwater (Wunder et al., 2008). It has been suggested recently that farmers should receive payments or 'green water credits' from downstream water users for good management practices that enhance green water (rainfall stored in soil moisture) retention as well as surface water and groundwater conservation (ISRIC, 2007).

23.6 CONCLUSIONS

This study replicates the distance function methodology for estimating scarcity rents that has been applied to the irrigated agricultural sector of the Kiti region of Cyprus employing data for a sample of farms situated in the Vozvozi River area, Thrace. In order to estimate the in situ shadow prices in a framework irrespectively of cost minimization restrictions, we opt for a methodology based on the input distance function, which does not require any behavioural assumptions. Documented failure of farmers to minimize costs, provides support for the use of the distance function and proves the potential for estimation inaccuracy should one wrongly choose to use the restricted distance function methodology. The suggested methodology could also be useful in estimating shadow prices for renewable resources such as groundwater, forest and fisheries. As has been mentioned, this shadow price is of vital importance to the implementation of the EU Water Framework Directive and EU Groundwater Directive, because it allows per farm estimation of the value of groundwater. It also allows the calculation of the difference between the current prices charged for groundwater, that is, the actual level of cost recovery.

In addition to the potential of this methodology as a demand management tool via pricing, technical inefficiency measures can be employed by the regulator for competitive benchmarking ('yardstick competition') in which taxes or subsidies granted to each farm are based on the costs of a similar (in terms of input mix) but more efficient firm. As indicated in the previous sections of the chapter, such a regulatory framework can spur managers toward efficiency, an admittedly difficult task when regulation of common-pool resources is at stake. Moreover, introducing competitive benchmarking could probably help to alleviate informational imbalance between the farmers and the regulators, an issue that calls for regulators' attention when it comes to the implementation of agricultural policies.

Results show that groundwater in our case study area is undervalued and economic instruments should provide incentives to use it more efficiently by agricultural sector, incorporating the notion of total economic value and therefore groundwater's indirect (ecosystem) value and non-use values in water management. However, in order to achieve this, as Kløve et al. (2011) note, integrated multidisciplinary knowledge on hydrology, geochemistry and biology from individual systems as well as on the scale of regional catchments and aquifers is needed. Therefore, it is important to clarify connections between groundwater processes, ecosystems and base stream flow and better define the extent to which changes in groundwater quality or quantity contribute to changes in ecologic values. Finally, other parameters of importance when designing policy are the finite nature of the resource needing a long-term view and the fact that any actions should consider avoiding an irreversible situation regarding groundwater.

NOTE

1. See Kupfersberger (2010).

REFERENCES

Aigner, D., C.A.K. Lovell and P.J. Schmidt (1977), 'Formulation and estimation of stochastic frontier production function models', *Journal of Economics*, **6**(1), 21–37.

Coelli, T.J. (1995), 'Recent developments in frontier modelling and efficiency measurement', *Australian Journal of Agricultural Economics*, **39**(03), 29–32.

Committee on Valuing Groundwater, National Research Council (1997), *Valuing Groundwater: Economic Concepts and Approaches*, Washington, DC: National Academies Press.

Halvorsen, R. and T.R. Smith (1984), 'On measuring natural resource scarcity', *Journal of Political Economy*, **92**(5), 954–64.

Halvorsen, R. and T.R. Smith (1991), 'Test of the theory of exhaustible resources', *Quarterly Journal of Economics*, **106**(1), 123–40.

Hasler, B., T. Lundhede, L. Martinsen, S. Neye and J.S. Schou (2005), *Valuation of Groundwater Protection Versus Water Treatment in Denmark by Choice Experiments and Contingent Valuation*, NERI Technical Report No. 543, Roskilde, Denmark: National Environmental Research Institute, Denmark.

ISRIC (2007), *Green Water Credits*, Wageningen: ISRIC – World Soil Information.
Jensen, P.R., A.B. Hallam, J. Hauch, E. Møllgaard and O.G. Petersen (1995), 'En velfærdsindikator for Danmark 1970–1990' [A welfare indicator of Denmark], Memorandum No. 8. Rockwool Foundation, Copenhagen.
Kemper, K., S. Foster, H. Garduño, M. Nanni and A. Tuinhof (2002–06), *Economic Instruments for Groundwater Management Using Incentives to Improve Sustainability*, Briefing Note Series No. 7, Sustainable Groundwater Management: Concepts and Tools, GW-MATE, accessed 15 February 2014 at http://www.limpoporak.com/_system/DMSStorage/3471en/GWMATEBN07.pdf.
Kløve, B., A. Allan, G. Bertrand, E. Druzynska, A. Ertürk and N. Goldscheider (2011), 'Groundwater dependent ecosystems. Part II Ecosystems services and management in Europe under risk of climate change and land use intensification', *Environmental Science and Policy*, **14**(7), 770–81.
Koundouri, P. (2000), 'Three approaches to measuring natural resource scarcity: theory and application to groundwater', PhD thesis, Faculty of Economics and Politics, University of Cambridge, accessed 15 February 2014 at http://www.dspace.cam.ac.uk/handle/1810/226658.
Koundouri, P. and C. Christou (2000), 'Dynamic adaptation to resource scarcity with backstop availability', paper presented at the 10th Annual Conference, European Association of Environmental and Resource Economists, Rethymnon, Greece, June.
Koundouri, P. and A. Xepapadeas (2004), 'Estimating accounting prices for common pool natural resources: a distance function approach', *Water Resources Research*, **40**(6), doi: 10.1029/2003WR002170.
Koundouri, P., E. Kougea, M. Stithou, P. Ala-aho, R. Eskelinen and T. Karjalainen et al. (2012), 'The value of scientific information on climate change: a choice experiment on Rokua esker, Finland', *Journal of Environmental Economics and Policy*, **1**, 82–102.
Kupfersberger, H. (2010), *Groundwater and Dependent Ecosystems: New Scientific and Technical Basis for Assessing Climate Change and Landuse Impacts on Groundwater Systems*, Deliverable 1.1 of Seventh Framework Programme ENV.2008.2.1.2.1, Groundwater Systems – GENESIS project.
Moutsopoulos, K., A. Gemitzi and V.A. Tsihrintzis (2008), 'Delineation of groundwater protection zones by the backward particle tracking method: theoretical background and GIS-based stochastic analysis', *Environmental Geology*, **54**(5), 1081–90.
Press, J. and T. Söderqvist (1998), 'On estimating the benefits of groundwater protection: a contingent valuation study in Milan', in T. Swanson and M. Vighi (eds), *Regulating Chemical Accumulation in the Environment*, Cambridge, UK: Cambridge University Press, pp. 121–82.
Rozan, A., A. Stenger and M. Willinger (1997, 'Valeur de préservation de la qualité de l'eau souterraine: une comparaison entre usagers et non-usagers' [Valuation of preserving the quality of underground water: a comparison between users and non-users], *Cahiers d' économie et sociologie rurales*, **45**, 61–92.
Shephard, R.W. ([1953] 1970), *Theory of Cost and Production Functions*, Princeton, NJ: Princeton University Press.
Wunder, S., S. Engel and S. Pagiola (2008), 'Taking stock: a comparative analysis of payments for environmental services programs in developed and developing countries', *Ecological Economics*, **65**(4), 834–52.

24. The economic feasibility of the creation of the Jardines de la Reina National Park

Tamara Figueredo Martín, Fabián Pina-Amargós and Jorge Angulo-Valdés

24.1 INTRODUCTION

24.1.1 Benefits of Marine Protected Areas

Marine protected areas (MPAs) have been regarded as passive management strategies, aiming at preserving and recovering marine resources. Lately, MPAs have become one of the favourite tools for management, conservation and recovery of marine resources (Roberts et al., 2001; Green and Donnelly, 2003; Christie and White, 2007; Sale, 2008).

MPAs provide a large array of benefits to both human and non-human components of marine ecosystems (Cesar, 2000). These benefits might be divided into two large groups: one related to extractive activities, like fishing, and the other one related to non-extractive activities, such as tourism. Many authors have reported that after the establishment of MPAs, fishing has increased outside of them (McClanahan and Mangi, 2000; Roberts et al., 2001), while others have observed an increase in fish density, biomass and richness inside and outside the MPAs limits (Pina-Amargós, 2008). Others, on the other hand, have expressed that achievement of such benefits is not as fast as the advocates of MPAs state (Hatcher, 1997; Brown et al., 2001).

Socioeconomic studies play a fundamental role among stakeholders to implement, maintain and approve MPAs (Angulo-Valdés and Hatcher, 2010). That is why involving the main actors in those studies regarding the direct or indirect uses of resources and environmental services of MPAs becomes crucial (Bunce et al., 1999; Brown et al., 2001).

24.1.2 Environmental Goods and Services in Marine Environments

Around 200 papers on environmental functions or environmental goods and services (EGS) of coral reefs have been published. However, only about a few dozen are related to MPAs. In most cases, articles focus more on the evaluation of environmental goods of extractive use than on sociocultural services (Brander et al., 2007). Besides, only a few examples (i.e., ten papers) have been found on cost–benefit analyses, as a tool to assess the economic feasibility of MPA design and implementation (e.g., Cesar et al., 2000; Hodgson and Dixon, 2000; Angulo-Valdéz, 2005).

In Cuba, only a single economic evaluation of the EGS of a marine protected area has been published (Angulo-Valdéz, 2005). This research includes a cost–benefit analysis, but does not take into account the cash flow for a certain period of time. Angulo-Valdéz (2005) disregards the fact that protection effects are ecologically variable. For that reason, an economic analysis must be carried out from mid- to long term. No comparison with another area regarding the value of the EGS is included in that paper, as no other investigation on the issue exists in the country. For this reason, it is very important to conduct economic feasibility studies prior to the establishment of MPAs, because of their sensibility and high management costs.

In the past ten years, a vast amount of scientific information has been gathered from the Jardines de la Reina Archipelago, Cuba. This information mainly focuses on ecological issues and not on the environmental benefits of ecosystems. When this study was carried out, this area was pending approval as a National Park, so an analysis of the area's economic benefits was a must to determine the goods and services resulting from the conservation of its natural resources (new management tools), and consequently the economic feasibility of the area. This survey determines and values the EGS of the Jardines de la Reina National Park and assesses different management alternatives to enhance the policy-making process.

24.1.3 Environmental Goods and Services Valuation Studies

Literature on EGS is abundant in the world, most of it produced and published in the last 20 years. Many authors assert that ecosystem management cannot be successfully implemented without an appropriate evaluation of the EGS (Barbier, 2008). As has been said before, these studies are infrequent in Cuba (Gómez-País, 2002; Angulo-Valdés, 2005; Zequeira-Álvarez, 2008), so this kind of study is critical to support decision-making.

24.2 MATERIALS AND METHODS

24.2.1 Study Area

The Jardines de la Reina archipelago stretches from Cay Bretón to Cay Cabeza del Este, off the south central coast of Cuba (Figure 24.1). It is about 135 km long and comprises 661 keys. Since 1996, about 950 km^2 were declared a Zone Under Special Regime of Protection and Use (Resolution 562/96 of the Ministry of Fisheries), equivalent to Marine Reserves. The Marine Reserve of Jardines de la Reina is the largest in the Caribbean region and has been regarded as one of the most successful examples of this kind of area (Appeldoorn and Lindeman, 2003).

The only kind of commercial fishing allowed in the area is that of spiny lobster (*Panulirus argus*), while the main tourist activities are fishing (mainly tarpon and bonefish) and diving. Jardines de la Reina is sold as an exclusive tourist destination for such activities (Figueredo-Martín et al., 2010a, 2010b). Currently, the main users of the area are the Ministry of Food (formerly Ministry of Fisheries), and the Ministry of Tourism (MINTUR), the National Enterprise for the Protection of the Flora and the Fauna (land tenant) and the Coastal Ecosystems Research Center (CIEC) of the Ministry of Science, Technology and Environment (research, monitoring and area management). From now on, the study area will be referred as the Jardines de la Reina National Park (JRNP).

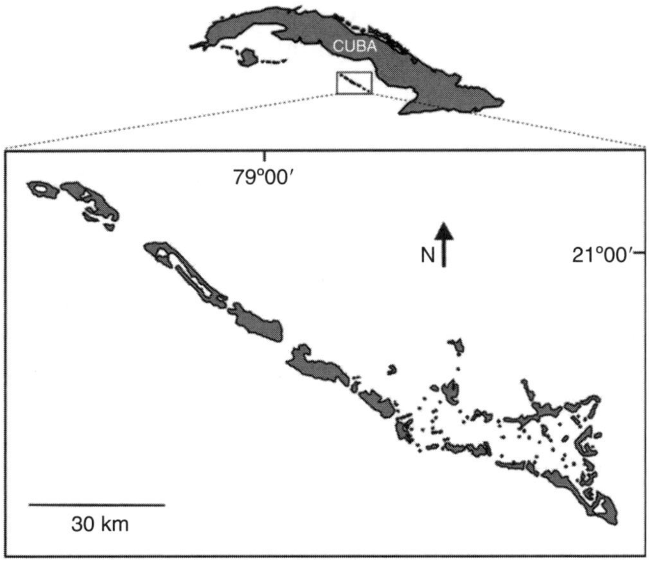

Figure 24.1 Study area, Jardines de la Reina Archipelago

24.2.2 Methods

Two management scenarios were created for the study area. In Scenario I, the Marine Reserve category has no formal administration and proper enforcement, and consequently has smaller protection costs. Scenario II, on the other hand, consists of a National Park, with associated administration and proper enforcement, therefore increasing protection costs. Scenario I is the current status of the area and Scenario II the future one. Cost–benefit methodology (Barzev and Ortiz-Frías, 1999; Barzev, 2002; Angulo-Valdéz, 2005) was used to evaluate both scenarios in order to provide technical information to the policy-makers. Scenario II aims for the implementation of new management and conservation tools and the improvement of tourism infrastructure in the JRNP. To do that, the net present value (NPV) (Zequeira-Álvarez, 2008) was used, through the projection of a cash flow (Barzev and Ortiz-Frías, 1999; Barzev, 2002) for 15 years, with a discount rate of 10 per cent. This technique is defined by the following equation:

$$NPV = \sum_{t=0}^{i} \frac{(B_t^F + B_t^{NF} + B_t^{ER} + B_t^C + B_t^{Pr} + B_t^{EP} + B_t^S) - C_t}{(1 + r)^t} - I_o$$

Where:

B_t^F = fishing benefits;
B_t^{NF} = non-fishing benefits;
B_t^{ER} = educational and research benefits;
B_t^C = cultural benefits;
B_t^{Pr} = process benefits;

B_t^{EP} = ecosystem and population benefits;
B_t^S = species benefits;
C_t = conservation costs
r = discount rate;
i = temporal horizon.

To calculate benefits for both scenarios, total economic value methods (Costanza et al., 1997; Gómez-País, 2002; Angulo-Valdés, 2005) were used, including a combination of market price valuation methods.

Various authors (Pina-Amargós et al., 2006; Pina-Amargós, 2008) were used to identify and select the EGS of the area. Then, EGS classification was done by adapting the methodology developed by different authors (Dixon and Sherman, 1990; Sobel, 1996; Costanza et al., 1997; Bohnsack, 1998; Cesar, 2000; National Academy of Science, 2001).

Different methods were used for the evaluation of the EGS: contingent valuation method (CVM) (Barzev, 2002; Hall et al., 2002; Brander et al., 2007; Asafu-Adjaye and Tapsuwan, 2008); travel cost method (TCM) (Barzev, 2002; Angulo-Valdés, 2005; Brander et al., 2007), transfer method (Angulo-Valdés, 2005; Brander et al., 2007) and the analysis of statistical and financial data from users of the area. Every calculation was made upon a year-period basis period and was converted to US dollars taking into account the official exchange rate in the country. An open and dichotomy format questionnaire was applied, using an interviewer, from May 2007 to July 2008 to a randomly selected sample of foreign visitors to the JRNP (Barzev and Ortiz-Frías, 1999; Babbie and Benaquisto, 2002).

Socioeconomic information on extracting fishing activities in the area was obtained from data gathered by Júcaro and Santa Cruz del Sur fishing cooperatives. Information about the tourist enterprise AZULMAR was collected through open interviews to workers and administrative officers and from the financial data of the enterprise. Semi-structured interviews were performed to complement the information on tourism incomes, particularly from tips.

All the values used were converted into US dollars (USD). Values referring to Cuban pesos (CUP) and Cuban convertible pesos (CUC) were converted using the current exchange rate of Cuba. Benefits were calculated upon a year-period basis.

Scenario I

Fishing benefits For Scenario I, the fishing economic benefit was first evaluated by taking into account finfish catches. Calculation of this benefit was made upon the basis of the estimated amount of catch reported for the area (Pina-Amargós, 2008) by spiny lobster boats, private fishers and the AZULMAR enterprise. The total annual catch of the area is 64 t: 60 t corresponding to illegal capture and 4 t to the AZULMAR catch. In the case of spiny lobster boats and private anglers, the price used to calculate the benefits was that of the underground market (22.00 CUP/kg). For the AZULMAR enterprise, the price used was 3.50 CUC/kg, corresponding to wholesale prices for enterprises (seafood market prices).

The second economic benefit of fishing comes from the catch of spiny lobster. For this benefit, the annual average catch of spiny lobster in the area (from 700 to 1000 t) was taken into account. To estimate this benefit, the annual mean catch (850 t) was multiplied by the average price of a ton (10000.00 CUC). This is the international market price. The total amount includes the benefits fishers directly receive.

The third economic fishing benefit is the spillover of species with high commercial value. This environmental service consists of fish movement from the reserve to areas outside the reserve limits, due to an abundance gradient (Pina-Amargós et al., 2010). To provide a value for this benefit, studies on fish movement were used. In this work, the exportation rate of species of high commercial value within 1 km was estimated for two months. This rate was applied to the number of individuals that move out of the Marine Reserve perimeter (225 km) and the average weight of fishes from the AZULMAR catch (Pina-Amargós, 2008), resulting in a biomass export of 86400 kg. The international market price was used in this case.

Non-fishing benefits Scuba diving and catch-and-release game fishing were identified in the case of non-fishing economic benefits. Income values from these activities were obtained from economic and financial data of the AZULMAR enterprise from 2001 to 2007.

Non-economic benefits For the non-economic benefits of scientific research, the most important projects in the zone were taken into account. Estimates were made upon the basis of the project average budget for a year.

Two environmental services ('Enhance aesthetic experiences and opportunities' and 'Enhance conservation appreciation' were chosen to value cultural non-economic benefits). For the former, a simple zonal travel cost approach that mostly uses secondary data (King and Mazzotta, 2001) and some simple data collected from visitors and from the AZULMAR enterprise were used. The average cost of an air ticket from and to the 16 countries that sent visitors to the area in 2007 was obtained from the travel agency related to AZULMAR. Data of the average package cost for anglers and divers in the area (accommodation, food, souvenirs, SCUBA diving or fishing gear, transfers, gratuities) were obtained from AZULMAR. The information about country of origin and number of visitors to the area corresponding to 2005, 2006 and 2007 was taken from the AZULMAR financial data. Values were calculated assuming the number of visitors to the area will not change, although AZULMAR predicts an increase of such figures.

In the case of 'Enhance conservation appreciation', fees for fishing licences were taken into account. This is the only charge to enjoy the environmental goods and services of the area ecosystems.

Abiotic components Four services were identified with abiotic components. The first one was that of mangroves as refuge and protection areas for fish and other species. To evaluate it, transfer methods were used based on a mangrove ecosystem study in Cuba (Gómez-País, 2002). In this study, the average value proposed for this service is 190.8 USD/ha/year. The mangrove area of the JRNP was estimated at 13500 ha.

The second benefit is the regulation of global climate, particularly through carbon sequestration. In this case, transfer methods were used as well, including services from mangrove and oceanic systems. In a study by Costanza et al. (1997), the authors proposed a value of 38.0 USD/ha/year for oceanic systems. The proposal of a local study (Gómez-

País, 2002) was taken into account for mangrove ecosystems. In this study, benefits from carbon sequestration were estimated in 310.5 USD/ha/year. The area of the JRNP oceanic system has been estimated at 200472 ha.

The third service is transformation, detoxification and the sequestration of pollutants. Costanza et al. (1997) proposed an index of 6.696 USD/ha/year for mangrove ecosystems.

The fourth service is nutrient cycles and the same methodology was followed. Costanza et al. (1997) proposed a 19.002 USD/ha/year index for seagrass contribution. For mangrove ecosystems, the authors proposed an index of 118.0 USD/ha/year.

Biotic components For ecosystem and population benefits, the maintenance of biodiversity was identified. A value was assigned on account of potential goods of unexploited mangroves. Gómez-País (2002) proposes a minimum economic estimate of 68.9 USD/ha/year for wood extraction and a minimum economic estimate of 90.8 USD/ha/year for apiculture.

Scenario II

Fishing benefits In Scenario II, the first economical benefit coincides with that of Scenario I (finfishing benefits). Estimates were made upon the basis of fish catch (4 t) by AZULMAR according to Pina-Amargós (2008). In this scenario we only took into account such a catch because surveillance and access to the protected area will make illegal catch impossible. The enterprise's wholesale price (3.50 CUC per kg) was used to estimate this benefit.

The second fishing economic benefit is that coming from spiny lobster catch. We took into account the fact that the conservation of the area and the new measures to protect the species (spiny lobster is the species of highest commercial value in the JRNP) will be maintained (1000 t). To estimate the international average market price, the amount of 10000.00 CUC/t was used. The total amount includes direct benefits to fishers.

The third fishing economic benefit for Scenario II is the spillover of species with high commercial value. Values for Scenario II were calculated taking into account future increase of spillover resulting from new protection measures in the area (catch decrease). Increase of spillover was calculated through the increase of fish abundance rate in the area, according to Alcolado et al. (2001) and Pina-Amargós et al. (2010). Assuming that mean weight of fishes will not change we calculated spillover increase by multiplying abundance rate by the exportation rate in a year.

Non-fishing benefits In the case of non-fishing economic benefits, SCUBA diving and catch-and-release game fishing were identified. The values of these are the same as in Scenario I.

Non-economic benefits For scientific research and non-economic education benefits, future opportunities from these services were taken into account. The proposal to use the future National Park as a field station for PhD and Master's dissertations and research by foreign students – common practice in protected areas of the world – was used in the case of educational services.

Two research projects to be undertaken in the zone in next few years were used in the case of research services. These projects will fill in the gaps of earlier investigations and research programmes, and will become one of the key priorities of the Coastal Ecosystems Research Center (the leading research institution in the JRNP).

To value cultural non-economic benefits, two environmental services (enhanced aesthetic experiences and opportunities and enhanced conservation appreciation) as in Scenario I, were assessed. In the case of the first service, the information obtained from CVM and TCM determining the average expenses of every visitor in the area were combined.

Through the CVM, the willingness of surveyed visitors to return to the area and the average number of times that they do so were obtained. This number is higher than the proposal from Scenario II, where we only took into account the increase of divers resulting from the increase of accommodation capacity of tourism facilities. Still, the number of visitors continues to be lower than the carrying capacity of the destination, at 320 divers per day (Pina-Amargós et al., 2006).

Abiotic components In the case of the abiotic components, four services were identified: mangroves as refuge and protection for fish and other species; the regulation of global climate, specifically through carbon sequestration; nutrient cycles and transformation; and detoxification and sequestration of pollutants. The value of these services was calculated as in Scenario I. The same process was followed for valuation of population and ecosystem benefits in Scenario II.

Biotic components For valuation of the loss of biotic components, four environmental services were identified: loss prevention of protected keystone and dominant species; loss prevention of vulnerable species; loss prevention of rare species; and protection of long-lived species. In all four cases, valuation was made according to sub-projects focusing on the study of species with certain features, to be undertaken in the next few years.

For protected keystone and dominant species, sub-projects for the study of the black urchin (*Diadema antillarum*) and hogfish (*Lachnolaimus maximus*) were included. The black urchin is the controller of algal abundance on coral reefs, and of great importance for ecosystem health. On the other hand, the hogfish is a fishery indicator, as it is one of the most susceptible species to overfishing due to its behaviour (Pina-Amargós, 2008).

Values of sub-projects of whale shark (*Rhincodon typus*) and Goliath grouper (*Epinephelus itajara*) were used to estimate loss prevention of vulnerable species. The whale shark is considered an endangered species and the Goliath grouper is classified as a critically endangered species according to the International Union for the Conservation of Nature (Hudson and Mace, 1996).

To prevent the loss of rare species, sub-projects for the study of insect species, subspecies of birds, and potential mollusc species, all of them new for Cuba, were proposed. Finally, for the protection of long-lived species, future investigations on tarpon (*Megalops atlanticus*) and sharks were taken into account. Tarpons live about 70 years (Andrews et al., 2001) and sharks about 50 years (Compagno, 1984).

Conservation cost

For the estimation of conservation cost, the classification of Dixon et al. (1995) and Pendleton (1995) was followed. They divided conservation costs into direct, indirect and opportunity costs. For Scenario I, tourism activities and research costs were included in the direct costs. For the opportunity cost, existing values calculated for Scenario II were taken, meaning that without conservation, all the environmental services could be lost. It is important to keep in mind that these values represent only a part of the truly existing value. In the case of indirect costs, none of the scenarios were included, because fishing is the activity affected with the establishment of the Zone Under Special Regimen of Use and Protection, and it is clear that the fishing effort has not decreased, but has moved to other fishing areas (sometimes even closer) that existed prior to restrictions. If there is any cost increase for some fishers it is not of high importance. For Scenario II, conservation costs and tourism costs were calculated as direct costs, as in Scenario I. Among conservation costs, management of resources, protection of resources, capacity building, environmental education, scientific research, administration, construction and maintenance, and cooperation and collaboration were taken into account. Every value of conservation cost was taken from the 2004 Executive Management Plan for the protected area (Plan de Manejo Operativo, 1995). In the case of opportunity cost, benefits not perceived by spiny lobster fishers and private fishers due to park regulations were taken into account.

24.3 RESULTS AND DISCUSSION

24.3.1 Economic Value of Scenario I and Scenario II

For the JRNP, 38 EGS were identified, 12 were evaluated for Scenario I and 17 for Scenario II (Table 24.1). Those 38 EGS identified support the fact that environmental functions of ecosystems are very diverse, indeed almost unlimited, because whenever we make an analysis of a certain natural resource, new functions or environmental services arise. The diversity of environmental functions strengthens the natural richness of the JRNP.

The number of benefits in scenarios with conservation and management tools is even higher. That is why conservation represents a desired scenario from the social and environmental points of view (De Groot, 2006). Besides, we proposed several alternatives for estimating existence of services, which could be used in further studies.

Figure 24.2 shows the values of the economic benefits for both scenarios. It is evident that finfishing benefits are the only ones higher in Scenario I. Scuba diving and sport fishing are the same for both scenarios. The benefits from the lobster catch and the spillover of high economic value fish are higher in Scenario II, particularly the lobster catch. Those are the benefits expected from a scenario with conservation.

Non-economic benefits are shown in Figure 24.3. In this case, only the 'Enhance conservation appreciation' benefit is the same for both scenarios. Scientific research, educational opportunities and 'Enhance aesthetic experiences and opportunities' are higher in Scenario II, particularly the latter.

Abiotic components (Figure 24.4) are similar in both scenarios, mainly because at the time the research was done, there was no evidence of degradation in ecosystem functions.

In Figure 24.5, biotic components from both scenarios are represented. Only in the case of maintenance of biodiversity protection, figures are the same for Scenarios I and

Table 24.1 Jardines de la Reina environmental goods and services

Total Economic Value – Benefits from the Jardines de la Reina Marine Reserve						
Use Value				No Use Value		
Direct use value Economic benefits		Indirect use value Non-economic benefits		Option value Abiotic components		Existence value Biotic components
Fishery	Non-fishery	Research and educational	Cultural	Process	Populations and ecosystems	Species
Finfishing	Tourism: scuba diving and sports fishing catch-and-release	Educational opportunities	Enhance aesthetic experiences and opportunities	Protection against coastal erosion	Preserve natural communities Composition and functioning	Protect keystone and dominant species
Lobster catch	Trekking	Research	Interpretative centre	Protection and refuge for migratory species and others	Ensure biodiversity protection	Prevent loss of vulnerable species
Spillover of high economic value fish	Bird watching		Enhance conservation appreciation	Global climate regulation. Carbon sequestration	Pharmaceutical potential	Sustain species presence and abundance
Raw materials for medicines	Whale sharks watching		Support cultural, religious and spirituals values	Avoid physical damage to habitats	Support community life	Prevent loss of rare species
Raw materials for souvenirs and handicrafts	Wildlife photography			Transformation, detoxification and sequestration of pollutants		Protection of long-lived species
Raw materials for construction				Allow for suitable nutrient cycles		Conservation of ecosystems
Live fishes and corals for aquarium				Sand production		Educative and recreational videos
				Exportation of organic materials and plankton to pelagic food network		Protect genetic resources and biodiversity

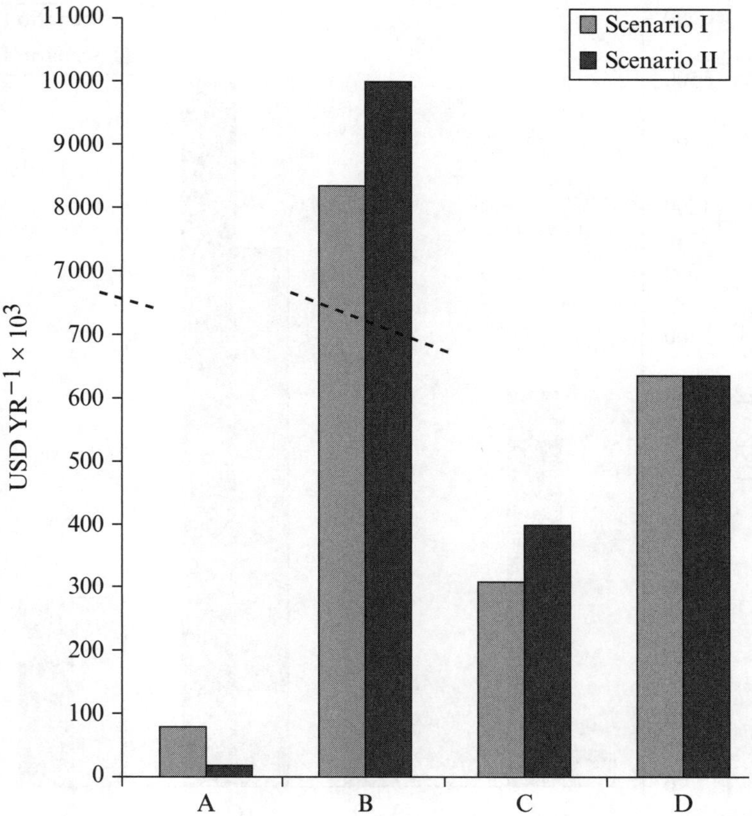

Figure 24.2 *Economic benefits from Scenario I and Scenario II. A: Finfishing. B: Lobster catch. C: Spillover of high economic value fish. D: Scuba diving and sport fishing catch-and-release*

II. Only in Scenario II were the other biotic components evaluated. Only with appropriate conservation measures, can loss prevention of protected keystone and dominant species, vulnerable species, rare species and long-lived species, be achieved. This is the reason why those environmental services were not taken into account for Scenario I.

Current benefits from Jardines de la Reina in Scenario I (Figure 24.6) and Scenario II (Figure 24.7), show that Scenario II has a total economic value of 59 532.6 USD yr^{-1} × 10^3, better than that of Scenario I of 55 357.2 USD yr^{-1} × 10^3. When comparing the benefits from both scenarios, fishing and cultural benefits are more prominent in Scenario II.

Conservation costs, following the classification by Dixon et al. (1995) and Pendleton (1995) were divided into direct, indirect and opportunity costs. For Scenario I, these costs were of 629.8 USD yr^{-1} × 10^3 (Table 24.2). For Scenario II (Table 24.3), conservation and opportunity costs were of 770.17 USD yr^{-1} × 10^3.

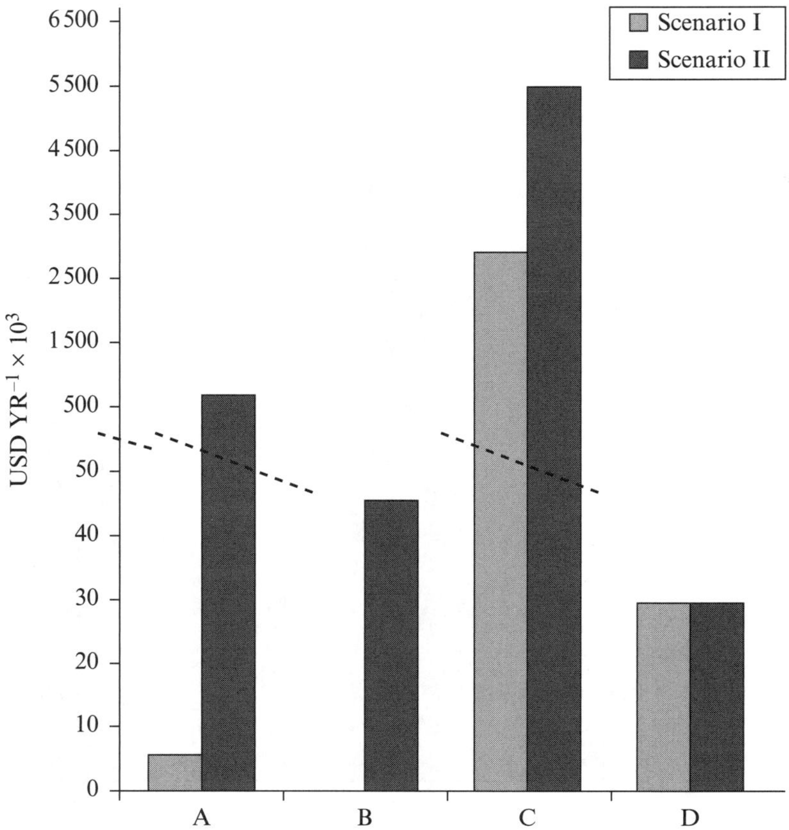

Figure 24.3 *Non-economic benefits from Scenario I and Scenario II. A: Scientific research. B: educational opportunities (just valued for Scenario II). C: Enhance aesthetic experiences and opportunities. D: Enhance conservation appreciation*

24.3.2 Financial Analysis

The cash flow of Scenario I yielded a net present value of 467.9 USD yr^{-1} × 10^3, and Scenario II 501.6 USD yr^{-1} × 10^3 (Figure 24.8). In both scenarios, NPV is above zero, so both are feasible from the financial point of view. Nonetheless, the differences of 4.1 USD yr^{-1} × 10^6 between benefit–cost and 33.8 USD yr^{-1} × 10^6 between NPV of Scenarios I and II indicate that under any circumstance, the establishment of a National Park in the Jardines de la Reina area is economically feasible, with all protection and conservation tools and the diversification of activities (tourism and research).

Results obtained are only a part of the total economic value of the Jardines de la Reina ecosystem, because value was only assigned to relevant EGS. Besides, the value assigned to each benefit only represents part of the benefit value.

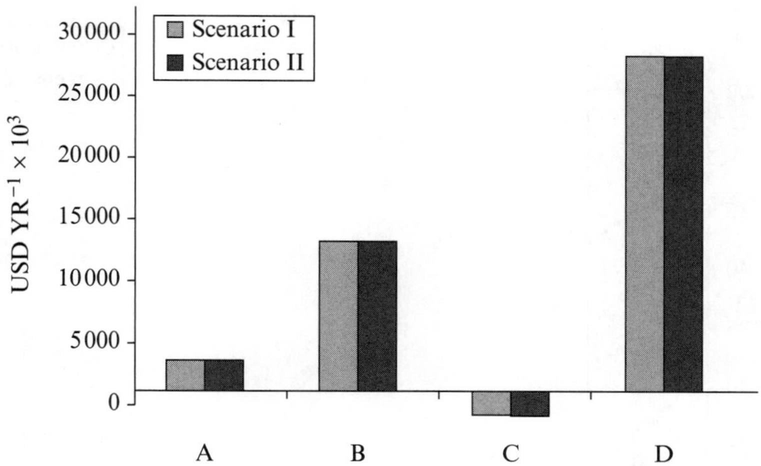

Figure 24.4 *Abiotic components from Scenario I and Scenario II. A: Protection and refuge for migratory species and others. B: Global climate regulation through carbon sequestration. C: Transformation, detoxification and sequestration of pollutants. D: Allow for suitable nutrient cycles*

24.3.3 Relation Between Economic and Non-economic Values

The sum of the biotic and abiotic components value in Scenario I represents 77 per cent of the total economic value. These results strengthen the fact that no-use value of EGS exceeds use value. Likewise, figures obtained from Scenario II confirm this, because no-use value represents 72 per cent of the total economic value. It is important to highlight that the number of benefits identified or calculated in both categories is not the same. Nonetheless, the difference between no-use value and use value of EGS is so high that the former will never be higher than the latter in any scenario. It is relevant that in Scenario II, use value exceeded by 4106.8 USD yr^{-1} × 10^3 that of Scenario I. This means that direct benefits of the protected area could be higher, with protecting tools and appropriate management decisions.

24.4 CONCLUSIONS

The main EGS of Jardines de la Reina were identified, evaluating those with higher market value and those with enough information available. In both scenarios, NPV and cost–benefit relation are above zero, so financially speaking, both are feasible. Nevertheless, the establishment of a National Park in the Jardines de la Reina area is more desirable. In both scenarios, it is evident that the non-use value of the EGS is higher than their use value.

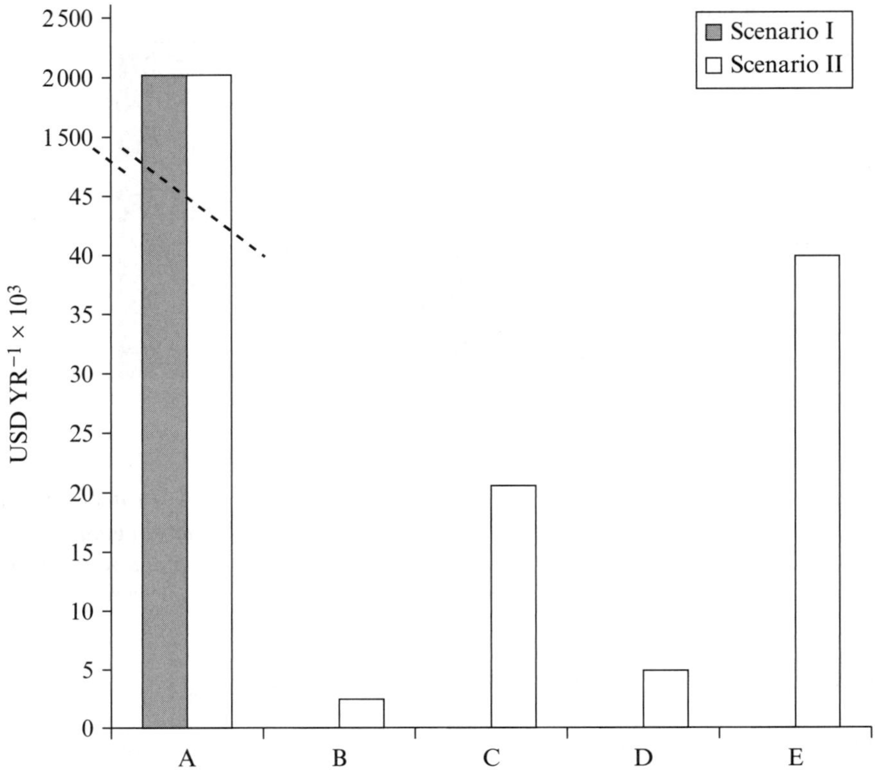

Figure 24.5 *Biotic components from Scenario I and Scenario II. A: Ensure biodiversity protection. B: Prevent loss of keystone and dominant species (just valued for Scenario II). C: Prevent loss of vulnerable species (just valued for Scenario II). D: Prevent loss of rare species (just valued for Scenario II). E: Protection of long-lived species (just valued for Scenario II)*

ACKNOWLEDGEMENTS

The authors are thankful to the staff of AZULMAR for logistical support on Jardines de la Reina, especially to Giuseppe Omegna (Pepe), Noel, Yuri and 'los primos', to WWF Canada for funding this research and to Idea Wild for providing us with equipment. We thank the Ministry of Science, Technology and Environment for funds and logistics. A million thanks to CIEC's staff for the institutional support to do our research in Jardines de la Reina, especially to V.O. Rodríguez Cárdenas for providing language help and to H. Salvat for helping with artwork. We are also grateful to C. Costello for his wise advice when reading the article and Matt for providing language help. We appreciate the reviews of Paulo A.L.D. Nunes, P. Kumar and T. Dedeurwaerdere for their critical comments that have improved this manuscript.

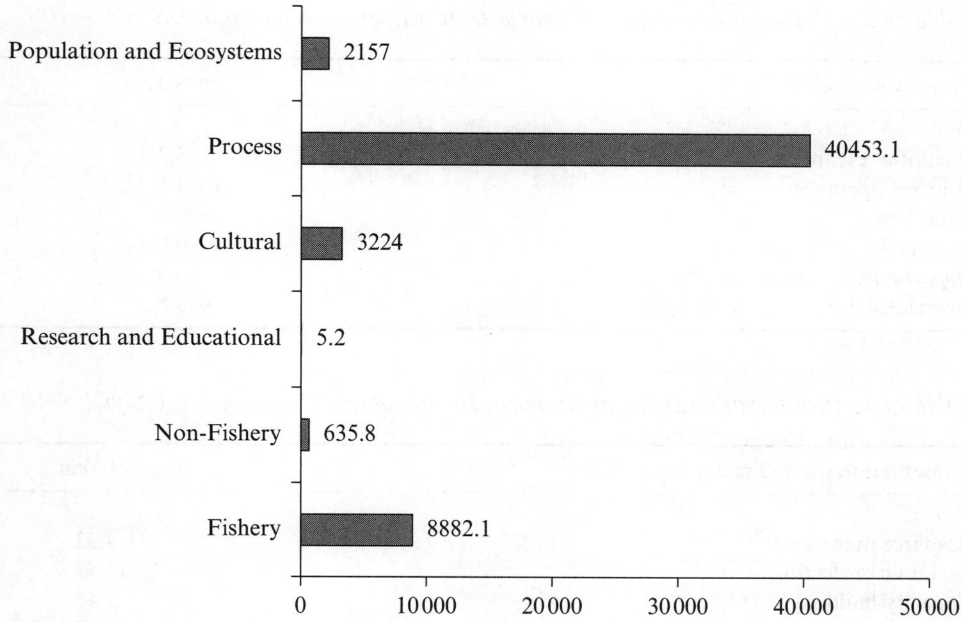

Figure 24.6 Current benefits from Jardines de la Reina, Scenario I (in USD $yr^{-1} \times 10^3$*)*

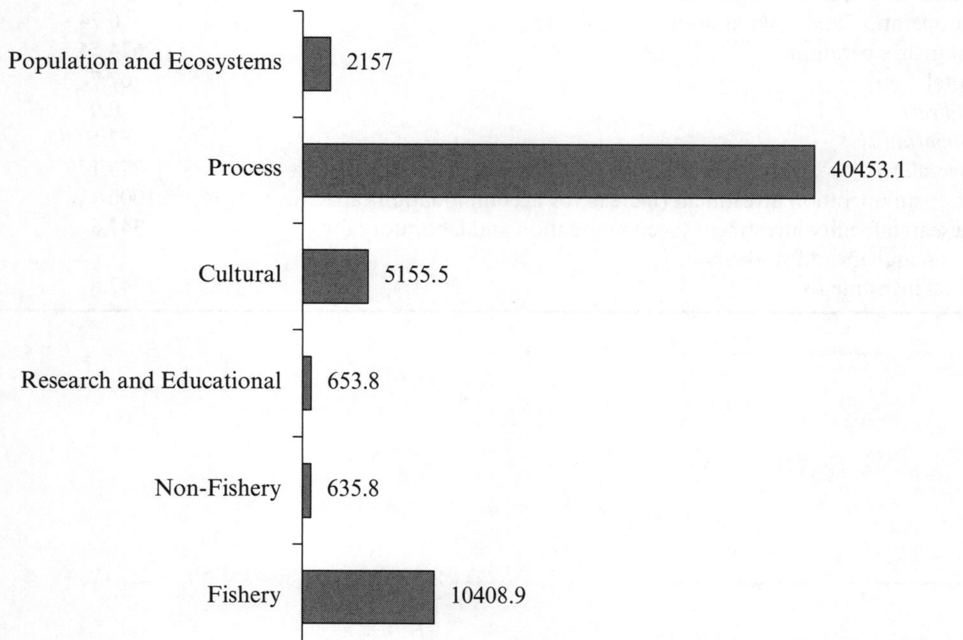

Figure 24.7 Current benefits from Jardines de la Reina, Scenario II (in USD $yr^{-1} \times 10^3$*)*

Table 24.2 Conservation cost for Scenario I. All values are refereed to US $yr^{-1} \times 10^3$

Conservation Cost	Figures/Year
Direct	
Scientific research	5.2
Tourism operation	624.5
Total	629.8
Indirect	0.0
Opportunity	0.0
Overall total	629.8

Table 24.3 Conservation costs for Scenario II. All values are refereed to US $yr^{-1} \times 10^3$

Conservation Cost Direct	Figures/Year
Direct	
Resource management	4.32
Resource protection	11.47
Capacity building	3.48
Environmental education	3.12
Scientific research	6.84
Administration	40.58
Construction and maintenance	12.57
Cooperation and collaboration	0.74
Tourism operation	624.54
Total	707.7
Indirect	0.0
Opportunity	62.5
Overall total	770.17
Tourism operation investment (increase of accommodation capacities)	1000
Research facility investment (accommodation and laboratory for researchers and students)	347.8
Total investments	1347.8

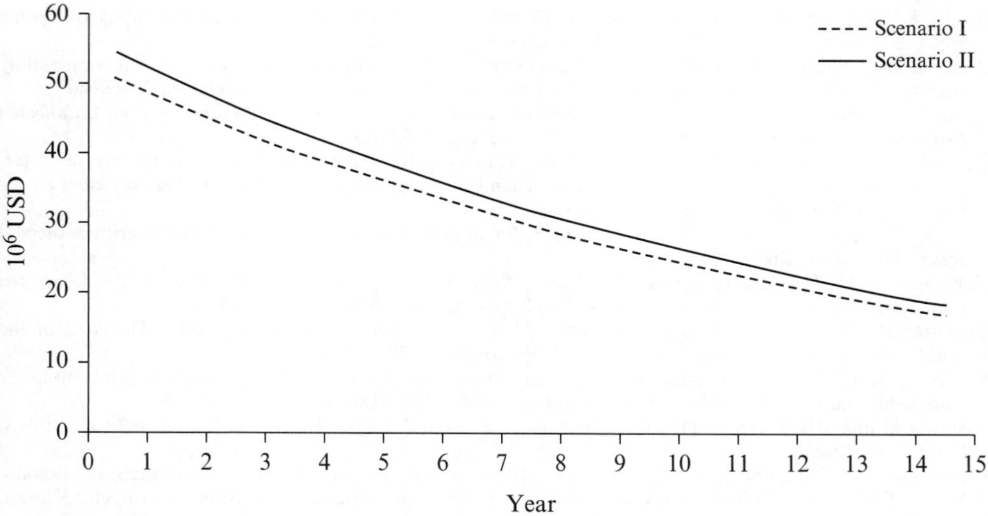

Figure 24.8 Discount cash flow for both scenarios. Calculations were made with a discount rate of 10%

REFERENCES

Alcolado, P.M., E. De la Guardia, F. Pina-Amargós, K. Cantelar, S. González-Ferrer and H. Caballero et al. (2001), 'Estado de salud de los arrecifes coralinos del Archipiélago Jardines de la Reina (SE de Cuba)' [Health status of coral reef of Jardines de la Reina Archipelago (SE of Cuba)], technical report of the project Assessment of a Relevant Ecological Area to be Included as Part of the National System of Protected Areas.

Andrews, A.H., E.J. Burton, K.H. Coale, G.M. Cailliet and R.E. Crabtree (2001), 'Radiometric age validation of Atlantic tarpon, *Megalops atlanticus*', *Fishery Bulletin*, **99**(3), 389–98.

Angulo-Valdés, J. (2005), 'Effectiveness of a Cuban marine protected area in meeting multiple management objectives', submitted in partial fulfilment of the requirements for the degree of Doctor of Philosophy, Dalhousie University Halifax, Nova Scotia, Canada.

Angulo-Valdés, J. and B. Hatcher (2010), 'A new typology of benefits derived from marine protected areas', *Marine Policy*, **34**(3), 635–44.

Appeldoorn, R.S. and K.C. Lindeman (2003), 'A Caribbean-wide survey of marine reserves: spatial coverage and attributes of effectiveness', *Gulf and Caribbean Research*, **14**(2), 139–54.

Asafu-Adjaye, J. and S. Tapsuwan (2008), 'A contingent valuation study of SCUBA diving benefits: case study in Mu Ko Similan Natural Park, Thailand', *Tourism Management*, **29**(6), 1122–30.

Babbie, E. and L. Benaquisto (2002), *The Basics of Social Research*, Canadian Edition, Toronto: Thomson Canada Limited.

Barbier, E.B., E.W. Koch, B.R. Silliman, S.D. Hacker, E. Wolanski and J. Primavera et al. (2008), 'Coastal ecosystem-based management with nonlinear ecological functions and values', *Science*, **319**(5861), 321–3.

Barzev, R. (2002), 'Guía metodológica de valoración económica de bienes, servicios e impactos ambientales' [Methodological guidance for economic valuation of environmental goods and services and environmental impact assessment], technical report of the Project for the Consolidation of Mesoamerican Biological Corridor, Technical Series No. 04.

Barzev, R. and J. Ortiz-Frías (1999), 'Valoración Económica del Potencial Turístico de las Reservas Naturales Miraflor y Moropotente' [Economic valuation of touristic potential of the Natural Reserves Miraflores and Moropotente], technical report of the Project PANIF, MARENA.

Bohnsack, J.A. (1998), 'Application of marine reserves to reef fisheries management', *Australian Journal of Ecology*, **23**(3), 298–304.

Brander, L.M., P. Van Beukering and H.S.J. Cesar (2007), 'The recreational value of coral reefs: a meta-analysis', *Ecological Economics*, **63**(1), 209–18.

Brown, K., W.N. Adger, E. Tompkins, P. Bacon, D. Shim and K. Young (2001), 'Trade-off analysis for marine protected area management', *Ecological Economics*, **37**(3), 417–34.
Bunce, L., K. Gustavson, J. Williams and M. Miller (1999), 'The human side of reef management: a case study analysis of the socioeconomic framework of Montego Bay Marine Park', *Coral Reefs*, **18**(4), 369–80.
Cesar, H.S.J. (2000), 'Coral reefs; their functions, threats and economic value', in H.S.J. Cesar (ed.), *Collected Essays on the Economics of Coral Reefs*, Kalmar, Sweden: CORDIO, pp. 14–39.
Cesar, H.S.J., M.C. Öhman, P. Espuet and M. Honkanen (2000), 'Economic valuation of an integrated terrestrial and marine protected area: Jamaica's Portland Bight', in H.S.J. Cesar (ed.), *Collected Essays on the Economics of Coral Reefs*, Kalmar, Sweden: CORDIO.
Christie, P. and A.T. White (2007), 'Best practices for improved governance of coral reef marine protected areas', *Coral Reefs*, **26**(4), 1047–56.
Compagno, L.J.V. (1984), *FAO Species Catalogue. Vol. 4. Sharks of the World. An Annotated and Illustrated Catalogue of Shark Species Known to Date. Part 2. Carcharhiniformes*, FAO, pp. 459–70.
Costanza, R., R. d'Arge, R. de Groot, S. Farber, M. Grasso and B. Hannon et al. (1997), 'The value of the world's ecosystem services and natural capital', *Nature*, **387**(6630), 253–60.
De Groot, R. (2006), 'Function-analysis and valuation as a tool to assess land use conflicts in planning for sustainable, multi-functional landscapes', *Landscape and Urban Planning*, **75**(3–4), 175–86.
Dixon, J.A. and P.B. Sherman (1990), *Economics of Protected Areas: A New Look at Benefits and Costs*, London: Earthscan.
Dixon, J.A., L.F. Scura and T. Vant-Hof (1995), 'Ecology and microeconomics as joint products: the Bonaire Marine Park in the Caribbean', in C.A. Perrings et al. (eds), *Biodiversity Conservation*, Dordrecht: Kluwer, pp. 127–45.
Figueredo-Martín, T., F. Pina-Amargós, J. Angulo-Valdés and R. Gómez-Fernández (2010a), 'Buceo contemplativo en Jardines de la Reina, Cuba: caracterización y percepción sobre el estado de conservación del área' [SCUBA diving in Jardines de la Reina: characteristics and perception about the conservation of the area], *Revista de Investigaciones Marinas*, **31**(1), 23–32.
Figueredo-Martín, T., F. Pina-Amargós, J. Angulo-Valdés and R. Gómez-Fernández (2010b), 'Pesca recreativa en Jardines de la Reina, Cuba: caracterización y percepción sobre el estado de conservación del área' [Recreational fisheries in Jardines de la Reina: characteristics and perception about the conservation of the area], *Revista de Investigaciones Marinas*, **31**(2), 141–8.
Green, E. and R. Donnelly (2003), 'Recreational SCUBA diving in Caribbean marine protected areas: do the users pay?' *Ambio*, **32**(2), 140–44.
Gómez-País, G. (2002), 'Análisis económico de las funciones ambientales del manglar en el Ecosistema Sabana Camagüey' [The economic analysis of environmental roles of the mangroves of Sabana Camagüey ecosystem], technical report of the project Priority Actions to Consolidate the Biodiversity Protection in Sabana Camagüey Ecosystem.
Hall, D., J. Hall and S. Murria (2002), 'Contingent valuation of marine protected areas: southern California rocky intertidal ecosystems', *Coastal Management*, **30**(2), 47–65.
Hatcher, B.G. (1997), 'Can marine protected areas optimize fishery production and biodiversity preservation in the same ecosystem?' in *Proceedings of the 50th Annual Gulf and Caribbean Fisheries Institute*, Merida, Mexico.
Hodgson, G. and J.A. Dixon (2000), 'El Nido revisited: ecotourism, logging and fisheries', in H.S.J. Cesar (ed.), *Collected Essays on the Economics of Coral Reefs*, Kalmar, Sweden: CORDIO.
Hudson, E. and G. Mace (1996), 'Marine fish and the IUCN Red List of threatened animals', in E. Hudson and G. Mace (eds), *Report of the Workshop Held in Collaboration with WWF and IUCN at the Zoological Society of London*, 29 April 29–1 May, Institute of Zoology, Regent's Park, London.
King, D.M. and M. Mazzotta (2000), 'Ecosystem valuation', website, accessed August 2001 at http://www.ecosystemvaluation.org.
McClanahan, T.R. and S. Mangi (2000), 'Spillover of exploitable fishes from a marine park and its effect on the adjacent fishery', *Ecological Applications*, **10**(6), 1792–805.
National Academy of Sciences (NAS) (2001), *Marine Protected Areas: Tools for Sustaining Ocean Ecosystems*, Washington, DC: National Academies Press.
Pendleton, L.H. (1995), 'Valuing coral reef protection', *Ocean and Coastal Management*, **26**(2), 119–31.
Pina-Amargós, F. (2008), 'Efectividad de la Reserva Marina de Jardines de la Reina para la conservación de la ictiofauna' [Effectiveness of Jardines de la Reina Marine Reserve on the conservation of the ichthyofauna], submitted in partial fulfilment of the requirements for the degree of Doctor of Philosophy, Havana University, Cuba.
Pina Amargós, F., L. Clero Alonso, F. Martín Blanco, L. Hernández Fernández, W. Acosta de la Red and L. Cabreja Ávila et al. (2006), 'Marine biota of the Jardines de la Reina ecosystem', in *Coastal Ecosystems: Biodiversity and Management of Natural Resources, Compilation for XV Anniversary of CIEC, 1991–2006*, pp. 396–449.

Pina-Amargós, F., G. González Sansón, A. Jiménez del Castillo, A. Zayas Fernández, F. Martín Blanco and W. Acosta de la Red (2010), 'An experiment of fish spillover from a Marine Reserve in Cuba', *Environmental Biology of Fishes*, **87**(4), 363–72.
Plan de Manejo Operativo (2004), Parque Nacional Jardines de la Reina.
Roberts, C.M., J.A. Bohnsack, F. Gell, J.P. Hawkins and R. Goodridge (2001), 'Effects of marine reserves on adjacent fisheries', *Science*, **294**(5548), 1920–23.
Sale, P.F. (2008), 'Management of coral reefs: where we have gone wrong and what we can do about it', *Marine Pollution Bulletin*, **56**, 805–9.
Sobel, J. (1996), 'Marine reserves: necessary tools for biodiversity conservation?' *Global Biodiversity*, **6**(1), 8–18.
Zequeira-Álvarez, M.E. (2008), 'Instrumento económico y metodológico para la gestión ambiental de humedales naturales cubanos con importancia internacional' [Economic and methodological tool for environmental management of Cuba's natural wetlands of international importance], submitted in partial fulfilment of the requirements for the degree of Doctor of Philosophy, Camagüey University, Camagüey, Cuba.

25. Valuation of ecosystem services provided by man-made wetlands

Nico B.P. Polman, Arianne T. de Blaeij, C. Martijn van der Heide, Vincent Linderhof and Stijn Reinhard

25.1 INTRODUCTION

Man-made wetlands are defined as commercial wetlands created by man and managed by profit-maximizing entrepreneurs. In the eastern part of the Netherlands near Haaksbergen a wetland of about 3 ha has been constructed by an entrepreneur. This wetland, called Waterpark *het Lankheet*, is constructed as a research project (Meerburg et al., 2010). It provides at least five different ecosystem services: (1) biomass production of reed (including capturing CO_2), (2) water quality improvement of surface water by growing reed, (3) improvement of biodiversity in the surrounding area by preventing desiccation problems, (4) water storage in times of flood risk, and (5) recreation. Also, the Dutch government recognizes that wetlands form a promising innovative option to improve water quality in combination with other functions and prefers voluntary types of governance structures (Minister of Agriculture, Nature and Food Quality, 2008).

The main objective of this chapter is to investigate the economic potential of man-made wetlands. In the ecological economics literature, most wetland valuation studies refer to natural wetlands, managed by non-profit organizations or governments. In the case of commercial (man-made) wetlands, the entrepreneur focuses on ecosystem services that generate the highest private revenues. Ecosystem services such as biomass production and recreation can be characterized as private services that potentially can be sold on a market. If a market exists, these private services will have a market price that coordinates supply and demand. The revenues of these private services are included in the profit-maximization function of the entrepreneur. Currently, the wetland entrepreneur is not rewarded for the provision of a combination of the ecosystem services, or for the private or public services. For instance, there is no demand for biomass reed.

To make commercial exploitation of wetlands feasible, a payment for ecosystem services (PES) mechanism as a governance structure is necessary. The basic reasoning underlying PES is that users of these services have a measurable value or willingness to pay (WTP) for those services (Engel et al., 2008; Jack et al., 2008; Daily et al., 2009). If demand for a particular ecosystem service is absent, the associated value is zero and PES arrangements will not emerge. For instance, if the quality of surface water is already sufficient according to society, extra water treatment service is not preferred and does not have any value. The benefits of the different types of wetland services are the values perceived by its consumers (Brown et al., 2007). Wetland valuation studies and meta-analyses enable the estimation of the value of (natural) wetlands, and the services these wetlands provide (see Brouwer et al., 1999; Woodward and Wui, 2001; Brander et al., 2006; Ghermandi et al., 2010). Ghermandi et al. (2010) conducted a meta-analysis on wetland value estimates to examine the value of ecosystem services provided by man-made wetlands. They conclude that biodiversity enhancement, water quality improvement and flood control are highly valued services.

This Chapter contributes to literature because it estimates values of ecosystem services derived from man-made wetland as a basis for economic governance. Following Dixit (2008, p. 1), economic governance 'consists of the processes that support economic activity and economic transactions by protecting property rights, enforcing contracts, and taking collective action to provide appropriate physical and organizational infrastructure' (see also Dixit, 2004, 2009). We focus on (potentially) commercial man-made wetlands to be located on Dutch agricultural land. We study the preferences for man-made wetland and its ecosystem services with a representative survey of a sample of the Dutch population. We use the contingent valuation method (CVM) to determine the total willingness to pay for the bundle of ecosystem services derived from man-made wetlands as a starting point for governance arrangements. The separate values for the different ecosystem services are obtained by the analytical hierarchical process (AHP) method to derive separate values for the different ecosystem services as a starting point for governance arrangements.

The chapter is organized as follows. The methodology section addresses the valuation method and the method to decompose the value over different ecosystem services. Then we present our valuation results of man-made wetlands, which is followed by a discussion on the estimated values and consequences for wetland governance.

25.2 METHODS AND DATA

We apply a CVM approach to determine the value of newly implemented man-made wetlands. Choice experiments (CE) are considered to be a more sophisticated method (e.g., Carlsson et al., 2003; Birol and Cox, 2007) but choice experiments use hypothetical changes in situations rather than examples from real-life situations like *Lankheet* Waterpark. Moreover, a number of the ecosystem services derived from man-made wetland are highly correlated, which makes it difficult to distinguish between individual services (Brouwer et al., 1999). For example, biomass production – when properly managed – contributes to improvements in surface water quality, while during flood events surface water quality is affected negatively due to a higher run-off of nutrients from water logging. We are particularly interested in the value of the separate services, and because those correlations make it difficult to separate these services a choice experiment is an inappropriate method for our analysis. Therefore, we apply CVM to obtain an aggregate WTP value for man-made wetlands, and subsequently the multi-criteria technique analytical hierarchical process (AHP) to decompose the aggregated WTP into values for the separate ecosystem services.

BOX 25.1 CVM VALUATION QUESTION

Do you want to pay for the construction of man-made wetlands, for example to compensate farmers?
1: yes
2: no
3: maybe

***[if yes or maybe] What is your extra willingness to pay to make the construction of man-made wetlands possible per year?**

indicate WTP by moving on a bar from €0 to €250, with an option more than €250, namely €. . . . . . .

The Value of Man-made Wetlands

CVM is a survey-based valuation technique used to determine the values of ecosystem services that do not have a market value (Barbier, 1997; Birol et al., 2006). The method is based on the economic welfare theory. It is frequently used to estimate the economic value of natural resources such as (man-made) wetlands. We asked respondents how much they are willing to pay for man-made wetlands (e.g., Bateman et al., 2003) taking into account the different services (Box 25.1).

As a very large share of the Dutch population is not familiar with man-made (commercially exploited) wetlands, its concept was described in detail in the questionnaire. Based on the explanation of the concept, respondents were asked how they feel about man-made wetlands (five categories ranging from negative to positive). Since the man-made wetlands provide services additional to current water management, we asked respondents about their expenditures on water services (drinking water, wastewater treatment, sewage system and water board levies). This question is also used to put their WTP bids into perspective; in this way we obtain more accurate WTP estimates from respondents. Since we used an internet questionnaire, respondents could indicate their WTP value by moving a bar from left (€0) to right (€250). The exact amount was shown on the screen while moving the bar. The WTP question elicits the maximum WTP indicated by the respondents for the bundle of ecosystem services provided by man-made wetlands.

Navrud (2002) indicated that the payment vehicle could differ according to different countries with heterogeneous institutional settings, cultures and preferences. Bergstrom et al. (2004) mentioned two ways of financing ecosystem services: financing with extra or marginal WTP and financing through reallocation of existing revenues. Usually, respondents that prefer reallocations of budget are classified as protest bidders, that is, respondents that object to payment questions. Protest bidders are usually excluded from the analysis, but in our study protest bidders are included because they might provide us with information on the institutional setting, which is important to use when interpreting the results.

In fact, we distinguish two types of protest bidders. In order to identify both protest bidders, we added a follow-up question for respondents with a zero bid to identify 'real' protest bidders (i.e., respondents that object to the concept of man-made wetlands) and 'budget allocation' bidders (i.e., respondents that do not fundamentally object, but object to the payment vehicle resulting in a zero WTP) (e.g., Hanley et al., 2002).

The Value of Man-made Wetland Services

We apply AHP to decompose the man-made wetland WTP value into separate values of the five different services. The AHP methodology was developed by Saaty in the late 1970s (Saaty, 1977, 1980) as a technique to support multi-criteria decision-making in marketing. The cognitive burden of respondents is reduced because AHP always uses two clear service comparisons (Kallas et al., 2007; Moran et al., 2007). The AHP method consists of a series of pairwise comparisons between the different services to assess the relative importance of each service on a nine-point scale (Kallas et al., 2007). From the choice questions and their nine-point scale, we derive the Saaty matrix A. The element a_{jk} of this matrix reflects the score of the importance of service j over service k. The Saaty matrix can be derived for all respondents. If the service j is extremely more important than service k, then $a_{jk} = 9$ and consequently $a_{kj} = 1/9$. If both services j and k are equally important, $a_{jk} = a_{kj} = 1$. By definition, $a_{jj} = 1$.

Following Kallas et al. (2007), we apply the practical approach to calculate the weights of the separate services for each respondent based on the Saaty matrix. The weights for the services w_{ij} are calculated based on Equation (25.1) for $k = 1, 2, 3, 4, 5$. The weights are calculated with a geometric function, because the scores of the elements of the Saaty matrix are non-linear:

$$w_{ij} = \left(\prod_{k} a_{ijk} \right)^{1/5} \tag{25.1}$$

The indices j and k reflect the different services, and i the respondents. For each individual, the service with the highest weight is the individual's most important service. Note that an individual might have similar scores for more than one service. In order to calculate the weights of the services (w_j) for the sample or sub-samples, we calculate the geometric mean over the (sub-)sample:

$$w_{j} = \left(\prod_{i} w_{ij} \right)^{1/N} \tag{25.2}$$

with N the number of respondents. From the weights of the ecosystem services (j), we derive the shares of the different services; see Table 25.1.

Data Description

We conducted an internet survey on the WTP for man-made wetland in the Netherlands. The data were collected in December 2008. Respondents were members of an internet panel of a specialized bureau (TNS-NIPO), and they were paid for filling out a questionnaire. Many socio-demographic characteristics of the respondents were known in advance, which made it possible to construct a representative sample based on education, age and gender. Furthermore, we oversampled in the Haaksbergen region (in

Table 25.1 Five separate ecosystem services of a man-made wetland

Services		Service as Described in the Survey
F_1	Water quality improvement	Water treatment in terms of nutrient emission reduction with helophyte-filters
F_2	Biomass production	Production of renewable energy from biomass
F_3	Desiccation prevention	Solving desiccation problems in nature conservation areas
F_4	Flood protection	Water storage to avoid flooding
F_5	Recreation	Recreation such as walking, cycling, picnicking, jogging, playing etc.

Table 25.2 Summary statistics for the sample surveys

Variable	Rest of the Netherlands (N = 826)		Haaksbergen Region (N = 134)	
	Mean	S.D.	Mean	S.D.
Wetland known concept	0.06	0.23	0.60	0.49
Income (*1000€)[a]	44.37	26.63	42.76	27.39
Household size	2.88	1.18	2.87	1.22
West	0.39	0.49		
North	0.09	0.28		
East	0.23	0.30		
South	0.29	0.45		
Age of respondent	51.45	14.47	48.28	13.65
Low education class	0.22	0.42	0.19	0.40
Sex (male = 1)	0.50	0.50	0.49	0.50
Water treatment most important	0.41	0.49	0.50	0.50
Biomass reed most important	0.05	0.21	0.05	0.22
Prevention of desiccation most important	0.06	0.24	0.08	0.28
Preventing waterlogging most important	0.40	0.49	0.30	0.46
Recreation most important	0.06	0.23	0.06	0.24
Necessary to compensate entrepreneur	0.79	0.41	0.80	0.40
Memberships of environmental organization	0.53	0.80	0.54	0.86
Recreation in nature	0.71	0.45	0.78	0.41
Preference for more nature in the Netherlands	0.66	0.48	0.72	0.45

Note: a. Household income is not available for all respondents. Household income is known for 656 respondents in the Rest of the Netherlands sample and 112 respondents in the Haaksbergen sample.

the vicinity of the *Lankheet*), so that respondents who were likely to be familiar with the *Lankheet* wetland can be analysed separately. The Haaksbergen sub-sample consisted of 134 respondents and the sub-sample of the Rest of the Netherlands (RoN) of 826 respondents. Table 25.2 reports summary statistics of these two samples. In the Netherlands, 48 per cent of the population lives in the west, 10 per cent in the north and 21 per cent in the south. The remaining 21 per cent lives in the eastern part of the country.

In our analysis, we test for the influence of the socio-demographic variables on the WTP. A priori, we do not expect any particular influence of the variables gender and age on the magnitude of their WTP bid. Most likely, income will have a positive relation to the decision whether to pay, and the magnitude of the bid. Other characteristics that can positively influence the decision to pay are whether respondents are a member of an environmental organization, and their preference with respect to natural environments in the Netherlands. Half of both sub-samples indicated that they are a member of an environmental organization, which was comparable to earlier findings for the Netherlands (e.g., Haile and Slangen, 2009). Furthermore, 71 per cent and 78 per cent of the respondents in the RoN and the Haaksbergen samples respectively conducts recreational activities in natural areas, while 66 per cent and 72 per cent of the respondents respectively reported the importance of developing more natural areas in the Netherlands.

25.3 RESULTS

Social Opinion on Man-made Wetlands and Contingent Valuation

In the Haaksbergen sample 60 per cent of the respondents was familiar with a man-made wetland, while in the RoN sample this was only 6 per cent. The results show that water treatment (41 per cent for RoN and 50 per cent for the Haaksbergen sample) and preventing waterlogging (40 per cent for RoN and 30 per cent for the Haaksbergen sample) are the most important services.

About 80 per cent of all the respondents agreed on the necessity to compensate farmers for their loss in income. About 72 per cent of the total sample is positive towards constructing commercially exploited man-made wetlands. However, more than half of the respondents in both samples is a protest bidder, that is, a respondent who refuses to pay any additional amount for a man-made wetland: 51.5 per cent and 50.4 per cent in the Haaksbergen and RoN sample respectively. For the RoN sample 79 per cent (328 out of 417) of the protest bidders would prefer the reallocation of the current budget (see Table 25.3). For the Haaksbergen sample 70 per cent (48 out of 69) of the protest bidders prefer budget reallocation. Those respondents were included in the analysis.

About 40 per cent of the respondents in the RoN sample indicated that man-made wetlands should be financed with existing revenues of taxes on water services prices. The share of zero WTP bids corresponds to earlier findings (e.g., Alberini et al., 2005; Jones et al., 2008). As explained earlier, we choose to include all zero bids for determining the WTP. Based on Table 25.3 we can conclude that excluding these bids will have a huge effect on the mean WTP.

Most of the respondents who are willing to pay for extra man-made wetlands in the Netherlands, had a willingness to pay between 0 and 100 euros. The average WTP was slightly but not significantly higher in the Haaksbergen sample than in the RoN sample. People living closer to the existing man-made wetland were not willing to pay more. The effect of excluding protest bidders and bidders that prefer budget reallocation was relatively large, due to the large amount of zero bidders in the sample.

Table 25.3 Frequency and average WTP for the Haaksbergen region and the Dutch sub-samples

	n	Haaksbergen Region	n	Rest of Netherlands
Protest bidders	69	51.5%	417	50.4%
• Zero bidders	21	15.7%	89	10.8%
• Reallocating budgets	48	35.8%	328	39.7%
WTP				
€0–25	20	14.9%	135	16.3%
€25–50	21	15.7%	131	15.9%
€50–100	18	13.4%	107	13.0%
€100–150	3	2.2%	25	3.0%
€150–200	2	1.5%	3	0.4%
€200–250	1	0.7%	8	1.0%
Total	134		826	
Average WTP (in €)	24.46		23.33	
Standard deviation WTP (in €)	39.54		37.99	

AHP Results

Water treatment and water storage were the highest rated services (see Table 25.1). In the Haaksbergen region, 'water treatment' had the highest weight, while in the rest of the Netherlands 'water storage' was ranked first. The man-made wetland was hardly associated with recreation. This was also the case for the Haaksbergen region in which a number of people did have experience with the man-made wetland.

Based on the weights derived by the AHP methodology, the shares of the WTP for the different functions in the WTP were calculated (see Table 25.4). The WTP for 'water quality improvement' was €6.17 for the Rest of the Netherlands and more than €7 for the Haaksbergen region. The WTP for 'flood protection' was €6.38 for the Netherlands and

Table 25.4 Weights and shares of the different functions

	Haaksbergen Region	Rest of Netherlands
Weights (w_j)		
Water quality improvement (nutrient reduction)	1.562	1.394
Biomass production (reed)	0.833	0.839
Desiccation prevention	1.156	1.019
Flood protection	1.323	1.441
Recreation	0.502	0.576
Percentages of functions j		
Water quality improvement (nutrient reduction)	29.1	26.5
Biomass production (reed)	15.5	15.9
Desiccation prevention	21.5	19.3
Flood protection	24.6	27.4
Recreation	9.3	10.9

slightly more than €6 for the Haaksbergen region. The WTP for 'recreation' was less than €3 for both samples. 'Biomass production', including CO_2 capturing, WTP was around €3.75 and 'desiccation prevention' ranged from €4.51 for the Netherlands to €5.26 for the Haaksbergen region.

Analysis of Heterogeneity in Preferences of Households

Next to the mean WTP, we investigated the composition of the social demand for man-made wetlands by analysing the relation between the respondents' socio-demographic characteristics and total WTP. Due to the zero bid respondents, a Tobit regression model (Tobin, 1958) was estimated to examine the heterogeneity in WTP values. The socio-demographic variables included in the analysis can be found in Table 25.2. Respondents with missing data on relevant explanatory variables, such as income, were excluded from this analysis.

Table 25.5 presents the Tobit regression results for the two sub-samples (Netherlands; $n = 656$; Haaksbergen; $n = 112$). Income per capita was positively related as expected and for the Rest of the Netherlands sample it was significant. The dummy variables of the residence of the respondents did not have a significant correlation with their WTP, nor the size of the household nor the gender of the respondent. The age of the respondent was positively correlated with WTP. The older the respondent, the higher the WTP for man-made wetlands. This effect persisted with income included in the analysis. A possible explanation is that older respondents are more aware of the problems that can be solved

Table 25.5 Estimation results for the WTP value from Tobit regression

Variables	Haaksbergen Region	Rest of Netherlands
Income per capita	0.0003	0.0002*
Household size	−4.84	−1.90
Western region (yes = 1, no = 0)		−0.26
Northern region (yes = 1, no = 0)		−3.83
Southern region (yes = 1, no = 0)		−5.44
Age of provider	0.36	0.62***
Sex (male = 1, female = 0)	−7.54	−2.60
Low education class (yes = 1, otherwise 0)	0.70	−12.37*
Water quality improvement most important (yes = 1; no = 0)	12.03	7.95
Flood protection most important (yes = 1; no = 0)	−8.73	0.82
Willingness to compensate farmer (yes = 1; no = 0)	20.56	27.80***
Number of memberships of environmental organization per household	14.30*	8.38***
Recreation regularly in nature (yes = 1; no = 0)	15.59	9.30
Preference for more nature in the Netherlands (yes = 1; no = 0)	0.66	28.84***
Intercept	−53.95	−85.28***
Σ	61.23***	59.32
Sample size	112	656

Note: Significance is indicated by ***, ** and *, referring respectively to the 1, 5 and 10% levels.

with man-made wetlands. Lower-educated people had a lower WTP for man-made wetlands. Possibly they value man-made wetlands because of lack of information or other preferences.

The water quality improvement and flood protection variables in the Tobit regression were dummy variables based on the question 'Which service is most important?' About 80 per cent of the respondents indicated that water quality improvement or flood protection were the most important service. These respondents did not have a significantly higher total WTP for man-made wetlands than respondents who choose another service as the most important service. So the WTP for man-made wetlands did not depend on the most important service of man-made wetlands as indicated by the respondents.

Two attitude variables were included in the analysis, the willingness to compensate farmers and the number of memberships of environmental organizations. Two groups of respondents share the opinion that farmers should be compensated but they do not want to pay these farmers themselves. The first group consists of respondents who do not have a preference for man-made wetlands at all; they prefer the status quo agricultural situation. If the government decides to construct man-made wetlands, the farmer should be compensated. The second group prefers man-made wetlands over the current situation, but are also not willing to pay to these farmers themselves. Respondents who agree to compensate farmers for managing man-made wetlands have a higher WTP than respondents who do not agree. An increase in the number of memberships of environmental organizations has a positive effect on the WTP for man-made wetlands. A similar effect is observed in earlier studies on multifunctional agriculture (Jongeneel et al., 2008; Haile and Slangen, 2009). Respondents who conduct recreational activities in natural areas do not have a higher WTP for man-made wetlands, while respondents who prefer the development of more nature conservation areas in the Netherlands have a significantly higher WTP. This is not the case for the Haaksbergen sub-sample.

Respondents do not appear to be interested in the recreational options of man-made wetlands. The recreation service is hardly mentioned as the most important function. An explanation for the lack of interest in the recreational service of man-made wetlands could be that the respondents are not interested in these recreational options. Moreover, the attractiveness of recreation also depends on the neighbouring alternative opportunities for recreation, which are not taken into account in our analysis.

The results show that about half of the Dutch population is willing to pay extra for the construction of man-made wetlands, for example to compensate farmers for investments and exploiting wetlands. The most appreciated services are water quality improvement and flood protection. Based on the average WTP of €23.33 and 7.2 million Dutch households, a simple aggregation method leads to an estimate of the total benefits for man-made wetlands in the Netherlands of approximately €170 million.

25.4 GOVERNANCE OF MAN-MADE WETLANDS AND VALUATION

With respect to the ecosystem services to be managed we can conclude from our empirical analysis that the main wetland services contributing to the social value are water quality improvement and flood protection, and that flood protection is valued more than water quality improvement, and water quality improvement more than desiccation prevention. This is comparable with the results of Brander et al. (2006) and Brouwer et al. (1999). The decomposition of this value shows that economic governance of man-made wetlands will have a multifaceted character linked to the different services provided. Because of this character, economic, ecological and institutional perspectives to create financial incentives reflecting the value of man-made ecosystem services have to be combined to determine adequate economic governance. Economic governance has to secure property rights of the different values giving incentives to invest in wetlands for private landowners. Constitutions at different jurisdictional levels attach property rights to the different ecosystem services because they specify land ownership, local governments, states, federal agencies and international organizations (de Blaeij et al., 2011). Economic governance will also have to enforce contracts in the case of voluntary transactions between an ecosystem service provider and an ecosystem service buyer. Finally, economic governance will need to secure collective action in managing the environment. This collective decision-making will become more complex where several stakeholders are involved.

Effective economic governance is a prerequisite for development of man-made wetlands considering key economic variables like private investment and supply and demand for ecosystem services. The different values of different ecosystem services are the result of economic activities and interactions and are essential for implementation of man-made wetlands in the Netherlands. Taking into account the specificities of public goods elements of the ecosystem services provided by man-made wetlands, hybrid governance arrangements are likely (see de Blaeij et al., 2011). Hybrid governance goes far beyond what the price system can provide and thus makes them distinct from pure market arrangements. Within these arrangements landowners maintain distinct and autonomous property rights and their associated decision rights. Man-made wetlands will not be exploited on the basis of private governance and payments only, unless the market price of reed increases to such a level that this service generates a profit for the entrepreneur. Consequently, public payments for social ecosystem services will remain essential but several challenges with respect to governance and scaling have to be overcome before realization (ibid.).

25.5 CONCLUSIONS

This chapter presents an original contribution to the current wetland valuation literature, as it concentrates on man-made wetlands exploited by profit-maximizing entrepreneurs and discusses consequences for economic governance. This chapter shows that respondents are interested in and have a demand for man-made wetlands, implying a social demand for man-made wetlands, despite the fact that most of the respondents indicated a zero WTP. They would prefer to allocate public payments for the construction of man-made wetlands, exploited on a commercial basis. It will be worthwhile to examine the effect of developing new economic governance arrangements, to determine the WTP for man-made wetlands taking into account the different social values of wetlands. Based on these conclusions, a next step to facilitate Dutch water management with efficient services would be to discuss with relevant Dutch stakeholders multifaceted governance mechanisms for the commercial provision of wetland services in the Netherlands.

REFERENCES

Alberini, A., P. Rosato, A. Longo and V. Zanatta (2005), 'Information and willingness to pay in a contingent valuation study: the value of S. Erasmo in the Lagoon of Venice', *Journal of Environmental Planning and Management*, **48**(2), 155–75.

Barbier, E.B., M. Acreman and D. Knowler (1997), *Economic Valuation of Wetlands: A Guide for Policy Makers and Planners*, Gland, Switzerland: Ramsar Convention Bureau.

Bateman, I.J., R.T. Carson, B. Day, W.M. Hanemann, N. Hanley and T. Hett et al. (2003), *Guidelines for the Use of Stated Preference Techniques for the Valuation of Preferences for Non-market Goods*, Cheltenham, UK and Northampton, MA, USA: Edward Elgar.

Bergstrom, J.C., K.J. Boyle and M. Yabe (2004), 'Trading taxes vs. paying taxes to value and finance public environmental goods', *Environmental and Resource Economics*, **28**(4), 533–49.

Birol, E. and V. Cox (2007), 'Using choice experiments to design wetland management programmes; the case of Severn Estuary Wetland, UK', *Journal of Environmental Planning and Management*, **50**(3), 363–80.

Birol, E., K. Karoukasis and P. Koundouri (2006), 'Using economic valuation techniques to inform water resources management: a survey and critical appraisal of available techniques and application', *Science of the Total Environment*, **365**(1–3), 105–22.

Brander, L.M., R.J.G. Florax and J.E. Vermaat (2006), 'The empirics of wetlands valuation: a comprehensive summary and a meta-analysis of the literature', *Environmental and Resource Economics*, **33**(2), 223–50.

Brouwer, R., I.H. Langford, I.J. Bateman, T.C. Crowards and R.K. Turner (1999), 'A meta analysis of wetland contingent valuation studies', *Regional Environmental Change*, **1**(1), 47–57.

Brown, T.C., J.C. Bergstrom and J.B. Loomis (2007), 'Defining, valuing and providing ecosystem goods and services', *Natural Resources Journal*, **47**(2), 329–76.

Carlsson, F., P. Frykblom and C. Liljenstolpe (2003), 'Valuing wetland attributes: an application of choice experiments', *Ecological Economics*, **47**(1), 95–103.

Daily, G.C., S. Polasky, J. Goldstein, P.M. Kareiva, H.A. Mooney and L. Pejchar et al. (2009), 'Ecosystem services in decision making: time to deliver', *Frontiers in Ecology and the Environment*, **7**(1), 21–8.

de Blaeij, A.T., N. Polman and S. Reinhard (2011), 'Economic governance to expand commercial wetlands: within- and cross-scale challenges', *Ecology and Society*, **16**(1), Article 33.

Dixit, A.K. (2004), *Lawlessness and Economics: Alternative Modes of Governance. Gorman Lectures in Economics*, Princeton, NJ: Princeton University Press.

Dixit, A.K. (2008), 'Economic governance', in S.N. Durlauf and L.E. Blume (eds), *The New Palgrave Dictionary of Economics*, online, 2nd edition, accessed 16 December 2012 at http://www.dictionaryofeconomics.com/article?id=pde2008_E000260.

Dixit, A.K. (2009), 'Governance institutions and economic activity', *American Economic Review*, **99**(1), 5–24.

Engel, S., S. Pagiola and W. Wunder (2008), 'Designing payments for environmental services in theory and practice: an overview of the issues', *Ecological Economics*, **65**, 663–74.

Ghermandi, A., J.C.J.M. van den Berg, L.M. Brander, H.L.F. de Groot and P.A.L.D. Nunes (2010), 'Values of natural land human-made wetlands: a meta-analysis', *Water Resources Research*, **46**(12), W12516.

Haile, D.R. and L. Slangen (2009), 'Estimating the willingness to pay for the benefit of AES using the contingent valuation method', *Journal of Natural Resources Policy Research*, **1**(2), 139–52.

Hanley, N., M. Ryan and R. Wright (2002), 'Estimating the monetary value of health care: lessons from environmental economics', *Health Economics Review*, **12**(1), 3–16.

Jack, B.K., R. Kousky and K.R.E. Sims (2008), 'Lessons from previous experience with incentive-based mechanisms', *PNAS*, **105**(28), 9465–70.

Jones, N., C.M. Sophoulis and C. Malesios (2008), 'Economic valuation of coastal water quality and protest response: a case study in Mitilini, Greece', *The Journal of Socio-Economics*, **37**(6), 2478–91.

Jongeneel, R., N.B.P. Polman and L.H.G. Slangen (2008), 'Why are Dutch farmers going multifunctional?' *Land Use Policy*, **25**, 81–94.

Kallas, Z., J.A. Gómez-Limón and J. Barreiro Hurlé (2007), 'Decomposing the value of agricultural multifunctionality: combining contingent valuation and the analytical hierarchy process', *Journal of Agricultural Economics*, **58**(2), 218–41.

Meerburg, B.G., P.H. Vereijken, W. de Visser, J. Verhagen, H. Korevaar and E.P. Querner et al. (2010)', 'Surface water sanitation and biomass production in a constructed wetland: pilot Lankheet', *Wetlands, Ecology and Management*, **18**(4), 463–70.

Minister of Agriculture, Nature and Food Quality, G. Verburg (2008), 'Zuiveringsmoerassen voor het verbeteren van waterkwaliteit' [Purification of marshes for improving water quality], letter to the Chairman of the House of Representatives, 6 November 2008.

Moran, D., A. McVittie, D.J. Allcroft and D.A. Elston (2007), 'Quantifying public preferences for agri-environmental policy in Scotland: a comparison of methods', *Ecological Economics*, **63**(1), 42–53.

Navrud, S. (2002), *The State-of-the-Art on Economic Valuation of Noise*, Final Report to European Commission DG Environment, 14 April 2002, Department of Economics and Social Sciences, Agricultural University of Norway.
Saaty, T. (1977), 'A scaling method for priorities in hierarchical structures', *Journal of Mathematical Psychology*, **15**(3), 234–81.
Saaty, T. (1980), *The Analytic Hierarchy Process*, New York: McGraw Hill.
Tobin, J. (1958), 'Estimation for relationships with limited dependent variables', *Econometrica*, **26**(1), 24–36.
Woodward, R.T. and Y.S. Wui (2001), 'The economic value of wetland services: a meta-analysis', *Ecological Economics*, **37**, 257–70.

26. The contribution of non-use values to inform the management of groundwater systems: the Rokua esker, Northern Finland

Phoebe Koundouri, Mavra Stithou, Eva Kougea, Pertti Ala-aho, Riku Eskelinen, Timo Karjalainen, Björn Klöve, Manuel Pulido-Velazquez, Kalle Reinikainen and Pekka M. Rossi

26.1 INTRODUCTION

This chapter focuses not only on the estimation of use but, as importantly, also on non-use values to inform the management of groundwater-dependent ecosystems (GDEs), using as a case study the Rokua esker in Northern Finland. GDEs are ecosystems of great importance because of the conservation, biodiversity, ecological, social and economic values they provide. Their basic characteristic is that they require access to groundwater to maintain their healthy condition. Following Evans and Clifton (2001) GDEs include: (1) terrestrial ecosystems that rely seasonally or episodically on groundwater; (2) river base-flow systems, including aquatic, hyporheic and riparian ecosystems that depend on groundwater input, especially during dry periods; (3) aquifer and cave ecosystems, often containing diverse and unique fauna; (4) wetlands dependent on groundwater influx for all or part of the time; and, (5) estuarine and nearshore marine ecosystems that rely on groundwater discharge.

As a result, a loss of groundwater resources is a major threat as ecosystems' functions and composition are reliant on the appropriate supply of groundwater. Consequently these ecosystems are very sensitive to climate change and natural variability. Across Europe, aquifers' resources are dramatically changing with groundwater resources to face increasing quantitative pressure mainly from land use issues and consumption pressures (Kløve et al., 2011). Also, all regions of the world show an overall net negative impact of climate change, freshwater resources and ecosystems and it is expected that many areas are likely to face a reduction in the value of the services provided by water resources (IPCC, 2007). Adaptation of measures and application of appropriate management practices have an important role in determining the impact of these pressures on water resources and on ecosystems.

As a response, policy-makers in Europe have developed the Water Framework Directive (Directive 2000/60/EC – WFD), which is probably the most ambitious piece of environmental legislation in the EU. While imposing environmental objectives to be achieved, the WFD also calls for the use of a set of instruments and procedures for analysing the socioeconomic and environmental impacts of water uses and at the same time provides guidance for the selection of measures for achieving these objectives. The WFD requires that Member States take the necessary measures for the protection of water bodies, promoting a sustainable water use based on a long-term protection of available water resources. The most cost-effective programme of measures should be selected in order to meet the WFD environmental objectives in all water bodies.

For groundwater bodies, along with WFD, the Groundwater Directive (GWD) requires the achievement of a 'good groundwater status', which is achieved when both its quantitative and chemical status are good. As emphasized in the new GWD (2006/118/EC), groundwater is characterized as particularly important for dependent ecosystems and for its use in water supply for human consumption. The value of GDEs such as wetlands or terrestrial ecosystems has long been recognized (Hynes, 1983). Therefore, according to the GWD, when establishing threshold values for groundwater pollutants, Member States need to consider the extent of interactions between groundwater and associated aquatic and dependent terrestrial ecosystems. However, although the WFD includes provisions to protect groundwater from pollution, and to ensure that groundwater abstraction does not threaten dependent terrestrial or wetland ecosystems, it is important that more emphasis is put on the fact that an aquifer has to be viewed as an ecosystem related to the surrounding environment (Danielopol et al., 2004). This is required in order to achieve an integrated and sustainable groundwater management that addresses the protection of ecologically valuable areas.

Economic valuation contributes to improved water management decisions by informing decision-makers about the full social cost of water use and full benefits of the goods and services that water provides. Many of the ecosystems' functions that water resources sustain, among others, recreation and aesthetic benefits, biodiversity benefits, research benefits and existence benefits, do not have a market price and as such are not recognized as having an economic value by the decision-makers (Bateman et al., 2003; Perman et al., 2003). Achieving a good water status necessarily requires the application of non-market economic valuation techniques, such as stated preference methods.

Stated preference methods, based on social survey techniques to elicit public preferences, have been used since the 1970s by environmental economists to value the non-market benefits of environmental changes. Of these, choice experiments (CE) are becoming a popular means of environmental valuation, where respondents are required to trade off changes in the levels of different attributes that describe the good against the cost of these changes. The contingent valuation (CV) method is another stated preference technique in which a hypothetical market is being created and respondents are asked directly to express their willingness to pay (WTP) for existing or potential environmental conditions not registered on any market (Mitchell and Carson, 1989).

In this study both techniques are employed to explore how people value groundwater quantity in an environment very sensitive to climate change and natural variability. The purpose of the CE is to investigate the local public's preferences for alternative management scenarios, defined by their impacts on water quantity, on the environment, recreation and total land income and by improved scientific information on climate change. Complementarily, a CV method was employed to investigate individuals' behaviour in a setting of uncertainty with respect to the damage level in the absence of a revised water management.

The rest of the chapter unfolds as follows. Section 26.2 offers an overview of the case study area, while Section 26.3 presents the survey design. Section 26.4 describes the models that were employed for the estimation, while Section 26.5 in its first subsection presents the results from the CE application and then results from the CV. Finally, Section 26.6 discusses and concludes the role of valuation results in policy design.

26.2 THE CASE STUDY AREA

The case study is Rokua esker located in Northern Finland. It is a part of a chain of esker ridges with small 'kettle' lakes situated within the esker area. Rokua is a dependent groundwater ecosystem. As such, the water level of most of the lakes in Rokua is a function of the level of the groundwater table of the esker; the latter is naturally recharged. However, during the last few decades, a significant reduction in the water level of many small lakes has been observed. Scientists have monitored groundwater quantity and observed that groundwater level tended to decline even in a period where the precipitation–evaporation ratio was increasing. Many reasons have been discussed for this decline in water quantity in groundwater and lakes such as climate change or land use and drainage. Forest drainage of the surrounding peatlands appears to disturb groundwater dynamics and thereby water level of lakes. However, there is yet a degree of uncertainty, as scientific knowledge is lacking on this complex ecosystem. The impacts of drainage and also the natural variability or impact of climate change on groundwater dynamics are not yet very clear. Even though more research is needed in order to better understand the extent and the nature of the problem, scientific observations provide sufficient evidence that a policy to mitigate a possible future environmental deterioration is needed.

Water resources in Rokua provide a diverse array of goods and services that can be

Table 26.1 Total economic value components of water resources in Rokua

Values	Goods/Services
Direct use values	Forestry
	Energy resources (peatland)
	Recreation
	Forestry
	Irrigation for agriculture/domestic water supply (to a lesser extent)
Indirect use values	Nutrient retention
	Pollution abatement
	External eco-system support
	Micro-climatic stabilization
	Reduced global warming
	Soil erosion control
Option values	Potential future uses of direct and indirect uses
	Future value of information on climate change
Non-use values (bequest, existence and altruistic values)	Biodiversity
	Cultural heritage

Source: Adopted and modified from Barbier et al. (1997).

translated into direct or indirect values, as presented in Table 26.1, for local society and visitors.

26.3 SURVEY DESIGN

The main goal of the questionnaire, as in stated preferences techniques, is to try to elicit information about environmental preferences from individuals through the construction of hypothetical but realistic scenarios of water management practices that involve an improvement in environmental aspects of water quantity. The questionnaire covered a number of topics and was divided into five sections – see Table 26.2 below. An accurate and clear description of attributes and their associated levels with the policy under consideration as well as an introduction to the study area were provided in the beginning of the questionnaire. To obtain more information on individuals' attributes, the questionnaire contained debriefing questions, questions that revealed environmental consciousness of the respondents and socioeconomic questions (age, gender, income categories, occupation and educational attainment).

The development of the survey instrument took place over a period of a year and initially involved focus group discussions, face-to-face interviews with local stakeholders and extensive discussions with experts. Discussions with local stakeholders revealed people's understanding of the issues related to the management of water resources and services in Rokua. Stakeholders (forestry, peatland industry, homeowners, local residents and service providers) were asked general questions, whether they are familiar with environmental conditions of water resources in Rokua and the issues related to it, that is, land use issues and climate change, and finally whether and how they value environmental goods or services that Rokua esker provides. After discussions with experts the valuation problem was refined. In particular, the exact attributes to be valued as well as the increase in their levels after the implementation of a policy option and the corresponding levels in case of deterioration were defined. The survey was administered with face-to-face interviews from

Table 26.2 Questionnaire structure

Section		
Section A	Presenting the problem	Site description Scientific facts Good to be valued Attributes to be valued Scenarios
Questions		
Section B	Choice experiment questions	Eight sets of choice cards
Section C	Debriefing questions	Questions to explain why respondents were or were not WTP
Section D	Environmental behaviour questions	Questions that reveal environmental consciousness
Section E	Contingent valuation question	Question to examine risk behaviour
Section F	Socioeconomic questions	Among others age, education, job and income

April to August 2011. A random sample of 170 respondents was collected from Oulu and around the municipalities of Utajärvi and Vaala where Rokua area is situated. The sample consisted of either local inhabitants of the area or recreational users of Rokua.

26.3.1 Choice Experiment Design

The good to be valued in the Rokua choice experiment is the revision of the water management in order to meet the objectives of the Water Framework and Groundwater Directives, which are the main pieces of European legislation in place to protect groundwater. A detailed description of choice experiment design can be found in Koundouri et al. (2012). Policy under consideration includes restriction of peatland drainage in the groundwater area, expansion of the conservation area and compensation when legally required. Implementation of the above would enable management of Rokua to comply with environmental, water and biodiversity legislation. The key requirements of the WFD in relation to groundwater-dependent ecosystems are to achieve and maintain 'good status' of these water bodies by 2015 and meet the overall environmental objectives for groundwater according to Article 4(b). Designations and actions should be implemented in order to maintain good water quantity in lakes, springs and aquifers and to sustain as many ecological and landscape functions as possible.

In this direction, proposed policy options will contribute to restore lakes' water levels and to avoid future possible deterioration. At the same time, environmental improvements would also lead to an increase in recreational values. Income opportunities of local people may also become affected. For example, environmental degradation that could occur in the absence of a holistic management could result in a decline in the popularity and the number of visitors to the Rokua area and as a result income from such activities would decline.

Policy under consideration is characterized by five different management attributes that are presented in Table 26.3. At this point it should be also noted that during testing the questionnaire as well as during the face-to-face interviews no indication that respondents experienced interrelationships between the attributes was revealed.

Experimental design techniques were employed in SPSS software to obtain an orthogonal design (Louviere et al., 2000; Hensher et al., 2005) consisting of main effects in 32 pair-wise comparisons of alternative wetland management scenarios randomly blocked to four different versions, each with eight choice sets. Each set contained two different wetland management scenarios and an option to select neither scenario considered as the status quo baseline alternative. The two wetland management scenarios are characterized by a change in attributes with respect to the status quo alternative.

26.3.2 Contingent Valuation

Complementary to the choice experiment a CV question following a split-sample approach was employed to investigate individuals' behaviour in a setting of uncertainty with respect to the damage level in the absence of the revised water management. Most of the studies in the water resources valuation literature employing a CV method aim to determine the WTP for water services considering single changes, that is, from a lower- to a medium- or higher-level water quality. Thus, there is a fundamental assump-

Table 26.3 Water management attributes and levels used in the CE

Attribute	Definition	Management Level
Water quantity	This attribute refers to the total quantity of water available in groundwater aquifers, lakes and springs	*Increased*: most of the lakes have restored their water level *Same as now*: some lakes have water quantity problems. Current state of water is sustained *Limited*: water quantity has considerably declined. The last alternative reflects what is expected to happen in the absence of revised management in the future (status quo level)
Recreation	This attribute refers to the sum of all values (direct and indirect) derived from recreational activities	*Increased*: environmental improvements result in an increase in recreational values *Same as now*: current levels of recreational values are sustained *Low*: this is the case where no measures are taken. As a result of environmental degradation in the absence of the revised management, recreational values are going to decline (status quo level)
Total land income	This attribute refers to the total income opportunities for the local people emerging from economic activities of logging, peat harvesting and tourism industry based in the Rokua area	*Same as now*: total income will remain unchanged *Restricted*: total income opportunities will get restricted (status quo level)
Investment in research	This attribute refers to the scientific research to better understand long-term environmental changes in Rokua	*High*: more resources *Medium*: current resources (status quo level) *Low*: stop current research
Price	One-off payment	0€, 10€, 20€, 50€, 100€

Source: Koundouri et al. (2012).

tion behind the survey design that all different scenarios presented can be achieved with certainty so respondents can reveal their underlying preferences upon certain outcomes (Roberts et al., 2008). Yet there are only few studies that have incorporated issues of uncertainty regarding the exact nature of damage under the status quo as well as the timing and the extent outcomes of the proposed environmental policies.

Johansson (1988), in a CV study in Sweden, presented some preliminary results on the consistency of WTP measure for public goods in an uncertain world. Respondents were asked about their willingness to pay for four different programmes that each would save all or some of the species. The result of one of the programmes was uncertain, that is, respondents were informed that there was a 50 per cent chance for the programme to save all species and 50 per cent to save every second species. The remaining programmes would save 50 per cent, 75 per cent and 100 per cent of the species respectively. Macmillan et al. (1996) employed a CV method to estimate WTP of the Scottish population for uncertain recovery and damage scenarios from reduced acid rain deposition in the semi-natural uplands of Scotland. In order to incorporate uncertainty with respect to the damage level in the absence of further reductions in emissions recovery, a split-sample survey format has been used that presented six alternative ecosystem recovery levels and damage levels scenarios.

Similarly in the Rokua case study, uncertainty was introduced through the use of subjective probabilities. That is, in the absence of any better information, equal probabilities to the mutually exclusive outcomes of water states (State A: High, State B: Good, State C: Moderate, State D: Poor, State E: Bad) have been assigned. Following the work of Macmillan et al. (1996), in the first sub-sample respondents were informed that in the absence of a revision of water management there is 50 per cent chance that water quantity will remain at current levels, that is, State (C), and 50 per cent chance for the level to go to State (E), implying an expected future damage State (D). In the other sub-samples respondents were informed that water quantity will be at State (D) with certainty. In both cases the revision of water management would result in a certain level of improvement. WTP bids (as a one-off payment) were elicited through a payment card contingent valuation (PCCV) mechanism. This method was first developed by Mitchell and Carson (1981, 1984) as an alternative to the bidding game. As the authors (Mitchell and Carson, 1989) noted, this approach does not require large samples compared to the referendum approach. However, although this method avoids the anchoring effects of dichotomous choice since respondents select their own WTP amount (Ariely et al., 2003) it is regarded that the chosen range of amounts can influence respondents' answers.[1] Respondents were asked to state the amount that best described their maximum WTP in a range of offered bids from 0 to > €100.[2] The maximum amount of €100 was derived from focus group interviews. An opportunity was also provided to state a higher WTP.

The aim of this set up is to test whether or not utility will differ between a management option according to which the expected future damage under status quo will be moderate but uncertain and a programme where future damage under status quo will be moderate but certain. Macmillan et al. (1996) have concluded that when individuals are faced with future environmental damage they appear to be risk-averse. As a result, the CV question is employed in order to investigate respondents' risk behaviour and hence to observe how and if the valuation result changes when respondents are aware of the uncertainty regarding environmental losses with respect to the status quo level.

26.4 MODEL SPECIFICATIONS

The models for both CE and CV we present in this study are the final equations selected following a specification search that tested all relevant explanatory variables in the data, and their natural logarithms, for significance, and kept only those that were found to be statistically significant at the 10 per cent level.

To contribute to an assessment of the validity of PCCV responses, we first examined descriptive cross-tabulations of sample WTP against variables in the dataset. We then conducted our econometric analysis beginning with a combination of explanatory variables, by picking the seemingly most important variables, and estimated two models using ordinary least squares (OLS) regressions. The first regression was run with WTP as the dependent variable; the second with *ln_WTP* as the dependent variable, where *ln_WTP* is the natural log of (1 + *WTP*). Findings from this analysis showed that the model with the 'log-linear' specification (*ln_WTP*) as the dependent variable fits the data substantially better than the model with *WTP* as the dependent variable. Subsequent analysis therefore focused on models with *ln_WTP* as the dependent variable. Furthermore, models were corrected for heteroscedasticity using the robust covariance estimator.

Based on the OLS regression, the WTP function for groundwater quantity is:

$$Ln\,(1 + WTP) = \beta_0 + \beta_1 * Gender + \beta_2 * Income + \beta_3 * Degree + \beta_4 * Group + \beta_5 * Visit \quad (26.1)$$

Then, according to the average selected sample and using Equation (26.1) the WTP is calculated.

Regarding the model specification for the analysis of CE data, Hausman and McFadden's (1984) tests led to the rejection of the IIA (irrelevance of independent alternatives) property, and therefore to the use of another model that relaxes the IIA assumption. As reported in Koundouri et al. (2012), where an overview of CE methodology can be found, the error components logit model (ECM) provided a better insight on preference heterogeneity and therefore was preferred. According to this model's specification the random part of utility[3] is decomposed to an individual unobserved effect and other variables that influence choice ($\varepsilon_{ij} = \alpha_i + k_{ij}$) and the possibility for error components in the combined change nest and the no change nest is examined. Because the probability function does not have a closed form solution, the model is estimated using simulated maximum likelihood methods (Train, 2003).

As the CE method is consistent with utility maximization and demand theory (Hanemann, 1984) the marginal value of change in water management programme attribute can be calculated as:

$$MWTP = -\frac{\beta_{attribute}}{\beta_{cost}} \quad (26.2)$$

This part-worth (or implicit price) formula represents the marginal rate of substitution between one-off payment and the water management programme attribute in question.

Before presenting model results, Table 26.4 describes the socioeconomic and attitudinal characteristics of the final usable sample for each method.

26.5 MODEL RESULTS

26.5.1 Error Component Model – CE Results

The error component models were estimated by simulated maximum likelihood using Halton draws with 500 replications (Train, 2003). All the choice attributes were

Table 26.4 Profile of respondents

Variable	Definition	CE Sample	CV Group 1	CV Group 2
Age	Average age of a person (in years)	41.58[a]	41.36	41.46
Gender	Dummy variable equals 1 if female, 0 if male (%)	40	48	34
Children	Dummy variable equals 1 if respondent has children, 0 otherwise (%)	43[a]	42	44
Degree	Dummy variable equals 1 if respondent has education with university degree and above, 0 otherwise (%)	35	38	30
Visited Rokua	Dummy variable equals 1 if individual has visited Rokua in the past, 0 otherwise (%)	78[b]	80	76
Income	Average annual gross household income (seven income bands from less than €10 000 to above €70 000)	3.74[c]	3.76	3.61
Group	Dummy variable equals 1 if respondent belongs to Group 1, 0 otherwise	54[a]	–	–
Sample size, *N*		166	91	79

Notes:
a. $N = 165$.
b. $N = 164$.
c. $N = 153$.

statistically significant. The models were estimated with NLOGIT 4.0 (Greene, 2002) and the full dataset of 1328 observations from 166 respondents. Initially, a full set of socioeconomic variables was entering the utility function either through interactions with the ASC (alternative specific constant) or as interaction terms with the choice attributes. Variables such as respondents' household size, gender or association with the farming or forestry community were not significant and are not included in the final model specifications. The first model reported in Table 26.5 includes only the choice attributes as explanatory variables in the utility function. All the estimated coefficients have the expected signs. Cost of new management is negative and significant, whereas an increase in seagrass area, riverside vegetation and rare species are positive and significant at the 5 per cent level. The ASC parameter is negative and significant, indicating that respondents generally prefer the 'new management' options over the no management scenario, ceteris paribus. The latent error term captures unobserved error correlations between the two new alternatives that deviate from the status quo option. The error component is significantly different from 0, indicating heterogeneity across the utilities that respondents derive from the new alternatives. It should be noted that results of this model are also reported in Koundouri et al. (2012).

The attribute-only model does not provide information about the sources of individual heterogeneity. In the second error component model reported in Table 26.5,

Table 26.5 ECM results

Variables	Attribute–only Model[a]		Model 1 with Interactions[a]		Model 2 with Interactions	
	Est.	*t*-ratio	Est.	*t*-ratio	Est.	*t*-ratio
Water Quantity	0.435***	4.699	0.400***	4.052	0.236**	2.017
Recreation	0.209***	2.761	0.171**	2.110	0.091	0.867
Research	0.551***	7.096	0.583***	7.182	0.383***	3.391
Total Land Income	0.158**	2.072	0.174**	2.183	0.236**	2.521
Cost	−0.016***	−9.846	−0.017***	−9.939	−0.017***	−9.991
SQ	−5.899***	−3.979	−8.778**	−2.385	−5.126**	−2.068
*Age*SQ*			0.070	1.205		
*Gender*SQ*			−4.437**	−2.517	−5.935***	−3.010
*Children*SQ*			−4.359**	−2.084		
*Visit*SQ*			4.958**	2.158		
*Income*SQ*			0.342	0.655	0.410	0.780
*Degree*SQ*			−3.835**	−2.083		
*Degree*Water Quantity*					0.465***	2.897
*Degree*Recreation*					0.278*	1.729
*Degree*Research*					0.504***	3.160
*Degree*Total Land Income*					−0.174	−1.150
St. Dev. of latent random effects						
No change	3.388	0.994	0.877	0.234	0.753	0.220
Change	7.802***	3.275	7.214***	5.031	7.552***	5.234
LL	−964.8493		−865.2540		−874.6694	
χ^2	988.2157		923.7392		940.0641	
Pseudo-R^2	0.34		0.35		0.35	
BIC	1.49641		1.51479		1.51052	
Observations	1328		1208		1224	
No. of respondents	166		151		153	

Notes:
a. Reported also in Koundouri et al. (2012).
(*) indicates significant at 10%; (**) indicates significant at 5%; (***) indicates significant at 1%.

socioeconomic variables were interacted with the ASC and the choice attributes. In addition, attributes were interacted with respondents' characteristics and out of a range of model specifications tested, the model that provided the best fit to our dataset included gender, income and interaction effects between education of respondents and the choice attributes in the utility function. Comparing the log-likelihoods and the pseudo-R^2 goodness-of-fit measures between models, the models with interactions that account for sources of preference heterogeneity provide a much better model fit than the attribute-only model.

Overall, the models are statistically significant and all attributes are significant determinants of choice, apart from recreation in Model 2 with interactions. The cost price is negative, indicating that an alternative is less likely to be chosen if the cost is higher, while other attributes' coefficients conform to theoretical expectation of increasing marginal utility. For both types of models, attribute-only and with interactions, respondents prefer water management practices that ensure higher water quantity, recreation, research potential and positive effect on total land income. The models also demonstrate a negative and significant coefficient for the status quo indicating that ceteris paribus, the status quo alternative is less desirable than the other options maintained also in both types of specifications.

Regarding models with interactions capturing individual observed heterogeneity, it is observed that in Model 1 respondents who are older and have visited Rokua in the past are more likely to choose the status quo than Option A or B, showing that familiarity with the site does not necessarily encourage change. An opposite effect is observed for female respondents, with children and a higher than secondary education. It is also noted that income has no effect on choice, which could be explained by the reluctance of participants to reveal their real income. Model 2 captures conditional heterogeneity by including in the utility function interactions of respondents' educational level with choice-specific attributes and interactions of income and gender with ASC. Results show that respondents with higher levels of education are likely to prefer management scenarios that assure and improve water quantity, research and recreation attributes. Furthermore, similarly to Model 1 with interactions female respondents are less likely to opt for the status quo scenario. Finally, the error component for the combined alternatives A and B is statistically significant, for all models, revealing alternative specific variance heterogeneity (heteroscedasticity) in the unobserved effects of these alternatives.

Using the Krinsky and Robb (1986) procedure with 1000 draws in LIMDEP 9.0 NLOGIT 4.0, respondents' valuation of water management programme attributes (following Equation 26.2) and 95 per cent confidence intervals were calculated for the attribute-only and with interactions ECM models and are reported in Table 26.6.

The estimated WTP values for all models indicate that overall an average household values the improvements positively. Specifically, in terms of research it is willing to pay from €21 to €33 to ensure that the scientific research to better understand long-term

Table 26.6 Implicit prices (per household, one-off payment) for water management attributes from NMNL and ECM and 95% confidence intervals

Attributes	Attribute-only Model	Model 1 with Interactions	Model 2 with Interactions
Water quantity	25.75	22.54	13.02
	(15.93, 35.73)	(13.44, 32.18)	(0.83, 25.05)
Recreation	12.46	9.71	0.00[a]
	(3.63, 22.15)	(0.57, 18.88)	
Research	33.05	33.50	21.41
	(24.22, 43.02)	(24.12, 43.34)	(9.36, 34.89)
Total land income	9.33	9.76	12.82
	(0.67, 17.51)	(1.50, 17.78)	(3.46, 22.73)

Note: a. WTP estimate was not found to be significantly different to zero and is expressed as zero.

Table 26.7 Estimation results for PCCV

	Log-linear Pooled		Log-linear Group 1		Log-linear Group 2	
	Est.	*t*-ratio	Est.	*t*-ratio	Est.	*t*-ratio
Gender	0.966***	3.893	1.131***	2.927	0.824**	2.263
Income	0.001***	6.979	0.001***	6.867	0.002**	2.073
Degree	0.749***	3.099	0.578	1.615	0.875**	2.309
Group	−0.174	−0.675				
Visit	0.0001	0.379	0.0001	0.145	−0.071	−0.169
Constant	2.387***	10.475	2.185***	7.128	2.460***	5.235
F-statistic	10.03		8.54		4.03	
R^2	23%		28%		18%	
Observations	170		91		79	

Note: (*) indicates significant at 10%; (**) indicates significant at 5%; (***) indicates significant at 1%.

environmental changes will not stop. Another important attribute for the households is improved water quantity, which varies from €13 to €26, while increased potential for recreation and total land income range from €9 to €13. Therefore, implicit prices clearly demonstrate the importance of water quantity for the respondents by supporting water management that will not allow the decline of total quantity of water available in groundwater aquifers, lakes and springs.

26.5.2 Log-linear Model – CV Results

Table 26.7 presents our main model estimation results for PCCV elicitation method estimated by OLS with $ln(1 + WTP)$ as the dependent variable.[4] To ensure the variance of coefficient estimates are consistently estimated, we use for all models the White standard errors employed in the LIMDEP heteroscedasticity command.[5]

Although the R^2 for the PCCV models are modest, significant relationships among variables such as gender, income and education have been revealed. The coefficients are almost uniformly in line with expectation, with WTP varying with income while use of the water environment was positive (log-linear pooled and Group 1) but not significant. Turning to other effects for which we held no clear prior prediction, we find that well-educated respondents and female respondents all gave higher values on average, and were more likely to accept the scenarios, than those with lower educational attainment and male respondents. However, the dummy variable for group although negative is not statistically significant and therefore doesn't pick up any differences between the groups. Based on the average sample and using Equation (26.1), the WTP from pooled data, Group 1 and Group 2 are reported in the third column of the Table 26.8. Findings show that respondents who faced uncertainty (Group 1) stated a smaller WTP compared to respondents of the second group who were willing to pay marginally more. Therefore, results of Table 26.8 are not conclusive regarding the effect of uncertainty on stated values. In the literature, risk behaviour has been observed in the context of uncertainty in financial losses. Kahneman and Tversky (1979) presented a body of empirical evidence that individuals are risk-seekers when financial losses are in prospect. Yet this result

Table 26.8 Mean WTP per household (one-off payment) for good status of water quantity and quality in five to ten years from now

	PCCV max WTP[a]	PCCV Regression Model
Pooled	41.55 (39.51)[b]	20.71
Group 1	40.42 (39.48)	15.20
Group 2	42.85 (39.76)	19.90

Notes:
a. All zero bidders are included.
b. Std. Dev. in parentheses.

opposes Macmillan et al.'s (1996) findings that individuals are risk-averse when faced with uncertain future environmental damage. It is believed that further research could shed more light on this respect.

Overall, Tables 26.6 and 26.8 demonstrate the importance of water quantity for the respondents, which is a prerequisite condition for healthy GDEs.

26.6 POLICY IMPLICATIONS AND CONCLUSIONS

The notion of the total economic value of groundwater-related ecosystems is very important for a holistic economic assessment that also considers that the functions performed by GDEs are an important component of the overall environmental services provided by a groundwater system. Therefore, decision-making needs to be informed by economic analysis that entails a relative assessment of the: (1) cost of protection in terms of the loss of alternative uses of groundwater and land, and the administration of the land use and groundwater control policy and (2) benefits of protection in terms of in situ value of groundwater and groundwater-related ecosystem services (Foster et al., 2006). In addition, findings of the current study confirm the significance of such benefits.

Overall, results show that respondents generally favour changes in water management and they prefer to deviate from the described status quo option by choosing a management scenario rather than a no management scenario. Both CE models, attribute-only and with interactions, revealed that individuals prefer water management practices that ensure not only higher water quantity but also recreation, research potential, as well as positive effects on total land income. Scientific research that reduces environmental uncertainty should be encouraged and promoted, since results indicate that an average household is willing to pay from €21 to €33 in order to ensure that the scientific research to better understand long-term environmental changes in Rokua will not stop. Another important attribute for the households is improved water quantity that varies from €13 to €26 (average value of €20).

CV models also revealed the importance of water quantity and quality for the respondents, with an average value of €19 per household (one-off payment). It is interesting that female respondents and those with higher education are most likely to accept changes in the management, revealing the higher environmental consciousness of those groups. On the other hand, results indicated that familiarity with the case study has an opposite effect, as older people who have already visited Rokua are the most likely to choose the status quo option. In addition, when applied the CV method respondents were presented with uncertainty regarding the losses conditional on a certain level of improvement. In the first subgroup respondents made a decision in a context of uncertainty that damage will occur, while in the second in a context of certain damage. So the CV question was employed in order to investigate respondents' risk behaviour and hence to observe if the valuation result changes when respondents are aware of the uncertainty regarding environmental losses. Both groups showed a moderate WTP for the services, with only a slightly lower WTP for the group confronted with only uncertain damage. However, interpretation of the above result does not provide a strong evidence of respondents' risk value under different degrees of uncertainty.

At this point it should be noted that while the economic benefits related, for example, to water supply may be easier to realize, non-use values of groundwater are often neglected. There are few studies that have estimated non-use values related to quality (Jensen et al., 1995; Rozan et al., 1997; Press and Söderqvist, 1998; Hasler et al., 2005) or quantity (Koundouri et al., 2012) of groundwater. Rozan et al. (1997) estimated €52 per household/year in 1995 of non-user households to protect the Alsatian aquifer (France). This value was considered as a proxy of its existence value and was used to assess the economic non-use value of the aquifer. Similarly, Press and Söderqvist (1998) employed the CV method to estimate the benefits of groundwater protection in the Milan area (Italy) in order to also consider non-use values directly. The authors elicited a value of ITL640 000 per household/year. Jensen et al. (1995), using the CV method, estimated the WTP for groundwater protection from pollution at DKK1000 per household/year elicited by an open-ended payment format, and at DKK2100 using the close-ended format. Furthermore, Hasler et al. (2005) employed a national CE study in order to assess the non-market benefits associated with increased protection of the groundwater resource and revealed an estimated WTP of DKK1899 per household/year for protected and naturally clean groundwater, not in the need for purification, a WTP for good conditions for flora and fauna in waterways and lakes of DKK1204 per household/year, and a WTP for purified water of DKK912 per household/year (all in 2005 prices). The authors also used a CV study to estimate the value of both naturally clean groundwater and very good conditions for plant and animal life (DKK711 per household/year) and purified water (DKK529 per household/year). It is noted that comparing values of the above studies with the current one, after accounting for differences across countries and years, show that the latter (CV and CE) has elicited values of lower magnitude with regard to water quantity.

Apart from stated preference methods, in situ values have been assessed using distance function methodology. Koundouri and Xepapadeas (2004) estimated the individual farmer's valuation of the marginal unit of groundwater in Kiti aquifer in Cyprus at £0.009 m^3 (in 1999 Cyprus pounds). Furthermore, the socially optimal shadow price of in situ groundwater (in Cyprus pounds) for the Kiti aquifer in 1999 was determined to be £0.2017 per m^3 of water, using an optimization model simulated under conditions of optimal groundwater extraction (Koundouri and Christou, 2000).

Furthermore, Koundouri (2000) reports the established in situ per cubic metre groundwater's total economic value. This total economic value is equal to the relevant backstop technology for water, which is, for example, the per cubic metre cost of desalination (at

€0.05). Divergence between this value and the estimated above shadow price of in situ groundwater points to the significant non-use values of groundwater, such as option value and ecosystem resilience value, as well as alternative use values of economic sectors other than agriculture.

From a policy perspective the findings of the current study provide an insight into the return value of the various services that groundwater-dependent ecosystems can provide. This result aims to inform policy-makers of how individuals value non-use and existence values of these ecosystems. Hence, it is regarded that results emphasize the need to broaden the policy options (e.g., related to the WFD implementation or future land use and ecosystem protection policies) beyond the consideration of only market and use values of groundwater systems and they contribute to justify decisions (ex ante and ex post) taken by government agencies (Bonnieux and Rainelli, 1999; Pearce and Ozdemiroglu, 2002).

As Boulton (2005) observes, mismanagement of the resource takes place and that happens for different reasons such as the difficulties of assessing groundwater volumes, recharge rates and sources, and groundwater quality, the relatively slow recognition of the linkages between groundwater and many surface water ecosystems and the lack of public visibility of groundwater's lag-time between changes in groundwater regime or quality and the response by surface GDEs.

Therefore, acknowledging and establishing vertical linkages between water bodies and exploring the relation between groundwater recharge and discharge is one of the most important aspects of the protection of ecologically valuable areas, especially when facing climate change uncertainty. Finally, given the likelihood of significant uncertainty over the impact on GDEs, decision-making strategy will normally have to embrace one or other of the following (Foster et al., 2006): (1) the precautionary principle of not authorizing any development until ecosystem risks are established and managed, (2) pragmatic initial development of groundwater resources with careful monitoring, evaluation and adaptation of development plans in the event of significant impacts, and (3) reserving specific environmental flows within the overall groundwater resource management strategy and planning to sustain key wetlands.

ACKNOWELDGEMENTS

We gratefully acknowledge financial support from the European Union via the 7th Framework Programme GENESIS: Groundwater and Dependent Ecosystems: New Scientific Basis on Climate Change and Land-use Impacts for the Update of the EU Groundwater Directive; WP-6 Groundwater Systems Management: Scenarios, Risk Assessment, Cost Efficient Measures and Legal Aspects. We finally thank two anonymous referees for constructive and insightful comments.

NOTES

1. Carson and Groves (2007) offer a discussion on these issues.
2. Offered bids were: €0, €2, €5, €10, €20, €50, €100, > €100.

3. The utility of a choice comprises a deterministic component (V) and an error component (ε), which is independent of the deterministic part and follows a predetermined distribution: $U_{ij} = V(z_{ij}, s_i) + \varepsilon(z_{ij}, s_i)$, where, for any individual i, a given level of utility will be associated with any alternative j. The researcher observes some attributes of the alternatives as faced by the individual, labelled $z_{ij}\ \forall j$, and some attributes of the individual, labelled s_i, and can then specify a function that relates these observed factors to the individual's utility.
4. Some respondents chose the option >100 among the payment options. This involves five respondents from Group 1 and five respondents from Group 2. These are included using the maximum bid as upper bound bid.
5. Although, the classic correction for heteroscedasticity is the HC0 estimator proposed by Huber (1967) and White (1980), MacKinnon and White (1985) discussed three improvements, HC1, HC2 and HC3 from which the latest is the best as suggested by Long and Ervin (2000), especially in small samples.

REFERENCES

Ariely, D., G. Loewenstein and D. Prelec (2003), 'Coherent arbitrariness: stable demand curves without stable preferences', *Quarterly Journal of Economics*, **118**(1), 73–105.

Barbier, E.B., M. Acreman and D. Knowler (1997), *Economic Valuation of Wetlands: A Guide for Policy Makers and Planners*, Gland, Switzerland: Ramsar Convention Bureau.

Bateman, I.J., R.T. Carson, B. Day, W.M. Hanemann, N. Hanley and T. Hett et al. (2003), *Guidelines for the Use of Stated Preferences Techniques for the Valuation of Preferences for Non-market Goods*, Cheltenham, UK and Northampton, MA, USA: Edward Elgar Publishing.

Bonnieux F. and P. Rainelli (1999), 'Contingent valuation methodology and the EU institutional framework', in I.J. Bateman and K.G. Willis (eds), *Valuing Environmental Preferences, Theory and Practice of the Contingent Valuation Method in the US, EU and Developing Countries*, New York: Oxford University Press, pp. 585–612.

Boulton, A. (2005), 'Chances and challenges in the conservation of groundwaters and their dependent ecosystems', *Aquatic Conservation: Marine and Freshwater Ecosystems*, **15**(4), 319–23.

Carson, R.T. and T. Groves (2007), 'Incentive and informational properties of preference questions', *Environmental Resource Economics*, **37**(1), 181–210.

Danielopol, D.L., J. Gibert, C. Griebler, A. Gunatilaka, H.J. Hahn and G. Messana et al. (2004), 'Incorporating ecological perspectives in European groundwater management policy', *Environmental Conservation*, **31**(3), 185–9.

Directive 2006/118/EC of the European Parliament and of the Council of 12 December 2006 on The Protection of Groundwater Against Pollution and Deterioration, accessed 20 February 2014 at: http://eurlex.europa.eu/LexUriServ/LexUriServ.do?uri=OJ:L:2006:372:0019:0031:EN:PDF.

Directive 2000/60/EC of the European Parliament and of the Council Establishing a Framework for the Community Action in the Field of Water Policy, accessed 20 February 2014 at http://eurlex.europa.eu/LexUriServ/LexUriServ.do?uri=CELEX:32000L0060:EN:NOT.

Evans, R. and C. Clifton (2001), *Environmental Water Requirements to Maintain Groundwater Dependent Ecosystems*, Canberra: Environment Australia.

Foster, S., P. Koundouri, A. Tuinhof, K. Kemper, M. Nanni and H. and Garduño (2006), *Groundwater Dependent Ecosystems: The Challenge of Balanced Assessment and Adequate Conservation*, Briefing Note Series No. 15, Sustainable Groundwater Management: Concepts and Tools, Washington, DC: World Bank.

Greene, W.H. (2002), *NLOGIT Version 4.0: User's Guide*, New York: Econometric Software.

Hanemann, W.M. (1984), 'Welfare evaluations in contingent valuation experiments with discrete responses', *American Journal of Agricultural Economics*, **66**(3), 332–41.

Hasler, B., T. Lundhede, L. Martinsen, S. Neye and J.S. Schou (2005), *Valuation of Groundwater Protection Versus Water Treatment in Denmark by Choice Experiments and Contingent Valuation*, NERI Technical Report No. 543, Roskilde, Denmark: National Environmental Research Institute.

Hausman, J. and D.L. McFadden (1984), 'Specification tests for the multinomial logit model', *Econometrica*, **52**(5), 1219–40.

Hensher, D., J. Rose and W. Green (2005), *Applied Choice Analysis: A Primer*, Cambridge, UK: Cambridge University Press.

Huber, P.J. (1967), 'The behavior of maximum likelihood estimates under non-standard conditions', *Proceedings of the Fifth Berkeley Symposium on Mathematical Statistics and Probability*, **1**, 221–33.

Hynes, H.B.N. (1983), 'Groundwater and stream ecology', *Hydrobiologia*, **100**(1), 93–9.

IPCC (2007), *Climate Change 2007: Impacts, Adaptation, and Vulnerability*, Contribution of Working Group

II to the Fourth Assessment Report of the Intergovernmental Panel on Climate Change, Cambridge, UK: Cambridge University Press.

Jensen, P.R., A.B. Hallam, J. Hauch, E. Møllgaard and O.G. Petersen (1995), 'En velfærdsindikator for Danmark 1970–1990' [A welfare indicator of Denmark from 1970 to 1990], Memorandum No. 8, Rockwool Foundation Research Unit, Copenhagen.

Johansson, P.-O. (1988), 'Valuing public goods in a risky world: an experiment', in H. Fomer and E. van Ierland (eds), *Valuation Methods and Policy Making in Environmental Economics, Studies in Environmental Science, Vol. 36*, Amsterdam: Elsevier.

Kahneman, D. and A. Tversky (1979), 'Prospect theory: an analysis of decision under risk', *Econometrica*, **47**(2), 263–91.

Kløve, B., A. Allan, G. Bertrand, E. Druzynska, A. Ertür and N. Goldscheider et al. (2011), 'Groundwater dependent ecosystems. Part II. Ecosystems services and management in Europe under risk of climate change and land use intensification', *Environmental Science and Policy*, **14**(7), 770–81.

Koundouri, P. (2000), 'Three approaches to measuring natural resource scarcity: theory and application to groundwater', PhD thesis, Faculty of Economics and Politics, University of Cambridge, Cambridge, UK.

Koundouri, P. and C. Christou (2000), 'Dynamic adaptation to resource scarcity with backstop availability', paper presented at the 10th Annual Conference, European Association of Environmental and Resource Economists, Rethymnon, Greece, June.

Koundouri, P. and A. Xepapadeas (2004), 'Estimating accounting prices for common pool natural resources: a distance function approach', *Water Resources Research*, **40**(6), W06S17.

Koundouri, P., E. Kougea, M. Stithou, P. Ala-aho, R. Eskelinen and T. Karjalainen et al. (2012), 'The value of scientific information on climate change: a choice experiment on Rokua esker, Finland', *Journal of Environmental Economics and Policy*, **1**(1), 85–102.

Krinsky, I. and R. Robb (1986), 'On approximating the statistical properties of elasticities', *Review of Economics and Statistics*, **68**(4), 715–19.

Long, J.S. and L.H. Ervin (2000), 'Using heteroscedasticity consistent standard errors in the linear regression model', *The American Statistician*, **54**(3), 217–24.

Louviere, J., D. Hensher, J. Swait and W. Adamowicz (2000), *Stated Choice Methods: Analysis and Application*, Cambridge, UK: Cambridge University Press.

MacKinnon, J.G. and H. White (1985), 'Some heteroskedasticity consistent covariance matrix estimators with improved finite sample properties', *Journal of Econometrics*, **29**(3), 305–25.

Macmillan, D., N. Hanley and S. Buckland (1996), 'A contingent valuation study of uncertain environmental gains', *Scottish Journal of Political Economy*, **43**(5), 519–33.

Mitchell, R.C. and R.T. Carson (1981), *An Experiment in Determining Willingness to Pay for National Water Quality Improvements*, draft report to the US Environmental Protection Agency, Washington, DC: Resources for the Future.

Mitchell, R.C. and R.T. Carson (1984), *A Contingent Valuation Estimate of National Freshwater Benefits*, technical report to the US Environmental Protection Agency, Washington, DC: Resources for the Future.

Mitchell, R.C. and R.T. Carson (1989), *Using Surveys to Value Public Goods: The Contingent Valuation Method*, Washington, DC: Resources for the Future.

Pearce, D.W. and E. Ozdemiroglu (2002), *Economic Valuation with Stated Preference Techniques. Summary Guide*, London: Department for Transport, Local Government and the Regions.

Perman, R.J., Y. Ma, J. McGilvray and M. Common (2003), *Natural Resource and Environmental Economics*, 3rd edition, Reading, MA: Addison Wesley Longman.

Press, J. and T. Söderqvist (1998), 'On estimating the benefits of groundwater protection: a contingent valuation study in Milan', in T. Swanson and M. Vighi (eds), *Regulating Chemical Accumulation in the Environment*, Cambridge, UK: Cambridge University Press, pp. 121–82.

Roberts, D.C., T.A. Boyer and J.L. Lusk (2008), 'Preferences for environmental quality under uncertainty', *Ecological Economics*, **66**(4), 584–93.

Rozan, A., A. Stenger and M. Willinger (1997), 'Valeur de préservation de la qualité de l'eau souterraine: une comparaison entre usagers et non-usagers' [Valuation of the preservation of underground water quality: a comparison between users and non-users], *Cahiers d'économie et sociologie rurales*, **45**, 61–92.

Train, K. (2003), *Discrete Choice Methods with Simulation*, Cambridge, UK: Cambridge University Press.

White, H. (1980), 'A heteroskedasticity-consistent covariance matrix estimator and a direct test for heteroskedasticity', *Econometrica*, **48**(4), 817–38.

PART V

THE ROLE OF GOVERNANCE AND SCIENCE–POLICY–BUSINESS INTERFACE IN BRINGING VISIBLE ECOSYSTEM VALUES

27. Governance is critical to managing coastal and marine resources: effects of marine management areas

Giselle Samonte, Daniel Suman, Juan Maté, Diego Quiroga, Carlos Mena, Adele Catzim-Sanchez, Patrick Fong and Xuanwen Wang

27.1 INTRODUCTION

Worldwide, coral reefs, mangroves, seagrass beds, and other highly diverse tropical marine ecosystems are under sharp decline. Anthropogenic impacts are degrading water quality, habitat configuration and the ecological structure of entire coastal systems. Consequently, most coastal marine fisheries are under an increasing threat of collapse. This global crisis poses an unprecedented challenge not only to marine biodiversity conservation, but also to the livelihood of millions of people who depend on healthy coastal ecosystems, especially in developing countries. Globally, almost 50 percent of fisheries are at maximum capacity, while more than 25 percent have been pushed beyond sustainable limits. Industrial fishing practices have depressed populations of large predatory fish to about 10 percent of pre-industrial levels throughout the global ocean. Recent assessments show that 20 percent of the world's coral reefs have been effectively destroyed, a further 24 percent are under imminent risk of collapse, and another 26 percent are under long-term threat from human-caused pressures.

Marine managed areas (MMAs) of various types are a form of resource management that regulates human activities in particular locations (area-based management strategy). There are many types and management regimes of MMAs, from multiple-use and community-managed areas to no-take reserves, but objectives generally converge at socio-economic (e.g., fisheries, tourism) and biodiversity conservation benefits. Due to their immense potential and cost-effectiveness, MMAs are being proposed as central coastal and marine management tools, and there has been increasing interest – particularly among international, non-governmental and multilateral development organizations – in evaluating and developing tools to increase MMAs' effectiveness (Orbach and Karrer, 2010). The current challenge, however, is to ensure that these commitments are transformed into meaningful actions.

Governance systems – those arrangements by which communities of people at different scales make common rules of behavior – occur in many different forms across nations and cultures (Figure 27.1). There is also a significant difference between governance structures on land and in the sea. On land, most property and many resources are subject to private ownership, as private property. In the sea, it is generally true that the water, seabed, and resources are common pool, or common property. That is, those environments and resources are held in trust by governments and managed for the

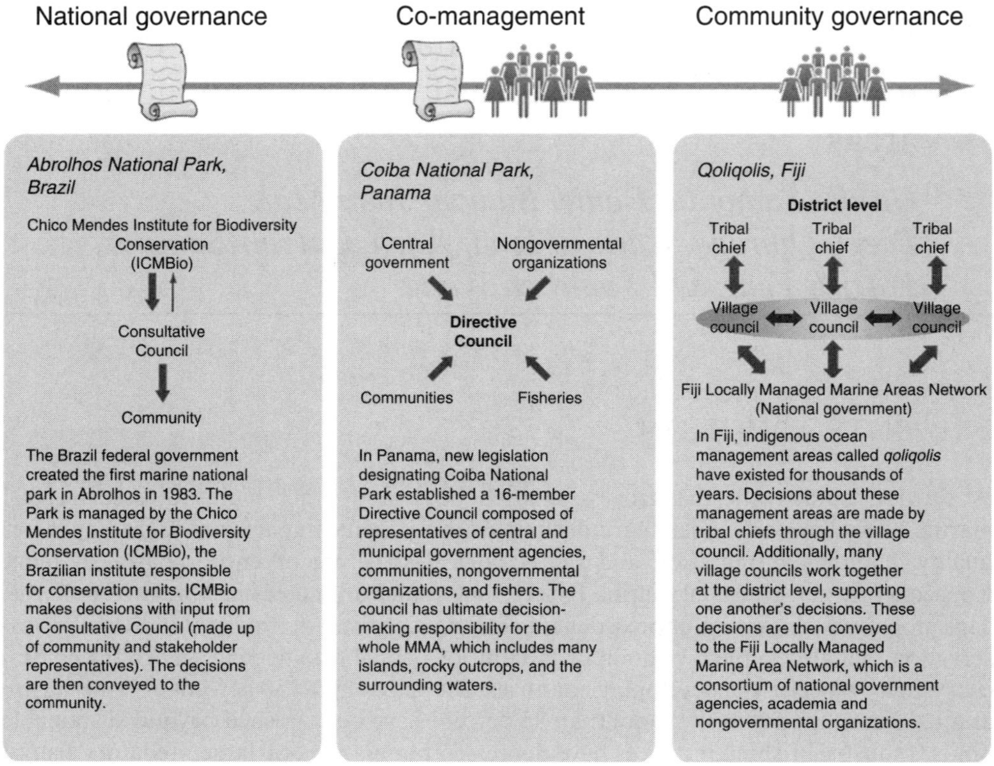

Source: Samonte et al. (2010).

Figure 27.1 Governance spectrum

benefit of all people. Within the common pool of marine environments and resources, there are many different governance arrangements, such as national governance, co-management, and community governance (Samonte et al., 2010). These arrangements can also extend across national boundaries for transboundary, international environments and resources. All, however, feature different scales of human communities with specific cultural values pertaining to the use of the marine environment. For example, a particular area may be valued by a society for spiritual or aesthetic reasons, and significant use or alteration may not be desirable. Another area may be valued for a particular extractive resource such as fish, and significant use may be desirable. The key is that the values of the culture and society are reflected in the marine managed area governance.

The importance of governance has been the focus and emphasis for the sustainability of socioecological systems (Ostrom, 2009). This includes studies on governance issues across geographical scales, that is from being able to simultaneously maintain access and use controls for the continuing sustainability of fishing grounds in community-based parks (Basutro, 2005) to the analysis of living marine resource governance in the large marine ecosystems in the Caribbean (Fanning et al., 2009). Some governance factors such as successful alternative income projects, high levels of participation in community decision-making, continuing advice from the implementing organization and inputs from local government (Pollnac et al., 2001) have been identified for effective management. Although a core set of guidelines for sustainable governance (responsibility, scale-matching, precaution, adaptive management, full cost allocation, and participation) has been developed (Costanza et al., 1999), common governance challenges, such as confused goals, conflict, and unrealistic attempts to scale up beyond institutional capacity (Christie and White, 2007) still exist.

This chapter contributes to the literature on marine conservation as it examines how marine managed areas have affected the resource use, dependence, and governance effects of coastal communities.

27.2 STUDY AREA AND DATA

Data for this study comes from the household surveys conducted in Belize (Caribbean), Fiji (Western Pacific), Ecuador (Eastern Pacific), and Panama (Eastern Pacific): five monitoring locations in Belize – Lighthouse Reef, South Water Caye, Laughing Bird/ Gladden Spit, Sapodilla Cayes, and Port Honduras; four locations in Fiji – Navakavu, Waitabu, Navatu, and Solevu; the Galapagos National Park in Ecuador and the Coiba National Park in Panama. Since the islands forming Coiba National Park are uninhabited and the local communities situated along the Panamanian coast and opposite Coiba National Park are historically dependent on the access to resources within and around Coiba National Park, nine communities out of these local communities were selected for the study (Figure 27.2).

All the households surveyed in these countries were randomly selected. The total sample size for Belize, Fiji, Ecuador, and Panama are 1341, 183, 365, and 497 respectively. The total sample size for cross-node analysis is 2386 (Table 27.1).

The indicators for resource use and dependence were income and livelihood. To test whether people's income and livelihoods have been increased or not after the establishment of MMAs, baseline data is needed to test the difference between the current situation and before the establishment of MMAs. However, it is very hard to get all the information before the establishment of MMAs. Therefore, we divide the whole sample into two groups, MMA beneficiaries and non-MMA beneficiaries, to test whether there is any significant difference between two groups. Non-MMA beneficiaries are treated as a baseline for those who use resources from MMAs. In total, there are 601 respondents who are non-MMA beneficiaries in Belize, while two villages and 33 respondents in Fiji take advantage of marine resources from non-MMA sites. Out of 497 respondents in Panama, 304 are non-Park users. Out of 363 respondents in Ecuador, 90 are Park users and others are non-Park users. Detailed information of MMA and non-MMA beneficiaries is summarized in Table 27.2.

A *t*-test with equal variance[1] is utilized to test the differences in the study. The null hypothesis is that the means of variables of the two groups are equal. The alternative hypothesis is that the mean of variable in MMA group is higher or lower than the mean of variable in the non-MMA group, which will depend on the effects we want to test. The *t*-test with equal variance is given by:

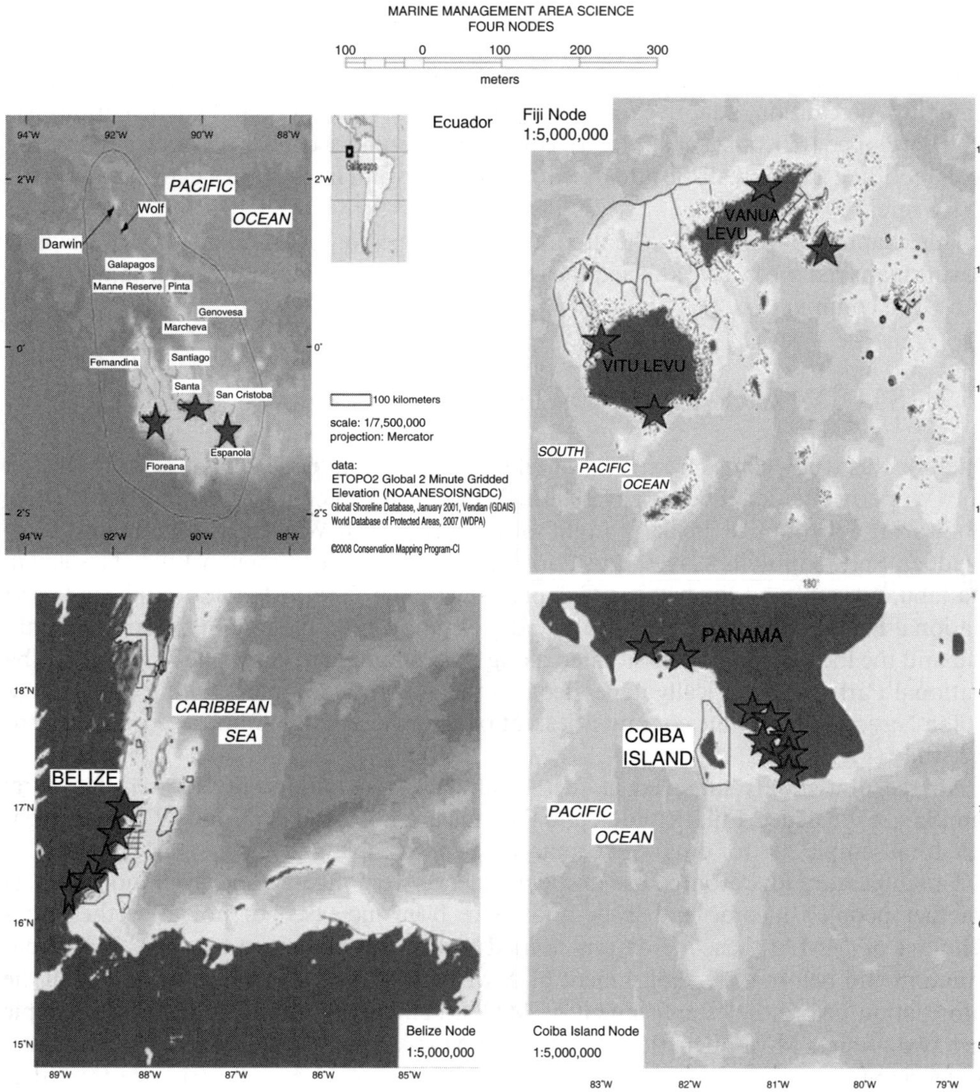

Source: Marine Management Area Science Program, Conservation International.

Figure 27.2 Coastal communities surveyed

$$t = \frac{\overline{X}_1 - \overline{X}_2}{S_{12}^* \sqrt{\dfrac{1}{N_1} + \dfrac{1}{N_2}}}, \; S_{12} = \sqrt{\frac{(N_1 - 1)S_1^2 + (N_2 - 1)S_2^2}{N_1 + N_2 - 1}}$$

Assume the variance is equal across countries.

Table 27.1 Sample sizes for selected MMAs and coastal communities

Country/Node	Marine Managed Areas	Coastal Communities	Respondents
Belize	Lighthouse Reef	Chunox	76
(n = 1341)	South Water Caye	Copper Bank	65
	Laughing Bird/Gladden Spit	Dangriga	276
	Sapodilla Cayes	Hopkins	105
	Port Honduras	Independence	187
		Monkey River	27
		Placencia	97
		Punta Negra	229
		Punta Gorda	7
		Sarteneja	141
		Seine Bight	104
		Sittee River	27
Fiji	Navakavu	Waiqanake	28
(n = 183)	Waitabu	Muaivuso	16
	Kubulau	Namakala	11
	Malolo	Nabaka	5
		Wai	5
		Waitabu	14
		Vurevure	6
		Navatu	15
		Raviravi	10
		Kiobo	10
		Namalata	7
		Solevu	23
Ecuador	Galapagos National Park	Santa Cristobal	113
(n = 365)		Isabela	63
		Santa Cruz	189
Panama	Coiba National Park	Bahia Honda	28
(n = 497)		El Puerto	76
		Gobernadora	15
		Hicaco	89
		Malena	26
		Pedregal	120
		Pixvae	69
		Puerto Mutis	15
		Santa Catalina	59
Cross-node (n = 2386)	11	36	2386

27.3 VALUATIONS (ECOLOGICAL VALUATION VERSUS ECONOMIC VALUATION)

27.3.1 Resource Use

The results of the *t*-test with equal variance are presented in Table 27.3. The null hypothesis is that the number of fishers among MMA resource users is equal to those

Table 27.2 MMA users vs non-MMA users

Country	Respondents Surveyed	MMA Beneficiaries		Non-MMA Beneficiaries	
Belize	1341	Total subsample	740	Total subsample	601
		Chunox		Chunox	
		Copper Bank		Copper Bank	
		Dangriga		Dangriga	
		Hopkins		Hopkins	
		Independence		Independence	
		Monkey River		Monkey River	
		Placencia		Placencia	
		Punta Negra		Punta Negra	
		Punta Gorda		Punta Gorda	
		Sarteneja		Sarteneja	
		Seine Bight		Seine Bight	
		Sittee River		Sittee River	
Fiji	183	Total subsample	150	Total subsample	33
		Waiqanake	28	Kalokolevu	23
		Muaivuso	16	Tavulomo	10
		Namakala	11		
		Nabaka	5		
		Wai	5		
		Waitabu	14		
		Vurevure	6		
		Navatu	15		
		Raviravi	10		
		Kiobo	10		
		Namalata	7		
		Solevu	23		
Ecuador	363	Total subsample	90	Total subsample	273
		Santa Cristobal	28	Santa Cristobal	85
		Isabela	21	Isabela	41
		Santa Cruz	41	Santa Cruz	147
Panama	497	Total subsample	193	Total subsample	304
		Bahia Honda	28	Hicaco	89
		El Puerto	76	Malena	26
		Gobernadora	15	Pedregal	120
		Puerto Mutis	15	Pixvae	69
		Santa Catalina	59		

Source: Household surveys conducted in these four countries.

among non-MMA users (H_0: $\mu_{MMA} = \mu_{Non-MMA}$), while the alternative hypothesis is that the number of fishers among MMA resource users is greater than those among non-MMA users (H_a: $\mu_{MMA} > \mu_{Non-MMA}$). The statistical results in Belize, Ecuador, Panama, and cross-node show that there is sufficient evidence to conclude that the null hypothesis is rejected, while there is no sufficient evidence in Fiji that there is sig-

Table 27.3 T-test results for resource use

Variable	Belize	Fiji	Ecuador	Panama	All Sites
Marine related					
MMA beneficiaries	0.35	0.67	1.00	0.60	0.41
Non-MMA beneficiaries	0.06	0.58	0.11	0.48	0.17
t-value	13.76	1.07	25.40	2.57	13.31
p-value	0.00***	0.14	0.00***	0.01***	0.00***
Fishing					
MMA beneficiaries	0.31	0.59	0.47	0.47	0.35
Non-MMA beneficiaries	0.01	0.58	0.00	0.45	0.14
t-value	15.83	0.12	14.75	0.38	12.43
p-value	0.00***	0.46	0.00***	0.35	0.00***
Tourism					
MMA beneficiaries	0.10	0.16	0.64	0.13	0.11
Non-MMA beneficiaries	0.05	0.00	0.11	0.03	0.03
t-value	3.24	2.49	11.30	4.37	6.87
p-value	0.00***	0.01***	0.00***	0.00***	0.00***
Boat drivers/divers					
MMA beneficiaries	0.08	0.05	0.00	0.00	0.06
Non-MMA beneficiaries	0.00	0.00	0.00	0.00	0.00
t-value	6.86	1.26	n.a.	n.a.	8.12
p-value	0.00***	0.10*	n.a.	n.a.	0.00***

Note: *, **, *** significant at 10%, 5%, and 1%, respectively; n.a. = not applicable.

nificant difference between MMA users and non-MMA users when it comes to fishing; however, there are more respondents who are involved in tourism among MMA users than those among non-MMA users. In-depth analysis of the benefits and challenges of 17 MMAs in Belize, Ecuador, Fiji, and Panama identified the following improvements: more diversified livelihoods – community members whose livelihood is directly tied to the MMA (MMA users) are much more likely to be engaged in both tourism and fishing than community members whose livelihood is marine based but not tied to the MMA (Figure 27.3).

27.3.2 Resource Dependence

One of the most important socioeconomic effects of MMAs is whether the income of coastal population increased or is maintained after the establishment of MMAs. Income variables in this study include average monthly household income and perception of economic situation. Average monthly household income is the total monthly income of all household members. The results of the *t*-test with equal variance are presented in Table 27.4. The null hypothesis is that the average monthly household income of

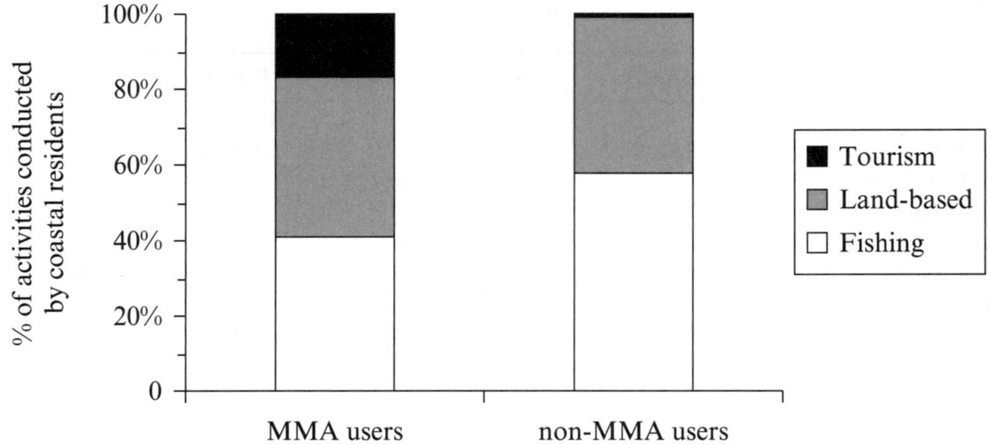

Figure 27.3 Resource use as reflected by more diversified livelihoods

Table 27.4 T-test results for dependence on marine resources, average monthly household income (US$)

Variable	Belize	Fiji	Ecuador	Panama	Cross-node
Total mean	1291	385	430	148	810
MMA beneficiaries	1378.53	370.63	3132.31	161.63	979.85
Non-MMA beneficiaries	1178.73	450.85	34.77	139.57	764.20
t-value	2.7643	−1.0765	8.4289	2.1276	4.1893
p-value	0.0029***	0.8584	0.0000***	0.0169**	0.0000***

Note: *, **, *** significant at 10%, 5%, and 1%, respectively.

respondents who use resources in MMA is equal to those who don't use resources in MMA (H_0: $\mu_{MMA} = \mu_{Non-MMA}$), while the alternative hypothesis is that average monthly household income of respondents who use resources in MMA is higher than those who don't use resources in MMA (H_a: $\mu_{MMA} > \mu_{Non-MMA}$). The statistical result in Belize, Ecuador, Panama, and cross-node show that there is sufficient evidence that the null hypothesis is rejected, while there is no sufficient evidence in the Fiji analysis that the null hypothesis is rejected. The reason is probably that about 51.5 percent of respondents' livelihood is non-marine related. The outcome is greater income – community members whose livelihood (e.g., fishing, tourism) is directly tied to the MMA (MMA users) have a higher average income than community members whose livelihood is marine based but not tied to the MMA (Figure 27.4).

27.3.3 Governance Effects

Management structures and strategies include local understanding of MMA rules and regulations and level of participation in the development of a management plan. The

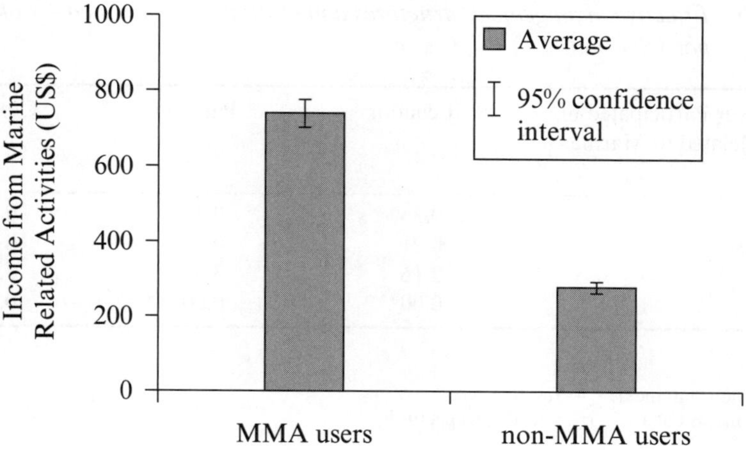

Figure 27.4 Resource dependence as indicated by greater income from marine managed areas

Table 27.5a Effective management structures and strategies maintained – rules and regulations

Variable	Belize	Fiji	Ecuador	Panama	Cross-node
MMA beneficiaries	0.48	0.91	0.65	0.26	0.51
Non-MMA beneficiaries	0.24	0.00	0.35	0.16	0.24
t-value	9.36	18.55	4.99	2.91	14.56
p-value	0.00***	0.00***	0.00***	0.00***	0.00**

Notes:
Answers for each statement: 1 = Yes; 0 = No.
*, **, *** significant at 10%, 5%, and 1%, respectively.

results of the *t*-test in Belize, Fiji, Ecuador, Panama, and cross-nodes are presented in Table 27.5a. The null hypothesis is that the awareness of regulation and rules of MMA users is equal to those who don't use resources in MMA (H_0: $\mu_{MMA} = \mu_{Non-MMA}$), while the alternative hypothesis is that the awareness of rules and regulations of MMA users is stronger than those who don't use resources in MMA (H_a: $\mu_{MMA} > \mu_{Non-MMA}$). The statistical results in all countries show that there is sufficient evidence that the null hypothesis is rejected. Therefore, we can say that marine users are more aware of the rules and regulations and more active in the management of MMAs. MMA users perceive that management in MMAs is more effective than non-MMAs compared to non-MMA users. Cross-node analysis shows that MMA users are more likely to know the rules and regulations than non-MMA users. Statistical results in Ecuador and Panama show that MMA users are more likely to be involved in MMA meetings and management plans (Table 27.5b).

Level of capacity building/training provided to stakeholders is used to measure the

Table 27.5b Effective management structures and strategies maintained – stakeholder participation

Have you Ever Participated in a Meeting Related to Marine Reserve?	Ecuador	Panama	Cross-node
Park users	0.36	0.28	0.27
Non-park users	0.21	0.16	0.16
t-value	3.16	3.31	3.93
p-value	0.00***	0.00***	0.00***

Notes:
Answers for each statement: 1 = Yes; 0 = No.
*, **, *** significant at 10%, 5%, and 1%, respectively.

effective stakeholder participation. This information was only collected in Belize and Fiji. The null hypothesis is that the level of capacity building/training provided to stakeholders in participation who use resources in MMA is equal to those who don't use resources in MMA (H_0: $\mu_{MMA} = \mu_{Non-MMA}$), while the alternative hypothesis is that the level of capacity building/training provided to stakeholders in participation who use resources in MMA is higher than those who don't use resources in MMA (H_a: $\mu_{MMA} < \mu_{Non-MMA}$). The statistical results show that there is sufficient evidence that the null hypothesis is rejected (Table 27.6).

Environmental awareness and knowledge includes six statements. Respondents were asked to choose among strongly agree, agree, neutral, disagree, and strongly disagree. These environmental awareness and knowledge variables were only collected in Belize, Fiji, and Panama. Therefore, cross-node analysis in this section is limited to these three countries. The results of the paired *t*-test are presented in Table 27.7. The null hypothesis is that the perception of environmental awareness and knowledge of respondents who use resources in MMA is equal to those who don't use resources in MMA (H_0: $\mu_{MMA} = \mu_{Non-MMA}$), while the alternative hypothesis is that the perception of environmental awareness and knowledge of respondents who use resources in MMA is higher than those who don't use resources in MMA (H_a: $\mu_{MMA} > \mu_{Non-MMA}$). The statistical results show that there is sufficient evidence that the null hypothesis is rejected. Respondents who use marine resources from MMAs have stronger environmental awareness and knowledge than their counterparts who don't use marine resources from MMAs. These perceptions of environmental awareness and knowledge of MMAs are only collected in Belize and Fiji.

The results of the paired *t*-test are presented in Table 27.8. The null hypothesis is that the information dissemination among respondents who use resources in MMA is equal to those who don't use resources in MMA (H_0: $\mu_{MMA} = \mu_{Non-MMA}$), while the alternative hypothesis is that the information dissemination among respondents who use resources in MMA is lower than those who don't use resources in MMA (H_a: $\mu_{MMA} < \mu_{Non-MMA}$). The statistical results show that there is sufficient evidence that the null hypothesis is rejected. Respondents who use marine resources from MMAs are more likely to get information and training from MMA bodies than respondents who don't use marine

Table 27.6 Level of capacity building and training

Statements	MMA beneficiaries (n = 890)	Non-MMA beneficiaries (n = 601)	t-value	p-value
1. Have you or anyone in your family ever received training in environmental education related to the MPA?	1.82	2.51	−8.5483	0.0000***
2. Have you or anyone in your family ever received any tour guide training as a result of the MPA?	1.77	2.13	−6.8197	0.0000***
3. Have you or anyone in your family ever received any arts and craft training that uses marine resources since the establishment of the MPA?	2.02	2.18	−2.9304	0.0017***
4. Have you or anyone in your family ever received a scholarship to attend formal schooling (primary or high school) as a result of assistance from the marine management body?	2.08	2.17	−1.7374	0.0413**
5. Have you or anyone in your family ever got a job related in some way to the MPA?	1.89	2.14	−4.7662	0.0000***

Notes:
Answers for each statement: 1 = Yes; 2 = No.
*, **, *** significant at 10%, 5%, and 1%, respectively.

resources from MMAs. Information dissemination is more efficient among MMA users than non-MMA users.

27.4 DISCUSSION AND CONCLUSIONS

Marine managed areas benefit coastal communities and are important in terms of resource use, dependences, and guaranteeing the possibility for local inhabitants to shift from an extractive economy based on the use of provisioning services to an economy based on the utilization of non-extractive cultural services. In the case of fisheries, marine managed areas have meant that their socioeconomic situation has improved owing to the restrictions on illegal fishing activities and the prohibition of the commercial vessels inside local fishing grounds. Users of marine managed areas are exploiting marine resources in a more sustainable way than they were ten years ago. Statistical results show, for example in Fiji, that current fishing techniques are more sustainable than those of ten years ago. In addition, many fishers are now hopeful that they will find alternative ways

Table 27.7 Environmental awareness and knowledge

Statements	MMA Beneficiaries (n = 1080)	Non-MMA Beneficiaries (n = 835)	t-value	p-value
1. Organizations that manage the resources are taking the bread out of people's mouths	3.09	2.85	3.5145	0.0002***
2. We do not have to worry about the sea and the fish. God will take care of it for us	3.66	3.49	2.5765	0.0050***
3. We should manage the sea to ensure that there are fish for our children and their children	1.48	1.60	−3.6723	0.0001***
4. We have to take care of the land and sea or they will not provide for us in the future	1.48	1.58	−3.1220	0.0009***
5. We want to protect the land and the sea but this is hard because we have economic needs now	2.33	2.05	4.8821	0.0000***
6. Protecting the land and the sea brings us more benefits than not protecting these resources	1.78	1.76	0.3148	0.6235

Notes:
Answers for each statement: 1 = Strongly agree; 2 = Agree; 3 = Neutral; 4 = Disagree; 5 = Strongly disagree.
*, **, *** significant at 10%, 5%, and 1%, respectively.

of making a living as they are already engaging in a wide range of activities that allows them to diversify their household economies and to lower the risk generated by natural and market fluctuations.

Contrary to fishing, the importance of tourism has been growing steadily. More people are increasingly getting involved in this sector, which has undoubtedly become the most important and dynamic sector of the local economy. In terms of the well-being of the general population, marine managed areas ensure an adequate level of income for an important part of the local population. The income of respondents who use marine managed areas have both higher household income and fishing income than respondents who do not use these resources. Coastal communities benefitting from marine managed areas perceive that they have benefited from tourism.

Marine managed areas can be considered as governance tools for managing the natural and cultural resources. This concept includes the rules that determine who has access to the area and when access is permitted; in essence, the legal framework and policies regarding resource use. Governance comprises more than the formal and informal rules of the game and also includes the structures created and adopted to reach decisions, the social actors that are responsible for the implementation of the actions that form part of the management of the marine area. A further aspect of governance embraces public participation and the degree of inclusion of the public, organized groups of users and other interested social groups.

Table 27.8 Information dissemination

Does the MPA Body Share Information with You or Your Family as it Relates to:	MMA Beneficiaries (n = 1080)	Non-MMA Beneficiaries (n = 834)	t-value	p-Value
1. The size and boundaries of the marine protected area?	1.66	1.96	−4.8522	0.0000***
2. The eco-system impact of having a marine protected area (e.g., the impact of having mangroves or the reef system)?	1.72	2.01	−4.2462	0.0000***
3. The biodiversity found within the marine protected area (e.g., give information on the kinds of animals and plants)?	1.76	2.03	−3.7833	0.0001***
4. The use of the natural resources within the marine protected area (e.g., use of the animals, plants, corals, beaches, mangroves)?	1.69	2.01	−4.8318	0.0000***
5. The social and economic benefits you can get from the marine protected area?	1.83	2.01	−2.6796	0.0037***
6. How you can participate in activities related to the marine protected area?	1.84	2.09	−3.4316	0.0003***

Notes:
Answers for each statement: 1 = Yes; 2 = No.
*, **, *** significant at 10%, 5%, and 1%, respectively.

The experience of the 11 marine managed areas is evolving, and all communities and management have had similar benefits. The process of the development of the management plan serves as a stimulus to all stakeholder groups to better define their roles and be more responsible in their approaches to marine conservation. A central principle of environmental management today is the promotion of public participation – involving the organized public in the park planning process. Nevertheless, we must ask whether the individuals who serve as members of these committees are really the best representatives of their social sectors that they apparently represent. Additionally, we must critically examine the transparency of the processes involved in the development of the management plans and search for mechanisms to more effectively include the social actors interested in and affected by the creation of marine managed areas.

Possible recommendations to improve this situation include public outreach campaigns targeted to distinct social sectors, as well as mechanisms to ensure that the public representatives who serve on the various committees are truly representative of their groups.

ACKNOWLEDGEMENTS

Funding and support was provided by The Gordon and Betty Moore Foundation through Conservation International's Marine Management Area Science Program.

NOTE

1. A *t*-test with unequal variance is also conducted in the study. The paired *t*-test with equal variance and unequal variance is given by $t = \frac{\bar{X}_1 - \bar{X}_2}{\sqrt{\frac{S_1^2}{N_1} + \frac{S_2^2}{N_2}}}$. Assume the variance is unequal across countries.

REFERENCES

Basutro, X. (2005), 'How locally designed access and use controls can prevent the tragedy of the commons in a Mexican small-scale fishing community', *Society and Natural Resources*, **18**(7), 643–59.
Christie, P. and A.T. White (2007), 'Best practices for improved governance of coral reef marine protected areas', *Coral Reefs*, **26**(4), 1047–56.
Costanza, R., F. Andrade and P. Antunes et al. (1999), 'Ecological economics and sustainable governance of the oceans', *Ecological Economics*, **31**(2), 171–87.
Fanning, L., R. Mahon and P. McConney (2009), 'Focusing on living marine resource governance: the Caribbean large marine ecosystem and adjacent areas project', *Coastal Management*, **37**(3–4), 219–34.
Orbach, M. and L. Karrer (2010), *Marine Managed Areas: What, Why, and Where*, Arlington, VA: Science and Knowledge Division, Conservation International.
Ostrom, E. (2009), 'A general framework for analyzing sustainability of social-ecological systems', *Science*, **325**(5939), 419–22.
Pollnac, R., B. Crawford and M. Gorospe (2001), 'Discovering factors that influence the success of community-based marine protected areas in the Visayas, Philippines', *Ocean & Coastal Management*, **44**(11), 683–710.
Samonte, G., L. Karrer and M. Orbach (2010), *People and Oceans*, Arlington, VA: Science and Knowledge Division, Conservation International.

28. Strengthening the science–policy interface: lessons from the Intergovernmental Platform on Biodiversity and Ecosystem Services

Anantha Kumar Duraiappah

28.1 CONTEXT

The six-year process to establish an intergovernmental science-policy interface for biodiversity and ecosystem services came to a culmination when more than 90 countries agreed on 21 April 2012 to establish the Intergovernmental Platform on Biodiversity and Ecosystem Services (IPBES). The growing recognition by the international community of the seriousness of the rapid change in biodiversity and ecosystem services and the impacts it will have on the well-being of present and future generations propelled countries to take this bold step (UNEP, 2012b).

The complex nature of the issue at hand and the need for credible and independent science to inform policy-making at multiple levels were primary reasons for going ahead with the establishment of IPBES. Another reason for establishing IPBES was the recognition of the large number of biodiversity and ecosystem-services-related multilateral environmental agreements (MEAs) and the strong and urgent need for a common science–policy platform (UNEP, 2009b). The platform is expected to strengthen these MEAs by providing highly credible science in a periodic and timely manner for policy-making at multiple levels of governance.

The platform is modeled after the Intergovernmental Panel on Climate Change (IPCC) but with some distinct differences. First, unlike climate change, which is predominantly a global change phenomenon, biodiversity and ecosystem services changes are more place based and interactions happen at multiple scales. Second, the close link with human well-being makes transdisciplinary knowledge and multi-knowledge necessary conditions for all activities of IPBES. Third, the platform has to inform not a single multilateral environmental agreement but multiple MEAs addressing the broad range of ecosystem services and biodiversity.

The main purpose of this chapter is to highlight a largely ignored but important component within the scientific and academic community. The science–policy interface is critical for the use of science in addressing and solving the increasingly complex and interdependent environmental problems we face today. Just doing science for science sake is no longer feasible. Policy-makers and the public at large who provide taxpayers' funds for science are expecting higher returns from their investments in scientific research.

The chapter is structured as follows. First, a brief historical account of the process by which IPBES was established is provided. This section will highlight the political sensitivity surrounding science and scientific findings. The Section 28.3 will provide an overview of the key objectives of the platform. This section will highlight the special niche IPBES will fill and the deliverables and outcomes expected of such a body. Section 28.4 will highlight the delicate practice of balancing the needs of multiple user constituencies. This section will also elaborate the complexities of negotiations in an intergovernmental context and the principle of decision-making by consensus. Section 28.5 will provide a brief overview of the science–policy cycle and the various activities IPBES will undertake to ensure efficient and effective achievement of its core mandate and objectives. Section 28.6 discusses the key principles and characteristics of IPBES. The chapter will end with some key lessons and conclusions in the last three sections.

28.2 THE IPBES STORY

The establishment of IPBES can be traced back to the following four major initiatives: (1) the Millennium Ecosystem Assessment (MA); (2) the International Mechanism of Scientific Expertise in Biodiversity (IMoSEB); (3) the Millennium Ecosystem Assessment Follow-up Global Strategy; and (4) the IPBES consultative process.

The MA was initiated in 2000 by a call from United Nations Secretary General Kofi Annan in his report to the UN General Assembly, *We the Peoples: The Role of the United Nations in the 21st Century* (UN, 2000). The objective of the MA was to assess the changes in ecosystem services and their consequences to human well-being. Although governments were not directly involved, their support was secured through decisions taken by a number of biodiversity and ecosystem-services-related MEAs. It is important to note that the MA was not an intergovernmental assessment. It had a science panel (led by Walt Reid, Harold Mooney, and Angela Crooper) and a multi-stakeholder board (led by Robert Watson and Hamid Zakri) to oversee its implementation. The MA was completed in 2005 with the involvement of over 1300 scientists from all major regions.

The IMoSEB was a process initiated by a call from French President Jacques Chirac in 2005 at the conclusion of the international conference organized by the government of France called, 'Biodiversity: Science and Governance' (Loreau et al., 2006). The objective of the IMoSEB process was to undertake a consultative process to explore the feasibility of establishing an IPCC-like body for biodiversity. The consultative process took two years, with regional consultation workshops held in Beijing (Asia), Montreal (Americas), Yaoundé (Africa), Geneva (Europe), Bariloche (South America) and Alotou (Pacific). This process was led by an international science committee (led by Michael Loreau and Alfred Oteng-Yeboah). The IMoSEB process concluded in 2007 with a final meeting held in Montpellier, and a statement requesting the UNEP executive director to take on board the outcomes of the IMoSEB process and to continue exploring the options of establishing an intergovernmental platform for biodiversity.

The MA Follow-Up Global Strategy was initiated in 2007 by UNEP (UNEP, 2008a). The decision to initiate an MA follow-up process was driven by the evaluations of the MA by the Global Environmental Facility (GEF) and the United Kingdom's House of Commons Environmental Audit Committee. Both evaluations suggested a strengthening of the sub-regional assessments undertaken by the MA, to establish a process whereby the global assessment on ecosystem services is conducted in a periodic and systematic manner and finally to support policy implementation of assessment findings. The UNEP MA Follow-Up Global Strategy (led by Anantha Duraiappah and Ibrahim Thiaw) held its first meeting Stockholm in October 2007 to develop the follow-up strategy and one key component was the establishment of a global assessment process.

The IPBES consultative process began in earnest in 2008. A small group of scientists and policy-makers convened in Paris in April 2008 by UNEP (led by Anantha Duraiappah and Ibrahim Thiaw) and the French Ministry of Environment (led by Laurence Tubiana and Didier Hoffshir) to develop a strategy to set in place an intergovernmental process to establish an Intergovernmental Platform on Biodiversity and Ecosystem Services. A concept note was developed for the Convention on Biological Diversity (CBD) in 2008 in Bonn, Germany (UNEP, 2008b). This was immediately followed by the first intergovernmental meeting hosted by the government of Malaysia in Putra Jaya. Although there was acknowledgement of the need to strengthen the science–policy interface on biodiversity and ecosystem services, many countries were still hesitant on the need for an intergovernmental platform (UNEP, 2008c). The reason for these hesitations will be provided in more detail in the following sections.

The second intergovernmental meeting took place in Nairobi in 2009. In this meeting governments were more receptive to the idea of an intergovernmental platform based on the information provided as requested from the first meeting (UNEP, 2009a). Governments requested UNEP to continue with the process and a third and final intergovernmental meeting was held in Busan, South Korea. At this meeting, the Busan outcome was declared and this document laid the building blocks and foundations for a proposed intergovernmental science–policy platform (UNEP, 2010).

Although IPBES was to be yet formally established, governments were fully on board for the establishment of such a body. The next two meetings took on a higher level of prominence and were called the first and second session of a plenary of an intergovernmental meeting. The mandate was clear for these meetings. The objective was to develop the governance structure for IPBES and to identify in broad terms the key components of the work program; in other words, to put in place the key structures needed to operationalize the intergovernmental platform. The first plenary was held in Nairobi 2011. Substantial advancement was made in terms of governance structures and program of work but no formal agreement was reached in establishing IPBES (UNEP, 2011). The second session of the plenary was held in Panama and at this meeting substantial agreement was reached on the governance structure, which included the establishment of a science panel (Multi-Expert Advisory Panel) and a bureau; a structure similar to the MA. The governments convened in Panama also agreed to formally establish the IPBES at this meeting (UNEP, 2012). IPBES was officially established on 12 April 2012.

28.3 IPBES: THE VALUE ADDED

One of the key reasons for countries hesitating to establish an Intergovernmental Platform on Biodiversity and Ecosystem Services at the Putra Jaya meeting was the existence of science subsidiary bodies of many MEAs who had similar tasks to those proposed for IPBES. Countries therefore requested at the end of the first intergovernmental meeting held in Putra Jaya a gap analysis on biodiversity and ecosystem services assessments. This document prepared by UNEP was submitted as an information document at the second intergovernmental meeting held in Nairobi (UNEP, 2009b). The gap analysis identified the following six key findings:

- *Finding 1: Multiple science–policy interfaces.* The gap analysis highlighted the presence of a wide range of science–policy interfaces already existing but many of them working in an ad hoc manner with little strategic thinking towards addressing the issue of biodiversity and ecosystem services in a comprehensive and holistic manner.
- *Finding 2: Effectiveness of science–policy interfaces.* The gap analysis highlighted the need for improving the scientific credibility, relevance, and legitimacy of many of the assessments undertaken by existing science–policy interfaces.
- *Finding 3: Common and shared knowledge base.* The gap analysis found that although there was an extensive knowledge base, there were no common conceptual frameworks, methodologies, and common understanding among the multitude of assessments undertaken. This prevented cross-scale and multi-scale comparisons and limited the degree of lessons learnt and duplicated across regions and ecosystems.
- *Finding 4: Policy impact.* A key gap identified in the analysis was the lack of support on acting on the assessment findings, in particular at the regional and national levels of governance.
- *Finding 5: Coordinated approach.* Closely related to Finding 1, this finding highlights the need for a more coordinated approach in the science–policy cycle. Therefore, not just focusing on assessments, which most science–policy interfaces tend to focus on, the finding suggests focusing on knowledge generation and research, knowledge brokering, modeling and scenarios, policy support, and capacity building.
- *Finding 6: Fundamental capacities.* The gap analysis identified the lack of capacity in undertaking multidisciplinary assessments as well as lack of capacity in undertaking assessments in many parts of the less industrialized countries, thus impeding their effective participation and engagement in the science–policy interface on biodiversity and ecosystem services.

The overarching benefit for a strengthened science–policy interface, drawing from the six findings of the gap analysis, is the increased support to multiple actors at multiple scales for mitigation and adaptation to unprecedented changes in biodiversity and ecosystem services from one common science–policy interface. IPBES would provide one common platform for biodiversity and ecosystem services with a common conceptual framework, methods, and tools to facilitate better coordination among the many biodiversity and ecosystem-related multilateral environmental agreements.

28.4 BALANCING NEEDS AND NEGOTIATION CHALLENGES

Establishing an intergovernmental science platform on biodiversity and ecosystem services is a non-trivial task. Although rationality plays a key role in identifying the gaps and needs for strengthening the science–policy interface, there are many other factors outside the realm of rationality that come into play in the negotiation process to establish such a platform.

First, there is a range of stakeholders who are directly or indirectly linked to the process with very different objectives and expectations. There are the governments who come at the top of the 'food chain' if one might call it such. Then come the international governmental organizations that include the United Nations and other multilateral organizations including international banks such as The World Bank among others. This group is then followed by the non-governmental organizations that have a special interest in biodiversity and ecosystem services; in this case, a host of environmental, scientific, and development organizations. Each group will have varying interests and therefore needs from the platform. Reconciling all of these in an open, transparent, and inclusive manner takes patience and commitment. It is an exercise that is new and in many ways foreign to the scientific community but one in which the community will need to invest resources and time to ensure the credibility of the science. In this chapter, I shall argue that unlike in the past, where the process to ensure scientific credibility is left to governments, it must be overseen by the scientific community itself right from the very start of the process as it is its reputation and accountability that is at stake.

Let me begin by addressing the stickiest points that emerged from the negotiation process to establish IPBES. The first comes from the single country versus regional grouping of countries. For example, countries in Africa form the African Block and will have one country from the region as the official voice of all countries in the block. This might seem at first glance to reduce the complexity of multiple positions from each country in the region but this is not necessarily the case. In the case of the African Block, individual countries still have the freedom to speak their position in addition to the position of the block and this is particularly true when no consensus can be reached on any particular point. This is the case in all other regional block groupings except in the case of the European Union. In the case of the EU, all countries speak with one voice and no individual country within the block speaks on its own accord.

Second, in the IPBES negotiations, most of the developing countries wanted capacity building to be a key and integral part of IPBES activities while the developed countries were keen to have IPBES acting purely as an assessment body with capacity building left to be addressed by existing capacity-building initiatives.

Third, one group of countries were of the strong opinion that IPBES should be part of the CBD while another group wanted IPBES to be truly independent of existing MEAs and UN bodies and should at the minimum be administered by one or a combination of these bodies.

Fourth, a group of countries wanted IPBES to be a truly scientific body with all officials being scientists of the highest caliber, while another group were more aligned to having a mix of policy-makers and scientists.

Fifth, some countries wanted IPBES to adopt a top-down approach similar to the IPCC and be focused on global assessments, while another group felt that IPBES should adopt a bottom-up approach and undertake sub-regional and regional assessments as the building blocks because of the place-based nature of biodiversity and ecosystem services.

I will not go into detail into the actual negotiations on each of the above points but will just highlight the need to come to consensus on all these and many more points in order to reach a final agreement on the actual establishment of the platform. Although there is scope for a majority vote option in final decision-making, in a majority of cases, countries prefer to choose the decision-making by consensus principle. This might mean that a single country with very strong objections might delay the whole process until some compromise is reached on the point or issue it has reservations about. Now imagine a decision-making process with over 90 countries with a large number of contestable points to deliberate and to reach consensus on every one of these points before a final decision can be made.

Some might ask, why did it take so many meetings and such a long time to come to what might have seemed from a rational view a simple decision to establish the platform? The principle of decision-making by consensus has come under increasing criticism, as the long gestation time taken to make decisions on important issues has caused much inertia in the global governance system and some countries have been arguing for a majority vote principle when impasses are reached. There are benefits and costs in taking either principle and which is superior is still left to be determined (Tomalin and Nicks, 2007; Heitzig and Simmons, 2012).

28.5 THE SCIENCE–POLICY CYCLE

One of the unique aspects of IPBES is its program of work. Instead of limiting IPBES to purely be an assessment body such as the IPCC, policy-makers and the scientific community to a large extent felt that the full range of activities, from knowledge generation, to assessments, to policy support, with a strong component of capacity building across all activities, would be needed to truly strengthen the science–policy interface for biodiversity and ecosystem services (Larigauderie and Mooney, 2010).

Knowledge generation as defined by the Busan outcome (UNEP, 2010) refers to: (1) identifying and prioritizing key scientific information needed for policy-making at appropriate scales; and (2) catalyzing efforts to generate new knowledge by engaging in dialogue with key scientific organizations, policy-makers, and funding organizations. But it must be emphasized that governments made it clear that IPBES should not undertake new research. The principal idea behind this work component was to ensure that knowledge gaps identified in an assessment cycle are brought to the attention of scientific research organizations who might then establish research programs as appropriate to fill those gaps before the next assessment cycle begins. An example of such an activity was the establishment of the Programme on Ecosystem Change and Society (PECs) project emerging from the MA (Carpenter et al., 2009).

All governments agreed that the assessments work component will be the key activity of IPBES. IPBES is expected to perform regular and timely assessments of knowledge on biodiversity and ecosystem services and their interlinkages with human well-being. These assessments will be taken at multiple scales ranging from regional, sub-regional and global. Governments agreed that IPBES will not undertake national assessments but requested that the UNEP-led Sub-Global Assessments under the Millennium Ecosystem Assessment Follow-up Global Strategy be focused on national assessments and work in close collaboration with IPBES. Governments also requested IPBES to undertake two additional types of assessments: one will be thematic and the other a sort of horizon-scanning exercise whereby the latest information from science but yet to gather policy attention should be assessed to determine if a full thematic assessment would need to be taken.

The third component within the science–policy cycle is policy support. This component focuses on providing decision-makers at all levels with appropriate tools and methodologies to turn knowledge into action. Providing the knowledge on biodiversity and ecosystem services changes and their impacts on human well-being is seen as only the first step in the policy-making process. Feedback from policy-makers on past assessments pointed towards a large amount of knowledge being generated by assessments but with very little follow-up support and in particular at the national level to act on the findings. However, there was emphasis for the support to be in the form of decision-making tools and methodologies and not actual decision-making.

A majority of developing countries felt strongly about the need for a strong capacity-building component within IPBES for strengthening institutional, scientific, and technical capacities for gathering, managing, and analyzing data and information on biodiversity and ecosystem services. There was consensus among all countries that capacity building should be an integral part of all IPBES activities and not be seen as a separate activity. The scientific community through a global survey (UNU-IHDP, 2011) identified a learning-by-doing approach whereby young scientists from developing countries are attached to assessments for a period of time under the mentorship of an experienced scientist. Many developing countries were also of the opinion that capacity building should also extend to the provision of knowledge infrastructure such as access to databases, access to literature, and access to computer tools and programs.

28.6 GUIDING PRINCIPLES FOR AN EFFECTIVE, EFFICIENT, AND EQUITABLE SCIENCE–POLICY PLATFORM

The IPBES's core mandate would be to provide authoritative, independent, credible, inclusive, internationally peer-reviewed, policy-relevant scientific information on changes in biodiversity and ecosystem services and their impacts on human well-being at regional and global levels. There are seven key overarching principles worth elaborating further, which are seen as essential for the efficient, effective, and equitable operation of the platform. The governance structure, together with the rules of procedures of the platform, needs to be designed accordingly to achieve these seven core principles.

Principle 1: Saliency

This principle relates to policy relevance. It is critical for a science–policy interface to focus on issues important for society. Undertaking an assessment purely out of science interests or importance is no longer sufficient. There is a growing need for science to provide the knowledge needed by decision-makers on challenges faced by society and the opportunities available for improving societal well-being (Mitchell et al., 2006). There are a number of ways to achieve this but key is having an open channel of dialogue between the scientific community and the decision-making community. Decision-making community here includes policy-makers, private sector as well as civil society. One possible mechanism for IPBES, which was suggested by a global survey of scientists on IPBES, was the establishment of informal regional consultative groups (UNU-IHDP, 2012a).

Principle 2: Scientific credibility

Ensuring the highest-quality science is a necessary condition for all assessment platforms including the IPBES (Miller and Edwards, 2001; Farrell et al., 2005). One of the reasons cited for the establishment of IPBES in some of the early documents was the lack of consistency in the quality of science across the assessment landscape. There are two key processes to ensure high standards of science. The first involves the selection of the scientists for the platform and the second revolves around the review process of the scientific outputs from the platform. In both cases, there needs to be a clear and open process whereby scientists can be nominated by all relevant parties including accredited scientific organizations and not just governments as in the case of most assessments. The final selection of the scientists has also been proposed by the scientific community to be overseen by an independent scientific body based on a set of criteria agreed by the plenary of the platform, with final approval by the plenary. The same level of openness and transparency is suggested for the review process, with an independent science body overseeing this (see Figure 28.1). These criteria and selection processes should ideally be reflected in the rules of procedure for the platform.

Principle 3: Scientific independence

Assessments made by IPBES must be independent of any political and/or special interest. This will require making the negotiation process separate from the scientific work undertaken by the platform. One way of ensuring the independence is to establish two subsidiary bodies within IPBES (UNU-IHDP, 2012b). The first subsidiary body (a bureau) will be focused on the administrative and political process within the platform, while the second (science panel) will work primarily on the science (see Figure 28.1). This would mean establishing a bureau that is tasked with overseeing the mandate of the platform on behalf of the plenary, while the science panel co-chaired by a prominent natural and social scientist respectively will focus solely on implementing the work program of the platform with the upmost scientific rigor and independence.

Principle 4: Inclusiveness

The high degree of interconnectivity and interdependence among biodiversity, ecosystem services and human well-being requires a transdisciplinary approach to be taken by IPBES. The platform needs to bring together a range of disciplines, from the natural sciences to the social sciences and humanities. But in addition to disciplines, there is also a need to include different knowledge systems. The important and relevant knowledge held by traditional and/or local knowledge on biodiversity and ecosystem services also requires IPBES to embrace these knowledge systems within its various work components (UNU-IHDP, 2011).

Principle 5: Bottom-up Approach

The place-based nature of biodiversity and ecosystem services and their corresponding impacts on human well-being suggests a bottom-up approach, with focus beginning at the regional and sub-regional scales. There was consensus among governments to build regional building blocks upon which the periodic global assessment would draw and synthesize. It was also felt that this approach will suit the needs of the respective regions more closely than global activities in all areas of IPBES activities such as knowledge

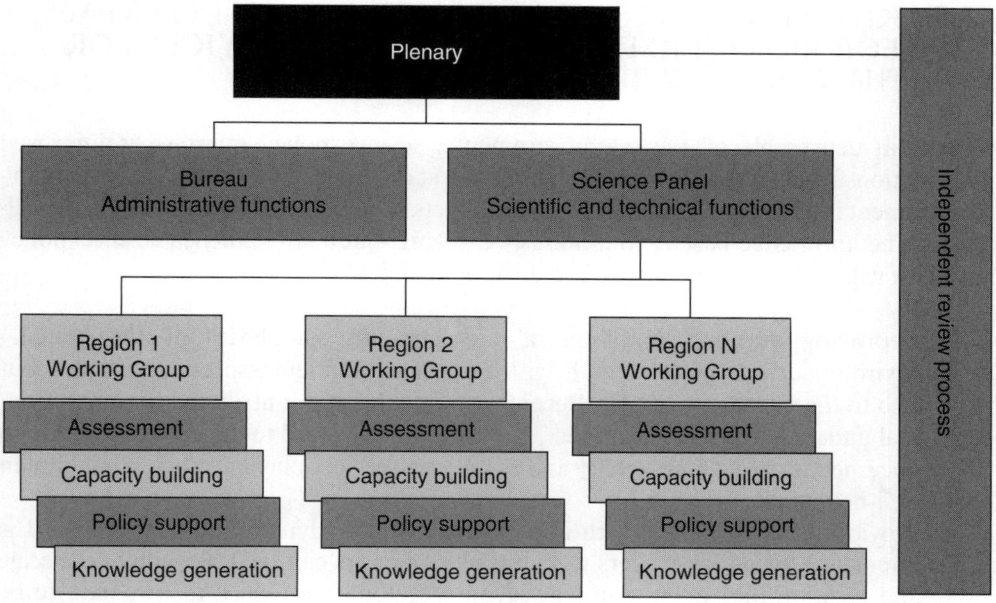

Source: UNU-IHDP (2012b).

Figure 28.1 Governance structure proposed for IPBES

generation, policy support, and capacity building where the needs can be site and context specific. The scientific community through the global survey suggested a regional and integrated approach – as illustrated in Figure 28.1 – to implement the IPBES work program.

Principle 6: Legitimacy

Legitimacy of the platform is critical if the results from the platform are to be acknowledged by decision-makers and adopted in their decision-making processes. This was one of the key reasons why the IPBES was designed to be an intergovernmental science–policy interface unlike other processes such as the Millennium Ecosystem Assessment and the Global Biodiversity Assessment. The uptake of findings and recommendations from such intergovernmental process as demonstrated by the IPCC highlights the utility of such mechanisms.

Principle 7: Equity

One of the primary concerns by many developing countries at the beginning of the deliberations on whether to establish an Intergovernmental Platform on Biodiversity and Ecosystem Services was possible lack of capacity within developing countries to participate in an equitable manner in the platform's activities. There was a fear that scientists from developed countries who to a large extent were not holders of biodiversity would dominate the process with their access to knowledge bases and financial resources. This was one of the key reasons why capacity building became one of the keystone activities of IPBES.

28.7 KEY ATTRIBUTES OF A SCIENCE–POLICY INTERFACE FOR BIODIVERSITY AND ECOYSTEM SERVICES FOR HUMAN WELL-BEING

The main deliverable of the intergovernmental science–policy interface is improved information from all relevant sources about the state, trends, and outlooks of human–environment interactions, with focus on the impacts of ecosystem change on human well-being. The knowledge base is continuously evolving, but a strengthened science–policy interface will strengthen our abilities to achieve the following:

- Improving our understanding of the immense complexity of the human–environment interactions, both with respect to different aspects and scales, but also to the system as a whole. Recent assessments have contributed to our conceptual understanding in this respect, but there is still a need to improve our ability to document the role biodiversity and ecosystem services play in determining human well-being.
- Providing policy-makers with information on scale dynamics. For example, it is important for policy-makers to know which factors can be dealt with at the scale and location they operate in (endogenous factors) and which factors have to be considered by other scales or other locations (exogenous factors). Furthermore, cross-scale interactions within both environmental and social processes may require mutually supportive responses at different scales to bring about desired results.
- Reflecting developments over time, be they trends or projections of environmental and social changes. Critical information for policy-makers includes information on emerging issues and time-lags – changes that continue even if the forces of change are slowed or reversed, or tipping points – beyond which there are abrupt, accelerating or potentially irreversible changes. There is a need to increase the ability at national, regional, and global level, to predict the consequences of current actions affecting biodiversity and ecosystem services and how they affect human well–being, to explore alternative scenarios, and to evaluate the effectiveness of measures taken (Perrings et al., 2011).

28.8 KEY MESSAGES

Strengthening the science–policy interface as demonstrated by the preceding paragraphs is not a trivial task. A number of key messages emerge in light of the experiences from the recent process of establishing the IPBES that are worth highlighting:

- *Key message 1*. Strengthening the science–policy interface on biodiversity and ecosystem services for human well-being involves multiple actors with different and in many cases conflicting objectives. These have to be reconciled through a consensus process involving all relevant stakeholders in an equitable manner.
- *Key Message 2*. The governance structure for a science–policy interface should be designed in a manner that minimizes the political aspects of the interface and

focuses on bringing together the 'best' science to guide policy-making at different levels of governance.

- *Key message 3*. A more active and equitable partnership of social sciences and humanities with the natural sciences can only strengthen the policy relevance of the science–policy interface.
- *Key message 4*. Regional representation in the science–policy interface can only strengthen the effectiveness of the science–policy interface in addressing the 'right' problems and providing the 'right' knowledge for policy-makers working at these levels.
- *Key message 5*. The science–policy interface should address the complete science–policy cycle beginning from knowledge scoping, assessments, policy support, and capacity building in order to increase its effectiveness.
- *Key Message 6*. Capacity building is critical for the success of the science–policy interface. Scientists from the respective regions should have the capacity to participate in an equitable manner in the interface and support policy-making across all levels.

28.9 CONCLUSIONS

Ultimately, the success of IPBES or any science–policy interface will depend on the willingness of the scientific community to engage in and give their time to the implementation of the platform's work program on a volunteer basis. This will only happen if there is confidence and trust in the mechanism. And this will only happen if the platform is open, transparent, and inclusive, with the seven principles highlighted in this chapter as core to the way it operates and delivers its findings.

But equally important is the active participation of the policy-making community in the dialogue process. The objectives and goals of the policy-making community will differ from those of the scientific community. These differences have to be acknowledged and respected by both communities and the challenge will be to find the common ground whereby both communities are happy with the partnership.

It must also be acknowledged that there will be strategic positioning among countries as well as the different scientific disciplines and/or ideologies. A successful science–policy interface will require these special interests to be minimized at the minimum and completely excluded at the optimal. Therefore, the success of IPBES will be to a large extent dependent on the governance structures put in place to oversee the implementation of its work program. This could be achieved by establishing a science panel co-chaired by prominent natural and social scientists from a developed and developing country working closely with a regionally balanced bureau comprised of government representatives but with a strong science background.

Finally, equally important would be the establishment of an independent review panel to oversee the review process of the outputs produced by the platform and to mediate any scientific disputes that might arise that the science panel might not be able to resolve due to conflicts of interest. Last but not least, the platform must adhere to being policy relevant not policy prescriptive.

REFERENCES

Carpenter, S.R., H.A. Mooney, J. Agard, D. Capistrano, R.S. Defiees and S. Diaz et al. (2009), 'Science for managing ecosystem services: beyond the Millennium Ecosystem Assessment', *Proceedings of the National Academy of Sciences of the United States of America*, **106**(5), 1305–12.

Farrell, A.E. and J. Jäger (eds) (2005), *Assessments of Regional and Global Environmental Risks: Designing Processes for the Effective Use of Science in Decision-making*, Washington, DC: RFF Press.

Heitzig, J. and F.W. Simmons (2012), 'Some chance for consensus: voting methods for which consensus is an equilibrium', *Social Choice and Welfare*, **38**(1), 43–57.

Larigauderie, A. and H. Mooney (2010), 'The Intergovernmental Science–Policy Platform on Biodiversity and Ecosystem Services: moving a step closer to an IPCC-like mechanism for biodiversity', *Current Opinion in Environmental Sustainability*, **2**(1–2), 9–14.

Loreau, M., A. Oteng-Yeboah, M.T.K. Arroyo, D. Babin, R. Barbault and M. Donoghue et al. (2006), 'Diversity without representation', *Nature*, **442**(7100), 245–6.

Miller, C. and P.N. Edwards (eds) (2001), *Changing the Atmosphere: Expert Knowledge and Environmental Governance*, Cambridge, MA: MIT Press.

Mitchell, R.B., C.C. William, D.W. Cash and N.M. Dickson (eds) (2006), *Global Environmental Assessments: Information and Influence*, Cambridge, MA: MIT Press.

Perrings, C., A.K. Duraiappah, A. Larigauderie and H. Mooney (2011), 'The biodiversity and ecosystem services science–policy interface', *Science*, **331**(6021), 1139–40.

Tomalin, B. and M. Nicks (2007), *The World's Business Cultures: And How to Unlock Them*, London: Thorogood Publishing.

UNEP (2008a), *The Millennium Ecosystem Assessment Follow-up: A Global Strategy for Turning Knowledge into Action*, Nairobi: UNEP/CBD/COP/9/INF/26.

UNEP (2008b), *An Intergovernmental and Multi-Stakeholder Approach to Strengthening the Science: Policy Interface on Biodiversity and Ecosystem Services*, Bonn: UNEP/CBD/COP/9/INF/37/Rev.1.

UNEP (2008c), *Report of the ad hoc Intergovernmental and Multi-Stakeholder Meeting on an Intergovernmental Science Policy Platform on Biodiversity and Ecosystem Services*, Putrajaya: UNEP/IPBES/1/6.

UNEP (2009a), *Report of the Second ad hoc Intergovernmental and Multi-Stakeholder Meeting on an Intergovernmental Science Policy Platform on Biodiversity and Ecosystem Services*, Nairobi: UNEP/IPBES/2/4/Rev.1.

UNEP (2009b), *Gap Analysis for the Purpose of Facilitating the Discussions on How to Improve and Strengthen the Science–Policy Interface on Biodiversity and Ecosystem Services*, Panama City: UNEP/IPBES/2/INF/12.

UNEP (2010), *Report of the Third ad hoc Intergovernmental and Multi-Stakeholder Meeting on an Intergovernmental Science Policy Platform on Biodiversity and Ecosystem Services*, Busan: UNEP/IPBES/3/3.

UNEP (2011), *Report of the First Session of the Plenary Meeting to Determine Modalities and Institutional Arrangements for an Intergovernmental Science Policy Platform on Biodiversity and Ecosystem Services*, Nairobi: UNEP/IPBES.MI/1/8.

UNEP (2012), *Report of the Second Session of the Plenary Meeting to Determine Modalities and Institutional Arrangements for an Intergovernmental Science Policy Platform on Biodiversity and Ecosystem Services*, Panama City: UNEP/IPBES.MI/2/9.

United Nations (2000), *We the Peoples: The Role of the United Nations in the 21st Century*, New York: United Nations Department of Public Information.

UNU-IHDP (2011), *Summary of Results from the International Science Workshop on Assessments for IPBES*, Nairobi: UNEP/IPBES.MI/1/INF/12.

UNU-IHDP (2012a), *Findings of the Global IPBES Assessment Survey*, Panama City: UNEP/IPBES.MI/2/INF/10.

UNU-IHDP (2012b), *Summary of Results from the Second International Science Workshop on Assessments for IPBES*, Panama City: UNEP/IPBES.MI/2/INF/10.

29. Governance of the transition to a green economy – responding to the values of nature

Patrick ten Brink, Leonardo Mazza, Tomáš Badura, Marianne Kettunen and Sirini Withana

29.1 INTRODUCTION

There have been increasing trends in the degradation and loss of biodiversity, ecosystems, and the services they provide. In the last century alone, 35 per cent of mangroves, 40 per cent of forests, and 50 per cent of wetlands have been lost (TEEB, 2011). Around 60 per cent of ecosystem services are estimated to have been degraded in the last 50 years (MA, 2005). Species are becoming extinct 100 to 1000 times faster than in geological times (Pimm et al., 1995). The unsustainable use of the world's natural resources is threatening to destabilize the functioning of many ecosystems, which, in turn, undermines their productive capacities. For example, 69 per cent of the world's fisheries are thought to be fully or over-exploited (FAO, 2009), leading to the fisheries sector underperforming by around US$50 billion annually in terms of lost economic benefits (World Bank and FAO, 2009). Such continued loss of ecosystems and biodiversity, which in many cases is expected to be amplified by the effects of climate change, is likely to have detrimental effects on human well-being as the functioning of the world's ecosystems underpin society and the economy (TEEB, 2011).

The past and on-going trends are expected to continue and the challenges we currently face are likely to escalate even further in the years to come. The world population is expected to increase to 9 billion by 2050. Life expectancy will continue to increase, as will the share of the world's population living in urban areas. The global economy is also expected to grow significantly – tripling or even quadrupling by 2050 (OECD, 2012). While some of these developments will result in benefits to an expanding middle class and may contribute to poverty alleviation around the globe, there are also a number of associated risks. The rising level of consumption and production will put increasing pressure on the planet's resources and ecosystems, accelerating the historic trends of pollution and the depletion of natural capital. As many ecosystems and landscapes continue to be used unsustainably and natural capital stocks and ecosystem service flows are further reduced, societal challenges associated with the loss of benefits from nature will rise – likely surpassing a range of critical ecological thresholds or 'tipping points' (Rockstrom et al., 2009). It is clear that humanity is consuming more than the regenerative capacity of the planet can supply. There is an urgent need for a fundamental change in these trends if major collapses are to be avoided (Randers, 2012).

There is growing recognition among policy-makers and private sector decision-makers that the current model of economic growth is socially, environmentally and economically unsustainable – in particular since the 2008 financial crisis. This has sparked a renewed focus on the need to change the current economic model towards a

BOX 29.1 TEEB DEVELOPMENTS AND OUTPUTS

The Economics of Ecosystems and Biodiversity (TEEB) was launched by the German government and the European Commission in response to a proposal by the G8+5 Environment Ministers in 2007 to develop a global study on the economics of biodiversity loss. This independent initiative, led by Pavan Sukhdev, is hosted by the United Nations Environment Programme with support from the European Commission, Germany, the UK, Belgium, Japan, Norway, the Netherlands, Switzerland, Sweden, and a growing number of other countries and organizations. It benefits from an international advisory board and input from around 500 authors and reviewers from all regions of the world in the fields of science, economics and policy.

Key outputs of the TEEB initiative include a range of reports focused at different end-users (see www.teebweb.org): *The Economics of Ecosystems and Biodiversity: Ecological and Economic Foundations* (TEEB, 2010a); *Mainstreaming the Economics of Nature: A Synthesis of the Approach, Conclusions and Recommendations of TEEB* (TEEB, 2010b); *The Economics of Ecosystems and Biodiversity in National and International Policy Making* (TEEB, 2011); *The Economics of Ecosystems and Biodiversity in Business and Enterprise* (TEEB, 2012a); and *The Economics of Ecosystems and Biodiversity in Local and Regional Policy and Management* (TEEB, 2012b).

To date, the TEEB initiative has supported a range of international and national policy processes, including the 10th and 11th Conference of Parties to the Convention on Biological Diversity in 2011 and 2012, the 11th Conference of Parties to the RAMSAR Convention in 2012 and the Rio+20 Conference on Sustainable Development in 2012. The initiative is currently in its third phase with a range of national (e.g., TEEB Netherlands, Germany, Finland, Belgium, India, Georgia, and Brazil) and regional assessments (e.g., TEEB Nordics; TEEB Arctic), targeted thematic reports (e.g., *Nature and the Transition to a Green Economy*; *TEEB for Water and Wetlands*) as well as guidelines (e.g., *TEEB for Cities Manual*; *Guidance Manual for TEEB Country Studies*) and case studies (see www.teebweb.org).

'green' economy that promotes social equity, poverty eradication and human well-being (UNEP, 2011; UN, 2012). This focus has been complemented by the increasing appreciation of biodiversity and ecosystem services (MA, 2005) and the economic value of nature, including its intrinsic value (TEEB, 2008, 2010a, 2010b, 2011, 2012a, 2012b; see Box 29.1). These two threads are closely interrelated as ecological resilience is necessary for long-term socio-economic development; efforts to build green economies should be based on a sound appreciation of the value and role of nature in this transition. This chapter highlights nature's role in the green economy, the multiple benefits to people and the economy of working with nature, and the building blocks of the transition to a green economy.

29.2 NATURE IN A GREEN ECONOMY

Nature underpins economic growth, human development and well-being. It is instrumental in fuelling today's economic system and is a core foundation in the transition to a green economy. In turn, the transition to a green economy will strengthen the foundations of nature by reducing the pressures of economic activities on biodiversity and ecosystems.

29.2.1 Nature and Natural Capital

Nature consists of ecosystems, landscapes, habitats, species, and genetic materials. It provides a number of ecosystem goods such as food, raw materials, medicine, and water, as well as a range of services such as regulating the climate, contributing to air and water quantity and quality, and mitigating natural hazards. Nature also offers a wide range of cultural benefits related to human health, recreation, tourism, scientific knowledge, and spiritual/cultural identity. These benefits depend, amongst other things, on the quantity, quality and diversity of species, genes and ecosystems. Interactions between the different components of nature (including living and non-living elements – i.e. biotic and abiotic), as well as those between nature and our societies and economies, such as through the flow of ecosystem services, are essential to the health and growth of economies, societies and individuals. Box 29.2 presents a simplified illustration of these interconnections.

In providing a series of services that benefit society and the economy, nature can be understood as providing natural assets, which are increasingly referred to as 'natural capital'. Natural capital stocks provide flows of ecosystem services. The analogy with other forms of capital (Box 29.3), such as manufactured and financial capital, has helped to highlight the role of nature in the economy. It has also been useful for underlining the loss of natural capital and in exploring the underlying causes of its unsustainable use and management (ten Brink et al., 2012a). While nature is understood to be more than 'natural capital', it is nonetheless a useful metaphor to communicate the value or benefits of nature to people and the wider economy (MA, 2005).

29.2.2 The Green Economy

UNEP defines a green economy as 'one that results in improved human well-being and social equity, while significantly reducing environmental risks and ecological scarcities. In its simplest expression, a green economy can be thought of as one which is low carbon, resource efficient and socially inclusive' (UNEP, 2011). The transition to a green economy takes us away from the traditional 'brown' economy approach that depends heavily on fossil fuels, unsustainable resource extraction, and environmental degradation towards an economy in which 'growth in income and employment are driven by public and private investments that reduce carbon emissions and pollution, enhance energy and resource efficiency, and prevent the loss of biodiversity and ecosystem services. These investments need to be catalysed and supported by targeted public expenditure, policy reforms and regulation change' (ibid.).

Thus, a green economy requires a change in economic planning across the economy as a whole, rather than investment in small 'green' subsectors. A central element of this is

BOX 29.2 FROM DRIVERS TO IMPACTS

Figure 29.1 illustrates the interconnections between the states of ecosystems; their functions and service flows; the drivers affecting those; the benefits that people, society and the economy have from nature; and the tools to value these benefits. The use of valuation tools (whether monetary or non-monetary) can help demonstrate values, which if taken into account, or 'captured', in decisions (whether policy, purchase or investment decisions) can lead to an improved state of biodiversity, better functioning of ecosystems, and flow of services.

Moving towards a green economy requires not only recognizing the benefits people derive from ecosystems and biodiversity, but also addressing the underlying causes of the loss and degradation of nature. In other words, to fully realize nature's contributions to development and prosperity, one needs to not only respond to the symptoms (degradation, loss of ecosystem functions and services) but also the underlying causes and drivers of the problems (e.g., production methods and consumption levels). Addressing these simultaneously will be essential to achieving lasting results.

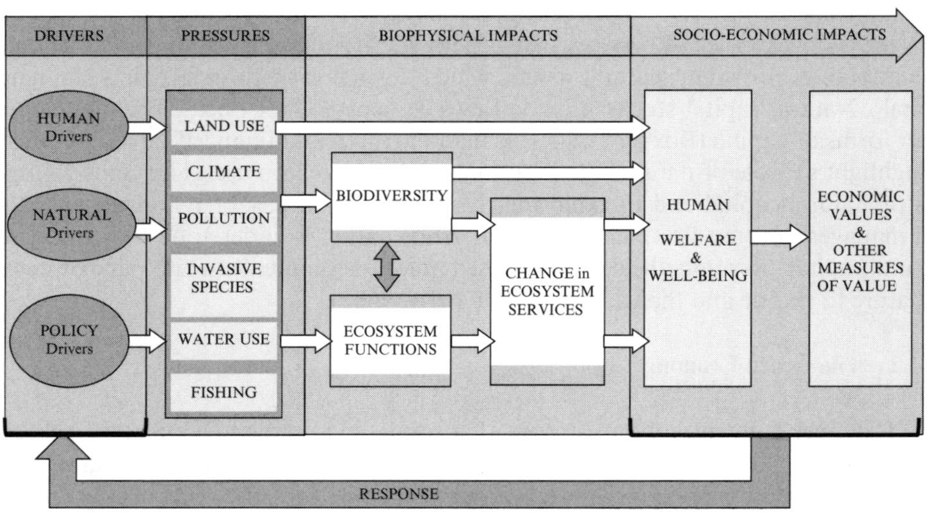

Source: Adapted from Braat and ten Brink et al. (2008).

Figure 29.1 The pathway from drivers to impacts

the integration of broader environmental and social criteria in public and private investments, policy frameworks, and governance.

The transition to a green economy aims to enable economic growth and investment while increasing environmental quality and social inclusiveness by inter alia:

BOX 29.3 THE DIFFERENT TYPES OF CAPITAL

Manufactured or 'man-made' capital

This includes produced assets that are used to produce other goods and services, such as machines, tools, buildings and infrastructure, that is, fixed assets. This category can also include money and other financial assets that are sometimes termed financial capital. 'Financial capital' is sometimes seen as a distinct category of capital (Aronson et al., 2007; Van Andel and Aronson, 2012).

Human capital

This generally refers to the health, well-being and productive potential of individual people and includes mental and physical health, education, motivation, and work skills. These elements not only contribute to a happy and healthy society, but also improve opportunities for economic development through a productive workforce.

Social capital

Like human capital, this is related to human well-being but on a societal rather than individual level. It consists of the social networks that support an efficient, cohesive society and facilitate social and intellectual interactions among its members. Social capital refers to those stocks of social trust: norms and networks that people can draw upon to solve common problems and create social cohesion, for example, neighbourhood associations, civic organizations and cooperatives. The political and legal structures that promote political stability, democracy, government efficiency, and social justice (all of which are good for productivity as well as being desirable in themselves) are also part of social capital.

Natural capital

In addition to natural resources like timber, fish, water and energy, and mineral reserves, this includes natural assets that are not easy to value monetarily (e.g., species diversity, ecosystems that perform ecological services like climate regulation and water provision) and can be considered as the components of nature linked directly or indirectly to human welfare. Forests, agricultural land and soil, grasslands, wetlands, rivers and coral reefs are examples of natural capital.

Source: TEEB (2011); GHK et al. (2005) building on Pearce et al. (1989) and Ekins (1992).

- improving human well-being: securing better healthcare, education and job security;
- increasing social equity: ending persistent poverty and ensuring social, economic, and financial inclusion;
- reducing environmental risks: addressing climate change, ocean acidification, the

release of hazardous chemicals and pollutants, and excessive or mismanaged waste; and
- reducing ecological scarcities: securing access to freshwater, natural resources, and improving soil fertility.

As economic development becomes less dependent on depleting natural capital and generating pollution, countries can achieve more sustainable economic development. Thus, the concept of a green economy does not replace sustainable development; rather it can be seen as a powerful tool (means) for achieving the end goal of sustainable development.

The broad focus of the green economy also means we will need to move away from measuring growth through narrowly defined indicators (such as GDP) to more holistic metrics to assess sustainable economic progress. A report by the UN Secretary's General High-Level Panel on Global Sustainability identifies the need to establish a common framework for measuring progress and calls for the development of a Sustainable Development Index or a set of indicators by 2014 (United Nations Secretary-General's High-Level Panel on Global Sustainability, 2012). In the outcome document of the June 2012 Rio+20 Conference, *The Future We Want*, governments also recognized the need for broader measures of progress to complement GDP and requested the UN Statistical Commission to launch a programme of work in this area (UN, 2012).

In February 2012, stakeholders gathered at the UNEP Governing Council to draft key principles of a green economy.[1] These became the focus of a wide international consultation and resulted in a set of principles that are being endorsed by an increasing number of organizations. The principles are considered necessary for developing a collective understanding and vision of what a green economy needs to deliver and serve as a reminder of the wider objectives of the transition to a green economy. The nine principles are:

1. The Sustainable Principle (a green, fair and inclusive economy delivers sustainable development).
2. The Justice Principle (it delivers equity).
3. The Dignity Principle (it creates genuine prosperity and well-being for all).
4. The Healthy Planet Principle (it invests in natural systems and rehabilitates those that are degraded).
5. The Inclusion Principle (it is inclusive and participatory in decision-making).
6. The Good Governance and Accountability Principle (it is accountable).
7. The Resilience Principle (it builds economic, social and environmental resilience).
8. The Efficiency and Sufficiency Principle (it delivers sustainable consumption and production).
9. The Generations Principle (it invests for the present and future).

29.2.3 Nature in the Transition to a Green Economy

Natural capital, together with other forms of capital, is a key input for a wide range of economic sectors and underpins much of our production and consumption patterns (see Figure 29.2). The conservation, restoration and sustainable use of natural capital

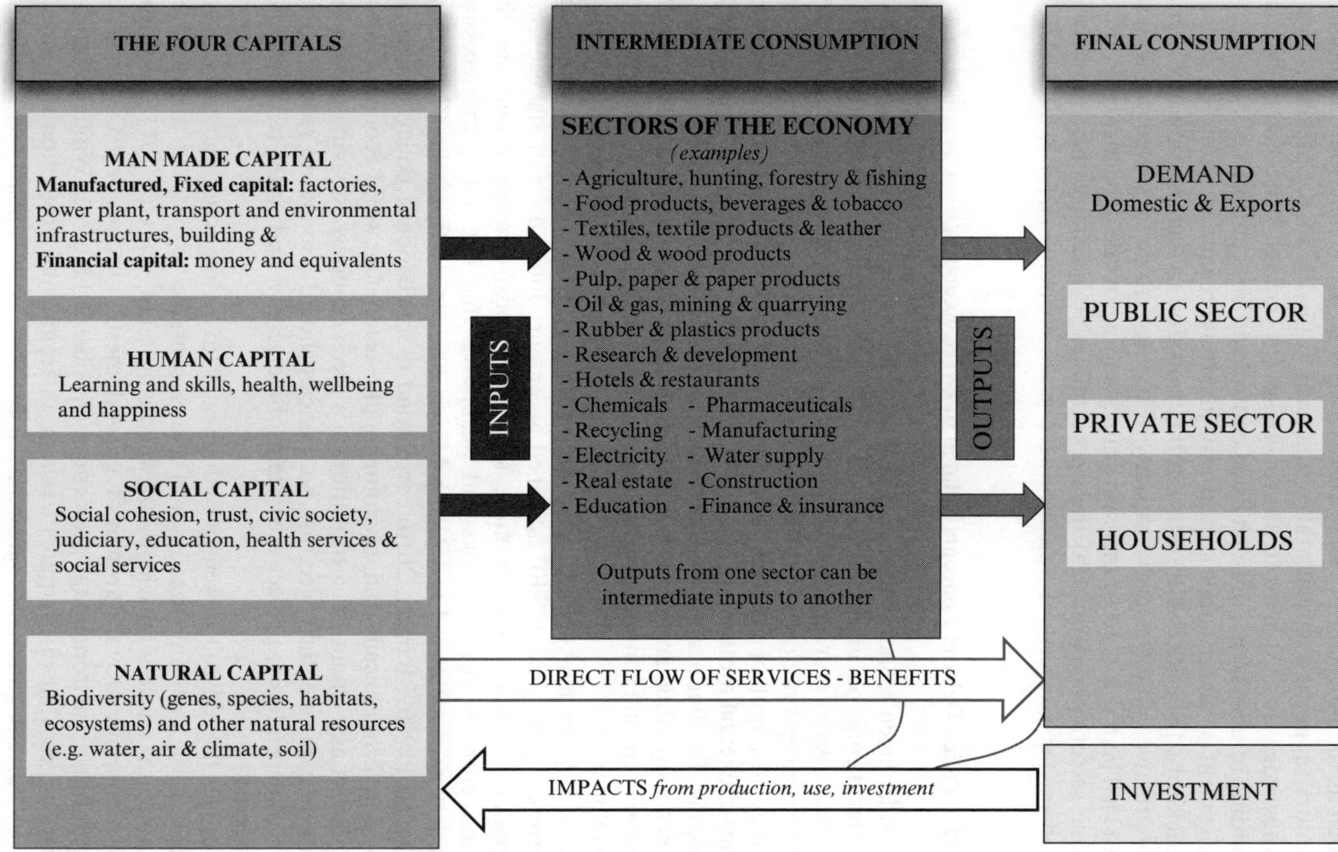

Source: Adapted from ten Brink et al. (2011a) in TEEB (2011).

Figure 29.2 Contribution of natural capital to human well-being and livelihoods

('working with nature') (1) is a key driver in the transition to a green economy as this provides what are often free, non-polluting, non-intrusive and low-carbon inputs to production (such as the provision of clean water by ecosystems, locally sourced products, wild pollinators supporting agricultural production); (2) is the source of materials and scientific knowledge, and (3) helps to attract investment through locational quality and the safeguarding of assets (e.g., through flood control and bio-control).

The engagement of all economic sectors (not just those reliant on nature's asset base such as agriculture, fisheries and forestry) in the transition to a green economy is critical if the productive and regenerative capacity of nature is to be preserved or increased (UNEP, 2011). There is a need to increase understanding of the dependence of different economic sectors on nature and the opportunities to minimize their impacts on the environment for a successful transition to a green economy (UNEP, 2011; TEEB, 2012a).

29.2.4 Wider Policy Debate: Concepts Related to Green Economy

Within the wider green economy debate, recent years have seen the emergence of a range of closely related concepts in the field of sustainable development, notably those of the 'green new deal', 'green growth' 'green economy' and more recently 'circular economy'. Though these concepts are used in different ways and arguably have different roles and meanings, they all seem to suggest that economic growth does not need to be incompatible with environmental protection (Jacobs, 2012). A brief explanation of these concepts is set out below. To offer a simplified picture green new deals can be seen as a catalyst to a green economy while green growth and circular economy as approaches or models contributing to a green economy, which in turn is an essential means of achieving 'sustainable development'.

The idea of a green new deal gained prominence following the 2008 economic and financial crises. At the time, UNEP launched an enquiry into how national economic recovery packages could result in more sustainable, post-recovery economies that would be less prone to the very risks and weaknesses that had led to the recession. Recommendations summarized in UNEP's March 2009 Policy Brief, the *Global Green New Deal* (UNEP, 2009; Barbier, 2010) correspond to an economic policy strategy for ensuring a more economically and environmentally sustainable world economic recovery that could act as a catalyst in a transition to a green economy. Reviving growth and creating jobs remain essential objectives, but policies should also aim to reduce carbon emissions and dependency; protect ecosystems, biodiversity, and water resources; and alleviate poverty.

According to the OECD, green growth means 'fostering economic growth and development, while ensuring that natural assets continue to provide the resources and environmental services on which our well-being relies' (OECD, 2011b). In other words, it is 'economic growth (growth of gross domestic product or GDP) that also achieves significant environmental protection' (Jacobs, 2012, p. 4). While definitions quite clearly suggest that this requires levels of environmental protection that are not being met by current patterns of growth, the concept leaves open how 'significant' improvements should be (ibid.). With the aim of helping governments devise country-specific strategies to successfully mainstream green strategies in national policies, the

OECD's Green Growth Strategy lists, inter alia, policy options to address barriers to green growth such as regulatory uncertainty, low research and development returns. It recommends, inter alia, the use of indicators in four main categories: environmental and resource productivity; economic and environmental assets; environmental quality of life; and economic opportunities and policy responses. Hence, the perspective taken by the OECD's Green Growth initiative provides a useful toolbox for delivering the measures and monitoring the progress towards achieving a green economy (ten Brink et al., 2012a).

The closed-loop or circular economy model emerged from the discipline of industrial ecology in which the functioning of ecosystems is used as an exemplar for industrial processes and systems. In contrast to a traditional linear economy, a circular economy is one which is 'restorative or regenerative by intention and design' (Ellen MacArthur Foundation, 2012). The concept covers issues of product design for durability, disassembly and refurbishment, cascading products and materials through different applications before their end of life; and eliminating the use of toxic elements in products to ensure low (or zero) impact disposal. Thus the circular economy concept offers another approach or model to progress towards achieving a green economy.

The Brundtland Report, *Our Common Future* (WCED, 1987), presented sustainable development as 'meeting present needs without compromising the ability of future generations to meet their own needs', linking two main concepts, namely 'needs' and 'limits', and thus integrating discussions about the environment and poverty reduction. The transition towards a green economy can be seen as a key vehicle to meeting the goals of sustainable development. Balancing developed and developing countries' interests, it stressed that while economic growth cannot stop, it needs to change course to account for the planet's ecological limits by giving primacy to the need for long-term preservation of living conditions following principles on intra- and inter-generational fairness. Needs should be met within these limits, defined through the environment (Fedrigo-Fazio and ten Brink, 2012). The 'ecological limits' issues was picked up again by the seminal Millennium Ecosystem Assessment (MA, 2005), which helped highlight the poor state of ecosystems and refocused attention on the environmental limits of a planet. The Rio+20 Outcome document (i.e., *The Future We Want* declaration), reaffirmed global commitments on sustainable development and eradication of poverty (UN, 2012). This declaration also recognized the green economy as one of the important tools available for achieving sustainable development.

29.3 NATURE, WELL-BEING, AND DEVELOPMENT

The transition to a green economy requires an understanding and appreciation of the role nature plays in human well-being, development, and poverty alleviation (see Box 29.4). Ecosystems and the services they provide underpin the well-being of people at different levels – from the local (e.g., supporting livelihoods), regional and national (e.g., water and food security), to the global level (e.g., nature-based solutions to climate change mitigation). It is of crucial importance to recognize the scale of the benefits people obtain from ecosystems as well as to understand and address the causes of their loss.

BOX 29.4 CONTRIBUTION OF NATURAL CAPITAL TO HUMAN WELL-BEING AND LIVELIHOODS

Building on the representation of the relationship between ecosystem services and human well-being developed in the context of the Millennium Ecosystem Assessment (MA, 2005), Figure 29.3 depicts the role of natural capital in this process. The flow of ecosystem services – provisioning, regulating, and cultural services – can provide direct and indirect support for livelihoods (food, materials, water, jobs), security (food, climate, and natural disasters), health (via clean water, disease control, and medicines), harmonious social relations and community well-being (MA, 2005; see Figure 29.3). Natural capital plays an essential role in the provision of these services as it underpins both the functioning of ecosystems, as well as other forms of capital.

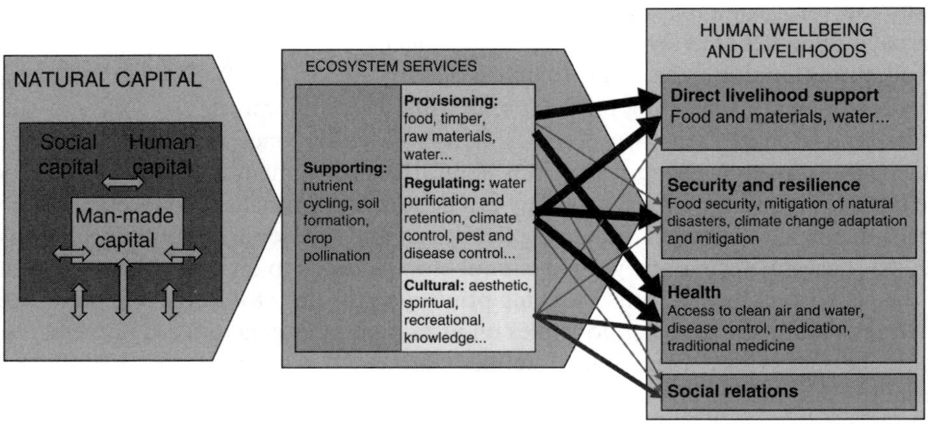

Note: As in the Millennium Ecosystem Assessment, the shading of the arrows presents the potential for mediation by socio-economic factors (i.e., substitutability): the darker the arrow the more opportunities for substitution. A light colour implies less potential for substitution. The arrow's width presents the intensity of linkages between ecosystem services and human well-being.

Source: Adapted from MA (2005) and TEEB (2011).

Figure 29.3 Contribution of natural capital to human well-being and livelihoods

29.3.1 Nature's Benefits to People and Communities

Human and societal well-being depends on nature. Where natural capital is degraded and lost, there is a risk that communities are undermined and humans will suffer. In contrast, efforts to conserve, restore and sustainably use natural capital can improve human well-being, support livelihoods and increase socio-economic and intergenerational equity (TEEB, 2011, 2012b). Increasing well-being and poverty alleviation are objectives that can be met by working with nature, especially since the poor are often more acutely and directly reliant on the flow of benefits from nature. For example, in South Africa, interventions to restore and improve conditions of wetlands under the Working for Water and Working for Wetlands programmes have increased their capacity to provide essential services to the poor including crop and reed production, water for domestic purposes and grazing for livestock, while at the same time creating important employment opportunities (TEEB, 2011).

Ecological resilience can be understood as the adaptive capacity of an ecosystem to withstand shocks and rebuild itself or persist on a given developmental trajectory. A resilient ecosystem can continue to provide ecosystem services to local communities under changing environmental conditions longer than degraded ecosystems and thus support community viability and livelihoods in the long term. The conservation, restoration and sustainable use of nature by communities, building on the innate strengths of their members and institutions, can in turn support ecosystem resilience. In other words, resilience in social or socio-ecological systems adds to and benefits from the capacity of people and society to anticipate and plan for the future. Healthy, functional, resilient ecosystems can be seen as a life insurance policy for many communities.

29.3.2 Nature's Contributions to Development and Prosperity

Healthy and resilient ecosystems may contribute to delivering development goals, especially those on poverty eradication, in a cost-effective way. The degradation and loss of natural capital, in most cases, undermines national development and long-term economic growth. Global commitments to improving well-being and eradicating poverty can only be achieved by recognizing and taking into account the value of natural capital and its associated benefits. This has all too often been overlooked and has led to a narrow focus on short-term gains at the expense of the viability and long-term prosperity of societies. In addition, private wealth and financial or manufactured capital are systemically prioritized over public welfare and natural capital, which then exacerbates the degradation and loss of natural capital. This is a case of capital misallocation (UNEP, 2011).

Healthy and resilient ecosystems are also crucial for the livelihoods and development of rural communities. For example, in the Indian village of Hiware Bazar, acute water shortages due to vegetation loss were undermining agricultural productivity. The subsequent regeneration of degraded forests and building of contour bunds around hills helped to conserve rainwater and recharge groundwater. This has increased agricultural production potential by several orders of magnitude and contributed to reducing poverty by 73 per cent in less than a decade (Singh, 2010; TEEB, 2012b, building on Tiwari et al., 2007). Another example is in the Shinyanga Region in Central Tanzania where the Nihili woodland was restored by utilizing traditional knowledge (*ngitili*). The resulting increase in the provision of ecosystem services from the woodland (fuel, fruits, building timber, honey, medicines, and fodder) led to a reduction in the time needed to collect fuel wood and non-timber forest products by several hours. The sale of tree products helped pay for children's schooling and allowed more time to be spent on education and productive work (Barrow and Shah, 2011; TEEB, 2012b).

Investments in the restoration of ecosystems and the designation of protected areas and associated conservation measures have demonstrated benefits from the local to the global level. For example, in Cambodia, the Ream National Park provides fish breeding grounds and other subsistence goods from mangroves and additional benefits such as storm protection and erosion control (Emerton et al., 2002). This is important for many local fisheries communities, both for livelihoods and food security, and is also of global importance as over 3 billion people worldwide rely on fish as a significant source of protein (FAO, 2009). At the city, regional and national levels, safeguarding and investing in natural resources can address environmental objectives, ecosystem degradation and loss, foster growth and development, and create employment opportunities. Nature in and around cities is often considered a core element of effective urban planning, investment and management. For example, managing and restoring an upstream watershed can be a cost-effective method for helping with water purification and ensuring its adequate supply (TEEB, 2011; 2012b; Kettunen and ten Brink, 2013).

Looking at the benefits of nature from a national perspective can also be important for long-term strategic planning. For example, the UK National Ecosystem Assessment explored future implications of different policy scenarios on the provision of various ecosystem services from 2000 to 2060. Those scenarios, which involved working with nature, resulted in significant gains in ecosystem services and led to the most important long-term economic gains to society (National Ecosystem Assessment, 2011).

29.3.3 Loss of Natural Capital and Implications for People and Communities

The deleterious impacts of environmental degradation and the associated loss of ecosystem services, which are borne disproportionately by the poor, are contributing to growing inequities and disparities across groups of people and are sometimes the principal factor causing poverty, social and political conflicts (MA, 2005). Those most immediately and directly affected by any loss of biodiversity are arguably the rural poor in developing countries, where the majority of populations depend directly on natural resources. Natural resources are a basic source of their income. Moreover, healthcare needs for the world's poor are mostly met by traditional medicines and treatments extracted from natural sources: loss of this biodiversity is particularly profound for this group as the cost of substitute treatment – modern medicines – is often prohibitive. The loss of ecosystem services will risk undermining progress towards achieving many of the Millennium Development Goals (MDGs) (UN, 2000; TEEB, 2008) and will play an integral role in reaching future global sustainable development goals (UN, 2012).

The livelihoods of many of the world's rural poor are furthermore intricately linked with exploited, fragile environments and vulnerable ecosystems (Barbier, 2005). Over 600 million of the rural poor currently live on lands prone to degradation and water stress and in upland areas, forest systems and drylands that are vulnerable to climatic and ecological disruptions (Comprehensive Assessment of Water Management in Agriculture, 2007). The tendency of rural populations to be clustered on marginal lands and in fragile environments is likely to be a continuing problem for the foreseeable future. This means that poorer households, particularly in rural areas, are likely to face disproportionate losses from the depletion of natural capital due to their relatively high dependence on certain ecosystem services for income and insurance against hard times.

Furthermore, the poor are particularly vulnerable to the climate-driven risks posed by rising sea levels, coastal erosion, glacier melting, and more frequent storms – in particular since these households have few means to cope with losses of critical ecosystem services; such as drinking water purification or protection from natural hazards. Around 14 per cent of the population and 21 per cent of urban dwellers in developing countries live in low elevation coastal zones that are exposed to these risks (McGranahan et al., 2007). The livelihoods of billions – from poor farmers to urban slum dwellers – are threatened by a wide range of climate-induced risks that affect food security, water availability, ecosystem and human health (OECD, 2008; UNDP, 2008). There is increasing evidence that their economic prospects are being adversely affected by the lack of efficient and sustainable management of their natural resource base (Barbier, 2002).

The transition to a green economy involves greening a range of key sectors that are particularly important for the poor (i.e., agriculture, forestry, fishery, and water management) and can contribute to eradicating poverty. Investing in greening these sectors, including through scaling up microfinance, is likely to benefit the poor by helping to secure livelihoods predominantly based on ecosystem services.

29.4 THE MULTIPLE BENEFITS OF VALUING NATURE

Working with nature forms a critical part of the transition to a green economy and can deliver multiple benefits that support economic growth and sustainability. In order to take full advantage of these opportunities, there must be a clear understanding of the value of nature and how to reflect this value in public and private decisions in light of the multiple benefits provided.

29.4.1 Valuing Nature and the Tools to Do So

As has been highlighted throughout this book, there is a strong case for valuing nature in both physical and monetary terms more systematically than at present. Historically, the lack of appreciation of the importance of nature, and the value of its contributions to society and the economy, have led to the degradation of nature and undermined the services that are central to well-being and prosperity. However, this has gradually started to change; there is a growing appreciation of the importance of biodiversity, ecosystem services (see e.g., MA, 2005), and the economic value of nature (see e.g., TEEB, 2008, 2010a, 2010b, 2011, 2012a, 2012b). The identification, assessment, estimation, and demonstration of the value of the services ecosystems provide are increasingly translated into policy responses. Some examples include water purification and supply via forest management, carbon sequestration via peatland restoration, climate change adaptation via floodplain conservation and green infrastructure in cities, and increased agricultural productivity via natural pollination and the bio-control of pests.

Assessing the value of biodiversity and ecosystem services can improve decision-making by helping to identify win-wins and trade-offs, where objectives can be met most cost-effectively and where there are multiple co-benefits. Economic valuation does not necessarily imply relying solely on economic solutions such as payments for ecosystem services (PES) and other market-based instruments (MBIs), which are currently receiving much attention (ten Brink et al., 2011b). Regulation and spatial planning are also fundamental ways of responding to the unsustainable use and management of biodiversity and ecosystems. There are a variety of other approaches, such as public–private partnerships, which can provide effective mechanisms to achieve multiple policy objectives including biodiversity conservation and job creation.

There is currently a real opportunity to transform decision-making so that it takes better account of nature, its intrinsic values, the wide range of public goods and services it provides, as well as private and collective benefits and values that are both market and non-market based (TEEB, 2011). There are a number of approaches to demonstrating the values derived from nature such as ecosystem indicators, maps demonstrating the flows of ecosystem benefits to local and downstream communities, the application of monetary valuation techniques, and wider multi-criteria approaches. Each tool has certain strengths and limitations, and decision-makers will need to rely on a mix of qualitative, quantitative, monetary, and spatial approaches and information.

29.4.2 Value for Money and Meeting Multiple Objectives

Investments in nature today can save money and promote economic growth in the long term and must therefore be seen as an integral part of the transition to and the foundation of a green economy. In the current context of austerity, it is worth taking a careful look at the role of nature and the benefits it provides as these can offer economic savings and opportunities for investments with real social and economic returns. Furthermore, sustainably using and managing natural capital can also support well-being, improve livelihoods, and create added value for both the public and private sectors. Investments in nature can, in many cases, be significantly more cost-effective than investments in other forms of capital or engineered solutions for delivering certain services or pursuing specific policy objectives – especially if the wider range of co-benefits delivered are factored into the equation. For example, investments in protected areas have led to benefits including increased visitor spending in Finnish protected areas, low cost water supply to the city of Dunedin in New Zealand, and avoided soil erosion and improved water supply for farmers in Venezuela (TEEB, 2011; Kettunen and ten Brink, 2013). Restoration has also been found to be a cost-effective solution. For example, the restoration of mangroves helped with flood and storm defences in Vietnam, the restoration of peatlands in Germany improved carbon storage, and the restoration and management of watershed ecosystems increased clean water provision to New York thus avoiding potentially significant increases in water prices (TEEB, 2011; see Box 29.5).

Ensuring the maintenance of healthy and resilient ecosystems can also contribute to meeting multiple policy objectives simultaneously, such as longer-term economic opportunities, the provision of food and clean water, floodwater management, combating desertification, health and recreation services, disaster prevention, and soil erosion control, as well as climate change mitigation and adaptation. Examples of pursuing various policy goals by working with nature include Mexican payments for ecosystem services programme (PSAH) or international negotiations on the Reduced Emissions from Deforestation and Forest Degradation (REDD+) (see Box 29.5).

BOX 29.5 PROVIDING VALUE FOR MONEY BY WORKING WITH NATURE

The benefits of protected areas

In Finland, the total annual revenue linked to visitor spending in national parks and key recreation areas (total of 45 areas) has been estimated at 87 million euros/year, generating 10 euros return for every 10 euros of public investment (Huhtala et al., 2010). In Venezuela, the national protected area system prevents erosion, flooding and water supply fluctuation that would, inter alia, reduce farm earnings by around US$4 million/year (based on data provided in Gutman, 2002, updated by the authors to account for inflation and increase in land under irrigated agriculture). In New Zealand, the Central Otago conservation area (Te Papanui Catchment) saved the city of Dunedin NZ$93 million (approximately US$65 million) in water supply costs (BPL, 2006).

Restoration as a cost-effective solution

During typhoon Wukong in Vietnam in 2000, areas planted with mangroves were relatively unharmed while neighbouring provinces without mangroves suffered significant losses of life and property (Brown et al., 2006). Also in Vietnam, mangrove restoration by volunteers cost US$1.1 million, but saved US$7.3 million annual expenditure on dyke maintenance and benefited the livelihoods of an estimated 7500 families (TEEB, 2011, building on IFRC, 2002).

In Mecklenburg-Vorpommern, Germany, 30 000 hectares of peatland were restored over the period 2000–08, leading to emission savings of up to 300 000 t CO_2-equivalent at an avoidance cost of CO_2 ~ 8–12 euros/t CO_2. If alternative land use options are realized (extensive grazing, reed production or alder forest growth), costs can decrease to 0 to 4 euros/t Co_2 (Förster, 2010).

A comprehensive PES programme for the 200km^2 Catskill Mountains watershed in the US cost around US$1 billion to US$1.5 billion over ten years. This was significantly less than the estimated cost of a water filtration plant (one-off costs of US$4 billion to US$6 billion and operational and maintenance costs of US$300 million to US$500 million). Nearly all (93 per cent) farmers in the region participate in the programme and water bills have been raised by 9 per cent instead of doubling, as would be the case with a new filtration system (Wunder and Wertz-Kanounnikoff, 2009; TEEB, 2011).

Meeting multiple policy objectives

The Mexican payments for ecosystem services scheme (PSAH) makes use of water charges earmarked to support selected community engagement in forest management for conservation and hydrological services (including aquifer recharge) and has led to slower deforestation rates and avoided greenhouse gas emissions. The scheme started in 2003 covering 127 000 hectares and has grown to 2.3 million hectares by 2010. By comparing statistically equivalent forests between 2000 and 2007, the PSAH reduced the rate of deforestation emissions; this equates to 3.2 million tonnes CO_2eq. (Muñoz et al., 2010; TEEB, 2011).

The proposals for the Reduced Emissions from Deforestation and Forest Degradation (REDD+) – a global initiative supporting the sustainable management of forests to enhance forest carbon stocks – represent a potentially cost-effective solution to addressing the challenges of climate change and biodiversity loss. It has been estimated that REDD+ could lead to a halving of deforestation rates by 2030, cutting emissions by 1.5–2.7Gt Co_2/year. It would require between US$17.2–33 billion/year and have an estimated long-term net benefit of US$3.7 trillion in present value terms (this accounts only for the benefits of reduced climate change) (Eliasch, 2009). Delaying action on REDD+ would reduce its benefits dramatically: waiting ten more years could reduce the net benefit of halving deforestation by US$500 billion (Hope and Castilla-Rubio, 2008; see also Hussain et al., 2012). Similarly, REDD+ can play a significant role in halting the biodiversity loss associated with deforestation. However, to deliver all the potential co-benefits, REDD+ needs to be well designed, coordinated, and implemented with a strong governance framework in place.

29.5 POLITICAL RESPONSE AND COMMITMENTS TO RESPOND TO THE CHALLENGES

Growing recognition of the urgent need for action to halt the degradation and loss of natural capital, avoid societal losses and safeguard future possibilities for sustainable growth and well-being has led to a range of recent international commitments. These include the call for the conservation, restoration, and sustainable use of biodiversity and ecosystem services in the Strategic Plan for Biodiversity 2011–2020 (see Box 29.6), support for the REDD+ mechanism and ecosystem-based mitigation and adaptation to climate change within the UN Framework Convention on Climate Change (UNFCCC), as well as a commitment to zero net land degradation and combatting global desertification under the UN Convention to Combat Desertification (UNCCD). There are also, of course, a number of other international commitments (via conventions, protocols, and institutions) that aim to respond to the challenges and risks currently faced. More specifically, the following commitments highlight the links between nature and the green economy:

- The reform of environmentally harmful subsidies, which has been identified by many governments as a priority for addressing the loss of biodiversity, the impacts of climate change, as well as making funds available for the transition to a green economy. Calls for subsidy reform have been highlighted under the Strategic Plan for Biodiversity 2011–2020 as well as in G20 statements (CBD, 2010; Lehmann et al., 2011; UNEP, 2011; ten Brink et al., 2012a, 2012b; Withana et al., 2012). Some commitments were also reiterated at the Rio+20 Conference in June 2012,

BOX 29.6 CBD STRATEGIC PLAN FOR BIODIVERSITY 2011–2020

The new Strategic Plan for Biodiversity 2011–2020 to implement the Convention on Biological Diversity was adopted at the tenth meeting of the Parties in October 2010 (Nagoya, Japan) and reaffirmed by Heads of State at the Rio+20 Conference on Sustainable Development in June 2012 (UN, 2012). The Plan represents a globally agreed framework for national implementation. It sets out five strategic goals and 20 headline targets for 2020 to guide national strategies (e.g., national biodiversity strategies and action plans, NBSAPs) and other efforts to preserve biodiversity and restore degraded ecosystems. While all 20 of the targets will be important, the following demonstrate the links between nature and green economy most clearly:

- *Target 2:* By 2020, at the latest, biodiversity values have been integrated into national and local development and poverty reduction strategies and planning processes and are being incorporated into national accounting, as appropriate, and reporting systems.
- *Target 3:* By 2020, at the latest, incentives, including subsidies, harmful to biodiversity are eliminated, phased out or reformed in order to minimize or avoid negative impacts, and positive incentives for the conservation and sustainable use of biodiversity are developed and applied, consistent and in harmony with the Convention and other relevant international obligations, taking into account national socio-economic conditions.
- *Target 4:* By 2020, at the latest, governments, business and stakeholders at all levels have taken steps to achieve or have implemented plans for sustainable production and consumption and have kept the impacts of use of natural resources well within safe ecological limits.
- *Target 6:* By 2020 all fish and invertebrate stocks and aquatic plants are managed and harvested sustainably, legally and applying ecosystem-based approaches, so that overfishing is avoided, recovery plans and measures are in place for all depleted species, fisheries have no significant adverse impacts on threatened species and vulnerable ecosystems and the impacts of fisheries on stocks, species, and ecosystems are within safe ecological limits.
- *Target 7:* By 2020 areas under agriculture, aquaculture and forestry are managed sustainably, ensuring conservation of biodiversity.
- *Target 14*: By 2020, ecosystems that provide essential services, including services related to water, and contribute to health, livelihoods and well-being, are restored and safeguarded, taking into account the needs of women, indigenous and local communities, and the poor and vulnerable.
- *Target 15:* By 2020, ecosystem resilience and the contribution of biodiversity to carbon stocks has been enhanced, through conservation and restoration, including restoration of at least 15 per cent of degraded eco

systems, thereby contributing to climate change mitigation and adaptation and to combating desertification.
- *Target 16:* By 2015, the Nagoya Protocol on Access to Genetic Resources and the Fair and Equitable Sharing of Benefits Arising from their Utilization is in force and operational, consistent with national legislation.

Target 2 is about good governance, which includes making appropriate links to green economy strategies and forms the basis of the commitment to natural capital accounting discussed above. Target 3 underlines Parties' commitments to 'getting the prices right' so as to improve economic signals for biodiversity and ecosystems and support the transition to a green economy. Targets 4, 6, and 7 relate to complementary aspects of sustainable consumption and production, a core foundation of the green economy. Targets 14 and 15 focus on the benefits from natural capital – with Target 14 contributing directly to the social dimension (also critical to the green economy) and Target 15 focusing on the contributions of natural capital to the UNFCCC and UNCCD objectives. Finally, Target 16 is focused on genetic resources, the basis for economy activity and livelihoods for many sectors and activities in the economy (see also IEEP et al., 2012).

in particular regarding fisheries subsidies and inefficient fossil fuel subsidies (UN, 2012; Kettunen and ten Brink, 2012; Oosterhuis and ten Brink, 2014).
- The integration of the value of ecosystem services into natural capital accounts and systems of economic and environmental accounts (SEEA) to ensure an improved and linked evidence base on nature and the economy. The SEEA was initiated by the United Nations Statistical Commission and is linked to the UN System of National Accounts (SNA) from which GDP is derived. The need for environmental accounting was recognised by its inclusion in Target 2 of the Strategic Plan for Biodiversity 2011–2020. The World Bank-led WAVES project and the Gaborone Declaration by ten African Nations (Gaborone Declaration, 2012) in support of green accounting are two further important commitments in this area. At the Rio+20 Conference in 2012, 57 countries and the European Commission supported a communiqué that called on governments, the UN system, international financial institutions, and other international organizations to strengthen the implementation of natural capital accounting around the world and factor the value of natural assets like clean air, clean water, forests, and other ecosystems into countries' systems of national accounting. In addition, 68 private companies also joined forces behind the move and committed to collaborating globally to integrate natural capital considerations into their decision-making processes (Russi et al., 2013).
- Mainstreaming conservation, restoration, and sustainable use of ecosystems and biodiversity as part of wider objectives, strategies, and plans for sustainable development, including national development plans, poverty alleviation strategies, and green economy strategies. This has been called for under the Strategic Plan for Biodiversity 2011–2020.

- The benefits of proactive investment in restoring degraded ecosystems, including Target 15 of the Strategic Plan for Biodiversity 2011–2020, suggested that opportunities for ecosystem-based adaptation should be systematically explored to support the objectives of the UNFCCC and commitments to addressing desertification and land degradation in drylands under the UNCCD.
- The need to further scientific knowledge and know-how and improve communications and cooperation across the science–policy interface has been recognized in the establishment of the Intergovernmental Platform on Biodiversity and Ecosystem Services (IPBES), which aims to reinforce the use of science in policy-making that affects or is affected by the status of biodiversity and ecosystems.
- The engagement of the business community in solutions – whether through improving accounting systems to include resource input dependency, risks and potential liabilities, commitments to carbon neutrality and no-net loss of biodiversity, codes of conduct and commitments to reporting, or improving research activities (McConville and ten Brink, 2012; TEEB, 2012a). Emerging developments include corporate sustainability reporting and accounting, such as PUMA's Environmental Profit and Loss Account (EP&L) and the Natural Capital Declaration of the financial sector (see above) (PUMA, 2011; Natural Capital Declaration, 2012).

The on-going global financial crisis should not slow down the transition to a green economy. On the contrary, the transition to a green economy should act as a catalyst to implement the above-listed commitments in order to achieve significant cost savings over time, exploit untapped opportunities to create jobs and growth, and finally to help society make the transition towards ecologically sustainable growth and, more broadly, a sustainable, desirable future.

29.6 ACHIEVING THE TRANSITION TO A GREEN ECONOMY

While different countries may opt for transition paths towards a green economy tailored to their national circumstances, it is likely that adopting a wider range of coherent and coordinated measures will be an integral part of successful transitions. The mix and emphasis of these measures will differ from one country to another. In most cases, a balanced approach will include both supply and demand measures, thereby greening the economy with production- and consumption-focused measures. As set out above, this approach should build on a sound appreciation of the value and role of nature, which will provide a core foundation of the development of a future green economy. If the transition is well managed, it will not only result in environmental sustainability but will also produce a paradigm shift to sufficiency, adequacy, equity, responsibility, and mutual respect within and across peoples, socio-economic classes, and generations.

29.6.1 Building Blocks in the Transition to a Green Economy

Over the years, a wealth of experience has been accumulated across countries on policies, approaches, and measures to reduce or avoid environmental damage, restore degraded ecosystems, and protect those that are intact and healthy. These measures have been a mix of traditional, 'business as usual' approaches to: (1) minimizing losses and avoiding inappropriate trade-offs, for example, the use of impact assessments, product life cycle assessments, project selection, and evaluation criteria; and (2) investing in environmental infrastructure to be compliant with legislation and regulation, for example, water supply and waste water infrastructure to meet water quality standards and waste infrastructure and air pollution control measures to meet emission and air quality standards.

They have also included, albeit less often, more 'active ecosystem management approaches' such as: (3) proactive approaches to risk management that build on a wider appreciation of risks, for example, risk mapping for flood control and taxonomy research for invasive species (e.g., the moth threat to Mexico's key economic and cultural icon the 'Nopal' cactus); and (4) investment in natural capital via restoration, conservation, and improved management practices through, for example, networks of protected areas (e.g., the EU's Natura 2000 network), the restoration of peatlands for carbon storage and other co-benefits (e.g., Mecklenburg-Vorpommern in Germany), the restoration of flood plains (e.g., Belgium's Scheldt estuary), or afforestation for flood control (China's Sloping Land Conversion Programme) (TEEB, 2011, 2012; ten Brink et al., 2011c, 2012a).

Finally, certain measures have focused on 'pursuing environmental sustainability' via: (5) measures for eco-efficiency and wider resource efficiency for example through water or other resource pricing and wider environmental fiscal reform to incentivize efficient resource use (e.g., fisheries and agricultural subsidy reform in New Zealand and Norway, water pricing reform in the Czech Republic) (OECD, 2011a; TEEB, 2011); and (6) decoupling the economy from resource use and its negative impacts through more radical innovation and changes in demand, for example, through new clean products and processes, building on (access to) genetic resources (e.g., pharmaceutical sector and plant-based cancer treatment) and biomimicry (e.g., floor tiles and waste, architecture and natural cooling), as well as consumption choice changes through information provision, civil society engagement, and the availability of near-zero impact alternatives (TEEB, 2011).

These six approaches, presented in Figure 29.4, together with good governance, are arguably the fundamental building blocks in the transition to a green economy. The mix and emphasis of measures will differ from one country to another depending on national circumstances and windows of opportunity for progress.

29.6.2 Putting the Building Blocks into Practice

The above building blocks represent the key steps to transforming our approach to natural capital and moving towards a development path that integrates economic, social and environmental concerns into a resource-efficient economy that works within the planet's ecological capacities.

To simplify and illustrate the challenge, Figure 29.5 presents two contrasting pathways for using and managing natural capital over time and identifies a range of contributing instruments and measures that can put the building blocks into practice. One pathway is based on the continuing erosion of natural capital from rising pressures of consumption and production of an increasingly rich and growing world population. The other leads to a slowing of and gradual halt in biodiversity loss through a wide range of polices and measures, followed by a reversal of such loss with additional efforts of restoration and investment in natural capital – thus driving a net positive gain for nature. This

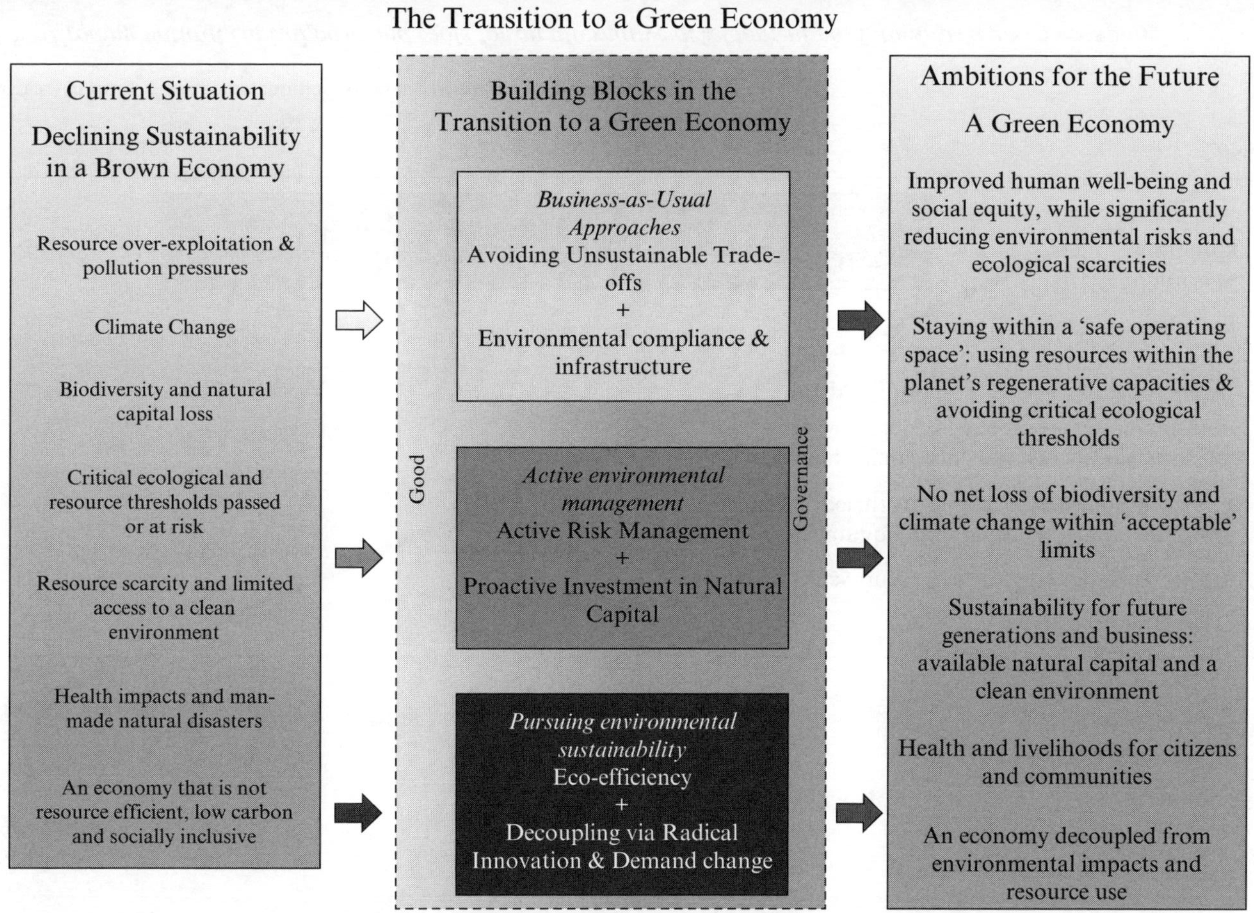

Source: ten Brink et al. (2012a).

Figure 29.4 Building blocks and instruments to enable the transition to a green economy

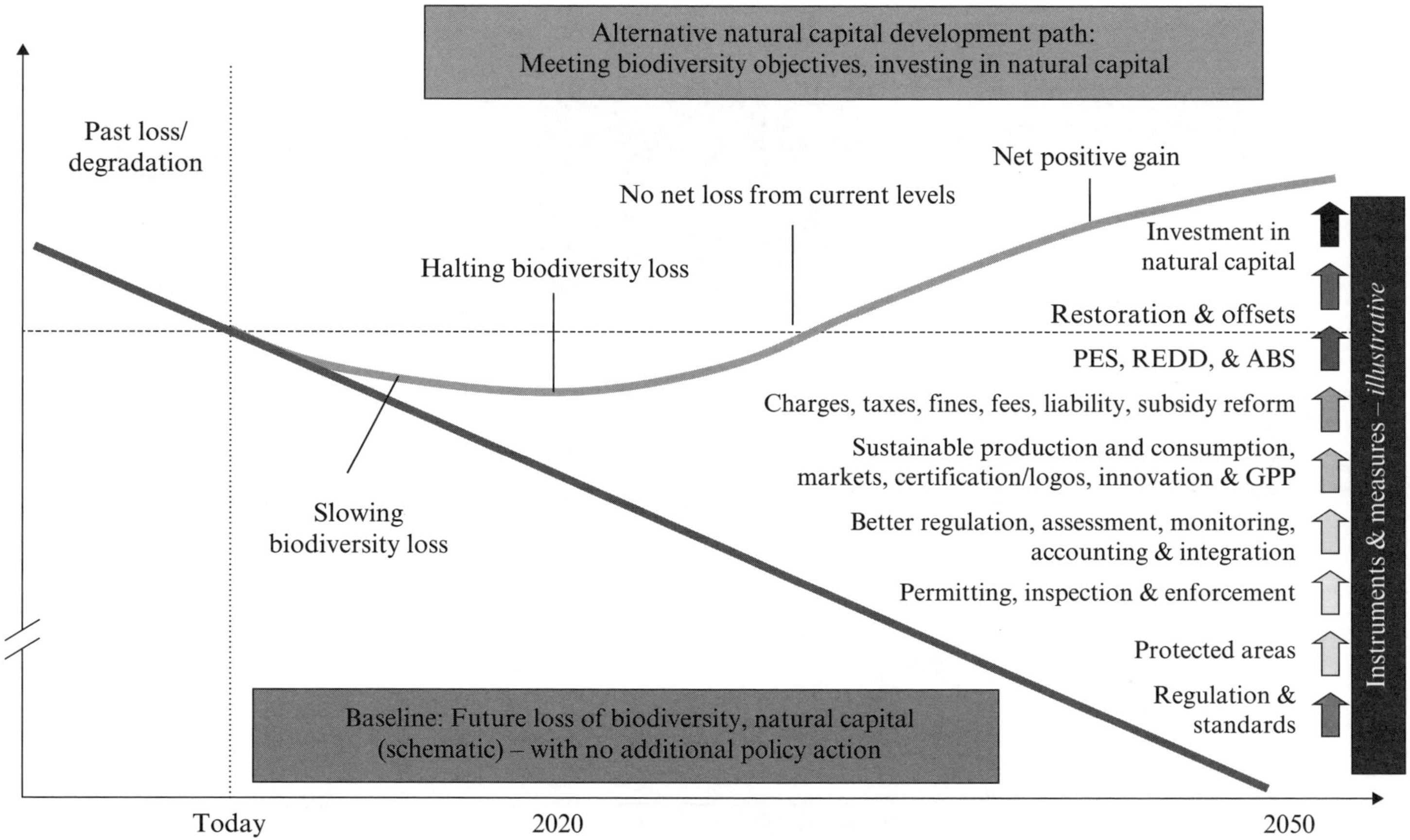

Source: Patrick ten Brink's own representation, from ten Brink et al. (2012a).

Figure 29.5 Eroding natural capital base and tools for an alternative development path, towards a green economy

latter pathway supports the transition to a green economy as recognition of the value of natural capital leads to improved stewardship, protection, and investment, and in turn to a net appreciation of the natural capital base.

The figure puts the milestone for halting the loss of biodiversity at 2020. It then envisages a gain so that no net loss relative to current levels is achieved by 2030 and a new paradigm of positive natural capital is entered thereafter. Unfortunately, without a significant increase in effort, this appears highly optimistic given the current rate of biodiversity loss and increasing (rather than decreasing) pressures from production and consumption. The lack of integration of biodiversity concerns and ecosystem service values into the market and sector policies also adds to this pessimism. Biodiversity policy alone will not be enough to halt biodiversity loss, never mind stimulate the necessary new paradigm of investments in natural capital and appreciation for the interconnections and interdependencies of economic and social systems with the world's ecosystems.

However, progress has been made in several countries and many tools are already proving their merits. Key challenges will be in increasing this effort, which requires 'mutual learning' (being inspired by each other), mainstreaming, good governance (see last section), and finding the finance to realize it.

29.6.3 Financing the Transition to a Green Economy

The transition to a green economy will require considerable financing that will need a wide range of tools and engagement across stakeholders and governance levels. Potential tools that simultaneously address certain environmental issues and raise funds include: subsidy reform (see Box 29.7); getting the prices right through the use of market-based instruments (for example, to encourage cost recovery and implement the polluter pays and user pays principles); better allocation of (e.g., through climate and biodiversity proofing public funds and more effective development cooperation funding); green public procurement and green private purchasing/consumption reducing the demand for

BOX 29.7 A TOOL TO SUPPORT THE REFORM OF ENVIRONMENTALLY HARMFUL SUBSIDIES

There are a number of potential benefits from reforming environmentally harmful subsidies including for the environment (by reducing negative impacts), for the economy (related to budgetary consolidation, competitiveness gains and promoting innovation by addressing technological lock-ins), and the social dimension (by improving the target/focus of subsidies, reducing health impacts of harmful activities). Thus, subsidy reform deserves priority attention.

The flowchart represented in Figure 29.6 sets out a checklist of questions to help develop an inventory of subsidies/incentives and identify clear steps for eliminating/redefining incentives that are harmful for biodiversity. A key issue for most countries is one of political priority; subsidy reform has to compete with many other priorities to attract ministers' attention, so the question of windows of opportunity may in reality be the first question for many countries.

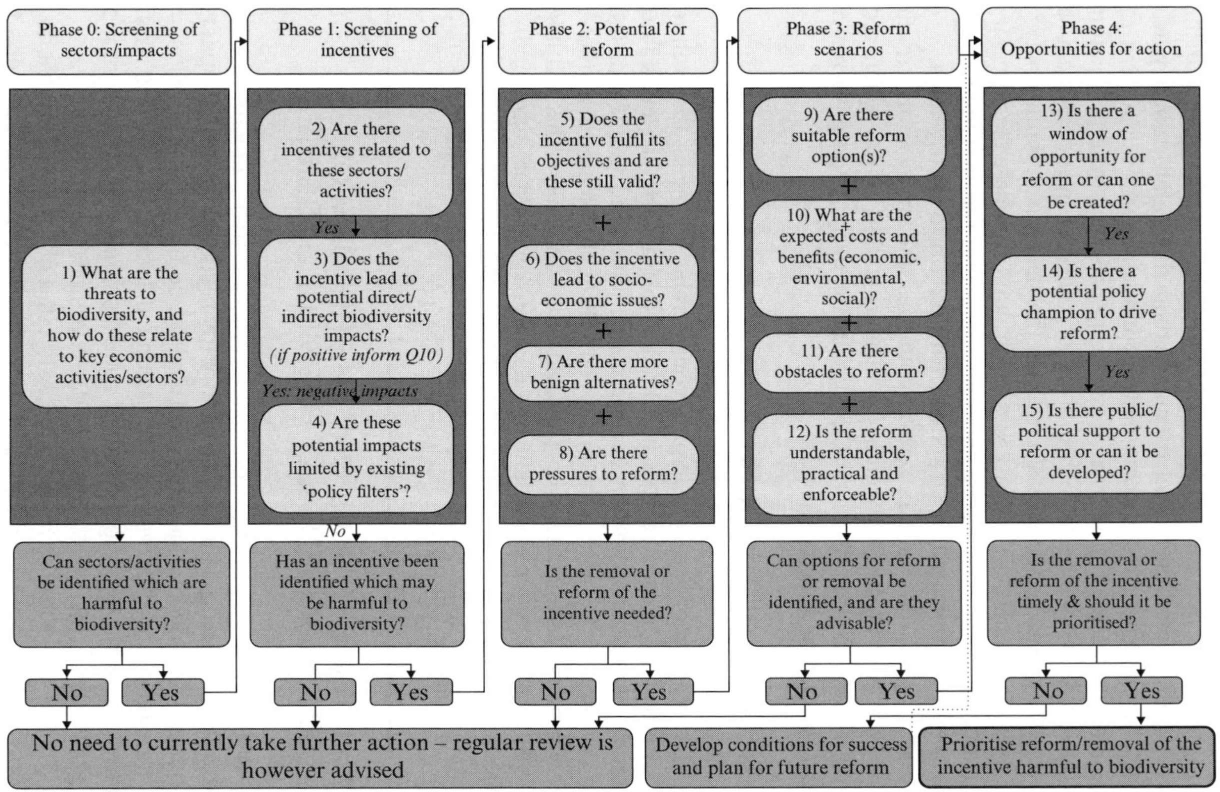

Source: ten Brink et al. (2012b), building on Valsecchi et al. (2009) and Lehmann et al. (2011).

Figure 29.6 Flowchart – the subsidy reform tool

environmentally damaging products while at the same time creating incentives for green production, products, and markets; sharing of benefits arising from the utilization of genetic resources; and other innovative financing tools (e.g., REDD+ and beyond).

There will also be a need to increase investment from the private sector. Ethical investment funds, insurance companies, banks, or indeed rating agencies, have not played a major role in financing nature's role in the transition to the green economy to date. There is, however, a potential for scaling up the contributions from this sector. This will in part be driven by an increased appreciation of nature's contribution to reducing risks related to increased resource scarcity and natural hazards exacerbated by climate change (TEEB, 2012a); it could be further leveraged through the effective use of financial instruments.

Non-economic measures that will help address the financing challenges include:

- *Raising the regulatory baseline* as this reduces risks and avoids costs associated with pollution and accidents. Making polluters pay helps avoid damages and increases industry's involvement and contribution in the transition to a green economy. It would also lessen investment needs to address the impacts of environmental pollution and damages.
- *Implementation and enforcement of regulation* to ensure resources for implementation and address non-compliance. This helps ensure that polluters pay, generally through investments to avoid pollution and, on occasion, via non-compliance fees and fines or compensation payments.
- *Clarifying property rights* (ownership, use, access, etc.), to give certainty and encourage action driven by a long-term perspective including investment in restoration, sustainable management practices, the setting up of PES schemes, and the leveraging of investments.

Those who pay and those who benefit are very often not the same people. Similarly, the costs and the benefits can occur at different timescales – degradation from activities today can cause costs for future generations, and investment in restoration can lead to benefits over many generations. Furthermore, action in one area can lead to benefits at different scales – from local to global. All of these issues are part of the governance challenge for the transition to a green economy. Work on innovative financing instruments is on-going and brings together a wider range of actors including national governments, the CBD, the OECD, and the Global Environment Facility (GEF) (ten Brink et al., 2012b).

29.6.4 Governance for a Green Economy

Actions at all governance levels involving the participation of relevant stakeholders are needed for a successful transition to a green economy. A culture of appreciation for the multiple values of nature can support good governance at many levels and can lead to increased use of a range of economic and non-economic approaches to realize associated benefits. These include the range of benefits to society and economy (the ecosystem services) as well as intrinsic values (TEEB, 2010a). Similarly a culture of evidence-based assessment, aiming to understand the full set of impacts of decisions (who are the winners and losers, what are the spatial impacts, the time profile of benefits and costs and trade-offs and synergies) is a critical aspect of good governance. The transition to a green economy will need to be an integrated multi-level governance process; recognizing the roles and responsibilities of all sectors and levels of governance and engaging a wide array of stakeholders. This can include public support for research and education, public funding for investments in natural capital, a regulatory framework and its enforcement, as well as public–private partnerships. Progress with the above will require engagement by the private sector, which will benefit from policy action to set up appropriate enabling conditions. Due participation, consultation, and engagement of civil society, communities including indigenous populations and citizens will also be of fundamental importance.

There are many potential champions of solutions at different geographic levels and in different groups. It is important to understand their respective interests, incentives, opportunities, and responsibilities in order to harness their potential. Good governance (at all levels from private, to community, to business, to the policy levels from local to national to global) can encourage these groups to contribute to the common goal of halting biodiversity loss and enable them to benefit from working sustainably with nature (TEEB, 2011).

Halting biodiversity loss will not be possible without the involvement of the business community. This is also often in a business's short- and long-term interests as it can serve to increase profitability, create new markets, and help avoid liabilities. The first steps for business are to identify impacts and dependencies, design appropriate responses, and find new ways to reduce biodiversity and ecosystem risks (TEEB, 2012a; see also Box 29.8). This is even the case for sectors that, at first sight, are not directly responsible for many of the pressures the economic system has on the natural environment.

NGOs are another key player. They can raise funds, purchase land, initiate payments for ecosystem service (PES) schemes, offer volunteers to help with monitoring, science and restoration, inform consumer choices and contribute arguments and evidence to policy formation as well as mobilise public support for policy change.

Scientists and academics can provide critical insights and discoveries to help the above groups benefit from nature's potential. This needs to be accelerated and backed by capacity building (e.g., national museums and laboratories on genetic resources) to make the most of new solutions, along with efforts to develop indicators for ecosystem services to better manage natural assets and support economic valuation and improved policy design. There is also a need for investment in both more fundamental science (e.g., functioning of ecosystems) of understanding ecology–economy–society links as well as in the area of economic science (e.g., valuation tools and their application), and investing in the science–policy interface (in order to strengthen evidence-based policy-making). The agreement to establish the Intergovernmental Science–Policy Platform on Biodiversity and Ecosystem Services (IPBES) is an opportunity for progress in this area, as are various national research programmes such as the National Institute of Biological Resources set up by Korea in 2007 to preserve indigenous biological resources that are national assets and provide support to the bio-industry with source materials and information on the properties of wild species.

This multitude of actors may engage in a wide range of processes at various governance levels. A combination of top-down and bottom-up processes will be needed. Decentralized experimentation has on multiple occasions made important contributions to testing and identifying solutions that were suitable for adoption at higher levels of gov-

BOX 29.8 REASONS FOR BUSINESS TO INTEGRATE BIODIVERSITY

There are also considerable incentives for businesses to take biodiversity into consideration in their decision-making. These refer to both risks of inaction and opportunities from proactive engagements, which can be categorized as follows:

Operational
Enhancing natural ecosystems can result in reduced cost of accessing resources, guarantee sustainable access to resources over the long term, and reduce the risk of disruption to resource base.

Access to new markets
Opportunities exist to develop new products that reduce the impact on natural ecosystems (as demonstrated by the expansion of certified biodiversity friendly products), capturing new revenues from company-owned natural assets or entering new markets such as watershed protection or carbon sequestration.

Reputation management
Companies may limit the risk of damage to corporate reputation and licence to operate by reducing impact on biodiversity or having a net positive impact; this in turn, may present an opportunity to increase brand recognition and improve market position.

Reduced exposure to regulation and legal action
Acting early to avoid negative impacts on biodiversity can reduce vulnerability to the risk of liability for damages and potential lawsuits by stakeholders and new regulatory frameworks, which might constrain business activities and reduce profitability.

Access to finance
As more banks and investors begin to adopt more rigorous lending and investment policies, reducing biodiversity impacts could result in more favourable financing terms and improved access to capital. Similarly, insurance coverage and cost is expected to become increasingly sensitive to environmental liabilities and hazards.

Source: Building on TEEB (2012b); WBCSD (2010); McConville and ten Brink (2012).

ernance at a later stage. In most countries, cities have competencies that make them key to producing and implementing models compatible with the green economy (see Box 29.9).

There has been a long and strong tradition of community-based management of natural resources, often highlighted in the literature with regard to fisheries, forests,

BOX 29.9 BOTTOM-UP APPROACHES: CITY ENGAGEMENT FOR SUSTAINABLE CITIES AND TRANSITION TOWNS

Over half the global population is now urban, and city authorities have both means and responsibilities for driving a transition to a green economy. There is growing recognition that nature plays a key role for cities and urban areas by offering water purification and supply, a range of recreation and health benefits, air pollution control, local climate effects (e.g., city cooling, mitigating the 'heat island effect'), as well as supporting tourism and the value of the housing stock. The level of benefits will depend on each city's green infrastructure and its management, and the interactions between the city, its citizens, and the surrounding ecosystems. Potential green economy initiatives at the city level include:

- designing for low impacts – transport, buildings, spatial planning (including densification to avoid urban sprawl);
- providing integrated public transport systems, restricting parking in urban areas, congestion/parking charges to discourage personal car use;
- creating and maintaining natural and semi-natural habitats – woods, parks, community gardens;
- investing in urban green infrastructures (green roofs, spaces, road verges);
- growing local food and supporting community-assisted agriculture schemes.

There are an increasing number of cities committing to becoming greener. The German city of Freiburg is an early example. In 1992, it was chosen as Germany's Environment Capital for its pioneering achievements such as the installation of an early warning system for smog and ozone pollution, pesticide bans, returnable packaging measures, its traffic and transport policy, and perhaps even its engaging 'feel-good' image. Almost every year since then, even more innovations in the field of environmental protection and solar engineering have been produced.

Another interesting example of bottom-up experimentation is the concept of 'transition towns', which are a grassroots network of communities working to build resilience in response to peak oil, climate change, and economic instability. Main aims generally include encouraging methods for reducing energy usage as well as reducing reliance on long supply chains, for example by sourcing food locally or encouraging community gardens to grow food. While the first transition towns were established in the United Kingdom, the concept has since spread across Europe, the USA, Australia, New Zealand, and South America.

Finally, a very wide range of experimentation around food and diet are being carried out by local authorities, which may help increase awareness and set trends that catalyse change on a larger scale. This ranges from organic food served in school canteens to initiatives like the one adopted by the City of Ghent (Belgium) in May 2009, which became the first city with 'official' weekly vegetarian days. Veggie Thursday (or 'Donderdag Veggiedag' in Dutch) was created by the Ethical Vegetarian Alternative, an organization partially funded by the Flemish government. Similarly, in April 2010, San Francisco became the first US city to officially declare Mondays to be 'meat free', calling it their Vegetarian Day.

Sources: Transition Network (www.transitionnetwork.org/), GreenCity Freiburg Brochure (www.freiburg.de/greencity), EVA vzw – Ethisch Vegetarisch Alternatief (www.evavzw.be/).

traditional knowledge and protected areas. As has been underlined, common property can be successfully managed by the groups who use it and, in some cases, it is actually better managed than if privatized or managed by government (Ostrom, 1990). Cooperation between the programmes, funds, and specialized agencies of the United Nations system, NGOs, and other local partners (in particular women's groups, indigenous and local communities) is essential for effective action at the national level (TEEB, 2011).

Finally, the transition to a green economy requires an active participation of citizens. This is already happening through different channels. Citizens, through their actions and purchasing decisions, can lower their consumption impacts out of a sense of responsibility or self-interest. Citizens are involved in various conservation and restoration activities or support local markets and initiatives, such as slow food or city gardening. Youth represent another distinctive group essential for successful transition to green economy, as they often play a direct role in many awareness raising and conservation activities globally and are actively involved in international negotiations. Focusing on youth via education and engagement will also be key, as supporting skills development, knowledge, and understanding in future generations is needed for ensuring the transition to a green economy. Social networks and new media tools are likely to play a substantial role in citizens' participation in the transition to a green economy and awareness raising. Many activities are initiated through social networks (e.g., Facebook, Twitter, LinkedIn), while at the same time major organizations increasingly communicate and disseminate knowledge using these channels.

Finally, a prerequisite to due engagement by the above stakeholders is to ensure good governance through improved transparency and opportunities for public participation (see Box 29.10).

29.7 CONCLUSION: ACCELERATING EFFORTS WHILE MANAGING THE TRANSITION

It is clear from the state of the environment and the on-going degradation that the transition to a green economy will not happen with only a marginal increase in existing greening and green activities. Current levels of ambition are unlikely to be enough to avoid high and increasingly volatile resource prices or do more than slow down the loss

BOX 29.10 TRANSPARENCY AND PARTICIPATION

The United Nations Economic Commission for Europe (UNECE) Aarhus Convention on Access to Information, Public Participation in Decision Making and Access to Justice in Environmental Matters establishes legally binding rights and obligations relating to government decision-making processes on matters concerning the local, national, and transboundary environment. The Convention has been signed by around 40 (primarily European and Central Asian) countries and the EU, and is an example of an international initiative with positive potential (UNECE, 1998, 2010). However, implementation of the Convention has been slow and a lack of transparency continues to hinder good governance. Lack of transparency acts as a brake to progress and can reduce/ hinder participation. In both cases, critical information is likely to be overlooked and not taken into account in decision-making.

Transparency is a prerequisite for progress in a number of issues discussed in this chapter. For example, it is a critical factor when reforming environmentally harmful subsidies. It also plays a key role in ensuring that the designs of payments for ecosystem service instruments are effective. For example, it will be essential in ensuring that the REDD+ instrument meets conditionality requirements (i.e., payment conditional of performance), offers added value (i.e., emissions reduction beyond what would have happened anyway), allows verifiability, and is an instrument worthy of trust and engagement. For biodiversity offsets, transparency is equally key: on transparency, the Business and Biodiversity Offsets Programme's (BBOP) principles state that 'the design and implementation of a biodiversity offset, and communication of its results to the public, should be undertaken in a transparent and timely manner (BBOP, 2008). In green public procurement schemes, transparency and clarity of processes and criteria are important for producers of all sizes to be able to adapt their offers. Similarly, transparency is critical in the area of access and benefit sharing (ABS) as regards sourcing of genetic material and traditional knowledge (IEEP et al., 2012).

National accounting and business accounting and reporting are key vehicles for improving transparency. Implementing the commitment to develop natural capital accounts (e.g., as per the commitment at the Rio+20 Conference, above) and the System of Environmental-Economic Accounting (SEEA) will be critical to ensuring that policy-makers have a more complete evidence base available for decisions to integrate insights on changes in the stock of natural capital and the implications for service flows and values. This is a long-term commitment that will not bring benefits immediately, but is a necessary foundation for a sustainable future. The European Union's Regulation on National Environmental Economic Accounts is an example of such a system, and efforts in this regard need to be further strengthened.

of biodiversity and other potentially irreversible environmental changes (e.g., as ecological thresholds are passed and ecosystems and climate systems move to new operations).

More conviction, commitments, and implementation of policy reforms and initiatives as well as a genuine acceleration in effort is needed to drive the transition to a truly sustainable future. There is a need to move from discrete cases of green economy transition to a fundamental systemic transition warranted by scientific findings (EEA, 2010; UNEP, 2011). This would not necessarily imply huge costs or indeed negative impacts on GDP. On the contrary, focused investment can lead to significant savings and efficiency gains.

While the transition to the green economy will lead to many win-wins, it may also mean losses for certain groups and trade-offs across sectors and capitals over time. These impacts need to be duly accounted for in transition plans. Managing the transition will thus be critical, as will the insurance of transparency and communication throughout the process. Transition management tools can include targeted education and skills training, the provision of early information, and the phased introduction of measures taking into account affordability (e.g., as regards moving towards cost recovery in pricing), spatial planning (e.g., zoning fisheries areas), investment in substitute products or services, and, in some cases, compensation for losses.

The follow-up to the Rio+20 Conference offers an important opportunity to commit to working with nature and driving the transition to a truly sustainable future that promotes social equity, poverty eradication, and human well-being. While the conference highlighted some of the limitations of multilateral processes it also gave more emphasis on the role that needs to be played by action cities, businesses, community and citizen initiatives, which can help to drive experimentation and progress at different levels. It is essential that these stakeholders take a proactive role in driving the transition to green economy forward. A multi-level governance response that builds on the strengths of various stakeholders is especially important now, given, on the one hand, the scale of challenge and urgency of action and, on the other, the limited policy appetite for ambitious measures in the current economic climate. The top-down policy measures will be far from sufficient and therefore the unique contributions and capacities of different stakeholders creating or making use of available windows of opportunity will be the key in achieving the needed transition to a green economy. Stakeholders can identify measures and processes that they can already initiate and/or proactively engage with, also creating opportunities for mutual learning, increasing joint motivation, and helping to identify practical solutions that can subsequently be picked up and promoted by other decision-makers at different levels. This helps to generate a system evolution, to address the systemic challenges that society faces, and achieve the needed transition to a green economy.

ACKNOWLEDGEMENTS

This article builds on the author's work for UNEP – *Nature and the Transition to a Green Economy* (2012). We would like to thank UNEP and the TEEB funders for the support for that work. We would also like to thank the following for valuable inputs and suggestions to that work that are also relevant for this chapter: Sasha Alexander, Maude Antoine, James Aronson, Ed Barbier, Nicholas Bertrand, James Blignaut, Allan Buckwell, Doreen Fedrigo-Fazio, Andreas Hauser, Marina Kosmus, Laure

Ledoux, Markus Lehmann, Dustin Miller, Alice Ruhweza, Bent-Arne Sather, Benjamin Simmons, Pavan Sukhdev, Nicola Tilche, James Vause, Axel Volkery, Heidi Wittmer, the wider TEEB Coordination Group and Advisory Board. It also benefits from research carried out within the EU 7FP research project, OPERAs (Operational Potential of Ecosystem Research Applications; Grant Agreement No. 308393), where green economy and the role of natural capital and ecosystem services is one of the emerging themes being researched. The contents and views contained in this chapter are those of the authors, and do not necessarily represent those of any of the contributors, reviewers or organizations supporting this work.

NOTE

1. See http://greeneconomycoalition.org/updates/sign-9-principles-green-economy.

REFERENCES

Aronson, J., S.J. Milton and J. Blignaut (eds) (2007), *Restoring Natural Capital: Science, Business and Practice*, Washington, DC: Island Press.

Barbier, E. (2002), 'The role of natural resources in economic development', 2002 Joseph Fisher Lecture, Adelaide University, Australia, 30 September 2002.

Barbier, E.B. (2005), *Natural Resources and Economic Development*, Cambridge, UK: Cambridge University Press.

Barbier, E.B. (2010), *A Global Green New Deal – Rethinking the Economic Recovery*, Cambridge, UK and New York: Cambridge University Press.

Barrow, E. and A. Shah (2011), 'TEEBcase: Traditional forest restoration in Tanzania', accessed 20 February 2014 at http://www.teebweb.org/wp-content/uploads/2013/01/TEEBcase-Traditional-forest-restoration-Tanzania-.pdf.

BBOP (Business and Biodiversity Offsets Programme) (2008), *Principles on Biodiversity Offsets Supported by the BBOP Advisory Committee*, accessed 20 February 2014 at http://www.forest-trends.org/bbop/guidelines/principles.pdf.

BPL (Butcher Partners Limited) (2006), *Evaluation Study of the Economic Benefits of Water in Te Papanui Conservation Park*, prepared for the New Zealand Department of Conservation of New Zealand.

Braat, L. and P. ten Brink (eds) with J. Bakkes, K. Bolt, I. Braeuer, B. ten Brink, A. Chiabai and H. Ding et al. (2008), *The Cost of Policy Inaction (COPI): The Case of Not Meeting the 2010 Biodiversity Target*, Brussels: European Commission.

Brown, O., A. Crawford and A. Hammill (2006), *Natural Disasters and Resource Rights: Building Resilience, Rebuilding Lives*, Manitoba, Canada: International Institute for Sustainable Development.

CBD (2010), *Decision X/2 on Updating and Revision of the Strategic Plan for the Post-2010 Period*, Tenth Meeting of the Conference of the Parties to the Convention on Biological Diversity.

Comprehensive Assessment of Water Management in Agriculture (2007), *Water for Food, Water for Life: A Comprehensive Assessment of Water Management in Agriculture*, London: Earthscan, and Colombo: International Water Management Institute.

EEA (2010) *The European Environment – State and Outlook 2010: Synthesis*, Copenhagen: European Environment Agency.

Ekins, P. (1992), 'A four-capital model of wealth creation', in P. Ekins and M. Max-Neef (eds), *Real-Life Economics: Understanding Wealth Creation*, London and New York: Routledge, pp. 147–55.

Ellen MacArthur Foundation (2012) *Towards the circular economy: Economic and business rationale for an accelerated transition*, Report Vol. 1

Eliasch, J. (2009). *Climate Change: Financing Global Forests*, London: UK Government.

Emerton, L., O. Philavong and K. Thanthatep (2002), *Nam Et-Phou Loei National Biodiversity Conservation Area, Lao PDR: A Case Study of Economic and Development Linkages*, Gland, Switzerland: IUCN.

Food and Agriculture Organization (FAO) (2009), *State of World Fisheries and Aquaculture 2008*, Rome: FAO.

Fedrigo-Fazio, D. and P. ten Brink (2012), *Green Economy: What Do We Mean by Green Economy?*, Briefing by

IEEP, UNEP and GLOBE Europe under Knossos 7FP, Nairobi, Kenya, accessed 20 February 2014 at http://www.unep.org/pdf/Main_briefing_2012.pdf.

Förster, J. (2010), 'TEEBcase: Peatlands restoration for carbon sequestration, Germany', accessed 20 February 2014 at http://www.teebweb.org/wp-content/uploads/2013/01/Peatlands-restoration-for-carbon-sequestration-Germany.pdf.

Gaborone Declaration (2012), *The Gaborone Declaration of the Summit for Sustainability in Africa*, 24–25 May 2012, Botswana, accessed 20 February 2014 at http://www.conservation.org/conferences/africa_sustainability_summit/Documents/Gaborone-Declaration-HoS-endorsed_5-25-2012_Govt-of-Botswana_CI_Summit-for-Sustainability-in-Africa.pdf

GHK, IEEP and PSI et al. (2005), 'SRDTOOLS: methods and tools for evaluating the impact of cohesion policies on sustainable regional development (SRD)', Contract No. 502485 Sixth Framework Programme Priority 8.3.1 Task 11 Regio Underpinning European Integration, Sustainable Development, Competitiveness and Trade Policies.

Gutman, P. (2002), 'Putting a price tag on conservation: cost benefit analysis of Venezuela's National Parks', *Journal of Latin American Studies*, **34**(1), 43–70.

Hope, C. and J.C. Castilla-Rubio (2008), 'A first cost benefit analysis of action to reduce deforestation', paper commissioned by the Office of Climate Change as background work to its report Climate Change, Cambridge: Judge Business School, University of Cambridge.

Huhtala, M., L. Kajala and E. Vatanen (2010), 'Local economic impacts of national park visitors' spending: the development process of an estimation method', Working Papers of the Finnish Forest Research Institute, accessed 20 April 2012 at http://www.metla.fi/julkaisut/workingpapers/2010/mwp149.htm.

Hussain, S.S., A. Markandya, L. Brander, A. McVittie, R. de Groot and O. Vardakoulias et al. (2012), 'The challenge of ecosystems and biodiversity', Copenhagen Consensus Paper.

IEEP, Ecologic and GHK (2012), *Study to Analyse Legal and Economic Aspects of Implementing the Nagoya Protocol on ABS in the European Union*, final report for the European Commission, DG Environment, Brussels and London: Institute for European Environmental Policy, accessed 20 February 2014 at http://ec.europa.eu/environment/biodiversity/international/abs/pdf/ABS%20FINAL%20REPORT.pdf.

IFRC (2002), *World Disasters Report 2002*, London: International Federation of Red Cross and Red Crescent Societies, Eurospan, accessed 20 February 2014 at http://www.ifrc.org/PageFiles/89755/2002/32600-WDR2002.pdf.

Jacobs, M. (2012), 'Green growth: economic theory and political discourse', Centre for Climate Change Economics and Policy Working Paper No. 108, accessed 20 February 2014 at www2.lse.ac.uk/GranthamInstitute/publications/WorkingPapers/Abstracts/90–99/green-growth-economic-theory-political-discourse.aspx.

Kettunen, M. and P. ten Brink (2012), 'Nature, green economy and sustainable development: the outcomes of UN Rio+20 Conference on Sustainable Development', *Nature Conservation*, **2**, 1–6.

Kettunen, M. and P. ten Brink (eds) (2013), *The Social and Economic Benefits of Protected Areas: An Assessment Guide*, London: Earthscan.

Lehmann M., P. ten Brink, S. Bassi, D. Cooper, A. Kenny, S. Kuppler, A. von Moltke and S. Withana (2011), 'Reforming subsidies', in *The Economics of Ecosystems and Biodiversity (TEEB) in National and International Policy Making: An Output of TEEB*, edited by Patrick ten Brink, IEEP, London: Earthscan.

McConville, A.J. and P. ten Brink (2012), 'Business and biodiversity: opportunities, conflicts and ways forward', an IEEP paper for JNCC, a contribution to the 11th ENCA Meeting, 23–25 September 2012, Brussels.

McGranahan, G., D. Balk and B. Anderson (2007), 'The rising tide: assessing the risks of climate change and human settlements in low elevation coastal zones', *Environment and Urbanization*, **19**(1), 17–37.

Millennium Ecosystem Assessment (MA) (2005), *Ecosystems and Human Well-being: Biodiversity Synthesis*, Washington, DC: World Resources Institute.

MLUV MV (2009), *Konzept zum Schutz und zur Nutzung der Moore* [*Concept for the Protection and Use of the Moore*], Ministerium für Landwirtschaft, Umwelt und Verbraucherschutz Mecklenburg-Vorpommern, Schwerin.

Muñoz, C., M. Rivera and A. Cisneros (2010), 'Estimated reduced emissions from deforestation under the Mexican Payment for Hydrological Environmental Services', INE Working Papers No. DGIPEA-0410, Mexico.

Natural Capital Declaration (2012), accessed 20 February 2014 at http://www.naturalcapitaldeclaration.org/the-declaration/#.

National Ecosystem Assessment (UK NEA) (2011), *The UK National Ecosystem Assessment*, Cambridge, UK: UNEP-WCMC, accessed 20 February 2014 at http://www.unep-wcmc.org/uk-national-ecosystem-assessment_646.html.

OECD (2008), *Costs of Inaction on Key Environmental Challenges*, Paris: OECD.

OECD (2011a), *Fisheries Policy Reform – National Experiences*, Paris: OECD, accessed 20 February 2014 at http://dx.doi.org/10.1787/9789264096813-en.

OECD (2011b), *Towards Green Growth: A Summary for Policy Makers*, Paris: OECD, accessed 20 February 2014 at http://www.oecd.org/dataoecd/32/49/48012345.pdf.
OECD (2012), *Environmental Outlook to 2050: The Consequences of Inaction*, Paris; OECD, accessed 20 February 2014 at http://www.oecd.org/document/11/0,3746,en_2649_37465_49036555_1_1_1_37465,00.html.
Oosterhuis, F. and ten P. Brink (eds) (2014), *Paying the Polluter. Environmentally Harmful Subsidies and their Reform*, Cheltenham, UK and Northampton, MA, USA: Edward Elgar Publishing.
Ostrom, E. (1990), *Governing the Commons: The Evolution of Institutions for Collective Action*, Cambridge, UK: Cambridge University Press.
Pearce, D.W., A. Markandya and E. Barbier (1989), *Blueprint for a Green Economy*, London: Earthscan.
Pimm, S.L., G.J. Russell, J.L. Gittleman and T.M. Brooks (1995), 'The future of biodiversity', *Science*, **269**(5222), 347–50.
PUMA (2011), PUMA's Environmental Profit & Loss Account for the year ended 31 December 2010, accessed 20 February 2014 at http://about.puma.com/wp-content/themes/aboutPUMA_theme/financial-report/pdf/EPL080212final.pdf.
Randers, J. (2012), *2052: A Global Forecast for the Next Forty Years*, White River Junction, VT: Chelsea Green Publishing.
Rockstrom, J., W. Steffen, K. Noone, A. Persson, F.S. Chapin and E.F. Lambin et al. (2009), 'A safe operating space for humanity', *Nature*, **461**, 472–5, accessed 20 February 2014 at http://www.nature.com/nature/journal/v461/n7263/full/461472a.html.
Russi, D. and P. ten Brink (2013), *Natural Capital Accounting and Water Quality: Commitments, Benefits, Needs and Progress. A Briefing Note*, The Economics of Ecosystems and Biodiversity (TEEB), available at: teebweb.org.
Schäfer, A. (2009), 'Moore und Euros – die vergessenen Millionen' [Moore and euros – the forgotten millions], *Archiv für Forstwesen und Landschaftsökologie*, **43**, 156–60.
Singh, S. (2010), 'TEEBcase: Enhancing agriculture by ecosystem management in Hiware Bazar, India', accessed 20 February 2014 at http://www.teebweb.org/wp-content/uploads/2013/01/Enhancing-agriculture-by-ecosystem-management-in-Hiware-Bazaar-India.pdf.
TEEB (2008) *TEEB Interim Report*, accessed 20 February 2014 at http://www.teebweb.org/publication/the-economics-of-ecosystems-and-biodiversity-an-interim-report/.
TEEB (2010a), *The Economics of Ecosystems and Biodiversity: Ecological and Economic Foundations*, edited by P. Kumar, London: Earthscan.
TEEB (2010b), *Mainstreaming the Economics of Nature: A Synthesis of the Approach, Conclusions and Recommendations of TEEB*, by P. Sukhdev, J. Bishop, H. Gundimeda, P. Kumar, C. Nesshöver and A. Neuville et al., Bonn/Brussels.
TEEB (2011), *The Economics of Ecosystems and Biodiversity in National and International Policy Making*, edited by P. ten Brink, London: Earthscan.
TEEB (2012a), *The Economics of Ecosystems and Biodiversity in Business and Enterprise*, edited by J. Bishop, London: Earthscan.
TEEB (2012b), *The Economics of Ecosystems and Biodiversity in Local and Regional Policy and Management*, edited by H. Wittmer and H. Gundimeda, Abingdon and New York: Earthscan from Routledge.
ten Brink P., Armstrong J., Kettunen M., Rayment M., Ruhweza A., Shine C., Vakrou A., and Wittmer H. (2011a). *The Global Biodiversity Crisis and related Policy Challenge*. In TEEB (2011) *The Economics of Ecosystems and Biodiversity (TEEB) in National and International Policy Making* edited by P. ten Brink. Earthscan, London.
ten Brink P., Bassi S., Bishop J., Harvey C.A., Karousakis K., Markandya A., Nunes P.A.L.D., McConville A.J., Ring I., Ruhweza A., Sukhdev P., Vakrou A., van der Esch S., Verma M., and Wertz-Kanounnikoff S. (2011b). *Rewarding Benefits through Payments and Markets*. In TEEB (2011) *The Economics of Ecosystems and Biodiversity (TEEB) in National and International Policy Making* edited by P. ten Brink. Earthscan, London.
ten Brink, P., T. Badura S. Bassi, E. Daly, I. Dickie and H. Ding et al. (2011c), *Estimating the Overall Economic Value of the Benefits Provided by the Natura 2000 Network*, Final Synthesis Report to the European Commission, DG Environment on Contract ENV.B.2/SER/2008/0038, Brussels: Institute for European Environmental Policy/GHK/Ecologic, accessed 20 February 2014 at www.ieep.eu/work-areas/biodiversity/nature-conservation-policy-and-its-implementation/2012/06/estimating-the-overall-economic-value-of-the-benefits-provided-by-the-natura-2000-network.
ten Brink, P., L. Mazza, T. Badura, M. Kettunen and S. Withana (2012a), *Nature and its Role in the Transition to a Green Economy*, TEEB, accessed 20 February 2014 at http://www.teebweb.org/wp-content/uploads/2012/10/Green-Economy-Report.pdf.
ten Brink, P., S. Bassi, T. Badura, K. Hart and M. Pieterse (2012b), *Incentive Measures and Biodiversity – A Rapid Review and Guidance Development. Volume 3 – Guidance to Identify and Address Incentives Which are Harmful to Biodiversity*, UK: Department for Environment, Food and Rural Affairs.

Tiwari, R. S. Phand and H.P.S. Arya (200), 'Dairy development through natural resource management: a success story of drought prone village in India', *Livestock Research for Rural Development*, **19**(8), accessed 15 July 2010 at www.lrrd.org/lrrd19/8/phan19112.htm.
UN (United Nations General Assembly) (2000), *United Nations Millennium Declaration*, accessed 20 February 2014 at http://www.un.org/millennium/declaration/ares552e.htm.
UN (United Nations) (2012), *The Future We Want – Outcomes of the Rio+20 Conference on Sustainable Development* (A/CONF.216/L.1), Rio de Janeiro 20–22 June 2012, accessed 20 February 2014 at http://www.uncsd2012.org/thefuturewewant.html.
UNDP (2008), *Human Development Report 2007/2008. Fighting Climate Change: Human Solidarity in a Divided World*, New York: United Nations Development Programme.
UNECE (1998), *Convention on Access to Information, Public Participation in Decision-Making and Access to Justice in Environmental Matters*, Aarhus, Denmark, accessed 21 February 2014 at http://www.unece.org/fileadmin/DAM/env/pp/documents/cep43e.pdf.
UNECE (2010), 'Introducing the Aarhus Convention: access to information, public participation in decision-making and access to justice in environmental matters', accessed 20 February 2014 at http://www.unece.org/env/pp/introduction.html.
UNEP (2009), *Global Green New Deal – Policy Brief*, accessed 20 February 2014 at http://www20.gencat.cat/docs/dmah/Home/Ambits%20dactuacio/Educacio%20i%20sostenibilitat/Desenvolupament%20sostenible/Conceptes%20clau/Economia%20ecoeficient/archivos/A_Global_Green_New_Deal_Policy_Brief.pdf.
UNEP (2011), *Towards a Green Economy – Pathways to Sustainable Development and Poverty Eradication, A Synthesis for Policy Makers*, accessed 20 February 2014 at http://www.unep.org/greeneconomy/Portals/88/documents/ger/GER_synthesis_en.pdf.
United Nations Secretary-General's High-Level Panel on Global Sustainability (2012), *Resilient People, Resilient Planet: A Future Worth Choosing, Overview*, New York: United Nations, accessed 20 February 2014 at www.un.org/gsp/sites/default/files/attachments/GSPReportOverview_Letter%20size.pdf.
Valsecchi, C., P. ten Brink, S. Bassi, S. Withana, M. Lewis and A. Best et al. (2009), *Environmentally Harmful Subsidies: Identification and Assessment*, final report for the European Commission's DG Environment, November 2009.
Van Andel, J. and J. Aronson (eds) (2012), *Restoration Ecology: The New Frontier*, 2nd edition, Oxford: Wiley-Blackwell.
WBCSD (2010), *Vision 2050: The New Agenda for Business*, World Business Council for Sustainable Development, accessed 21 February 2014 at http://www.wbcsd.org/vision2050.aspx.
Withana, S., P. ten Brink, L. Franckx, M. Hirschnitz-Garbers, I. Mayeres, F. Oosterhuis and L. Porsch (2012), *Study Supporting the Phasing Out of Environmentally Harmful Subsidies*, a report by the Institute for European Environmental Policy (IEEP), Institute for Environmental Studies – Vrije Universiteit (IVM), Ecologic Institute and VITO for the European Commission – DG Environment, Final Report, Brussels: IEEP.
World Commission on Environment and Development (WCED) (1987), *Our Common Future*, Oxford: Oxford University Press, Oxford, p. 43.
World Bank and FAO (2009), *The Sunken Billions: The Economic Justification for Fisheries Reform*, Washington, DC: Agriculture and Rural Development Department, The World Bank, accessed 20 February 2014 at http://siteresources.worldbank.org/EXTARD/Resources/336681–1224775570533/SunkenBillionsFinal.pdf.
Wunder, S. and S. Wertz-Kanounnikoff (2009), 'Payments for ecosystem services: a new way of conserving biodiversity in forests', *Journal for Sustainable Forestry*, **28**(3), 576–96

30. New business decision-making aids in an era of complexity, scrutiny, and uncertainty: tools for identifying, assessing, and valuing ecosystem services

Sissel Waage

30.1 INTRODUCTION

Imagine a corporate manager under pressure to continually add metrics to track and analyze business impacts and performance. The number of parameters, complexity of processes, and cost to the company would constantly rise. Global companies face this dynamic. In response, business leaders seek integrated approaches to track a multitude of metrics concurrently within the broader system of performance. As part of continual corporate improvement processes, a range of companies in the early twenty-first century began to consider whether the concept of ecosystem services might be a useful lens through which to view environmental and social performance of their businesses.[1] This exploratory work was driven by a growing number of players who noted that current corporate environmental performance tracking methods were inadequate for assessing many issues related to maintenance of natural capital, as well as ecological structure and function, and the flow of nature's benefits.

The expanding set of environmental issues is clear when viewed in terms of activity on ecosystem services, which is on the rise within:

- *the financial services sector*, including the International Finance Corporation (IFC), which in turn affected mainstream financial services institutions through the Equator Principles, which serve as international best practice on due diligence processes;
- *academic institutions*, such as Stanford University[2] in the United States, Wageningen University[3] in the Netherlands, and Fundação Getulio Vargas[4] in Brazil, which all have active research and teaching initiatives underway related to ecosystem services;
- *environmental organizations*, such as Flora and Fauna International, the International Union for Conservation of Nature, The Nature Conservancy, World Resources Institute, and the World Wildlife Fund, among many others around the world;
- *national governments* (such as India, Colombia, Mexico, the Philippines, the UK, and Norway) and *subnational governments* (such as Brazil's state of Acre)[5] through independent activities as well as, at the national level, through engagement with the Global Partnership for Wealth Accounting and the Valuation of Ecosystem Services (WAVES).[6]

Efforts focused on natural capital and ecosystem services will likely continue to build. If they do, companies will face expanded corporate management and performance expectations that include ecosystem services parameters. For example, companies may need to employ new ecosystem-services-related risk and impact assessment protocols to identify potential effects of new projects and possible disruptions to supply chains based on changes in ecosystem services flows. At the same time, businesses will have new opportunities to create, measure, and capture value by investing in activities that enhance ecosystem services.

In response to this emerging area of corporate performance, and in collaboration with the USGS (US Geological Survey) and US BLM (Bureau of Land Management), BSR's Ecosystem Services Working Group convened and facilitated a roundtable discussion – which included industry, tool developers, and government – focused on preliminary findings from a comparative tool application of emerging ecosystem-services-focused tools. This application aimed to help corporate decision-makers, tool developers, government officials, and other interested entities to better understand how to apply these tools in decision-making processes, at what cost, and with what added value, particularly in terms of new insights on impacts and dependencies that are associated with activities on the landscape. This comparative tool assessment sought to compare the tools through applications at a common study site, using the same datasets and the same technical analyst to oversee the process.

This chapter lays out the comparative tool assessment's process and findings. It also discusses potential applications of tools within corporate contexts.

30.2 COMPARATIVE ECOSYSTEM SERVICES TOOL ASSESSMENT PROCESS

The objective of the tool assessment was to compare and assess seven ecosystem-services-focused tools through a test application within the US San Pedro Watershed in Arizona, and the question of where to site a hypothetical residential housing project. The study focused on four parameters: water provisioning, carbon sequestration, cultural services, and biodiversity. (Even though biodiversity is not an ecosystem service, its role in ecosystem structure and function makes it a significant parameter.) The study also tracked the time, cost, level of expertise, and other factors associated with applying each tool within this case study. The intent was to gather information that would help assess the feasibility of tool applications within corporate settings.

The study area was initially chosen by the US Bureau of Land Management and the US Geological Survey for an Ecosystem Services Valuation Pilot study, which included, among others, the ARIES and InVEST tools (described below). The site was chosen because it is a data-rich, ecologically diverse area with a history of mining activity and cultural significance for 11 Native American tribes. The BLM and USGS agreed to engage with BSR in their review in order to expand the comparative assessment to include five additional tools, which the tool developers applied using the same datasets, insofar as those datasets meshed with the tools' data demands. Technical analyst Dr. Ken Bagstad[7] served as a contractor to the BLM and USGS, as well as BSR, to collect data and feed it into these tools.

The comparative study including the following tools:

ARIES (ARtificial Intelligence for Ecosystem Services) is:

> a web-based technology offered to users worldwide to assist rapid ecosystem service assessment and valuation. Its purpose is to make environmental decisions easier and more effective. ARIES helps discover, understand, and quantify environmental assets and what factors influence their values, in a geographical area and according to needs and priorities set by its users. ARIES can accommodate a range of different use scenarios, including spatial assessments and economic valuations of ecosystem services, optimization of payment schemes for ecosystem services, and spatial policy planning.[8]

InVEST (Integrated Valuation of Ecosystem Services and Trade-offs) is:

> designed to help local, regional, and national decision-makers incorporate ecosystem services into a range of policy and planning contexts for terrestrial, freshwater, and marine ecosystems, including spatial planning, strategic environmental assessments, and environmental impact assessments. InVEST models are based on production functions that define how an ecosystem's structure and function affect the flows and values of ecosystem services. The models account for both service supply (for example, living habitats as buffers for storm waves) and the location and activities of people who benefit from services (for instance, location of people and infrastructure potentially affected by coastal storms). Since data are often scarce, the first version of InVEST offers relatively simple models with few input requirements. These models are best suited for identifying patterns in the provision and value of ecosystem services. With validation, these models can also provide useful estimates of the magnitude and value of services provided.[9]

EcoAIM (Ecological Asset Inventory and Management) is a tool:

> to (1) inventory ecological services and help in making decisions regarding development, transactions, and ecological restoration; (2) develop specific estimates of ecosystem services in a geographically relevant context, and (3) offer the means for evaluating trade-offs of ecosystem services resulting from different land or resource management decisions.[10]

It is intended to be applied at a watershed level down through a site-specific level. The approach is one of GIS optimization model analysis of rare species with a risk analysis basis, including metric weightings of stakeholder preferences.

EcoMetrix is:

> an environmental measurement and modeling tool that supports sustainable infrastructure, restoration projects, and enterprise-level program decision-making. EcoMetrix models and quantifies changes within an ecosystem, enabling users to evaluate the positive or negative effects of different scenarios and alternative designs on ecosystem services.[11]

This is a site-level tool that relies on ecological field site data collection on presence and status of ecosystem services at a particular location.

ESR (Ecosystem Services Review) is:

> a structured methodology for corporate managers to proactively develop strategies for managing business risks and opportunities arising from their company's dependence and impact on ecosystems.[12]

ESValue is:

> a strategic decision support tool that integrates scientific and economic information to show the impact and value of alternative environmental management strategies on ecosystem services. The objective of the tool is to integrate existing information and expert opinion with stakeholder values to efficiently and effectively identify the key site-specific ecological effects and resulting change in economic value for different management strategies.[13]

NAIS (The Natural Assets™ Information System) was developed:

> to estimate Ecosystem Service Values (ESV) using 'state of the art' value transfer methods and geospatial science. Value transfer involves the adaptation of existing valuation information to new policy contexts where valuation data is absent or limited. For ESVs, this involves searching the literature for valuation studies on ecosystem services associated with ecological resource types (for example, forests, wetlands, etc.) present at the policy site. Value estimates are then transferred from the original study site to the policy site based on the similarity of ecological resources at the policy site. Value transfer is a 'second-best' approach for gathering information about the value to humanity of ecosystem goods and services. However, the alternative, primary valuation research is extremely costly and is rarely feasible in the context of the policy and planning process. Therefore, value transfer integrated with geospatial science has proven to be a critical tool in decision making and planning.[14]

Developed by a range of specialists with distinct skills and approaches, the tools span the gamut in terms of approach, analytical architecture, as well as outputs created. Therefore, despite a common set of overarching questions, the applications of the tools to the San Pedro assessment were distinct due to significant differences among tools, as well as other considerations. For example:

- ARIES and InVEST are GIS-enabled computer simulation tools.
- The ESValue tool incorporates stakeholder preferences and ecological analysis in the GIS impact analysis.
- The ESR is a structured approach to setting priorities among experts (and/or stakeholders).
- NAIS was not applied because of time constraints, a pro bono budget, and a lack of primary economic valuation studies to apply to arid and semi-arid ecosystems. However, NAIS's developers presented their tool's approach and participated in the roundtable discussions.

The two least comparable approaches were the EcoAIM and the EcoMetrix tools. Their non-comparability was due to the decision to scope tool application to the San Pedro Watershed differently, as well as to the inherent nature of the tools. Specifically, the EcoAIM tool team focused only on one ecological parameter: minimizing biodiversity impacts, as a proxy for habitat-provisioning services of the landscape. The EcoAIM team also narrowed the geographic scope to built-up areas with impervious surfaces to select the site of the hypothetical housing development. Since the EcoMetrix tool is designed for parcels or sites rather than landscapes, it is most useful when applied to the sites selected by landscape-level assessment tools, such as ARIES and InVEST, and then applied for more granular site assessment. However, at the time of the BSR assessment, the BLM and USGS assessment was not yet complete. The EcoMetrix team instead selected field sites with the objective of ensuring a diverse set of sites and accessibility for fieldwork teams, rather than in the more ideal 'hand in glove' approach with other landscape assessment tools.

Finally, the process of applying the ecosystem service tools to the San Pedro Watershed varied due to the reality that tool developers engaged in this comparative test voluntarily, often without direct project funding. Only Ken Bagstad, the technical analyst and ARIES tool team developer, was paid directly; he applied both ARIES and InVEST in the BLM-USGS project. For these reasons, we defined each tool's application to ensure that it was appropriate, as well as feasible within the pro bono budget.

Within these comparative study parameters, the assessment process asked each tool developer to apply their tool to the hypothetical scenario to answer the same set of questions, specifically:

- *New project siting and project development*. Where would be the ideal site for a new residential project that would have the least impact on ecosystem services? Why?
- *Existing infrastructural and project expansion*. Where (and if possible how) would you expand growth of residential units on the US side of the border? Why?
- *Land management*. In what areas would focused ecosystem services-related investments offer potential benefits? What are the recommended investments? What return on investment (ROI), quantitative or qualitative, would be realized and on what time frame (for example, payments from environmental market transactions, real estate sales, etc.)? How might developers avoid regulatory exposure in light of: endangered species habitat sites, indigenous peoples, Native American claims, and other concerns?

All of the tools except EcoAIM, due to time and budget constraints, and EcoMetrix, due to the nature of the tool, looked at the San Pedro Watershed as a whole to identify the site (or sites) that would have the least impact on all four parameters (water provisioning, carbon sequestration, cultural services, and biodiversity).

Tables 30.1 and 30.2 summarize how each tool's team applied it within the San Pedro comparative application with what time implications.

30.3 ASSESSMENT RESULTS

The tools were assessed in terms of relevance and value added to:

- new project siting;
- existing infrastructure project expansion;
- land management.

Each of these application areas are discussed below with specific insights for each tool.

Table 30.1 Tool parameters and boundaries of the San Pedro application

Tools	All Four Parameters Examined?	Other Boundaries Placed on Analysis
ARIES	Yes	None. It examined the whole landscape for hypothetical development
EcoAIM	No, biodiversity only	This assessment focused on privately owned parcels with existing infrastructure (for example, road access) as constraints to development. Note that the tool is not limited to biodiversity; rather this parameter was selected as the best for illustrating the tool
EcoMetrix	Yes	This tool selected five study sites – each 20 acres, projected to 500 acres – in consultation with technical analyst Dr. Ken Bagstad who oversaw all tool applications and based on a set of heterogeneous sites accessible for field research. The study's timing did not allow for the preferable process of examining sites identified by other tools, since those findings were not available in time for the fieldwork period prior to the roundtable
ESValue	Yes	None. This tool examined the whole landscape for hypothetical residential development within context of ecological impact and stakeholder values
ESR	Yes	This tool conducted the strategic priority-setting exercise (the tool's focus) with one technical expert, Ken Bagstad, in consultation with WRI tool developers
InVEST	Yes	None. Technical analyst Dr. Ken Bagstad examined the whole landscape for hypothetical development

New Project Siting

The *ARIES* tool produced maps of potential and actual service provision based on each service's spatial dynamics, to answer questions, including maps of:

- provisioning services where an ecosystem provides a direct benefit to beneficiaries (for example, to identify areas of high carbon sequestration, high biodiversity for recreational values, high-quality views, and areas of high precipitation, infiltration, and groundwater recharge);
- detrimental sinks of ecosystem services where landscape features deplete the quality of an ecosystem service benefit for beneficiaries, such as highways, visual blight, and areas of increased high-intensity fire risk;
- preventive services where an ecosystem mitigates a negative carrier's access to beneficiaries, such as areas absorbing floodwater, absorbing detrimental nutrients, or promoting sediment deposition.

Overall, model results identified sites that minimize impacts to: (1) sources of key ecosystem services (for example, areas of high carbon sequestration, high biodiversity

Table 30.2 Time required for San Pedro Watershed case study[15]

Tool	Time Required for Application
ARIES	200–300 hours of senior technical expert with GIS capabilities Note: Time noted is to develop and parametrize a new case study, which currently can only be done by working with the ARIES team. In the future, applications in areas where models have already been developed will require substantially less time
InVEST	160–260 hours of senior technical expert with GIS capabilities Note: The time it takes to use InVEST varies dramatically by site and according to the technician's level of expertise. The bulk of the time needed to run InVEST is to review literature and parameterize the models. Time used can be substantially reduced if literature is assembled beforehand
ESValue	Approximately 200 hours of a company's staff time, including: 60 hours gathering input from stakeholders (not including about eight hours of each stakeholder's time individually) 100 hours preparing the GIS data, meeting with scientists, and collecting expert opinion, as well as setting up the ecological relationships 40 hours running the tool and analyzing the results
EcoAIM	25 hours reviewing, identifying, downloading, converting, and uploading data, with administrative staff spending eight hours downloading and scientists' work accounting for the remainder Note: for an application limited to biodiversity
ESR	Fewer than 40 hours were needed to complete the ESR worksheet and document assumptions, strengths, and weaknesses of the approach A 'real-world' ESR application would require more time. It would bring together corporate representatives from different business units, inform them about ecosystem services, and seek their input and then have an analyst synthesize it. The time requirements may not be trivial, depending on the scope of the analysis and baseline knowledge about ecosystem services within the organization
EcoMetrix	Field data collection, data entry, and data verification can range from 15 to 60 minutes per acre, depending on the site's complexity

for recreational values, high-quality views, and areas of high precipitation, infiltration, and groundwater recharge); (2) beneficial sinks of ecosystem services (areas of infiltration and groundwater recharge) and avoid new detrimental sinks of ecosystem services. These sites generally include avoiding disturbance to areas of high carbon storage, avoiding creation of additional road infrastructure, and minimizing additional water demand (that is, requiring as few new wells and groundwater extraction requirements as possible).

The *EcoAim* tool's application to this case study focused only on biodiversity impacts – as defined by metric weightings generated through stakeholder preferences – and generated maps for four metrics, including species richness and special protection status, distance to impervious surfaces, distance to wildlife corridors, and vegetation type. Results showed a combined biodiversity score for a region. The average biodiversity score of four potential development sites could then guide developer choice. Overall, the tool enables users to drill down into each metric for information by parcel (for instance, number of endangered species with known distribution in the parcel).

The *EcoMetrix* tool results identified whether or not potential actions will make a contribution to meeting larger goals at a landscape level, which led to selection of a proposed site for a hypothetical residential development. The tool assumed standardized lot design for all sites and identified alternative design features to improve performance and impact minimization (for example, cattle removal, lot design changes, pavement material, roof type, existence of pool, etc.). The approach allows for cost–benefit evaluation of alternatives (for instance, effect of removing swimming pools at each hypothetical residence for water provisioning vs marketability of lots).

The *ESValue* model results indicated sites would be best for development based on the ecosystem services that stakeholders valued and on the ecological impacts of development. Drivers of differences in scores and ecological impacts across alternatives were given.

The *ESR* relied on expert advisors to answer questions where to best site the hypothetical project and to follow general guidelines that minimize site impact (for example, reducing project footprint, acres of land disturbed, acre-feet of water withdrawn, degree of site fragmentation, avoidance of more valuable land cover types, etc.).

The *InVEST* tool provided outputs indicating that an ideal project site would have the following attributes.[16] (1) Carbon: seek to avoid areas of high carbon storage (for instance, forests, riparian areas, oak woodland, and mesquite). Focus on ecosystems, like grasslands and desert scrub, that store less carbon. (2) Biodiversity: new development should be accompanied by *minimal additional road infrastructure* (that is, be located near roads) and *minimal additional water demand* (that is, requiring as few new wells and groundwater extraction requirements as possible). In terms of water, for arid areas, instead of predicting water availability, this model at the time of application combined surface and groundwater into a single flow ('water yield'), which, along with some other factors, raised some problems with model outputs, worth noting here. First, increases in water yield may not always be beneficial, and may actually be detrimental in systems with the attributes of the San Pedro Watershed. Results for each scenario show that water yield increases with new development because impervious surfaces are added. But, in arid systems where evapotranspiration is high, additional impervious surfaces increase runoff speed and quantity, which can lead to problems with erosion and lower dry-season flows. So, development should minimize the increase in water yield from pre- to post-development conditions. (It is noteworthy, however, that this interpretation of the model outputs (increase in water yield is detrimental) may, however, not be accurate in other arid applications because it includes expert information about locally important hydrological processes that are not included in the model parameters (erosion, flooding, and infiltration zones). For example, if the area drained into a reservoir (not present in the San Pedro Watershed), erosion was not a problem, and additional water demand from the new development could be met, an increase in water yield may be beneficial.

While it might at first appear counterintuitive to minimize increases in water yield, this application provides an important illustration: water yield model outputs must be examined carefully in the local context and along with other hydrologic ecosystem services to assess whether water yield changes are beneficial, detrimental, or if more information is needed. This assessment requires local expert knowledge, will be related to the timing and location of runoff and presence or absence of reservoirs, and may require additional information about trade-offs or feedback loops with other processes and services or disservices (erosion, flooding, etc.).

Existing Infrastructure Project Expansion

In assessing the tools applications of where to (hypothetically) expand existing infrastructure, the findings were as follows:

- *ARIES*: answers to this question are nearly identical to those for the development of new projects (above). Although not explicitly considered in the ARIES models, existing expansion near currently developed areas would reduce the landscape and habitat fragmentation resulting from highly dispersed development.
- *EcoAIM*: restoration siting assumptions focused on preservation (for example, preserve corridors, join parcels, explore vegetation type, and consider species richness). Results, in the form of tables, identified seven restoration projects, from which stakeholders can choose the optimal one. Return on investment (ROI) calculations was based on restoration costs, land acquisition costs, license to operate benefits, and credit market benefits.
- *EcoMetrix* did not answer this question directly because it is not a project-level question – the level on which it focuses. Yet, the tool could contribute to assessing incremental impact to cumulative impacts questions.
- *ESValue* did not consider this issue within the analysis, due to time and budget constraints.
- *ESR*: the answers to this question were identical to new project siting (above).
- *InVEST*: the answers to this question were also nearly identical to those for the development of new projects. Although not explicitly considered in the InVEST models, siting existing projects near currently developed areas would reduce the landscape and habitat fragmentation resulting from highly dispersed development.

Land Management

In considering the land management questions, the tool applications yielded the following results:

- *ARIES*: investments should be considered in terms of their ability to: (1) maximize carbon sequestration and storage; (2) minimize water demand in the watershed, which reduces groundwater pumping and can affect biodiversity and associated ecosystem services. Regulatory exposure may be reduced by reconsidering water withdrawals from the San Pedro's riparian ecosystem, increasing water conservation, and bringing the San Pedro's water budget into balance. Given the complex nature of the San Pedro's groundwater and the way ARIES deals with groundwater, its value to support decision-making about groundwater will probably remain limited until it can incorporate existing groundwater models (that is, local applications of the MODFLOW model). Finally, ARIES can couple with other tools, in particular those that scope ecosystem services-based decisions (for instance, ESR), to estimate economic values (for example, Defenders of Wildlife's Wildlife Habitat Benefits Estimation Toolkit) or to map potential impacts to biodiversity (for instance, IBAT).
- *EcoAIM*: specific evaluation factors depend on stakeholder preferences.

Stakeholders choose the best sites for restoration from the seven sites the tool identified. Architecture allows user-driven sensitivity analysis to reveal the relative effect of stakeholder preferences on modeled outcomes. ROI was based on restoration costs, land acquisition costs, license to operate benefits, and credit market benefits.

- *EcoMetrix*: site-specific benefits include: decreased impacts and facilitated project approval; decreased supply chain dependencies (in this case, water); decreased site management costs; strengthened social license to operate, and opportunities to create PES. Future mitigation needs were identified in terms of opportunities for off-site habitat restoration (and possible eco-asset credits for banking, markets, etc.). The tool also enables understanding where investments get the biggest ecosystem return. Expected ROI would likely come from: improved project delivery; decreased infrastructure costs; decreased residential utility bills; increased residential quality of life, and credit generation.
- *ESValue*: clearly demonstrated that stakeholders favored strong land management practices that protect key ecosystems services. Developments that protect aquatic habitat, stream flow, water quality, and recreational access are much more likely to receive stakeholder approval.
- *ESR*: not applicable.
- *InVEST*: similar to the earlier answers, investments might be made to maximize carbon sequestration and storage, minimize threats to biodiversity, and reduce water demand in the watershed (which also reduces the threat brought on by excessive groundwater pumping). A key issue is to consider dewatering of the San Pedro's riparian ecosystem, and thus consider increasing water conservation and bringing the San Pedro's water budget into balance. Given the complex nature of the San Pedro's groundwater and the simplistic way InVEST deals with this matter, this tool will have limited value until it can incorporate existing groundwater models (that is, local applications of the MODFLOW model). Perhaps the InVEST Tier 2 models will address these concerns.

30.4 KEY INSIGHTS FROM TOOL ASSESSMENT

A number of insights emerged from the comparative assessment of ecosystem services tools.

Insight 1: Side-by-side ecosystem services tool comparisons are difficult, given the tools' very different definitions of the term and approaches to measurement and assessment.

The San Pedro Watershed comparative tool assessment used a diverse set of analytical measures and generated an equally diverse set of results. The methods and metrics rarely overlapped across the tools. Interestingly, however, the overall conclusions of some tools, such as ARIES and InVEST, generally agreed. Other tools, such as ESValue, reached distinctly different conclusions.

This finding highlights the ongoing debate over how to translate ecosystem services concepts into clear, commonly accepted measures that can inform decision-making.

Without agreed-upon metrics and assessment methodologies for ecosystem services, corporate decision-makers will have to carefully justify their selection of any one tool over another. Independent examination and a common set of measures, or metrics, as well as a methodology, will hasten uptake by providing key credibility for, and validation of, a particular approach.

Insight 2: The ecosystem services tools offer insights that can be relevant to corporate decision-making processes. However, none readily mesh with key existing corporate processes. Thus, they do not appear to be ready for immediate, widespread, off-the-shelf business application.

Analysis of tools during and following the 2010 BSR roundtable led to the conclusion that none are ready for broad-scale implementation in the corporate context. At the time, all of the tools would either require assistance with interpreting findings within a corporate setting or would need to be tailored to fit particular corporate decision-making contexts.

One key issue is the gap between what tool developers offer and what corporate decision-makers need. Specifically, many of these tools have been developed for use with expert support to provide detailed assessments using powerful modeling and scenario development for forecasting. Several of the tools require complex validation and are research driven. Some tools incorporate stakeholder input, but others do not allow for this input.

Many corporate decision-makers are looking for a flexible, modular toolbox. They commonly want help making more immediate, practical decisions, including exploring options for fast-track action, especially at the project and site level. Business decision-makers also need tools to help understand how stakeholders depend on, value, and approach trade-offs among ecosystem services in specific contexts. Consideration of stakeholder needs and priorities is integral to many key corporate decision processes.

Given this gap, both ecosystem services tools, as well as business decision-makers, could benefit from pilot test applications and refinement in business settings. Overlapping areas among tools and collaborative application opportunities are other areas of opportunity.

Insight 3: Because tool applications are limited within corporate decision-making processes, it is not yet clear what additional value ecosystem services tools will add when compared to the existing approaches companies use to assess performance.

Private-sector test applications of tools will be essential to understanding the value these tools add and for building the business case within companies. In particular, the way that tools may mesh with existing corporate decision-making processes, particularly as related to environmental and social impacts of projects, remains to be seen. Ultimately, ecosystem services tools will need to demonstrate key benefits not otherwise obtainable if their use is to be justified.

Overall, business managers need clarity on how, when, and why to apply tools to particular business activities and issues. At present, the diversity of both tools and business settings present a significant challenge. Pilot applications will have to consider both issues around which tools are most appropriate for a certain decision-making context, as well as how tools could link to, or augment, existing processes and protocols, most notably including environmental/social impact assessments (E/SIAs) and life cycle analyses (LCAs).

Looking forward, business managers will need to learn from a robust set of new private-sector applications of tools. These pilot applications will likely require some industry-specific, as well as industry-initiated, work, because of unique issues and assessment processes. Testing tools in multiple private sector contexts will help clarify whether and how ecosystem services metrics and tools can interface with existing corporate processes for undertaking environmental and social assessments. Ideally, pilot tests will also bring greater clarity on when, where, and how to integrate new metrics around ecosystem services, or even help identify the need for new tool development.

Discussions among corporate members of BSR's Ecosystem Services Working Group indicate that there are some promising potential applications of ecosystem services tools within corporate settings, which could include:

- new project planning and development, particularly in terms of impact assessment and permitting processes, to show companies, governments, and other stakeholders where and how impacts or co-benefits may result;
- real estate strategy and management;
- property portfolio priority-setting exercises to assess relative risk and opportunity for property retention, disposition, remediation, restoration, and other options;
- ongoing management and decommissioning of operations;
- valuations of ecosystem services impacts or benefits;
- corporate performance and communication dashboard or scorecard, in terms of measuring performance and progress toward a corporate-level ecosystem services goal (including key performance indicators, baseline, impacts, progress, and monitoring);
- providing a more complete picture of corporate environmental performance, using ecosystem services concepts to integrate currently discrete natural resource parameters;
- scenario planning and modeling, such as linking to corporate climate change adaptation strategy development;
- project planning within a landscape-level context, in terms of natural resources uses, beneficiaries, and minimum ecological parameters for continued flow of ecosystem services;
- operational assessment of E/SIAs, such as by providing additional baseline data, integrating existing baseline data, assessing significant potential for future ecosystem service changes, and identifying necessary mitigation or enhancement measures;
- operational assessment of life cycle impacts of products, such as in terms of additional parameters and bounding analyses;
- selection of potential building sites in terms of optimization of benefits and minimization of impacts;
- understanding ecosystem service functions at facility scale;

Engage with Stakeholders	Step	Description	Phase	Tool
↑	Step 1	Identify ecosystems and ecosystem services of concern *(primarily a qualitative assessment)*	SCREEN	Tool: Ecosystem Services Review (ESR)
	Step 2	Prioritize and rank ecosystem services of most concern, including eliciting preferences from stakeholders *(If needed or desired)*	SCOPE	Tool: EcoAIM ESR, ESValue
	Step 3	Given a specific geographic area (or set of scenarios), what will be the change in ecosystem services? Who will be affected? How will they be affected?	ASSESS	Tool: ARIES, INVEST EcoAIM, EcoMetrix
	Step 4	What is the value of changes in ecosystem services? Value to whom? Quantified how? *(if needed or desired. Note that value can be intrinsic or monetary)*		Tool: EcoAIM, ESValue NAIS
↓	Step 5	Implement and Monitor	DEVELOP MANAGEMENT PLAN	Tool: EcoAIM, EcoMetrix

Figure 30.1 Steps and tools for applying ecosystem services in corporate settings

- identifying corporate dependencies on ecosystem services at various geographical and supply chain levels;
- exploring new strategies and scenarios;
- optimizing the sourcing of natural resources;
- engaging stakeholders in at least some of the above-mentioned contexts.

Figure 30.1 suggests a potential mapping of tools included in the assessment to steps in an illustrative corporate decision-making process.

Overall, ecosystem services metrics and tool uptake by companies will be based on a strong business case that demonstrates the value that an ecosystem services lens adds relative to current approaches to considering performance (for instance, E/SIAs, LCAs, etc.). Specifically, a side-by-side assessment of 'business as usual' (in terms of E/SIAs and environmental management approaches) versus an ecosystem services tool-based approach would be extremely useful, particularly in clarifying the need for additional data or new processes, as well as insights that companies could gain.

Insight 4: Independent, third-party comparative review of analytical and modeling frameworks underlying ecosystem services tools will build and ensure credibility.[17]

The ecosystem services tool domain is relatively crowded and complex. Multiple definitions, frameworks, and approaches are used. Tools often lack transparency, and they are difficult to compare.

Looking forward, it will be essential to have clear evidence of tool credibility and widespread support to justify applying ecosystem services concepts and tools to their activities. Key areas of need are to explore harmonization of ecosystem services definition and metrics used by tools, to conduct rigorous comparative assessments of multiple tools, and to assess data needs, as well as provide quality control.

Insight 5: A 'taxonomy' for the emerging ecosystem services tool domain would assist with selecting tools that are best suited for specific applications.

This San Pedro Watershed comparative tool assessment highlighted areas of potential complementarities between tools. For example, ESR could offer a structure for priority setting prior to doing a landscape-level assessment using either ARIES or InVEST. The EcoMetrix tool could then assist with site-level analysis. However, many questions remain, such as:

- How would connections among tools work in practice? Would there be linked software or online web portals?
- Would it be time- or cost-effective to use multiple tools?
- Would new insights, relative to current corporate environmental assessments, be gleaned?

In addition, it would be useful to be able to more systematically compare tools to one another and understand them in this context and in terms of their ideal applications. Table 30.3 outlines a possible approach to a tool taxonomy.

30.5 CONCLUSIONS AND A LOOK FORWARD

Overall, the comparative test uncovered important distinctions between the tools and ideal contexts for their application (see also Figure 30.2), including the following:

- ESR provides a structured but relatively high-level approach to understanding and prioritizing the relevance of ecosystem services in the business context defined by a tool user.
- ARIES and InVEST are landscape-level computer models for conducting detailed, spatially explicit analyses and simulations. ARIES is an online modeling platform that integrates data with probabilistic and deterministic models to map ecosystem service flows, and InVEST is based on land use land cover (LULC) data, corresponding ecological attribute data for each LULC type, and generally accepted ecological-process models, applied in an ArcGIS environment.
- EcoAIM and ESValue integrate stakeholder preferences in consideration of ecosystem services impacts. EcoAIM in this instance focused on a modified risk analysis approach. ESValue used stakeholder input to assess the relative social value of ecosystem services and scientific input to assess the ability of the ecological landscape to provide those values.
- EcoMetrix, a field site tool, is intended to be used once a broader landscape-level assessment has identified the parcels where ecosystem services are least likely to be affected. It helps define an approach to design that minimizes impact.

The findings of this comparative assessment leave many open questions. However, discussions during the process indicated that ecosystem services approaches and tools could offer value to companies if the tools can help companies more effectively and efficiently:

Table 30.3 Potential ecosystem services tool taxonomy

Target User	Policy-maker	Corporate	Academic and NGO
User Motivation	New policy designs or elimination of subsidies Regulatory enforcement Mapping of new protected areas Education Seeding of new environmental markets	Risk mapping for ecosystem decline Strategy and policy design Location screening Footprint measurement Liability transfer New revenue-generating transactions Social license-to-operate	Advancement of conservation science techniques Recommendations for delineation of protected areas Integration of datasets with other organizations
Desired outputs	Spatially-explicit maps Valuation analysis ROI prediction Sensitivity-analysis results for scenario planning	Spatially-explicit maps Valuation analysis ROI prediction Sensitivity-analysis results for scenario planning	Spatially-explicit maps Valuation analysis ROI prediction Sensitivity-analysis results for scenario planning
Primary ecosystem services of interest	Supporting services (from MEA) Provisioning services Regulating services Cultural services	Regulating services Cultural services	Supporting services (from MEA) Provisioning services Regulating services Cultural services
Quality of input data	High quality[a] Medium quality[b] Low quality[c]	High quality Medium quality Low quality	High quality Medium quality Low quality

Notes:
a. Recommended suite of tools: X, Y, Z.
b. Recommended point of application.
c. Recommended roles and responsibilities.

- compare the trade-offs various projects or initiatives would involve;
- broaden the benefits for the local populations where they operate;
- retire or decommission a project in a way that maximizes benefits and is cost-effective and timely;
- take a landscape-level perspective, but also see facility-level effects along the supply chain;
- understand potential effects and dependencies on ecological functions;
- understand how local populations may affect and depend on ecological functions;
- make informed decisions based on sound science and stakeholder input;
- collaborate and communicate with regulators and communities in a transparent process.

Figure 30.2 Distinctions and complementarities among tools

This list is ambitious, but not impossible to achieve. As ecosystem services tools continue to develop, they may complement one another or be able to be integrated into internal processes many companies have already undertaken, and so ultimately deliver on many desired benefits.

Looking forward, a set of robust ecosystem services tools that have been well-vetted, peer reviewed, and ultimately approved by regulators, stakeholders, and companies for use at different geographic scales in a wide variety of business activities will eventually emerge. Ideally, the tools will allow companies to carefully examine various decisions' trade-offs in ways that mesh with current decision-making protocols in a cost- and time-effective manner.

To be most useful to companies, tools will need to have certain key attributes that may be challenging to deliver, such as:

- scalability and adaptability for different locations, conditions, and types of company activity;
- ability to generate and compare scenarios;
- ease of use (related to time and resources);
- generation of spatially explicit displays of information (for example, maps);
- transparency (no 'black boxes') – easy to understand and communicate tool inputs, operation, and outputs;
- avoidance of new corporate-level metrics (unless there is a corporate-wide policy that names ecosystem services);

- levels of (un)certainty;
- 'roll up' and 'roll down' findings;
- highlight trends;
- benchmarks;
- maps, charts, and tables – all of which should be able to be exported into PowerPoint presentations or documents;
- development of internal corporate plug-and-play applications or augment existing tools.

In addition, the universe of tools is likely to proliferate in the coming years. It will encompass a range of tools specialized for different scales, climates, and applications. Within this context, private-sector players will increasingly need to develop their own understanding of, not only of using ecosystem services concepts and tools, but also of costs and resources needed to integrate these issues within corporate governance, strategy, and operations.

The trends are clear, even if specific outcomes are currently uncertain. Ecosystem services concepts and tools appear increasingly in public policy and business operational arenas globally. All stakeholders have an opportunity to engage in robust, constructive discussion around how ecosystem services concepts and tools can best support efforts to improve the structure and functioning of ecosystems, and their ability to deliver the services on which society and business relies.

ACKNOWLEDGEMENTS

This chapter is based on work both prior to, during, and following an October 2010 roundtable that focused on emerging ecosystem services tools and included analytical work, narrative text, as well as comments on drafts, by (in alphabetical order): Ken Bagstad, University of Vermont, Gund Institute for Ecological Economics; Kevin Benck, Parametrix; Pieter Booth, Exponent; Chris Colvin, The Natural Capital Project; Kenna Halsey, Parametrix; Kevin Halsey, Parametrix; John Finisdore, World Resources Institute; Gary Johnson, University of Vermont, Gund Institute for Ecological Economics; Sheryl Law, Exponent; Doug MacNair, Cardno ENTRIX; Mary Ruckelshaus, The Natural Capital Project; David Saah, Spatial Informatics Group; and Austin Troy, University of Vermont and Spatial Informatics Group. We are grateful to all of these individuals for the time they invested.

This report draws upon discussions about emerging ecosystem services tools during the October 2010 roundtable with the individuals listed above as well as (in alphabetical order): Brian Bellew, US Bureau of Land Management (BLM); Greg Biddinger, ExxonMobil; Roberto Bossi, Eni; Sarah Connick, Chevron; Amanda DeSantis, DuPont; Craig Duxbury, The Walt Disney Company; Sofie Gudum Faaborg, DONG Energy; Ann George, Freeport-McMoRan Copper & Gold; Sachin Kapila, Shell; Jasper Lament, BC Hydro; John Lin, US Environmental Protection Agency; Mark Nechodom, US Department of Agriculture; Anne Neville, Rio Tinto; David Norris, Freeport-McMoRan Copper and Gold; Shirley Olivera, BP; Malka Pattison, US Department of the Interior; Mark Rekshynskyj, BLM; Darius Semmens, US Geological Survey (USGS); Carl D.

Shapiro, USGS; Manuel Winograd, European Environment Agency; Robert Winthrop, BLM; and Richard Wood, ExxonMobil. We are deeply appreciative for the strong participation and level of engagement during the roundtable discussion.

Finally, we would like to thank the corporate members of BSR's Ecosystem Services Working Group, including BC Hydro, BP, Chevron, DONG Energy, DuPont, Eni, Exxon Mobil Corporation, Freeport-McMoRan Copper & Gold, GlaxoSmithKline, Merck & Co., Rio Tinto, Shell, and The Walt Disney Company. Any errors in the report are those of the authors alone.

NOTES

1. For an introduction to ecosystem services for businesses, please see BSR's publications on the topic at: http://www.bsr.org/consulting/working-groups/environmental-markets.cfm. For further scientific background on the topic, please see the Millennium Ecosystem Assessment (MA) at: http://www.maweb.org/en/index.aspx.
2. See http://www.naturalcapitalproject.org/about.html.
3. See http://www.fsd.nl/everyone.
4. See http://www.fgv.br/ces.
5. See Hwang et al. (2012).
6. See http://www.wavespartnership.org/.
7. Dr. Ken Bagstad did his doctoral work at the University of Vermont and participated on the ARIES development team.
8. Excerpt from http://ebmtoolsdatabase.org/tool/aries-artificial-intelligence-ecosystem-services.
9. Excerpt from http://www.naturalcapitalproject.org/pubs/NatCap_InVEST_and_Case_Study_Summary_TEEB_2010.pdf.
10. Excerpt from Exponent materials for the 2010 BSR roundtable.
11. Excerpt from Parametrix materials for the 2010 BSR roundtable.
12. Excerpt from http://www.wri.org/publication/corporate-ecosystem-services-review.
13. Excerpt from Cardno ENTRIX materials for the 2010 BSR roundtable
14. D. Saah and A. Troy (2010), personal communication (for more information on NAIS, please see: http://www.sig-gis.com/services/ecosystem-services).
15. Times are based on this single case study and may not be broadly applicable, but this first concurrent application of the tools allows us to begin to compare them.
16. This material on InVEST was jointly written by Dr. Ken Bagstad, who applied the InVEST tool to the San Pedro Watershed case study, and Chris Colvin of the InVEST tool team and the Natural Capital Project.
17. Some individual tools have been vetted through peer-reviewed publications. For example, see: Nelson et al. (2009, 2010); Tallis and Polasky (2009); Kareiva et al. (2011); Polasky et al. (2011).
18. Developers of the Habitat Estimation Toolkit and MEASURES tool were approached for inclusion in this study, but they declined due to time and cost constraints.

REFERENCES

Hwang, L., S. Waage and K. Armstrong (2012), *Global Public Sector Trends in Ecosystem Services: 2009–2011 Summary*, San Francisco: BSR, accessed 20 February 2014 at https://www.bsr.org/reports/BSR_Ecosystem_Services_Policy_Synthesis_09–11.pdf.

Kareiva, P.M., T.H. Ricketts, G.C. Daily, H. Tallis and S. Polasky (eds) (2011), *Natural Capital: Theory and Practice of Mapping Ecosystem Services*, Oxford: Oxford University Press.

Nelson, E., G. Mendoza, J. Regetz, S. Polasky, H. Tallis, D.R. Cameron et al. (2009), 'Modeling multiple ecosystem services, biodiversity conservation, commodity production, and trade-offs at landscape scales', *Frontiers in Ecology and the Environment*, **7**(1), 4–11.

Nelson, E., H. Sander, P. Hawthorne, M. Conte, D. Ennaanay and S. Wolny et al. (2010), 'Projecting global land-use change *and* its effect on ecosystem service provision and biodiversity with simple models', *PLoS ONE*, **5**(12),

e14327, accessed 21 February 2014 at http://www.plosone.org/article/info%3Adoi%2F10.1371%2Fjournal.pone.0014327.
Polasky, S., E. Nelson, D. Pennington and K. Johnson (2011), 'The impact of land-use change on ecosystem services, biodiversity and returns to landowners: a case study in the State of Minnesota', *Environmental and Resource Economics*, **48**(2), 219–42.
Tallis, H. and S. Polasky (2009), 'Mapping and valuing ecosystem services as an approach for conservation and natural-resource management', *Annals of the New York Academy of Sciences*, **1162**, 265–83.

APPENDIX

Table 30A.1 Ecosystem services tools included in assessment[18]

Tool	Developers and Websites
ARIES (ARtificial Intelligence for Ecosystem Services)	University of Vermont's Gund Institute and Ecoinformatics Collaboratory (USA) Basque Centre for Climate Change (Spain) Conservation International (USA) Earth Economics (USA) Instituto de Ecología (Mexico) http://www.ariesonline.org/
InVEST (Integrated Valuation of Ecosystem Services and Tradeoffs)	The Natural Capital Project, including: Stanford University (USA) University of Minnesota WWF (World Wildlife Fund) The Nature Conservancy http://www.naturalcapitalproject.org/
EcoAIM (Ecological Asset Inventory and Management)	Exponent http://www.conference.ifas.ufl.edu/aces10/Presentations/Wednesday/Coyote-B-E/PM/Yes/0135%20P%20Booth.pdf
EcoMetrix	EcoMetrix Solutions Group http://www.ecometrixsolutions.com/index.html
ESR (Ecosystem Services Review)	World Resources Institute (WRI) Meridian Institute World Business Council on Sustainable http://www.wri.org/publication/corporate-ecosystem-services-review
ESValue	Cardno ENTRIX http://www.cardnoentrix.com/documents/San_Pedro_Watershed.pdf
NAIS (Natural Assets Information System)	Spatial Informatics Group http://www.sig-gis.com/services/ecosystem-services

NIA →

Index